Franco Laeri, Ferdi Schüth,
Ulrich Simon, Michael
Wark (Eds.)

Host-Guest-Systems
Based on Nanoporous
Crystals

Related Titles from WILEY-VCH

F. Schüth, K.S.W. Sing, J. Weitkamp (Eds.)

Handbook of Porous Solids

Five volumes
2002, ISBN 3-527-30246-8

G. Meyer, D. Naumann, L. Wesemann (Eds.)

Highlights in Inorganic Chemistry

2001, ISBN 3-527-30265-4

U. Schubert, N. Hüsing

Synthesis of Inorganic Materials

2000, ISBN 3-527-29550-X

J.S. Miller, M. Drillon (Eds.)

Magnetism: Molecules to Materials

Vol. I: Molecules to Materials
2001, ISBN 3-527-29772-3

Vol. II: Molecule-Based Materials
2001, ISBN 3-527-30301-4

Vol. III: Nanosized Magnetic Materials
2001, ISBN 3-527-30302-2

Vol. IV: Molecules to Materials IV
2002, ISBN 3-527-30429-0

Host-Guest-Systems Based on Nanoporous Crystals

Franco Laeri, Ferdi Schüth, Ulrich Simon, Michael Wark (Eds.)

WILEY-VCH

WILEY-VCH GmbH & Co. KGaA

Dr. Franco Laeri
Institute of Applied Physics
Technical University Darmstadt
Schloßgartenstr. 7
64289 Darmstadt
Germany

Prof. Dr. Ferdi Schüth
Max-Planck-Institute of Coal Research
Kaiser-Wilhelm-Platz 1
45470 Mülheim an der Ruhr
Germany

Prof. Dr. Ulrich Simon
Institute of Inorganic Chemistry
RWTH Aachen
Professor-Pirlet-Str. 1
52074 Aachen
Germany

Dr. Michael Wark
Institute of Physical Chemistry and
Electrochemistry
Hannover University
Callinstr. 3-3a
30167 Hannover
Germany

Library of Congress Card No.: applied for
A catalogue record for this book is available from the British Library.
Bibliographic information published by Die Deutsche Bibliothek
Die Deutsche Bibliothek lists this publication in the Deutsche Nationalbibliografie; detailed bibliographic data is available in the Internet at http://dnb.ddb.de

Printed in the Federal Republic of Germany.
Printed on acid-free paper.

Typesetting Asco Typesetters, Hong Kong
Printing betz-druck gmbh, Darmstadt
Bookbinding Litges & Dopf Buchbinderei GmbH, Heppenheim

ISBN 3-527-30501-7

Contents

List of Contributors

Institute of Applied and Physical Chemistry
Fachbereich 2
University of Bremen
PF 330 440
28334 Bremen
Germany
altindag@uni-bremen.de

Carsten Baehtz
Institute of Materials Science
Darmstadt University of Technology
Petersenstraße 23
64287 Darmstadt
Germany
carsten.baehtz@desy.de

Monika Bauer
Fraunhofer-Institut für Zuverlässigkeit und Mikrointegration
Kantstraße 55
14513 Teltow
mbauer@epc.izm.fhg.de

Martin Beneke
Institute of Applied and Physical Chemistry
Fachbereich 2
University of Bremen
PF 330 440
28334 Bremen
Germany
Present adress:
Airbus Germany
Hienefeldstraße 1-5
28199 Bremen
Germany
Martin.Beneke@airbus.com

Peter Behrens
Institute of Inorganic Chemistry
University of Hannover
Callinstraße 9
30167 Hannover
Germany
peter.behrens@mbox.acb.uni-hannover.de

Thomas Bein
Department of Chemistry
University of Munich
Butenandtstraße 5-13 (E)
81377 Munich
Germany
tbein@cup.uni-muenchen.de

Martin Beneke
Institute of Applied and Physical Chemistry
Fachbereich 2
University of Bremen
PO Box 330 440
28334 Bremen
Germany
jse@uni-bremen.de

Lhoucine Benmohammadi
Darmstadt University of Technology
Petersenstraße 23
64287 Darmstadt
Germany

Christoph Bräuchle
Department of Chemistry and Center of Nanoscience
Ludwig-Maximilians-Universität München
Butenandtstraße 11
81377 Munich
Germany
christoph.braeuchle@cup.uni-muenchen.de

Jürgen Brickmann
Department of Chemistry
Darmstadt University of Technology
64287 Darmstadt
Germany
brick@pc.chemie.tu-darmstadt.de

Ingo Braun
Institute of Applied and Physical Chemistry
University of Bremen
Bibliotheksstr. 1
28359 Bremen
Germany

Gion Calzaferri
Department of Chemistry and Biochemistry
University of Bern
Freiestraße 3
3000 Bern 9
Switzerland
gion.calzaferri@iac.cmibe.ch

Fred-Walter Deeg
Carl BAASEL Lasertechnik GmbH & Co. KG
Petersbrunner Straße 1b
82319 Starnberg
Germany

Wulf Depmeier
Institute of Geological Science
Christian-Albrechts University at Kiel
Olshausenstraße 40
24098 Kiel
Germany
wd@min.uni-kiel.de

Rolf Eckhardt
ATMI Sensoric
Justus-von-Liebig-Straße 22
53121 Bonn
Germany
reckhardt@sensoric.de

A. Erodabasi
Darmstadt University of Technology
Petersenstraße 23
64287 Darmstadt
Germany

Detlef Fanter
Fraunhofer-Institut für Zuverlässingkeit
und Mikrointegration
Außenstelle Polymermaterialien
Kantstraße 55
14513 Teltow
Germany
fanter@epc.izm.fhg.de

R. Dieter Fischer
Institute of Inorganic and Applied Chemistry
University of Hamburg
Martin-Luther-King-Platz 6
20146 Hamburg
Germany
dieter.fischer@chemie.uni-hamburg.de

Rolf Fricke
Institute of Applied Chemistry Berlin-Adlershof e. V.
PO Box 96 11 56
12474 Berlin
Germany
fricke@aca-berlin.de

Michael Fröba
Institute of Inorganic and Analytical Chemistry
Justus-Liebig University, Gießen
Heinrich-Buff-Ring 58
35392 Gießen
Germany
michael.froeba@anorg.chemie.uni-giessen.de

Ligia Frunza
Institute of Applied Chemistry in Berlin Adlershof e.V.
Postfach 961156
12474 Berlin
Germany
Permanent address:
National Institute of Materials Physics
PO Box Mg 07
76900 Bucharest-Magurele
Romania
frunza@alpha1.infim.ro

Matthias Ganschow
Institute of Applied and Physical Chemistry
University of Bremen
Bibliotheksstraße 1
28359 Bremen
Germany
ganschow@uni-bremen.de

Stephan Glaus
Department of Chemistry and Biochemistry
Universtity of Bern
Freiestraße 3
3012 Bern
Switzerland
stephan.glaus@laengasse.ch

Annett Gräser
Infineon Technologies AG
R &D Lithography
MH E FE
PO Box: 80 09 49
81609 München
annett.graeser@infineon.com

Gerd Grubert
Institute for Applied Chemistry
Berlin-Adlershof e. V.

PO Box 96 11 56
12474 Berlin
Germany
grubert@ACA-Berlin.de

Hilka Hanika-Heidl
Institute of Inorganic and Applied Chemistry
University of Hamburg
Martin-Luther-King-Platz 6
20146 Hamburg
Germany
hanika@chemie.uni-hamburg.de

Christian Hellriegel
Department of Chemistry and Center of Nanoscience
Ludwig-Maximilians-Universität München
Butenandtstraße 11
81377 Munich
Germany

G. Ihlein
Max-Planck-Institut für Kohlenforschung
Kaiser-Wilhelm-Platz
45470 Mülheim
Germany

Nils I. Jaeger
Institute of Applied and Physical Chemistry
Fachbereich 2
PF 330 440
28334 Bremen
Germany
jse@uni-bremen.de

Christian Jäger
Labor 1331
Magnetische Resonanzspektroskopie
Eichard-Willstätter Straße 11
12489 Berlin-Adlershof
Germany
christian.jaeger@bam.de

Jörg Kärger
Faculty of Physics and Geological Sciences
University of Leipzig
Linnéstraße 5
04103 Leipzig
Germany
kaerger@physik.uni-leipzig.de

Michael Kant
Institute of Applied Chemistry Berlin-Adlershof e. V.
Richard-Willstätter Straße 12
12489 Berlin
Germany
mkant@aca-berlin.de

K. Koch
Darmstadt University of Technology
Petersenstraße 23
64287 Darmstadt
Germany

Jan Kornatowski
Department of Chemical Technology
University Technology of Munich
Lichtenbergstraße 4
85747 Garching
Germany
jan.Kornatowski@ch.tum.de

Hendrik Kosslick
Institute of Applied Chemistry Berlin-Adlershof e. V.
PO Box 96 11 56
12474 Berlin
Germany
kosslick@aca-berlin.de

Friedrich Kremer
Universität Leipzig
Fakultät für Physik und Geowissenschaften,
Linnéstraße 5
04103 Leipzig
Germany
kremer@physik.uni-leipzig.de

Ken'Ichi Kuge
Faculty of Engeneering
Chiba University
1-33 Yayoi-cho
Inage-ku
Chiba263
Japan
kuge@image.tp.chiba-u.ac.jp

Ulrich Kynast
University of Applied Sciences/ Fachhochschule Münster
Stegerwaldstraße 39,
48565 Steinfurt
Germany
UK@FH-muenster.de

Franco Laeri
Darmstadt University of Technology
Institut für Angewandte Physik
Schloßgartenstraße 7
64289 Darmstadt
Germany
franco.laeri@physik.tu-darmstadt.de

Claudia Leiggener
Department of Chemistry and Biochemistry

Universtity of Bern
Freiestraße 3
3012 Bern
Switzerland
claudia.leiggener@iac.unibe.ch

Min Ling
Guangxi University
Industrial Testing Centre
Nanning 53004
P.R. China
lingmin@gxu.edu.cn

Frank Marlow
Max-Planck-Institut für Kohlenforschung
Kaiser-Wilhelm-Platz 1
45470 Mülheim an der Ruhr-
Germany
marlow@mpi-muelheim.mpg.de

Peter Neumann
Institute of Applied Chemistry Berlin-Adlershof e. V.
Richard-Willstätter Straße 12
12489 Berlin
Germany
neum@aca-berlin.de

Manfred Noack
Institute for Applied Chemistry
Berlin-Adlershof e.V.
Richard-Willstätter-Straße 12
12489 Berlin
Germany
noack@aca-Berlin.de

J. U. Nöckel
University of Oregon
Eugene. OG 97403-1274
USA

N. Owschimikow
Darmstadt University of Technology
Petersenstraße 23
64287 Darmstadt
Germany

Notker Rösch
Institute of Physical and Theoretical Chemistry
Technical University of Munich
Lichtenbergstr. 4
85747 Garching
Germany
roesch@theochem.tu-muenchen.de

Joachim Sauer
Institute of Chemistry
Humboldt University Berlin
Unter den Linden 6
10099 Berlin
Germany
js@chemie.hu-berlin.de

Jürgen Schneider
Fraunhofer-Institut für
Zuverlässigkeit und Mikrointegration
Außenstelle Polymermaterialien und Composite
Kantstraße 55
14513 Teltow
Germany
juergen.schneider@epc.izm.fhg.de

Sigurd Schrader
Institute of Physics
University of Potsdam
Am Neuen Palais 10
14469 Potsdam
Germany
schrader@rz.uni-potsdam.de

Ulrich Schülke
Michael Kant
Institute of Applied Chemistry Berlin-Adlershof e. V.
Richard-Willstätter Straße 12
12489 Berlin
Germany
schue@aca-berlin.de

Ferdi Schüth
Max-Planck-Institut für Kohlenforschung
Kaiser-Wilhelm-Platz
45470 Mülheim
Germany
schueth@mpi-muelheim.mpg.de

Günter Schulz-Eckloff
Institute of Applied and Physical Chemistry
University of Bremen
PF 330 440
28334 Bremen
Germany
jse@uni-bremen.de

Christian Seebacher
Department of Chemistry and Center of Nanoscience
Ludwig-Maximilians-Universität München
Butenandtstraße 11
81377 Munich
Germany

Dorota Sendor
Institut for Anorg. Chemistry

RWTH Aachen
ProfessorPirlet-Straße 1
52074 Aachen
dorota.sendor@ac.rwth-aachen.de

Ulrich Simon
Institut für Anorganische Chemie
RWTH Aachen
Professor-Pirlet-Straße 1
52064 Aachen
Germany
ulrich.simon@ac.rwth-aachen.de

Stefan Spange
Polymer Chemistry
Department of Chemistry
Faculty of Natural Science
Chemnitz University of Technology
Straße der Nationen 62
09111 Chemnitz
Germany
stefan.spange@chemie.tu-chemnitz.de

Frank Starrost
Institut für Theoretische Physik und Astrophysik
Christian-Albrechts-Universität Kiel
Leibnizstraße 15
24118 Kiel
Germany
Starrost@tp.cau.de

Sabine Stockhause
Institute of Physics
University of Potsdam
Am Neuen palais 10
14469 Potsdam
Germany
stockhau@rz.uni-potsdam.de

Michael Tiemann
Institute of Inorganic and Analytical Chemistry
Justus-Liebig University, Gießen
Heinrich-Buff-Ring 58
35392 Gießen
Germany
michael.tiemann@anorg.chemie.uni-giessen.de

U. Vietze
Darmstadt University of Technology
Petersenstraße 23
64287 Darmstadt
Germany

Michael Wark
Institute of Physical Chemistry and Electrochemistry
University of Hannover
Callinstr. 3-3A
30167 Hannover
Germany
Michael.Wark@pci.uni-hannover

Kornelia Weh
Institute for Applied Chemistry
Berlin-Adlershof e.V.
Richard-Willstätter-Straße 12
12489 Berlin
Germany
weh@ACA-Berlin.de

Özlem Weiß
Hämeenkatu 30 E39
20700 Turku
Finnland
oweiss@abo.fi
ab 1. April
Kalkofenstraße 26
66125 Saarbrücken
fuddah@imail.de

J. Wiersig
Max-Planck-Institut für Pyysik komplexer Systeme
D-01187 Dresden
Germany

Dieter Wöhrle
Institute of Organic and Macromolecular Chemistry
University of Bremen
PO Box 330 440
28334 Bremen
Germany
woehrle@chemie.uni-bremen.de

Gabriela Zadrozna
Department of Chemical Technology
University of Technology of Munich
Lichtenbergstraße 43
85747 Garching
Germany
Gabriela.Zadrozna@ch.tum.de

Part 1
Synthesis Routes for Functional Composites Based on Nanoporous Materials

Synthesis Routes for Functional Composites Based on Nanoporous Materials

Michael Wark

Molecular engineering is reaching highly elaborate levels of sophistication. The analysis of the cooperative behavior of single molecules or clusters of molecules within controlled spatial assemblies is a field undergoing continuous progress. The most common inorganic matrices for the construction of inorganic/inorganic or inorganic/organic host–guest composites are zeolites, aluminum phosphates, and mesoporous silicates or aluminum silicates. An overview of their synthesis procedures was recently published by van Bekkum, Flanigan, Jacobs, and Jansen [1]. Over the past 20 years, there has been a dramatic increase in the literature of design, synthesis, characterization, and property evaluation of zeolites and molecular-sieve based composites for catalysis and optical applications. In addition to metal and metal oxide clusters embedded in the regular pore systems of the host materials, the encapsulation of organic dye molecules and metal organic compounds has gained particular attention. A summary of novel composite materials based on zeolites and related structures, including pigments, phosphors, optical hole burning materials, nonlinear optical materials, quantum size effect materials, molecular wires, membranes, and sensors, is given by Behrens and Stucky [2].

Reviews summarizing the synthesis procedures leading to the formation of metal clusters or metal nanoparticles in the pore systems have been written by Kawi and Gates [3] and by Schulz-Ekloff [4]. Principles important for the introduction of metal oxide or metal sulfide clusters were reviewed by Weitkamp et al. [5].

Bioinorganic chemistry is profiting from a more and more developed design of molecular systems and nanoscale mechanisms. For example, bio-inorganic structural motifs can potentially model metalloenzyme structures and functions in terms of steric effects imposed by the inorganic edifice. One aim of such model systems is the mimicking of enzymatic systems. Overviews regarding synthesis routes and properties of zeolite-based supramolecular assemblies of metal organic compounds, such as salens or phthalocyanines, are given by De Vos and Jacobs [6], or very recently by Wark [7]. The preparation and the optical properties of all kinds of chromophores in zeolites, porous silica, and are described by Schulz-Ekloff et al. [8].

The chapters in this section highlight some recent and detailed developments in the synthesis and construction of host–guest composites with novel optical properties and high potential for applications such as miniaturized optical switches, optical gas sensors, or highly effective light emitters.

The first four chapters concentrate on organic dye molecules as guests, mainly on microporous zeolites or aluminophosphates as matrices providing pores with diameters less than 2 nm. In the subsequent chapters mesoporous materials with channel diameters between 2 and 10 nm are mainly used. The synthesis of these hosts is based on long-chain alkyl amine surfactants [9], block copolymers [10], or even expanded block-copolymers [11] as structure-directing agents. Recently, polymer-templated ordered silicas with cage-like mesostructure have been developed [12].

In the first chapter (Chapter 1.1) Behrens et al. present methods for the preparation of functional composites based on zeotypes. They incorporated different chromophors. As synthesis routes they used either an unspecific co-occlusion, where the guest species is just added to the zeolite synthesis gel containing an additional structure-directing agent (SDA), or a direct method, in which the modified functional guest species directly acts as SDA. The incorporated functional units obtained are arranged and protected by the inorganic framework leading to altered optical properties. These first examples concentrated on rather stable guest molecules, however, the development of milder synthesis methods, to introduce species with new magnetic properties for example, seems to be imminent.

A real "ship-in-the-bottle" synthesis of organic dyes in the cages of faujasite-type zeolites was carried out by Wöhrle et al. (Chapter 1.2). The developed methods use the fixation of a first educt with the host by acid–base interactions. Then the synthesis of the chromophore is achieved by reaction of the second educt, also introduced into the pores. The obtained loadings were as high as 10^{-4} mol dye per gram zeolite. The host–guest interactions were studied for the encapsulated photochromic spiropyran as an example. Compared with organic polymer hosts in the matrix of a dealuminated zeolite Y, a dramatically improved stability of the switched state against thermal relaxation and an extreme high stability during photoinduced switching were found.

Ganschow et al. (Chapter 1.3) established a one-step procedure for the covalent anchorage of dyes at the pore walls of the mesoporous Si-MCM-41 and they achieved the stable crystallization inclusion of highly fluorescing dye molecules during the synthesis of microporous $AlPO_4$-5 by using microwave radiation. It turned out that during the rapid microwave-assisted crystallization, a preferential accommodation of smaller chromophores takes place. Larger dye molecules enter later. Such accommodation enables directed energy transfer between the hosted dye molecules. The dye accommodation in porous minerals can be analyzed by bifocal microscopy (Chapter 4.3 by Seebacher et al.). In order to obtain optimized crystal geometries for micro-lasing (Chapter 4.6 by Benmohammadi et al.) the synthesis conditions were varied so that $AlPO_4$-5 crystals with low length-to-width aspect rations were formed.

The chapter of Kornatowski and Zadrozna (Chapter 1.4) deals also with the control of the crystal morphology of the $AlPO_4$-5 molecular sieve and its derivatives. Their growth can be controlled to a high extent and extremely flat crystals with length-to-width aspect ratios reduced to about 0.1 and the crystal width enlarged to about 120 μm were obtained for the first time for CrAPO-5. The crystal length is reduced owing to the adsorption of organic and inorganic additional components/co-templates on the growing crystals.

Nanoporous crystals can also be used for the confinement of liquid crystals. This is demonstrated by Frunza et al. (Chapter 1.5) who studied the influence of the molecular sieve pore/cavity system on the phase transition characteristic and the host–guest interactions that stabilize the cyanobiphenyl liquid crystal molecules inside the pores. It has been found that size as well as shape and interconnectivity of the pores play an important role for the modification of properties of liquid crystals. Phase transitions characteristic of liquid crystals were only observed if the nanoporous hosts provide interconnected pores larger than 3 nm as they exist in extra large pore SBA-15 material.

Hybrid materials with adjustable content and molecular weight of the loaded organic polymer fraction can be synthesized by cationic host–guest polymerization of vinyl ether monomers within MCM-41 materials. The synthesis routes to reach this goal are discussed by Spange et al. in Chapter 1.6. The structures of the polymer chains in MCM-41 are identical to the pure, bulk polymers, whereas the glass-transition temperature is significantly different from those of the bulk fraction. The given synthesis procedures are suitable for producing flexible polymer chains within pores of inorganic materials to study their dynamics in confined geometry (compared to chapter 3.2 by Kremer et al.).

The next chapter by Behrens et al. (Chapter 1.7) report that it is possible to obtain functional mesostructured organic/inorganic hybrid materials directly by a self-assembly process in which the functional organic molecules act themselves as amphiphilic SDAs in a synthesis approach analogous to the preparation of M41S mesophases. Special structure-directing effects that cannot be observed with non-functional amphipihiles become apparent: aggregation tendencies between the functional amphiphiles can lead to a clear preference for only one type of mesostructure and the possibility of forming aggregates of different type can give rise to different mesostructures for different surfactants with similar lengths. The aggregation phenomena are influenced by interactions between the aromatic systems of the chromophore amphipihiles.

Besides organic dye molecules, various inorganic guest species also can be arranged and stabilized by encapsulation in nanoporous materials. The next two chapters give some examples of the development of composite materials with prospective new physical and especially optical properties.

In Chapter 1.8. Wark et al. discuss the arrangement of metal oxide species in the pores of molecular sieves either in mononuclear dispersion or as clusters or nanoparticles. The encapsulation was predominately achieved by post-synthetic treatment using chemical vapor deposition (CVD), ion exchange, and impregnation. The stabilized differently sized metal oxide species differ drastically in their behavior against reductive gases. The composites can be used for a sensing of gases based on optical detection. The optical changes are correlated to the number of oxygen vacancies formed in the clusters or nanoparticles. By use of Ti^{IV} oxide/molecular sieve and Sn^{IV} oxide/molecular sieve composites concentrations of H_2 and CO in air down to 10 ppm as well as changes in the ratio of CO/air mixtures could easily be monitored with very fast response times.

Beneke et al. (Chapter 1.9) describe a route to the formation of stable sub-nanometer platinum clusters within the cages of supporting zeolites. The sub-nanometer platinum clusters formed via direct carbonylation of $[Pt(NH_3)_4]^{2+}$ exchanged zeolites and decomposition in oxygen or vacuum correspond in size to the skeleton of a platinum carbonyl precursor complex. This could be inferred from the observation of a size quantization effect and from the rapid and almost quantitative recarbonylation of the cluster to the initial carbonyl complex. The metal clusters obtained after vacuum decomposition show a surprisingly high thermostability. These stable noble metal clusters of uniform subnanometer size appear to be very promising for the development of new devices with prospective electronic and catalytic behavior.

Mesoporous metal oxides as powders [13,14] or thin films [15], periodic mesoporous organosilicas [16], and mesostructured aluminum phosphates are attracting more and more attention as host materials.

In Chapter 1.10 Tiemann and Fröba report some new nonaqueous synthesis routes to prepare mesoporous aluminum phosphates. With *n*-dodecyl phosphate as a structure director, a composite with an inverted hexagonal structure with strict 1:1 molar ratio of Al and P is obtained. The utilization of primary alkyl amines leads to materials with randomly ordered tubular mesopores.

Stockhause et al. (Chapter 1.11) use bisphosphonic acids to form functional multi-layers by self-assembly. For this a chemical reaction between the bisphosphonic acid and a transition metal is necessary. For application in electronic devices the bisphosphonic acid layers can be anchored on conducting substrates such as indium–tin oxide (ITO). Within the obtained film organic moieties can be inclined forming domains with different directions. Incorporating zirconium bisphosphonate films in LED structures with aluminum as top electrode leads to devices emitting in the blue region of the spectrum.

A further trend in the development of supramolecularly assembled materials with ordered porous structure focuses on the use of metal/organic building blocks. For example, the formation of a zeolite-like structure consisting of porphyrin building blocks has been reported [17]. Also carboxylates and bis-pyridyls were used as organic linkers to obtain highly porous nanostructured materials [18]. The chapter by Fischer et al. (Chapter 1.12) fits into this research topic. Syntheses of Prussian-blue-derived organometallic coordination polymers with nanometer-sized cavities are reported. The structural properties, the crosslinking, and the resulting porous structures of different guest-free and guest-containing super Prussian-blue derivatives are discussed. Controlled thermolysis of numerous nanostructured Prussian-blue assemblies under oxidative and reductive conditions has turned out to afford amorphous and crystalline, oxidic, or intermetallic phases of promising interest for applications such as heterogeneous catalysts.

References

1 H. Van Bekkum, E.M. Flanigan, P.A. Jacobs, J.C. Jansen (eds.), Studies in Surface Science and Catalysis, Vol. 137: Introduction to Zeolite Science and Practice, Elsevier, Amsterdam **2001**.

2 P. Behrens, G.D. Stucky, in Comprehensive Supramolecular Chemistry, Vol. 7: Solid-State Supramolecular Chemistry: Two- and Three Dimensional Inorganic Networks, G. Alberti, T. Bein (eds.), Pergamon, Oxford **1996**, p. 721.

3 S. Kawi, B.C. Gates, in Clusters and Colloids, G. Schmid (ed.), VCH, Weinheim **1994**, p. 299.

4 G. Schulz-Ekloff, in Comprehensive Supramolecular Chemistry, Vol. 7: Solid-State Supramolecular Chemistry: Two- and Three Dimensional Inorganic Networks, G. Alberti, T. Bein (eds.), Pergamon, Oxford **1996**, p. 549.

5 J. Weitkamp, U. Rymsa, M. Wark, G. Schulz-Ekloff, in Molecular Sieves–Science and Technology, Vol. 3: Modification, H.G. Karge, J. Weitkamp (eds.), Springer, Berlin **2002**, p. 339.

6 D.E. De Vos, P.A. Jacobs, in Studies in Surface Science and Catalysis, Vol. 137: Introduction to Zeolite Science and Practice, H. Van Bekkum, E.M Flanigan, P.A. Jacobs, J.C. Jansen (eds.), Elsevier, Amsterdam **2001**, p. 957.

7 M. Wark, in The Porphyrin Handbook, Vol. 17: Phthalocyanines: Properties and Materials, K. Kadish, K.M. Smith, R. Guilard (eds.), Academic Press, St. Louis **2003**, p. 247.

8 G. Schulz-Ekloff, D. Wöhrle, B. van Duffel, R.A. Schoonheydt, *Microp. Mesop. Mater.* **2002**, *51*, 91.

9 J.S. Beck, J.C. Vartuli, W.J. Roth, M.E. Leonowicz, C.T. Kresge, K.D. Schmitt, C.T.W. Chu, D.H. Olson, E.W. Sheppard, S.B. McCullen, J.B. Higgins, J.L. Schlenker, *J. Am. Chem. Soc.* **1992**, *114*, 10 835.

10 D. Zhao, J. Feng, Q. Huo, N. Melosh, G.H. Fredricksson, B.F. Chmelka, G.D. Stucky, *Science* **1998**, *279*, 548.

11 J.H. Sun, J.A. Moullin, J.C. Jansen, T. Maschmeyer, M.O. Coppens, *Adv. Mater.* **2001**, *13*, 327.

12 J.R. Matos, L.P. Mercuri, M. Kruk, M. Jaroniec, *Langmuir* **2002**, *18*, 884.

13 U. Ciesla, F. Schüth, *Microp. Mesop. Mater.* **1999**, *27*, 131.

14 D.M. Antonielli, *Angew. Chem. Int. Ed.* **2002**, *41*, 214.

15 G.A. Ozin, *Chem. Comm.* **2000**, 419.

16 T. Asefa, M. Kruk, M.J. MacLachlan, N. Coombs, H. Grondey, M. Jaroniec, G.A. Ozin, *J. Am. Chem. Soc.* **2001**, *123*, 8520.

17 K.J. Lin, *Angew. Chem. Int. Ed.* **1999**, *38*, 2730.

18 H. Li, M. Eddaoudi, M.O'Keeffe, O.M. Yaghi, *Nature* **1999**, *402*, 276.

1
Guest Functionalized Crystalline Organic/ Inorganic Nanohybrid Materials

Peter Behrens, Christian Panz, Clemens Kühn, Bernd M. Pillep, and Andreas M. Schneider*

1.1
Introduction

Zeolites and related compounds (zeotypes) can act as organizing and protecting media for organic molecules and metal complexes introduced into their voids [1–3]. The resulting substances can possess interesting properties if the guest molecules carry a specific function. Apart from catalytic reactivity, such functions can for instance include that of a chromophore, a luminophore, or a magnetic moment. The specific properties of zeotype frameworks and the strict spatial organization they impose on the arrangement of the guest species can lead to interesting material properties and possible applications. Examples of this novel class of nanostructured materials include the insertion of *p*-nitroaniline molecules into the linear channels of $AlPO_4$-5 yielding an efficient material for second harmonic generation (SHG) [4–6], the formation of a nonasil composite containing an organometallic complex that exhibits electric-field induced second harmonic generation (EFISH) [7], the inclusion of laser dyes into $AlPO_4$-5 crystals leading to micrometer-sized lasing crystals [8–11], the construction of a light-harvesting complex in zeolite L in an attempt to mimic photosynthetic processes [12,13], and the incorporation of switchable organic molecules into zeotypes that can control diffusion within the pore system [14–16]. It is remarkable that these examples of novel zeotype-based materials rely mainly on optical functionalities. This is because zeotype frameworks are especially suited for such functionalities, as they usually possess high optical transparency extending into the UV region. Apart from the more sophisticated applications mentioned above, optically transparent zeotype frameworks can also be used to construct pigments [17,18] by loading organic dyes into the porous hosts, thus rendering them insoluble and protecting them against photochemical or photophysical damage. The protecting influence of zeotype frameworks on their guest species against photochemical [18,19] and thermal attack [20] has been studied in some detail. Recently, an overview about chromophores in porous silicas and zeotypes has been published [21].

Before the exciting properties of chromophore–zeotype composites can be studied and possibly exploited, such materials have to be synthesized. The synthesis of zeotypes generally follows the recipe of structure-directed synthesis [22–24] in which organic molecules (or organometallic complexes) are added to the synthesis gel as structure-directing agents (SDAs): They become incorporated into the growing crystals and thus influence the structure of the inorganic framework. So, there is at least a basic compatibility of the synthesis system with organic molecules, although the SDAs normally do not contain any specific functions.

There are several methods for constructing functionalized guest–host assemblies based on zeotypes (Fig. 1) [21].

- The microporous inorganic framework can be synthesized according to the general principles of structure-directed synthesis; the SDA molecules are then removed, typically by calcination. The now empty pores of the host can then be loaded either from the vapor phase or from solution (Fig. 1a). These processes are designated as insertion (for neutral molecules) or ion-exchange (for cationic molecules). High and homogeneous loadings can be achieved and, interestingly, the insertion process itself can induce the formation of ordered arrangements of the functional molecules, leading, for example, to well-ordered dipole chains of *para*-nitroaniline [6]. As in molecular-sieving applications and in shape-selective catalysis, the size of the pores determines which molecules can be sorbed, and, as a caveat to this method, desorption is often as easy as loading.
- Precursors of a functional molecule can be sorbed into the empty zeotype framework, which are then induced to form a larger entity within a pore. As an advantage of this method, the newly formed molecule is typically larger than the surrounding pore windows and cannot escape anymore from the zeotype framework. Therefore, this method is called "ship-in-the-bottle" synthesis (Fig. 1b). It is, however, an expeditious multi-step technique. High and homogeneous loadings are often difficult to achieve, which can be a disadvantage in certain applications, but lower loadings can also be preferred in some cases, such as catalysis.
- The functional guest molecules can also be introduced into the host zeotype during its formation. As was stated above, there is a general compatibility of the synthesis systems used for structure-directed synthesis with molecular species. For this purpose, the chemical properties of the functional molecules (for example their solubilities) have to be adapted to the synthesis system and they have to be stable enough to withstand the synthesis procedure, an important and not easily fulfilled condition, as will be detailed below. Two variants of this occlusion procedure are known:
  In the unspecific co-inclusion method the functional guest molecule is added to a typical zeotype synthesis gel that contains among the other necessary ingredients also an SDA controlling the formation of a specific structure type. The SDA as well as the functional molecule then become occluded within the pores of the zeotype host (Fig. 1c). Owing to the necessary presence of at least some SDA molecules within the pores, no full loading can be achieved in this way. However, this method even offers the possibility of introducing guest molecules into zeotypes that are larger than the pores generated by the presence of the SDA; in

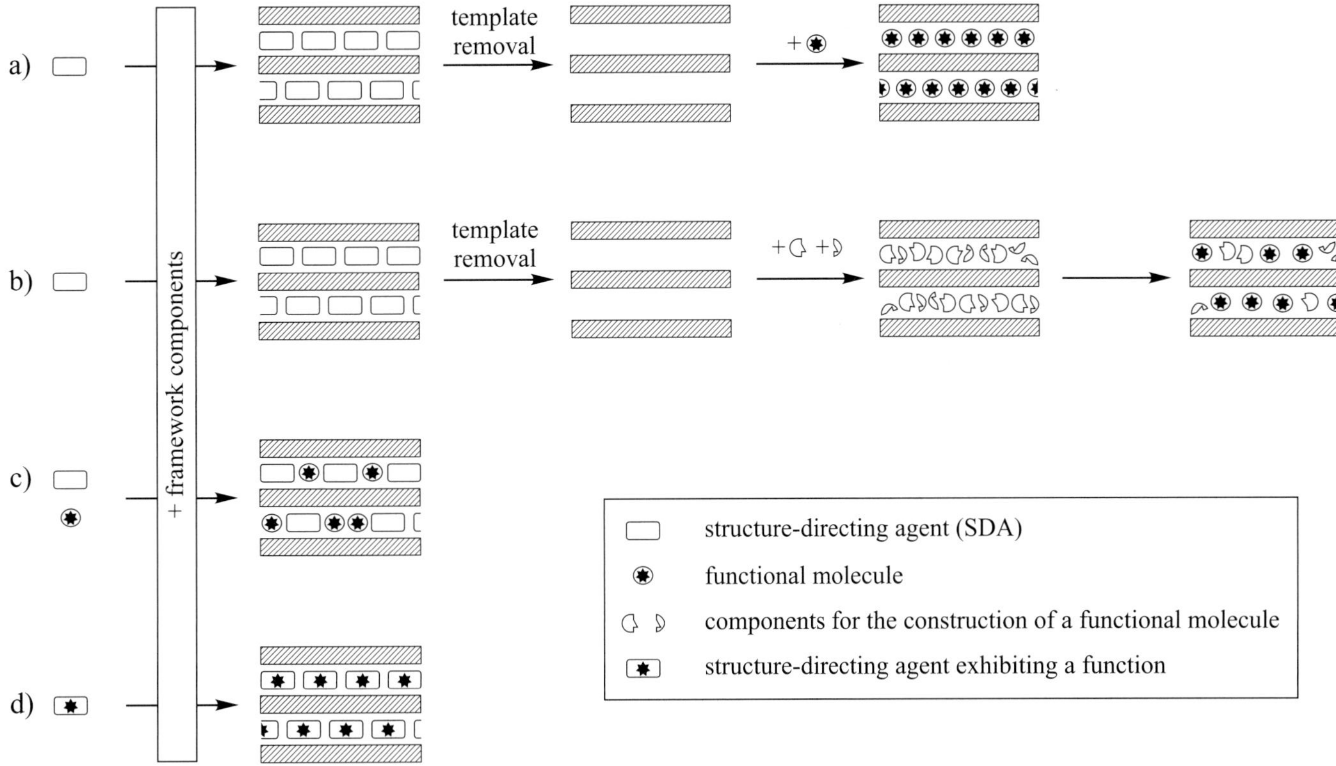

Fig. 1. Synthesis pathways for the construction of functionalized guest–host assemblies based on zeotypes: (a) standard synthesis of zeotype followed by removal of the SDA and subsequent loading of the functional species; (b) "ship-in-the-bottle" synthesis of functional molecules inside the pores; (c) occlusion of functional molecules during synthesis; (d) direct synthesis using functionalized SDAs.

such cases, the formation of the inorganic framework is locally hindered and the functional molecule resides in an enlarged defect pore that it has created during its occlusion.

In the other variant of the occlusion method, the functional molecule itself acts an SDA. This direct synthesis of functional organic/inorganic host–guest systems (Fig. 1d) puts several high demands on the compatibility and stability of the molecule: the molecule must be equipped to function as an SDA and it must contain a functionality. When these requirements are fulfilled, highly ordered composites with optimum loading can be produced in a one-step direct synthesis [25].

1.2 Direct Construction of Functional Host–Guest Compounds: Synthesis Between Scylla and Charybdis

As discussed above, the preparation of functionalized zeotypes puts strict requirements on the organic functional molecules: they must withstand the harsh conditions of zeotype synthesis, and in the last example, they also have to act as an SDA. A way to make these requirements less strict is of course to soften the reaction conditions, for example by lowering the synthesis temperatures, decreasing the reaction times, or switching to more moderate pH values. Then, on the other hand, elaborated synthesis procedures might not work anymore, and navigating between the stability of functional organic molecules and less severe reaction conditions becomes similar to the attempt to cross the famous narrow path between Scylla and Charybdis [26].

This chapter is organized into three main sections. When no special allowances are made with regard to the stability of the SDA, that is, when the synthesis system is not especially adapted with regard to, for example, lower temperatures or shorter reaction times, then only very stable functional molecules can be used as functional SDAs. An example is the use of organometallic cations in the synthesis of porosils, which is described in Section 1.3. A special synthesis for the aluminophosphate $AlPO_4$-5 was developed in order to reduce the synthesis time and the amount of water present in the synthesis. This method, described in Section 1.4, allows the introduction of sensitive organic dye molecules into this host. Finally, in Section 1.5, we switch to easily crystallizing zincophosphates. In these syntheses, cobalt-amine complexes act as SDAs.

1.3 Stable Functional Structure-Directing Agents in the Synthesis of Porosils

Porosils are microporous compounds with a pure silica framework. They can be subdivided into clathrasils with cage-like voids and zeosils with channel-like voids [27]. The typical conditions for the synthesis of porosils are among the most severe in the preparation of zeotypes, typically involving long synthesis times (weeks to months) and high temperatures (160–200 °C) [28–30]. Therefore, SDAs for the synthesis of porosils must be very stable. Typically, aliphatic amines are used, but few molecules that carry a functionality are stable enough to withstand such synthesis conditions. It was shown that the most stable organometallic complexes, which are colored and thus carry the functionality of a chromophore, can act as effective SDAs for the synthesis of porosils [20,25,31–47].

Figure 2 summarizes the results of successful syntheses of porosils using organometallic SDAs. The syntheses can be carried out in a basic solution or, with fluoride as a mineralizer, in a neutral or weakly acidic medium. Some of the preparation procedures require comments.

Owing to its framework topology, which features fourteen-membered rings and thus the largest pore size available among the zeolites and porosils [37], UTD-1 is probably the most famous of microporous material synthesized using an organometallic SDA. According to our experiences [46,47], in the hydroxide system UTD-1 can be synthesized only starting from a solution of $[Co(cp^*)_2]^+OH^-$ [32], but not from the chloride or the hexafluorophosphate salt of the SDA. This is one of the cases in which (somewhat unexpectedly) the anion of a cationic SDA influences the structure formation. The synthesis in the hydroxide system yields microcrystalline powders of several polymorphs of UTD-1. In contrast, in the fluoride system, one of the polymorphs (framework type DON) of UTD-1 is formed selectively as needle-like crystals [45]. On a textured powder sample of this compound, the crystal structure of UTD-1 was determined based on X-ray diffraction data [44,45]. We were recently able to confirm this structural analysis on the basis of single-crystal X-ray data [46,47]. There is an interesting difference in the behavior of UTD-1 samples synthesized by the hydroxide or the fluoride route during template removal that is performed by calcination and subsequent washing with hydrochloric acid. Whereas the hydroxide-derived UTD-1 samples yield a porous solid by this procedure [32,38,46–48], UTD-1 samples prepared in the fluoride system are nonporous and do not even allow the insertion of iodine molecules. The reasons for this diverging behavior have recently been elucidated by X-ray absorption spectroscopic investigations of the calcined samples [49].

The 1,1′-dimethylcobalticinium cation and the benzol-cyclopentadienyliron cation form the clathrasil dodecasil 1H only under special conditions. The fluoride-based dry-gel synthesis method [30,50–52], although not really a nonaqueous technique, allows the preparation of microporous solids using only minimum amounts of water (which is introduced by water-containing silica sources and released from the reaction $SiO_2 + 4\,NH_4F \rightarrow SiF_4 + 2\,H_2O$). The decreased water content appears to be essential for the iron complex, which is destroyed in conventional water-rich synthesis attempts [34]. But even when the dry-gel method is used, the synthesis is not easily reproduced (large autoclave volumes appear to be of advantage), and, whereas benzol-cyclopentadienyliron-DOH is the only compound that gives reflections in the powder X-ray diffractograms obtained on successful synthesis attempts, Mössbauer spectroscopy shows that the synthesis product consists of more than one iron-containing species. The 1,1′-dimethylco-

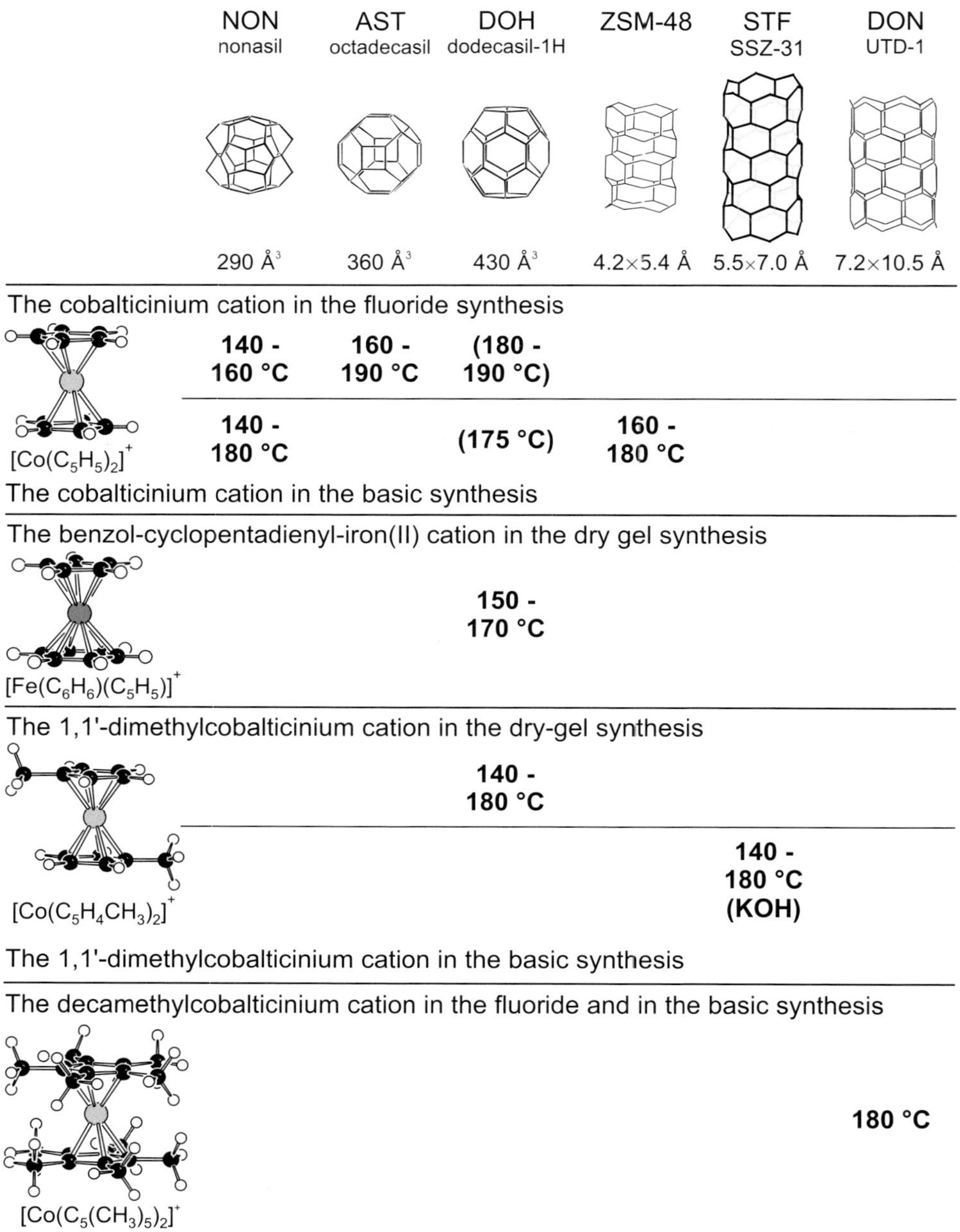

Fig. 2. Overview over synthesis of porosils with organometallic SDAs. The syntheses can be carried out in a basic medium or with fluoride as a mineralizer. The temperature regions in which the corresponding compound forms is indicated.

balticinium complex is stable under normal aqueous synthesis conditions, but it does not act as an SDA and does not generate a porosil. The fact that it does form a DOH compound from a dry gel was ascribed to the strongly increased concentration of this SDA under these conditions [34].

Using the fluoride synthesis system, the unsubstituted cobalticinium cation can form the NON framework (at low synthesis temperatures: 150–170 °C) and the AST framework (at higher temperatures: 170–190 °C). At the higher temperatures, a DOH compound is formed as a by-product with a fraction of about 5 % of the yield [20]. This sequence of framework types formed with increasing temperature is typical and corresponds to the increasing volumes of the main clathrasil cages. This possibly reflects the increasing space requirements of the SDA with increasing thermal motion [20,29]. By another variation of the synthesis method, namely the application of a high pressure (about 400 bar) of a noble gas (Ar, Kr, Xe) during the hydrothermal synthesis, it is possible to enforce the formation of a pure DOH clathrasil, irrespective of the synthesis temperature [53–55]. This can be rationalized as follows. The noble gas has a strong tendency to become occluded within the crystalline compound that forms during a porosil synthesis, and, in fact, as shown by a crystal structure analysis on $|[Co(cp)_2]^+F^-Ar|$-DOH, occupies the smaller $[5^{12}]$ and $[4^3 5^6 6^3]$ cages of the DOH structure (the large $[5^{12}6^8]$ cage contains the cobalticinium cation). The NON and the AST framework are not formed under these conditions, as the smaller cages of these frameworks are not large enough to host noble gas atoms. The presence of a high pressure of a noble gas usually improves the quality and the size of the crystals produced [53]. Similar co-structure-directing effects for so-called "help gases", when applied at high pressures, were reported before [30].

Owing to the fact that the organometallic complexes are colored and that for these sandwich complexes the color-giving electronic transitions are polarized along their principal axes, first insights on the structure of the porosils, namely on the arrangement of the SDAs, can already be obtained by simple polarization microscopy. Figure 3 shows corresponding photographs of some of the compounds listed in Fig. 2. These show that the metal complexes are aligned in the $|[Co(cp)_2]^+F^-|$-NON and the $|[Co(cp)_2]^+F^-|$-DOH crystals. Their respective orientations are in agreement with the results from X-ray structural analyses shown below. On the other hand, for $|[Co(cp)_2]^+F^-|$-AST, such a preferred orientation is not obvious, possibly due to rotational disorder of the cobalticinium cation within the nearly spherical large $[4^6 6^{12}]$ cage of the pseudo-cubic AST framework.

In principle, orientation-dependent absorption behavior can make some of these compounds useful as polarizers. In any case, these results show the strong organizing power of porosil frameworks that possess a clearly distinct principal axis. The dichroitic absorption behavior can be quantified by UV/vis spectroscopy and similar orientation-dependent absorption behavior was also be detected by IR spectroscopy [36].

X-ray structural analyses on some of the organometal-porosil nano-hybrids yield further insight into the properties of these compounds. In the $|[Co(cp)_2]^+F^-|$-NON (Fig. 4), the cobalticinium cation is fixed and does not exhibit orientational

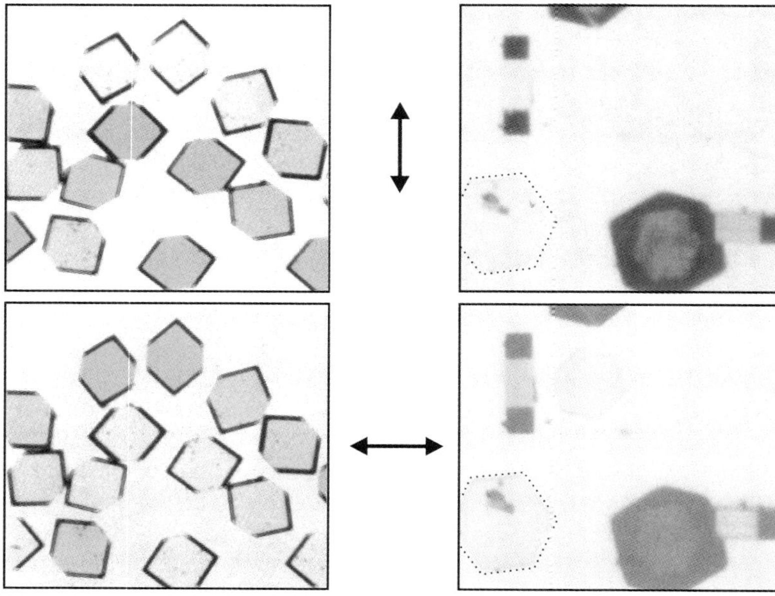

Fig. 3. Investigation of cobalticinium-containing clathrasils by polarization microscopy. The polarization of the light is indicated by the arrows. Left: $[Co(cp)_2]^+F^-$|-NON crystals that appear yellow or colorless in dependence of the orientation of the crystals with regard to the polarization. Right: crystals of |$[Co(cp)_2]^+F^-$|-DOH and of |$[Co(cp)_2]^+F^-$|-AST. Two DOH crystals (above left and below right) are standing on their prism faces and appear yellow or colorless in dependence of the orientation of the crystals with regard to the polarization. Another DOH crystal is lying on its basal face; it appears colorless under all polarizations and is therefore surrounded by a dotted line. The isometric crystals of AST (below center) are yellow and do not exhibit any dichroism.

disorder within the nonasil cage up to a temperature of 200 °C [20,33]. A single-crystal structural analysis for cobalticinium-containing DOH was only possible for the compound synthesized under a high pressure of argon gas [53,54,56]. The result is in qualitative agreement with the findings from polarization microscopy, but there is strong rotational disorder of the complex within the cage. The structure of |$[Co(mecp)_2]^+F^-$|-DOH (mecp: methylcyclopentadienyl) could only be derived by a combination of structural modeling and the Rietveld refinement of powder X-ray diffraction (PXRD) data [35,43]. Due to the increased size of the organometallic SDA, the complex is in a tilted orientation with regard to the *c* axis of the DOH framework (see Fig. 5a, which corresponds to a snapshot picture and does not show the disorder). Modeling also results in a reasonable model for |$[Fe(bz)(cp)]^+F^-$|-DOH (bz: benzene) [56]. The powder X-ray diffractogram calculated on the basis of this structure is in good agreement with the experimental one (snapshot picture given in Fig. 5b). The principal axis of the iron complex is aligned with that of the $[5^{12}6^8]$ cage.

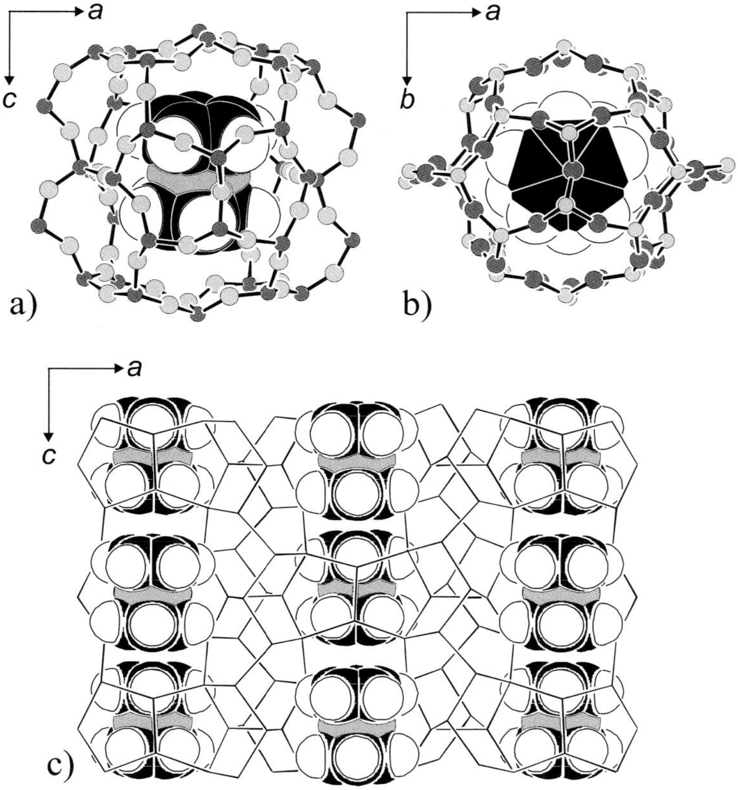

Fig. 4. Crystal structure of $|[Co(cp)_2]^+F^-|$-NON from single-crystal XRD structural analysis. The cobalticinium cation is fixed and does not exhibit orientational or rotational disorder within the nonasil cage up to a temperature of 200 °C [20,33]. (a, b) Different views of the structure; (c) larger excerpt of the structure showing the alignment of the cobalticinium cations (oxygen atoms omitted in c).

As an example of the possible functionality of such composite structures, we carried out an investigation on $|[Co(cp)_2]^+F^-|$-NON with regard to the possible occurrence of an EFISH effect (EFISH: electric-field induced second harmonic generation) [7]. It clearly shows the favorable interplay between a silica host structure and a functional organometallic guest species. The generation of the second harmonic of laser light is a nonlinear optical effect of second order, which can only occur in noncentrosymmetric structures. As $|[Co(cp)_2]^+F^-|$-NON crystallizes in the centrosymmetric space group *Pccn*, an SHG effect cannot be expected. However, it is possible to induce a noncentrosymmetric electron distribution by the application of an electrical field, which can induce the polarization of easily polarizable electrons. This EFISH effect can be considered as a nonlinear optical effect of third order, for which there are no symmetry restrictions. The experimental setup for our study is shown in Fig. 6 (above), together with the results (below), which

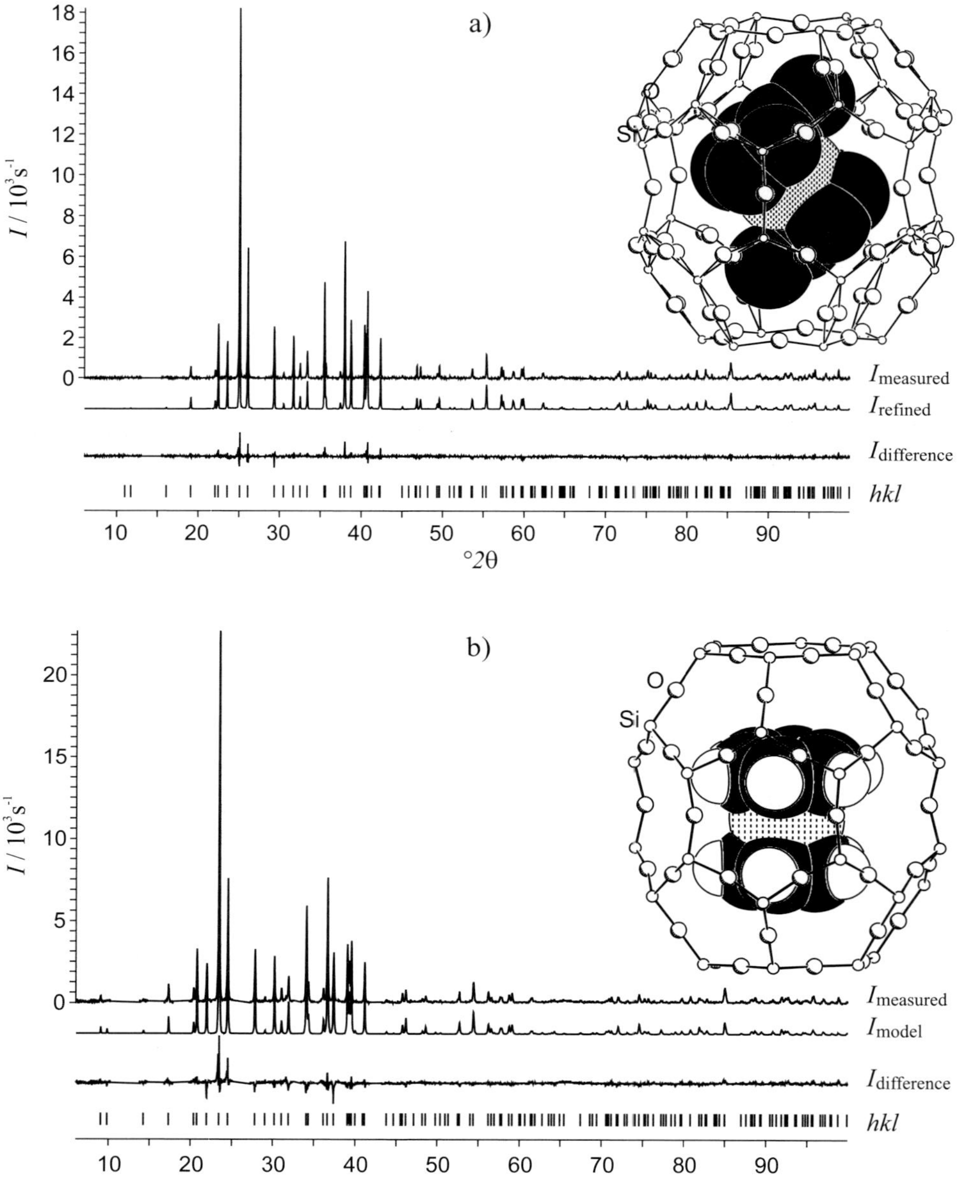

Fig. 5. (a) Crystal structure of $|[Co(mecp)_2]^+ F^-|$-DOH from structural modeling and Rietfeld refinement of powder X-ray diffraction data [35,43]. Only one position of the dimethylcobalticinium cation, which exhibits pronounced rotational disorder, is shown. The positional disorder of some of the oxygen atoms of the framework could be resolved. The Rietfeld plot of the refinement is also shown. (b) Crystal structure of $|[Fe(bz)(cp)]^+F^-|$-DOH from structural modeling and comparison with powder X-ray diffraction data [56]. Only one position of the benzolcyclopentadienyl iron cation, which exhibits pronounced rotational disorder, is shown. A comparison between the experimental powder X-ray diffraction pattern and that calculated based on the modeled structure is also shown.

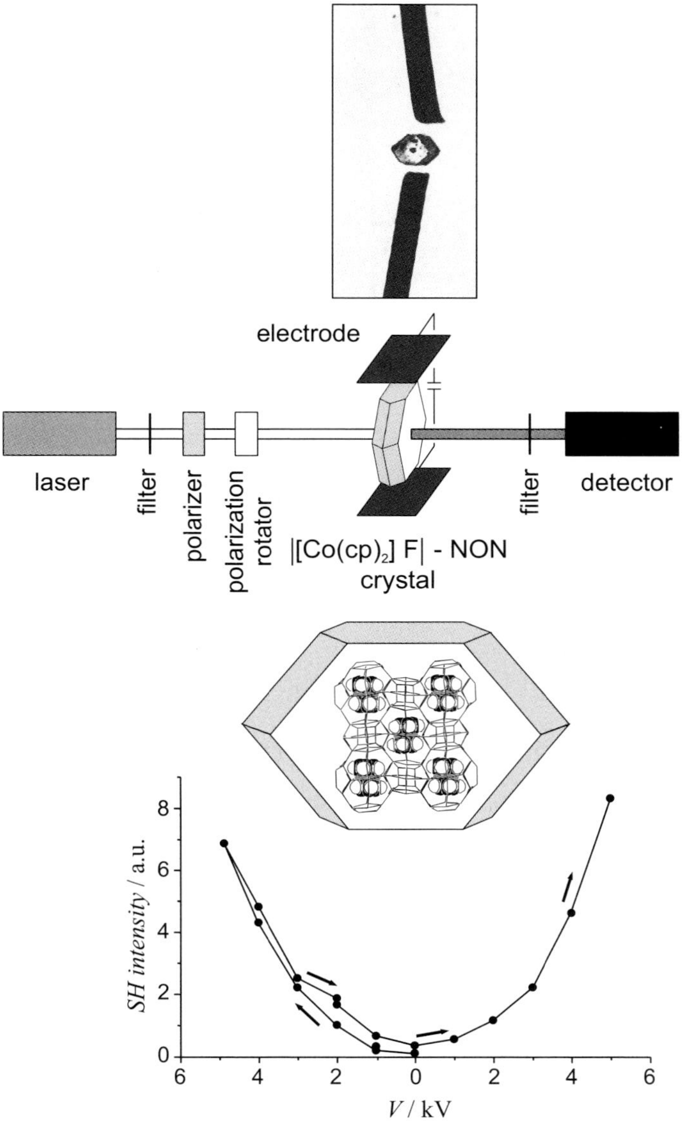

Fig. 6. EFISH effect on $|[Co(cp)_2]^+F^-|$-NON [7]. Center: schematic depiction of the experimental set-up: Infrared laser light (the polarization of which can be rotated) is frequency-doubled by a crystal of $|[Co(cp)_2]^+ F^-|$-NON in the orientation shown. Above: light micrograph of the actual experimental set-up. Below: results of the EFISH experiment. The parabolic dependence of the frequency-doubled light on the applied voltage is expected from theory.

show a parabolic increase of the intensity of the second harmonic light with voltage. This dependence is expected from the theory of third-order nonlinear optical effects. Furthermore, it was found that the intensity of the frequency-doubled light depends upon the angle between the polarization vector of the laser light and the orientation of the crystal [7].

The EFISH effect found on $|[Co(cp)_2]^+F^-|$-NON could make this substance an important material for electro-optical applications, such as for controlling the flow of light by electrical signals. More importantly, it shows the favorable interplay between the properties of the silicon dioxide host framework and of its functional guest molecules. In their cyclopentadienyl units, these molecules possess easily polarizable π electrons, which give rise to the polarization responsible for the EFISH effect. The organizing forces of the framework align these molecules so that the effect is maximized. Owing to the strong bonds within the silicon dioxide host, it is optically transparent and can also serve as a stable dielectric medium: The application of similarly high fields on simple salts of the cobalticinium cation would probably result in electrical discharges. The porosil framework also stabilizes its organometallic guest species. For example, the thermal stability of the cobalticinium cation in air (as deduced from thermogravimetric measurements) increases from 375 °C in the simple hexafluorophosphate salt to about 650 °C in the nonasil compound [20]. $|[Co(cp)_2]^+F^-|$-NON can thus be considered to be the most stable organometallic compound.

1.4
The Glycol Method for the Fast Synthesis of Aluminophosphates and the Occlusion of Organic Dye Molecules

$AlPO_4$-5 with the AFI framework type is a very prominent microporous host material, especially for the construction of advanced zeolite-based materials [3]. For example, $AlPO_4$-5 loaded with *para*-nitroaniline exhibits a strong SHG effect [4–6], $AlPO_4$-5 containing laser dyes that were enclosed during synthesis acts as a novel laser material [8–10] and $AlPO_4$-5 loaded with azobenzene represents an interesting photoresponsive material [14]. Therefore, the synthesis of $AlPO_4$-5, and especially the inclusion of functional organic molecules during the synthesis has been studied extensively, not only with regard to conventional hydrothermal crystallization procedures in standard autoclaves [57–60], but also with respect to milder synthesis conditions. Special attention has been paid to the synthesis of $AlPO_4$-5 using microwaves as a heating source [61–68]. The use of microwaves allows to accelerate the preparation procedure drastically (by a factor of 100) with respect to the conventional technique and so allows the direct inclusion of functional organic molecules, even when they are sensitive to higher temperatures, as for example laser-active dyes [67]. Also, the crystal shape and size can be tailored by adapting the synthesis conditions [68]. On the other hand, a fast synthesis (on the timescale of minutes) can also be achieved using an open system and very high temperatures and heating rates [69].

We have developed a novel synthesis method for the aluminophosphate $AlPO_4$-5 [26,70]. It makes use of ethylene glycol as a solvent with a high boiling point ($T_b = 198$ °C), thus allowing to maintain high reaction temperatures without the need to use closed reaction vessels. In fact, the synthesis is routinely carried out in a simple glass beaker containing boiling ethylene glycol, to which aqueous solutions of the reagents (solution A: containing for example triethylamine as an SDA, water, H_3PO_4, hydrolyzed aluminum triisopropylate, and hydrofluoric acid; solution B: containing the sensitive chromophore molecule) are added. The water is evaporated immediately and nucleation is thus induced instantaneously. The reaction can be terminated as soon as the addition of the reactant solutions is finished. It is also possible to terminate the reaction by quenching (i.e., by pouring the synthesis batch from the open beaker into cold water). Typically, reactions are completed within minutes. This method thus has the advantages of short reaction times and minimum water contents, both of which can serve to prevent the destruction of sensitive organic molecules. In addition, the open synthesis system allows visual control of the reaction and a part of the sample can be removed and investigated, for example, by light microscopy. If appears to be necessary, further ingredients can be added. Both is not given with either the conventional procedures nor the microwave synthesis. In this way for example the destruction of organic chromophores can be detected and the reaction can be stopped, if necessary. Syntheses of zeotypes in nonaqueous solvents, and especially in ethylene glycol, have been described before [30,71–74], but these employed standard autoclave techniques and did not use open systems as is the case here. As is to be expected from a reaction system in which nucleation is triggered in a crude and uncontrolled manner, the crystals obtained from the glycol synthesis are small (5–10 μm), so that for the investigation of certain optical properties, confocal microscopy techniques have to be used [75,76]. However, the size distribution of the crystals is narrow and they possess well-defined morphologies (Fig. 7).

Employing the glycol method, we were able to include various organic dye molecules and inorganic complex molecules into $AlPO_4$-5. As an example, the cationic dye 4-(4-dimethylaminostyryl)-1-methyl-pyridinium was occluded within $AlPO_4$-5 to give a fluorescent solid. Optical investigations using confocal microscopy, which are further described in the contribution of Bräuchle and coworkers in this volume [75], show that the dye molecules are incorporated in an oriented fashion and are distributed homogeneously throughout the crystal.

We also introduced amphiphilic azobenzene molecules (as they were also used as structure-directing agents in the synthesis of mesostructured solids [77], Fig. 7) into $AlPO_4$-5 syntheses via the glycol method and obtained interesting products [26,70,78]. The mass of crystals has a greenish-yellow color and exhibits a strong luminescence under UV light (Fig. 7, center). Investigations using the confocal microscope (carried out by Bräuchle, Deeg, and coworkers at the Ludwig-Maximilians University, Munich), gave interesting results. As can be seen in Fig. 7, the luminescence stems only from the tips of the crystals. By investigating broken crystals, it was shown that this is not a waveguide effect (such crystals fluoresce only at one tip). As can also be seen, the luminescence is orientation-dependent:

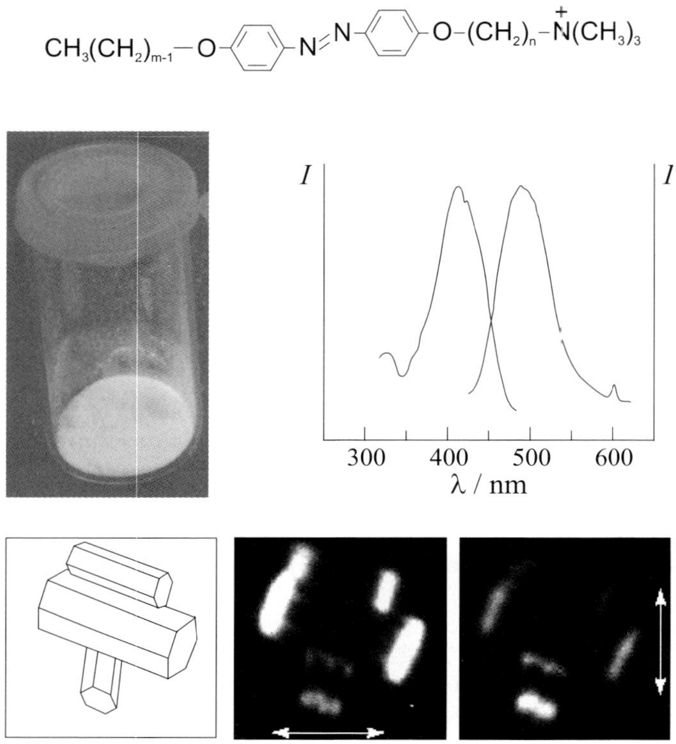

Fig. 7. Luminescence from $AlPO_4$-5 crystals prepared by the glycol method in the presence of surfactants containing azobenzene units (above). Center left: luminescent sample with a greenish-yellow color under UV light. Center right: fluorescence and fluorescence excitation spectra of this sample. Below left: schematic depiction of the $AlPO_4$-5 crystals from which pictures (below center and below right) were obtained by confocal microscopy under polarization with the polarization vectors indicated (these investigations were carried out by Bräuchle, Deeg, and coworkers from the Ludwig-Maximilians University, Munich). The luminescence stems only from the tips of the crystals and is polarization-dependent.

The fact that the luminescence can only be excited when the vector of polarization of the exciting light is perpendicular to the channels of the crystals gives a strong indication that the luminescent species are encapsulated within and aligned along the channels of the $AlPO_4$-5 crystals.

In further investigations it became clear that is was not the intact dye molecules that cause the luminescence, but either a by-product remaining from the synthesis of the azo surfactants or a product of the possible destruction of these molecules, which might have taken place during the $AlPO_4$-5 synthesis. Determining the true identity of the luminescent species is difficult and has not yet been achieved. The luminescent molecules cannot be obtained by dissolution of the $AlPO_4$-5 crystals and subsequent analysis of the formerly enclosed organic material due to the fact that the amount of the luminescent species is very small (only the tips of the crystals exhibit luminescence). Traces of luminescent species were isolated from the synthesis mixtures of the azo surfactants and characterized by luminescence spectroscopy, and there appears to be some resemblance to the luminescence spectra of those $AlPO_4$-5 crystals. A steady decay in luminescence intensity was observed for the species isolated from the synthesis solution, but not for the $AlPO_4$-5 crystals, providing evidence for a protecting action of the surrounding aluminophosphate framework and further proof of encapsulation [78].

The research on these "crystals with glowing tips" continues. The work appears worthwhile in view of the intriguing fact that the luminescent species are incorporated only at the tips of the crystals. The underlying mechanism of formation must be very interesting, not only with regard to the synthesis of luminescent zeolite-based materials, but also in view of the possibility of preparing organic/inorganic composite structures with a spatially modulated contents of organic guest molecules.

The glycol method cannot only be used for the unspecific co-inclusion of functionalized organic guest species, but also allows the use of sensitive SDAs. As an example, we have used *N,N*-dimethylaminoferrocene as an SDA for the synthesis of a mixture of $AlPO_4$-5 and $AlPO_4$-11. Other attempts to use this organometallic complex as SDA, such as in conventional syntheses or in microwave syntheses, had failed. With prolonged reaction times, the complex also deteriorates in the glycol synthesis, but this process can be easily observed and counter-measures can be taken (addition of more complex, early termination of the reaction) [70].

1.5 Easily Crystallizing Inorganic Frameworks: Zincophosphates

Numerous microporous zincophosphates have been synthesized during the last 20 years via the structure-directed synthesis approach [79]. Mostly, organic amino compounds have been used as SDAs, but it was shown that microporous frameworks can also be obtained in purely inorganic compounds [80]. Zincophosphates can be crystallized at very mild conditions. In some cases even reaction temperatures below ambient temperature have been used successfully [80,81], and, more typically, crystallizations are carried out at temperatures below 100 °C. Unlike to the synthesis of porosils and aluminosilicates, an increased pressure is not necessary, and crystallization has even been achieved simply by grinding the reaction mixture [82]. Owing to these exceptionally mild and simple synthesis conditions, zincophosphates have been used as examples for the preparation of crystalline microporous materials under special conditions. For example, microporous zincophosphates were synthesized in reverse micelles [83–86], and small [Zn-P-O]-FAU crystallites were attached to specially prepared surfaces to form thin films [87].

Such a mild and versatile synthesis system of course represents a favorable basis for the preparation of host–guest systems with sensitive functional guest species. For our basic studies in this direction, we chose amine complexes of cobalt(III) $[CoL_6]^{3+}$ as SDAs [70,88–94]. These complexes are among the most stable members of the large class of metal amine coordination compounds, which can possess a variety of interesting properties.

The cobalt(III)-amine complexes we have so far used as SDAs are displayed in Fig. 8 (left). In spite of the extensive work on the structure-directed synthesis of zincophosphates, our synthesis system poses novel unprecedented problems. This is due to the high charge of the SDA entities, which easily form weakly soluble precipitates with a variety of anionic species. Therefore, we have developed a novel synthesis procedure with a specific sequence for the addition of the reagents, including intermittent aging times [70,88]. This procedure is depicted in Fig. 8 (right). Our investigations have shown that it is of prime importance to obtain a well-defined precipitate of the cobalt(III)-amine complex with tetrachlorozincate as an anion. For this purpose, the zinc chloride used must be strictly anhydrous. This necessary requirement appears to be in contradiction to the fact that the synthesis is carried out in an aqueous system anyway. However, $ZnCl_2$ is strongly hygroscopic, and if this substance has been exposed to water, -(OH)- or -O- bridges are formed between the Zn atoms, which cannot be disrupted in the following course of the reaction, so that the crystallization may fail [70,89].

We have synthesized five different zincophosphates using the cobalt(III) amine complexes shown in Fig. 8 [70,88–94]. Their compositions are given in Table 1, and their structures will be described in detail in forthcoming publications. A compound that appears to be similar to LMU-6 has recently been described; however, the authors give their compound another formula, ignoring the water contents and assigning a 2+ charge to the metal complexes, implying a reduction of the cobalt ions [95]. We have found by XANES (X-ray absorption near edge spectroscopy) and by magnetic measurements (indicating diamagnetism) that the complex cations in LMU-6 carry a 3+ charge in the zincophosphate [93].

The most remarkable property of this synthesis system is the one-to-one relationship between the SDA and the zincophosphate structure. Even the two complexes *mer*-$[Co(dien)_2]^{3+}$ and *s-fac*-$[Co(dien)_2]^{3+}$, two isomers that are very similar to one another, induce strikingly different zincophosphate structures. This points to strong host–guest interactions, which also lead to a transfer of at least some of the point symmetry elements of the metal complex to the space group symmetry of the composite (Table 1). For example, the fourfold symmetry axis of the hexamine cobalt(III) complex with its O_h symmetry is transferred to the point and the space symmetry of LMU-4. On the other hand, threefold axes are not transferred from the SDAs to the zincophosphate structures (in LMU-6 and UH-1); possibly, threefold symmetry is not compatible with the zincophosphate structures. Centers of symmetry (in LMU-7 and UH-1) and twofold symmetry axes (in LMU-6) are again transferred.

As the physical properties of crystals depend strongly on their symmetry, with many physical effects arising only in polar or in low-symmetry crystals, the transfer of symmetry elements from an SDA to a composite structure is very interesting. For example, the *mer*-$[Co(dien)_2]^{3+}$ complex does not contain a center of symmetry. Correspondingly, the zincophosphate LMU-5 derived from this SDA crystallizes in a noncentrosymmetric structure with space group *Fdd*2. With this polar struc-

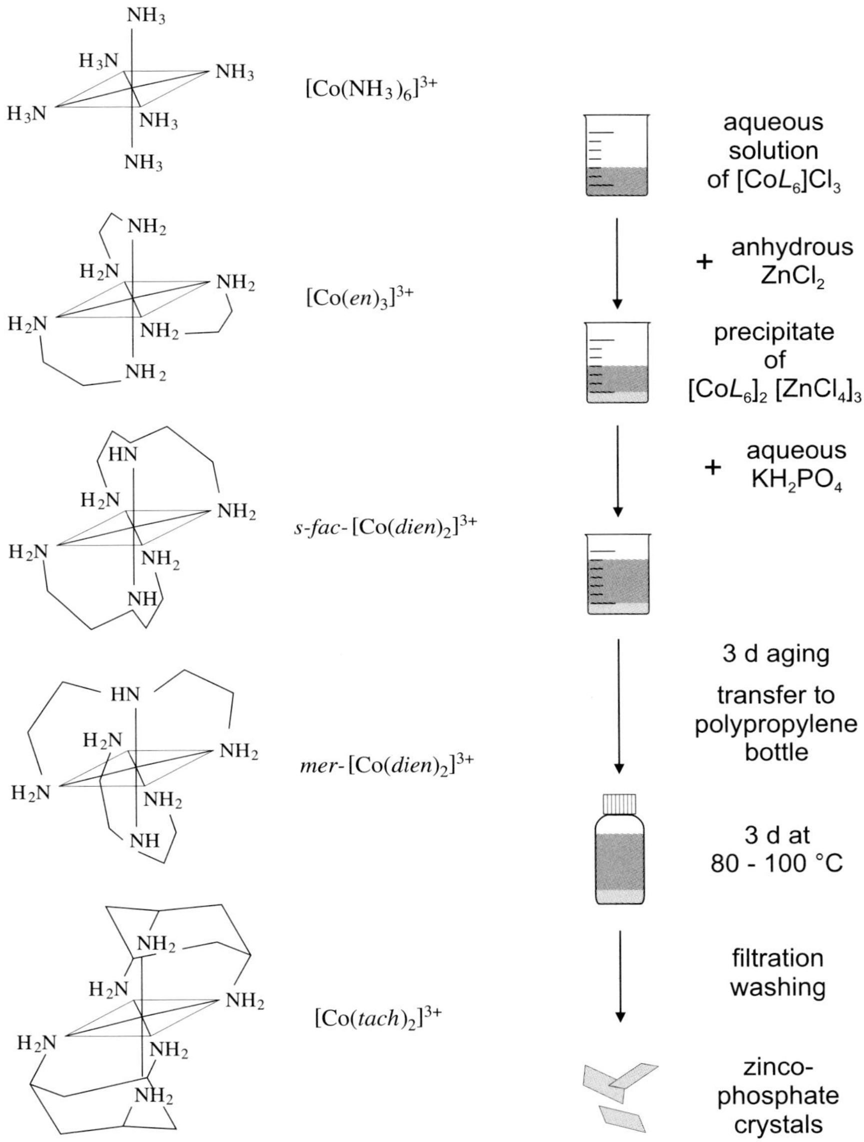

Fig. 8. Left: the cobalt(III)-amine complexes we have so far used as SDAs in the synthesis of zincophosphates. Right: The synthesis procedure that was developed for the preparation of zincophosphates with cobalt(III)-amine complexes as SDAs.

Tab. 1. Designations, formula and symmetry data on zincophosphates prepared using cobalt(III) complexes as SDAs.

Compound	*SDA*	*Formula*	*Point group symmetry of complex*	*Point group symmetry of composite*	*Space group symmetry of composite*
LMU-4	$[Co(NH_3)_6]^{3+}$	$[Co(NH_3)_6]_3\ [H_8Zn_8P_{10}O_{40}]\ PO_4$	$O_h \equiv 4/m\ -3\ 2/m$	$4/m$	$I\ 4/m$
LMU-5	*mer*-$[Co(dien)_2]^{3+}$	$[Co(C_4N_3H_{13})_2]\ [H_5Zn_2P_4O_{16}]\cdot H_2O$	$D_{2d} \equiv -4\ 2\ d$	$m\ m\ 2$	$F\ dd2$
LMU-6	$[Co(en)_3]^{3+}$	$[Co(C_2N_2H_8)_3]_2\ [H_6Zn_6P_8O_{32}]\cdot 1/2\ H_2O$	$D_3 \equiv 3\ 2$	$2/m$	$C\ 2/c$
LMU-7	*s-fac*-$[Co(dien)_2]^{3+}$	$K\ [Co(C_4N_3H_{13})_2]\ [H_6Zn_4P_6O_{24}]$	$C_i \equiv -1$	-1	$P-1$
UH-1	$[Co(tach)_2]^{3+}$	$[Co(C_6N_3H_{15})_2]_3\ [H_{15}Zn_{15}P_{30}O_{72}]\cdot 4.5\ H_2O$	$D_{3d} \equiv -3/m$	-1	$P-1$

ture, the compound exhibits a significant SHG effect [70]. The strong structure-directing effect that exists within the system Co(III) complexes–zincophosphate allows the pre-determination of certain symmetric features and thus of certain physical properties of the composite.

1.6
Conclusions

The preparation of functional composites based on zeotypes by either the unspecific co-occlusion or the direct method can yield solids with interesting physical properties. This was shown here by using chromophores as functional units, which are arranged and protected by the inorganic framework. So far, however, only rather stable molecules have been used as functional species. Further work will involve the introduction of other functions, for example magnetism, and for this purpose it will be furthermore necessary to develop milder synthesis methods.

Acknowledgements

This work was in part carried out at the Institut für Anorganische Chemie of the Ludwig-Maximilians Universität, Munich. It was supported by the Deutsche Forschungsgemeinschaft in the framework of the Schwerpunktprogramm "Nanoporöse Wirt-Gast-Systeme" (Be1664/6) and by the Fonds der Chemischen Industrie. We like to thank all our cooperation partners from the Schwerpunktprogramm, especially Ferdi Schüth and Frank Marlow from the Max-Planck-Institut für Kohlenforschung in Mülheim, Fred-Walter Deeg and Christoph Bräuchle from the Chemistry Department, Ludwig-Maximilians Universität, Munich, Günter Engelhardt, now retired from the Institut für Technische Chemie I of the Universität of Stuttgart, and Franco Laeri from the Institut für Angewandte Physik of the Technische Universität, Darmstadt, as well as all their coworkers who participated in cooperations with us.

References

1 G.D. Stucky, J.E. MacDougall, *Science* **1990**, *247*, 669.
2 G.A. Ozin, *Adv. Mater.* **1992**, *4*, 612.
3 P. Behrens, G.D. Stucky, in Comprehensive Supramolecular Chemistry, J.L. Atwood, D.D. MacNicol, J.E.D. Davies, F. Vögtle F (eds.), Vol. 7, G. Alberti, T. Bein (Vol. eds.), Pergamon Press, Oxford **1996, p. 721.**
4 L. Werner, J. Caro, G. Finger, J. Kornatowski, *Zeolites* **1992**, *12*, 658.
5 F. Marlow, J. Caro, J. Kornatowski, S. Dähne, *J. Phys. Chem.* **1993**, *97*, 13 685.
6 F. Marlow, M. Wübbenhorst, J. Caro, *J. Phys. Chem.* **1994**, *98*, 12 315.
7 F. Marlow, G. van de Goor, P. Behrens, *Adv. Mater.* **1999**, *11*, 238.

8 G. Ihlein, F. Schüth, O. Krauß, U. Vietze, F. Laeri, *Adv. Mater.* **1998**, *10*, 1117.

9 U. Vietze, O. Krauß, F. Laeri, G. Ihlein, F. Schüth, B. Limburg, M. Abraham, *Phys. Rev. Lett.* **1998**, *81*, 4628.

10 I. Braun, G. Ihlein, F. Laeri, J.U. Nöckel, G. Schulz-Ekloff, F. Schüth, U. Vietze, *Appl. Phys. B* **2000**, *70*, 335.

11 M. Ganschow, G. Schulz-Ekloff, M. Wark, M. Wendschuh-Josties, D. Wöhrle, *J. Mater. Chem.* **2001**, *11*, 1823.

12 N. Gfeller, S. Megelski, G. Calzaferri, *J. Phys. Chem. B* **1998**, *102*, 2434.

13 G. Calzaferri, M. Pauchard, H. Maas, S. Huber, A. Khatyr, T. Schaafsma, *J. Mater. Chem.* **2002**, *12*, 1.

14 K. Hoffmann, F. Marlow, J. Caro, *Adv. Mater.* **1997**, *9*, 567.

15 F. Marlow, K. Hoffmann, *Ber. Bunsenges. Phys. Chem.* **1997**, *101*, 1731.

16 K. Hoffmann, U. Resch-Genger, F. Marlow, *Microporous Mesoporous Mater.* **2000**, *41*, 99.

17 D. Wöhrle, A. Sobbi, O. Franke, G. Schulz-Ekloff, *Zeolites* **1995**, *15*, 540.

18 C. Schomburg, D. Wöhrle, G. Schulz-Ekloff, *Zeolites* **1996**, *17*, 232.

19 R. Hoppe, G. Schulz-Ekloff, D. Wöhrle, Ch. Kirschhock, H. Fuess, *Stud. Surf. Sci. Catal.* **1994**, *84*, 821.

20 G. van de Goor, C.C. Freyhardt, P. Behrens, *Z. Anorg. Allg. Chem.* **1995**, *621*, 311.

21 G. Schulz-Ekloff, D. Wöhrle, B. van Duffel, R.A. Schoonheydt, *Microporous Mesoporous Mater.* **2002**, *51*, 91.

22 M.E. Davis, R.F. Lobo, *Chem. Mater.* **1992**, *4*, 756.

23 R. Szostak, Handbook of Molecular Sieves, Van Nostrand-Reinhold, New York **1992**.

24 R.F. Lobo, S.I. Zones, M.E. Davis, *J. Incl. Phenom. Molec. Recogn. Chem.* **1995**, *21*, 47.

25 P. Behrens, Functionalized Structure-Directing Agents for the Direct Synthesis of Nanostructured Materials, in Organosilicon Chemistry III, N. Auner, J. Weis (eds.), Wiley-VCH, Weinheim **1997**, p. 649.

26 Ch. Panz, P. Behrens, Synthesis between Scylla and Charybdis: Encapsulating labile structure-directing molecules into microporous materials during synthesis, in *Proc. 10th German Zeolite Conf.*, Bremen, 9–11 March **1998**.

27 R.P. Gunawardane, H. Gies, F. Liebau, *Z. Anorg. Allg. Chem.* **1987**, *546*, 189.

28 H. Gies, in Inclusion Compounds, Vol. 5, Academic Press, London **1995**, **p. 1.**

29 H. Gies, B. Marler, *Zeolites*, **1992**, *12*, 42.

30 H. Gies, B. Marler, U. Werthmann, in Molecular Sieves: Science and Technology, H.G. Karge, J. Weitkamp (eds.), Vol. 1: Synthesis, Springer, Berlin **1998**, p. 35.

31 K.J. Balkus, S. Shepelev, *Microporous Mater.* **1993**, *1*, 383.

32 K.J. Balkus, A.G. Gabrielov, N. Sandler, *Mater. Res. Soc. Symp. Proc.* **1995**, *368*, 369.

33 P. Behrens, G. van de Goor, C.C. Freyhardt, *Angew. Chem.* **1995**, *107*, 2895; *Angew. Chem. Int. Ed. Engl.* **1995**, *34*, 2680.

34 G. van de Goor, B. Lindlar, J. Felsche, P. Behrens, *Chem. Commun.* **1995**, 2559.

35 A.M. Schneider, P. Behrens, *J. Mol. Graphics* **1996**, *14*, 381.

36 G. van de Goor, K. Hoffmann, S. Kallus, F. Marlow, F. Schüth, P. Behrenst, *Adv. Mater.* **1996**, *8*, 65.

37 C.C. Freyhardt, M. Tsapatsis, R.F. Lobo, K.J. Balkus, M.E. Davis, *Nature* **1996**, *381*, 295.

38 K.J. Balkus, M. Biscotto, A.G. Gabrielov, *Stud. Surf. Sci. Catal.* **1997**, *105*, 415.

39 P. Behrens, Ch. Panz, V. Hufnagel, B. Lindlar, C.C. Freyhardt, G. van de Goor, *Solid State Ionics* **1997**, *101–103*, 229.

40 R.F. Lobo, M. Tsapatsis, C.C. Freyhardt, S. Khodabandeh, P. Wagner, C.Y. Chen, K.J. Balkus, S.I. Zones, M.E. Davis, *J. Am. Chem. Soc.* **1997**, *119*, 8474.

41 A.R. Ramsaran, K.J. Balkus, M.A. Biscotto, J.E. Sheats, K.T. Micai, S. Furyk, A. Hamilton, *Preprints Am. Chem. Soc., Div. Petrol. Chem.* **1998**, *43*, 289.

42 A.M. Schneider, P. Behrens, *Chem. Mater.* **1998**, *10*, 679.

43 P. Behrens, A.M. Schneider, in *Proc. Silica '98*, Mulhouse, France **1998**, p. 25.

44 T. Wessels, C. Baerlocher, L.B. McCusker, *Science* **1999**, *284*, 477.

45 T. Wessels, C. Baerlocher, L.B. McCusker, E.J. Creyghton, *J. Am. Chem. Soc.* **1999**, *121*, 6242.

46 R. Jäger, Doctoral Thesis, University of Hannover, **2002**.

47 R. Jäger, P. Behrens, in preparation.

48 R. Jäger, A. M. Schneider, P. Behrens, B. Henkelmann, K.-W. Schramm, D. Lenoir, A. Kettrup, in preparation.

49 R. Jäger, M. Hartl, Ch. Seebacher, Ch. Bräuchle, P. Behrens, in preparation.

50 W. Xu, J. Dong, J. Li, J. Li, F. Wu, *J. Chem. Soc. Chem Commun.* **1990**, 755.

51 R. Althoff, K. Unger, F. Schüth, *Microporous Mater.* **1994**, *1*, 191.

52 U. Deforth, K.K. Unger, F. Schüth, *Microporous Mater.* **1997**, *9*, 287.

53 B.M. Pillep, Doctoral Thesis, Ludwig-Maximilians University, Munich, **1998**.

54 B.M. Pillep, M. Kämper, R. Wartchow, P. Behrens, *Chem. Eur. J.*, in preparation.

55 M.A. Hartl, Doctoral Thesis, University of Hannover, **2002**.

56 A.M. Schneider, Doctoral Thesis, Ludwig-Maximilians University, Munich, **1998**.

57 S.T. Wilson, *Stud. Surf. Sci. Catal.* **1991**, *58*, 137.

58 S. Qin, W. Pang, H. Kessler, J.L. Guth, *Zeolites* **1989**, *9*, 440.

59 H. Kessler, J. Patarin, C. Schott-Darie, *Stud. Surf. Sci. Catal.* **1994**, *85*, 75.

60 G. Finger, J. Richter-Mendau, M. Bülow, J. Kornatowski, *Zeolites* **1991**, *11*, 443.

61 J. Caro, F. Marlow, K. Hoffmann, C. Striebel, J. Kornatowski, I. Girnus, M. Noack, P. Kölsch, *Stud. Surf. Sci. Catal.* **1997**, *105*, 2171.

62 I. Braun, G. Schulz-Ekloff, M. Bockstette, D. Wöhrle, *Zeolites* **1997**, *19*, 128.

63 M. Bockstette, D. Wöhrle, I. Braun, G. Schulz-Ekloff, *Microporous Mesoporous Mater.* **1998**, *23*, 83.

64 I. Braun, G. Schulz-Ekloff, D. Woehrle, W. Lautenschläger, *Microporous Mesoporous Mater.* **1998**, *23*, 79.

65 T. Kodaira, K. Miyazawa, T. Ikeda, Y. Kiyozumi, *Microporous Mesoporous Mater.* **1999**, *29*, 329.

66 K. Tetsuya, M. Kohji, I. Takuji, K. Yoshimichi, *Microporous Mesoporous Mater.* **1999**, *29*, 329.

67 M. Ganschow, G. Schulz-Ekloff, M. Wark, M. Wendschuh-Josties, D. Wöhrle, *J. Mater. Chem.* **2001**, *11*, 1823.

68 O. Weiss, G. Ihlein, F. Schüth, *Microporous Mesoporous Mater.* **2000**, *35–36*, 617.

69 B.L. Newalkar, R.V. Jasra, V. Kamath, T.S.G. Bhat, *J. Chem. Soc. Chem. Commun.* **1994**, 1041.

70 Ch. Panz, Doctoral Thesis, Ludwig-Maximilians University, Munich, **1998**.

71 D.M. Bibby, M.P.Dale, *Nature* **1985**, *317*, 157.

72 H. Qisheng, X. Ruren, *J. Chem. Soc. Chem. Commun.* **1990**, 783.

73 H. Qisheng, X. Ruren, *J. Chem. Soc. Chem. Commun.* **1992**, 168.

74 C.M. Braunbarth, P. Behrens, J. Felsche, G. van de Goor, G. Wildermuth, G. Engelhardt, *Zeolites* **1996**, *16*, 207.

75 C. Seebacher, C. Hellriegel, F.W. Deeg, C. Bräuchle, Confocal microscopy and spectroscopy for the characterization of host–guest materials, this volume.

76 C. Seebacher, J. Rau, F.-W. Deeg, C. Bräuchle, S. Altmaier, R. Jäger, P. Behrens, *Adv. Mater.* **2001**, *13*, 1374.

77 P. Behrens, A.M. Glaue, O. Oellrich, Direct synthesis and selected properties of organic-inorganic hybrid mesostructures, this volume.

78 C. Kühn, O. Oellrich, P. Behrens, Ch. Seebacher, J.P. Rau, F.-W. Deeg, Ch. Bräuchle, contribution to the *14th Deutsche Zeolith-Tagung*, Frankfurt am Main, 6–8 March **2002**.

79 A.K. Cheetham, G. Ferey, T. Loiseau, *Angew. Chem. Int. Ed. Engl. 1999*, *38*, 3268.

80 T.E. Gier, G.D. Stucky, *Nature* **1991**, *349*, 508.

81 P. Reinert, A. Khatyr, J. Patarin, B. Marler, in *Proc. 12th Int. Zeolite Conf.*, M.M.J. Treacy, B.K. Marcus, M.E. Bisher, J.B. Higgins (eds.), MRS, Warrendale (PA) **1999**, p. 1757.

82 S. Kowalak, A. Jankowska, E. Baran, *Chem. Commun.* **2001**, 575.

83 P.K. Dutta, K.S.N. Reddy, L. Salvati, M. Jacupka, *Nature* **1995**, *374*, 44.

84 K.S.N. Reddy, L.M. Salvati, P.K. Dutta, P.B. Abel, K.I. Suh, R.R. Ansari, *J. Phys. Chem.* **1996**, *100*, 9870.

85 M.J. Castognola, P.K. Dutta, *Microporous Mesoporous Mater.* **2000**, *34*, 61.

86 R. Singh, P.K. Dutta, *Langmuir* **2000**, *16*, 4148.

87 S. Feng, T. Bein, *Nature* **1994**, *368*, 834.

88 P. Behrens, Ch. Panz, G. Nuspl, K. Polborn, Ein neues Zincophosphat mit Schichtstruktur durch milde Hydrothermalsynthese: $[Co(NH_3)_6]_3$ $[H_8Zn_8P_{10}O_{40}]$ PO_4 (LMU-4), contribution to the *9th Vortragstagung Festkörperchemie of the GDCh*, Saarbrücken, 23–25 Sept **1998**.

89 C. Kühn, diploma thesis, University of Hannover, **2000**.

90 C. Kühn, P. Behrens, UH-1, a new layered zincophosphate structure, contribution to the *12th Deutsche Zeolith-Tagung*, Munich, 1–3 March **2000**.

91 C. Kühn, Ch. Panz, P. Behrens, Vorgabe der Symmetrieeigenschaften von Festkörpern durch strukturdirigierte Synthese, contribution to the *10th Vortragstagung of the Fachgruppe Festkörperchemie und Materialforschung of the GDCh*, Munich, 26–29 Sept **2000**.

92 C. Kühn, C. Panz, P. Behrens, Zincophosphates as host structures for inorganic complexes, contribution to the *13th Deutsche Zeolith-Tagung*, Erlangen, 7–9 March **2001**.

93 C. Panz, C. Kühn, P. Behrens, to be published.

94 C. Kühn, P. Behrens, to be published.

95 J. Yu, Y. Wang, Z. Shi, R. Xu, *Stud. Surf. Sci. Catal.* **2001**, *135*, 05-O-04.

2
In Situ Synthesis of Azo Dyes and Spiropyran Dyes in Faujasites and their Photochromic Properties

Dieter Wöhrle, Carsten Schomburg, Yven Rohlfing, Michael Wark, and Günter Schulz-Ekloff*

2.1
Introduction

Mineral-hosted dyes can exhibit unique properties for practical applications as well as for fundamental research. In comparison with polymers, crystalline porous mineral hosts offer advantages like higher thermal, mechanical, or chemical stability. In contrast to glasses, their uniform void structure enables the study of the influences of host–guest interaction and of spatial constraints on the dynamics and chemistry of accommodated chromophores. Recently, preparation methods, properties, and possible applications of chromophores in porous silicas, molecular sieves, and minerals have been summarized by Schulz-Ekloff et al. [1]. Practical applications can be expected (1) as inclusion pigments because of improved migration stability by stable anchoring and enhanced photo stability of encaged dye monomers [2], (2) for frequency doubling of laser radiation because of an organization of molecular dipoles resulting in a macroscopic hyperpolarization [3–7], high luminescence and also lasing from the excited state of chromophores [8–11], and as photosensitized reactions, stimulated by visible light [12–16].

Four different methods have been applied for the incorporation of dye molecules in molecular sieves [1]: (1) ion exchange of cationic chromophores [12–14,17,18]; (2) deposition from the vapor phase [3–7]; (3) crystallization inclusion [2,8–10,19]; (4) *in situ* synthesis of metal complexes [20–23]. Small dye molecules fitting into the zeolite cavities can be accommodated on crystallographically defined positions, as revealed by Rietveld refinement of X-ray diffractograms [18], and can exhibit preferred orientations, such as pearl-string-like arrangement of molecular dipoles in channel structures as detected by optical spectroscopy with polarized light [24,25]. Phthalocyanine derivatives with a molecule diameter exceeding the channel diameter of the molecular sieve $AlPO_4$-5 are incorporated monomolecularly in mesopores [2,26].

In this chapter three aspects of organic chromophores in the pores of faujasites (the microporous structure-type of zeolite Y) are summarized [1,27–30]:

- *In situ synthesis (also called "ship-in-the-bottle" technique)* of azo dyes and spiropyran dyes in faujasites NaY, HY, and DAY. This method for the embedding of pure organic dye molecules into the zeolite pores was mainly introduced by our group [27–30]. Prior to that work, only one paper describing the *in situ* synthesis of the yellow-colored, strong oxidizing 2,4,6-triphenylpyrylium cation in faujasite HY was available [31]. The *in situ* synthesis makes it possible to include polar dyes and to build-up chromophore molecules with diameters exceeding the pore width in the zeolite cages. After their formation the dyes cannot be removed by solvent extraction.
- *Photochromic properties of encapsulated spiropyran.* A broader application of organic photochromic compounds for fluid-flow visualization, optical switching, and information storage [32] is mainly restricted by their limited photostability (photodegradation) and quick relaxation of the energy-rich photoinduced state. It is expected that the polar faujasite matrix will stabilize the optical switching properties due to distinct host–guest interactions.
- *Embedding of the chromophore-loaded faujasites in a polymer matrix.* For a number of potential practical application the use of zeolite powders is disadvantageous. To overcome this situation the molecular sieve crystals can be dispersed in a polymer matrix, which would allow the preparation of larger samples for optical applications, for example. In order to avoid light scattering of crystals with a size of >1 μm, based on differences in refractive indices between the zeolite and the polymer, index matching has to be carried out.

2.2 *In Situ* Synthesis of Azo Dyes in Faujasites

The applied faujasite-type Y-zeolites (Si/Al ratio: 2.9, crystal size: 1–2 μm; overall surface area: $\approx 850\ \mathrm{cm^2\ g^{-1}}$; supercage diameter: 1.3 nm; pore size 0.74 nm) are based on hydrothermally synthesized NaY [33]. The faujasite HY was obtained from the NaY via ion exchange with NH_4^+ in aqueous solution and subsequent calcination [28]. A DAY zeolite was prepared from HY via isomorphous substitution of Al by Si (Si/Al ratio: 100).

The developed method for the synthesis of pure organic chromophores in the supercages of faujasites follows the simple scheme $A + B \rightarrow C$. In a first step a precursor for the dye has to be fixed by interaction with the zeolite lattice on adsorption sites. For example, a basic precursor can be bound by acid–base interaction with the host (Step 1). Therefore, HY is preferable compared to NaY and DAY. After careful cleaning of the molecular sieve from the excess of noninteracting basic precursors, the chromophore synthesis is achieved by diffusing the second educt into the porous zeolite structure and subsequent reaction (Step 2). Whereas the small educt molecules can enter the pore system, the larger size of the formed dye molecules inhibits their diffusion out of the supercages (size of synthesized azo and spiropyran dyes: $1.7 \times 0.85 \times 0.35\ \mathrm{nm^3}$, $1.4 \times 0.74 \times 0.35\ \mathrm{nm^3}$, respectively) [27,28,34].

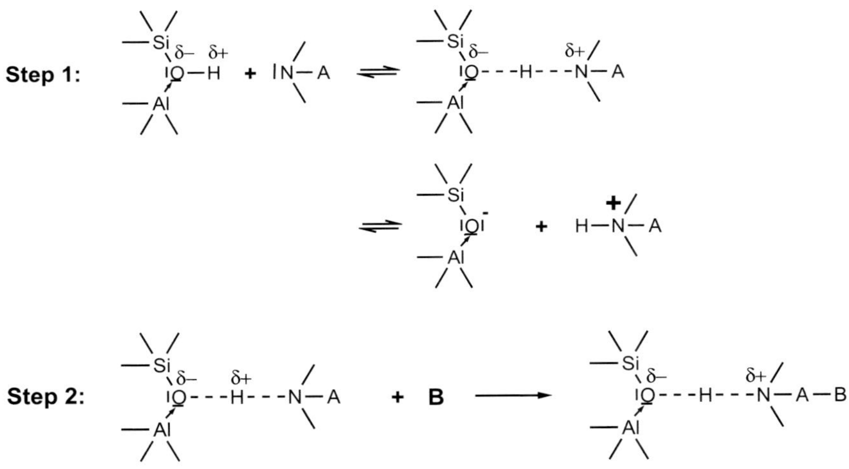

For the syntheses of the azo dyes **1a–g** (Eq. 3) each 3 g of dried faujasite NaY or HY was treated with 10 mL N,N-dialkylaniline without additional solvent for 24 h followed by filtration and intensive washing with ethanol and acetone [27]. HY absorbs 240 mg g^{-1} and NaY 100 mg g^{-1} of N,N-dimethylaniline. Then 3.5 g of the molecular sieve were reacted with 4 mmol of a diazonium zinc double salt in 40 mL aqueous HCl at pH 3 for 24 h. Treatment with NaOH and several days of Soxhlet extraction with ethanol confirmed the diffusion-stable incorporation of the differently colored azo dye substituents **1a–g** in the zeolite pores. Alternatively, the basic precursor was dissolved in an organic solvent like ethanol, toluene, or *n*-hexane [34]. For example, 7.5×10^{-4} mol dialkylaniline in 15 mL

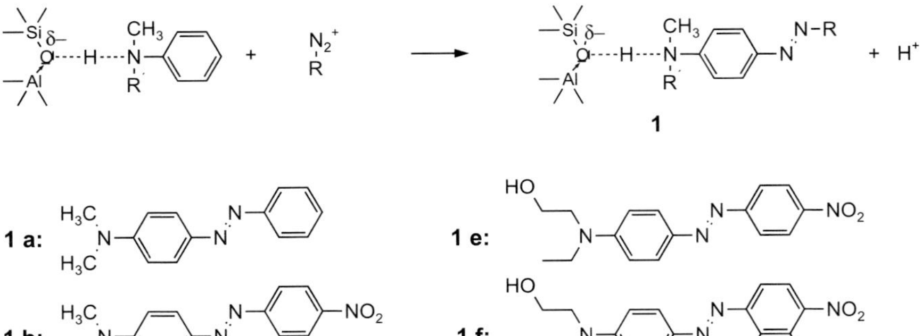

1

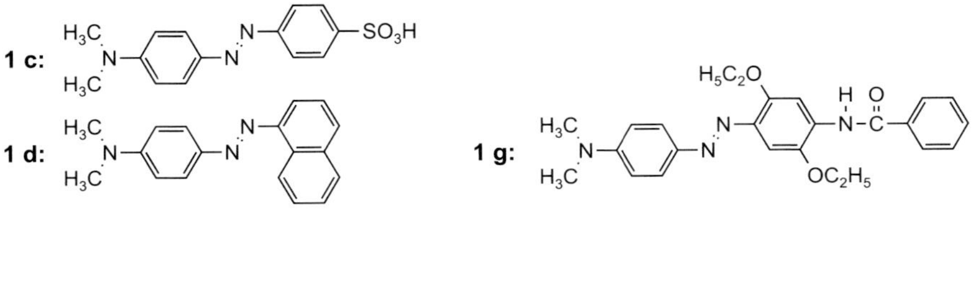

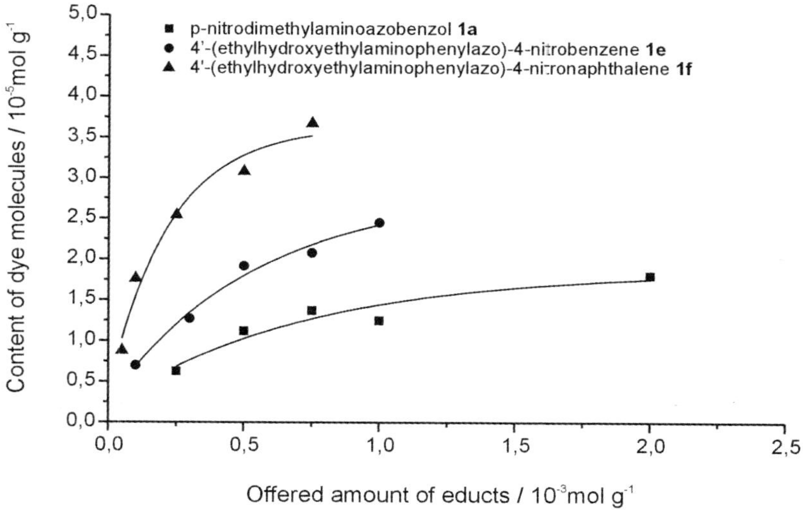

Fig. 1. Loading of the faujasite HY with azo dyes as a function of the offered amount of educts.

solvent were treated for 72 h with 1 g NaY or HY (1 g faujasite contains 5.1×10^{-4} mol supercages). After washing with organic solvents, 1 g of this pretreated faujasite was reacted with 7.5×10^{-4} mol of a diazonium tetrafluoroborate in aqueous HCl at pH 3.5.

The loading of the faujasite with the azo dyes was estimated semiquantitatively from the reflectivities of the diffuse reflectance UV/vis spectra by comparison with azo dyes deposited on the external surface of the faujasites from solution and subsequent evaporation of the solvent. With NaY loadings of about 10^{-6}–10^{-5} mol dye g^{-1} could be achieved whereas the loading of HY is higher (10^{-5}–10^{-4} mol g^{-1}, i.e., about 0.25–2 wt.-%). The alternative method was tested for the synthesis of **1e** leading to a maximum loading of 4.1×10^{-5} mol g^{-1} or 1.45 wt.-%. In this case 9.1% of the supercages are occupied and the reaction yield for the azo coupling is 6.2%.

Figure 1 shows the dependence of the degree of loading from the concentrations of both educts. It becomes obvious that the loadings reach different maximum loadings in dependence of the kind of educts for the synthesis of the azo dyes. On one hand it was found that the reactivity of the educts for the azo dye synthesis in the cages is different [34], on the other hand the amount of solvent extraction of the dye depends on its size. Whereas for example the dye **1a** (size: 1.36 nm × 0.57 nm) is very slowly extracted by ethanol over several weeks, dyes **1e** and **1f** (sizes: **1e**: 1.56 nm × 0.71 nm; **1f**: 1.56 nm × 0.79 nm) are really fixed without loss by solvent extraction. The UV/vis reflectance spectra of the monomolecular encapsulated azo dyes differ from those in solution and indicate an intensive interaction with the host system. The azo dye **1a** shows absorption maxima in water of pH 7 at $\lambda = 457$ nm and in acidic solution of pH 2 at $\lambda = 505$ nm, in HY it exhibits a

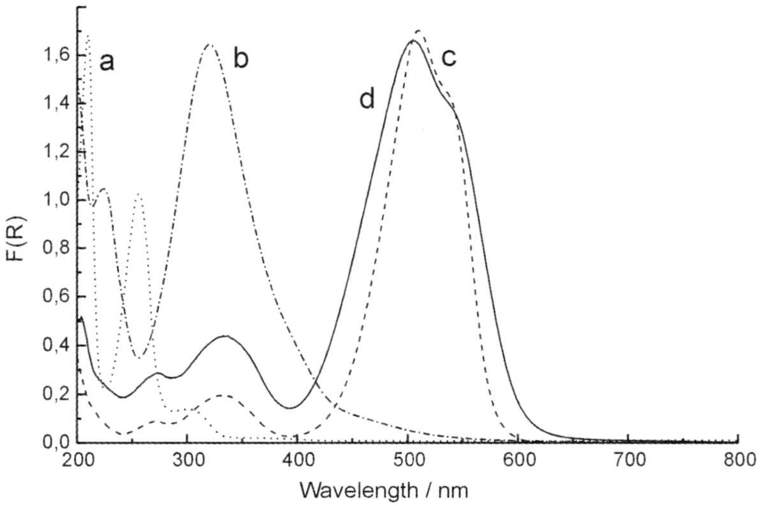

Fig. 2. Diffuse reflectance UV/vis spectra of: (a) N-ethyl-N-hydroxyethylaniline; (b) 1-diazonium-4-nitrobenzene; (c) 4′-(ethylhydroxyethylaminophenylazo)-4-nitrobenzene **1e** adsorbed on HY; (d) **1e** *in situ* synthesized in HY.

maximum at $\lambda = 487$ nm with an intensive shoulder at $\lambda = 530$ nm [34]. A weak absorption in the UV region signifies that the excess of nonreacted starting materials is washed out. This is exemplarily shown for **1e** and its educts in Fig. 2. The spectrum of the adsorbed dye is included for comparison. The obtained colors of the HY encapsulated azo dyes are: **1a** intensively pink, **1b** intensively pink, **1c** intensively orange, **1d** beige, **1e** dark red, **1f** dark brown-red, **1g** beige. In case of **1c** (methyl orange) the color is, as expected, pH dependent.

2.3
In Situ Synthesis of Spiropyran Dyes in Faujasites

Also the ship-in-the-bottle synthesis of photochromic spiropyran dyes in faujasites was successfully carried out in NaY, HY, and DAY [28,29,34]. The size of 6-nitro-1′,3′,3′-trimethyl-indolospiro[2H-1]-benzopyran **2** fits, as mentioned before, with the diameter of the supercage and pore opening. After the synthesis no leaching of **2** was observed. As precursor molecules for the synthesis of **2** 1,3,3-methyl-2-methylene-indoline and 5-nitro-salicylaldehyde were used (vide infra); 1 g Zeolite was suspended in a solution of 5×10^{-4} mol of 1,3,3-methyl-2-methylene-indoline in 15 mL ethanol. The slurry was agitated for 3 d, filtered, and thoroughly washed with ethanol. Subsequently, a solution of 4×10^{-3} mol of 5-nitrosalicylaldehyde in 15 mL ethanol was added and the suspension was stirred at room temperature for 3 h and heated under reflux for 70 h. Finally, the hot slurry was filtered and extracted in a Soxhlet apparatus with ethanol for 70 h.

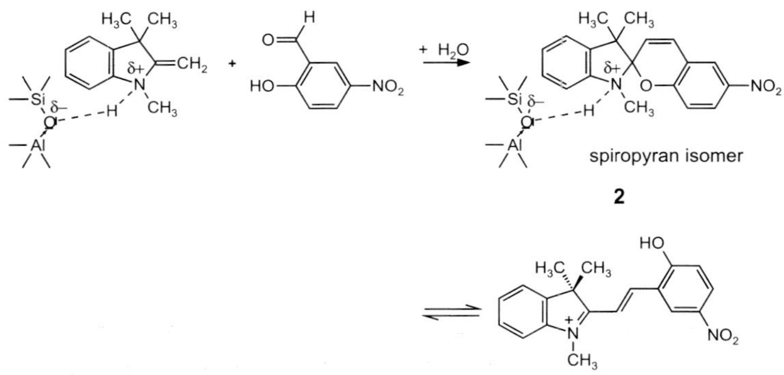

merocyanine isomer

2

The *in situ* synthesis of **2** in the faujasite structures NaY, HY, or DAY yielded loadings, which can be expressed as percent of filled supercages, such as NaY (0.2%) < DAY (4%) < HY (34%). In the case of HY this corresponds to 5.6 wt.-% or 1.7×10^{-4} mol g^{-1}. More than one dye molecule is hosted per unit cell, so neighboring supercages are filled with spiropyran molecules. The degrees of loading from *in situ* synthesis were estimated from UV/vis spectra taken in diffuse reflectance using calibrations with series of dye-impregnated faujasites, and dilution with the parent zeolite material to keep the Kubleka–Munk function in the range 0.1–3.0 in which linear correlation with the dye concentration is valid.

The nature of the thermodynamically stable constitutional isomer of a spiropyran **2** incorporated in a host matrix depends on the host–guest interactions, as elucidated for various mineral matrices [35–37]. If the host exhibits significant Brønsted acidity, as in the case of molecular sieves, the protonated merocyanine isomers of the spiropyrans have to be considered, too (Fig. 3) [38,39]. The different spectral patterns obtained for zeolite-encapsulated spiropyran isomers (Fig. 4) can be assigned to the following species, which can exist in acidic environments [28]. For dye incorporated in NaY the pattern of a mixture appears containing predominately the neutral, closed spiropyran (SP) and the protonated, open merocyanine form (BH^+), gleaned from the maxima around $\lambda = 230$, 270, 350, and 400 nm. In the more acidic DAY host the protonated transoid merocyanine form BH^+, with maxima around $\lambda = 300$ and 420 nm, exists predominantly. For the most acidic HY matrix the superposition of the BH^+ form and a protonated merocyanine *cis*-form Y with a maximum at $\lambda = 320$ nm dominates the spectral patterns (Fig. 4).

All the samples loaded with **2** (even the highly loaded HY) exhibit a broad fluorescence peak centered around $\lambda = 510$ nm. This indicates the presence of the merocyanine in the open form, since the closed spiropyran (SP) is not fluorescent, and in a nonaggregated state excluding self-quenching.

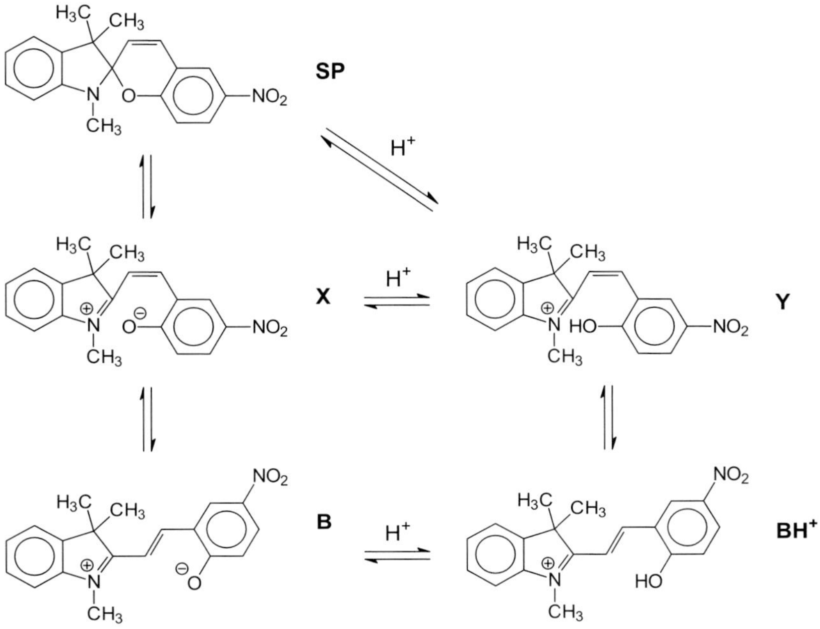

Fig. 3. Constitutional isomers of zwitterionic (X, B) and protonated (Y, BH^+) merocyanine forms of the spiropyran **2**.

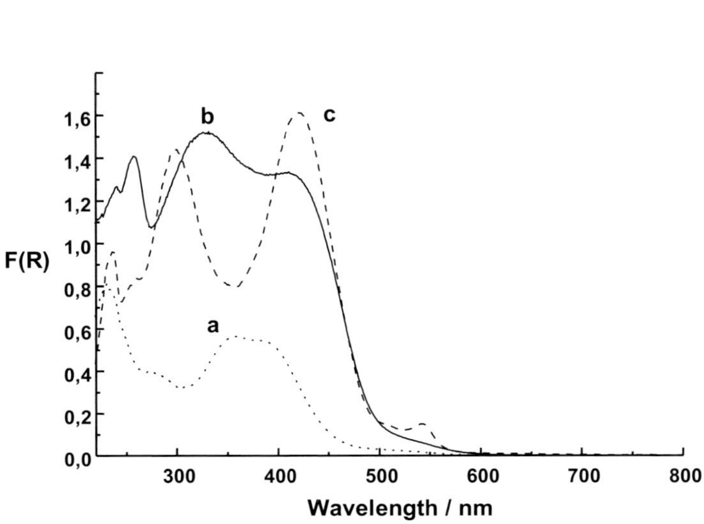

Fig. 4. Diffuse reflectance UV/vis spectra of constitutional isomers of the spiropyran **2** incorporated in zeolites via *in situ* synthesis: (a) NaY; (b) HY; (c) DAY. Inset: (a) emission spectrum of **2** in zeolite HY embedded in a refractive index matched copolymer (methylmethacrylate-tetrafluoroacrylate) [40]; the copolymer shows no fluorescence.

Incorporation of spiropyran in Si-MCM-41 via impregnation at lower loadings results in characteristic spectra of the BH^+ form. At higher loadings ($> 5 \times 10^{-5}$ mol dye per g Si-MCM-41) broader maxima appear, indicating that the spiropyran is also adsorbed on nonacidic sites in the zwitterionic open merocyanine form [28]. The reference sample, polyvinylacetate(PVA)-hosted spiropyran, exhibits the spectrum of the closed spiropyran form SP.

2.4
Optical Switching of Azo and a Spiropyran Dyes in Molecular Sieves

Photochromism experiments were carried out using dye laser pulses (Coumarin 307, $\lambda = 465$–550 nm or RDC 360-NEU, $\lambda = 340$–370 nm) of high irradiation power, which were focused to 1 cm spots and directed to the sample, which was fixed in a sample holder [28,34]. Local overheating or hot spots, respectively, were avoided by the application of short laser pulses (100 ns) and irradiation in the flanks of the absorption signals. All the samples have been diluted with parent matrix material to an overall dye concentration 1.0×10^{-5} mol g^{-1}. 50 mg of irradiated sample in all cases contain 3×10^{17} dye molecules. The diffuse reflectance of the samples was immediately measured after laser irradiation with an UV/vis spectrometer equipped with a praying mantis cell.

Azo dyes belong to the most intensively investigated photochromic systems due to distinct absorptions of the *E*- and *Z*-isomers at different wavelengths [40–42]. The optical switching was investigated in different host materials [7,32,43,44]. The switching ability was also shown for the unsubstituted azobenzene in the pores of $AlPO_4$-5, ZSM-5 and Silicalite-1 molecular sieves [7,45]. Also in faujasites optical switching of photochromic molecules is possible as investigated in detail for *trans*- and *cis*-thioindigo [18]. Surprisingly, the tested azo dyes **1a**, **1e**, and **1f** exhibit no significant changes in the UV/vis reflectance spectra after laser light irradiation of the *E*-isomers [34]. This could be due to space limitations of the host for the substituted azo dyes and fast relaxations from the *Z*- to the *E*-isomers [17]. In order to check this and to designate the principal possibility of *Z*/*E*-isomerization of azo dyes in faujasite, a sample prepared by impregnation and diffusion of the unsubstituted azobenzene in HY was investigated [34]. Irradiation with laser light at $\lambda = 347$ nm of 1 s duration results in a decrease of the π–π^* transition at $\lambda = 320$ nm and an increase of the n–π^*transition at $\lambda = 420$ nm for the *E*- to *Z*-isomerization (Fig. 5). A high quantum yield of 0.13 was calculated [34]. The back isomerization from the *Z*- to the *E*-isomer is possible by irradiation at 467 nm. This documents that in the cases of the larger azo dyes **1a**, **1e**, and **1f** the spatial constraint prevents the switching.

The photochromism of spiropyrans in solution, in films of organic polymers and on non- or low-structured inorganic solids had been investigated in detail before [32,40–42]. For applications like information storage, a long-term stability of the switched state against thermal relaxation is required. To compare and to judge on different photochromic systems, data on the relaxation kinetics, such as the time in

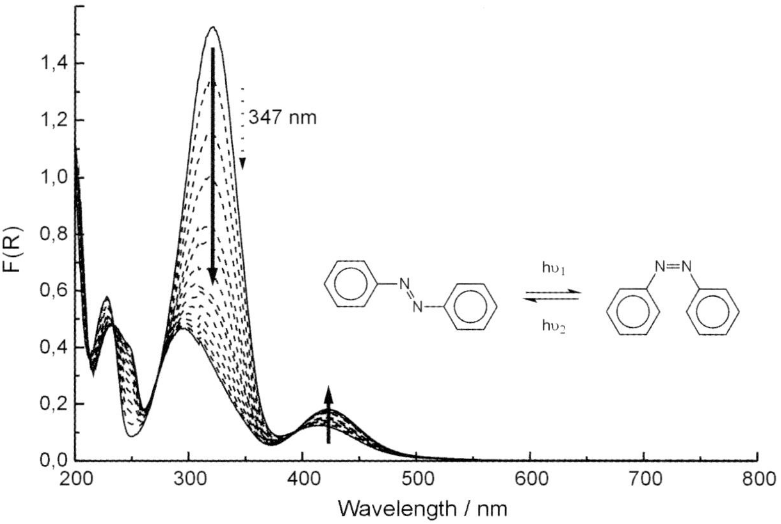

Fig. 5. Diffuse reflectance UV/vis spectra of 1.5×10^{-5} mol g^{-1} azobenzene in faujasite HY. Irradiation in steps of 1 s at $\lambda = 347$ nm.

which half of the molecules relax thermally into the original state ($t_{1/2}$), are necessary. Recently, it was shown that spiropyrans after incorporation into a wide-pore mesoporous molecular sieves of the SBA-15 type containing hydrophobic parts exist in the SP form (Fig. 3). After switching to the zwitterionic open merocyanine form B, relative rapid thermal relaxation to the SP form occurs with $t_{1/2} \approx 1.25$ h [45].

In the following the switching procedure for the spiropyran **2** in faujasites is presented exemplarily [28,29]. The as-prepared state of **2** in the supercages of DAY, the BH^+ form, (Fig. 4, spectrum c) is switched using the excitation wavelength $\lambda_{exc} = 467$ nm. The switched state (Fig. 6, spectrum a) is assigned to the protonated merocyanine *cis*-form Y. This switching process from the *trans*- to the *cis*-form is called "reverse photochromism". The slow thermal relaxation of the *cis*-form results in spectrum b in Fig. 6 after 65 h. The attainment of the final state would have required several weeks. However, the expected spectrum of this final state was produced by photoexcitation (re-switching), using $\lambda_{exc} = 347$ nm. The extinction of the band at $\lambda = 420$ nm in this final state (Fig. 6, spectrum c) is strongly reduced compared to the initial state after the *in situ* synthesis (compare Fig. 4, spectrum c). This points out that the initial states for the as-prepared samples are metastable and are converted into photostationary states after irradiation.

The photochromism experiments of **2** in HY displayed changes of the spectrum similar to that on DAY. However, faster thermal relaxation is observed (Fig. 7). In Si-MCM-41 thermal relaxation appears much faster than in all the faujasites (Fig. 7). The initial state is attained by thermal relaxation but not by photoexcited re-

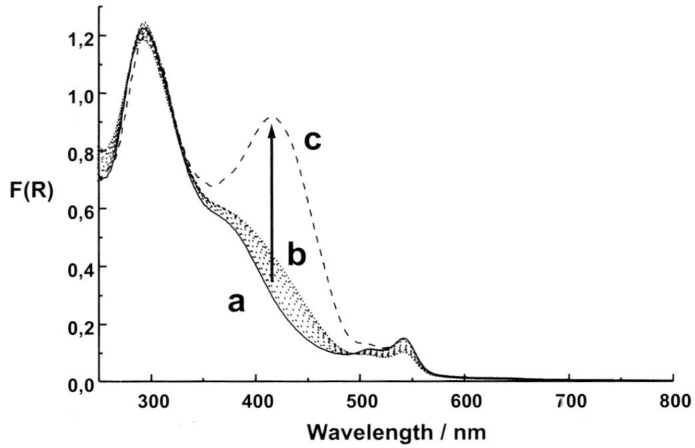

Fig. 6. Diffuse reflectance UV/vis spectra: (a) of the protonated merocyanine *cis*-form Y in DAY (1×10^{-5} mol g^{-1}) formed after irradiation ($\lambda_{exc} = 467$ nm) of the incorporated merocyanine *trans*-form BH^+; (b) after thermal relaxation at room temperature for 65 h; (c) after irradiation ($\lambda_{exc} = 347$ nm).

switching, since both photoexcited and initial state absorb at the applied wavelength $\lambda = 347$ nm. The spiropyran in polyvinylacetate exhibits the "normal photochromism" behavior from the *cis*-spiropyran SP to the zwitterionic merocyanine *trans*-form B. The photoexcited state shows a very rapid thermal relaxation (Fig. 7).

The thermal relaxation kinetics could be fitted by a first-order rate equation. Only in the case of the spiropyran-DAY composite a bi-exponential rate equation results

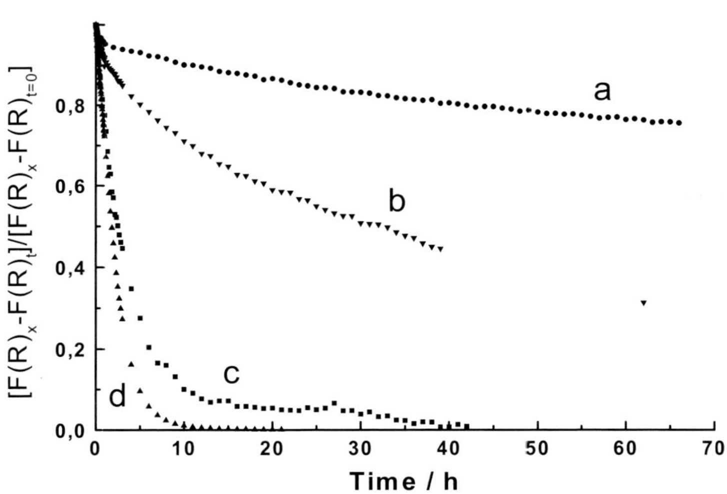

Fig. 7. Normalized kinetic plot of thermal relaxation of N-BIPS **2** vs. *t* in different host materials: (a) DAY; (b) HY; (c) Si-MCM-41; (d) PVA.

in a better fit [28,29]. The evaluated data for the lifetimes of the photoinduced states ($t_{1/2}$), indicating the stability of a photoinduced switching, are 1.5 h for the PVA-hosted spiropyran, approximately 2 h in Si-MCM-41, approximately 40 h in HY, and almost 400 h for the merocyanine in the DAY zeolite, and are thus by far best for the faujasite composite. In the pores of Si-MCM-41 and especially in the PVA-matrix, where only weak interactions exist, the dye can readily undergo all rotational movements, which are necessary for the attainment of the thermodynamic equilibrium: the thermal relaxation to the initial form. In the faujasites, however, a strong host-guest interaction for the merocyanine molecules in the supercages is obvious. Since the large spiropyran molecule ($1.4 \times 0.74 \times 0.7$ nm) fills most of the free space in the supercage (diameter: 1.2 nm), its free relaxation is impeded. The faster fading rate in HY compared to DAY might be an effect of the different acidities, since an activation of the thermal relaxation process has also been observed for dyes in acidified silica [35,46].

The analysis of the reversibility of the photoinduced switching is usually described via $N_{1/2}$, so the number of photoinduced switching cycles N after which the extinction difference ΔA_0, that is the difference between initial and final state extinctions at a selected observation wavelength, is decreased to $\Delta A_0/2$. The results of corresponding experiments are represented in Fig. 8. High repetition numbers for the reversible photoinduced switching without degradation indicate a photostability needed for possible application. In case of spiropyran derivatives hosted in molecular sieves due to the metastable status of the isomers in the as-synthesized samples, the changes of extinction after a first complete photoinduced switching cycle were taken as initial ΔA_0. For spiropyran or merocyanine, respectively, the obtained values for $N_{1/2}$ were 7 (in PVA), 20 (in Si-MCM-41), 45 (in HY), and 80 (in DAY). The higher $N_{1/2}$ values for all mineral-hosted merocyanines can be

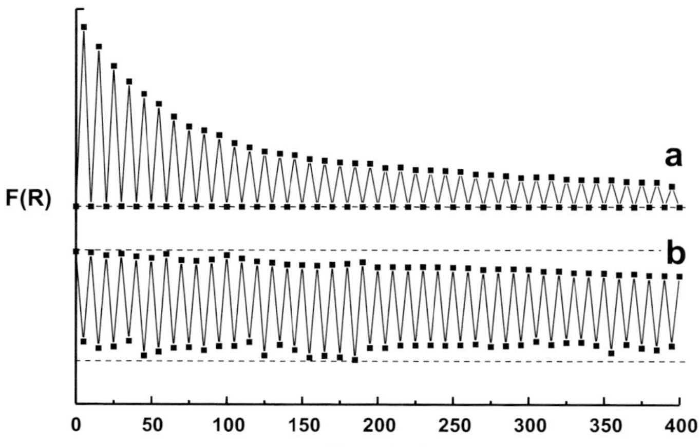

Fig. 8. Photoinduced switching cycles for (a) **2** in PVA and (b) the merocyanine conformers in the DAY zeolite at $\lambda = 420$ nm ($\lambda_{exc} = 467$ nm and 347 nm).

attributed to the fact that here the photoinduced processes change only the conformations of the participating molecules. No electrocyclic binding or bond cleavage take place, as in the case of participating closed spiropyran in PVA.

In order to prepare compact samples, which are in practical applications often necessary for better handling, the molecular sieves were dispersed in an organic polymer. To achieve optical homogeneity and to avoid light scattering the refractive indices of the organic polymers and the molecular sieves were matched. At first the refractive indices of the faujasites NaX, NaY, and HY were determined [30]. Photometric measurements of suspended zeolites (without dye molecules) in different toluene/ethanol mixture resulted in the wavelength region of $\lambda = 300–850$ nm in a turbidity minimum at which the $n(\lambda)$ values of the zeolite and the solvent mixture correspond. Thus, refractive indices $n(\lambda)$ of the faujasite in the range of 1.45 and 1.49 were determined. In a second step dispersive refractive indices of differently composed copolymers of 2,2,2-trifluoroethyl methacrylate and methyl methacrylate (TFM-MMA) were estimated by variable angle spectroscopic reflection ellipsometry. The higher the wavelength of light and the higher the amount of the fluorinated component the lower is the refractive index of the resulting copolymer. Finally, HY containing the encapsulated azo dye **1e** was dispersed in a suitable TFM–MMA copolymer [30]. The transmittance of light and the refractive index were chosen as criteria for the success of the index matching. A copolymer of the composition 10 wt.-% TFM and 90 wt.-% MMA was found to be best to disperse the faujasite crystals (size about 3 µm). Fig. 9 displays the differences in transparency of the samples investigated. Pure MMA-TFM copolymer shows about 90%

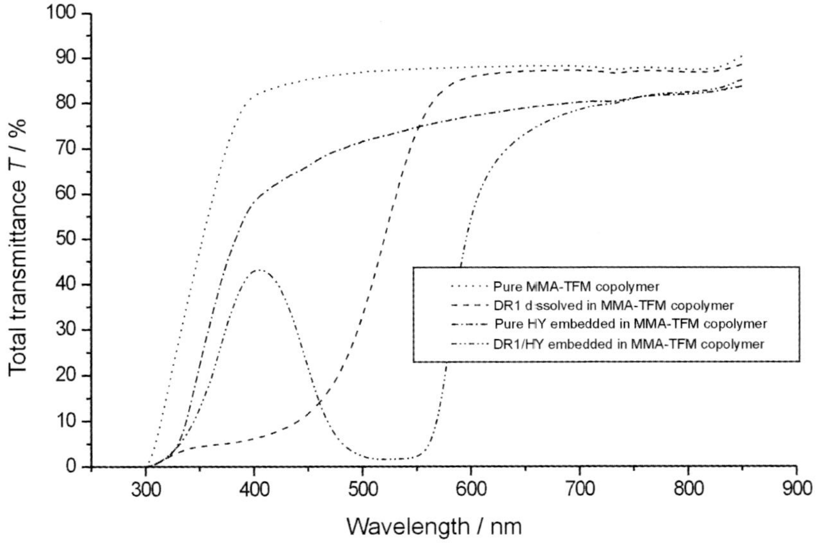

Fig. 9. Dispersive total transmittance $T(\lambda)$ of TFM–MMA copolymers unfilled and differently filled with faujasite HY and/or azo dye **1e**.

transmittance at a wavelength $\lambda > 400$ nm. After addition of 2 wt.-% unloaded zeolite an optimum transparency of 75–80% could be reached. The embedding of HY faujasite loaded with azo dye led to a similar behavior in light transmittance in comparison to the embedded parent material with exception of the absorption peak of the azo dye Disperse Red 1 (DR1, **1e**). It should be mentioned that during the polymerization reaction for the preparation of the composite samples no influence on the loaded zeolites could be observed. The color remained red. However, if the pure dye DR1 (without zeolite) was dissolved in the monomer mixture during the polymerization the color changed to yellow and a blue shift of the absorption peak of about 100 nm occurred. The reason for this effect is a reaction between free radicals of the polymerizing system and the amino and/or nitro groups of the dye.

The embedding of dye-loaded zeolites is explained in more detail in the chapter by J. Schneider et al. [47].

2.5 Conclusions

For the first time *in situ* syntheses of pure organic dyes in cages of zeolites were carried out. The developed methods use the fixation of a first educt with the host by acid–base interactions. Then the synthesis of the chromophor is achieved by reaction of the second educt, also introduced into the pores. The synthesis of azo dyes and a spiropyran shows that the amounts of embedded dye increase in dependence on the host NaY < DAY < HY up to a loading 10^{-4} mol g^{-1}. The method developed is also suitable for the *in situ* synthesis of other organic dyes in the void structures of molecular sieves.

The host–guest interactions were studied for the encapsulated spiropyran dyes. For the photochromic spiropyran a dramatically improved stability of the switched state against thermal relaxation and an extreme high stability during photoinduced switching were found in the faujasite DAY in comparison to non-structured matrices or such with wider pores.

Acknowledgments

Financial support from the Deutsche Forschungsgemeinschaft (Wo 237/16) is gratefully acknowledged.

References

1 G. Schulz-Ekloff, D. Wöhrle, B. van Duffel, R.A. Schoonheydt, *Microporous Mesoporous Mater.* **2002**, *51*, 91.

2 D. Wöhrle, A. Sobbi, O. Franke, G. Schulz-Ekloff, *Zeolites* **1995**, *15*, 540.

3 J. Caro, F. Marlow, M. Wübbenhorst, *Adv. Mater.* **1994**, *6*, 413.

4 J. Caro, G. Finger, J. Kornatowski, J. Richter-Mendau, L. Werner, B. Zibrowious, *Adv. Mater.* **1992**, *4*, 273.
5 K. Hoffmann, F. Marlow, J. Caro, *Zeolites* **1996**, *16*, 281.
6 F. Marlow, J. Caro, L. Werner, J. Kornatowski, S. Dähne, *J. Phys. Chem.* **1993**, *97*, 11 286.
7 K. Hoffmann, F. Marlow, J. Caro, *Adv. Mater.* **1997**, *9*, 567.
8 I. Braun, M. Bockstette, G. Schulz-Ekloff, D. Wöhrle, *Zeolites* **1997**, *19*, 128.
9 M. Bockstette, D. Wöhrle, I. Braun, G. Schulz-Ekloff, *Microporous Mesoporous Mater.* **1998**, *23*, 83.
10 I. Braun, G. Ihlein, F. Laeri, J.U. Nöckel, G. Schulz-Ekloff, F. Schüth, U. Vietze, O. Weiß, D. Wöhrle, *J. Appl. Phys. B* **2000**, *70*, 335.
11 F. Marlow, M.D. Gehee, D. Zhao, B.E. Chmelka, G.D. Stucky, *Adv. Mater.* **1999**, *11*, 632.
12 A. Kunzmann, R. Seifert, G. Calzaferri, *J. Phys. Chem. B* **1999**, *103*, 18.
13 G. Calzaferri, *Chimia* **1998**, *52*, 525.
14 N. Gfeller, S. Megelski, G. Calzaferri, *J. Phys. Chem. B* **1999**, *103*, 1250.
15 T. Bein, in Comprehensive Supramoleuclar Chemistry, Vol. 7, Solid State Supramoleuclar Chemistry: Two- and Three-Dimensional Networks, G. Alberti, T. Bein (eds.), Pergamon, Oxford **1996**, p. 579.
16 O. Bartels, M. Wark, D. Wöhrle, unpublished results.
17 R. Hoppe, D. Wöhrle, G. Schulz-Ekloff, E.S. Shpiro, O.P. Tkachenko, *Zeolites* **1993**, *13*, 222.
18 R. Hoppe, G. Schulz-Ekloff, D. Wöhrle, C. Kirschhock, H. Fuess, *Adv. Mater.* **1995**, *7*,61.
19 S. Wohlrab, R. Hoppe, G. Schulz-Ekloff, D. Wöhrle, *Zeolites 1992*, *12*, 862.
20 D.E. DeVos, F. Thibault-Starzyk, P.P. Knops-Gerrits, R.F. Parton, P.A. Jacobs, *Macromol. Symp.* **1994**, *80*, 157.
21 B.V. Romanowsky, *Macromol. Symp.* **1994**, *80*, 185.
22 G. Meyer, M. Mohl, D. Wöhrle, G. Schulz-Ekloff, *Zeolites 1984*, *4*, 30.
23 K.J. Balkus, in Phthalocyanines–Properties and Applications, C.C. Leznoff, A.B.P. Lever (eds.), Vol. 4, VCH Publishers, New York **1996**, p. 287.
24 F. Marlow, W. Hill, J. Caro, G. Finger, *J. Raman Spectrosc.* **1993**, *24*, 603.
25 F. Marlow, M. Wübbenhorst, J. Caro, *J. Phys. Chem.* **1994**, *98*, 12 315.
26 M. Ehrl, F.W. Deeg, C. Bräuchle, O. Franke, A. Sobbi, G. Schulz-Ekloff, D. Wöhrle, *J. Phys. Chem.* **1994**, *98*, 47.
27 C. Schomburg, D. Wöhrle, G. Schulz-Ekloff, *Zeolites* **1996**, *17*, 232.
28 C. Schomburg, M. Wark, Y. Rohlfing, G. Schulz-Ekloff, D. Wöhrle, *J. Mater. Chem.* **2001**, *11*, 2014.
29 C. Schomburg, D. Wöhrle, G. Schulz-Ekloff, M. Wark, in Zeolites and Mesoporous Materials at the dawn of the 21st centrury, A. Galarneau, F. Di Renzo, F. Fajula; J. Vedrine (eds.), Studies in Surface Science and Catalysis, Vol. 135, Elsevier, Amsterdam **2001**, p. 359 (22-P-07).
30 J. Schneider, D. Fanter, M. Bauer, C. Schomburg, D. Wöhrle, G. Schulz-Ekloff, *Microporous Mesoporous Mater.* **2000**, *39*, 257.
31 A. Corma, V. Fornes, H. Garcia, M A. Miranda, J. Primo, M.-J. Sabater, *J. Am. Chem. Soc.* **1994**, *116*, 2276.
32 J.C. Crano, R.J. Guglielmetti, Organic Photochromic and Thermochromic Compounds, Vol. 2, Kluwer/Plenum, New York **1999**.
33 H Kacirek, H. Lechert, *J. Phys. Chem.* **1975**, *79*, 1589.
34 C. Schomburg, PhD Thesis, University of Bremen, **2000**.
35 H.H. Tagaya, T. Nagaoka, T. Kuwahara, M. Karasu, J. Kadokawa, K. Chiba, *Microporous Mesoporous Mater.* **1998**, *21*, 395.

36 X. Sun, M. Fan, E.T. Knobbe, *Mol. Cryst. Liq. Cryst.* **1997**, *297*, 57.

37 M. Ueda, K. Kudo, K. Ichimura, *J. Mater. Chem.* **1995**, *5*, 1007.

38 R. Heiligman-Rim, Y. Hirschberg, E. Fischer, *J. Phys. Chem.* **1992**, *96*, 2465.

39 T. Bercovici, R. Heiligman-Rim, E. Fischer, *Mol. Photochem.* **1969**, *1*, 189.

40 H. Dürr, H. Bonas-Laurent (eds.), Photochromism, Elsevier, Amsterdam **1990**.

41 H. Dürr, *Angew. Chem.* **1989**, *101*, 427.

42 F. Ghebremichael, M.G. Kucy, *J. Appl. Phys.* **1995**, *77*, 2896.

43 J.O. Morley, R.M. Morley, R. Docherty, M.H. Charlton, *J. Am. Chem. Soc.* **1997**, *119*, 10 192.

44 F. Marlow, K. Hoffmann, in *Proc. 12th Int. Zeolite Conf.*, M.M.J. Treacy, B.K. Marcus, M.E. Bisher, J.B. Higgins (eds.) Materials Research Society, Pennyslvania **1999**, Vol. III, p. 2121.

45 G. Wirnsberger, B.J. Scott, B.F. Chmelka, G.D. Stucky, *Adv. Mater.* **2000**, *12*, 1450.

46 C.J. Drummond, D.N. Farlong, *J. Chem. Soc. Faraday Trans. B* **1990**, *8*, 3613.

47 J. Schneider, D. Fanter, M. Bauer, this volume, Chapter 4.8.

3
Microwave-Assisted Crystallization Inclusion of Dyes in Microporous $AlPO_4$-5 and Mesoporous Si-MCM-41 Molecular Sieves

Matthias Ganschow, Ingo Braun, Günter Schulz-Ekloff, and Dieter Wöhrle*

3.1
Introduction

The structural peculiarities of molecular-sieves enable the incorporation of optically active guest molecules in crystallographically defined positions or highly organized arrangements [1–4]. The advantages of mineral hosts, which are of high thermal and mechanical stability as well as optical transparency in the visible region and of stability towards ultraviolet radiation, are also valid for the molecular sieves, used for the incorporation of organic chromophores and metal complexes. The encapsulation leads to composite materials with novel optical properties, which give potential applications as pigments or as materials for optical data storage, frequency doubling, microlasing, gas sensing, catalysis, or photocatalysis [5].

Dye molecules can be incorporated in molecular sieves by four different methods: (1) ion exchange in aqueous solution, (2) deposition from the vapor phase into the molecular sieve, (3) *in situ* synthesis in cavities or channels of a molecular sieve, or (4) crystallization inclusion during the hydrothermal synthesis of the molecular sieve [6,7]. The last method, in particular, enables stable encapsulation of dyes, such as phthalocyanines with a diameter larger than the channel opening, in mesopores [8,9]. Smaller organic chromophores, such as methylene blue or thioindigo, which fit approximately into the diameter of the channel openings, can also be stably fixed in zeolites and $AlPO_4$-5 due to host–guest or guest–guest interactions [10]. Dyes incorporated by the method of crystallization inclusion exhibit increased photooxidative stability [8], and enhanced properties of photochromic molecules for optical switching and storage [11].

The drastic reaction conditions of the hydrothermal synthesis (for example faujasites: pH =13–14, $T = 80$ °C, $t = 24$–720 h or $AlPO_4$-5: pH = 2–4, $T = 150$–210 °C, $t = 6$–24 h) can result in hydrolytic or thermal degradation of dye molecules. However, unstable dyes can be successfully incorporated by application of microwave (MW)-assisted synthesis of the molecular sieve. Since the decomposition of the chromophores increases with the crystallization time for the host sys-

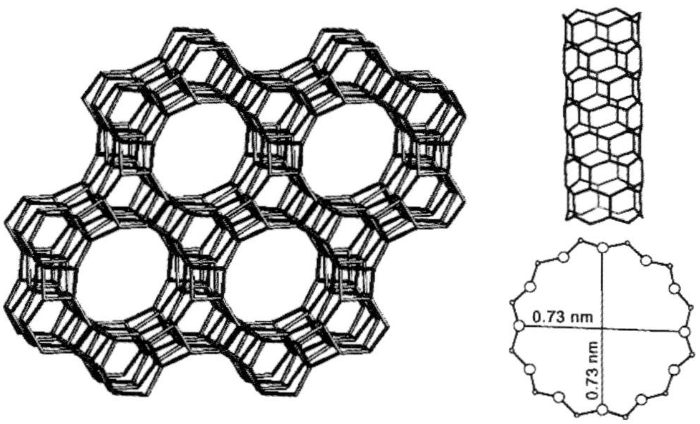

Fig. 1. Channel structure of $AlPO_4$-5.

tem, during which the dye molecules are exposed to the hydrolytic reaction conditions, being up to several days, the much faster (15–45 min) crystallization of $AlPO_4$-5, Si-MCM-41, and faujasites by MW heating represents a prospective new method for undestroyed encapsulation of chromophores [12].

3.2 Dyes in the Microporous Molecular Sieve $AlPO_4$-5

The molecular sieve $AlPO_4$-5 exhibits a hexagonal, anisotropic, one-dimensional channel structure (Fig. 1) [13].

Composites based on $AlPO_4$-5 molecular sieve crystals exhibit peculiar advantages for optical functions. Dipolar chromophores fitting in the channels of calcined $AlPO_4$-5, can be included via adsorption from the gas phase. Steric restrictions and chemical interactions result in aligned orientations of the guest molecules along the channel axis, exhibiting optical second-harmonic generation, such as for *p*-nitroaniline or *p*-dimethyl-aminobenzonitrile [14].

In the conventional (CV) hydrothermal synthesis from a standard gel composition (1.0 Pr_3N/1.0 Al_2O_3/1.0 P_2O_5/35 H_2O) after a crystallization period of 6–24 h at 160 °C a pure $AlPO_4$-5 phase is obtained. In the MW-assisted preparation only dense phases and no $AlPO_4$-5 are obtained from this standard gel composition. An increase in the amount of template Pr_3N as well as of the water fraction is required to obtain a pure $AlPO_4$-5 phase. The modified gel composition under MW conditions is 2.0 Pr_3N/1.0 Al_2O_3/1.0 P_2O_5/150 H_2O. Compared with the CV synthesis with the reaction time of 6–24 h, under MW conditions the reaction time is reduced to <45 min. Furthermore, it is necessary to heat the starting gel to 160–170 °C very quickly (within 1 min), otherwise no successful synthesis of a pure $AlPO_4$-5 phase is achieved [15–17].

The drastically shortened reaction time for the MW-assisted synthesis can be related to the different heating process using MW radiation at 2450 MHz, which nearly matches the rotation frequency of the water dipoles. Three predominant variables influence the phase and morphology of the molecular sieve formed during the heating period of the reaction gel under MW conditions. These variables are the starting gel volume, the water content, and the MW power. When MW energy penetrates the gel mixture, the energy is absorbed by the sample at a rate that depends on the dissipation factor for absorptive samples. Since the frequency of the applied MW radiation (2450 MHz) corresponds with the reciprocal of the dielectric relaxation time for water dipole rotation, an aqueous solution exhibits a high dissipation factor and, consequently, rapid heating [18]. This explains the faster process of molecular sieve crystallization for the MW-assisted synthesis compared with that using a CV heating system. The larger fraction of water, required for the former synthesis method, is responsible for the necessary larger fraction of template for a more diluted synthesis batch. The selective energy input guarantees not only the short heating time but also the superfluid state of the sol-gel suspension, characterized by a lower degree of clustering of the water molecules by hydrogen bonding, of which the influence on (1) the processes of fragmentation and dissolution of the gel, (2) transport of the secondary building units through the solution, and (3) crystallization of the molecular sieve is not yet fully understood. The changes in the crystallization process are reflected in the required modification of the batch composition (the increase of the fractions of template and water) for an optimal product: pure $AlPO_4$-5 phase.

In addition, for the first time the synthesis of $AlPO_4$-5 in a MW-heated, continuous-flow, high-pressure tube reactor is described, based on recipes optimized for the MW-heated batch reactor [19].

3.2.1
Crystallization Inclusion of Dyes in $AlPO_4$-5

The crystallization inclusion of the unstable Coumarin dye Basic Yellow 40 (formula in Fig. 6) using the CV method and the corresponding optimal gel composition 1.0 Pr_3N/1.0 Al_2O_3/1.0 P_2O_5/35 H_2O, leads after UV/vis to a decomposition of the chromophore molecules. The characteristic absorption at $\lambda = 436$ nm disappears and a new absorption at $\lambda = 362$ nm is observed [15]. The degradation of Basic Yellow 40 is, presumably, initiated by the hydrolysis of the lactone bond. The intermediate product, a substituted (Z)-*o*-hydroxycinammic acid, is hydrolyzed after hydration of the aliphatic double bond in a retroaldol reaction down to *o*-hydroxybenzaldehyde.

If, however, the dyes are added to the batches for MW-assisted crystallization (gel composition: 2.0 Pr_3N/1.0 Al_2O_3/1.0 P_2O_5/150 H_2O) a yellow powder is obtained. The corresponding UV/vis reflectance spectra show absorption bands similar with the spectra in solution. An expected inhomogeneous band-broadening appears, indicating the host–guest interactions of the accommodated chromophore molecules. After destruction of the $AlPO_4$-5 with hydrochloric acid, the dye contents were determined by photometry. Between 5.1 and 13.5 % of the offered dyes were incorporated, corresponding to a loading between 8.0×10^{-6} and 2.3×10^{-5} mol per gram molecular sieve. These quantities correspond to about one dye molecule per 33–88 unit cells (u.c.). Up to this loading, the diffractograms of the dye-loaded $AlPO_4$-5 exhibit no significant difference from the pure $AlPO_4$-5 phase [15,17].

The dye is stably incorporated and not absorbed on the surface because it could not be extracted by intensive treatment with boiling ethanol. Deposition of the dye at the external surface by stirring of pure $AlPO_4$-5 with solutions of the dyes led to colored products from which the dyes could be removed totally by washing with hot ethanol. Molecular modeling of the dimension of the dye with the Hyperchem program (size of Basic Yellow 40 is 0.6 nm × 1.6 nm) indicate that the chromophores fit in the $AlPO_4$-5 channel (diameter 0.73 nm). The dye molecules are fixed in the channels by host–guest interactions as well as by the large amount of incorporated template molecules acting as barriers to the removal of the dye.

$AlPO_4$-5 molecular sieves can also accommodate chromophore molecules, being larger than the channel dimensions of the mineral host, using the methods of CV or MW-assisted crystallization inclusion. Little is known about the parameters limiting the amount of dye that can be incorporated in the $AlPO_4$-5 framework. A positive charge at the chromophore is, in any case, favorable for crystallization inclusion [4]. Larger amounts of dye in the synthesis mixture affect the morphology of the $AlPO_4$-5 crystals and favor the formation of aggregates. We investigated the influence of the dye structure on the extent of incorporation in detail by using substituted rhodamine derivates (size 0.85 nm × 1.6 nm) (Fig. 2) [16].

The UV/vis spectra of the encapsulated rhodamines exhibit agreements as well as deviations from those in aqueous solution (Fig. 3). The positions of the principal absorption bands at $\lambda \approx 555$ nm (Rh B $\lambda = 551$ nm, Rh 3B $\lambda = 557$ nm, Rh BE50 $\lambda = 559$ nm), originating from the π–π^* transitions from binding HOMO (highest occupied molecular orbitals) to antibinding LUMO (lowest unoccupied molecular orbitals) along the longest dimension of the conjugated system [20], are largely unchanged compared with the aqueous solution. However, the blue shoulder at $\lambda \approx 520$ nm, usually ascribed to a dimer [21], exhibits a strongly increased extinction in all cases and a most marked blue shift for the samples obtained by CV synthesis (Fig. 3). The widths of the double-band, the principal band and shoulder, are strongly broadened. The bands at $\lambda < 450$ nm originate from π–π^* transitions from the NHOMO (next highest occupied molecular orbitals) to the LUMO [22].

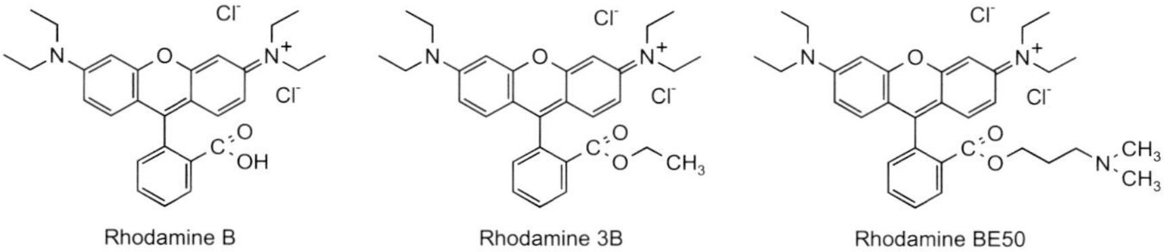

Fig. 2. The structures of the incorporated rhodamine derivates Rh B, Rh 3B, and Rh BE50.

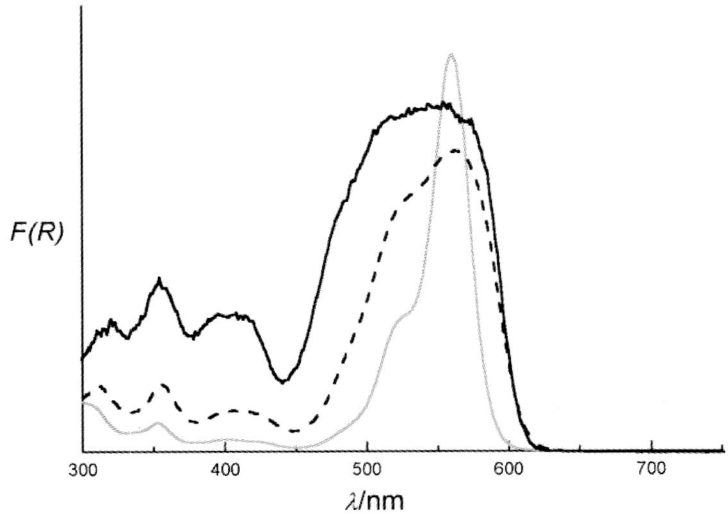

Fig. 3. Absorption spectrum of rhodamine BE50 in H_2O (gray), reflectance spectrum of rhodamine BE50 in $AlPO_4$-5 obtained by MW synthesis (dashes, $\lambda_{max} = 559$ nm) and reflectance spectrum of rhodamine BE50 in $AlPO_4$-5 obtainec by CV synthesis (black).

These transitions, which are usually forbidden by symmetry, are strongly enhanced and exhibit longitudinal as well as transverse polarizations.

The maximal dye uptake by crystallization inclusion in $AlPO_4$-5 expressed as number of unit cells accommodating one dye molecule, #Q# increases in the sequences (i) Rh B (140) < Rh 3B (75) < Rh BE50 (35) for the CV method of crystallization inclusion and (ii) Rh B (180) < Rh 3B (100) < Rh BE50 (30) for the MW-assisted one (Fig. 4) [16]. The explanation has to be found in the structure of the different chromophores. Rh B is characterized by a free carboxyl group. From its pK_a value of 3.1–3.5 [23], it follows that this dye exists as zwitterion during the synthesis in which the pH value rises from 3–4 at the start to 8–9 at the end. The dye Rh 3B, however, has an esterified carboxyl group, eliminating the zwitterionic character of Rh B and increasing its ability to compete with the template molecules for accommodation. This ability is strongly increased for Rh BE50, being esterified with 3-dimethylamino-1-propanol. This new derivative dye molecule exhibits an additional positive charge by protonation of the aliphatic amino group and improved competition to the template Pr_3N. Presumably, the localized positive charge at the aliphatic amino group results in a better interaction with the aluminophosphate framework than the delocalized positive charge distributed over the conjugated electron system of the xanthene chromophore. By rationalization of this effect one comes to the conclusion that higher rates of dye inclusion will be achieved, generally, with template analogue moieties substituted to the chromophore molecule via spacer groups protecting the localization of the positive charge at the amino group.

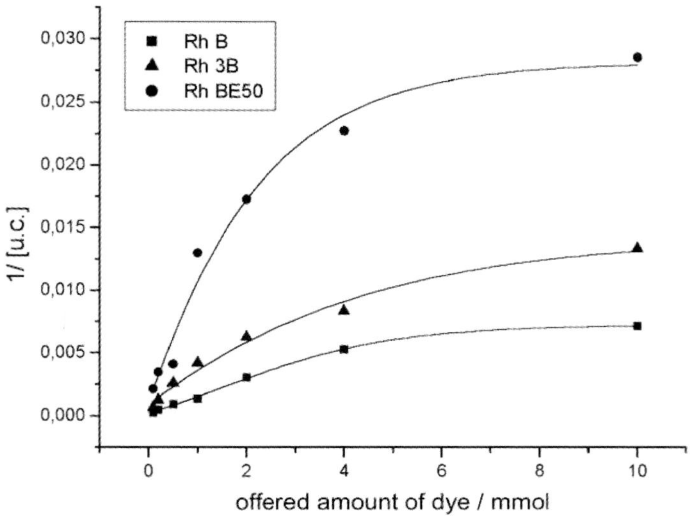

Fig. 4. Dye uptake (molecules per unit cell) against dye offered for samples prepared by the MW-assisted method.

For all chromophores, such as rhodamine and coumarin encapsulated during hydrothermal synthesis in the neutral channel-structured $AlPO_4$-5, the energy transfer via the Förster mechanism was studied [16,17]. With increasing loadings the maxima of the fluorescence of Rh BE50 in the range $\lambda = 570–615$ nm are increasingly shifted towards the red region of the spectrum. The fluorescence intensities depend on the dye loading and pass through a maximum at loadings of $\approx$400–200 u.c. per dye molecule. The dye concentration has been converted into average distances between dye molecules. The highest fluorescence intensities of rhodamines and Basic Yellow 40 are obtained at a distance of $\approx$10–12 nm (Fig. 5) [16,17].

The decrease in the fluorescence intensity with dye loadings lower than 1 molecule per 200–400 u.c. is explained by the too large average distances, and the decrease with dye loadings higher than 1 molecule per 200 u.c. is attributed to the long-range resonance coupling between excited dipolar chromophores and non-excited ones. Based on Förster's theory of radiationless energy transfer [24,25] with $(R_o/r)^6$ (R_o is the critical distance for a 50 % decrease of fluorescence intensity; r is the actual average Förster distance of dye molecules), values of $r \approx 8$ nm for rhodamines and of $r \approx 6$ nm for Basic Yellow 40 were calculated [16].

Both effects, the blue shift of the center of the double-band width at half maximum in the UV/vis spectrum and the red shift of the maximum of the fluorescence spectrum, have been combined in the representations of the Stokes shift, being much larger for the samples prepared by the CV synthesis. It is found that the Stokes shift is nearly independent of the kind of rhodamine dye.

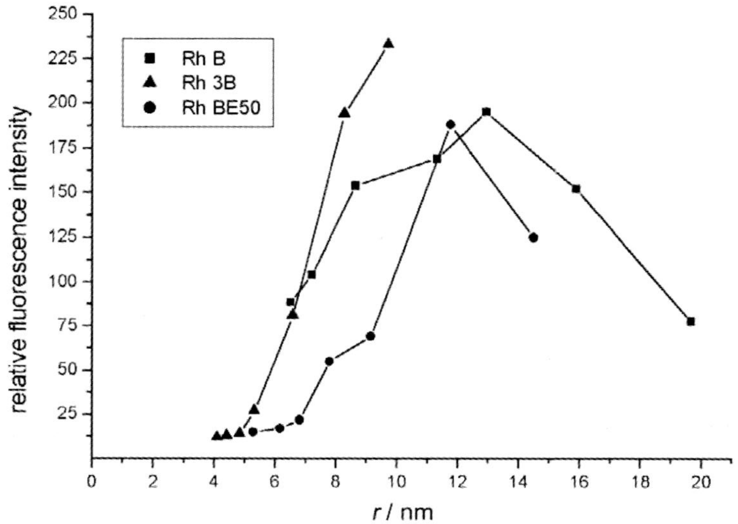

Fig. 5. Relative fluorescence intensity against the average separation of chromophores in $AlPO_4$-5.

The strong host–guest interaction in the dye-loaded $AlPO_4$-5 is obvious from the marked shifts in the absorption and fluorescence spectra [16]. The strong increase of the blue shoulder of the absorption band is especially indicative of such an interaction. The blue shoulder cannot be attributed to the presence of a high fraction of dimers, since the samples exhibit strong fluorescence. It has been clearly demonstrated that rhodamine dimers do not fluoresce but exhibit strong self-quenching by efficient intersystem crossing [25]. This means that the strong blue shoulder in the principal absorption band for the encaged dye originates from the increased probability of transitions from the vibrational ground state of the electronic ground state to an excited vibrational state of the excited electronic state of a monomer, 0–1 transitions according to the Franck–Condon principle. The maximal Stokes shifts found for the $AlPO_4$-5-encaged dyes (50–70 nm) are much larger than those for the dissolved ones (about 25 nm) or those reported for silica-encaged rhodamine dyes (about 25 nm) [26]. The strong shifts or strong host–guest interactions, might be attributed to the peculiar structure of the $AlPO_4$-5 framework. This type of molecular sieve has a polar nature originating from a partial negative charge at the Al and a partial positive one at the P atoms of the framework. This results in dipoles aligned along the *c*-axis of the $AlPO_4$-5 crystal with translational symmetry. The alignment of the permanent framework dipoles in a superposition leads to a macroscopic dielectric susceptibility or dipolar crystal. This effect explains the unusually strong host–guest interactions, by a polarization of the π-electron system of the chromophores by the framework dipoles. This polarization (1) increases the HOMO–LUMO gap in the ground state, explaining the blue shift in the absorption spectra and (2) decreases the energy of the electronic states in the dipolar excited state due to dipole–dipole interaction, explaining the red shift in fluorescence spectra. The increasing red shift with increasing dye loading indicates participation of the dye molecules in the dipole–dipole interaction and, thus, a locally oriented arrangement of the chromophores in the $AlPO_4$-5 framework enabling their mutual interaction. These strong host–guest interactions cause marked symmetry distortions in the orbital and bond structure of the chromophores, resulting in the strong increase in the blue shoulder of the absorption spectra or the increased probability of 0–1 transitions, respectively. Furthermore, the symmetry distortions increase the probability of transitions in the $\lambda < 450$ nm region that are normally forbidden by symmetry. Symmetry operations for rhodamine dyes are not affected by the carboxyphenyl-group, not participating in the conjugated chromophore system.

The larger Stokes shift for the samples prepared by the CV method has to be attributed to the much slower rate of crystallization. Presumably, the slower rate of crystallization yields a host crystal exhibiting either a lower density of defect sites (with a less flexible framework for the accommodation of dye molecules exceeding the dimensions of the $AlPO_4$-5 channels) or a more perfect orientation of the framework dipoles (resulting in a stronger electrostatic interaction between host and guest) [17].

The slim coumarin dye is orientedly encapsulated in the $AlPO_4$-5 host. The alignment of the chromophores is shown by UV/vis absorption spectroscopy using polarized radiation. Strong absorptions for the coumarin chromophores in $AlPO_4$-5 appear only if the plane of polarization of the radiation is parallel to the *c*-axis of the host crystal (Fig. 6). The geometric restrictions by the size matching pores and the polarity of the $AlPO_4$-5 structure aligns the chromophores so that their principal transition dipole moment is parallel to the *c*-axis of the host crystal [17,27]. The

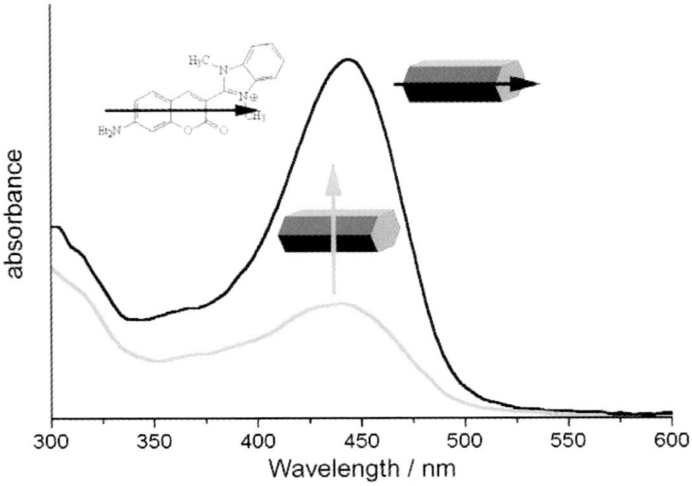

Fig. 6. UV/vis spectra of Basic Yellow 40 using polarized radiation.

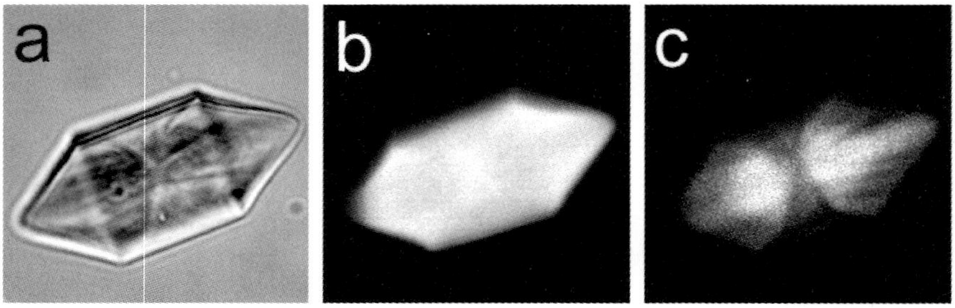

Fig. 7. The typical appearance of Basic Yellow 40 and Rhodamine BE50 in an $AlPO_4$-5 crystal in transmission (a) and the corresponding fluorescence images (b and c). The fluorescence images are taken in the focal plane of the objective (*xy*-plane) with an excitation of 442 nm and an detection wavelength of (b) 470 nm (Basic Yellow 40) and (c) 550 nm (Rhodamine BE50).

bulky Rhodamine BE50 is not orientedly encapsulated in the $AlPO_4$-5 host, so that their principal transition dipole moment is randomly oriented to the *c*-axis of the host crystal.

If both types of chromophores are added to the batch for the hydrothermal synthesis, then a selective accommodation of the different dyes is observed by us [28]. The slim coumarin is encaged in the center of the crystals (Fig. 7b), whereas the rhodamine is preferentially accommodated close to the pyramidal end of the crystals (Fig. 7c). This selective dye inclusion indicates that the smaller chromophores are accommodated first (fast crystal growth), whereas the more bulky rhodamine, which requires the formation of defects in the host, is incorporated at a later stage (slow crystal growth) of the crystal growth.

Single molecule spectroscopy (SMS) is a valuable tool for characterizing materials in a straightforward manner. SMS completely eliminates ensemble averaging, gives direct access to the heterogeneities of the system and direct information about the near environment of the molecule [29], in our case the orientation of the single molecules against the crystal framework (channels). An orientational distribution of a dye molecule accommodated in the channels of a molecular sieve has been observed showing a preferential orientation depending on the size of the molecule [30]. Three different size variations of oxazine dyes were studied (Fig. 8) [28,31].

The orientations of the single fluorophores with respect to the *c*-axis were evaluated by detecting the fluorescence response signal of the chromophores after rotary modulation of the excitation radiation. Many orientations are measured for every probe. A histogram over the angles generates the distributions shown in Fig. 9. They are the direct forms of the angular distribution.

In the case of the smaller Oxazine 1 molecules, a narrow distribution of angles aligned to the main crystal axis is found, whereas the larger Oxazine 170 distribution is broader, suggesting a more intense distortion of the channel structures. The largest molecule studied, Oxazine 750, shows no preferential orientation at all.

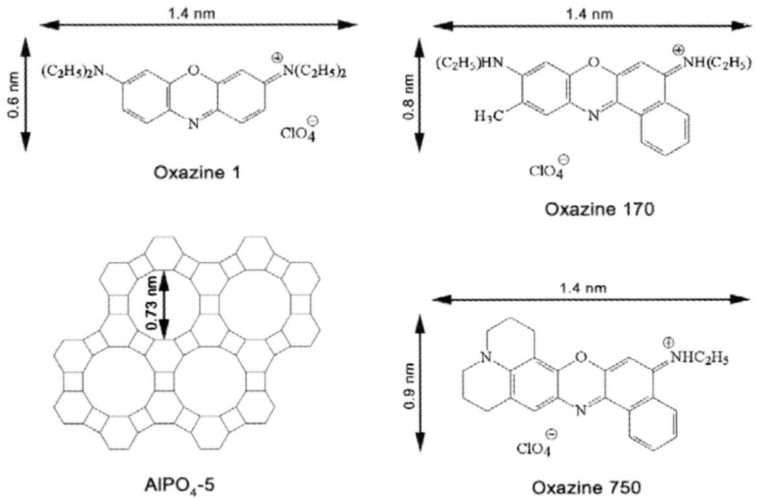

Fig. 8. The three oxazine dyes of different size and the $AlPO_4$-5 structure.

In summary it may be said that (1) small molecules are aligned and cause minor defects, (2) medium molecules are aligned and cause bigger defects, and (3) huge molecules have no alignment at all [28,31].

3.2.2
Crystal Morphology of $AlPO_4$-5

Remarkably, up to concentrations of 1 molecule per 75 unit cells there are no negative consequences for the crystal morphology, as shown in Fig. 10a. But the crystal morphologies change with higher dye content from single crystallinity (Fig. 10a) to bunched aggregation (Fig. 10b) (1.5×10^{-5} mol g^{-1} Rhodamine BE50) [15,17].

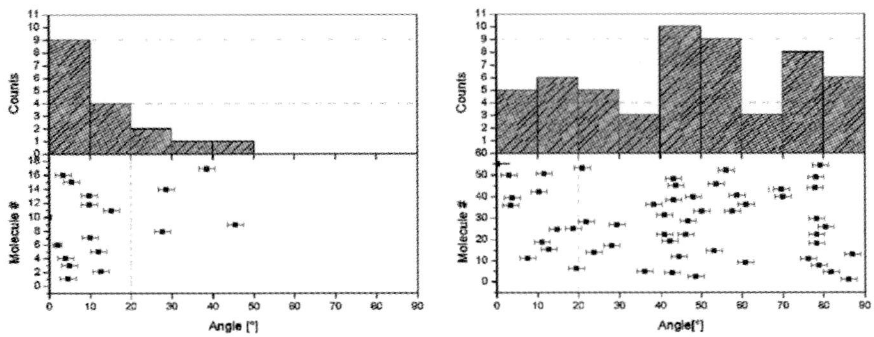

Fig. 9. Histogram/distribution for Oxazine 1 (left) and Oxazine 170 (right).

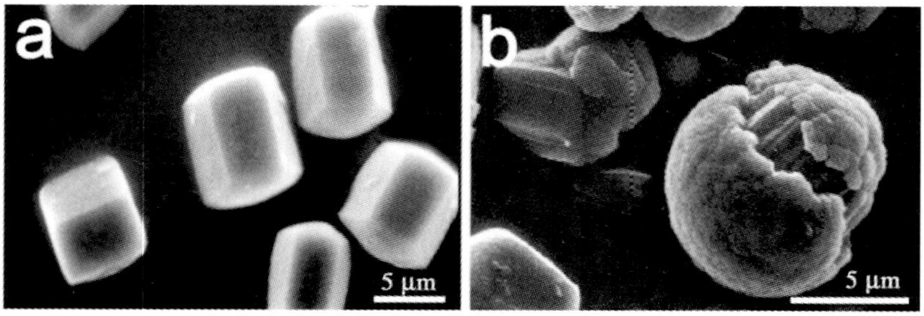

Fig. 10. Morphology of Rhodamine BE50/$AlPO_4$-5 crystals: (a) 4.8×10^{-6} mol g^{-1}, (b) 1.5×10^{-5} mol g^{-1} (loadings: mol Rhodamine BE50 per g $AlPO_4$-5).

The control of the $AlPO_4$-5 crystallization by the CV heating method was studied repeatedly regarding crystal growth and morphology [32]. The charges of $AlPO_4$-5 single crystals, obtained by CV, often exhibit a wide size distribution, different kinds of crystal aggregation and by-products. During the faster MW-heated crystallization very small crystals can be formed since many more crystal nuclei are generated simultaneously during the rapid increase of the gel temperature and the growth of these nuclei is much more homogeneous [33]. Multiple morphologies are often present in the synthesis products [16]. For several applications it is necessary to get $AlPO_4$-5 crystals with a defined and uniform morphology [4].

We studied the formation of different phases and morphologies of aluminum phosphate molecular sieves by the MW-assisted reaction versus the synthesis temperature and the reaction time [34]. After MW heating for 15 min at 145 °C, a solid product is isolated consisting of a mixture of molecular sieve phases including about 20 mass-% of nonreacted pseudoboehmite Al_2O_3, according to X-ray diffraction. The crystalline fraction consists of $AlPO_4$-8 [35] and VPI-5 [36] needles and a small amount of the characteristic hexagonal $AlPO_4$-5 crystals (Fig. 11a).

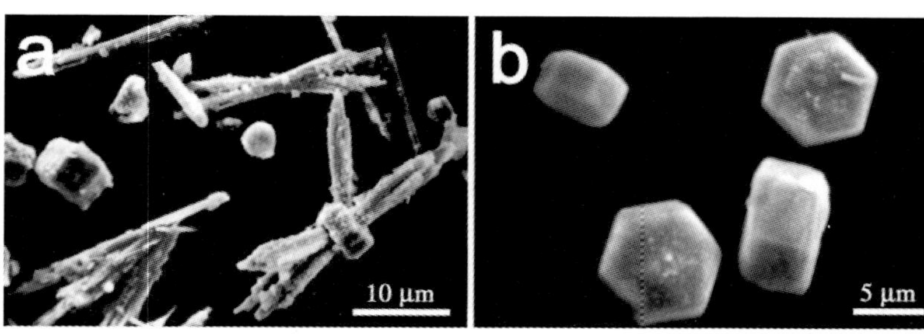

Fig. 11. (a) Scanning electron micrograph of $AlPO_4$-5 and $AlPO_4$-8/VPI-5 crystals (reaction time 15 min, reaction temperature 145 °C); (b) $AlPO_4$-5 crystals 4×9 µm (reaction time 10 min, reaction temperature 150 °C).

A prolongation of the reaction time results in an increase of the fraction of $AlPO_4$-5 and $AlPO_4$-8 and a decrease in the portion of VPI-5 phase. After 40 min reaction time the VPI-5 is no longer observable. The VPI-5, formed within the first 15 min, is energetically unstable due to its 18-ring channels, and, thus, recrystallizes into the more stable molecular sieve $AlPO_4$-8 possessing only 14-membered rings [37]. Both, VPI-5 and $AlPO_4$-8 belong to the class of $AlPO_4$ hydrates, which possess six-fold coordinated Al atoms and are therefore less stable than the aluminum phosphates with only four-fold coordinated Al, of which $AlPO_4$-5 is a member [38].

If MW-assisted crystallization is carried out at 150 °C, a solid product can be isolated after a short reaction time of about 10 min. In addition to minor amounts of nonreacted pseudoboehmite, VPI-5 crystals with needle-like morphology and $AlPO_4$-5 crystals with hexagonal plate-like morphology are formed (Fig. 11b). The length of the hexagonal plate-like crystals (*c*-direction, parallel to the channel system) is only about 4 μm and, thus, only about 50 % of the diameter in *a*–*b*-direction of 8 μm (plane perpendicular to the channels).

After reaction times longer than about 15 min, $AlPO_4$-8 (originating from the first precipitated VPI-5) and $AlPO_4$-5 crystals are preferentially formed. Pure $AlPO_4$-5 without any subordinary phases is obtained after 40 min reaction time. Scanning electron micrographs show hexagonal barrel-shaped crystals with a length of about 8 μm in the *c*-direction (Fig. 10a). This means that the plates have changed into hexagonal barrels, in which the hexagonal morphology and expansion in the *a*–*b* plane are kept and only a growth in *c*-direction took place [28,33] This kind of behavior has also been reported for several kinds of zeolites using the CV heating method [39].

A further prolongation of the reaction time to more than 50 min leads to an obviously oriented, probably epitaxial overgrowth of appendices out of the $AlPO_4$-5 bodies in the *c*-direction, so the crystals possess the barrel-like hexagonal body, typical of $AlPO_4$-5 with antenna-like appendices attached at its ends in the direction of the *c*-axis (Fig. 12a). If rhodamine BE50 is added to the gel before the synthesis, an analysis of the mixed crystals by optical microscopy shows that only the hexagonal bodies are intensively colored, whereas the antennae are colorless, the pink-colored rhodamine BE50 is incorporated only in $AlPO_4$-5 crystals, whereas $AlPO_4$-8 crystals do not host any dye (Fig. 12b).

The synthesis of $AlPO_4$-5 has no template specificity, since it can be prepared with many different templates [38]. In the $AlPO_4$-5 organic templates are acting as pore fillers with only weak van der Waals interactions to the framework atoms [40]. In contrast, in the formation of $AlPO_4$ hydrates, such as VPI-5 and $AlPO_4$-8, water molecules takes over the function as micropore filler. The introduction of charged molecules, such as organic templates or dyes, in the micropores is expected to disturb the specific water structures, such as the triple helix structure of water molecules present in the micropores of VPI-5 [41]. This explains (1) why no incorporation of organic molecules in the micropores during crystallization was found for the VPI-5 and (2) why the $AlPO_4$-8 antennae are free of dye molecules.

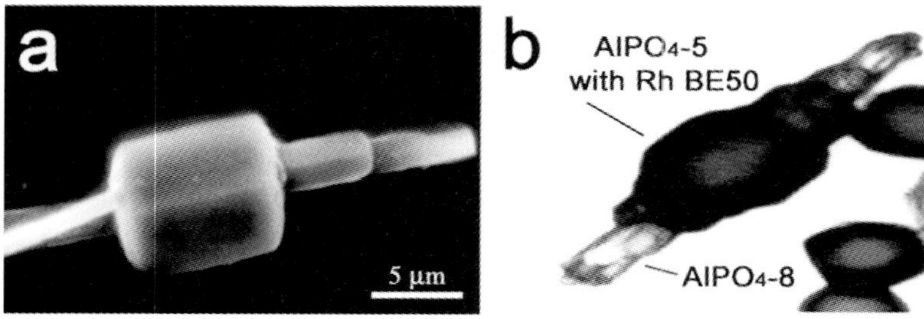

Fig. 12. (a) Epitaxial overgrowth of $AlPO_4$-5/-8 20 × 9 µm, after 60 min and 150 °C reaction temperature; (b) the optical microscope picture of an epitaxial overgrowth of $AlPO_4$-5/-8 with embedded Rhodamine BE50.

At a reaction temperature of 155 °C transient phases as well as the final formation of the dense phase T-$AlPO_4$, a tridymite analogue of aluminum phosphate are observed [42]. After 5 min reaction time hexagonal discs of $AlPO_4$-5 crystals are detected besides pseudoboehmite. Subsequently, these discs grow preferentially in the *c*-direction. Only the T-$AlPO_4$ phase appears after a prolonged reaction time (15 min) at 155 °C or dominates after a reaction time of 15 min above 160 °C. The T-$AlPO_4$ covers previously formed $AlPO_4$-5 crystals or appears exclusively at 170 °C as small spherical crystals with a diameter of 2–5 µm.

The temporal development of the morphology, from hexagonal discs (Fig. 11b) to hexahedral prisms (Fig. 10a), occurs as reported previously [33,34]. The oriented growth of the orthorhombic $AlPO_4$-8 on the hexagonal $AlPO_4$-5 might be favored by (1) the similar topology of the channels of both structure types and (2) the nearly identical framework densities. The time-dependent phenomena of transient phase formation as studied in detail are summarized in a temperature–time diagram (Fig. 13) [34].

Further, pencil- or stick-like morphologies can be received by adding different fluoride salts to the gel before the MW-synthesis (Fig. 14) [28].

3.3
Dyes in the Mesoporous Molecular Sieve Si-MCM-41

In molecular sieves with cage diameters less than 1.5 nm (such as $AlPO_4$-5 described above), dye molecules can be incorporated that are stable against extraction using *in situ* synthesis (ship in the bottle) or crystallization inclusion [4]. The mesoporous host Si-MCM-41 is a member of the class of M41S materials with hexagonally arranged cylindrical pores [43], which is formed by a template-assisted liquid crystal process. The structure-directing micelles are build by cationic organic surfactant molecules, such as cetyltrimethylammoniumbromide (CTAB) [44]. The

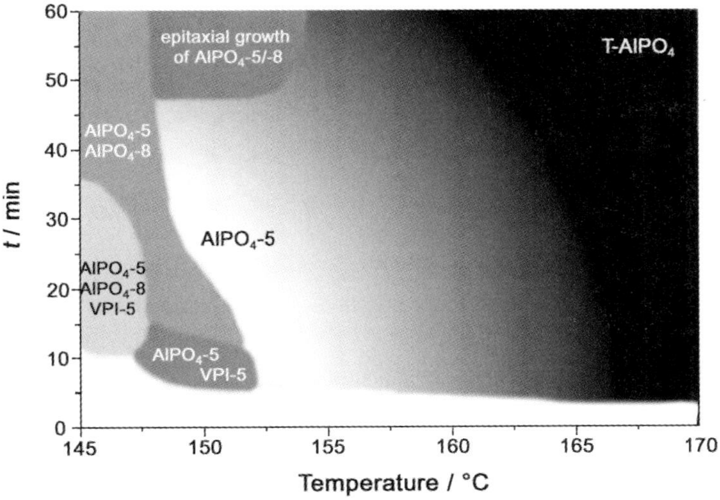

Fig. 13. Formation of different phases against reaction temperature and reaction time.

surfactant molecules inside the pore structure can be burnt-off by calcination at 873 K or can be removed by extraction [45]. The wide channel-like pores of MCM-41 (about 3 nm in diameter) enable the inclusion of relatively large chromophores; these are, however, either extracted or destroyed during removal of the template [46].

Applications in photocatalysis or optical sensing, however, particularly require stability against extraction, which can only be achieved by covalent anchoring of the dyes onto the walls of Si-MCM-41. For this purpose a suitable new method for covalent anchoring of chromophores at the walls of molecular sieves was recently developed by us [47]. The idea is that the chromophore is covalently bond to one of the building blocks of the host lattice. By co-condensation, covalent bonding in the molecular sieve is achieved. In order to avoid splitting of the silane–chromophore bond during crystallization, MW-assisted synthesis is advantageous. For the anchoring of dyes at MCM-41, 4-[4-(dimethyl-amino)phenylazo]benzoic acid

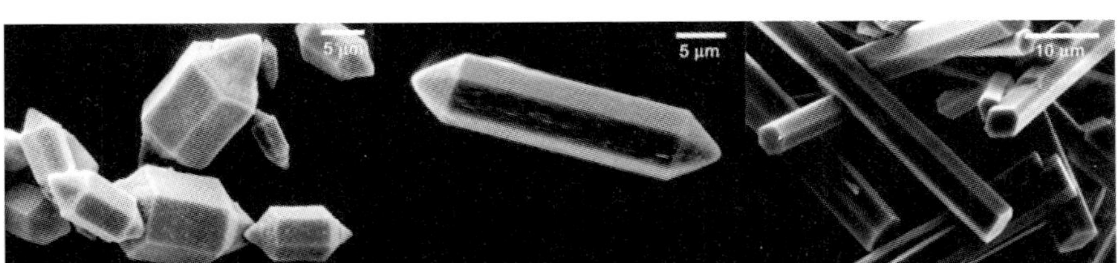

Fig. 14. Pencil- and stick-like morphologies from adding fluoride salts to the gel.

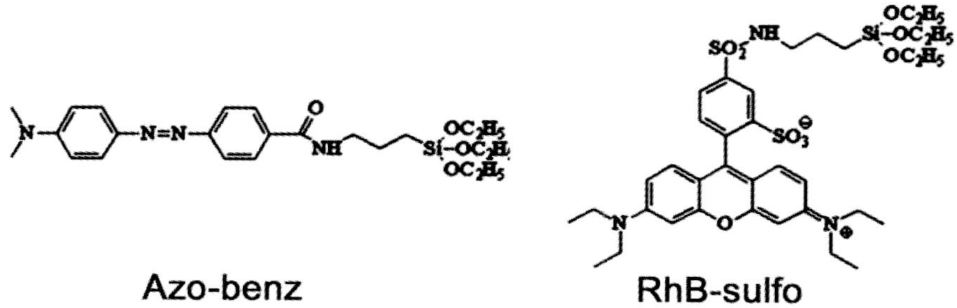

Fig. 15. Modified chromophores for covalent anchoring inside the channels of MCM-41.

(Azo-benz) or the sulfonic acid of Rhodamine B (RhB-sulfo) were first bonded to 3-aminotriethoxysilane (Fig. 15).

A mixture of tetraethoxysilane and modified dye was employed with cetyltrimethylammonium bromide in the synthesis of Si-MCM-41, which was finished under MW conditions after 30 min at 130 °C [47]. This ensures that no dye degrades during the hydrothermal synthesis of the Si-MCM-41 host. X-ray data confirm the Si-MCM-41 phase formation. On scanning electron micrographs (SEM) Si-MCM-41 particles with a uniform morphology and a narrow particle size distribution were observed independent of the loading with dye molecules. The template was removed with aqueous HCl.

With the azo dye, powders colored from slightly yellow to dark orange are obtained, depending on the amount of dye added to the synthesis gel. The dye cannot be washed out of the pores either with water or by extraction with hot ethanol for several days. In contrast, dye molecules introduced by impregnation, which are not covalently anchored, are removed within a few minutes. Owing to the high stability of the anchored dye molecules, the template molecules can be removed with an aqueous HCl solution (1 molar) at 343 K. Even from template-free samples no dye could be removed by extraction with ethanol, which indicates an acid stability of the anchoring amide bonds.

The removal of the template leads to a decrease in the electron density in the pores, which results in an increase of the tinctorial strength of the samples due to changes in the refractive index. In diffuse reflectance UV/vis spectra an increase of the Kubelka–Munk ($F(R)$) values by a factor of about 2 and a red shift of the maximum by 30 nm is observed [47]. In analogy to a change of the solvent, the removal of the surfactant molecules alters the chemical environment of the dye molecules. The resulting strong influence on the reflectance of the samples demonstrates the location of dye molecules inside the pore system and excludes an exclusive anchoring on the external surface, which represents only 5% of the total surface.

The rhodamine is monomolecularly encapsulated. No great differences between the absorption and emission spectra of, for example, Rh B-sulfo in ethanol and covalently bonded at MCM-41 are seen. The Rh B/Si-MCM-41 is highly fluorescent, which indicates a monomeric incorporation of the dye molecules.

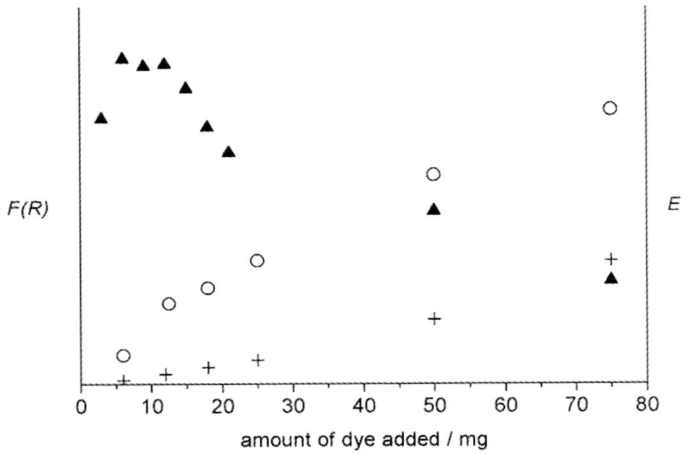

Fig. 16. $F(R)$ values (○, dye **1**; +, dye **2**) and fluorescence E (▲, dye **2**) against the amount of dye added to the synthesis gel.

Contrary to the situation in $AlPO_4$-5, no limit for inclusion was observed in the investigated concentration range for both dyes. The amount of encapsulated dye increases linearly with the offered concentration (Fig. 16).

Owing to the monomeric incorporation of the rhodamine dye in Si-MCM-41, an intense orange fluorescence is found. The emission maximum is continuously red-shifted with increasing dye concentration; the Stokes shift amounts to 31–39 nm for Si-MCM-41 with template and 32–47 nm for template-free samples. In analogy to the situation in solution, the fluorescence intensity passes through a maximum with increasing dye loading (Fig. 16). There are no indications for the formation of dimers in the diffuse reflectance spectra. Therefore it is assumed that red-shift and decrease in fluorescence intensity result from a radiationless energy transfer, which is known as Förster quenching [48]. The critical Förster distance between chromophores, the distance that leads to a 50% decrease in the maximal fluorescence intensity, is about 8 nm. From Fig. 16 it is obvious that the fluorescence decreases to 50% if 50 mg (0.07 mmol) of dye are added to 2.4 g TEOS in the synthesis gel. Considering the observed monomeric anchoring onto the inner surface of Si-MCM-41 (BET value: $\approx$1050 $m^2\ g^{-1}$), an average distance between the dye molecules of 6–7 nm is valid, which is in fair agreement with the literature data demonstrating again the uniformity of the incorporation [28,47].

The short synthesis time (minutes) facilitates the inclusion of dyes that will degrade under CV hydrothermal conditions. An uniform monomeric anchoring of the dye molecules on the inner surface of the Si-MCM-41 host of high quality results from X-ray diffraction, SEM, diffuse reflectance UV/vis, and fluorescence data.

In addition, an alternative synthetic method was developed by us to anchor azo and rhodamine dyes at the surface of Si-MCM-41 [49]. At first the external surface of the Si-MCM-41 was blocked by dichlorodiphenylsilan. Then in a second step the internal surface was activated by 3-aminopropyltriethoxysilan, followed in a third step by the covalent binding of the dyes. This method exhibits the advantage of only binding at the internal surface. But the method discussed before has the advantage of only one reaction step to get the modified molecular sieve.

3.4 Outlook

Stable crystallization inclusion of dyes is advantageous for the accommodation of various suitably substituted chromophores, being sensitive toward the conditions of hydrothermal synthesis. Moreover, it has turned out that selective inclusion of smaller dyes occurs during rapid MW-assisted crystallization; a preferential accommodation of smaller chromophores takes place first. Larger dye molecules enter later, in the further growing sections of the $AlPO_4$-5 crystal. The practical aspects of such sandwich-like accommodation for directed energy transfer have to be investigated [28,30].

The dye accommodation in porous minerals can be analyzed by bifocal microscopy with respect (1) to orientation-relationships via the "orientation" of the emitting fluorescence and (2) to dynamics of the chromophores in the channels and cages, giving further information about the nature and strength of host–guest interactions in porous matrices. Fluorescing chromophores represent helpful probes for following steps of the hydrothermal crystallization, which are only poorly understood.

A one step procedure for the covalent anchorage of dyes at the pore walls of the mesoporous Si-MCM-41 was developed. This method enables the synthesis of highly fluorescing materials [47].

The synthesized chromophore/molecular sieve materials have several potential applications: new kinds of hybrid pigments, fluorescent pigments, micrometer-sized lasers, and optical sensors [4,5d].

Acknowledgements

The financial support of the Deutsche Forschungsgemeinschaft under grant Schu 426/10-3 is gratefully acknowledged. The authors are grateful for fruitful collaboration with C. Bräuchle and coworkers (LMU Munich), F. Laeri and coworkers (University of Darmstadt), and K. Hoffmann (Berlin).

References

1 a) S.D. Cox, T.E. Gier, G.D. Stucky, J. Bierlein, *J. Am. Chem. Soc.* **1988**, *110*, 2986; b) S.D. Cox, T.E. Gier, G.D. Stucky, *J. Solid State Ionics* **1990**, *32/33*, 514; c) S.D. Cox, T.E. Gier, G.D. Stucky, *Chem. Mater.* **1990**, *2*, 609.

2 a) J. Caro, G. Finger, J. Kornatowski, J. Richter-Mendau, L. Werner, B. Zibrowius, *Adv. Mater.* **1992**, *4*, 273; b) F. Marlow, J. Caro, L. Werner, J. Kornatowski, S. Dähne, *J. Phys. Chem.* **1993**, *97*, 11 286.

3 G. Schulz-Ekloff, in Advanced Zeolite Science and Applications, J.C. Jansen, M. Stöcker, H.G. Karge, J. Weitkamp (eds.), Studies in Surface Science and Catalysis, Vol. 85, Elsevier, Amsterdam **1994**, p. 145.

4 G. Schulz-Ekloff, D. Wöhrle, R.A. Schoonheydt, B. van Duffel, *Micropor. Mesopor. Mater.* **2002**, *51*, 91.

5 a) M. Ehrl, F.W. Deeg, C. Bräuchle, O. Franke, A. Sobbi, G. Schulz-Ekloff, D. Wöhrle, *J. Phys. Chem.* **1994**, *98*, 47; b) G. Ihlein, F. Schüth, U. Vietze, O. Krauß, F. Laeri, *Adv. Mater.* **1998**, *10*, 1117; c) D.E. de Vos, F. Thibeault-Starzyk, P.P. Knops-Gerrits, R.F. Parton, P.A. Jacobs, *Macromol. Symp.* **1994**, *80*, 157; d) I. Braun, G. Ihlein, F. Laeri, J.U. Nöckel, G. Schulz-Ekloff, F. Schüth, U. Vietze, Ö. Weiß, D. Wöhrle, *Appl. Phys. B* **2000**, *70*, 335.

6 C. Schomburg, D. Wöhrle, G. Schulz-Ekloff, *Zeolites* **1996**, *17*, 233.

7 D. Wöhrle, G. Schulz-Ekloff, *Adv. Mater.* **1994**, *6*, 875.

8 D. Wöhrle, A. Sobbi, O. Franke, G. Schulz-Ekloff, *Zeolites* **1995**, *15*, 293.

9 a) S. Kowalak, K.J. Balkus, *Collect. Czech. Chem. Commun.* **1992**, *57*, 774; b) K.J. Balkus, S. Kowalak, K.T. Ly, D.C. Hargis, in Zeolite Chemistry and Catalysis, P.A. Jacobs, N.I. Jaeger, L. Kubelkova, B. Wichterlova (eds.), Studies in Surface Science and Catalysis, Vol. 69, Elsevier, Amsterdam **1991**, p. 93.

10 a) R. Hoppe, G. Schulz-Ekloff, D. Wöhrle, C. Kirschhock, H. Fuess, *Langmuir* **1994**, *10*, 1517; b) R. Hoppe, G. Schulz-Ekloff, D. Wöhrle, M. Ehrl, C. Bräuchle, in Zeolite Chemistry and Catalysis, P.A. Jacobs, N.I. Jaeger, L. Kubelkova, B. Wichterlova (eds.), Studies in Surface Science and Catalysis, Vol. 69, Elsevier, Amsterdam **1991**, p. 199; c) R. Hoppe, G. Schulz-Ekloff, D. Wöhrle, C. Kirschhock, H. Fuess, L. Uytterhoeven, R.A. Schoonheydt, *Adv. Mater.* **1995**, *7*, 61.

11 M. Ehrl, W. Kindervater, F.W. Deeg, C. Bräuchle, R. Hoppe, *J. Phys. Chem.* **1994**, *98*, 11 756.

12 a) I. Girnus, K. Jancke, R. Vetter, J. Richter-Mendau, J. Caro, *Zeolites* **1995**, *15*, 33; b) I. Girnus, M. Pohl, J. Richter-Mendau, M. Schneider, M. Noack, D. Venzke, J. Caro, *Adv. Mater.* **1995**, *8*, 711; c) B.I. Whittington, N.B. Milestone, *Zeolites* **1992**, *12*, 815; d) A. Arafat, J.C. Jansen, A.R. Ebaid, H. van Bekkum, *Zeolites* **1993**, *13*, 162.

13 a) J.M. Bennett, J.P. Cohen, E.M. Flanigen, J.J. Pluth, J.V. Smith, in Intrazeolite Chemistry, G.D. Stucky, F.G. Dwyer (eds.), *ACS Symp. Series*, **1983**, *218*, 109; b) W.M. Meier, D.H. Olson, Ch. Baerlocher, Atlas of Zeolites Structure Types, Elsevier, London **1996**, p. 26.

14 F. Marlow, W. Wübbenhorst, J. Caro, *J. Phys. Chem.* **1994**, *98*, 12 315.

15 I. Braun, G. Schulz-Ekloff, M. Bockstette, *Zeolites* **1997**, *19*, 128.

16 M. Bockstette, D. Wöhrle, I. Braun, G. Schulz-Ekloff, *Micropor. Mesopor. Mater.* **1998**, *23*, 83.

17 I. Braun, Doctoral Thesis, University of Bremen **1999**.

18 H.M. Kingston, L.B. Jassie, Introduction to Microwave Sample Preparation, Am. Chem. Soc., Washington **1988**, pp. 7–27.

19 I. Braun, G. Schulz-Ekloff, D. Wöhrle, W. Lautenschläger, *Micropor. Mesopor. Mater.* **1998**, *23*, 79.

20 a) G.N. Lewis, J. Bigeleisen, *J. Am. Chem. Soc.* **1943**, *65*, 2102; b) J. Fabian, H. Hartmann in Light Absorption of Organic Colorants, Reactivity and Structure Concepts in Organic Chemistry, Vol. 12, Springer, Berlin **1980**, p. 137 and references therein.

21 a) E. König, T. Förster, *Z. Elektrochem., Ber. Bunsenges. Physik. Chem.* **1957**, *61*, 347; b) H. Jakobi, H. Kuhn, *Z. Elektrochem., Ber. Bunsenges. Physik. Chem.* **1962**, *66*, 46.

22 a) H. Suzuki, Electronic Absorption Spectra and Geometry of Organic Molecules, Academic Press, New York **1967** and references therein; b) J. Griffith, K. J. Pender, Dyes and Pigments **1981**, *2*, 37 and references therein; c) R. Zahradnik, A. Kruhn, J. Panci, *Coll. Czech. Chem. Commun.* **1965**, *34*, 2553.

23 a) R.W. Ramette, E.B. Sandell, *J. Am. Chem. Soc.* **1956**, *78*, 4872; b) I. López Arbeloa, P. Ruiz Ojeda, *Chem. Phys. Lett.* **1981**, *79*, 347; c) M. El Baraka, M. Deumié, P. Viallet, *J. Photochem. Photobiol. A: Chem.* **1991**, *62*, 195.

24 a) E. Gaviola, P. Pringsheim, *Z. Physik* **1924**, *24*, 24; b) F. Perrin, *Ann. Chim. Phys.* **1932**, *17*, 283; c) T. Förster, *Ann. Phys.* **1948**, *2*, 55; d) T. Förster, *Z. Naturforsch. A* **1949**, *4*, 321.

25 a) A.K. Chibisov, T.D. Slarnova, *J. Photochem.* **1978**, *8*, 285; b) V.L. Levshin, L.V. Krotova, *Opt. Spectrosk.* **1962**, *11*, 457; c) K.H. Drexhage, in Dye Lasers, F.P. Schäfer (eds.), Springer, Berlin **1977**, Chapter 4.

26 D. Avnir, D. Levy, R. Reisfeld, *J. Phys. Chem.* **1984**, *88*, 5956.

27 I. Braun, C. Schomburg, M. Bockstette, G. Schulz-Ekloff, D. Wöhrle, *Proc. 12th Int. Zeolite Conf.*, Baltimore, 3–10 July 98, Vol. III, p. 2233.

28 M. Ganschow, Doctoral Thesis, University of Bremen **2001**.

29 a) Ph. Tamarat, A. Maali, B. Lounis, M. Orrit, *J. Phys. Chem. A* **2000**, *104*, 1; b) Th. Basché, W.E. Moerner, M. Orrit, U.P. Wild, eds., Single Molecule Optical Detection, Imaging and Spectroscopy, VCH, Weinheim **1997**.

30 S. Megelski, A. Lieb, M. Pauchard, A. Drechsler, S. Glaus, C. Debus, A.J. Meixner, G. Calzaferri, *J. Phy. Chem. B* **2001**, *105*, 25.

31 C.F. Seebacher, C. Hellriegel, C. Bräuchle, M. Ganschow, M. Wark, D. Wöhrle, *J. Phys. Chem.* in preparation.

32 a) S.T. Wilson, in Introduction to Zeolite Science and Practice, H. van Bekkum, E.M. Flanigen, J.C. Jansen (eds.), Studies in Surface Science and Catalysis, Vol. 58, Elsevier, Amsterdam **1991**, p. 137; b) S. Qin, W. Pang, H. Kessler, J.L. Guth, *Zeolites* **1989**, *9*, 440; c) H. Kessler, J. Patarin, C. Schott-Darie, in Advanced Zeolite Science and Applications, J.C. Jansen, M. Stöcker, H.G. Karge, J. Weitkamp (eds.), Studies in Surface Science and Catalysis, Vol. 85, Elsevier, Amsterdam **1994**, p. 75.

33 T. Kodaira, K. Miyazawa, T. Ikeda, Y. Kiyozumi, *Micropor. Mesopor. Mater.* **1999**, *29*, 329.

34 M. Ganschow, G. Schulz-Ekloff, M. Wark, M. Wendschuh-Josties, D. Wöhrle, *J. Mater. Chem.* **2001**, *11*, 1823.

35 J.W. Richardson, E.T.C. Vogt, *Zeolites* **1992**, *12*, 13.

36 M.E. Davis, C. Montes, P.E. Hathaway, J.M. Garces, in Zeolites: Facts, Figures, Future, P.A. Jacobs, R.A. Santen (eds.), Studies in Surface Science and Catalysis, Vol. 49A, Elsevier, Amsterdam **1989**, p. 199.

37 E.T.C. Vogt, J.W. Richardson, *J. Solid State Chem.* **1990**, *87*, 469.

38 J.A. Martens, P.A. Jacobs, in Advanced Zeolite Science and Applications, J.C. Jansen, M. Stöcker, H.G. Karge, J. Weitkamp (eds.), Vol. 85, Elsevier, Amsterdam **1994**, p. 653.

39 F. Fajula, in Guidelines for Mastering the Properties of Molecular Sieves: Relationship between the Physicochemical Properties of Zeolitic Systems and their Low Dimensionality, D. Barthomeuf, E.G. Derouane, W. Hölderich (eds.), Plenum, New York **1990**, p. 53.

40 J.J. Pluth and J.V. Smith, in Zeolites: Facts, Figures, Future, P.A. Jacobs, R.A. Santen (eds.), Studies in Surface Science and Catalysis, Vol. 49B, Elsevier, Amsterdam **1989**, p. 835.

41 L.B. McCusker, Ch. Baerlocher, E. Jahn, M. Bülow, *Zeolites* **1991**, *11*, 308.

42 O.W. Flörke, *Z. Kristallogr.* **1967**, *125*, 134.

43 C.T. Kresge, M.E. Leonowicz, W.J. Roth, J.C. Vartuli, J.S. Beck, *Nature* **1992**, *359*, 710.

44 W. Zhou, J. Klinowski, *Chem. Phys. Lett.* **1998**, *292*, 207.

45 O. Groeger, G. Engelhardt in *Proc. 11th German Zeolite Meeting*, **1999**, PO35.

46 a) M. Wark, A. Ortlam, M. Ganschow, G. Schulz-Ekloff, D. Wöhrle, *Ber. Bunsenges. Phys. Chem.* **1998**, *102*, 1548; b) S. Ernst, R. Gläser, M. Selle, *Stud. Surf. Sci.* **1997**, *105*, 1021; c) I. Honma, H.S. Zhou, *Adv. Mater.* **1998**, *10*, 1532; d) M. Ganschow, D. Wöhrle, G. Schulz-Ekloff, *J. Porphyrins Phthalocyanines* **1999**, *3*, 299.

47 M. Ganschow, M. Wark, D. Wöhrle, G. Schulz-Ekloff, *Angew. Chem.* **2000**, *39*, 161.

48 H.G.O. Becker, *Einführung in die Photochemie*, 1. Aufl., Deutscher Verlag der Wissenschaften, Berlin **1991**.

49 Y. Rohlfing, D. Wöhrle, M. Wark, G. Schulz-Ekloff, J. Rathousky, A. Zukal, *Stud. Surf. Sci. Catal.* **2000**, *129*, 295.

4
Large and Perfect, Optically Transparent Crystals of an Unusual Habitus

Jan Kornatowski and Gabriela Zadrozna*

4.1
Introduction

Control of crystal morphology is still a new direction in the investigation of molecular sieves. The significant role of morphology has been known since the first large-scale application of zeolite A in the production of phosphate-free washing powders: the cubooctahedral crystals appeared to be ideal carriers for surfactants as opposed to the cubic crystals, which were useless for this task. The investigations of morphology reported in the literature [1–21], describe either the growth of large crystals of random morphology or the synthesis of layer-like aggregates (quasi-membranes), but do not deal with control of crystal growth in particular crystallographic directions. This is probably because of poor knowledge about the growth processes. Thus, synthesis methods for growing zeolitic materials with particular morphologies need to be researched. An interesting habitus is that of large and flat, plate- or even pane-like, optically transparent, and possibly perfect crystals. Such crystals are promising materials for various nonchemical applications.

The controlled introduction of guest compounds into the pores of molecular-sieve hosts is a new way of creating ensembles of ordered molecules [22–25]. During adsorption, the nonsymmetric molecules form chains of head–tail orientation. These organized systems can have properties that are not observed for nonordered free molecules, such as doubling of frequency, which is an effect of second harmonic generation (SHG) occurring as the result of symmetry brake [26–28]. Other applications comprise laser, magnetic, electric, and fluorescence effects. All of them are based on anisotropic properties of adsorbed molecules and cumulative amplification of the effects [29–36].

The modified zeolitic materials are potentially useful for two other new ideas.

1. Creation of wave-conductor structures by lateral defined adsorption of guest molecules in the pores; depending on the type of the guest molecules, such structures can also reveal nonlinear optic (NLO) properties [37–39].

2. Control of NLO properties by *in situ* transformation of *trans*-isomers into hyperpolarizable *cis* ones.

Kinetic investigations have been limited to general effects [40–43] and crystal orientation [44–48]. The new crystals are expected to offer the possibility of kinetic measurements into particular crystallographic directions [49].

Especially noteworthy is that knowledge of parameters which decide and make possible the control of crystal growth leads to progress in understanding the crystallization processes and potentially to new horizons in the synthesis of zeolitic materials. Examinations of sorption properties of large crystals as model adsorbents, as well as studies on sorption mechanisms [9,50–52], are powerful tools coinciding with the synthesis experiments.

4.1.1
Synthesis of Molecular Sieve Crystals of Tailored Dimensions and Habitus

The syntheses of various zeolitic materials in the form of large crystals [1–21], as well as possible single crystals [4,6,8–12], are the only results so far. The so-called tailoring of a defined crystal habitus [4,5,17,18] is practically limited to the crystal size. Papers in this subject are still reporting only new synthesis routes, yielding crystals larger than those grown previously [53]. Optical properties have hardly been reported [21]. Syntheses of plate-like crystals of molecular sieves have not been reported yet. As far as we know, the papers on MeAPO-5 materials by the authors of this Chapter are the first detailed studies in this new research field, with one exception [54]. A number of parameters that influence the synthesis and play a deciding role in the control of crystal habitus are discussed in Section 4.2.

Layers or films of molecular sieves (AFI and MFI type) are another direction in morphology control [55–58]. However, as polycrystalline materials, they are not presented here.

For the nonclassical applications described here, the crystals have to reveal the following attributes.

- Good quality for optics: a special form (habitus) and structure, preferably free of defects of morphology, chemical composition, and optical purity. The morphology (crystal sizes and the ratios of dimensions) should be adjustable for desired applications. This means control and/or tailoring the crystal growth in particular crystallographic directions.
- Good sorption properties: high effective loading with compounds such as dyes and other polar or polarizable molecules. Sorption and diffusion in the pores should not be hindered, which is directly connected with proper morphology and perfect internal order. The present contribution is focused on molecular sieves of AFI structure type: aluminiumphosphates with a unidimensional system of tubular parallel pores.

4.2
Results and Discussion

4.2.1
General Remarks and Synthesis Procedure

The goals described above require synthesis investigations in two main directions: determination of the parameters that decide the crystal growth in particular crystallographic directions, and synthesis control *via* optimization of these parameters. Of particular importance are: induction of co-templating effects by additional components in the reaction gel, significance of various Al compounds, influence of metal (Me) compounds on the synthesis reaction and products, and determination of the role and significance of pH and its variations during the reaction.

The reaction gels for preparation of the materials had the following molar ratios of components as oxides: a Al_2O_3: 1.00–1.05 P_2O_5: b (Me_2O_3) or (2 MeO): c TEA: d H_2O: e HA: f HX, where: $a + b = 1.00$; $b = 0.05$–0.2 or 0 for the nonsubstituted material; $c = 1.5$–2.5; $d = 200$–300; $e = 0.05$–2.50; $f = 0.15$–0.60 and 0.05–0.45 for nonsubstituted $AlPO_4$-5. TEA (triethylamine) was the standard template, HA represented compounds tested as potential co-templates (Sections 4.2.2–4.2.7.), and HX the second, supporting co-templates, mainly HCl.

The significance of Al sources was tested by using crystalline hydroxides Pural SB and Catapal B (73.6 % Al_2O_3) of pseudoboehmite structure, pseudoboehmite-like Al oxide hydrate sol, Al sulfate, acetate, isopropoxide, and phosphate salts (all Merck), and amorphous Al hydroxide sols/gels (Giulini Chemie). The main metal tested was Cr (see below), used in the form of $Cr(NO_3)_3 \cdot 9\,H_2O$, $CrCl_3 \cdot 6\,H_2O$, or Cr acetate (all Aldrich).

The synthesis procedure is as follows [2–8,50–52]. The initial mixture A is formed by dissolving/dispensing an Al compound, a Me salt, and co-template(s) in water under vigorous stirring. The mixture B is formed by reacting H_3PO_4 (85 wt.-%, Merck) diluted about 1:1 in water with triethylamine (Merck) added when stirred and cooled. After adding mixture B to A, the whole is homogenized by stirring at room temperature, then placed in PTFE-lined autoclaves, and heated at 443–463 K, depending on the gel composition, for 4–72 h. The resulting products are decanted in water until free of the remains of unreacted gel, then filtered, washed, dried at 380 K overnight, and calcined under air at 773 K for 2 days.

The crystalline phases are formed in all batches with yields typical of the applied procedure, which is 60–100 % depending on the gel composition, amount(s) of co-template(s), crystallization period, and the Al compound used [2–9,59]. Occasionally occurring by-products form mostly aggregates with primary crystallite size significantly smaller than that of the AFI type crystals. The classic composition of the reaction gel (without additional components) influences only the yield and rate of crystallization as well as, to some extent, the amount of incorporated Me.

The modified synthesis procedure has been developed from the method [2–8,50–52] originally developed for growing $AlPO_4$-5 and its derivatives in the form of large crystals as elongated hexagonal prisms. Such crystals can have lengths up to about 1 mm but their width is limited to about 40 μm. A specially tailored composition of the reaction mixture can significantly influence the crystal morphology. The largest effects exert additional components introduced into the reaction gel. These components are usually not necessary for the formation of the AFI type structure, that is, for the synthesis of the $AlPO_4$-5 phase. Such a procedure can yield the crystals with aspect ratios *l*:*d* (length:width) of about 1:1, though with a similarly limited width. Further progress in tailoring the morphology, that is shortening the length and broadening the width of the crystals, has been achieved by the following operations.

- Testing a wide range of additional components.
- Optimization of the amounts and the ratios between the components.
- Testing the influence of pH values of the reaction mixture on the synthesis results.
- Testing the indirect influence of the pH on the other reaction parameters.
- Determination of the influence of Al compounds and mutual interaction/influence of the used Al species and the additional components.

4.2.2
Inorganic Acids and Salts of Alkaline Metals as Additional Components

The inorganic acids HCl, H_2SO_4, HNO_3, and $HClO_3$ taken into the gels require an adjustment of the initial pH of the reaction mixtures with use of additional amounts of the template (amine). The salts of alkaline metals do not cause such problems. Small amounts of all these additives give hardly visible effects, but greater amounts commonly cause the collapse of the crystallization. In effect, no AFI type phase is formed, especially with the salts, which indicates clearly an unfavorable influence of the alkaline cations. The relatively best influence can usually be observed for low HCl contents as the Cl^- ions seem to play a primary role (see 4.2.7).

4.2.3
Inorganic Salts of 2+ and Higher Valence Metal Ions as Additional Components

The test syntheses with metals such as Be, Mg, Co, Ni, Cu, Zn, Mn, Cd, Cr, Fe, Ti, Zr, Si, V, Mo, W, and Sb, which were added into the reaction gels in several different amounts each, mainly in the form of chlorides, sulfates, nitrates, or acetates, reveal a positive influence of numerous metals on the crystal morphology, that is, a shortening of the length of the hexagonal prisms. The kind of chemical compound in which the metal is used also plays a significant role. However, a clear relationship between the metal ions or their compounds and the effect on the morphology cannot be concluded from the syntheses performed. It is probably connected with the complexity of the system and the fact that any change made in the reaction gel composition is followed by consequent changes in several other parameters. This makes systematic investigations of particular parameters impossible.

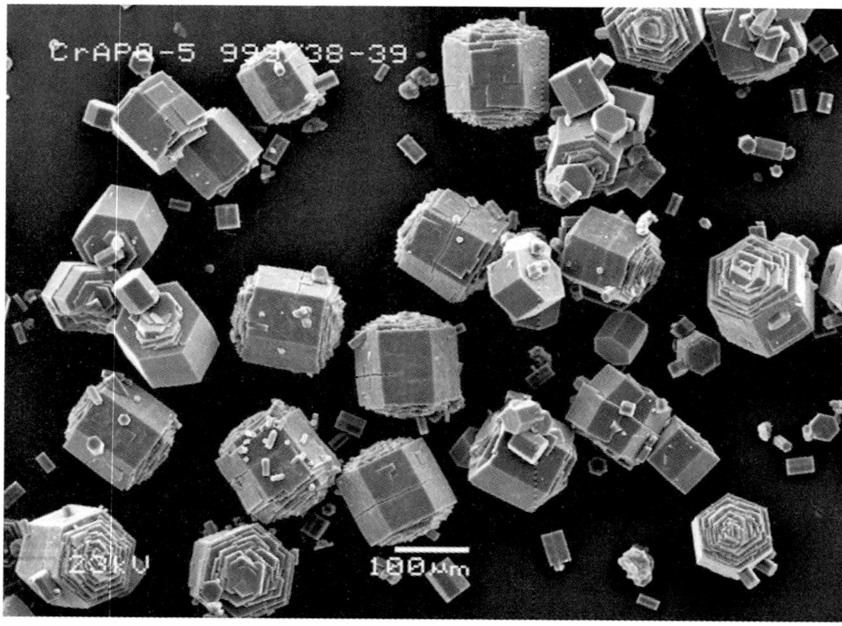

Fig. 1. An example of the CrAPO-5 phase synthesized without co-template.

The best results with the lowest *l:d* ratios arise from the use of Cr^{3+}, Fe^{3+}, Si^{4+}, and Co^{2+} ions. The latter two ions cause formation of the crystals with *l:d* of about 1:1 in the whole batch and with a good reproducibility. The 3+ ions, especially Cr^{3+}, can yield shorter, flatter crystals. Depending on the compounds used, the CrAPO-5 crystals grow as plates of *l:d* ratios below 1 (Fig. 1). These crystals can have flat and smooth hexagonal walls (110) and reveal good optic properties [59]. The aspect ratio of about 0.25, desired for many applications, is a limit which could not be improved by changing the reaction parameters. On the other hand, some syntheses yield only elongated hexagonal prisms. The cause of this discrepancy appears to be fundamental: flat crystals grow only when the Cr^{3+} ions are substituted for Al^{3+} in the framework [8,9,59–63]. In this way, the synthesis method described enabled successful incorporation of the Cr^{3+} ions into framework positions for the first time: the first preparation of stable CrAPO-5 materials [8,9,59–63]. Nonincorporated Cr^{3+} ions or compounds with Cr in other oxidation states do not shorten the hexagonal prisms [8,9,62,63] and additionally clog the pores [8,9,59].

The presence of the Cr^{3+} ions results in green crystals, which may cause some limitations in optical applications. However, the Cr^{3+} ions substituted for Al^{3+} do not cause any framework charge and thus create no strong sorption centers. Therefore, the CrAPO-5 crystals have good sorption properties, especially for nonpolar molecules, and reveal weaker interactions with the sorbate molecules than other derivatives with charged frameworks [8,9,59]. Thus, homogeneous dye uptake can be expected.

Fig. 2. FeAPO-5 material synthesized with acetic acid as co-template.

The stable framework position of Cr also causes the stability of its 3+ oxidation state [59,61]. The latter is not favorable for catalysis [60], but promising for all the required physical applications. This situation is quite different from the behavior of the Fe^{3+} ions that can easily undergo reversible redox processes. It is obvious that the framework iron ions are sorption centers similar in behavior to the Cr ones, so they may cause some limitations for sorption of polar molecules. For the above reasons, the syntheses of FeAPO-5 (Fig. 2) have not been optimized. However, the similarity in the influence of the both metals on morphology clearly demonstrates a deciding role of the metal centers for controlling the crystal growth.

4.2.4
Other Organic Templates as Additional Components and/or Co-Templates

The syntheses investigated are based on the standard TEA template. Other templates, tested for a possible influence on crystal morphology and growth, include tripropylamine, triethyltetramine, methyldiethanolamine, triethanolamine, diphenylamine, dipropylamine, ethylenediamine, and tetramethylammonium and tetrapropylammonium hydroxides, used alone or in a mixture with TEA under standard conditions for the TEA synthesis. They lead either to smaller crystals or to crystallization of by-products and contaminated mixed phases. No required influence on the crystal morphology can be achieved.

4.2.5
Organic Acids as Additional Components and Co-Templates

Acetic acid and acetate ions show an almost revolutionary influence. Such synthesis batches contain CrAPO-5 crystals with the length reduced below the width and with a perfect shape in the sense of flat and smooth external wall surfaces. Thus, extensive optimization experiments and broad investigations [62] have been performed with several groups of aliphatic acids of various length of carbon chains. The groups contain: (a) saturated formic HCOOH, acetic CH_3COOH, propionic CH_3CH_2COOH, butyric $CH_3(CH_2)_2COOH$, pentanoic $CH_3(CH_2)_3COOH$, and hexanoic $CH_3(CH_2)_4COOH$; (b) unsaturated acrylic $CH_2{=}CHCOOH$, propiolic $CH{\equiv}CCOOH$, vinylacetic $CH_2{=}CHCH_2COOH$, methacrylic $CH_2{=}C(CH_3)COOH$, crotonic $CH_3CH{=}CHCOOH$, 4-pentenoic $CH_2{=}CH(CH_2)_2COOH$, and sorbonic $CH_3CH{=}CHCH{=}CHCOOH$; and (c) bifunctional oxalic HOOCCOOH, malonic $HOOCCH_2COOH$, succinic $HOOC(CH_2)_2COOH$, maleinic $HOOCCH{=}CHCOOH$, 3-hydroxybutyric $CH_3CH(OH)CH_2COOH$, glutaric $HOOC(CH_2)_3COOH$, and adipic $HOOC(CH_2)_4COOH$ acids. The length of hydrocarbon chains, the number and the distribution of functional groups and/or multiple bonds, as well as the amount of the acids in the reaction gels significantly influence the properties and morphology of the crystals. The optimization experiments allowed the determination of favorable amounts of co-templates, their ratios to the other components, and relations to the other parameters such as gel preparation and aging, as well as crystallization temperature and time [59–63].

A simultaneous use of metals and co-templates makes possible to grow crystals with the aspect ratio $l{:}d$ down to about 0.05 and even less for single pieces. Moreover, the crystals can easily be grown wider than the limit of 40–50 μm (Section 4.2.1) and the largest ones are about 120 μm wide [59,62]. The bifunctional acids allow the introduction of the highest amounts of Cr, which, interestingly, are not necessarily accompanied by a perfect morphology or further shifted aspect ratios of the crystals [62,63]. The crystals with the best morphology can be synthesized with a co-template from the group of C_2 to C_4 acids (Fig. 3).

Among the products prepared with the saturated acids, the most regular and flat crystal morphology is observed for the C_1 to C_4 acids. Pentanoic acid causes a slight, and hexanoic acid a drastic, decrease in the crystals shape and morphology, despite their good crystallinity and sorption properties. The group of unsaturated acids yields products with excellent regular morphology of the crystals that are as good (acrylic, vinylacetic, 4-pentenoic) as those with acetic acid or better (the other acids except propiolic). Metacrylic acid leads to the most regular morphology among all products, followed by crotonic and sorbonic acids. The latter is a surprise, considering its long C_6 chain. However, it contains two double bonds in contrast to the others. Another surprise is propiolic acid with a triple bond, which produces clearly worse ($l > d$) and highly aggregated crystals. Within the group of bifunctional acids, a good morphology is observed only for unsaturated bicarboxyl maleinic acid (C_4). The comparative synthesis with saturated C_4 succinic acid gives much worse results. Surprisingly, the best crystals within this group result from

Fig. 3. SEM micrographs of the CrAPO-5 materials showing the best morphology of the crystals; two examples for each group of the co-templating acids used: saturated (top), unsaturated (middle), bifunctional (bottom).

the use of 3-hydroxybutyric acid. Especially unsatisfying morphology is obtained with oxalic (C_2, elongated crystals as in the case of a simple uncontrolled synthesis) and adipic (C_6) acids.

The group composed of acetic, propionic, acrylic, propiolic, vinylacetic, and crotonic acids was tested as co-templates also for nonsubstituted $AlPO_4$-5 [64]. These acids show the required influence and appear to be the first components of the reaction that are able to control the synthesis of "pure" $AlPO_4$-5. The crystals have aspect ratios *l:d* mostly between 1 and 2, while the whole range is about 0.5–3 [64]. These values depend not only on the type and amount of the co-templating acid but also on other synthesis parameters. The favorable tendency to shortening the crystal length is, however, hindered by an undesirable affinity to formation of convex external hexagonal walls (110) that look apparently like aggregates (Fig. 4). Only the smallest crystals seem to form flat hexagonal walls. The maximum width of the $AlPO_4$-5 crystals synthesized reaches about 80 μm (Fig. 4) [64].

Despite numerous syntheses with various acids, the number of both the acids and the experiments appears to be too low to conclude defined tendencies or dependences between the structure of the co-template molecules and its effect on the morphology. The only clear dependences resulting from the experiments are a type of mutual interaction between the template and co-templates [62–65] and a synergetic effect of heterometals and co-templates. The latter, together with the high stability of Cr in framework positions, strongly suggests that the co-templates form complex species with Cr that are more favorable for the framework incorporation than simple hydrated ions. The role, significance, and interaction of the co-templates with the particular components of the reaction gel as well as the synergetic effects observed between the co-templates and metals being introduced into the framework positions [59] are under further intensive investigation [65].

4.2.6 Alcohols as Additional Components and Co-Templates

Methanol, ethanol, propanol, ethylenglycol, and glycerin were also tested as co-templates, when taken in various amounts aside from the main template, TEA. After some optimization procedures, the experiments showed that the alcohols also shorten the crystals, however, to a lower extent than the acids [59,63].

4.2.7 Mixed Organic/Inorganic Additional Components as Co-Templates

The favorably acting acetic acid, used in mixtures with small amounts of acetates of alkaline and alkaline earth metals, results only in strongly overgrown aggregates composed of small crystals and in no desired effects on morphology and growth.

Small amounts of inorganic acids used together with acetic acid give clearly advantageous results, especially with respect to formation of highly smooth and transparent external crystal walls in the case of CrAPO-5. Of all inorganic acids examined, HCl works best in connection with all the organic acids tested

Fig. 4. Examples of $AlPO_4$-5 synthesized from pseudo-boehmite with co-templating acids: (top left) acetic, (top right) acetic + ethanol, (middle left) propionic, (middle right) acrylic, (bottom left) acetic [amorphous $Al(OH)_3$ gel], (bottom right) propiolic. The latter are the largest in *a–b* plane $AlPO_4$-5 crystals synthesized.

[59,62,63]. However, the inorganic acids, combined with the unsaturated organic ones, yield more aggregated crystals than the organic acids alone, which is an undesired effect [62]. These observations strongly imply that the synergetic effect of HCl or Cl^- ions is connected with their interaction in two directions: complexing of the introduced metals, and buffering effects (pH control) in the reaction gel during the whole crystallization period [59,62–65]. The investigations continue.

In the case of nonsubstituted $AlPO_4$-5, the effect of HCl or other inorganic acids is much weaker and similar to the effect of the organic acids alone. These observations strongly support the above suggestions about the role of HCl and Cl^- ions. An addition of HCl seems to result in less convex hexagonal walls of the crystals, but definitely does not lead to significantly improved crystals with flat and smooth external walls [63]. A large number of syntheses in such systems implies that the introduction of HCl is probably not sufficient to achieve a regular growth of flat walls. Thus, the regular growth of smooth flat walls might directly be connected with the presence of heterocenters in the framework and dependent on it [59,62–65].

The Cr content of the samples synthesized with the acetic acid varies from 0.5 to 0.06 Cr per unit cell (uc), or 2–16.5 unit cells per Cr atom (ICP) and does not directly depend on the Cr amount added to the synthesis gel. It differs for crystals synthesized from apparently similar gels with the same amount of Cr by a factor of up to three, even when the same Al compounds are used. The differences follow the ratios of particular components and especially the amount of co-template and type of Cr^{3+} compound.

The saturated acids reveal a favorable influence on the amount of substituted chromium, which increases systematically with the chain length from 0.25 Cr/uc (C_2) to 0.63 Cr/uc (C_5). Formic acid is an exception yielding an apparently high Cr content of about 0.50 Cr/uc while hexanoic (C_6) acid reveals a drastic decrease to 0.36 Cr/uc The unsaturated acids can be divided in two groups. For all the C_3 and C_4 acids, the amount of Cr is approximately constant (about 0.25 Cr/uc) and similar to that achieved with acetic (C_2) acid [59]. For the C_5, C_6, and branched methacrylic (C_4) acids, the contents between 0.30 and 0.37 Cr/uc correspond to those of the saturated C_3 and C_4 acids. This implies different, most likely stronger, interactions of Cr with the unsaturated acids than with the saturated ones. This tendency seems to be confirmed by the group of bifunctional acids that allow the introduction of the greatest amounts between 0.4 and 0.74 Cr/uc Exceptions are C_2 and C_6 acids which, however, show strong deviations in the morphology and sorption properties of the crystals.

4.2.8 Aluminum Source as Directing Agent

The kind of aluminum compound used plays a deciding role in the syntheses of numerous molecular sieves. The new CrAPO-5 material is the best example. Its synthesis is generally controlled by the kind of the Al compound in the reaction gel [8,59]. In particular, the Al compound, such as $Al(OH)_3$, $AlO(OH)$, Al_2O_3, $[Al(H_2O)_6]^{3+}$, and the degree of agglomeration of the Al species are of primary importance for the reaction in all aspects: formation of AFI type structure, framework incorporation of Cr, nucleation rate, crystal growth rate, and also morphology and dimensions of the crystals [8,59,62,63].

Among numerous Al compounds tested [59,62,63], the most successful results yield those with Pural SB, Catapal, and aluminum hydroxide sols, that is the pseudoboehmite-like phases. This finding confirms that the synthesis of CrAPO-5 and control of the crystal morphology are connected with a synergetic effect of at least three parameters: co-template(s), heterometal, and appropriate Al compound [59,62,63]. One may assume that intermediate compounds, which have to be formed as building units for the crystal framework, as well as the controlled growth of the crystals require very specific conditions in the reaction mixture. These conditions can probably be satisfied only with defined chemical compounds. Moreover, each system shows its own properties and behavior that have to be tested and optimized separately. These observations support the conclusion that it is not possible to change only one parameter in such a complicated system, which enormously hinders real systematic studies on the role and significance of particular synthesis factors.

4.2.9
Preparation of the Reaction Gel as a Control Tool

The commonly known parameters and conditions important for the preparation of the reaction gel are: compounds of particular elements, composition of the gel with respect to the overall concentration and the ratios between particular components, and the sequence of mixing of the components. There are two methods of gel preparation. In the classic one by Flanigen *et al.* [66], all components are mixed together in one vessel. The main components Al_2O_3, P_2O_5, and template are added in the ratios 1:1:1 with relatively little water, and no defined sequence of the components. The method by Kornatowski and Finger [2–7], developed for growing large crystals, requires more specific conditions: more template by about 50–80 %, about 20 times more water (lower concentration of the gel), and, first of all, a separate preparation of two initial mixtures: one (A), combined by suspending an Al compound and possible other metals or additional components in a large amount of water and the other (B), by reacting about 1:1 diluted phosphoric acid with the template amine. For the fine tuning of the synthesis with the aim of morphology control, several other parameters appear to have a fundamental meaning.

- The type of Al compound.
- Preparation of the reaction gel not only in the classic meaning of the mixing sequence, but also introduction of the additional components into A or B mixture, initial concentration of particular components, and possible aging.
- Ratios of particular components not only in the classic meaning of the main components, but also template to water, template to co-template(s), and co-templates one to another.
- Concentration of the reaction gel.
- pH of the reaction gel and its control in all particular stages.

The pH value of the reaction gel generally decides the formation of $AlPO_4$-5 or other aluminophosphate phases. A simple synthesis without special requirements for the dimensions or morphology of the crystals is possible within a relatively broad pH range of 2.5–7. The final pH value after the synthesis is hardly important. However, control of the morphology requires a possibly exact adjustment of pH in all particular stages of the reaction: in the A and B mixtures, in the reaction gel, and after the crystallization. The latter equals to the control and/or stabilization of pH during the synthesis reaction. As found experimentally, the pH values of the mixtures A and B as well as the final pH value should be possibly equal to each other within the range 2.5–4.0, which means that this pH should be maintained during the reaction [59,63]. Large differences in pH before and after synthesis counteract the expected morphology effects. The stability of pH is probably responsible for the control of nucleation rates.

The reaction gels prepared from amorphous Al compounds change their pH value immediately after the initial mixtures are combined and ΔpH can reach 2 and more after 20 min stirring [59,63]. In the case of pseudoboehmite and pseudoboehmite-like Al compounds, the ΔpH is lower than 0.5 even after several hours stirring. The pH values after the hydrothermal crystallization are higher by 2 or more for the amorphous Al compounds. Thus, the two types of Al compounds react in different ways, yielding various products [59,63].

4.2.10
Sorption Characteristics of the Tailored Crystals

Sorption measurements are the most sensitive tests for the crystal and the pore structure as well as for the pore accessibility to sorbate molecules and thus for the presence of extra framework species in the pores. The sorption isotherms for adsorbates of various dimensions, shape, dipole moment, and electron structure (H_2O, C_6H_6, N_2) supply complex information about these crystal features [9,50–52,59–63]. In the case of CrAPO-5, the synthesized products could be divided into two groups. Amorphous Al compounds yield samples with a very low sorption capacity for benzene and nitrogen while those synthesized from pseudoboehmite or pseudoboehmite-like Al compounds exhibit much higher sorption capacities [8,9,59]. The products synthesized with co-templating organic acids show full sorption capacity, higher even than that for pure $AlPO_4$-5 synthesized without co-templates [9,59]. It is almost independent of the Cr content. Therefore, the materials are referred to as LS for low- and HS for high sorption capacity (Figs. 5 and 6). The high sorption capacities for benzene and nitrogen show open pore systems of the HS samples. Limited sorption capacities of the LS samples indicate that the pores are hardly accessible to either sorbate.

The samples of both groups reveal the sorption capacity for water almost identical with that of pure $AlPO_4$-5. The capacity for water cannot be used as an indication of open or clogged pores, as these small molecules can pass the 6-ring windows of the structure and penetrate the crystals not only along the channels but also in other directions, apparently "through the walls" of the pores. However, the low-pressure step of the isotherms for water is sensitive to the content of heteroatoms substituted into the framework [9,59] (Fig. 7). The HS samples show a clear shift of the step to lower relative pressure with growing content of Cr (Fig. 8), which indicates that the Cr centers influence the sorption process [9,59]. A stable

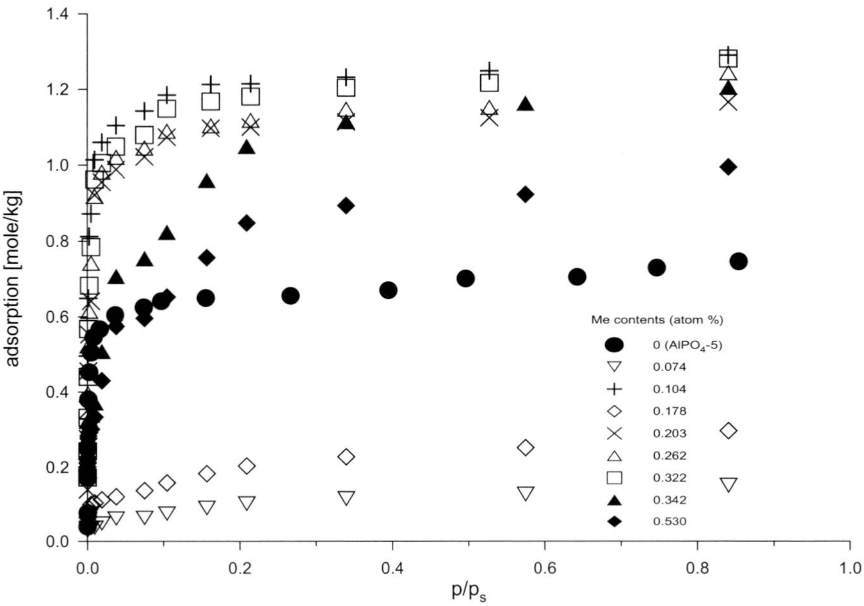

Fig. 5. Isotherms of benzene sorption on the CrAPO-5 materials with various Cr contents. The HS samples (upper curves) were synthesized with acetic acid. Two the lowest curves illustrate sorption on the LS materials.

position of the step for the LS samples reveals the lack of interactions between the adsorbent and water and thus the absence of substituted chromium [59].

All the materials synthesized with the saturated and unsaturated acids exhibit high sorption capacities for benzene and nitrogen with the surprising exception of the sample synthesized with acrylic acid [62]. This sample exhibits the capacities reduced to about a half of the values observed for the other materials, in spite of an extremely high crystallinity and excellent crystal morphology, and a standard sorption capacity is observed only for water [62]. In the case of the bifunctional acids, the sorption capacities are also high except oxalic (C_2) and adipic (C_6) acids. The former shows all values, also for water, reduced to about 1/3 of the usual level, which indicates substantial structural problems seen also in XRD [62]. Surprisingly, high sorption capacities for benzene and nitrogen are exhibited by almost all products of the worst crystal morphology [62]. These apparent inconsistencies of the results strongly suggest that the acid molecules take part in the synthesis process *via* formation of some species with the Cr ions and thus play a role of real co-templates controlling the features of the growing crystals [62,63]. This might also explain the minor differences between the highly sorbing materials, although these features depend probably on the distribution of Cr centers, which probably vary between the samples.

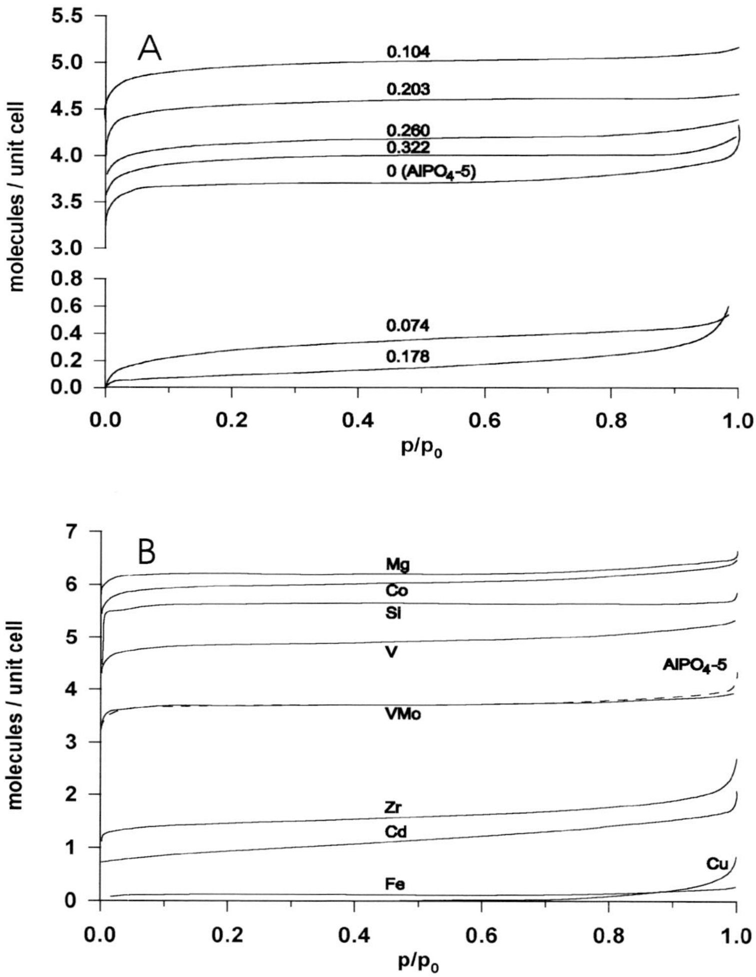

Fig. 6. Isotherms of nitrogen sorption (77 K): (A) CrAPO-5 with various Cr contents (upper part HS samples synthesized with acetic acid; lower part LS samples; numbers = Cr contents in atom %); (B) (Me)APO-5 samples with various Me contents (syntheses without co-templates).

4.3
Conclusions

The crystal morphology of the AFI type molecular sieves ($AlPO_4$-5 and derivatives) can be controlled to a great extent. The crystals can be grown as large and flat, even pane-like hexagonal plates with the shortest *c* axis (direction of the channels). The best flat crystals are synthesized with the Cr^{3+} ions. This is the first CrAPO-5 mate-

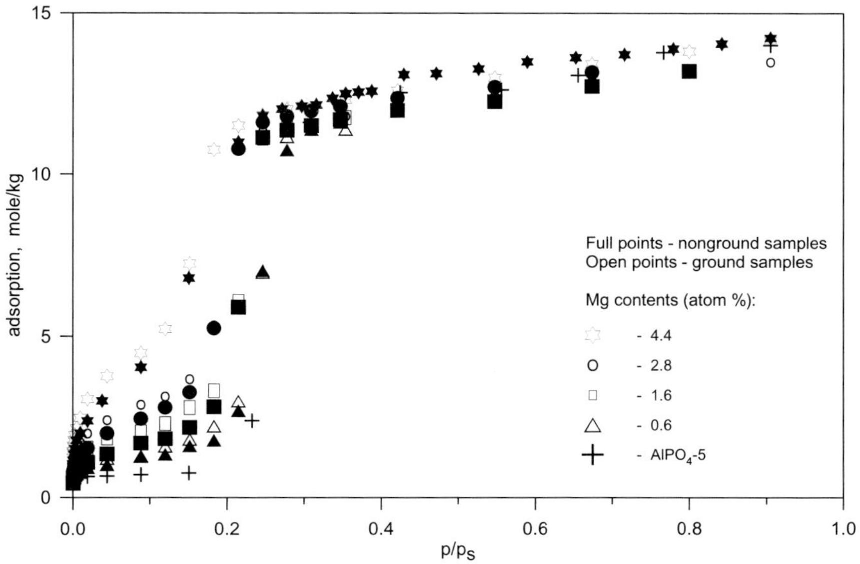

Fig. 7. Isotherms of water sorption on large nonground and ground crystals of MgAPO-5.

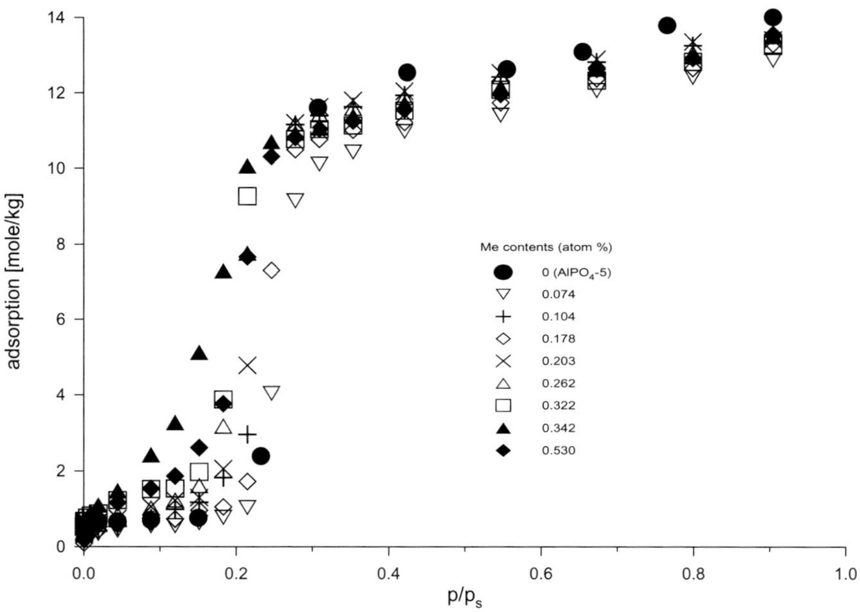

Fig. 8. Isotherms of water sorption on the CrAPO-5 materials with various Cr contents (synthesized with acetic acid).

rial with chromium stably substituted into the framework positions [8,9,59–62]. The crystal length can be reduced by the use of organic and inorganic additional components being co-templates in the reaction gel. The aspect ratios *l*:*d* of the CrAPO-5 crystals can be reduced to 0.1–0.05 and the crystal width enlarged to about 120 μm. Such flat crystals have been synthesized for the AFI structure type for the first time. These new materials are promising for various physical applications.

The crystal dimensions and their distribution are broad and depend, similarly to the morphology, on the gel composition, the use of co-templates, the preparation method, and the resulting substitution of Cr. As these parameters also influence the nucleation process, control of the size distribution is most complicated. Up to now, the best results have been attained by application of energy control (microwave heating) [54,67–69].

The Al source as well as pH and its stability play a deciding role in morphology control. The type and amount of the acidic co-templates exert significant effects on the crystallinity, dimensions and morphology of the crystals as well as on the content of substituted Cr. The highest crystallinity can be obtained with acetic, acrylic, methacrylic, or crotonic acids while the most perfect morphology with use of the unsaturated acids with the C_3 to C_6 chains. The highest substitution of Cr can be achieved with the bifunctional acids of the C_3 to C_5 chains. Only methacrylic acid seems to satisfy all these requirements to a relatively high extent.

We postulate that the co-templating acid molecules are included in the pore system of the crystals at the Cr centers, and Cr^{3+} substituted for Al^{3+} in the framework strongly restrains the growth of the crystals along the *c*-axis. With the help of the synergetic effect of the both factors, the crystals can be grown as flat hexagonal plates. The co-templates also exert a favorable buffering influence and control the synthesis by adjusting the pH value of the reaction gel. The co-templated synthesis gives rise to a reaction mechanism that results in a very high crystallinity, perfect morphology, and high sorption capacities of the products.

High crystallinity is not necessarily equivalent to perfect morphology and high sorption capacities or directly dependent on the Cr content of the crystals and *vice versa*. The co-templating acid has to be chosen with respect to the required properties of the product, specifically important for the planned application of the crystals.

The best optical properties are shown by the metal-containing crystals which, however, may potentially show a reduced sorption of dyes. The Cr^{3+} ions substituted for Al^{3+} in small amounts do not hinder loading the crystal pores with nonpolar dyes or other sorbates, required for optical or kinetic investigations. The large and "short" metal free crystals of $AlPO_4$-5 show commonly no flat and perfectly transparent hexagonal external walls (110), which are observed only in the small crystals.

In general, synthesis with morphology control is based on an equilibrium between parameters causing acceleration and slowing down of the crystal growth. The nucleation phase influences the morphology only indirectly *via* the crystal size and size distribution.

The following features of the crystals are independent of each other to a great extent and have to be controlled separately.

- Ratio of the crystal dimensions *l*:*d*, or the horizontal and vertical growth of the crystals.
- The width of the *a*–*b* plane: the hexagonal wall (110) perpendicular to the pores.
- Perfectly flat and transparent hexagonal external walls (110).
- Twinning and aggregation of the crystals.
- Average size of the crystals and distribution of sizes, both connected with nucleation.

Further studies are required to give an improvement in the synthesis of non-substituted $AlPO_4$-5 crystals, an enlargement of the plate-like crystals with high aspect ratio, a better homogeneity of crystal dimensions, and syntheses of other structure types, especially high quality MFI type crystals (ZSM-5).

Acknowledgements

Thanks are due to the Deutsche Forschungsgemeinschaft (DFG, Key-area "Nano-structured Host–guest Systems", Project No Ko 1641/1-3) for financial support of the investigations.

References

1 J. Kornatowski, *Zeolites* **1988**, *8*, 77.
2 G. Finger, J. Kornatowski, *Zeolites* **1990**, *10*, 615.
3 J. Kornatowski, G. Finger, *Bull. Soc. Chim. Belg.* **1990**, *99*, 857.
4 G. Finger, J. Kornatowski, J. Richter-Mendau, K. Jancke, M. Bülow, M. Rozwadowski, in Catalysis and Adsorption by Zeolites, G. Öhlmann, H. Pfeifer, R. Fricke (eds.), Studies in Surface Science and Catalysis, Vol. 65, Elsevier, Amsterdam **1991**, p. 501.
5 G. Finger, J. Richter-Mendau, M. Bülow, J. Kornatowski, *Zeolites* **1991**, *11*, 443.
6 J. Kornatowski, M. Sychev, W.H. Baur, G. Finger, *Coll. Czech. Chem. Commun.* **1992**, *57*, 767.
7 W.H. Baur, W. Joswig, D. Kassner, J. Kornatowski, G. Finger, *Acta Cryst.* **1994**, *B50*, 290.
8 J. Kornatowski, G. Zadrozna, in *Proc. 12th Int. Zeolite Conf.*, M.M.J. Treacy, B.K. Marcus, M.E. Bisher, J.B. Higgins (eds.), Vol. III, MRS, Warrendale, Pennsylvania **1999**, p. 1577.
9 J. Kornatowski, G. Zadrozna, J. Wloch, M. Rozwadowski, *Langmuir* **1999**, *15*, 5863.
10 E.R. Geus, J.C. Jansen, H. van Bekkum, *Zeolites* **1994**, *14*, 82.
11 R. de Ruiter, A.P.M. Kentgens, J. Grootendorst, J.C. Jansen, H. van Bekkum, *Zeolites* **1993**, *13*, 128.
12 R. de Ruiter, J.C. Jansen, H. van Bekkum, in Synthesis of Microporous Materials, M.L. Occelli *et al.* (eds.), Vol. 1, Van Nostrand, New York **1992**, p. 167.
13 H. Lermer, M. Draeger, J. Steffen, K.K. Unger, *Zeolites* **1985**, *5*, 131.

14 R. Mostowicz, L.B. Sand, *Zeolites* **1992**, *2*, 143.
15 M. Ghamami, L.B. Sand, *Zeolites* **1993**, *3*, 155.
16 S.Z. Chen, K. Huddersman, D. Keir, L.V.C. Rees, *Zeolites* **1988**, *8*, 106.
17 D.T. Hayhurst, J.C. Lee, in New Developements in Zeolite Science and Technology, Y. Murakami, A. Iijima, J.W. Ward (eds.), Studies in Surface Science and Catalysis, Vol. 28, Kodansha, Tokyo **1986**, p. 113.
18 J. Patarin, H. Kessler, J.L. Guth, *Zeolites* **1990**, *10*, 674.
19 A. Kuperman, S. Nadimi, S. Oliver, G.A. Ozin, J.M. Garces, M.M. Olken, *Nature* **1993**, *365*, 239.
20 U. Müller, K.K. Unger, *Zeolites* **1988**, *8*, 154.
21 C. Weidenthaler, R.X. Fisher, R.D. Shannon, O. Medenbach, *J. Phys. Chem.* **1994**, *98*, 12 687.
22 G.A. Ozin, A. Kuperman, A. Stein, *Angew. Chemie (Int. Ed.)* **1989**, *28*, 359.
23 G.A. Ozin, *Adv. Mat.* **1992**, *4*, 612.
24 G.D. Stucky, J.E. McDougall, *Science* **1990**, *247*, 669.
25 T. Bein (ed.), Supramolecular Architecture, ACS Symp. Ser., Vol. 499, **1992**.
26 S.D. Cox, T.E. Gier, G.D. Stucky, J. Bierlein, *J. Am. Chem. Soc.* **1988**, *110, 2986.*
27 S.D. Cox, T.E. Gier, G.D. Stucky, *Chem. Mater.* **1990**, *2*, 609.
28 S.D. Cox, T.E. Gier, G.D. Stucky, *Solid State Ionics* **1990**, *32/33*, 514.
29 F. Marlow, J. Caro, L. Werner, J. Kornatowski, S. Dähne, *J. Phys. Chem.* **1993**, *97*, 11 286
30 F. Marlow, J. Caro, *Mol. Cryst. Liq. Cryst.* **1994**, *240*, 175.
31 J. Caro, F. Marlow, M. Wübbenhorst, *Adv. Mater.* **1994**, *6*, 413.
32 F. Marlow, M. Wübbenhorst, J. Caro, *J. Phys. Chem.* **1994**, *98*, 12 315.
33 F. Marlow, J. Caro, *Zeolites* **1992**, *12*, 433.
34 K. Hoffmann, F. Marlow, J. Caro, *J. Fluoresc.* **1994**, *4*, 74.
35 F. Marlow, W. Hill, J. Caro, *J. Raman Spectr.* **1993**, *24*, 603.
36 W. Hill, F. Marlow, J. Kornatowski, *Appl. Spectr.* **1994**, *48*, 224.
37 L. Werner, J. Caro, G. Finger, J. Kornatowski, *Zeolites* **1992**, *12*, 658.
38 J. Caro, G. Finger, J. Kornatowski, J. Richter-Mendau, L. Werner, B. Zibrowius, *Adv. Mater.* **1992**, *4*, 273.
39 J. Caro, G. Finger, E. Jahn, J. Kornatowski, F. Marlow, M. Noack, L. Werner, B. Zibrowius, in *Proc. 9th Int. Zeolite Conference*, R. von Ballmoos, J.B. Higgins, M.M.J. Treacy (eds.), Vol. II, Butterworth-Heinemann, Boston **1993**, p. 683.
40 P.B. Weisz, *Chemtech* **1973**, *3*, 498.
41 M.F.M. Post, in Introduction to Zeolite Science and Practice, H. van Bekkum, E.M. Flanigen, J.C. Jansen (eds.), Studies in Surface Science and Catalysis, Vol. 58, Elsevier, Amsterdam **1991**, p. 392.
42 J. Kärger, D.M. Ruthven, Diffusion in Zeolites and other Microporous Solids, Wiley, New York **1992**.
43 J. Caro, M. Bülow, H. Jobic, J. Kärger, B. Zibrowius, *Adv. Catal.* **1992**, *39*, 351.
44 E.R. Geus, A.E. Jansen, J. Schoonman, H. van Bekkum, in Catalysis and Adsorption by Zeolites, G. Öhlmann, H. Pfeifer, R. Fricke (eds.), Studies in Surface Science and Catalysis, Vol. 65, Elsevier, Amsterdam **1991**, p. 327.
45 A.R. Paravar, D.T. Hayhurst, *Proc. 6th Int. Zeolite Conf.*, D. Olson, A. Bisio (eds.), Butterworths, Guildford **1984**, p. 217.
46 V.V. Nesariker, D.B. Shah, D.T. Hayhurst, poster at the *13th Ann. Mtg Br. Zeolite Assoc.*, Chislehurst, UK, 10–15 July **1990**.
47 D.L. Wernick, E.J. Osterhuber, *Proc. 6th Int. Zeolite Conf.*, D. Olson, A. Bisio (eds.), Butterworths, Guildford **1984**, p. 122.
48 D.L. Wernick, E.J. Osterhuber, *J. Membr. Sci.* **1985**, *22*, 137.
49 V. Kukla, J. Kornatowski, D. Demuth, I. Girnus, H. Pfeifer, L.V.C. Rees, S. Schunk, K.K. Unger, J. Kärger, *Science* **1996**, *272*, 702.

50 J. Kornatowski, M. Rozwadowski, in *Proc. 12th Int. Zeolite Conf.*, M.M.J. Treacy, B.K. Marcus, M.E. Bisher, J.B. Higgins (eds.), Vol. I, MRS, Warrendale **1999**, p. 285.

51 M. Rozwadowski, J. Kornatowski, R. Golembiewski, K. Erdmann, *Langmuir* **1999**, *15*, 5857.

52 J. Kornatowski, M. Rozwadowski, J. Wloch, J.A. Lercher, in Porous Materials in Environmentally Friendly Processes, I. Kiricsi, G. Pál-Borbély, J.B. Nagy, H.G. Karge (eds.), Studies in Surface Science and Catalysis, Vol. 125, Elsevier, Amsterdam **1999**, p. 675.

53 S. Ahn, H. Chon, *Micropor. Mater.* **1997**, *8*, 113.

54 M. Ganschow, G. Schulz-Ekloff, M. Wark, M. Wendschuh-Josties, D. Wöhrle, *J. Mater. Chem.* **2001**, *11*, 1823.

55 J.H. Koegler, A. Arafat, H. van Bekkum, J.C. Jansen, in Progress in Zeolite and Microporous Materials, H. Chon, S.-K. Ihm, Y.S. Uh (eds.), Studies in Surface Science and Catalysis, Vol. 105, Elsevier, Amsterdam **1997**, p. 2163.

56 K.J. Chao, C.N. Wu, H.C. Shin, T.G. Tsai, Y.H. Chiou, in Progress in Zeolite and Microporous Materials, H. Chon, S.-K. Ihm, Y.S. Uh (eds.), Studies in Surface Science and Catalysis, Vol. 105, Elsevier, Amsterdam **1997**, p. 2187.

57 J. Hedlund, B.J. Schoeman, J. Sterte, in Progress in Zeolite and Microporous Materials, H. Chon, S.-K. Ihm, Y.S. Uh (eds.), Studies in Surface Science and Catalysis, Vol. 105, Elsevier, Amsterdam **1997**, p. 2203.

58 Lixiong Zhang, Mengdong Jia, Enze Min, Progress in Zeolite and Microporous Materials, H. Chon, S.-K. Ihm, Y.S. Uh (eds.), Studies in Surface Science and Catalysis, Vol. 105, Elsevier, Amsterdam **1997**, p. 2211.

59 J. Kornatowski, G. Zadrozna, M. Rozwadowski, B. Zibrowius, F. Marlow, J.A. Lercher, *Chem. Mater.* **2001**, *13*, 4447.

60 G. Zadrozna, E. Souvage, J. Kornatowski, *J. Catal.* **2002**, *208*, 270.

61 B. Padlyak, J. Kornatowski, G. Zadrozna, M. Rozwadowski, A. Gutsze, *J. Phys. Chem. A* **2000**, *104*, 11 837.

62 J. Kornatowski, G. Zadrozna, J.A. Lercher, M. Rozwadowski, in Zeolites and Mesoporous Materials at the Dawn of the 21st Century, A. Galarneau, F. Di Renzo, F. Fajula, J. Vedrine (eds.), Studies in Surface Science and Catalysis, Vol. 135, Elsevier, Amsterdam **2001**, p. 04-P-12.

63 J. Kornatowski, G. Zadrozna, J.A. Lercher, in Impact of Zeolites and Other Porous Materials on the New Technologies at the Beginning of the New Millenium, R. Aiello, G. Giordano, F. Testa (eds.), Studies in Surface Science and Catalysis, Vol. 142A, Elsevier, Amsterdam **2002**, p. 399.

64 G. Zadrozna, J. Kornatowski, J.A. Lercher, in preparation.

65 J. Kornatowski, G. Zadrozna, J.A. Lercher, in preparation.

66 S.T. Wilson, B.M. Lok, C.A. Messina, T.R. Cannan, E.M. Flanigen, in Intrazeolite Chemistry, G.D. Stucky, F.G. Dwyer (eds.), *Am. Chem. Soc. Symposium Ser.*, Vol. 218, Am. Chem. Soc., Washington **1983**, p. 79.

67 I. Girnus, K. Hoffmann, F. Marlow, J. Caro, G. Döring, *Micropor. Mater.* **1994**, *2*, 537.

68 J. Caro, F. Marlow, K. Hoffmann, J. Kornatowski, I. Girnus, M. Noack, P. Kölsch, in *Proc. 2nd Polish–German Zeolite Colloquium*, Torun, Poland **1995**, M. Rozwadowski (ed.), N. Copernicus University Press, Torun, **1995**, p. 186.

69 I. Girnus, K. Jancke, R. Vetter, J. Richter-Mendau, J. Caro, *Zeolites* **1995**, *15*, 33.

5
Nanoporous Crystals as Host Matrices for Mesomorphous Phases

Ligia Frunza, Hendrik Kosslick, and Rolf Fricke*

5.1
Introduction

Thermotropic liquid crystals (LCs) are matter that, between crystalline (C) and isotropic (I) liquid states, possess well-defined phases such as smectic A (S_A) and nematic (N) with long-range orientational order. The transition from one phase to another (e.g., crystalline to smectic A (C–S_A), nematic to isotropic (N–I)) takes place at a fixed temperature, characteristic of a given material. Nematic LCs have the simplest known structure among the LCs: the elongated molecules orient on average in a parallel way to each other. The preferred alignment direction is given by a so-called director. The liquid crystalline phases are also called mesomorphous phases.

The macroscopic effect of a solid substrate on the nematic LCs consists in fixing the mean orientation taken by the molecules with respect to this surface [1]. This effect, known as anchoring, is the result of complex mechanisms involving short-range liquid crystal–substrate interactions and long-range fluid–fluid interactions that are specific to LCs [2]. LC anchoring depends on the properties of both the surface and the LC.

Porous materials as host matrices for liquid crystals offer an outstanding opportunity to study the effects caused by the contact of liquid crystals with the internal and external surfaces on the characteristic properties of LCs. Therefore, the confinement of liquid crystals to restricted geometries has recently received considerable attention [3]. Only simple restricted geometries or porous media with large pores (2.5–350 nm in diameter) such as silicagel, Vycor glass, Anopore, Nuclepore, controlled porous glass (CPG) were employed as host materials. In addition to the finite size and surface effects, these complex matrices introduce randomness, imposed by pore geometry, namely a disorder.

Generally, such a disorder lowers the transition temperature, it may broaden the transition range, lower the heat capacity peak, and may also change the order of the transition(s) [4]. It is now well known [5] that the pore surface leads to an orientational order near the pore wall while disordering effects exist at higher distances.

A variety of experimental techniques among which are high-resolution calorimetry [6,7], deuterium NMR [2,8–12], static or dynamic light scattering [6,13,14], X-ray/neutron scattering [15], time-resolved grating optical Kerr effect [16], and FTIR [17,18] experiments give information on the molecular alignment of the LC, on its orientational order and phase transitions, on the dimensionalities of the pore voids and shapes, and on the anchoring and wetting properties of the internal surface. Studies of the dynamic behavior of such composites have been carried out by dielectric measurements [12,14,19–25].

The following general characteristics were observed. The bulk N–I transition temperature can be shifted either up or down depending on the cavity size and on the anchoring properties of the cavity surface. The N–I phase transition can be also replaced by a gradual evolution of the nematic order for very small cavities.

5.2 Liquid Crystals Confined in Molecular Sieves

By using molecular sieves as hosts for LCs it was possible to decrease the limit size of the confining pores below 2.5 nm. Under such conditions, different framework topologies such as MFI, FAU, CLO, MCM-41, and SBA-15 were applied for hosting. The pore systems of these crystalline nanoporous materials, either zeolites or mesoporous ordered materials, have a definite size and shape, contrary to silicagels and other porous materials with pores not well defined. Besides, these materials possess a high internal surface area, and chemical as well as mechanical stability. A list of the nanoporous materials used for confinement studies of LC is given in Table 1.

Some AlSBA-15 samples were additionally exchanged with metallic cations having different charges and radii such as Na^+, K^+, Rb^+, Ca^{2+}, La^{3+}, and Co^{2+} [30,31]. We used monocomponent LC from the alkyl-cyanobiphenyl series, namely 4-n-pentyl-4′-cyanobiphenyl (5CB) and 4-n-octyl-4′-cyanobiphenyl (8CB). In Fig. 1,

Tab. 1. Samples of molecular sieves used for LC confinement

Sample	*Framework oxide composition*	*Pore diameter [nm]*	*Pore volume [$cm^3\ g^{-1}$]*	*BET surface area [$m^2\ g^{-1}$]*	*Source*
Si-MFI	Si	0.54 × 0.56	0.14	321	[26]
NaY	Al, Si	0.74	0.44	928	[Aldrich]
CLO	Ga, P	1.32	0.30	800	[17,27]
Si-CLO	Si, Ga, P	1.32	0.23	704	[17,27]
AlMCM-41	Al, Si	2.0	0.69	678	[25]
Si-MCM-41	Si	≈4	0.70	724	[28]
SBA-15	Si	10.2	1.584	794	[29]
AlSBA-15	Al, Si	7.5	1.086	638	[29]

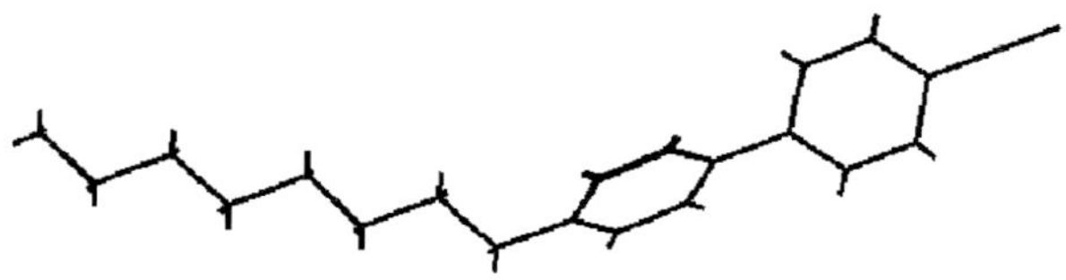

Fig. 1. Molecular structure of 8CB molecule.

a sketch of the molecular structure of 8CB as resulted from DFT calculations [32] is shown. These bulk LCs have the following sequence of phase transitions.

$$\text{5CB: C} \xrightarrow{295.5\ \text{K}} \text{N} \xrightarrow{308.3\ \text{K}} \text{I}$$

$$\text{8CB: C} \xrightarrow{294.1\ \text{K}} \text{S}_\text{A} \xrightarrow{306.5\ \text{K}} \text{N} \xrightarrow{313.9\ \text{K}} \text{I}$$

The cyan group determines the dipolar moment and the polarization of the 5CB and 8CB molecules. The dimension and the shape of these molecules are also important for confinement experiments: The molecular length of the 8CB is about 2 nm, whereas its width (geometrical diameter) is only 0.67 nm. One has to notice also that biphenyl is not planar; the two rings are twisted with a dihedral angle of about 50°. 5CB has a shorter alkyl chain and therefore the molecule length is a little smaller than for 8CB.

The loaded samples were routinely characterized by X-ray diffraction pattern, nitrogen sorption measurements, scanning electron microscopy, MAS NMR (^{29}Si, ^{27}Al, ^{1}H, ^{71}Ga), TG-DTA measurements, FTIR spectroscopy, diffuse reflectance spectroscopy in the UV/vis range as already described in detail [17,25,26]. DSC, TG-DTA, and (*in situ*) FTIR methods have given additional information on phase transition behavior, LC loading and host–guest interactions [25].

We have also made use of broadband dielectric spectroscopy as a convenient tool for probing the dynamics of LC loaded in molecular sieves [25]. The methods of recording the dielectric loss spectra and processing the data have been described [33,34].

5.3 Methods of Loading Molecular Sieves with Liquid Crystals

Several loading methods have been investigated for studying the degree of loading and its results in connection with the location of the LCs in the molecular sieves.

A Contact of LC in the isotropic state with the porous material [25]. The loaded samples have a high LC content.

B The fraction of the LC adsorbed on the outer surface by method A was removed by evacuation [25] at a temperature higher than N–I transition of the bulk.

C Loading of LC from the vapor state [17].

Tab. 2. Examples LC loadings by different methods

Sample	*Loading method*	*LC amount (TG meas.) [%]*	*LC amount (pore volume)* [%]*
5CB/Si-MFI	A	36	12
5CB/Si-MFI	D	24	12
5CB/NaY	A	26	30
5CB/NaY	D	4	30
5CB/CLO	C	38	23
5CB/Si-CLO	C	45	19
8CB/AlMCM-41	A	56	40
8CB/AlMCM-41	B	41	40
8CB/SBA-15	D	62	61
8CB/AlSBA-15	D	37	52

* Estimated values by considering the complete filling of the pores with LC.

D Confinement was achieved by using impregnation of the solid materials with a solution of LC in an easily evaporating solvent, such as acetone [35].

The samples thus prepared were labeled with nCB/STR(L) in which $n = 5{,}8$ enters in the LC acronym, STR is the framework topology and L = A–D is the loading method. Table 2 gives some examples of the quantitative efficiency of loading obtained by different methods compared with the estimated theoretical maximum possible values.

In this connection an "excess of LC" is identified when the amount of sorbed LC exceeds the estimated maximum loading capacity of the pores, so there is additional amount of LC on the external surface. Evidence for excess loading of LC has already been obtained in some cases by TG measurements. However, if the loading thus found is lower than the estimated maximum value, location of LC both on the inner as well as on the outer surface of the molecular sieve has to be considered.

As information on the location of the LC on the outer surface or inside the pores/cavities of the host is concerned, some comments on the methods investigating the loading are necessary.

Thus, careful examination of the TG-DTA curves and their changes with loading parameters gave a first rough indication of LC location. A main exothermic peak in the heat flow at about 500 K (which is not observed either for bulk LC or for the unloaded molecular sieve) was correlated with the re-arrangements of LC molecules on the external surface of the molecular sieve (Section 5.4.1). In addition, a second main peak appeared at about 818 K, which is due to the oxidation of LC. Compared with the bulk LC the temperature is shifted to lower temperatures, which might indicate that the LC is located preferentially on the external surface of the molecular sieve grain and destabilized by this re-arrangements. For a location of the LC inside the pores, a "protective" effect of the pores is expected (as in the case of other guest–host interactions [36]); therefore the decomposition/oxidation processes are expected to occur at higher temperature than in the bulk state.

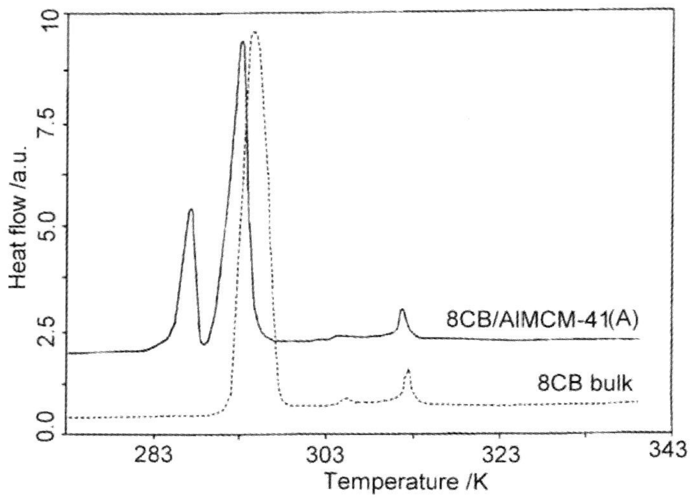

Fig. 2. DSC measurements of the samples 8CB/AlMCM-41 (A) and of 8CB bulk.

DSC measurements have shown that samples with LC on the external surface might present bulk-like phase transitions, shifted, however, towards lower temperatures [25,37]. An example is given in Fig. 2. At the same time, the samples having the LC only inside the pores do not show any phase transitions but a very broad low-intensity DSC curve (not shown here).

Dielectric measurements can also give evidence of the LC location: the amount located on the outer surface or in the large pores leads to the appearance of a bulk-like loss peak in the dielectric spectra, while the surface layer has a slow dynamics (Section 5.4.4 and 5.4.5).

With respect to the location of LC on the internal or external surface of the host, the experiments have shown that none of the above mentioned loading methods allows a decisive conclusion *a priori*. Method A ensures an adsorbed LC amount higher than predicted on the basis of pore volume and LC density, while method B ensures that almost the estimated amount is hosted inside the pores. Method C leads to samples having the LC located on the outer surface as well as inner surface, while method D seems to facilitate LC uptake and the amount of the LC given for loading and can be in accordance with the value estimated from the pore volume.

5.4
Nanoporous Composites Based on Different Molecular Sieves

The results of confinement with LCs located inside the pores or cavities of some representative molecular sieves are further discussed below in order of increasing pore or cavity size.

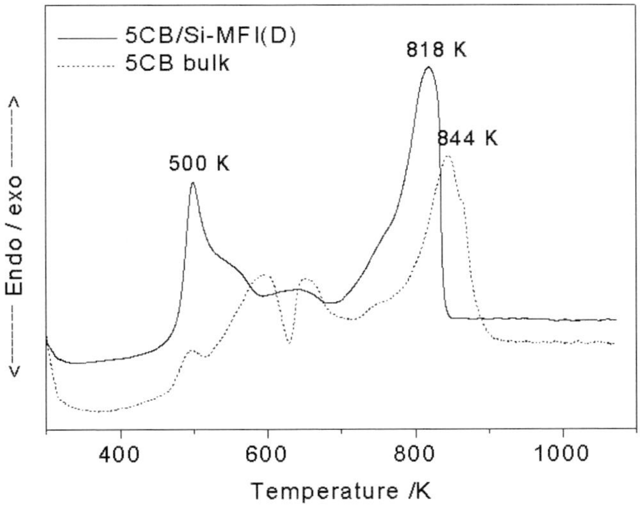

Fig. 3. DTA curves for 5CB/Si-MFI(D) sample and bulk 5CB.

5.4.1
MFI Type Molecular Sieves

Elliptical 10-membered rings of oxygen atoms control the entrance to the straight and sinusoidal channels of MFI topology. Therefore, entering of 5CB or 8CB molecules inside the pore system of MFI zeolites seems rather unlikely. Even though the channel intersections are the largest spaces ("cavities" of 0.87 nm diameter), they are reached with difficulty by the guest LC molecules due to small access windows (Table 1).

However, samples with mesopores might show some uptake of LCs. Si-MFI samples were specially synthesized [27] to have large crystals of about 50 μm, with an apparently perfect shape but with hidden voids (nanopores of about 2 nm diameter).

In Fig. 3, the corresponding DTA curves of 5CB/bulk and 5CB/MFI(D) are shown, indicating a main exothermic peak at about 500 K, which is present neither in the bulk curve nor in that of the unloaded MFI zeolite. This peak is not accompanied by a significant weight loss. The temperature of this process is too high for a LC phase transition and too low for an oxidation process, since bulk 5CB is thermally stable up to about 600 K. *In situ* FTIR spectra indicate changes in the guest–host interaction around 500 K [38]. Therefore, it was concluded that the DTA exothermic peak at 500 K might be due to a transition involving some guest–guest and especially guest–host interactions, leading to the re-arrangement of LC molecules on the surface of Si-MFI. In fact, further studies indicated that a catalytic dealkylation and oxidation take place.

A second intense DTA peak appears at 818 K (Fig. 3). This is due to the oxidative decomposition of the LC, at a temperature shifted 26 K lower than for the bulk LC.

This shift corresponds to a similar shift in the TG curves. This behavior might indicate that the LC is located mainly on the external surface of the molecular sieve grains.

Therefore, such mesoporous silicalite samples allowed the study of LC located on the external surface of the molecular sieve and probably at the mesopore mouth as well.

5.4.2
Faujasite

The faujasite structure contains truncated octahedra arranged in a tetrahedral coordination forming a network of large cages 2.37 nm in length and 1.3 nm in diameter, with the bottlenecks being only 0.74 nm in diameter. The positions of sodium ions are mostly in the single or double six-membered rings as well as in the large cages [39].

The efficiency of the loading of faujasite with LC has been proven by TG-DTA measurements, although the amount of LC in the faujasite material was found to be low (Table 2). However, further evidence came from dielectric measurements, because these indicated the hindering of sodium ion mobility by the presence of LC, which is in agreement with the behavior of other polar molecules confined in nanopores [12].

Figure 4 presents the dielectric loss of some faujasite samples loaded with 5CB by the method A (leading to LC excess) and B, as well as of bulk LC and of unloaded samples. Toward low frequencies, a contribution from the dc conductivity of faujasite is observed (process I). Besides, there are three further relaxation pro-

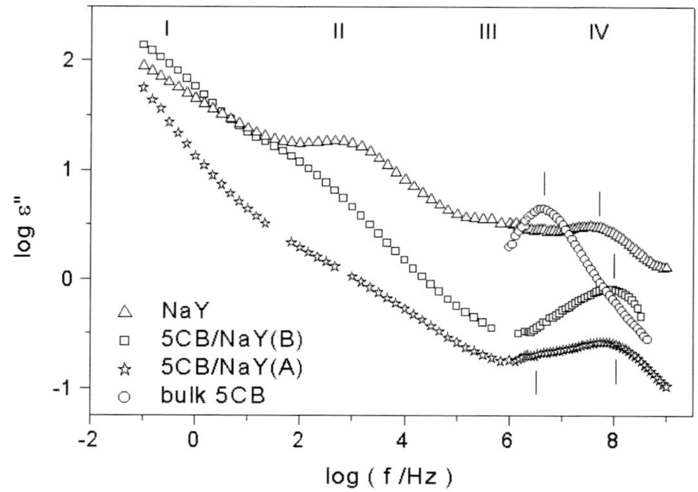

Fig. 4. Dielectric loss of two loaded NaY samples, at 300 K, and also of unloaded NaY and bulk 5CB. Roman numerals refer to the assignment of relaxation processes in empty zeolite.

cesses (II-IV), which are mostly contributions of the empty zeolite. The assignment of these processes is known or still under discussion [40,41]. The low-frequency relaxation processes II was assigned to a cation relaxation, whereas the relaxation processes III was assigned to movements of correlated Na^+ ions (or to the cation motion in the sodalite cages). Process IV was assigned to water molecules/OH groups in sodalite cages [42]. The comparison with the bulk 5CB indicates, however, that there is also a distinct overlap of contributions coming from the presence of LC in zeolite NaY.

The dielectric strength of process II for the sample 5CB(excess)/NaY is strongly decreased. This favors the interpretation that this process is in fact a Maxwell–Wagner–Sillars effect taking part at these intergrains. The Maxwell–Wagner–Sillars effect is due to interfacial phenomena appearing between the grains within the sample or between the bulk sample and the electrodes. It consists in a strong dispersion towards low frequencies. A striking fact is the decrease in the dielectric intensity of process III for the 5CB/NaY samples compared with the unloaded pores [41]. Assuming that this process is due to the movement of Na^+ ions in the supercages, these movements of sodium ions are strongly depressed now due to the presence of 5CB molecules.

This assignment is supported also by the values of the activation energy as estimated by representing the relaxation rates as function of the inverse temperature (Fig. 5). The (slight) increase of the activation energy (of 19.5 and 5.1 kJ mol^{-1}, respectively) for processes II and III [41] indicates differences in the electrostatic interactions between the sodium ions and the oxygen lattice, owing to the loading with LC. As long as the number and the type of cations is the same in the loaded samples as in the unloaded one, one can infer that some of the sodium ions are blocked by the presence of 5CB molecules in the supercages. As a consequence,

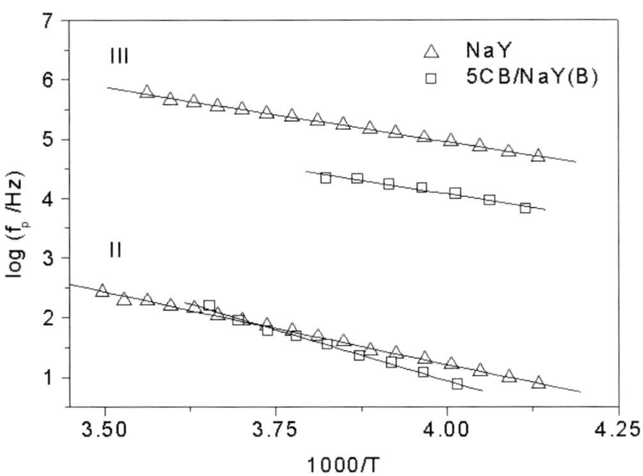

Fig. 5. Relaxation rates versus $1/T$ for the composite system 5CB/NaY(B). The linear dependence is described by the Arrhenius law.

the loaded systems behave as if having a lower concentration of sodium ions or their mobility being hindered.

In the high-frequency region, the dielectric loss spectrum of 5CB/NaY(A) sample is complex (Fig. 4). However, one can notice a shift of the bulk-like 5CB when compared with the bulk 5CB and of relaxation process IV in loaded zeolite when compared with the unloaded zeolite, as marked by vertical lines. These shifts are additional evidence for the reciprocal influence of the two system components: the LC and the molecular sieve.

5.4.3 Cloverite

The gallophosphate cloverite (CLO) molecular sieve [43] has large pores of 1.3 nm diameter and supercages of 3 nm at the pore intersections.

The loading of cloverite with 5CB or 8CB was demonstrated by several methods such as XRD, IR, and dielectric spectroscopy. Thus, in the XRD patterns (Fig. 6), the peaks at high 2θ have a decreased intensity when compared with the unloaded cloverite, owing to a distorted structure caused by the LC loading. Moreover, XRD temperature programmed experiments [17] have shown that when temperature increases, the main peaks at (200) and (222) are shifted toward lower values owing to a shrinkage of the lattice by removing the loaded material, showing that at least part of the LC is inside the pores/cavities of cloverite.

In addition, the IR band at 2262 cm^{-1} of 5CB/CLO(C) indicated that the CN group interacts with Ga ions and OH groups located on the internal surface (Fig. 7). Besides, the LCs molecules inside the supercages of the cloverite are thermally stabilized against the bulk [44], demonstrated by *in situ* experiments.

Dielectric loss spectra of 8CB/CLO(B) have a complex form, difficult to fit with model functions [37]. However, two processes were revealed, one bulk-like in the high-frequency region and a second one at frequency two orders of magnitude less. The variation with temperature of bulk-like relaxation obeys the Arrhenius law. The corresponding activation energy is 38.8 kJ mol^{-1} in the isotropic state and

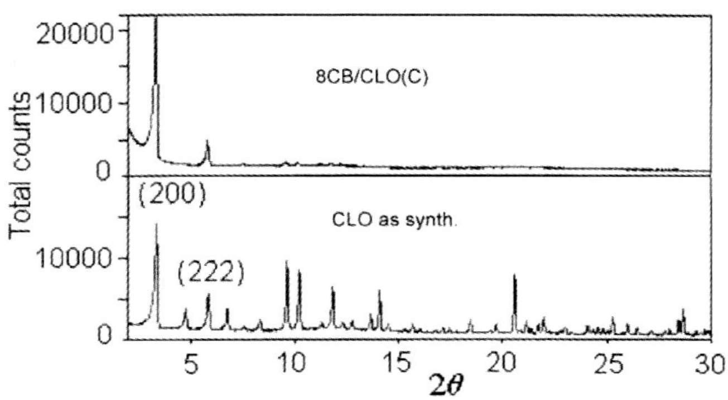

Fig. 6. XRD patterns of cloverite samples, in the form as-synthesized or loaded with 8CB.

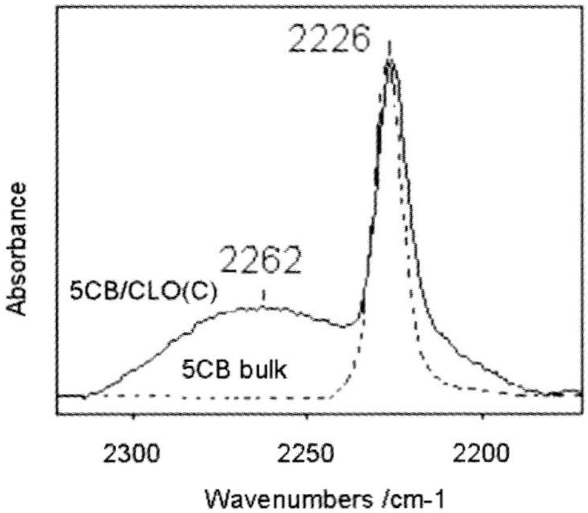

Fig. 7. FTIR spectra in the CN stretching vibration of 5CB and 5CB/CLO(C) sample.

68.5 kJ mol^{-1} in the nematic–smectic range [37], whereas the corresponding bulk values are lower: 35.6 and 59.8 kJ mol^{-1}, respectively [45]. The higher activation energies of the confined 8CB seem to indicate that the fluctuations responsible for the relaxation are more hindered inside the cloverite cages than in the bulk.

In conclusion, cloverite has been shown to be a good host for the investigated LC, since it allows observation of characteristic phase transitions. However, the instability of the framework to humidity is a drawback for further applications.

5.4.4
MCM-41 Molecular Sieves

MCM-41 samples used for confinement have a rather narrow pore radii distribution centered at 2 nm and a high volume fraction (Table 1). These nanoporous materials consist of randomly oriented grains having cylindrically shaped unidirectionally oriented pores. The inherent randomness is imposed by the grain orientation.

Evidence for a successful loading process is given by TG-DTA, DSC, IR, and dielectric measurements [25]. Thus, the dielectric loss against frequency for 8CB confined to AlMCM-41(A and B) is compared with the bulk material in the isotropic state (Fig. 8). Two relaxation processes can be detected for 8CB/AlMCM-41(A): The relaxation process at higher frequency (I) corresponds obviously to the bulk; it does not appear, however, for the sample loaded by method B. The process at lower frequencies (II) is not present in the bulk 8CB [46]. However, more rigorous considerations show that the molecular motions of these "bulk-like" molecules are not really identical to those of the corresponding bulk phases (Fig. 9), since the changes of the relaxation time at N–I transition in the intergrain voids are not so sharp and not so strong as in the bulk state. Moreover, the temperature dependence, even in the range corresponding to a certain mesogenic phase, is not exactly linear, as expected, but seems to be slightly curved.

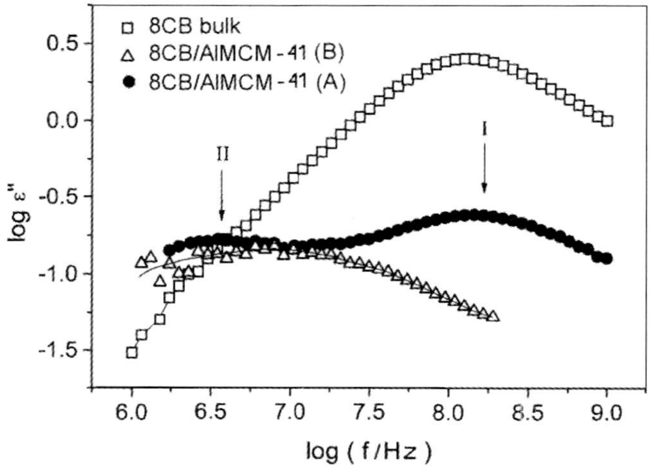

Fig. 8. Comparison of the dielectric loss of 8CB in the bulk isotropic state (360 K) and loaded to the pores of AlCM-41 molecular sieve, by method A or B.

The relaxation rate of the low-frequency process is almost two orders of magnitude lower than the slowest bulk relaxation process and its temperature dependence shows a curved trace in a plot against $1/T$ (Fig. 9). This process also takes place for samples having the LC molecules located only in the inner surface layer. Figure 10 gives the temperature dependence of this slow relaxation process for

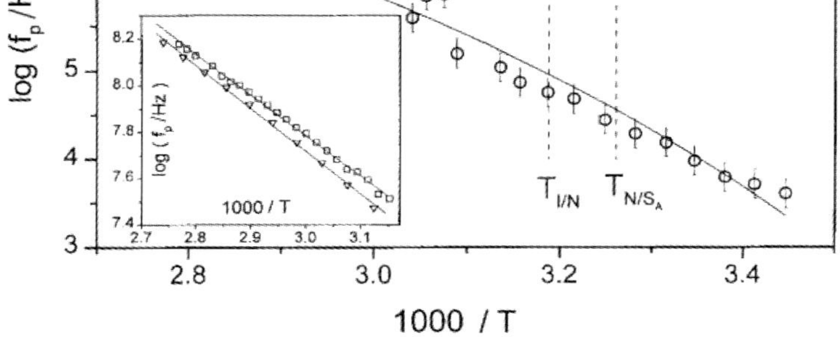

Fig. 9. Relaxation rates against $1/T$ for 8CB/AlMCM-41 (A) and bulk 8CB.

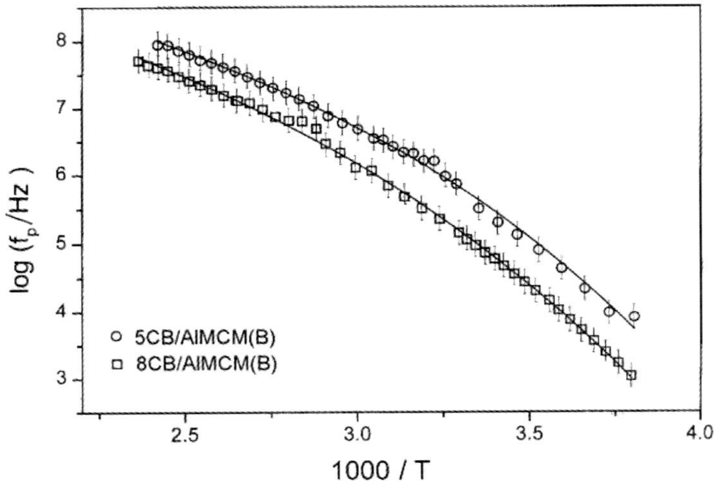

Fig. 10. Relaxation rates against $1/T$ for 8CB/AlMCM-41 (B) and 5CB/AlMCM-41 (B).

both 5CB and 8CB loaded into molecular sieve material by method B. Both these LCs have a similar behavior.

Any bulk-like behavior of the LC molecules inside the unidirectional pores of MCM-41 materials was revealed, despite the close dimension of the pores and of the cloverite supercages. This might be an indication that the interconnectivity (present for cloverite with large access windows) of the pore system plays an important role in the appearance of phase transitions characteristic of LC behavior.

5.4.5
SBA-15 Materials

Recently, the synthesis of novel mesoporous molecular sieves of SBA-15 type has been reported [47,48]. These have larger pores (7.5–10 nm), thicker walls, and (consequently) higher stability than MCM-41.

The dielectric behavior of 8CB confined in these SBA-15 materials [38] shows not only the surface layer, as for MCM-41 materials [25], but also a bulk-like relaxation process. Besides, there are differences between the two loaded SBA adsorbents, probably related to particular host–guest interactions.

The interaction between the LC molecules and the SBA hosts results in changes in spectral parameters of some fundamental bands of both interacting components in the FTIR spectra. In additions to the changes in the CN groups, already discussed for other molecular sieves, bands appeared between 1800 and 2000 cm^{-1}, which are due to out-of-plane deformation vibrations, show [49] the interaction of the π system of the aromatic rings with the (Al) cations.

Two representative pictures of the dielectric behavior of the loaded SBA materials are shown in Fig. 11. In the frequency range in which the bulk 8CB does not

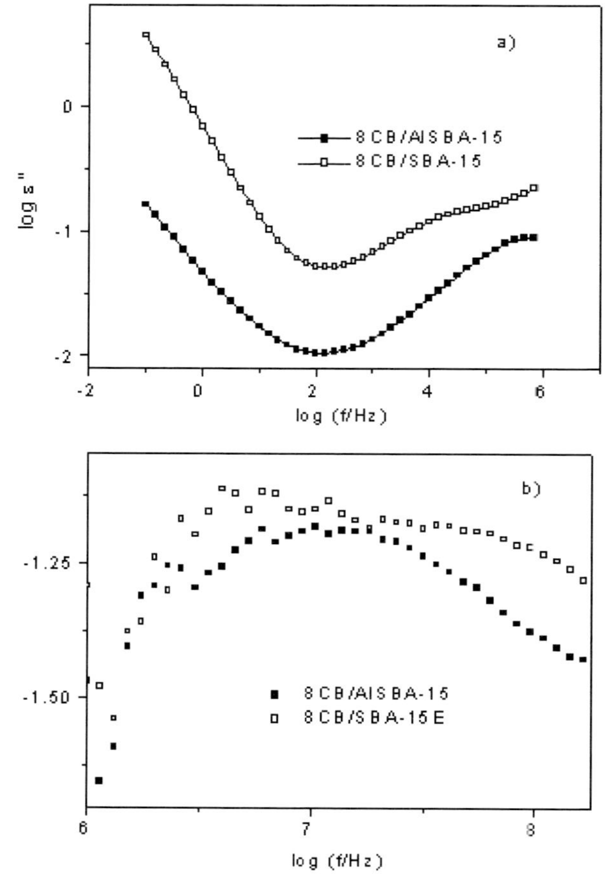

Fig. 11. Dielectric loss of 8CB/SBA-15 and 8CB/AlSBA-15 samples at 301 K: (a) low-frequency region; (b) high-frequency region.

present appreciable losses (Fig. 11a) silica SBA-15 loaded with 8CB indicates only a small absorption centered at about 100 Hz. This absorption might be due to the surface layer with a slow dynamics, as it was found for MCM-41 materials [25]. A conductivity contribution to the dielectric loss appears at low frequency but only for higher temperatures. At the same time, the spectra of the 8CB/AlSBA-15 sample are much more complex than for the appropriate Al-free sample. The conductivity contribution is present for all the investigated temperatures, probably due to some extra-framework Al species. The surface layer contribution also seems to be much more important than in the case of silica SBA-15 material.

Figure 11b illustrates the behavior of the loaded samples in the high-frequency domain in which the bulk LC also shows a dielectric loss [45]. While the loaded SBA sample clearly shows absorptions at temperatures higher than room temperature, when the bulk LC is not in a solid (crystalline) state, the loaded AlSBA-15 sample shows more complex spectra. However, the parameters characterizing the bulk-like relaxation process are somehow different from those of the bulk LC, probably due to the influence of confinement substrate. Similar behavior was found also for LC molecules located in the intergrain void in 8CB/AlMCM-41 samples [25].

It is noteworthy that the pores of these SBA-15 adsorbents are of cylindrical shape and arranged parallel in a honeycomb-like lattice. The absence of the pore channel intersections guarantees that the pore networking effects are negligibly small [50]. Therefore, the observed dynamic is attributed to the movements of the LC molecules inside singular pores.

5.4.6 Exchanged Nanoporous Materials

Cation exchanged AlSBA-15 samples with different charge compensation cations have shown particular host–guest interactions due to these cations by a two-fold effect. First, because it changes the chemical composition of the surface of the molecular sieve surface. Second, because the cations modify the hydrophilic features of the surface. In addition, some of the cations might hinder the entrance of the LC molecules inside the pores of the molecular sieves.

TG measurements performed on the exchanged samples (without a previous equilibration with water), gave first information on the water content of these samples and allow us to better interpret the TG measurements on the samples loaded with 8CB. Under similar conditions of preparation and registering, the samples are arranged in the following sequence according to their hydrophilicity: parent AlSBA-15 < K-AlSBA-15 < Rb-AlSBA-15 < Na-AlSBA-15 < Co-AlSBA-15 < Ca-AlSBA-15 < La-AlSBA-15. This order shows that cations with higher charge lead to higher water retention.

The loading of the exchanged samples was performed by using 8CB solution (method D). The samples can be arranged in the order of increasing the weight loss as follows: 8CB/La-AlSBA-15 < 8CB/Na-AlSBA-15 < 8CB/K-AlSBA-15 < 8CB/Co-AlSBA-15 < 8CB/Rb-AlSBA-15 < 8CB/Ca-AlSBA-15. This order differs from that of unloaded samples, indicating different interactions between cations and the guest molecules. Besides, all these loaded exchanged samples have a weight loss less than that of the unexchanged samples. To the host–guest interactions mentioned above, the exchanged cations add expected interactions with π electrons of the aromatic rings. In this connection, the high shield of OH groups/water molecules still present around the La cation might explain the position of the sample 8CB/La-AlSBA-15 in the previous series.

DTA measurements for the loaded exchanged samples can be seen in Fig. 12; these samples do not have the low-temperature DTA peak at 500–560 K that was found to be due to the LC on the outer surface or in larger intergranular voids. In addition, DSC measurements did not reveal phase transitions.

IR spectra of loaded exchanged samples show that the cyan groups and benzene

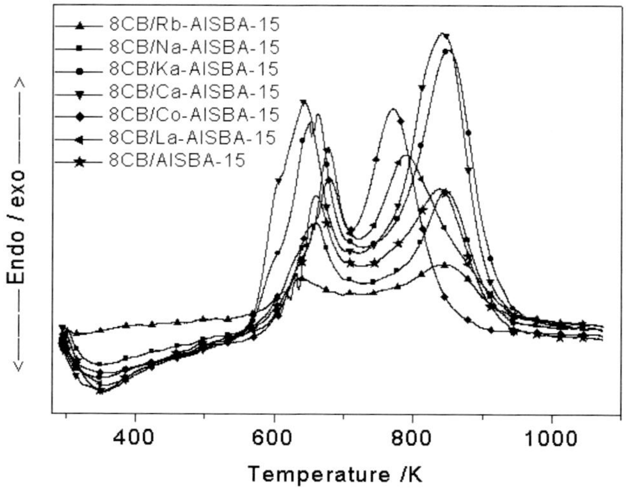

Fig. 12. DTA curves for loaded exchanged AlSBA-15 samples.

rings of LCs are involved in host–guest interactions by their π electrons (to the charge compensating cations) and ring H (to the network oxygen atoms), in agreement with the literature [17,18,51].

Therefore, we presume that he host–guest interactions play a special role in the appearance of the low-temperature DTA peak and that these exchanged samples are not good hosts for nematic LCs.

5.5
On the Location of Liquid Crystals Inside the Pores or Cavities of Molecular Sieves

When discussing the location of the LC molecules inside the pores or cavities of the molecular sieves, the structures of all components and their mutual interactions have to be considered. The location and the position of cyanobiphenyl in FAU supercages or in the straight channels of MFI structures are already known [52,53]. One phenyl group is coordinated to the extra-framework cations (if present) and might be oriented parallel to the surface wall.

As far as guest–guest interactions are concerned, the molecules are arranged anti-parallel to each other, forming dimers in which the biphenyl cores overlap significantly and the alkyl chains point roughly in opposite directions. Evidence for such an arrangement has been obtained from X-ray diffraction studies [54], and a quadrupolar ordering of the dimers [55] was investigated by experiments using second harmonic generation. However, the pores cause the large dipolar molecules to arrange themselves in a highly ordered manner [56,57].

When the pore diameter approaches the size of the CB molecules, only a low loading of the nanoporous materials is possible. The LC molecules lie with their

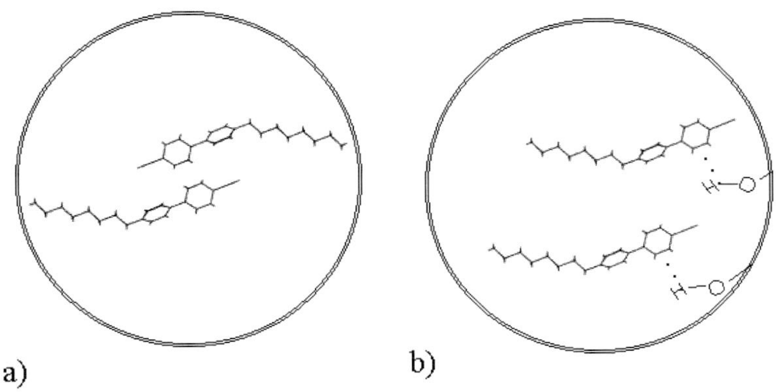

Fig. 13. Cross section through pores of mesoporous ordered materials loaded by CBs. (a) antiparallel arrangement of two LC molecules, the LC molecules are perpendicular to the pore walls; (b) arrangement allowing interaction of the rings with the surface OH groups (unscaled).

long axis either perpendicular (Fig. 13) or parallel (Fig. 14) to the pores or cavity walls. In these figures the pores are seen in cross section as cylinders with their axis longer than the diameter (for which 3.5 nm is assumed). The 8CB molecules are drawn as in the DFT image (Fig. 1) or as cylinders that include the biphenyl part and the alkyl chain (Fig. 14). Even though these models have not yet been confirmed by energy minimization calculations, they are still suggestive. It is important to note that in the models sketched in Figs. 13 and 14, surface Lewis acid centers can appear instead of surface OH groups. It is easy to imagine that in a parallel arrangement the pores are better filled than in a perpendicular one. Moreover, if the pores or cavities are much larger than the molecule size, the LC mole-

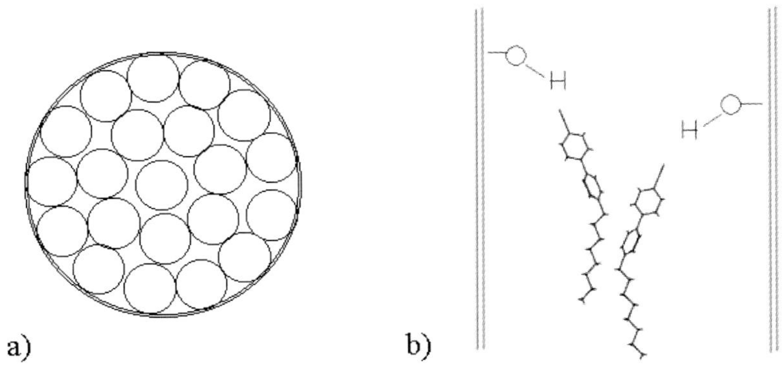

Fig. 14. Schematic representation of the pores of mesoporous ordered materials loaded with CBs parallel to the pore walls: (a) cross section, (b) longitudinal section.

cules can be packed as usually in the bulk. If the size of a LC molecule is, however, comparable with the channel/pore/cavity size, it is quite reasonable to assume that the translational degree of freedom is restricted to motion parallel to the channels [57] and, therefore, molecules are arranged parallel to the pore walls (Fig. 14). Such a parallel arrangement is wholly supported by the dynamic behavior of the molecules as given by dielectric measurements: a planar alignment of the director and rotational fluctuations around the short molecular axis.

5.6 Conclusions

The confinement of 5CB and 8CB liquid crystals inside and outside the pores or cavities of MFI, FAU, CLO, MCM-41, and SBA-15 type molecular sieves has been studied. We looked for the influence of the molecular sieve pore/cavity system on the phase transition(s) characteristic to a given LC, for host–guest interactions that stabilize the LC molecules inside the pores, and even for new composite materials.

According to these goals, dielectric spectroscopy was used to investigate the molecular dynamics, while DSC, DTA-TG, and FTIR were used for static properties of the composite systems. These methods were also used to discriminate between the parts of LCs that interact with the outer surface of the grains and which are located inside the pores.

Several loading methods were adapted for very restrictive confinement. In general, the suitability of each method depends on the aim of investigation. If the key point of these investigations is to obtain samples containing the LC mostly inside the pores then method D that uses a solution of the LC seems to be the recommended method. In other cases the removal of LCs from the outer surface by vacuum outgassing may also be applied for samples prepared by other methods (A–C).

It has been found that size, shape, and interconnectivity of the pores play an important role in the modification of properties of liquid crystals. Thus, the dynamics of the cyanobiphenyl molecules (5CB and 8CB) inside the small pores of molecular sieves strongly depends on various factors; in the case of faujasite by the presence of sodium ions in the sodalite and faujasite supercages the mobility of which is hindered by the adsorbed LC. Instead, the larger supercages of cloverite and its 3D pore system allow unhindered location and orientation of a sufficient concentration of LC to perform bulk-like phase transition(s). For similar-sized pore systems but having only singular cylindrical pores (MCM-41 materials), such phase transitions cannot be observed. In the extra large pores of SBA-15 materials, which are distinctly larger than those of MCM-41, liquid crystalline behavior can be observed.

The investigations have shown that nanoporous crystalline molecular sieves are interesting hosts for molecules of liquid crystals. However, phase transitions characteristic of LCs are only observed if the nanoporous hosts provide pore systems of appropriate size and interconnectivity. It seems that the size of the interconnected pores must be larger than 3 nm.

Since nanosized crystalline molecular sieves with defined shape, size, and chemical composition are now available, it should be interesting to create new inverse molecular sieve–LC nanocomposites by embedding the nanoslabs in the bulk LC. In fact, such studies already started for related systems such as aerosil-LC [35] and clay-LC [58] systems.

Acknowledgements

The authors gratefully acknowledge DFG financial support (Project Ko 1639/2-3). They are also grateful for excellent collaboration and constant interest from Dr. H.-L. Zubowa (ACA) and Dr. A. Schönhals (BAM).

References

1 C. Maugin, *Compt. Rend. Acad. Sci.* **1913**, *156*, 1246.

2 G.P. Crawford, R. Stannarius, J.W. Doane, *Phys. Rev. A* **1991**, *44*, 2558.

3 G.P. Crawford, S. Zumer, eds., Liquid Crystals in Complex Geometries Formed by Polymer and Porous Networks, Taylor & Francis, London **1996**, p. 606.

4 M.D. Dadmun, M. Muthukumar, *J.Chem. Phys.* **1993**, *98*, 4850.

5 G.S. Iannacchione, D. Finotello, *Phys.Rev. E* **1994**, *50*, 4780.

6 T. Bellini, N.A. Clark, C.D. Muzny, L. Wu, C.W. Garland, D.W. Schaefer, B.J. Oliver, *Phys. Rev. Lett.* **1992**, *69*, 788.

7 L. Wu, B. Zhou, C.W. Garland, T. Bellini, D.W. Schaefer, *Phys. Rev. E* **1995**, *51*, 2157.

8 S. Kralj, G. Lahanjar, A. Zidansek, N. Vrbancic-Kopac, M. Vilfan, R. Blinc, M. Kosec, *Phys. Rev. E* **1993**, *48*, 340.

9 G.S. Iannacchione, G.P. Crawford, S. Zumer, J.W. Doane, D. Finotello, *Phys.Rev. Lett.* **1993**, *71*, 2595.

10 A. Zidansek, S. Kralj, G. Lahanjar, R. Blinc, *Phys. Rev. E* **1995**, *51*, 3352.

11 S. Kralj, A. Zidansek, G. Lahanjar, I. Musevic, S. Zumer, R. Blinc, M.M. Pintar, *Phys. Rev. E* **1996**, *53*, 3629.

12 C. Cramer, T. Cramer, M. Arndt, F. Kremer, L. Naji, R. Stannarius, *Mol. Cryst. Liq. Cryst. A* **1997**, *304*, 209.

13 X. Wu, W.I. Goldburg, M.X. Liu, J.Z. Xue, *Phys. Rev. Lett.* **1995**, *69*, 470.

14 F.M. Aliev, M.N. Breganov, *Zh. Exper. Teor. Fiz.* **1989**, *95*, 122.

15 N.A. Clark, T. Bellini, R.M. Malzbender, B.N. Thomas, A.G. Rappaport, C.D. Muzny, D.W. Schaefer, L. Hrubesh, *Phys. Rev. Lett.* **1995**, *71*, 3505.

16 G. Schwalb, F.W. Deeg, *Phys. Rev. Lett.* **1995**, *74*, 1383.

17 H.-L. Zubowa, H. Kosslick, E. Carius, S. Frunza, L. Frunza, H. Landmesser, M. Richter, E. Schreier, R. Fricke, *Micropor. Mesopor. Mater.* **1998**, *21*, 467.

18 I. Gnatyuk, G. Puchkovska, O. Yaroshchuk, K. Otto, G. Pelzl, T. Morawska-Kowal, *J. Mol. Struct.* **1995**, *563–564*, 498.

19 W. Gorbatschow, M. Arndt, R. Stannarius, F. Kremer, *Europhys. Lett.* **1996**, *35*, 71.

20 C. Cramer, T. Cramer, F. Kremer, R. Stannarius, *J. Chem. Phys.* **1996**, *106*, 3730.

21 S.A. Rozanski, R. Stannarius, H. Groothues, F. Kremer, *Liq. Cryst.* **1995**, *20*, 59.

22 G.P.Sinha, F.M. Aliev, *Mol. Cryst. Liq. Cryst. A* **1997**, *304*, 309.

23 G.P.Sinha, F.M. Aliev, *Phys. Rev. E* **1998**, *58*, 2001.

24 S.A. Rozanski, R. Stannarius, F. Kremer, *Z. Phys. Chem.* **1999**, *211*, 159.

25 S. Frunza, A. Schönhals, L. Frunza, H.-L. Zubowa, H. Kosslick, R. Fricke, H. Carius, *Chem. Phys. Lett.* **1999**, *307*, 167.

26 H.-L. Zubowa, U. Bentrup, H. Kosslick, R. Fricke, *Stud. Surf. Sci. Catal.* **1999**, *125*, 321.

27 R. Fricke, H.-L. Zubowa, J. Richter-Mendau, E. Schreier, U. Steinicke, *Proc. Int. Symp. Zeolites in China*, Jiangsu Petroleum Society, Nanjing **1995**, p. 2–19.

28 H. Landmesser, H. Kosslick, W. Storek, R. Fricke, *Solid State Ionics* **1997**, *101*, 771.

29 H. Kosslick, I. Mönnich, E. Paetzold, G. Oehme, R. Fricke, *Micropor. Mesopor. Mater.* **2001**, *44*, 537.

30 H. Berndt, A. Martin, A. Brückner, E. Schreier, D. Müller, H. Kosslick, G.-U. Wolf, B. Lücke, *J. Catal.* **2000**, *191*, 384.

31 C.W. Lee, D.H. Ahn, B. Wang, J.S. Hwang, S.E. Park, *Micropor. Mesopor. Mater.* **2001**, *44–45*, 587.

32 A. Hofmann, unpublished results.

33 F. Kremer, D. Boesse, G. Meier, E.W. Fischer, *Prog. Colloid Polym. Sci.* **1989**, *80*, 129.

34 E. Schlosser, A. Schoenhals, *Colloid Polym. Sci.* **1989**, *267*, 963.

35 S. Frunza, L. Frunza, H. Goering, H. Sturm, A. Schönhals, *Europhys. Lett.* **2001**, *56*, 801 and references herein.

36 L. Frunza, H. Kosslick, H. Landmesser, E. Höft, R. Fricke, *J. Mol. Catal.* **1997**, *123*, 179.

37 L. Frunza, S. Frunza, A. Schönhals, H.-L. Zubowa H. Kosslick, R. Fricke, *J. Mol. Struct.* **2000**, *563–564*, 491.

38 L. Frunza, S. Frunza, A. Schönhals, U. Bentrup, R. Fricke, I. Pitsch, H. Kosslick, *Stud. Surf. Sci. Catal.* **2002**, *142*, 1323.

39 W.H. Baur, *Am. Mineral.* **1964**, *49*, 697.

40 U. Simon, M.E. Franke, *Micropor. Mesopor. Mater.* **2000**, *41*, 1.

41 L. Frunza, H. Kosslick, S. Frunza, R. Fricke, A. Schönhals, *J. Non-Cryst. Solids* **2002**, *307–310*, 503.

42 L. Frunza, H. Kosslick, S. Frunza, A. Schönhals, *J. Phys. Chem. B* **2002**, *106*, 9191.

43 M. Estermann, L.B. McCusker, C. Baerlocher, A. Merrouche, H. Kessler, *Nature* **1991**, *353*, 320.

44 S. Frunza, L. Frunza, A. Schönhals, H.-L. Zubowa, H. Kosslick, R. Fricke, *Stud. Surf. Sci. Catal.* **2001**, *135*, A21P14.

45 A. Schönhals, H.-L. Zubowa, R. Fricke, S. Frunza, L. Frunza, R. Moldovan, *Cryst. Res. Technol.* **1999**, *34*, 1309 and references cited herein.

46 S. Frunza, L. Frunza, A. Schönhals, *J.Phys. IV France* **2000**, *10*, Pr7-115.

47 D. Zhao, Q. Huo, J. Feng, B.F. Chmelka, G.D. Stucky, *J. Am. Chem. Soc.* **1998**, *120*, 6024.

48 P. Yang, D. Zhao, D. Margolese, G.D. Stucky, *Nature* **1998**, *396*, 152.

49 L. Frunza, H. Kosslick, U. Bentrup, I. Pitsch, S. Frunza, A. Schönhals, *J. Mol. Struct.* **2003**, in press.

50 K. Morishige, K. Kawano, *J. Phys. IV France*, **2000**, *10*, Pr7-91.

51 D. Barthomeuf, *Catal. Rev. Sci. Eng.* **1996**, *38*, 521.

52 I. Gener, G. Ginestet, G. Buntinx, C. Bremard, *J. Phys. Chem.* **2000**, *104*, 11656.

53 I. Gener, G. Buntinx, C. Bremard, *Micropor. Mesopor. Mater.* **2000**, *41*, 253.

54 A.J. Leadbetter, J.C. Frost, J.P. Gaughan, G.W. Gray, A. Mosley, *J. Phys. France* **1979**, *40*, 375.

55 B. Jerome, J. O'Brien, Y. Ouchi, C. Stanners, Y.R. Shen, *Phys. Rev. Lett.* **1993**, *71*, 758.

56 I. Leike, F. Marlow, *Zeolites* **1996**, *16*, 65.

57 F. Marlow, D. Demuth, G. Stucky, F. Schüth, *J. Phys. Chem.* **1995**, *99*, 1306.

58 M. Kawasumi, N. Hasegawa, A. Usuki, A. Okada, *Liq. Cryst.* **1996**, *21*, 769.

6
Cationic Host–Guest Polymerization of Vinyl Monomers in MCM-41

Stefan Spange, Annett Gräser, Friedrich Kremer, Andreas Huwe, and Christian Jäger*

6.1
Introduction

The synthesis of inorganic oxide/organic polymer hybrid materials has been intensively studied during the last decade [1–11]. In this context the study of flexible polymers under conditions of constricted geometry, as in cavities or nanopores of inorganic or organic solid materials, is a special experimental challenge. The dynamics of the embedded polymers is found to be determined by the counterbalance between surface and confinement effects [12]. The former result in a decrease in molecular dynamics and hence an increase in the calorimetric glass transition temperature (T_g), while the later are characterized by increased mobility causing a decrease in T_g [13]. As the thickness of the polymer layer decreases, the glass transition temperature (T_g) should decrease [13]. Various experimental procedures are available for the preparation of individual polymer chains within solid inorganic materials. Individual polymer chains can be enclosed by sol-gel processes in a hybrid material, deposited in 2D in layered silicates, or adsorbed in nanopores of solid porous materials [14–18].

The threading of linear flexible chains in highly ordered pore systems of HY zeolites or MCM-41 is experimentally very difficult, as direct threading appears to be relatively unsuccessful because of the associated loss of entropy (Fig. 1). A promising option involves synthesizing the polymer directly in the pore system of a mesoporous silicate or other host.

The linking of mobile monomers to polymers within cavities of solid or soft macromolecular materials has been termed host–guest polymerization [19]. In the broadest sense there are three possibilities for the association of guest and monomer in host–guest polymerization.

A The host is flexible and fits in the geometry of the monomer structure exactly. The polymer is produced from specific inner sites in this active cavity and then liberated. This example essentially corresponds to the enzymatic synthesis of all-cis polyisoprene (natural rubber) and other biopolymers [19].
B The host is rigid and the pore geometry fixed. The monomer is mobile and dif-

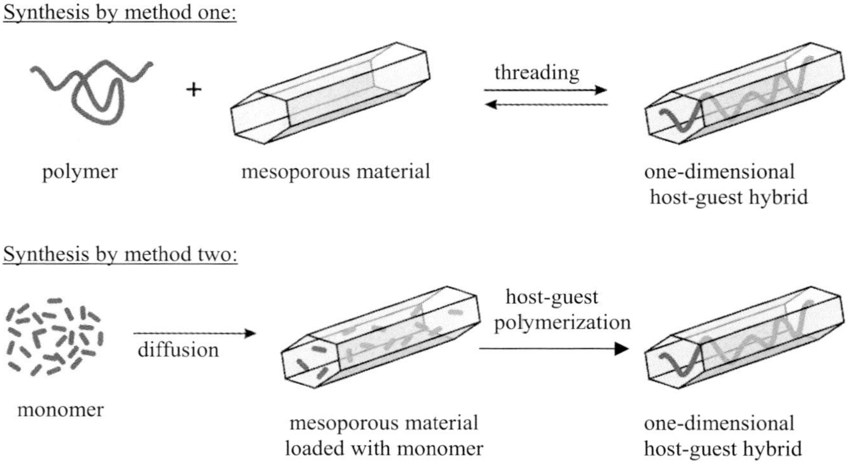

Fig. 1. Syntheses of 1D polymer/oxide hybrids.

fuses into the pores of the host system. The polymerization process can then be initiated by internal (immobilized catalysts) or external (light, radiation) energy sources. Polymer fractions are formed in this process, which are distributed statistically within the host framework. Defects in the pore structure or geometry of the host can have a drastic effect on the polymerization process (see below).

C A flexible mobile host building block with a capacity for self-assembly and the polymerizable monomer form an aggregate that forms a supramolecular system, in which the monomer can be polymerized to give a flexible chain. The rigid host is formed at the same time. These types of complementary syntheses have so far only been described as laboratory procedures [20–22].

In this review we concentrate on option B, whereby the hosts are inorganic oxides and the monomers 1-olefins, producing flexible polymer chains directly in the channels of mesoporous and nanoporous inorganic oxide materials. Various methods have been developed for these host–guest polymerizations. In pioneering work Bein et al. [23–29] have successfully studied the electropolymerization of pyrrole in HY zeolites and the radical polymerization of methacrylates in MCM-41. In these processes the initiation of the polymerization occurs by chance, on or within the solid material. If the initiator is immobilized by covalent bonding to the inner surface of a mesoporous silicate or/and aluminosilicate (a method used successfully with transition metal complexes for ethylene polymerization) polyethylene fibers can be synthesized directly inside the mesopores of, for example, MCM-41 [31–35]. A general feature of all host–guest polymerizations is the inner surface chemistry of the host. It should interfere as little as possible (if at all) with the mechanism of the polymerization reaction, for example by inducing transfer and degradation reactions.

Tab. 1. Examples of guest polymerizations of organic monomers in one dimensional inorganic host materials. For a review see also Ref. [19].

Guest monomer	*Inorganic host*	*Catalysis*	*Reference*
acrylonitrile	MCM-41	radical	28
aniline	Y zeolite	redox	23
pyrrole	Y zeolite	redox	24, 30
pyrrole	Cu mordenite	redox	36
thiophene	Cu mordenite	redox	36
styrene	13X zeolite	radical	5
ethylacrylate	13X zeolite	radical	37
methylmethacrylate	MCM-41	radical	25
methylmethacrylate	MCM-41	radical	28, 29
methylmethacrylate	Y zeolite, zeolite β	radical	26
methylmethacrylate	ZSM-5, MCM 48	radical	26
vinylether	HY zeolite	cationic	38
ethylene	MCM-41	metallocene	31, 32
ethylene	MCM-41	metallocene	33
ethylene	MCM-41	metallocene	35
vinylether	MCM-41	cationic	39, 40, 41, 42
N-vinylcarbazole	MCM-41	cationic	40, 41, 42

A brief literature review of host–guest polymerizations in 1D directed inorganic materials is given in Table 1 [36–42]. Tajima and Aida [19] reviewed controlled polymerization with constrained geometries.

For HY zeolites, direct cationic, radical, and redox polymerizations have been used mainly, because immobilization of surface initiators on HY zeolites affords no advantage. The direct linking of polymers on the internal surface of MCM-41 is possible by using immobilized transition-metal complex initiators. Also, the polymerization of freely mobile vinyl monomers by radical polymerization within MCM-41 has been done. Owing to the acidic nature of the inorganic materials considered, anionic polymerizations have not been used till now.

6.2 Concept

Cationic polymerization as a method for surface functionalization has been applied to supports such as silica, carbon black, and aluminosilicates (montmorillonite clay). A detailed review of cationic surface polymerizations of organic monomers on inorganic materials is given elsewhere [43,44].

Proton (H^+) surface initiation has been observed for aluminosilicates [14] and when protic acids are adsorbed on silica [44]. Pure silicate materials are usually unable to initiate directly the cationic surface polymerization of vinyl ethers (VE), *N*-vinylcarbazole (NVC), styrene, or other vinyl monomers, even in suspension of solvents that are established for cationic polymerization, for example dichloromethane, toluene, or hexane.

Suitable initiators for cationic surface polymerization are halogeno arylmethanes ($R^1R^2R^3C$–X, X = Cl or Br) that become cationically active on acidic surfaces [44–53]. However, on zeolites (HY or ZSM-5) only the external surface of the material is of relevance because $R^1R^2R^3C$–Cl is too large to enter [54–56]. Therefore, initiation of cationic polymerization of vinyl ethers within HY zeolites is achieved only with the mobile protons of the HY zeolite. Based on own knowledge and experience of the mechanism of cationic surface polymerization we have defined reaction conditions enabling preferential polymerization in the silicate channels. Eq. (1) gives an expression for the rate of total monomer consumption taking into consideration the reaction steps of the cationic propagation reactions in the channels (host–guest, HG), those on the outer surface of the MCM-41 (OS), and in the surrounding solution (S). If there is exclusive cationic host–guest polymerization terms 2 and 3 in Eq. (1) vanish.

$$-\mathrm{d}[\mathrm{M}]/\mathrm{d}t = \underbrace{k_{\mathrm{pHG}}[\mathrm{R}^+]_{\mathrm{HG}}[\mathrm{M}]_{\mathrm{HG}}}_{1} + \underbrace{k_{\mathrm{pOS}}[\mathrm{R}^+]_{\mathrm{OS}}[\mathrm{M}]_{\mathrm{OS}}}_{2} + \underbrace{k_{\mathrm{pS}}[\mathrm{R}^+]_{\mathrm{S}}[\mathrm{M}]_{\mathrm{S}}}_{3} \qquad (1)$$

This can be achieved best experimentally by keeping the concentration of the carbenium ion in the surrounding solution $[R^+]_S$ or on the outer surface $[R^+]_{OS}$ as small as possible. Halogeno arylmethanes [X-$CR^1R^2R^3$] are used as cationically inactive initiators for this host–guest polymerization [39,41,42]. They are essentially inactive in solution but are activated specifically on the inner surface of, for example, porous siliceous materials. For cationic vinyl ether polymerization on silica, chloro triphenylmethane (TCl) and chloro bis-(4-methoxyphenyl)methane (MeO) (Table 2) have been selected as suitable cationically active surface initiators. Both compounds have specific advantage as surface initiators on silica for cationic polymerization. Both corresponding carbenium ions $(C_6H_5)_3C^+$ and $(4\text{-}CH_3OC_6H_4)_2CH^+$ have approximately the same electrophilicity as expressed by their pK_{R+} values [46,47] and similar size. They also fit into the channel of MCM-41 with an average pore radius of 1.82 nm (Table 2).

Tab. 2. Largest dimension of the monomers and initiator molecules used in this work.

Guest molecule		*Size [nm]*
2,3-dihydrofuran	DHF	0.47
ethylvinylether	EVE	0.65
isobutylvinylether	IBVE	0.78
cyclohexylvinylether	CHVE	0.76 (equatorial) / 0.80 (axial)
2-chlorethylvinylether	ClEVE	0.70
N-vinylcarbazole	NVC	0.90
chloro triphenylmethane	TCl	0.93 (0.91)
chloro bis-(4-methoxyphenyl)methane	MeO	1.34

Other carbenium precursors that involve strong electrophilic carbenium ions such as $(C_6H_5)_2CHCl$ and $(4\text{-}CH_3C_6H_4)_2CHCl$ are very weakly cationically active on silica [48] and MCM-41. With silica as catalyst, the largest apparent rate constant k' (maximum point of the curve lg k' versus pK_{R+}) for the surface mediated hydride transfer reaction of 1,4-cyclohexadiene with carbenium ions has been found for MeO with $pK_{R+} = -5.6$ [44,49].

In this chapter the use of cationic host–guest polymerizations of substituted vinyl ethers and *N*-vinylcarbazole in MCM-41 and two related materials MCM-48 (average pore radius 1.28 nm) and a porous glass Gelsil (average pore radius 2.50 nm) is reported. Table 2 contains the monomers and initiators used in this study, their abbreviations, and optimum sizes calculated with molecular modeling. Detailed descriptions of the experiments and analytical equipment used have been published recently [38,41,42].

6.3 Results and Discussion

A particularly useful initiator for cationic host–guest polymerizations on MCM-41 is $(4\text{-}CH_3OC_6H_4)_2CHCl$ [50–56]. When it is adsorbed on the surface of the silicate, heterolytic bond cleavage occurs at the central carbon–halogen bond: the ion formed, $(4\text{-}CH_3OC_6H_4)_2CH^+$, can be detected directly using UV/vis transmission spectroscopy in MCM-41, measured in the suspension using light-conducting optics (Fig. 2) [45]. The transmission technique allows the carbenium ion fraction to be determined within the mesopores, because as a result of the high transparency of MCM-41 suspended in CH_2Cl_2, these UV/vis absorptions can be measured cleanly. The influence of light scattering on the position of UV/vis absorption can be neglected.

Figure 2 shows the increase in the UV/vis absorption of the carbenium ions with time during the adsorption of $(4\text{-}CH_3OC_6H_4)_2CHCl$ in MCM-41. The formation of $(4\text{-}CH_3OC_6H_4)_2CH^+$ can be clearly recognized using the characteristic UV/vis absorptions at $\lambda = 511$ nm [46]. The largest fraction of $(4\text{-}CH_3OC_6H_4)_2CH^+$ in terms of quantity is fixed in the channels of the MCM-41, because on nonporous carriers, such as Aerosil 300, no time-dependence is observed, but instead complete adsorption occurs within seconds [45,52]. The temperature influence on the adsorption ionization equilibrium of $(4\text{-}CH_3OC_6H_4)_2CHCl$ on MCM-41 is reversible. With decreasing temperature the carbenium ion UV/vis absorption band at $\lambda = 511$ nm increases, indicating an exothermic process for the adsorption-ionization of $(4\text{-}CH_3OC_6H_4)_2CHCl$ on MCM-41 [42]. Furthermore, it reacts very rapidly with vinyl ethers or *N*-vinylcarbazole, so that the conditions for rapid initiation and thus molecular mass control of the polymer fraction formed are given. The initiation of the host–guest polymerization is achieved by rapid addition of the diarylmethyl carbenium ion to the double bond of the vinyl monomer [39,42]. After addition of the monomer to the $(4\text{-}CH_3OC_6H_4)_2CH^+$/MCM-41 composite about 3 min of diffusion time are required until the initiator (MeO) or (TCl) is consumed.

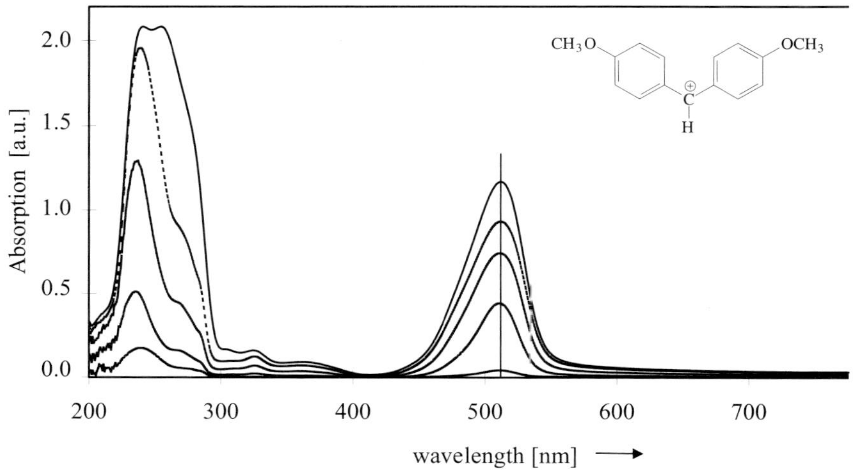

Fig. 2. Formation of $(4\text{-}CH_3OC_6H_4)_2CH^+$ as a function of time during the adsorption of $(4\text{-}CH_3OC_6H_4)_2CHCl$ in MCM-41 measured using UV/vis transmission spectroscopy in dichloromethane at room temperature (22 °C) after 4 s (- -), 17 s (- - -), 45 s (- - - - -), 140 s (. . . .), and 250 s (___).

Figure 3 shows the initiation of a vinylether with triphenylmethylium ion inside MCM-41. For this experiment, $(C_6H_5)_3CCl$ is adsorbed on MCM-41 from dichloromethane solution. The corresponding UV/vis absorption of $(C_6H_5)_3C^+$ at $\lambda = 410/435$ nm on MCM-41 has been measured by means of the immersion cuvette. Its intensity remains constant after 40 min adsorption time (Fig. 4). After EVE is added, the decrease of intensity of the UV/vis absorption of $(C_6H_5)_3C^+$ at $\lambda = 410/435$ nm on MCM-41 is easily seen (Fig. 4).

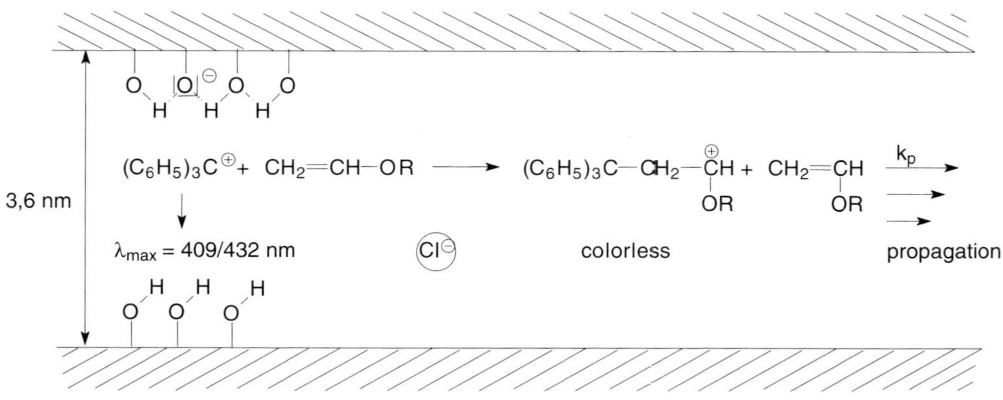

Fig. 3. Cationic initiation of substituted VE with triphenylmethylium inside MCM-41.

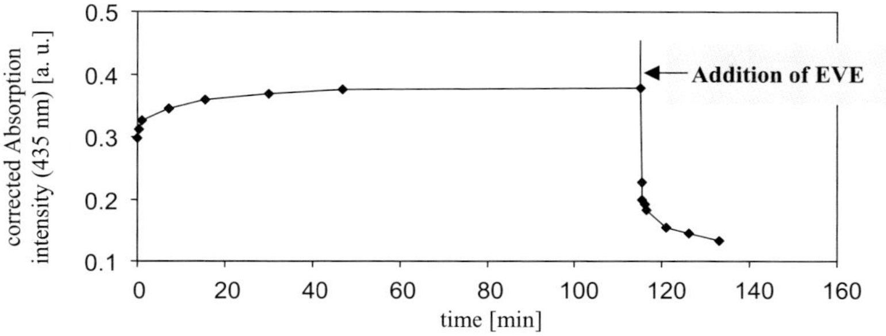

Fig. 4. Intensity of the UV/vis absorption of $(C_6H_5)_3C^+$ at $\lambda = 435$ nm as a function of time and after addition of ethylvinylether. Experimental conditions: 15.6 mg (56 μmol) TCl of 67.2 mg MCM-41; solvent: 10 mL dichloromethane; monomer: 200^ sμL (2.091 mmol; $M/I = 37.4$) EVE; temperature: 293 K.

However, initiation can also be achieved by very low concentrations of particularly acidic protons on the inner surface of the MCM-41 skeleton [40]. The propagation reaction proceeds through addition of further monomer molecules to the active chain, whereby the chain itself does not move within the channel, but monomer diffusion towards the chain determines the propagation rate, since the counterion can migrate as a result of very rapid proton exchange between the silanol groups on the inner surface area of MCM-41 [43]; in other words, the exchange of silanol protons on the inner MCM-41 surface proceeds more rapidly than the propagation of the polymer. We presume the overall entropy loss, which has to be overcome during direct threading, is distributed over the individual steps of the propagation reaction and is compensated by the reaction energy liberated in each addition step. Figure 5 shows the mechanistic principles of the cationically induced host–guest propagation reaction.

In this way many cationically polymerizable vinyl monomers can be converted to the corresponding polymer in the cavities of inorganic materials.

The solid hybrid materials obtained by cationic host–guest polymerization contain up to 30 wt.-% carbon as determined by quantitative elemental analysis [41,42]. The covalently linked part of the polymer fraction cannot be removed from the MCM-41 by simple extraction with an organic solvent. Approximately 300–400 mg polymer can be generated per 1 g MCM-41 by this procedure. Typical plots for polymer generated inside MCM-41 as a function of the monomer/initiator ratio are shown in Fig. 6.

Similar curves are obtained when the amount of generated polymer inside MCM-41 is plotted against the [monomer]/[MCM-41] mass ratio used [42].

Despite the absence of internal surface initiators $(4\text{-}CH_3OC_6H_4)_2CH^+$ or $(C_6H_5)_3C^+$ for some experiments (unfilled points in Fig. 6), the polymer mass generated per gram MCM-41 fits well in the curves. For these experiments the polymers formed have a narrower MWD (molecular weight distribution) than obtained with $(4\text{-}CH_3OC_6H_4)_2CHCl$ or $(C_6H_5)_3CCl$ as internal surface initiators, because the proton initiation occurs only inside the channel. The lowest cat-

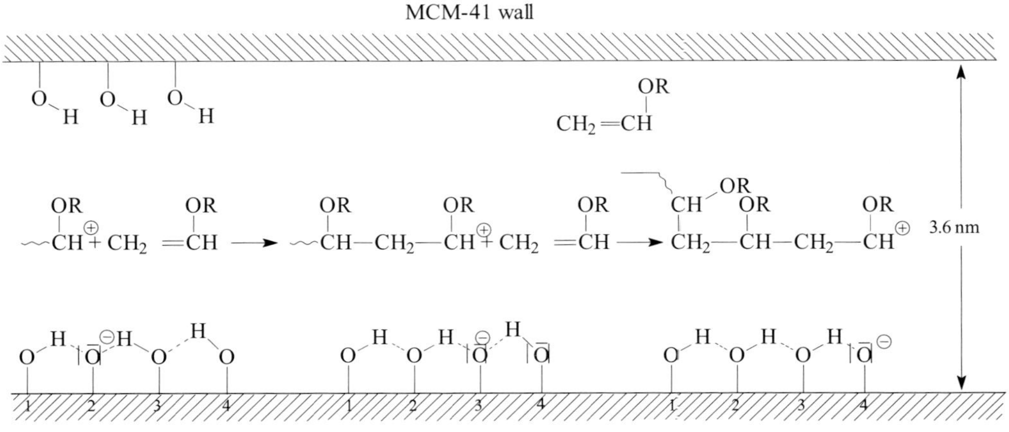

Fig. 5. Suggestion for the source of cationic propagation reaction of vinyl ethers in an MCM-41 channel. The silanol group density is not actual.

ionic reactivity for guest polymerization in MCM-41 shows 2-chlorethylvinylether. Mainly oligomeric fractions with low molecular weight are obtained for this monomer. To characterize the molecular mass distribution and the structures of the enclosed polymers more precisely it was therefore necessary to dissolve the PVE/MCM-41 hybrid in aqueous KOH.

In this way we established that the number average molecular weight (M_n) of the guest-poly(vinyl ether) fraction rarely exceeds 4000 g mol^{-1} in the case of IBVE or 2-chloroethyl vinyl ether, independent of the starting monomer concentration or the temperature. Representative results for guest polymerizations of vinylethers inside MCM-41 are compiled in Table 3.

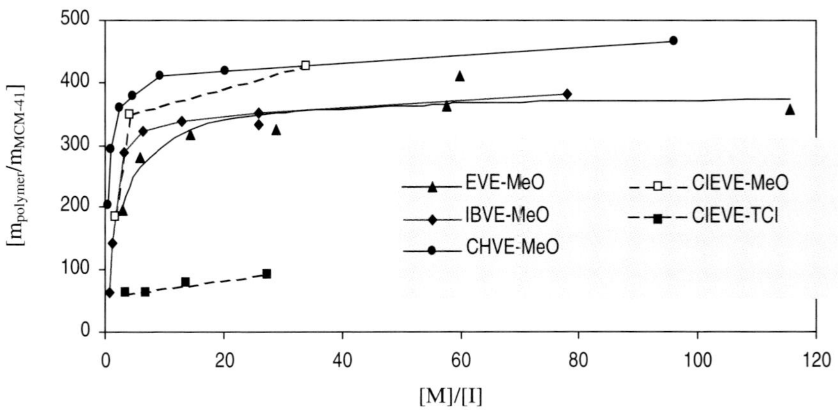

Fig. 6. Polymer generated inside MCM-41 as a function of the initial monomer/initiator ratio.

Tab. 3. Results for guest polymerization of vinylethers inside MCM-41 compared to the extractable fraction, T = 249 K, in 10 ml dichloromethane.

Monomer (batch)		*Initiator*		*MCM-41*	M_n/g mol^{-1}		M_w/M_n	
	mol × 10⁴		*mol × 10⁴*	*[g]*	*inside MCM-41*	*extractable*	*inside MCM-41*	*extractable*
IBVE-4	38	TCl	5.9	0.1	3070	1470	1.6	3.0
IBVE-8	153	TCl	2.0	0.2	8200	23400	3.4	2.3
IBVE-15	19.2	—	—	0.11	1750	3700	1.9	2.0
IBVE-17	76	—	—	0.21	4700	7200	2.0	2.2
DHF-2	265	MeO	3.2	0.31	4300	15000	2.5	1.7
DHF-4	265	—	—	0.3	4100	8000	2.5	1.6
EVE-6	313	MeO	5.2	0.31	4300	6400	1.8	2.2
CHVE-3	17.7	MeO	7.56	0.09	2100	3400	1.7	1.4
CHVE-7	141.2	MeO	1.5	0.17	6270	10500	2.0	1.3
CHVE-11	19.2	—	—	0.12	2300	6500	1.7	3.0

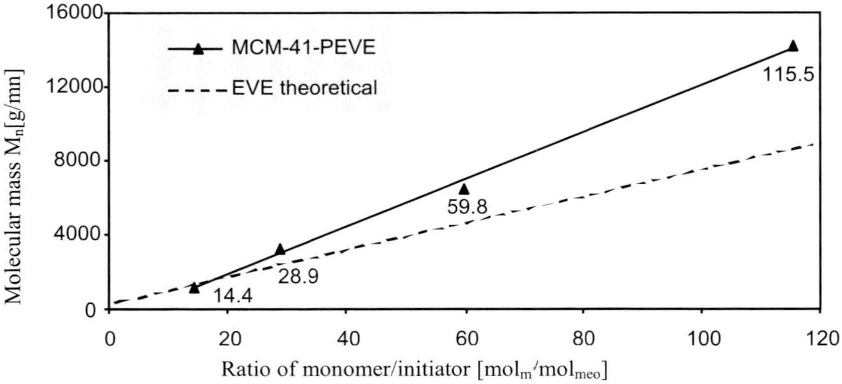

Fig. 7. Molecular mass of extracted PEVE as a function of the ratio of the quantity of monomer to the quantity of initiator MeO with exact details of the M/I ratio in the diagram. Curve (dashed) was calculated as follows: $M_n = M_0\,[M]/[I] + M_{MeO}$, M_0 = molecular mass of EVE; [M] = quantity of EVE; [I] = quantity of initiator bis-(4-methoxyphenyl)methyl chloride.

The comparison of the two methods clearly indicate that an additional initiator is not essential. However, the polymer fraction that remains inside the MCM-41 pores always possesses a lower molecular weight than those of the soluble fractions. It is still not clear that part of the extractable fraction is formed inside the pores, because the critical section in the pore is near the window of the MCM-41.

The molecular mass of the polymer fraction inside the channel can be controlled in the range from $M_n = 1000$ g mol^{-1} to about $M_n = 6000$ g mol^{-1} (Table 3). For the EVE polymerization, a good relationship between the molecular mass and the initial monomer/initiator (I) ratio [M]/[MeO] has been established (Fig. 7).

The deviation of the experimental curve from the theoretical one (see legend to Fig. 7) can be attributed to the fact that not all the $(4\text{-}CH_3OC_6H_4)_2CHCl$ [R–Cl] is available and therefore consumed during the initiation. This means that the actual concentration of $[R^+]$ is lower than the quantity of RCl used. For the other monomers, the theoretical plot [M]/[I] is satisfactory.

The PVE/MCM-41 hybrid materials obtained by means of initiation with the bare MCM-41 contain well-defined polymers without structural defects, as shown by solid state MAS CP $^{13}C\{^1H\}$ NMR spectroscopy [39,42]. The solid state MAS CP $^{13}C\{^1H\}$ NMR spectra of PVE/MCM-41 hybrid materials show no significant signal at about $\delta = 10$ ppm indicating methyl head groups. This is a further indication that proton transfer reactions to the monomer inside the channel are suppressed even at 249 K. Theoretical considerations with regard to the effective channel length in MCM-41 (l_{MCM-41}) and the contour size (length of polymer chains) of the whole guest-polymer fraction (r_{cont}) show that the ratio of r_{cont}/l_{MCM-41}, in the case of smaller monomers, such as DHF or EVE, is significantly larger than for NVC or CHVE [41].

Tab. 4. Comparison of the calculated contour length sum of the polymers generated corresponding to the channels available in MCM-41.

Hybrid Sample no.	*Mass P/mass C* $m_{Polymer}/m_{MCM-41}$ $[g_P/g_{MCM-41}]$	*Σ Contour lengths* r_{cont} $[10^{10}$ m$]$	*Σ Contour lengths/ channel lengths MCM-41* r_{cont}/l_{MCM-41} [m/m]
C-DHF-2	0.493	110	16.3
C-EVE-7	0.358	76	11.5
C-EVE-1	0.196	40	6.3
C-ClEVE-3	0.487	67	10.6
C-IBVE-6	0.333	51	7.7
C-CHVE-6	0.412	51	7.6
C-NVC-6	0.532	40	6.4
C-Styrene-2	0.100	14	2.2

Since the contour length is a theoretical parameter (the actual chain length of the enclosed polymer in MCM-41 is significantly smaller) a maximum of 4–5 chains should be deposited together (estimation) (Table 4).

Residual monomer or unreacted initiator are extracted completely. Accordingly, the pore volume of the PVE/MCM-41 hybrid materials decreases continuously with increasing polymer loading in all investigated samples. The pore radius distribution in the hybrid material becomes correspondingly wider, as can be seen, for example, in the case of MCM-41/PEVE hybrids (Fig. 8). In PEVE/MCM-41 pores with a smaller radius ($r = 1.1$–1.8 nm) can be observed alongside pores of unloaded MCM-41 regions ($r = 1.82$ nm).

For the PNVC/MCM-41 hybrid materials, only a decrease of the whole BET pore volume is observed in increasing polymer content without narrowing the pores. The pore diameter of the former material is not decreased. Accordingly, PNVC penetrates only in the window section of the pores. Then it polymerizes rapidly and the pores become plugged. The rigid PNVC fills this section of the pores completely because PNVC is restricted in its thermal motion owing to its larger glass transition temperature compared to PVE (see later).

Altogether, these results clearly show that the pores of the MCM-41 are filled with the polymers. The polymers are strongly bonded to the internal surface of the MCM-41 wall, because they cannot be extracted with solvents suitable for dissolution of PVE or PNVC. The strong bonding of the polymer to the MCM-41 wall can be attributed either to a covalent Si–O–C bond that can be formed by the reaction of the growing cationically active chain with the silanol groups, which nucleophilically trapping the chain end, or to strong adsorption of the polymer on the MCM-41 wall.

The Si–O–C bond on the surface can be easily identified by solid state MAS CP $^{13}C\{^1H\}$ NMR spectroscopy [43,44]. The ^{13}C NMR signal for the relevant carbon atom is expected at about $\delta = 58 \pm 2$ ppm [57,58]. In MAS CP $^{13}C\{^1H\}$ NMR

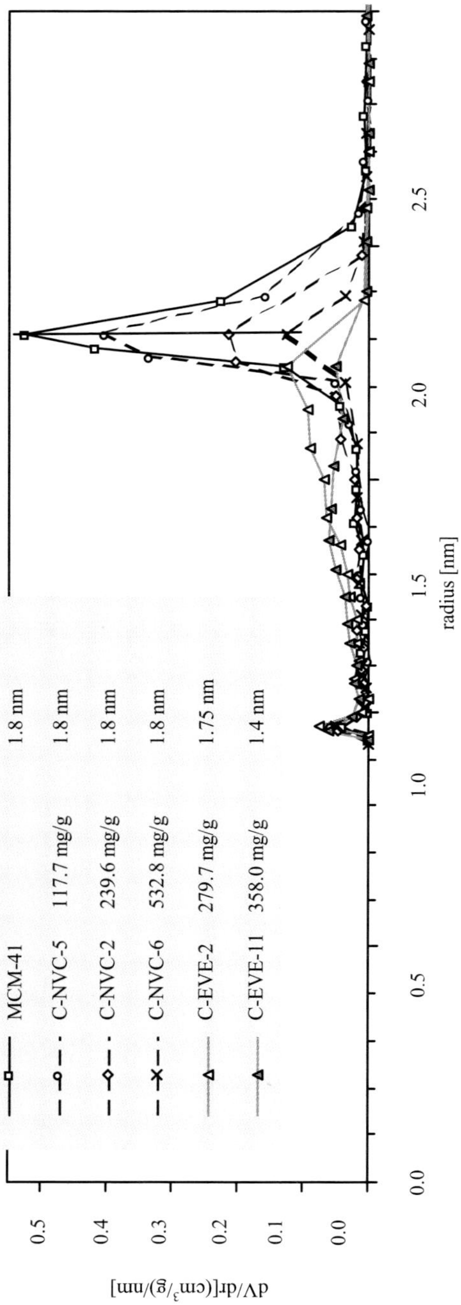

Fig. 8. Adsorption of nitrogen per volume [dV] as a function of the pore radius [dr] for MCM 41, PNVC/MCM-41, and PEVE/MCM-41 hybrid materials with different loadings of polymer. The maximum peak value of each batch is given.

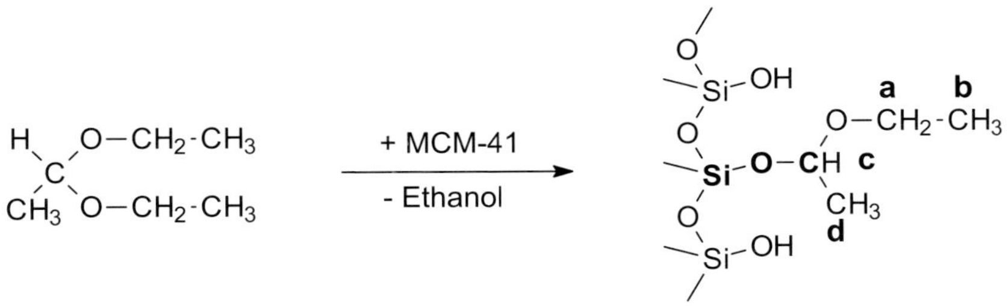

Fig. 9. Reaction of acetaldehyde diethylacetal with MCM-41 and assignment of the NMR signal to surface groups NMR signals found. 1H MAS: δ(ppm) = 0.5–1.6 (6H, Hb–Hd), 3.0–4.0 (3H, Ha + Hc) and protons of the residual silanol groups at δ(ppm) = 1.7–3.2. 13C{1H}CP MAS δ(ppm) = 63 (Cc or Cd), 59 (Ca or Cc), 21 (Cd), and 15 (Cb). 29Si MAS: δ(ppm): −110 (Q4) and −100 (Q3).

spectra of PVE/MCM-41 hybrid materials measured, a weak signal in this section is observed. In addition, solid state NMR investigations (^{13}C and ^{29}Si) have so far not given any unambiguous indication as to whether the poly(vinylether) formed is actually covalently bonded to the inner MCM-41 wall by an Si–O–CHR^{1}OR bond. It is very difficult to differentiate between the ^{13}C NMR signal of the C-atom of the Si–O–C bond and that of the end group of the poly(vinyl ether) chains [42].

Acetals or ketals react readily with silanol groups forming Si–O–CHR^{1}OR bonds [59]. This option is very likely, especially as acetaldehyde diethylacetal reacts smoothly with MCM-41 with loss of ethanol (Fig. 9).

In the MAS CP ^{13}C{^{1}H} solid state NMR spectrum of the product of the reaction between $CH_3CH(OC_2H_5)_2$ and MCM-41, two signals at $\delta = 58.2$ ppm and $\delta = 58.6$ ppm are clearly detectable. They relate to the expected Si–O–C bond and the ether bond of the acetal [59]. An assignment of the two bonds is difficult. However, the reaction of $CH_3CH(OC_2H_5)_2$ with MCM-41 also occurs rapidly at room temperature and, therefore, evidently supports the formation of Si–O–C bonds during the host–guest polymerization of the vinyl ethers. As long as water or acidic and basic impurities are excluded, this bond remains stable. This prevents extraction of the polymer by organic solvents. The strong bonding of the PVE to the internal MCM-41 wall is supported by DRIFT spectroscopy and solid state ^{29}Si NMR spectroscopy. Data are given elsewhere [42]. The -CH_2- vibration of the polymer inside MCM-41 shifts to higher energy compared to that of the bulk polymer. This is a clear indication that the polymer is adsorbed on the internal MCM-41 surface. It is assumed that the oxygen atoms of the absorbed PVE chain interact with the silanol groups (Fig. 10). As a consequence, the energy required for inducing the -CH_2- vibration must be larger.

The results of ^{29}Si NMR spectroscopy of the PVE/MCM-41 hybrid materials do not allow a detailed interpretation concerning the bonding of the polymer to the MCM-41 material [42], because the spectrum in the section of the Q^3 (O_3Si–OH) and Q^4 (O_3Si–O–) signals of the silica framework is poorly resolved, because of adsorbed water traces that amount differs dependent on polymer loading.

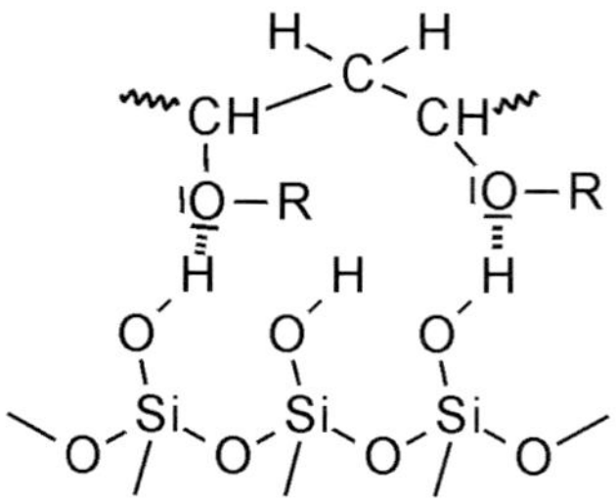

Fig. 10. Proposed adsorption of a polyvinylether segment on the internal silanol groups of MCM-41.

Broadband dielectric spectroscopy is a suitable tool for investigating the molecular dynamics in polymer/MCM hybrid systems. In measuring the dynamic glass transition (α-relaxation) the glass transition temperature can be determined as the temperature with a mean relaxation rate of 10^{-2} Hz [60]. For the dielectric investigations we chose poly(ethylvinylether) (PEVE)/ and poly(isobutyl vinyl ether) (PIBVE)/MCM-41 hybrid materials, as in both cases the corresponding organic polymer fractions have a T_g of about -40 °C in the pure bulk and thus can be easily investigated within a temperature range of about -150 °C to 100 °C.

Dielectric loss spectra of the poly(isobutylvinylether) (PIBVE)/MCM-41 hybrid have been measured at 150 and 160 K [41]. Two relaxation processes can be recognized: the fast β-relaxation progress is attributed to movement of the ether groups [41]. The dynamic glass transition (α-relaxation) corresponds to the relaxation of the main chain between structural substrates.

The logarithmic relaxation rate as a function of the reciprocal temperature for PIBVE in the melt and in MCM-41 has been determined [41]. The relaxation rate of the local β-relaxation of PIBVE in MCM-41 does not differ from that of the free polymer melt (bulk phase). Further details are available in this book [64].

The calorimetric glass transition corresponds to the temperature at which the relaxation rate is about 0.01 s^{-1} for dynamic measurements [60–62]. The results of the T_g determination of PVE in geometrically different states are shown in Table 5.

The fact that the relaxation rate of the polymer in the hybrids is many orders of magnitude larger than in the free melt can be attributed to the constricting geometry of the channels in the porous silicates. If, on the other hand, a polymer melt is mixed with unloaded MCM-41 powder, a slight lowering of the relaxation rate is observed because of surface effects. For the calorimetric glass transition this would cause a shift to higher temperatures.

The increase in the relaxation rate of PIBVE in the constricted geometry of the MCM materials is more pronounced when the pore radius is smaller. This is shown by the comparison of PIBVE in MCM-41 (pore diameter 3.6 nm) and MCM-48 (pore diameter 2.4 nm) (Table 5). Such a confinement effect has been investigated in detail for low molecular weight systems [61–63]. This effect has its molecular basis in the inherent length scale of the dynamic glass transition, which can increase to values of a few nanometers with decreasing temperature [60–63].

Tab. 5. Representative glass transition temperatures (T_g) of the extractable polymer (**E**), polymer fractions enclosed in the hybrid (**H**), and polymers physically adsorbed on MCM-41 (**A**) determined by dielectric spectroscopy or DSC.

Hybrid Material	*E*	*H*	*A*
MCM-41/PCHVE	311	no dynamic T_g	320
MCM-41/PEVE	239	131	245
MCM-41/PDHF	335 (DSC)	128	225 (DSC)
MCM-41/PIBVE	244 ± 1	135	225 ± 1
MCM-48/PIBVE	244 ± 1	134	225 ± 1
Gelsil/PIBVE	244 ± 1	155	225 ± 1
MCM-41/PNVC	370 (DSC)	n.d.	n.d.

The PIBVE polymer chain in the porous channels of the MCM-41 and MCM-48 is surrounded by solvent molecules facilitating its segmental fluctuations. As result of annealing (at 350 K) a certain fraction of these solvent molecules (acting in some sense as a plasticizer) is removed. This leads to a pronounced decrease of the mobility of fluctuating polymer segments. Furthermore, a considerable fraction of the chains will be immobilized. The effect is fully reversible as shown for several solvents (dichloromethane, cyclohexane, and water) [41].

Since heterogeneously induced polymerization in the pores of siliceous materials can be supported by immobilized catalysts (propylene polymerization initiated by immobilized transition metal catalysts), the destruction of the siliceous support during the polymerization process is desired. However, this effect is detrimental to the production of nanostructural hybrid materials by host–guest polymerizations.

The nondefect channels of the MCM-41 material can be clearly shown in a TEM image (Fig. 11). They have a length of about 100–150 nm. Note that the polymers have no contrast and cannot be detected by this method.

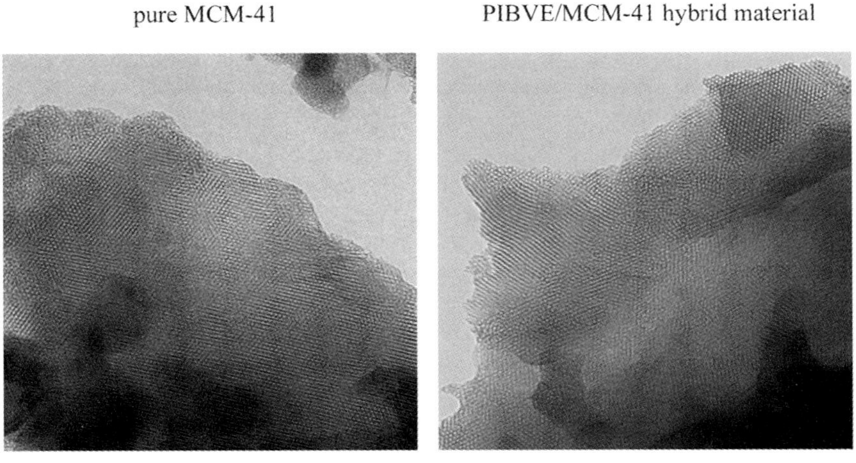

Fig. 11. TEM images of pure MCM-41 and a PIBVE/MCM-41 hybrid material.

6.4 Conclusions and Outlook

PVE/MCM-41 hybrid materials with adjustable polymer content and molecular weight of the loaded polymer fraction can be synthesized by cationic host–guest polymerization of vinyl ether monomers within MCM-41 materials. For this the inorganic materials are used in form as particle powders in a slurry of a solvent suitable for cationic polymerization. The structures of the polymer chains in MCM-41 have properties identical to the pure, bulk polymers, whereas the glass transition temperature is significantly different from those of the bulk fraction. The procedure presented is, therefore, suitable for producing flexible polymer chains within pores of inorganic materials to study their dynamics in confined geometry.

Acknowledgements

This work was supported by the DFG within the framework of the programs "Nanoporous Host Guest Systems". For scientific cooperation within this program we thank Prof.Dr. P. Behrens, Dr. C. Tintemann, Prof.Dr. H. Fuess, Dr. C. Baehtz, Prof.Dr. C. von Borczyskowski, Dr. U. Rempel, and Prof.Dr. R. Holze. We also thank the SFB 294 "Molecules interacting with interfaces" and the Fonds der Chemischen Industrie. We also thank Prof.Dr. M. Antonietti, MPI of Colloids and Interfaces, for the support in the morphological studies.

References

1 H. Eckert, M. Ward (eds.) Organic–Inorganic Nanocomposite Materials, special issue of *Chem. Mater.* **2001**, *13*, 3059–3809. DSNN 0897-4756.

2 Z. Pu, J.E. Mark, G. Beauge, *Rubber Chem. Techn.* **1999**, *72*, 138–151.

3 J.E. Mark, C.Y.-C. Lee, P.A. Bianconi (eds), Hybrid Inorganic–Organic Composites, American Chemical Society, Washington **1995**, Vol. 585.

4 L.L. Beecroft, C.K. Ober, *Chem. Mater.* **1997**, *9*, 1302–1317.

5 H.L. Frisch, J.E. Mark, *Chem. Mater.* **1996**, *8*, 1735–1738.

6 J. Wen, G.L. Wilkes, *Chem. Mater.* **1996**, *8*, 1667–1681.

7 J. Shi, C.J. Seliskar, *Chem. Mater.* **1997**, *9*, 821–829.

8 Y. Ikeda, S. Kohjiya, *Polymer* **1997**, *17*, 4417–4423.

9 R. Tamaki, K. Naka, Y. Chujo, *Polym. Bull.* **1997**, *39*, 303–310.

10 R. Tamaki, T. Horiguchi, Y. Chujo, *Bull. Chem. Soc. Jpn.* **1998**, *71*, 2749–2756.

11 J.V. Crivello, Z. Mao, *Chem. Mater.* **1997**, *9*, 1554–1561, 1562–1569.

12 R. Canalini, D. Fioretto, D. Livi, M. Luccheri, P.A. Rollan, *Phys. Rev. B* **1997**, *56*, 3016–3021.

13 P.G. de Gennes, *Euro. Phys. J. E* **2000**, *2*, 201–206.

14 M. Biswas, S.S. Ray, *Polymer* **1998**, *39*, 6423–6428.

15 S.S. Ray, M. Biswas, *Mater. Res. Bull.* **1999**, *34*, 1187–1194.

16 A. Matsumoto, T. Kitajima, K. Tsutsumi, *Langmuir* **1999**, *15*, 7626–7631.

17 F. Leroux, J.-B. Besse, *Chem. Mater.* **2001**, *13*, 3507.

18 A.-C. Franville, B. Dunn, J.I. Zink, *J. Phys. Chem. B* **2001**, *105*, 10 335–10 339.

19 K. Tajima, T. Aida, *Chem. Commun.* **2000**, 2399.

20 J.K. Whitesell (ed.), Organized Molecular Assemblies in the Solid State, John Wiley **1999**.

21 C.M. Paleos (ed.), Polymerization in Organized Media, Gordon & Breach, New York **1992**.

22 L.J. Prins, D.N. Reinhoudt, P. Timmermann, *Angew. Chem. Int. Ed.* **2001**, *113*, 2446–2492.

23 T. Bein, P. Enzel, *Angew. Chem.* **1989**, *101*, 1737–1738; *Angew. Chem. Int. Ed.* **1989**, *28*, 1692–1694.

24 P. Enzel, T. Bein, *J. Phys. Chem.* **1989**, *93*, 6270–6272.

25 K. Möller, T. Bein, R. X. Fischer, *Chem. Mater.* **1998**, *10*, 1841–1852.

26 K. Möller, T. Bein, *Chem. Mater.* **1998**, *10*, 2950–2963.

27 C.-G. Wu, T. Bein, *Science* **1994**, *264*, 1757–1759.

28 C.-G. Wu, T. Bein, *Chem. Mater.* **1994**, *6*, 1109–1112.

29 J.L. Meinershagen, T. Bein, *J. Am. Chem. Soc.* **1999**, *121*, 448–449.

30 A. Mutsumoto, T. Kitajima, K. Tsutsumi, *Langmuir* **1999**, *15*, 7626–7631.

31 S.M. Ng, S. Ogino, T. Aida, K.A. Koyano, T. Tatsumi, *Macromol. Rapid Commun.* **1997**, *18*, 991–996.

32 P. Lehmus, B. Rieger, *Science* **1999**, *285*, 2081–2082.

33 K. Kageyama, J.I. Tamazawa, T. Aida, *Science* **1999**, *285*, 2113–2115.

34 R. Ramachandra Rao, B.M. Weckhuysen, R.A.Schoonheydt, *Chem. Commun.* **1999**, 445–446.

35 M. Weckhuysen, R. Ramachandra Rao, J. Pelgrims, R.A. Schoonheydt, P. Bodart, G. Debras, O. Collart, P. Van Der Voort, E.F. Vansant, *Chem. Eur. J.* **2000**, *6*, 2960–2970.

36 J.G. Millar, G.F. McCamm, C.M. Hobbis, G.A. Bowmaker, R.P. Conney, *J. Chem. Soc. Faraday Trans.* **1994**, *90*, 2579.

37 H.L. Frisch, J.M. West, C.G. Göltner, G.S. Attard, *J. Appl. Polym. Sci.* **1996**, *34*, 1823.

38 A. Gräser, S. Spange, *Chem. Mater.* **1998**, *10*, 1814–1819.

39 S. Spange, A. Gräser, P. Rehak, C. Jäger, M. Schulze, *Macromol. Rapid Commun.* **2000**, *21*, 146–150.

40 S. Spange, Y. Zimmermann, A. Gräser, *Chem. Mater.* **1999**, *11*, 3245–3251.

41 S. Spange, A. Gräser, A. Huwe, F. Kremer, C. Tintemann, P. Behrens, *Chem. Eur. J.* **2001**, *7*, 3722–3728.

42 S. Spange, A. Gräser, H. Müller, Y. Zimmermann, P. Rehak, C. Jäger, H. Fuess, C. Baethz, *Chem. Mater.* **2001**, *13*, 3698–3708.

43 S. Spange, U. Eismann, S. Höhne, E. Langhammer, *Macromol. Symp.* **1997**, *126*, 223–236.

44 S. Spange, *Prog. Polym. Sci.* **2000**, *25*, 781–849.

45 U. Eismann, S. Spange, *Macromolecules* **1997**, *30*, 3439–3446.

46 S. Schneider, H. Mayr, P.H. Plesch, *Ber. Bunsenges. Phys. Chem.* **1987**, *91*, 1369.

47 H. Mayr, in Cationic Polymerization: Mechanisms, Synthesis, and Applications, K. Matyjaszewski (ed.), Marcel Dekker: New York **1996**, pp. 51–136.

48 S. Spange, A. Fährmann, A. Reuter, R. Walther, Y. Zimmermann, *J. Phys. Org. Chem.* **2001**, *14*, 271.

49 S. Spange, S. Adolph, R. Walther, Y. Zimmermann, *J. Phys. Chem.* **2002**, *107*, 298–305.

50 H.P. Leftin, in Carbonium Ions, G.A. Olah, P.R. Schleyer (eds.), John Wiley & Sons, **1968**, *1*, 363.

51 H.G. Karge, *Surf. Sci.* **1971**, *27*, 419.

52 S. Spange, D. Fandrei, F. Simon, H.-J. Jacobasch, *Coll. Polym. Sci.* **1994**, *272*, 99.

53 S. Adolph, S. Spange, Y. Zimmermann, *J. Phys. Chem. B* **2000**, *104*, 6417.

54 J.C. Scaiano, H. Garcia, *Acc. Chem. Res.* **1999**, *32*, 783–793.

55 M.L. Cano, A. Corma, V. Fornes, H. Garcia, M.A. Miranda, C. Baerlocher, C. Lengauer, *J. Am. Chem. Soc.* **1996**, *118*, 11 006–11 113.

56 T. Tao, G.E. Maciel, *J. Am. Chem. Soc.* **1995**, *117*, 12 889–12 890.

57 D. Hoebbel, M. Nacken, H. Schmidt, *J. Sol-Gel Sci. Technol.* **1998**, *12*, 169.

58 C. Wies, K. Meise-Gresch, W. Müller-Warmuth, W. Beier, A.A. Göktas, G.H. Frischat, *Ber. Bunsenges. Phys. Chem.* **1988**, *92*, 689.

59 B.R. Guidotti, E. Herzog, F. Bangerter, W.R. Caseri, U.W. Suter, *J. Coll. Interf. Sci.* **1997**, *191*, 209–215.

60 E. Donth, Glasübergang, Akademie Verlag, Berlin **1981**.

61 M. Arndt, R. Stannarius, H. Groothues, E. Hempel, F. Kremer, *Phys. Rev. Lett.* **1997**, *79*, 2077–2080.

62 A. Huwe, F. Kremer, P. Behrens, W Schwieger, *Phys. Rev. Lett.* **1999**, *82*, 2338–2341.

63 F. Kremer, A. Huwe, M. Arndt, P. Behrens, W. Schwieger, *J. Phys. Condens. Matter* **1999**, *11*, A175–A188.

64 F. Kremer, A. Huwe, A. Gräser, S. Stange, P. Behrens, this book.

7
Direct Synthesis of Functional Organic/Inorganic Hybrid Mesostructures

Peter Behrens, Andreas M. Glaue, and Olaf Oellrich*

7.1
Introduction

The proposal to use the ordered pore systems of molecular sieves as organizing media for the design of materials was made in the beginning of the 1990s [1,2] and by the middle of that decade, important progress had been made towards this goal [3], with a strong impetus carrying on until today. In addition to crystalline zeolite-type materials [4], mesostructured host matrices can also play an important role in the construction of such materials.

The synthesis of ordered regular mesostructures from simple amphiphiles such as alkyltrimethylammonium cations C_nTMA^+ and silica solutions as well as their calcination to give mesoporous materials with uniform pore width distribution, so called M41S materials, has been a most important topic in materials synthesis during the last ten years [5–11]. The synthesis of the mesostructures is based on the idea of combining the ability of amphiphilic organic molecules to self-assemble with the formation of an inorganic polymer [12–14]. The supramolecular structures that the surfactants form in these materials are in most cases similar to the arrangements in simple lyotropic liquid-crystal phases; in M41S materials, however, these arrangements are embedded in a silica matrix. Interestingly, it was shown that the presence of silica can strongly enhance the tendency of amphiphilic molecules to form mesostructures [14,15] compared with the formation of lyotropic phases in simple surfactant–water systems, that is, the presence of inorganic polyelectrolyte systems can strengthen ordering tendencies of amphiphilic molecules. The amorphous inorganic host materials of mesostructures can provide protection to the organic guest molecules. Mesostructures have also been shown to be especially versatile with regard to the design of special morphologies (films or fibers) [16,17].

As for the generation of host–guest compounds based on zeolites and related solids [4,18], there are several synthesis pathways for the generation of functional mesocomposites based on a silica host (Fig. 1).

- The host structure is prepared using a simple amphiphile, such as an alkyltrimethylammonium ion, as structure-directing agent (SDA); the organic mate-

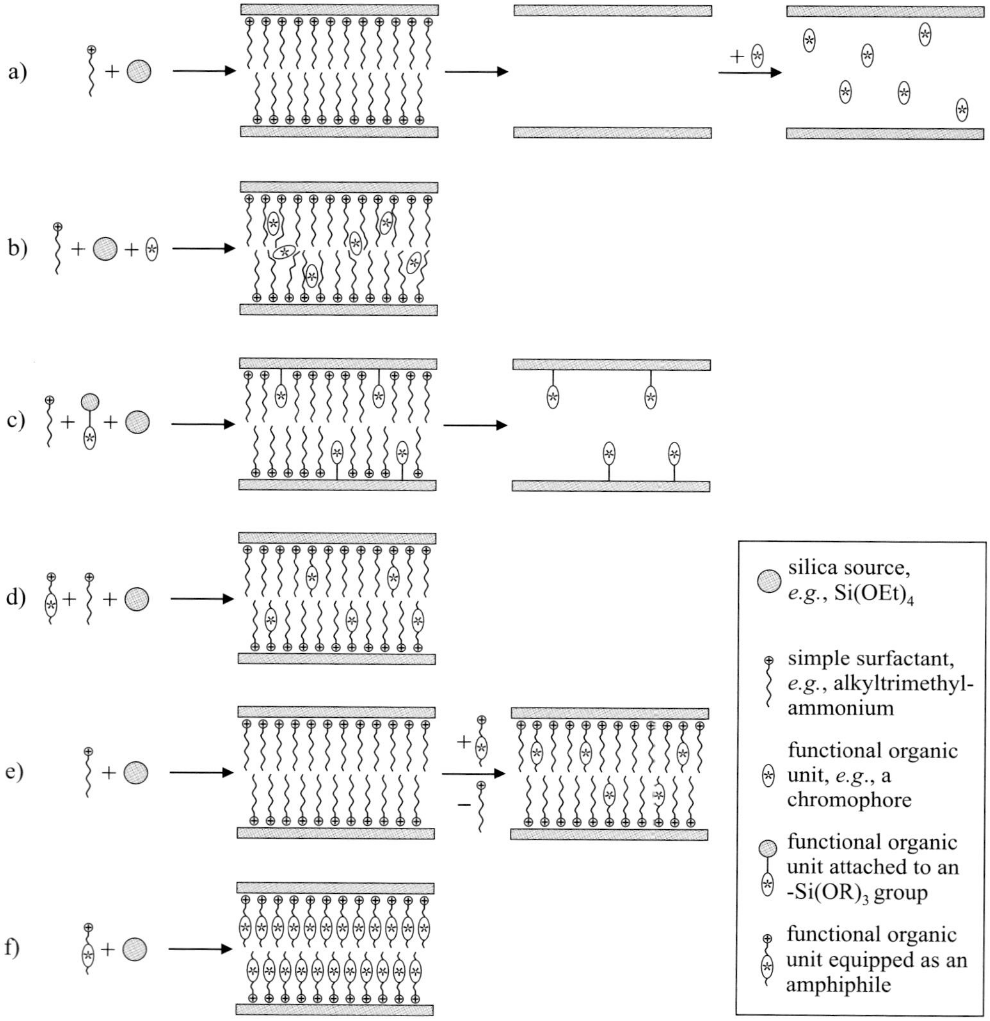

Fig. 1. Different synthesis pathways for the generation of functionalized mesostructured materials. For further explanation see the text.

rial is then removed by extraction or calcination and the functional organic entities are inserted into the mesoporous host, for example by loading via the vapor-phase or from a solution (Fig. 1a). With larger molecules, a full loading of the pores is usually not achieved.

- Unpolar organic molecules can be introduced into mesostructures by occupying the inner space of the micels built from a simple surfactant (Fig. 1b). While this is a simple method working for a large variety of organic compounds, the degree of loading is usually small (otherwise the lyotropic structure of the simple surfactant is destabilized) and it is not possible to selectively remove the surfactant molecules alone.

- When functional organic molecules are equipped with a $-Si(OR)_3$ group, they can take part in the condensation process of the silica host framework; the simple surfactant that controls the formation of the mesostructure can then be removed by extraction, leaving behind the empty mesopores with attached functional organic units (Fig. 1c). Typically, the degree of loading of the pores is small.
- When the functional molecule is itself amphiphilic, it can take part actively in the micel formation and thus in the construction of the mesostructure, which is controlled, however, by a simple surfactant being present in a greater amount (Fig. 1d). The number of functional molecules that can be integrated into the mesostructure in this way is small (again due to the fact that otherwise the lyotropic structure of the simple surfactant is destroyed), but for special bola amphiphiles (which interact more strongly with the host than simple surfactants) it was shown that it is possible selectively to extract the molecules of the simple surfactant after the synthesis [19].
- With special amphiphiles, a post-synthetic ion-exchange procedure can also be successful (exchange of simple surfactant against functional amphiphile, Fig. 1e) [19].
- Finally, functional molecules equipped with amphiphilic headgroups can themselves be used as SDAs, controlling the formation of the host structure without addition of additional simple surfactant and directly introducing functionality (Fig. 1f). This direct synthesis path to functional mesostructures allows for a full and homogeneous filling of the pores [20–33].

For many purposes, a full or a high degree of loading with the functional molecules is not necessary. This is true, for example, in sensing applications or for catalysis, in which isolated and easily accessible reaction centers are preferred. On the other hand, some special functionalities can be attained only by aggregates of molecules. This is especially true for chromophores. Here we show how the last-mentioned of the described synthesis pathways can be used to obtain organic/inorganic composite materials with special properties relying on aggregates of chromophores within the organic part of the mesostructures [22–33].

From progress in the organic supramolecular chemistry of dye molecules it has become clear that organized assemblies of chromophore-containing molecules have great potential for the design of functional materials in a variety of application areas, but especially in optics and photonics [34–37]. Many of the typical methods of generating supramolecular arrangements of chromophores rely on the self-assembly of amphiphilic dye molecules into organized aggregates like micels, membranes, or lyotropic phases. The Langmuir–Blodgett technique is a powerful tool for transferring self-assembled monolayers onto a solid substrate, providing the opportunity to construct thin films with specifically designed architectures. Nonamphiphilic dye molecules equipped with thiol or trialkoxysilyl groups can organize themselves to form self-assembled monolayers (SAMs) on gold or silica substrates, respectively, and thus provide another opportunity to exploit collective effects resulting from chromophore ensembles.

Organic/inorganic composite mesostructures are mechanically more stable than aqueous lyotropic phases or Langmuir–Blodgett type aggregates. When chromophores are embedded in inorganic matrices, they are protected by their surrounding, a fact that can be of special importance when photophysical or photochemical applications are envisaged [3,11,38,39]. Compared with chromophore arrangements constructed at 2D surfaces as SAMs or Langmuir–Blodgett films, the photophysical and photochemical properties of the chromophore arrangements within the mesostructures will be enhanced owing to the 3D nature of these systems. Photophysical and photochemical applications of such materials especially will benefit from the optical transparency of the silica framework. Therefore, it is not surprising that a number of chromophore–silica mesostructures [3,11,39,40] with interesting optical properties (pH sensing [41], photochromicity [42], thermochromicity [43], laser action [44–47]) have recently been described.

Here we present our work on the structure-directing properties of amphiphilic azo dyes [22–32] and on the properties of the resulting surfactant–silica composites. We also investigated bola amphiphiles based on a porphyrin core [22,33], but these will not be described here. It is worth pointing out once more that these amphiphilic dyes were used as the only SDA in the synthesis (pathway f in Fig. 1), so they were not mixed with other surfactants, such as simple alkyltrimethylammonium cations. Similar work on a direct synthesis of functional mesostructures has, to the best of our knowledge, been restricted to that performed by Zhou and co-workers, who have shown that hexagonal silica mesostructures can be formed with an 11-ferrocenylundecyltrimethylammonium surfactant [20,21]. However, this amphiphile behaved largely like a simple alkyltrimethylammonium surfactant, that is, no special effects in the synthesis caused by the ferrocenyl units were observed and no remarkable photophysical or photochemical properties caused by the occlusion of these chromophore units within the composite mesostructure were reported.

We will show that for silica mesostructures based on amphiphilic azo dyes, special properties resulting from the aggregation of the chromophores arise [22–33]. However, apart from the interesting properties that amphiphilic dyes may lend to the mesostructured products, the presence of chromophore units within the hydrophobic tail of the SDAs can also yield additional insight into the structure of the micels of the mesostructure and on the structuring process, both of which can then be investigated spectrometrically. Even more interestingly, the chromophore entities, located in the hydrophobic part of the amphiphile, may exert additional structure-directing influences, for example because of specific interactions (hydrogen bonds or π–π interactions). In the field of the chemistry of M41S and related materials, many efforts were directed towards the modification of the headgroup of the surfactant molecules (yielding new structural topologies) [13,48–54] or to the composition of the inorganic part. Only minor attention, however, was paid to manipulations of the hydrophobic part of the surfactant units.

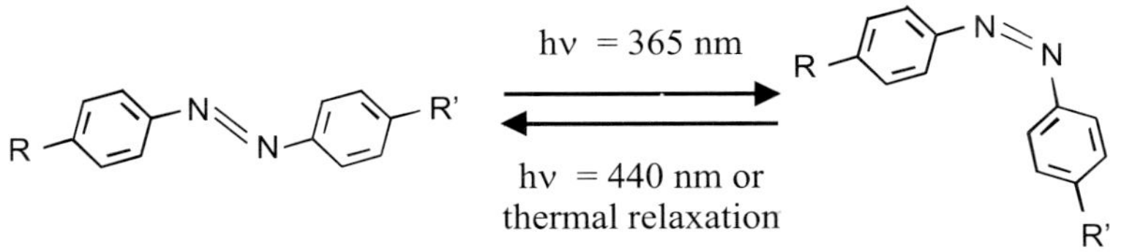

Fig. 2. Photoinduced switching of azobenzene chromophors due to *cis–trans* isomerization.

7.2
Mesostructured Composites of Azobenzene Surfactants and Silica

The azobenzene unit is one of the most widely used functional entities in organic supramolecular and materials chemistry, owing to comparatively simple synthesis procedures and to the possibility of photochemically switching between the *trans* and the *cis* isomer [34,36,55–57]. This photoisomerization process of azobenzene compounds promises interesting material applications, and so it has been investigated thoroughly. Switching from the thermodynamically stable *trans* to the *cis* isomer is achieved by ultraviolet light with a wavelength of about 365 nm. The back reaction can be triggered with light of longer wavelength (about 440 nm) or can occur thermally in a slower dark reaction (Fig. 2). As the *cis–trans* isomerization of an azo moiety involves significant changes in the shape, dipole moment, polarizability, and spectral characteristics of the molecule, the photoreactivity of materials based on azo dyes is of special interest for applications in the field of optoelectronics, optical switches, optical data storage, and photoactive membranes.

Azobenzene units have been introduced into zeotype molecular sieves [58–62]. The *cis–trans* isomerization of azobenzene was investigated in $AlPO_4$-5 und ZSM-5. It leads to a strong and reversible change of the refractive index, a remarkable photosensitivity that could give rise to new materials for optoelectronic computing [60–62].

Azobenzene derivatives were also intercalated into inorganic layered materials as clays [63–66] and zirconium phosphates [67]. Ogawa and coworkers described some interesting observations on clays intercalated with cationic "azo amphiphiles" [63,64], that is, molecules of the type $H_3C–(CH_2)_{m-1}$-O-C_6H_4-N=N-C_6H_4-O-$(CH_2)_n–N(CH_3)_3^+$ (Fig. 3). In the following, these azo amphiphiles are designated as C_mAzoC_nTMA$^+$, in which m is the number of carbon atoms in the alkyl tail, "Azo" represents the azobenzene group, n is the number of methylene groups in the alkyl spacer and "TMA$^+$" stands for the trimethylammonium headgroup; alkyl tail and spacer are attached to the aromatic rings of the azobenzene group via oxygen ether linkages. Inspired by the fact that such azo amphiphiles can form lyotropic phases in aqueous solutions [68,69], we have used them for the direct preparation of azo amphiphile–silica mesostructures. A variety of azo amphiphiles with different tail and spacer lengths ($m = 1, 6, 8, 10, 12$; $n = 3, 6, 8, 10$; not all combinations of m and n) were employed as SDAs in the synthesis of silica mesostructures [22–32].

Fig. 3. Chemical formula and symbol of azo amphiphiles.

Ogawa and coworkers also intercalated a related amphiphilic azo chromophor, C_6H_4-N=N-C_6H_4-O-$(CH_2)_2$-$N(CH_3)_2(CH_2CH_2OH)^+$, into magadiite and observed changes in the basal spacing upon irradiation with ultraviolet light [65]. The same group described the attempt to intercalate this amphiphile into lamellar MCM-50 materials that were prepared with trimethyloctadecylammonium cations [70]. However, the results do not appear to be conclusive with regard to a full occupation of the interlayer.

7.2.1
Synthesis and Structural Characterization of Azobenzene Surfactants in the Synthesis of Silica Mesostructures

The azo amphiphiles used in this work were prepared by slight variations [22,32] of a literature procedure [69]. For the preparation of the dye–silica composites, synthesis gels were prepared from SiO_2, $C_mAzoC_nTMA^+Br^-$, KOH, and H_2O and treated hydrothermally for 3 days in Teflon-lined steel autoclaves at temperatures between 110 and 160 °C. Yellow powders were obtained. Calcination of the samples was performed by heating in air at 600 °C for 2 h. The results presented in this section refer to samples synthesized at 110 °C.

Table 1 lists the different $C_mAzoC_nTMA^+$ amphiphiles that were used as SDAs and characteristic data of their mesostructured silica composites. The mass losses

Tab. 1. Structural properties of lamellar $C_mAzoC_nTMA^+$-silica composites synthesized at 110 °C: basal spacing c as calculated from the 00l peaks of the PXRD patterns; mass loss Δm upon calcination; λ_{max} of the excitonic absorption in the UV-Vis spectra; type of aggregation as deducted from the basal spacing c and from λ_{max}.

SDA	*c [nm]*	*Δm [%]*	*λ_{max} [nm]*	*Type of aggregate*
$C_{12}AzoC_6TMA^+Br^-$	5.6	58	394	*J*
$C_{10}AzoC_8TMA^+Br^-$	5.8	—	381	*J*
$C_{10}AzoC_6TMA^+Br^-$	5.3	67	390	*J*
$C_8AzoC_6TMA^+Br^-$	4.9	69	388	*J*
$C_6AzoC_6TMA^+Br^-$	4.6	63	380	*J*
$C_6AzoC_{10}TMA^+Br^-$	3.8	78	310	*H*
$C_1AzoC_{10}TMA^+Br^-$	3.5	62	350	*H*
$C_1AzoC_3TMA^+Br^-$	2.7	48	340	*H*

upon calcination as given there represent the sum of the mass fraction of the organic component and of a portion of water liberated from the condensation of silanol groups in the composite. For comparison, for lamellar M41S-type materials synthesized with the simple surfactant hexadecyltrimethylammonium cation, the average mass loss is about 60%.

Powder X-ray diffraction (PXRD) patterns of the mesostructures obtained are shown in Fig. 4. The diffraction patterns typically show three peaks at 2θ of 1–10°, which can be indexed according to a lamellar structure with basal spacings as given in Table 1 for the different azo amphiphiles. Upon calcination, these peaks disappear in most of the cases due to structural collapse, as is typical of truly lamellar mesostructures (see, however, Section 7.2.2).

The fact that only lamellar materials are obtained must be due to a special structure-directing effect of the tail-functionalized amphiphiles used here. For example, the normal amphiphile trimethyltetradecylammonium ($C_{14}TMA^+$) is of comparable length to the $C_1AzoC_3TMA^+$ amphiphile. The $C_{14}TMA^+$ amphiphile can give rise to different M41S materials: hexagonal MCM-41, cubic MCM-48, LMU-1 [15] (or KIT-1 [71]) or lamellar materials [15]. The $C_1AzoC_3TMA^+$ azobenzene amphiphile, however, prefers an arrangement that leads to lamellar composite structures only. This is because of special arrangements of the hydrophobic tails induced by the azo groups of this special amphiphile.

There are obviously two series of samples for which the relative intensities of the reflections are different. For composites synthesized with amphiphiles of the type $C_mAzoC_nTMA^+$ with $m \geq n$, the first reflection 001 is very strong, 002 is typically very weak, and 003 is weak. Composites synthesized with amphiphiles $C_mAzoC_nTMA^+$ in which $m < n$, have 001 and 002 reflections with both stronger intensities (except for $C_1AzoC_3TMA^+$), whereas 003 is very weak. Within each series, the basal spacing increases with increasing length of dye molecule, clearly showing its influence on the inorganic structure and testifying its role as SDA. However, in relation to the total lengths of the amphiphiles, the basal spacings of the compounds of the second group are always smaller than the repeat distances of the first. This becomes clear when the basal spacings of composites produced using $C_{10}AzoC_6TMA^+$ and $C_6AzoC_{10}TMA^+$ surfactants are compared. These molecules have similar lengths, but their silica mesostructures have clearly different d values. This fact is due to an additional structure-directing effect exerted by the functionalized hydrophobic tails, which can assume different arrangements in dependence of the relative lengths m and n of the alkyl chains (see below). It provides an instructive example of how delicate variations of the hydrophobic part of the SDAs can lead to substantially different mesostructures and also represents the first approach to mesostructure control by variation of this part of the amphiphile, thus complementing the numerous attempts to control mesostructures by variation of the headgroup of the chain [13,48–54].

The above-mentioned differences in the intensity distributions of the $00l$ reflections in the PXRD diagrams of the two groups of composites of course reflect differences in the electron density distribution along the direction normal to the layers (c axis). Assuming that the electron densities of the silica sheets within these

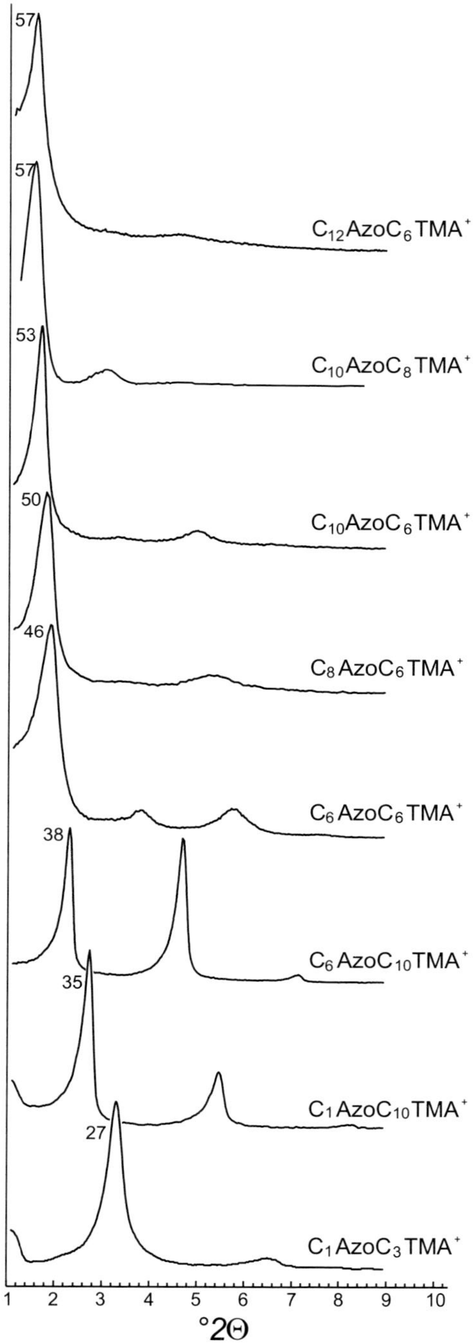

Fig. 4. Powder X-ray diffraction patterns of lamellar composites synthesized with different amphiphilic azo dyes. The *d* value of the first peak is indicated.

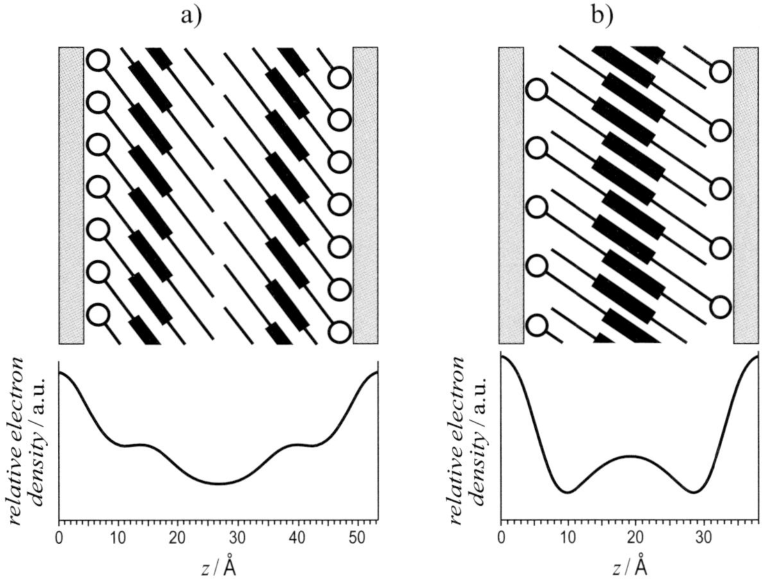

Fig. 5. Projections of the electron density of lamellar azo amphiphile–silica composites as calculated on the basis of PXRD patterns: (a) $C_{10}AzoC_6TMA^+$-based composite; (b) $C_6AzoC_{10}TMA^+$-based composite. The aggregation forms of the azo amphiphiles are also shown: (a) *J* aggregate; (b) *H* aggregate. Regions of high electron density correspond either to the silica layers or to the azobenzene moieties.

two types of mesostructures are similar and that the electron density in the azobenzene unit is significantly higher (about twice as high) as in the alkyl chains of the surfactant, we conclude that it must be the relative locations of the aromatic segments (in a projection of the structure onto the *c* axis) of the two types of composites that gives rise to the differences in diffraction intensities. Even when only three or four 00*l* reflections are available, it is possible to calculate a Fourier transform, which then represents a projection of the electron density along the *c* axis (a rough estimate of the resolution of such projections is given by dividing the basal spacing by the number of reflections that were used in the calculation). Figure 5 shows by examples, that these projections are distinctly different for the two groups of compounds. The underlying structural model to explain these projections was substantiated by UV/vis spectroscopy.

Owing to the fact that the surfactants used in this work contain chromophores in their hydrophobic tails, UV/vis spectra can be used to analyze further the structures of the lamellar composites. In Fig. 6, a typical UV/vis spectrum from a member of each of the two groups of composites is shown, namely those of the samples containing $C_{10}AzoC_6TMA^+$ and $C_6AzoC_{10}TMA^+$, respectively. The absorption spectrum of a diluted solution of the corresponding dye in ethanol is included for comparison. The intensive absorption band at 360 nm in the solution

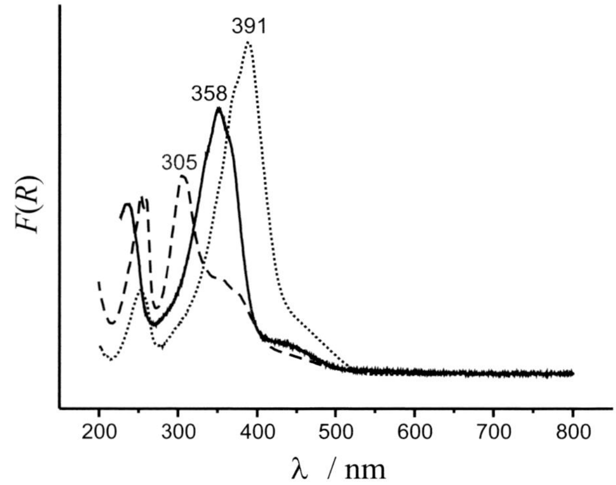

Fig. 6. Diffuse reflectance UV/vis spectra (as the Kubelka – Munk function [72]) of the $C_{10}AzoC_6TMA^+$-based composite (*J* aggregate, dotted line) and the $C_6AzoC_{10}TMA^+$-based composite (*H* aggregate, dashed line) at a temperature of 20 °C. The absorption spectrum of an azo amphiphile in a diluted solution in ethanol is included for comparison (solid line). The wavelengths of the absorption maxima of the excitonic transition or the π–π^* transition are indicated.

spectrum is ascribed to the π–π^* transition [73]. The reflectance spectra of the composites obtained at 20 °C show remarkable shifts of this band (which are collected for all composites under study here in Table 1). Shimomura et al. have thoroughly investigated the spectral properties of aqueous solutions and cast films of azobenzene amphiphiles [68,69] and found that the absorption spectra are strongly affected by the aggregation state of the amphiphiles. Since the isolated azobenzene chromophore as present in diluted ethanol solutions does not show strong solvent effects, variations in the spectra are assigned to intermolecular interaction within molecular assemblies. In line with this, spectra of our silica composites measured at 90 °C (not shown here) do not exhibit these shifts. Obviously, the amphiphilic dye molecules in these composites are aggregated at 20 °C, but the aggregation is destroyed by increasing temperature.

Kasha [74] has developed a rather simple model to explain shifts of the absorption bands that are due to aggregation. According to his molecular exciton theory a strong coupling of transition dipoles delocalizes a photonic excitation over a number of molecules, which leads to a splitting of the excited $^1[\pi$–$\pi^*]$ state (S_2 state) into Frenkel-excitonic states. Optical selection rules allow transitions from the ground state only to the highest or to the lowest state of the exciton band, in dependence on the orientation of the transition dipoles to each other within the assembly. A hypsochromic shift of the absorption maximum (with respect to the spectrum of the isolated molecule) is assigned to a parallel side-by-side orientation of the transition dipole moments of the azobenzene units (*H* aggregate as depicted

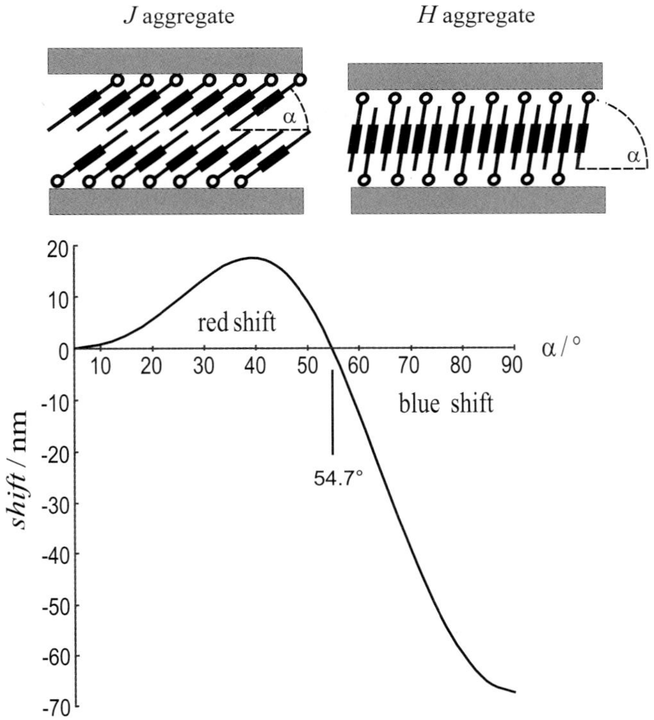

Fig. 7. Top: idealized schematic drawing of the arrangements of amphiphilic azo dye molecules between silica layers in the composites. Bottom: Relationship between $\Delta E_{\text{aggregate}}$ and α according to Eq. (1).

in the top of Fig. 7), whereas a bathochromic shift is caused from a tilted bilayer arrangement of the molecules with head-to-tail orientation of the transition dipole moments (J aggregate as depicted in Fig. 7). A quantitative treatment of the shifts is given by McRae and Kasha [75,76]. According to their theory, the position of the excitonic absorption can be estimated by

$$\Delta E_{\text{aggregate}} \approx \Delta E_{\text{monomer}} + 2((N-1)/N)(\mu^2/(DR^3))(1 - 3\cos^2\alpha) \qquad (1)$$

with

$\delta E_{\text{monomer}}$ = transition energy of the monomer

μ = transition dipole moment of the monomer

D = dielectric constant of the vacuum

R = center-to-center distance of the molecules in the aggregate

α = tilt angle between the transition dipoles and the line connecting the centers of the molecules (Fig. 7)

N = number of molecules contributing to the excitonic state.

For the calculations presented in the following, μ was derived from the integrated extinction of the π–π^* band of the corresponding azo dye in a diluted ethanol solution. R is given by $R = r/\sin\alpha$, where r is the distance of closest contact found in molecular crystals of amphiphilic azo dyes; r is about 3.6 Å [77]. For a large number of molecules in the aggregate ($N = 100$), Eq. (1) gives the curve $\Delta E_{aggregate} = f(\alpha)$ shown in Fig. 7. According to this relation, the spectrum of an aggregate shows a red or a blue shift depending on whether α is smaller or larger than the "magic angle" (54.7°), respectively.

On the basis of this simple model, the tilt angle of the chromophores can be estimated. For C_mAzoC_nTAB–silica composites with $m \geq n$, bathochromic shifts of 20–30 nm are observed (Table 1). This is consistent with a tilt angle of around 40° when a large aggregation number N is assumed. Tilt angles as small as this occur only in *J*-type aggregates. The bilayers characteristic of this type of aggregate lead to a large thickness of the surfactant layers and correspondingly, the composites possess a large extension normal to the layers. The aggregate absorption band of the $C_6AzoC_{10}TAB$ composite, where $m < n$, is blue-shifted by 50 nm, corresponding to a tilt angle of about 75°. Therefore, a side-by-side arrangement of the chromophores can be assumed, consistent with the structural model of an arrangement with interdigitating dye molecules. Such an *H*-type aggregate is consistent with the smaller basal spacings observed for composites based on C_mAzoC_nTAB surfactants with $m < n$. The proposed structural arrangements of the composites agree well with the projections of the electron density shown in Fig. 5, with high electron densities within the silica layers and one or two maxima within the organic layer corresponding to the positions of the azobenzene units that have a higher electron density than the alkyl chains. For the different composites under study here, the aggregation form is given in Table 1.

Similar differences in the aggregation behavior of azo surfactants with different tail and spacer lengths m and n were observed by Shimomura and Aiba [69] on azo surfactant films cast on quartz substrates. They found that *H* aggregates were formed for azo surfactants with $m \leq n + 2$, whereas *J*-type arrangements were found for $m > n + 2$. These authors reasoned that favorable π–π interactions between the azo moieties in *H* aggregates can act only for azo surfactants with $m \leq n + 2$; for molecules with large m, no interactions of this type would take place in a *J*-type arrangement.

Ogawa and Ishikawa prepared lamellar structures by the intercalation of the azo amphiphiles $C_8AzoC_{10}TMA^+$ and $C_{12}AzoC_5TMA^+$ into montmorillonite, which bear some resemblance to the materials under discussion here. For both azo surfactants, the authors assume a *J*-type arrangement [64]. This is unlike our layered silica materials, in which we would expect the $C_{12}AzoC_5TMA^+$ dye to form an *H*-type aggregate structure. This difference is possibly because the arrangement of the azo surfactants within the montmorillonite composite is influenced by the charge density of the clay layers, which is pre-determined before the intercalation, whereas in the directly synthesized mesostructure, the charge of the cationic azo surfactants is balanced by depronated silanol groups of the layers, the number of which can be adapted to the requirements of the packing of the surfactant molecules.

7.2.2
Mesoporous Materials from Templating with Azobenzene Amphiphiles

Several points of evidence have been brought forward in Section 7.2.1 for a lamellar structure of the azobenzene surfactant–silica composites. Upon calcination, truly lamellar composites lose their mesostructure due to a collapse when the surfactant layers are removed. However, noncollapsing lamellar phases synthesized with simple C_nTMA^+ surfactants have been described and named MCM-50 [78]. The authors ascribe the persistence of porosity after surfactant removal to the presence of pillars between the layers, which, when irregularly spaced, cannot be observed by X-ray diffraction.

In fact, among our composites, we find some samples, which after calcination yield mesoporous solids with pore characteristics similar to those of M41S materials. Such materials are typically obtained at low synthesis temperatures (110–130 °C). Figure 8 shows as an example the results of a sorption measurement performed on a $C_8AzoC_6TMA^+$–silica composite prepared at 130 °C and then calcined at 600 °C for 2 h. The steep incline of the adsorption and the desorption isotherms indicate a rather narrow pore size distribution (although not as narrow as in MCM-41 materials of good quality). The specific surface area of this sample (1138 $m^2\ g^{-1}$) and its pore volume (1.39 $cm^3\ g^{-1}$) compare favorably with the values of typical M41S materials. After calcination, this and similar materials typically exhibit a strong 001 reflection and one further, weak and broad, reflection. These

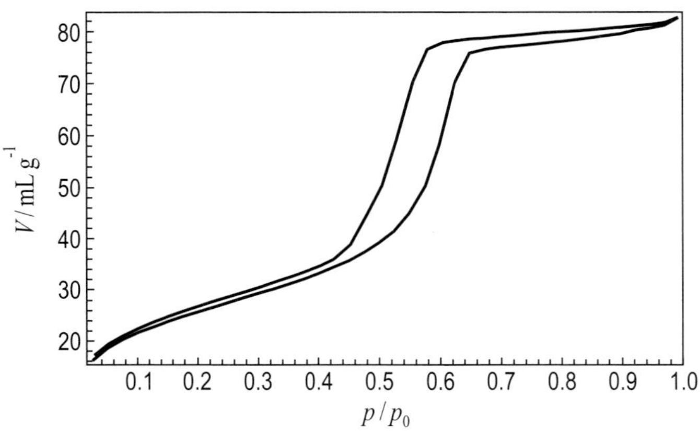

Fig. 8. Nitrogen sorption measurement performed on a $C_8AzoC_6TMA^+$–silica composite prepared at 130 °C.

mesoporous materials are of interest, as large pores with diameters up to 50 Å and more can be synthesized using surfactants with the simple trimethylammonium headgroup and without the need to use swelling agents. Further details, especially on the mesostructure of these silicas as revealed by TEM will be given in a forthcoming publication [22,79].

7.2.3
Photoisomerization in Azo Amphiphile–Silica Composites

The photoinduced *cis-trans* isomerization of azo dyes is the basis for the interest these molecules find currently in the field of advanced functional materials and other areas of science. The isomerization of an azo moiety in a dye molecule changes its shape. Therefore, the isomerization process depends on the space available to the molecule and is correspondingly strongly affected by the environment of the dye. Azo dyes incorporated into polymer networks (either bonded to side-chains or inserted into the polymer backbone) have been discussed as probe molecules for the existence of free volume in the network and the possibility of dynamic mobility [80,81]. Ueda et al. entrapped azo dyes in silica-based sol-gel materials and derived structural characteristics of these materials from the isomerization properties [82]. The sterical effects upon which such investigations are based will be especially strong when the azobenzene group carries substituents, like in the amphiphilic dyes used here, in which the sterical demands of the *trans* and the *cis* isomers differ remarkably. They might be even more pronounced in the more ordered aggregated state of the composite materials under study, in which the chromophore molecules are arranged in layers separated by silica sheets. Within the layers, the microenviroment of an azo dye surfactant is governed by the packing of the arrangement. Investigations on the isomerization processes will therefore also give interesting additional information on the structure of the chromophore assembly.

We examined the photoisomerization of $C_{12}AzoC_6TMA^+$ and $C_6AzoC_{10}TMA^+$ in lamellar composites with silica. The experiments for the photoinduced *trans*-to-*cis* isomerization were carried out by irradiation of the samples with the 365 nm line of a mercury lamp (power = 0.05 mW cm^{-2}). After a certain time of irradiation (30 min in our experiments), a photostationary state (PSS) is reached where the part of azo dye molecules that were isomerized from the *trans* to the *cis* configuration does not increase anymore. In the PSS, the maximal *cis–trans* ratio of a certain system has thus been obtained and this state can be maintained by ongoing irradiation. If the irradiation is stopped, a thermal *cis–trans* relaxation occurs within hours. Absorption spectra corresponding to the PSS, obtained at a temperature of 12 °C, are shown in Figs. 9a and 9b for the composites based on $C_{12}AzoC_6TMA^+$ and $C_6AzoC_{10}TMA^+$, respectively. The mole fraction of the *cis*-isomer α_{cis} was estimated from the absorption band of the n–π* transition using Eq. (2).

$$\alpha_{cis} = [(A_t/A_0) - 1]/[(\varepsilon_{cis}/\varepsilon_{trans}) - 1] \tag{2}$$

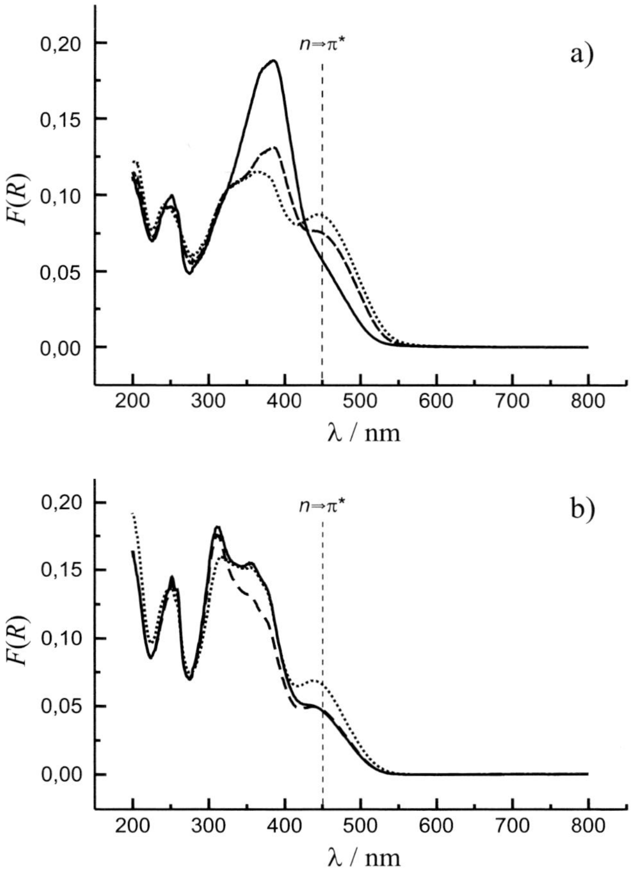

Fig. 9. Diffuse reflectance UV/vis spectra (as the Kubelka–Munk function [72]) for the investigation of the photoisomerization of azo amphipihiles in silica composites: (a) the $C_{12}AzoC_6TAB$-based composite (*J*-type aggregate) before irradiation (solid line), in the PSS at a temperature of 12 °C (dashed line) and in the PSS at 40 °C (dotted line); (b) the $C_6AzoC_{10}TAB$-based composite (*H*-type aggregate) before irradiation (solid line), in the PSS at a temperature of 13 °C (dashed line) and in the PSS at 38 °C (dotted line).

with

A_0 and A_t = absorbances at $\lambda = 450$ nm before and after irradiation, respectively

ε_{cis} and ε_{trans} = molar extinction coefficients of the two isomers at $\lambda = 450$ nm

The extinction coefficients were determined from absorption spectra of diluted methanol solutions before and after irradiation. For monitoring α_{cis}, we chose to use the intensity of the n–π^* band rather than the π–π^* band usually employed.

Owing to the excitonic shifts, the π–π^* band undergoes strong thermochromic effects making a determination of the *cis*/*trans* ratio at different temperatures problematic. Since the excitonic shift depends on the square of the transition dipole moment, the effects of aggregation on the weaker n–π^* band can be neglected. An overlap with another band at the observed wavelength was checked for and excluded by means of least-squares fit analyses using Gaussian lines.

In the PSS generated at 12 °C, the $C_6AzoC_{10}TAB$ composite contains only 2% of the *cis* isomer whereas in the $C_{12}AzoC_6TAB$ composite the fraction of the *cis* isomer is 40%. According to the structural model described above, the dye molecules in the $C_6AzoC_{10}TAB$ composite are arranged in *H*-aggregate-like assemblies with interdigitating hydrophobic chains. The lower *cis* content observed in its PSS can be explained by a smaller motional freedom in this tight-packed arrangement. Sato et al. have shown that in Langmuir–Blodgett films of azobenzene-containing long-chain fatty acids photoisomerization is hindered to a high extent by the formation of *H* aggregates [83]. The higher *cis* fraction in the $C_{12}AzoC_6TAB$ composite supports the structural model of *J*-type noninterdigitating bilayers with a less dense and less ordered arrangement of the azo chromophors. Upon increasing the temperature, the *cis* fraction in the PSS also increases and reaches about 60% at 40 °C in the *J*-type $C_{12}AzoC_6TAB$ composite. For the *H*-type $C_6AzoC_{10}TAB$ composite, the evaluation of the *cis* fraction is less reliable owing to changing spectral shapes caused by thermochromic effects. The *cis* fraction can be estimated to be smaller than 50% at 38 °C. The thermally induced increase of the efficiency of the *cis–trans* photoisomerization can be ascribed either to the increase in dynamic motion offering additional space for the associated shape change within the aggregate or by a liberation of chromophore molecules from the aggregate.

Ogawa and Ishikawa [64] described investigations on the *cis–trans* photoisomerization of $C_{12}AzoC_5TMA^+$ intercalated into montmorillonite. In the PSS at room temperature, the part of the *cis* isomer was estimated to 60%. This is in agreement with the assignment of *J*-type aggregate structures to these composites. However, unlike these authors we do not consider such isomerization ratios as effective.

7.2.4
Chemical Switching of Azobenzene Surfactant–Silica Composites: Basis for a "Nanoscale Elevator"?

The fact that the *trans–cis* isomerization does not proceed effectively in our composites is disappointing. However, we found another most remarkable property of our composites [22,28–32]. This feature is observed in mesostructures synthesized in a similar way to that described above (Section 7.2.1), but with the difference that the hydrothermal treatment is carried out at 160 °C. Materials prepared under these conditions always show a collapse of the layered mesostructure upon calcination and do not become mesoporous, unlike the materials described in Section 7.2.2.

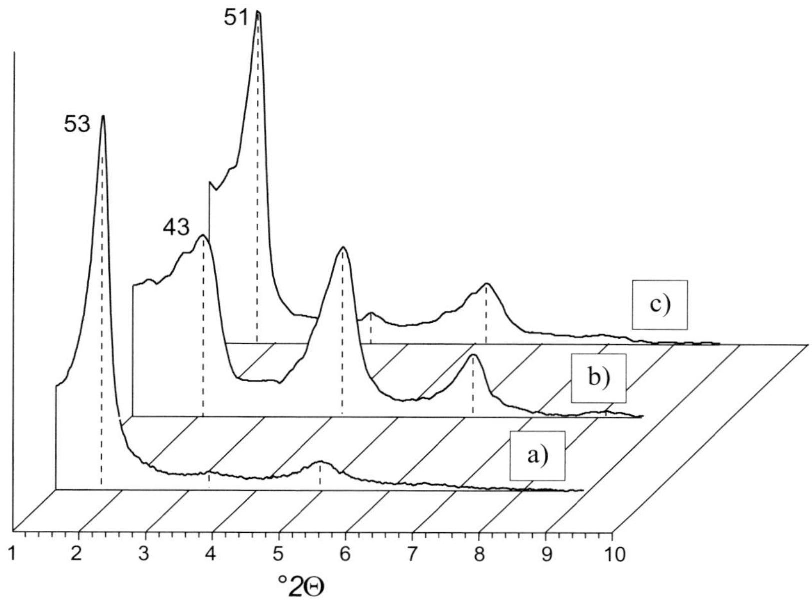

Fig. 10. PXRD patterns of a $C_8AzoC_6TMA^+$-based composite: (a) after synthesis, (b) after exposure to methanol vapor, (c) after exposure to water vapor. The *d* value of the first peak is indicated.

Figure 10 shows the PXRD of a mesostructure based on $C_8AzoC_6TMA^+$ (a) after synthesis, (b) after exposure to methanol vapor, and (c) after exposure to water vapor. Figure 11 gives the UV/vis spectra of the same sample (a) after synthesis, and (b) after exposure to methanol vapor. Obviously, drastical changes in the properties of this mesocomposite occur during these processes. These can be rationalized as follows.

The exposure of the sample to methanol vapor causes a transition from a *J* arrangement to an *H*-type aggregate structure. The evidence for this lies in the decrease of the basal spacing from 53 to 43 Å, in the changes in the intensity distribution (decrease of the intensity of the 001 reflection, increase of the intensity of the 002 reflection), and in the shift of the excitonic absorption from 391 to 328 nm. Upon exposure to water vapor, the changes are largely reversible, leading back to a mesostructure with a *J* arrangement, albeit with a somewhat smaller basal spacing (51 Å). Figure 12 shows projected electron density distributions calculated on the basis of the PXRD patterns of an azo amphiphile–silica composite, which confirm the assignments of the different aggregation types. Similar *J*–*H* switching processes can be observed using linear alcohols (C1–C7), *tert*-butanol, or cyclopentanol. Only water has so far been found to induce the back-switching process from *H* to *J* aggregation.

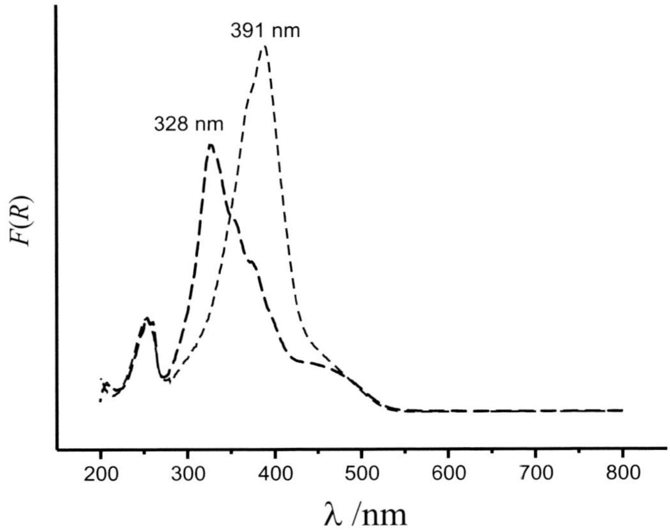

Fig. 11. Diffuse reflectance UV/vis spectra (as the Kubelka–Munk function [72]) of a $C_8AzoC_6TMA^+$-based composite: (a) as-synthesized (solid line), (b) after exposure to methanol vapor (dashed line).

For the $C_8AzoC_6TMA^+$ composite, good reproducibility of the switching process (investigated for up to ten cycles) was observed, also with respect to the basal spacings of the lamellar mesostructure. The switching process has been observed on a number of other composites, too. In all cases, the switching is complete according to the results of UV/vis spectroscopy, but upon switching back from an *H*-type to a *J*-type aggregate, the basal spacing does not increase to its original value in many cases. ^{13}C MAS NMR spectroscopy (Fig. 13) shows that this is possibly due to an ordering process within the surfactant layers. After synthesis (Fig. 13a), the signals are relatively broad and unstructured. Upon sorption of methanol, the peaks become significantly sharper (Fig. 13b, in this spectrum, the signal due to the methanol molecules is indicated by an arrow). This can be associated with a better order of the azo amphiphiles within the organic layers (as already stated above, the denser *H* aggregate structure implies a higher degree of ordering of the surfactant molecules). Upon exposure to water vapor, the signals remain nearly as sharp (Fig. 13c). Obviously, the *J*-type aggregate obtained by chemically switching back from an *H*-type state retains part of the higher order of the latter arrangement. With ^{13}C MAS NMR spectroscopic measurements conducted in the high-power proton decoupling (HPDEC) mode (not shown here), a quantitative evaluation of the carbon-containing species becomes possible, so the ratio between methanol and amphiphile molecules in the switched state can be estimated. A methanol:surfactant ratio of about 1:3 is obtained.

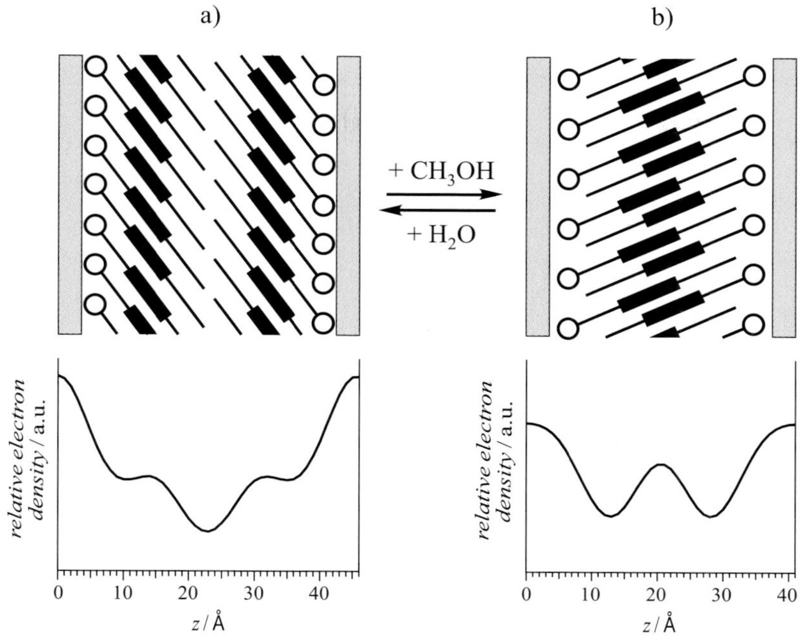

Fig. 12. Projections of the electron density of a lamellar azo amphiphile–silica composite as calculated on the basis of PXRD patterns: (a) after synthesis, (b) after exposure to methanol vapor.

The basal spacing of layered materials can easily be changed by swelling the interlamellar space or by intercalation. In these cases, the interlayer distance is usually a direct function of the volume of the material added to (or subtracted from) the interlayer galleries. It should be clearly stated that the effect observed here is different. By the exchange of a relatively small amount of alcohol, the interlayer distance can be strongly decreased, and the arrangement of the organic species present within the interlayer space can be changed drastically. Ogawa et al. [65] have shown that the interlayer distance of an azobenzene amphiphile-magadiite composite can be varied to a small amount by light. Again, the mechanism is different and involves *cis–trans* isomer switching instead of *J–H* aggregate switching.

Although the exact mechanism of the chemical switching processes is not clear yet, we have indications that the sorbed alcohol resides in the palisade region near to the polar headgroups of the amphiphiles [22]. The fact that only such compounds that can act as hydrogen bond donors can induce the *J–H* switching points to hydrogen bonding interactions within that region. A strong influence of the humidity on the spectral properties of a Langmuir–Blodgett layer on quartz, com-

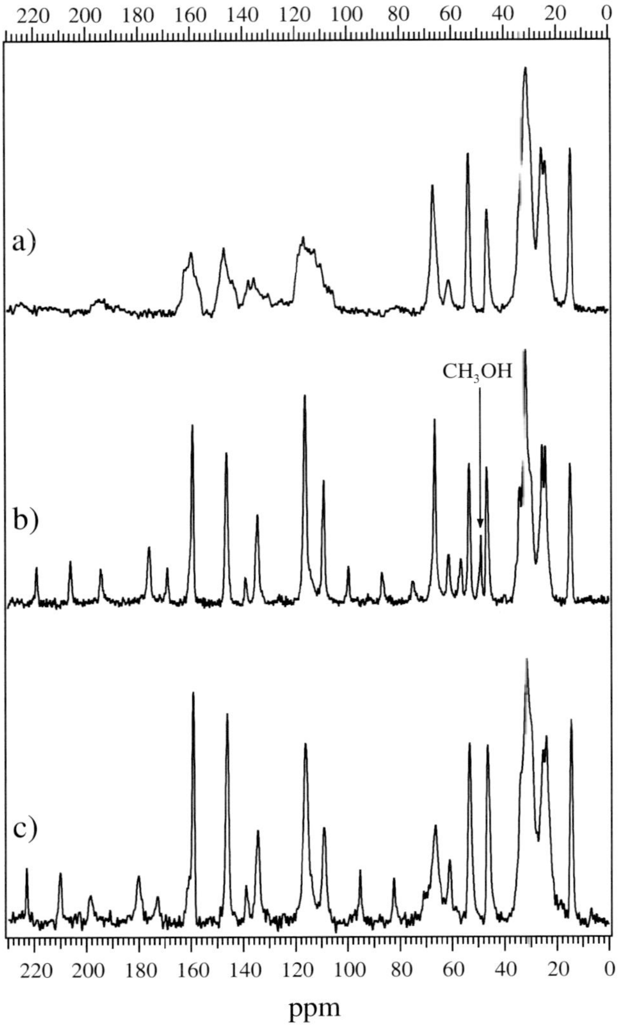

Fig. 13. ^{13}C CPMAS NMR spectra of a $C_8AzoC_6TMA^+$-based composite: (a) after synthesis, (b) after exposure to methanol vapor (the peak caused by the presence of methanol is indicated), (c) after exposure to water vapor.

posed of an azobenzene amphiphile carrying an urea headgroup, was noticed before. However, no changes in the layer spacing could be observed by X-ray diffraction [84].

The switching process observed could possibly find applications in sensing devices or for chemically induced motions on the Å to nanometer scale ("nanoscale elevator"). For these purposes, shaping the morphology of the composites is a prime requisite. Currently, attempts are underway in our group to prepare films of these materials.

7.4 Conclusions

This work has shown that it is possible to obtain functional mesostructured organic/inorganic hybrid materials directly by a self-assembly process in which the functional organic molecule acts itself as an amphiphilic SDA in a synthesis approach analogous to the preparation of M41S mesophases. By inclusion within the mesostructure, the arrangements of and the organic molecules themselves are stabilized. The composites show interesting expected (variations in the electronic spectra, *cis–trans* switching of azo dyes) and unexpected (chemical switching of aggregate type of azo dyes) properties.

In addition to these interesting properties, special structure-directing effects that cannot be observed with nonfunctional amphipihiles, become apparent. The simple fact of special aggregation tendencies between the functional amphiphiles can lead to a clear preference for only one type of mesostructure (lamellar in this case) and the possibility of forming aggregates of different type can give rise to different mesostructures for different surfactants with similar lengths. The aggregation phenomena are influenced by interactions between the aromatic systems of the chromophore amphipihiles, so that the preparative work described here may be considered as the first example of actively controlling mesostructure formation via the hydrophobic part of the amphiphile.

A large number of functional organic molecules are nowadays routinely equipped with specific side-groups (such as cationic headgroups, thiol groups, trialkoxysilyl groups) in order to be able to form aggregated states (lyotropic phases, Langmuir–Blodgett films, SAMs) with interesting properties. The concerted formation of mesostructured silica composites, as presented here, opens a new possibility of obtaining not only increased mechanical stability, but also to easily extend the aggregates to the third dimension: to bulk phases.

Acknowledgements

This work was in part carried out at the Institut für Anorganische Chemie of the Ludwig-Maximilians Universität, Munich. It was supported by the Deutsche Forschungsgemeinschaft in the framework of the Schwerpunktprogramm "Nanoporöse Wirt-Gast-Systeme" (Be1664/3) and by the Fonds der Chemischen Industrie. We like to thank the colleagues within the program, especially Katrin Hoffmann, Frank Marlow, Michael Wark, and Dieter Wöhrle, for interesting discussions.

References

1 G.D. Stucky, J.E. MacDougall, *Science* **1990**, *247*, 669.

2 G.A. Ozin, *Adv. Mater.* **1992**, *4*, 612.

3 P. Behrens, G.D. Stucky, in Comprehensive Supramolecular Chemistry, J.L. Atwood, D.D. MacNicol, J.E.D. Davies, F. Vögtle (eds.), Vol. 7, G. Alberti, T. Bein (eds.), Pergamon Press, Oxford **1996**, p. 721.

4 P. Behrens, Ch. Panz, C. Kühn, B.M. Pillep, A.M. Schneider, Guest Functionalized Crystalline Organic/Inorganic Nanohybrid Materials, in Guest–Host Systems on Nanoporous Crystals, F. Schüth (ed.), Wiley-VCH, Weinheim **2003**.

5 C.T. Kresge, M.E. Leonowitcz, W.J. Roth, J.C. Vartuli, J.S. Beck, *Nature* **1992**, *359*, 710.

6 J.S. Beck, J.C. Vartuli, W.J. Roth, M.E. Leonowitcz, C.T. Kresge, K.D. Schmitt, C.T.W. Chu, D.H. Olson, E.W. Sheppard, S.B. McCullen, J.B. Higgins, J.L. Schlenker, *J. Am. Chem. Soc.* **1992**, *114*, 10 834.

7 P. Behrens, G.D. Stucky, *Angew. Chem. Intern. Ed. Engl.* **1993**, *32*, 696.

8 J.Y. Ying, C.P. Mehnert, M.S. Wong, *Angew. Chem. Intern. Ed. Engl.* **1999**, *38*, 56.

9 U. Ciesla, F. Schüth, *Micropor. Mesopor. Mater.* **1999**, *27*, 131.

10 G. Oye, J. Sjöblom, M. Stöcker, *Adv. Coll. Interface Sci.* **2001**, *89–90*, 493.

11 F. Schüth, *Stud. Surf. Sci. Catal.* **2001**, *135*, 1.

12 A. Monnier, F. Schüth, Q. Huo, D. Kumar, D. Margolese, R.S. Maxwell, G.D. Stucky, M. Krishnamurty, P. Petroff, A. Firouzi, M. Janicke, B.F. Chmelka, *Science* **1993**, *261*, 1299–1303.

13 Q. Huo, D.I. Margolese, U. Ciesla, P. Feng, T.E. Gier, P. Sieger, R. Leon, P.M. Petroff, F. Schüth, G.D. Stucky, *Nature* **1994**, *368*, 317–321.

14 G.D. Stucky, A. Monnier, F. Schüth, Q. Huo, D. Margolese, D. Kumar, M. Krishnamurty, P. Petroff, A. Firouzi, M. Janicke, B.F. Chmelka, *Mol. Cryst. Liq. Cryst.* **1994**, *240*, 187.

15 P. Behrens, A. Glaue, Ch. Haggenmüller, G. Schechner, *Solid State Ionics* **1997**, *101–103*, 255.

16 F. Schüth, *Curr. Opin. Coll. Interf. Sci.* **1998**, *3*, 174.

17 F. Schüth, U. Ciesla, S. Schacht, M. Thieme, Q. Huo, G. Stucky, *Mater. Res. Soc. Bull.* **1999**, *34*, 483.

18 M. Wark, M. Ganschow, Y. Rohlfing, G. Schulz-Ekloff, D. Wöhrle, *Stud. Surf. Sci. Catal.* **2001**, *135*, 21-O-02.

19 B.T. Holland, C. Walkup, A. Stein, *J. Phys. Chem. B* **1998**, *102*, 4301.

20 H.S. Zhou, H. Sasabe, I. Honma, *J. Mater. Chem.* **1998**, *8*, 515.

21 I. Honma, H.S. Zhou, *Chem. Mater.* **1998**, 10, 103.

22 A.M. Glaue, Ph.D. Thesis, Ludwig-Maximilians University, Munich, Logos Verlag, Berlin **1999**, ISBN: 3-89722-3331-7.

23 A. Glaue, P. Behrens, contribution to 9th Deutsche Zeolithtagung, Halle, 3–5 March **1997**.

24 P. Behrens, Ch. Eich, A. Glaue, Ch. Haggenmüller, G. Schechner, contribution to the VIth Europ. Conf. Solid State Chemistry, Zürich, 17–20 September **1997**.

25 A. Glaue, P. Behrens, contribution to 10th Deutsche Zeolithtagung, Bremen, 9–11 March **1998**.

26 U. Mennicke, T. Salditt, G. Goerigk, A. Glaue, P. Behrens, HASYLAB Annual Report **1998**, HASYLAB/DESY, Hamburg **1999**, p. 581.

27 O. Oellrich, A.M. Glaue, P. Behrens, contribution to 12th Deutsche Zeolithtagung, Munich, 1–3 March **2000**.

28 O. Oellrich, A.M. Glaue, P. Behrens, contribution to 10th Vortragstagung of the Fachgruppe "Festkörperchemie und Materialforschung" of the GDCh, Münster, 26–29 September **2000**.

29 O. Oellrich, A. Glaue, P. Behrens,

contribution to 13th Deutsche Zeolithtagung, Erlangen, 7–9 March **2001**.

30 O. Oellrich, P. Behrens, contribution to 14th Deutsche Zeolithtagung, Frankfurt am Main, 6–8 March **2002**.

32 A.M. Glaue, O. Oellrich, P. Behrens, submitted to *Adv. Funct. Mater.*

33 P. Behrens, A.M. Glaue, *Chem. Monthly – Monatsh. Chem.* **2002**, *133*, 1405.

34 V. Balzani, F. Scandolo, Supramolecular Chemistry, Horwood, Chichester **1991**.

35 V. Ramamurthy, Photochemistry in Organized and Constrained Media, VCH, New York **1991**.

36 H. Dürr, H. Bouas-Laurent, Photochromism–Molecules and Systems, Elsevier, Amsterdam **1990**.

37 I. Willner, *Acc. Chem. Res.* **1997**, *30*, 347.

38 D. Wöhrle, G. Schulz-Ekloff, *Adv. Mater.* **1994**, *6*, 875.

39 G. Schulz-Ekloff, D. Wöhrle, B. van Duffel, R.A. Schoonheydt, *Micropor. Mesopor. Mater.* **2002**, *51*, 91.

40 G. Wirnsberger, G.D. Stucky, *Chem. Phys. Chem.* **2000**, *1*, 89.

41 G. Wirnsberger, B.J. Scott, G.D. Stucky, *Chem. Commun.* **2001**, 119.

42 G. Wirnsberger, B.J. Scott, B. Chmelka, G.D. Stucky, *Adv. Mater.* **2000**, *12*, 1450.

43 Y.F. Lu, Y. Yang, A. Sellinger, M.C. Lu, J.M. Huang, H.Y. Fan, R. Haddad, G. Lopez, A.R. Burns, D.Y. Sasaki, J. Shelnutt, C.J. Brinker, *Nature* **2001**, *410*, 913.

44 F. Marlow, M.D. McGehee, D.Y. Zhao, B.F. Chmelka, G.D. Stucky, *Adv. Mater.* **1999**, *11*, 632.

45 P.D. Yang, G. Wirnsberger, H.C. Huang, S.R. Cordero, M.D. McGehee, B. Scott, T. Deng, G.M. Whitesides, B.F. Chmelka, S.K. Buratto, G.D. Stucky, *Science* **2000**, *287*, 465.

46 G. Wirnsberger, G.D. Stucky, *Chem. Mater.* **2000**, *12*, 2525.

47 B.J. Scott, G. Wirnsberger, M.D. McGehee, G.D. Stucky, *Adv. Mater.* **2001**, *13*, 1231.

48 Q. Huo, D. Margolese, U. Ciesla, D. G. Demuth, P. Feng, T. E. Gier, P. Sieger, A. Firouzi, B. F. Chmelka, F. Schüth, G. D. Stucky, *Chem. Mater.* **1994**, *6*, 1176.

49 Q. Huo, D.I. Margolese, G.D. Stucky, *Chem. Mater.* **1996**, *8*, 1147.

50 G.D. Stucky, Q. Huo, A. Firouzi, B.F. Chmelka, S. Schacht, I.G. Voigt-Martin, F. Schüth, *Stud. Surf. Sci. Catal.* **1997**, *105*, 3.

51 P.T. Tanev, T.J. Pinnavaia, *Science* **1995**, *267*, 865.

52 S.A. Bagshaw, E. Prouzet, T.J. Pinnavaia, *Science* **1995**, *269*, 1242.

53 P.T. Tanev, T.J. Pinnavaia, *Chem. Mater.* **1996**, *8*, 2068.

54 Q. Huo, R. Leon, R.M. Petroff, G.D. Stucky, *Science* **1995**, *268*, 1324.

55 H. Rau, in Photochemistry and Photophysics, Vol. 4, J.F. Rabek (ed.), CRC Press, Florida **1990**, p. 110.

56 G.S. Kumar, D.C. Neckers, *Chem. Rev.* **1989**, *89*, 1915.

57 J.I. Anzai, T. Osa, *Tetrahedron* **1994**, *50*, 4039.

58 C. Schomburg, D. Wöhrle, G. Schulz-Ekloff, *Zeolites* **1996**, *17*, 232.

59 L.T. Yuen, J.S. Geilfuss, S.I. Zones, *Micropor. Mater.* **1997**, *12*, 229.

60 K. Hoffmann, F. Marlow, J. Caro, *Adv. Mater.* **1997**, *9*, 567.

61 F. Marlow, K. Hoffmann, *Ber. Bunsenges. Phys. Chem.* **1997**, *101*, 1731.

62 K. Hoffmann, U. Resch-Genger, F. Marlow, *Micropor. Mesopor. Mater.* **2000**, *41*, 99.

63 M. Ogawa, *Chem. Mater.* **1996**, *8*, 1347.

64 M. Ogawa, A. Ishikawa, *J. Mater. Chem.* **1998**, *8*, 463.

65 M. Ogawa, T. Ishii, N. Miyamoto, K. Kuroda, *Adv. Mater.* **2001**, *13*, 1107.

66 T. Fujita, N. Iyi, Z. Klapyta, *Mater. Res. Bull.* **2001**, *36*, 557.

67 S.K. Doughty, G.J. Simpson, K.L. Rowlen, *J. Am. Chem. Soc.* **1998**, *120*, 7997.

68 M. Shimomura, R. Ando, T. Kunitake, *Ber. Bunsenges. Phys. Chem.* **1983**, *87*, 1134.

69 M. Shimomura, S. Aiba, *Langmuir* **1995**, *11*, 969.

70 M. Ogawa, J. Mori, K. Kuroda, *Stud. Surf. Sci. Catal.* **2000**, *129*, 865.

71 R. Ryoo, J.M. Kim, C.H. Shin, *J. Phys. Chem.* **1996**, *100*, 17 713.

72 G. Kortüm, Reflektionsspektroskopie, Springer, Berlin **1969**, p. 91.

73 J. Griffiths, *Chem. Soc. Rev.* **1972**, *1*, 481.

74 M. Kasha, *Radiat. Res.* **1963**, *20*, 55.

75 E.G. McRae, M. Kasha, *J. Chem. Phys.* **1958, 721.**

76 E.G. McRae, M. Kasha, in Physical Processes in Radiation Biology, A.G. Augenstein, R. Mason, R. Rosenberg (eds.), Academic Press, New York **1964**, p. 17.

77 K. Okuyama, H. Watanabe, M. Shimomura, K. Hirabayashi, T. Kunitake, T. Kajiyama, N. Yasuoka, *Bull. Chem. Soc. Japan* **1986**, *59*, 3351.

78 J.C. Vartuli, C.T. Kresge, W.J. Roth, S.B. McCullen, J.S. Beck, K.D. Schmitt, M.E. Leonowicz, J.D. Lutner, E.W. Sheppard, in Advanced Catalysts and Nanostructured Materials, W.R. Moser (ed.), Academic Press, San Diego **1996**, p. 1.

79 P. Behrens, O. Oellrich, A.M. Glaue, N. Pfänder, R. Schlögl, in preparation.

80 C.D. Eisenbach, *Makromol. Chem.* **1978**, *179*, 2489.

81 I. Mita, K. Horie, K. Hirao, *Macromolecules* **1989**, *22*, 558.

82 M. Ueda, H. Kim, K. Ichimura, *Chem. Mater.* **1994**, *6*, 1771.

83 T. Sato, Y. Ozaki, K. Iriyama, *Langmuir* **1994**, *10*, 2363.

84 T. Seki, T. Fukuchi, K. Ichimura, *Langmuir* **2000**, *16*, 3564.

8
Metal-Oxide Species in Molecular Sieves: Materials for Optical Sensing of Reductive Gas Atmospheres

Michael Wark, Yücel Altindag, Gerd Grubert, Nils I. Jaeger, and Günter Schulz-Ekloff*

8.1
Introduction

Since air pollution is a serious problem and for protection against explosions, the detection of flammable gases in air is a matter of considerable current interest and importance [1,2]. Since most of the reductive (oxygen consuming) gases, in particular H_2, CO, or nitrous oxides, are in addition to their flammable nature also harmful at very low concentrations, several techniques for their sensing are used in various fields, such as chemical and clinical analysis and environmental monitoring [3,4]. With regard to the development of gas-sensitive layers, metal-oxide layers are widely employed as active compounds in resistivity sensors for reducing gases [5,6], for example, ZrO_2 in the λ probe for exhaust emission control [7], thin films of TiO_2 [8,9], or SnO_2 in the Taguchi sensor [10]. To achieve further miniaturization of sensing systems, research is focussed on the development of metal-oxide based materials in which the surface to volume ratio is drastically increased [11]. Several attempts to develop TiO_2 phases of high porosity (>600 m^2/g) by template-assisted self-organization [12,13] or electrodeposition [14] have been reported. The best sensing, however, was obtained with highly dispersed nanometer-sized metal-oxide particles [15]. For example, SnO_2 particles with diameters less than about 5 nm are able to monitor flammable gases in low ppm concentration [16]. Several routes exist for producing highly effective nanocrystalline SnO_2 including vacuum sputtering [17,18] and sol-gel synthesis [19,20].

Sol-gel processing is a very cheap and straightforward method for synthesizing metal-oxide nanoparticles with high dispersion and narrow size distribution in mono- and polycomponent systems [21,22]. However, although more expensive in their syntheses, since template molecules or supramolecular arrangements of surfactant are involved [23,24], molecular sieves (zeolites and mesoporous SiO_2 of the Si-MCM-41 type) are more suitable hosts for the storage of well-defined dispersions of very small metal-oxide clusters or nanoparticles [15,16]. At the same time their regular pore structures provide an excellent accessibility for the gases to be detected. Li and Kawi obtained H_2 selectivities over 90% for simple mechanical mixtures of SnO_2 and Si-Al-MCM-41 at 673 K by measuring the sensor activity by resistance measurements [27].

During the last decade various clusters of different metal oxides, such as ZnO [28], Fe_2O_3 [29,30], MoO_3 [31], WO_3 [32], have been encapsulated in the pore systems of molecular sieves. Most popular were TiO_2 [33–35] and V_2O_5 [36,37] due to applications in the catalysis of oxidation reactions [38,39]. The state of the art has recently been summarized by Weitkamp et al. [40].

However, if the metal-oxide particles are really encapsulated in the pores of molecular sieves, a serious problem had to be overcome regarding the detection of gases. The insulating properties of molecular sieves prevent the usual recording of changes in the resistivity as a function of the surrounding gas atmosphere. Optical detection has turned out to be a potential alternative method. It has been demonstrated that changes in the oxidation state of TiO_2 clusters stabilized in the pores of zeolites due to the presence of H_2 can be very quickly recorded by diffuse reflectance spectroscopy [41]. In a recent paper the use of optical detection with Pd-coated WO_3 sensors was reported for H_2 sensing in the high-throughput screening of hydrogen producing materials [42].

In the following the changes in the optical behavior of TiO_2, V_2O_5, and SnO_2 clusters and nanoparticles, encapsulated in the pores of faujasite-type zeolite Y or the channels of Si-MCM-41, are studied in dependence on the surrounding gas atmosphere and in relation to their structure and to changes in their average stoichiometry.

8.2 Titanium Oxide Clusters

The loading of the zeolite NaY (Si/Al = 2.7) and the mesoporous molecular sieve Si-MCM-41 was performed by chemical vapor deposition (CVD). For this the molecular sieves were dehydrated at 673 K for 12 h, loaded at 373–673 K for 15–90 min in a N_2 stream saturated with $TiCl_4$, hydrolyzed at 373 K in a N_2 stream saturated with water and finally calcined in a dry O_2 stream at 673 K for 4 h. Details of the CVD procedure are given elsewhere [26,41]. For some samples this treatment was repeated several times. Samples containing 4 nm TiO_2 particles, purchased from the Sachtleben Co. (Germany), a TS-1 zeolite [43,44] and NaY zeolite ion-exchanged with $(NH_4)_2TiO(C_2O_4)_2$ in aqueous solution [45–47] were used as reference materials.

Initially $TiCl_4$ binds to one or two OH groups of the molecular sieve, either in the supercages (pores with a diameter of 1.3 nm with tetrahedrally arranged 0.7 nm wide pore openings) of the NaY zeolite or in the 3.5 nm wide channels of Si-MCM-41. After hydrolysis mononuclear six-fold coordinated $Ti^{IV}O_x$ species with one or two oxygen bridges to the matrix are formed [26,41]. The index x signifies that OH groups, or framework oxygen in Si–O–Si bridges of the zeolite complete the coordination sphere of the titanium.

If loaded only once with $TiCl_4$, the zeolite Y samples exhibited only slight (10–15%) decreases of the BET values indicating a negligible influence on the porosity of the zeolite. Samples loaded by applying two or three CVD cycles, however, showed BET values around 400 $m^2 g^{-1}$ pointing to a damaged zeolite structure or a partial blockage of zeolite cages by titanium oxide species [41].

In X-ray diffractograms of singly loaded samples the relative intensities of the zeolite Y reflections remain constant, but the absolute intensities decrease slightly indicating a random distribution of the $Ti^{IV}O_x$ species. Reflections of titanium oxide around $2\theta = 25.1°$ (anatase reflection with highest intensity) could not be observed, so crystalline TiO_2 particles exceeding a diameter of about 3 nm are absent even on the external surface of the zeolite crystallites. This is consistent with the absence of a Raman signal at 144 cm^{-1}, typical for crystalline anatase particles of diameter > 2 nm.

After ion exchange with $(NH_4)_2TiO(C_2O_4)_2$ TiO^{2+} ions were found on cation sites of the zeolites [46]. The Ti/Si ratios determined by XPS were nearly equal to that of TS-1, in which Ti is located on framework positions of the zeolite, so no enrichment of titanium species at the outer surfaces of the NaY zeolite crystals was found [45]. For TiNaY samples loaded with $TiCl_4$ by CVD, however, a distinct enrichment of the titanium species on the outer surface of the zeolite crystals was detected [41].

In the pores of Si-MCM-41 a tailored generation of titanium oxide species of uniform size without a substantial enrichment on the external surface of the host is possible by a repeated addition and hydrolysis of the titanium compound in consecutive steps [48]. Mononuclear Ti(IV) oxide species, Ti(IV) oxide oligomers, and anatase nanoparticles of a well-defined size up to 3 nm were generated.

Figure 1 represents normalized diffuse reflectance (DR) UV/vis spectra of TiNaY and Ti-MCM-41 loaded singly or triply by CVD, and 4 nm TiO_2 particles for com-

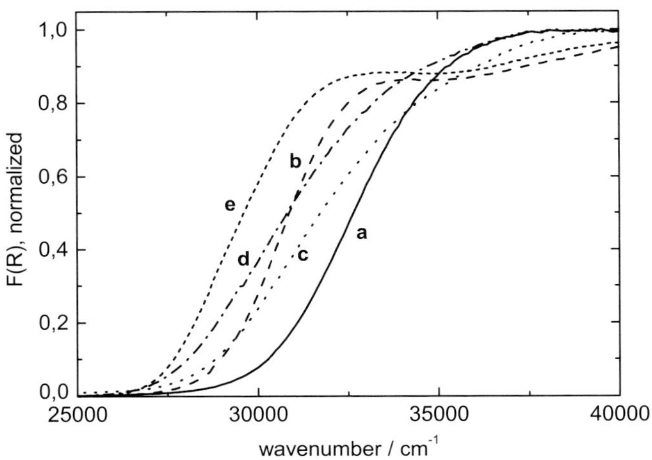

Fig. 1. Normalized DR-UV/vis spectra of singly loaded TiNaY (a), triply loaded TiNaY (b), singly loaded Ti-MCM-41 (c), triply loaded Ti-MCM-41 (d), and 4 nm TiO_2 nanoparticles (e). All CVD loadings were performed at 373 K. With permission from [58].

parison. The reflectance spectra were recorded at room temperature and were converted to Kubelka–Munk values ($F(R_\infty)$ [49,50] using a Teflon standard and the parent molecular sieves as reference materials [41,46]. For all the samples the absorption is significantly blue-shifted compared to that containing 4 nm TiO_2 particles. The most pronounced shift occurs for the singly loaded TiNaY zeolite in which the onset of the absorption is shifted by about 3000 cm^{-1}. The onset is defined as the intersection of the tangent through the point of inflection in the absorption edge with the abscissa. The absorption originates from electron charge transfer transitions from oxygen 2p levels to titanium 3d levels [51]. The $\pi O \rightarrow TiCT$ electronic transitions of more or less octahedrally coordinated species exhibit more than one absorption band between 34 000 and 50 000 cm^{-1} [52]. The position of the bands with lowest energy depends strongly on the distortion of the oxygen coordination sphere. The more pronounced the distortion of the octahedral coordination sphere, the higher the energy of the $\pi O \rightarrow TiCT$ transition [51]. Thus, the electronic absorption in the region around 34 000 cm^{-1} can be assigned to $\pi O \rightarrow TiCT$ in five- or six-fold coordinated $Ti^{IV}O_x$ species, which are monofunctionally bound to single isolated OH group in the zeolite. The absorption region around 45 000 cm^{-1} originates from to electronic transitions in $Ti^{IV}O_x$ species that are bifunctionally attached to vicinal silanol groups possessing a more distorted coordination sphere [26]. The relative amount of these species is increasing with the temperature of the CVD loading with $TiCl_4$.

With increasing number of CVD cycles the onset of the absorption edge is red-shifted. The shift is more pronounced for the TiNaY samples. This indicates a growth of the initially mononuclear $Ti^{IV}O_x$ species to $Ti^{IV}{}_yO_x$ clusters ($y > 1$). The clusters are smaller than 2 nm, since the absorption edge resulting from these species remained blue-shifted relative to that of the 4 nm TiO_2 particles, so their optical behavior is still ruled by size quantization effects [15,53]. Three-dimensional (3D) $Ti^{IV}{}_yO_x$ clusters are formed by a bonding of $TiCl_4$ to already anchored $Ti^{IV}O_x$ species. In the mesoporous Si-MCM-41 polynuclear $Ti_y{}^{IV}O_x$ ($y > 1$) species are found even after the first CVD loading.

The DR-UV/vis spectra of ion-exchanged TiNaY samples are comparable to that of TS-1 with a slightly red-shifted absorption maximum at 48 000 cm^{-1}. The oxygen coordination sphere around the Ti is significantly deviant from octahedral symmetry, owing to the interaction of TiO^{2+} on cation sites with the zeolite matrix.

For all the TiNaY samples binding energies for the Ti $2p_{3/2}$ electrons around 459.2 eV were measured, which is in between the values found for tetrahedrally coordinated framework Ti in TS-1 (460.0 eV) and for octahedrally coordinated Ti in crystalline anatase (458.6 eV) [43].

Further information regarding the coordination spheres of the different $Ti^{IV}O_x$ species was obtained from X-ray absorption near edge spectroscopy (XANES) and extended X-ray absorption fine structure (EXAFS) spectra. The pre-edge peaks in Ti K-edge XANES measurements are sensitive to the environment of the titanium atoms [54,55]. Octahedrally coordinated titanium species, such as those in anatase, exhibit multiple pre-edge peaks of low intensity between 4960 and 4979 eV; tetrahedrally coordinated titanium atoms such as in TS-1 show only one pre-edge peak of high intensity [56]. The TiNaY samples prepared by CVD exhibited peak positions, peak heights, and FWHM, which are again in between the values of TS-1 and anatase. This indicates again a deviant octahedral coordination. Samples containing $Ti_y^{IV}O_x$ clusters exhibited parameters of the center pre-edge peak that correspond closely to anatase pointing to a less distorted octahedral oxygen coordination sphere [46].

Ion exchanged samples exhibit unusual pre-edge parameters and an increased pre-edge peak. The significantly different chemical environment of TiO^{2+} ions on cationic sites near the tetrahedral Al centers in the zeolite framework leads to p–d orbital mixing thus allowing the otherwise-forbidden 1s–3d transition [46].

In EXAFS for all the Ti-oxide loaded zeolite samples the Ti–O distances and coordination numbers N are significantly decreased compared with anatase. The decrease is most pronounced for CVD samples loaded at 673 K. From this a proposal for the structures of the zeolite-hosted titanium oxide species prepared via the CVD method was deduced (Sketch 1). Mononuclear $Ti^{IV}O_x$ species, bifunctionally bound to two OH groups of the zeolite, mainly formed at high loading temperatures (higher than 573 K), reveal five-fold oxygen coordination, whereas samples loaded at lower temperatures contain predominately monofunctionally bound $Ti^{IV}O_x$ species with more octahedral like oxygen coordination sphere (Sketch 1A and 1B). Samples containing $Ti^{IV}{}_yO_x$ clusters, obtained by multiple loading at moderate temperatures, exhibit short Ti–Ti distances of 3.06 Å and show coordination numbers slightly more reduced compared with anatase [46]. This can be related to a high number of Ti atoms at the surface of the $Ti^{IV}{}_yO_x$ clusters, which are close to the pore walls of the growth restricting zeolite cages. Oxygen atoms of the zeolite framework form dative bonds to the Ti (Sketch 1C). The dative binding is favored, since it avoids a cleavage of Si–O–Si or Si–O–Al bonds in the zeolite. For the ion exchanged samples an observed short average Ti–O distance of about

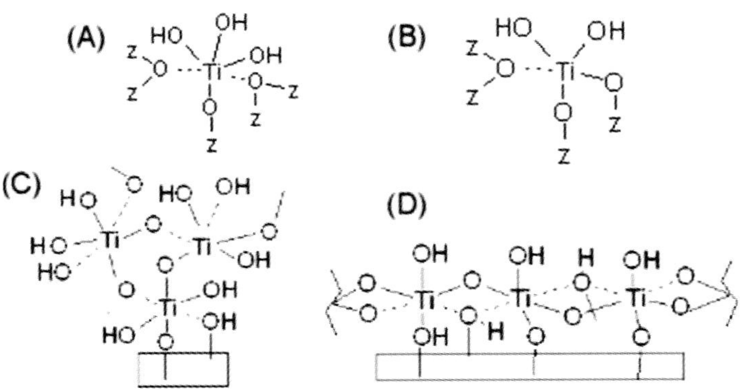

Sketch 1. Illustration of possible structures of mononuclear $Ti^{IV}O_x$ species monofunctionally bound with six-fold oxygen coordination (A) and, bifunctionally bound with five-fold coordination (B), $Ti^{IV}{}_yO_x$ cluster ($y > 1$) in the pores of the NaY zeolites (C), and 2D polynuclear $Ti^{IV}{}_yO_x$ species in the channels of Si-MCM-41 (D). The dashed lines assign dative bonds to the Ti atoms. Z stands for a Si or Al of the zeolite framework.

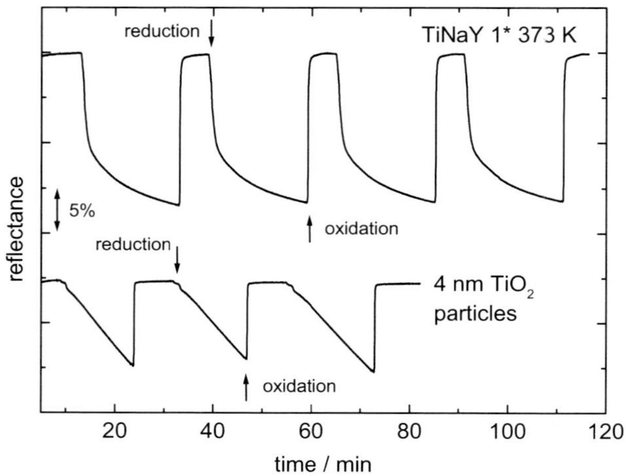

Fig. 2. Reflectance at 16 200 cm^{-1} with time during alternating exposure to reductive (10 vol.-% H_2 in Ar) or oxidative (10 vol.-% O_2 in Ar) atmospheres at 773 K for a TiNaY singly loaded by CVD at 373 K compared with 4 nm TiO_2 particles. The arrows mark changes of the gas atmosphere. With permission from [58].

1.77 Å indicates the formation of Ti=O bonds. The ammonium titanyl oxalate precursor interacts with cation sites of the zeolite and after its decomposition the titanyl bond is retained.

8.2.1 Redox Properties

For all the TiNaY or Ti-MCM-41 samples the evolution of the reflectance at a fixed wavenumber (16 200 cm^{-1}) with time during reduction with H_2 or CO and oxidation with O_2 (25 vol.-% each), performed at temperatures between 573 and 773 K, is completely reversible. This is demonstrated for singly loaded TiNaY in Fig. 2. The decreasing reflectance during reduction indicates an increasing absorption, which is proportional to the concentration of Ti^{III} formed in the samples [41]. Compared with the 4 nm TiO_2 particles, the changes in the reflectance of the mononuclear $Ti^{IV}O_x$ species occur much faster during reduction and the signal can be clearly distinguished from the noise after 5–10 s. The re-oxidation (healing of oxygen vacancies) is completed for all samples within about 15 s and, thus, occurs faster than the reduction. Owing to the bigger cross section of oxygen compared with hydrogen the presence of barriers for gas diffusion in the zeolite pore system can be ruled out.

The reduction kinetics of different samples were analyzed by closer examination of single reduction steps [41]. Since highly distorted mononuclear $Ti^{IV}O_x$ species can easily be reduced by H_2, the fastest response of the absorption is found for TiNaY samples singly loaded at high temperatures.

The Si-MCM-41 matrix possesses a relative high density of silanol groups: 2.5–3 OH groups exist per square nanometer of the inner surface of Si-MCM-41 [57]. Therefore, the $Ti^{IV}O_x$ species anchored on the inner pore surface can interact and form two-dimensional (2D) polynuclear $Ti_y{}^{IV}O_x$ ($y > 1$) species (Sketch 1D). Owing to the formation of Ti–O–Ti bridges the Ti atoms in these clusters are more protected against reduction by H_2. During multiple loading the number of 2D clusters increases because Ti species are anchored deeper in the channels (the anchoring starts presumably at the pore mouths), but growth into the third dimension does not occur owing to strong interactions with the Si-MCM-41 matrix. This opposite to the situation in the pores of zeolite Y in which growth into the third dimension is most probable, because an optimal interaction with the matrix can be achieved this way. In the 3D clusters the Ti^{IV} is shielded against reduction by H_2. Thus, an induction period is found for the reduction such as in free 4 nm TiO_2 particles [58]. In the channels of Si-MCM-41 a growth of 3D TiO_2 clusters is only possible, if the samples are hydrated and calcined after every CVD loading [48].

TS-1 and ion-exchanged TiNaY samples, however, show no significant differences in the DR-UV/vis spectra between the oxidized material and following exposure to H_2, demonstrating that the tetrahedrally coordinated titanium framework species in TS-1 and TiO^{2+} ions are not reducible.

For the NaY zeolites, loaded by CVD, three rate constants for the reduction kinetics for three different $Ti^{IV}O_x$ species were determined based on pseudo-first order kinetics [41]. The $Ti^{IV}O_x$ species bifunctionally bound to the zeolite reveal the highest reduction rate constant (k_1), followed by the $Ti^{IV}O_x$ species on the external surface of the zeolite crystallites (k_2). The lowest reducibility was found for $Ti^{IV}O_x$ species, which are monofunctionally bound to the zeolite (k_3). The reduction constants are listed in Table 1. The temperature dependence of the reduction constants results in straight lines in an Arrhenius plot. The activation energies for the reduction of the three different $Ti^{IV}O_x$ species and the pre-exponential factors are given in Table 1.

The activation of molecular hydrogen can be assumed to be the rate-limiting step. This is supported by the activation energies, which are of the same order of magnitude compared with energies found for the dissociative adsorption of hydrogen on TiO_2 with a defect structure (about 80 kJ mol^{-1}) [59,60]. For the bifunctionally bound $Ti^{IV}O_x$ species both the pre-exponential factor and the activation energy are responsible for the high k_1 value. This leads to the conclusion that the structure of the distorted bifunctionally bound species, with five-fold oxygen coordination, enhances the activation of hydrogen compared to the more octahedral-

Tab. 1. Reduction rate constants (k_i), activation energies (ΔE_a) and pre-exponential factors (A) for mononuclear TiO_x-species. The errors in ΔE_a and A are about $\pm$ 10%.

k_i [s^1]	773 K	748 K	723 K	698 K	673 K	ΔE_a [kJ mol^{-1}]	A [s^{-1}]
k_1	1.6	1	0.77	0.43	0.24	80	4.5×10^5
k_2	0.13	0.11	0.07	0.04	0.018	87	1.2×10^5
k_3	0.03	0.017	0.01	0.005	0.002	100	1.8×10^5

like monofunctionally bound species. The reduction rate of the latter can be drastically increased, if the H_2 molecules are split into H atoms at 3–5 nm Pt particles encapsulated in mesopores in a PtNaY zeolite [41]. In this case two zeolites pellets are used. Owing to a spillover effect the H atoms are able to migrate from the pellet containing Pt nanoparticles to that hosting Ti^{IV} oxide clusters.

8.2.2
Sensing Properties

The response times after which the reflectance decreases by 15% during reduction at 773 K in a H_2 atmosphere were found to be around 50 s for singly loaded TiNaY, 110 s for triply loaded TiNaY, and 80 s for Ti-MCM-41 samples. They are drastically longer (more than 500 s) for 4 nm TiO_2 particles [46,58]. In combination with a PtNaY zeolite the response time for the singly loaded TiNaY can be further decreased to 25 s [41]. An extrapolation of the response times to temperatures more relevant for sensing of exhaust gases and achievable without degradation of the composite material (1073 K) results in times between 0.3 and 0.7 s [45,46], which are similar to those found for solid state electrolyte sensors [61].

The detection of alterations in mixtures of hydrocarbons and O_2 is important for lean burning in vehicle motors (λ value). In this process CO is a intermediate, and thus the optical registration of deviations of CO:O_2 mixtures with zeolite-hosted $Ti^{IV}O_x$ species was tested at 773 K. For pulsing the composition of the mixtures, the concentrations of either CO or O_2 were varied for periods of 5 s from the lambda ratio of O_2:CO = 1:2, used as starting composition. Figure 3 demonstrates the possibility of monitoring deviations in the λ ratio by measuring the changes in reflectance. The response time is only a few seconds, and the changes in the reflectance are approximately proportional to the alterations in the concentrations. The sensitivity appears to be low. This arises, however, mainly due to a relatively large dead volume in the test apparatus leading to a broadening of the gas inlet pulse.

8.3
Tin Oxide Clusters

8.3.1
Tin Oxide Nanoparticles in Zeolites

In the preparation of SnO_2 nanoparticles in NaY zeolites by CVD with $SnCl_4$, no loss of zeolite crystallinity has been observed either by X-ray diffraction or by N_2 adsorption at 77 K up to tin oxide contents of about 2 wt.-% Sn. The inner surface area of the SnO_2-loaded zeolites around 800 $\pm$ 30 m^2g^{-1} is very close to that of the parent NaY zeolite (820 m^2g^{-1}). For higher loading, however, the crystallinity is strongly decreased and for tin contents beyond 3 wt.-% the zeolite matrix is destroyed [62].

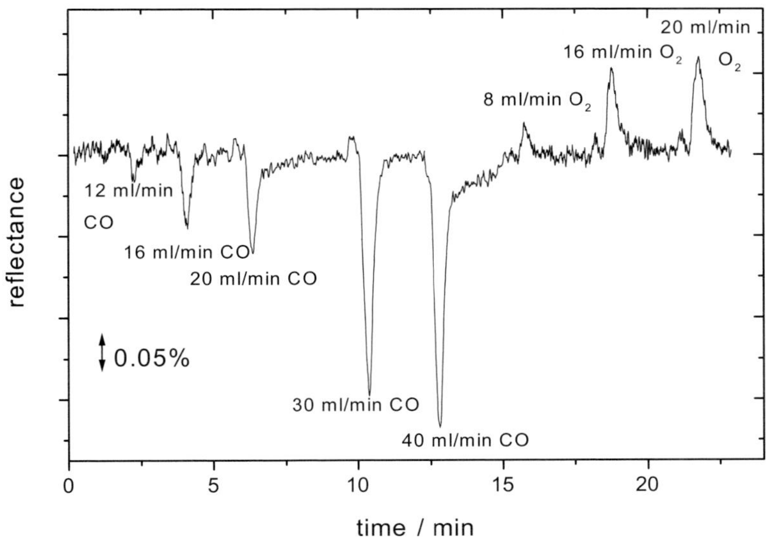

Fig. 3. Monitoring of pulsed deviations from a stoichiometric mixture for the combustion of CO (10 mL/min CO; 5 mL/min O_2 in Ar) over a pellet of TiNaY singly loaded by CVD at 673 K by changes in the reflectance at a temperature of 673 K. After [46].

Higher loading of NaY zeolites with SnO_2 nanoparticles without noticeable destruction of the zeolite matrix can be achieved by impregnation, for example with $Sn(OAc)_4$, and subsequent calcination with dry oxygen at 673 K [63]. No distinct loss of crystallinity was detected up to 10 wt.-% Sn by XRD and N_2 physisorption. However, adsorption/desorption isotherms with N_2 indicate the formation of mesopores. Although the introduced acetate anions possess a higher basicity than chloride ions, for example, the protons formed in low quantity during the calcination step are not fully neutralized and attack the zeolite framework [64].

The sizes of the embedded SnO_2 particles, deduced from transmission electron microscopy (TEM) and from the blue-shift of the absorption edge observed in DR-UV/vis spectra, depend strongly on the preparation method. The DR-UV/vis spectra of samples prepared by CVD show the most distinct blue-shifts compared with bulk SnO_2, indicating the presence of very small particles of less than 1 nm in diameter. Also no particles were observable by TEM after calcination, which led to the conclusion that $SnCl_4$ in the adsorption step is bound to silanol groups of the zeolite in molecular dispersion. During calcination in O_2 the chloride is removed from these SnO_xCl_y units and molecularly dispersed SnO_2 units are formed.

After introduction via ion exchange most of the particles have diameters of around 5 nm; impregnation leads to particles 3–5 nm in diameter. A typical TEM micrograph is shown in Fig. 4. X-ray photoelectron spectroscopy (XPS) gave Sn/Si ratios at the surface of the zeolite crystals, which are increased by less than 10%

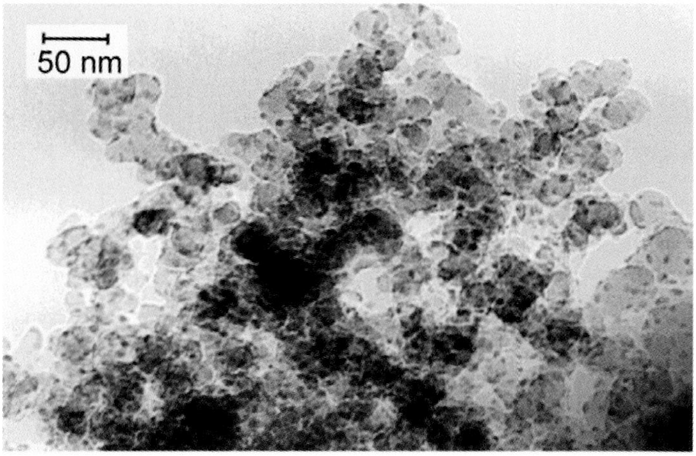

Fig. 4. TEM micrograph of a NaY zeolite loaded with SnO_2 nanoparticles (8 wt.-%) by impregnation. The embedded SnO_2 particles are mainly 3–5 nm. With permission from [63].

compared with ratios found by atomic absorption spectroscopy after complete dissolution of the Sn-loaded zeolite crystallites in acidic (HF/HNO_3) solution [65], that is, tin oxides have not been precipitated on the outer surface of the zeolite crystals. Thus, SnO_2 particles with diameters up to 5 nm are stabilized inside the zeolite framework in the mesopores formed during calcination [63]. As found by impedance spectroscopy, electron hopping is possible between SnO_2 particles hosted in neighboring mesopores [66].

The presence of reducing gases can be monitored via the reflectance of the samples in the wavenumber range 16 000–17 000 cm^{-1}. The time-dependent decreases in the reflectance, which reflect the formation of oxygen vacancies in the SnO_2 nanoparticles, were found to be most pronounced for the samples prepared by impregnation [63]. These samples also showed the shortest induction period (the time passing between exposure to CO atmosphere and the start in the decrease of the reflectance) as well as the highest rate of the decrease in reflectance. Presumably, the SnO_2 nanoparticles in mesopores are mainly located in the outer shell of the zeolite crystals [62]. Therefore, a fast diffusion of the gases to the SnO_2 nanoparticles occurs, which ensures that a large fraction of the nanoparticles reacts.

In contrast, samples prepared by CVD showed a very slow initial decrease and a relatively small total change of the reflectance as well as incomplete reversibility. Although the SnO_2 nanoparticles formed by CVD are at less than 1 nm much smaller than that formed by impregnation, they cannot be reduced very effectively, since they interact strongly with the zeolite matrix due to a large number of intrinsic defects [63].

If the measurements are performed in air or if different partial pressures of O_2 are applied during the *in situ* DR-UV/vis studies, less pronounced decreases in the

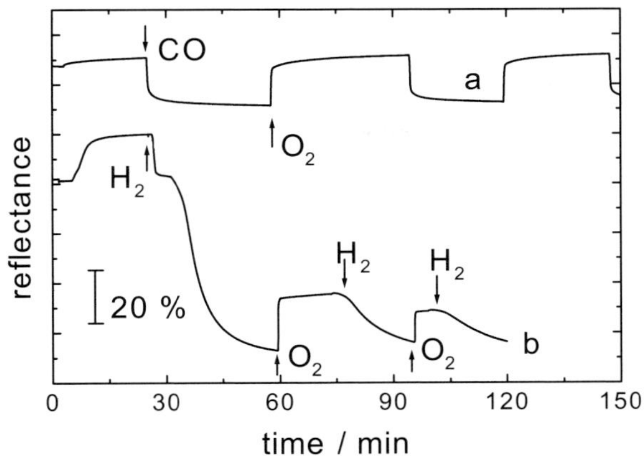

Fig. 5. Time dependent reflectance at 16 200 cm^{-1} of SnO_2 nanoparticles in zeolite Y loaded by impregnation (8 wt.-% Sn) during alternating exposure of CO and O_2 (a) and H_2 and O_2 (b) at 673 K. The arrows mark changes of the gas atmosphere. With permission from [63].

reflectance have been found since an equilibrium between the formation of oxygen vacancies, depending on the partial pressure of CO, and the healing of oxygen vacancies, depending on the partial pressure of O_2, is established [67]. In air, concentrations down to 50 ppm can be detected unambiguously [62].

For different reducing gases strong alterations concerning the reversibility of the reduction were found. While the effect of the exposure to CO can be completely reversed (Fig. 5), re-oxidation of SnO_2 nanoparticles reduced with H_2 is not complete and only about 30% of the initial reflectance can be restored by re-oxidation after three cycles. This indicates that a distinct amount of the tin remains in an oxidation state lower than Sn^{IV}.

The changes of the oxidation state of the tin atoms in the SnO_2 nanoparticles after reduction and re-oxidation were monitored by Mössbauer spectroscopy [63]. Mössbauer spectra of a sample prepared by impregnation (8 wt.-% Sn) in contact with various gases are presented in Fig. 6.

In the as-synthesized state and after the complete redox cycles with CO, isomer shifts very close to SnO_2 were found. After treatment with CO besides a remaining intensity of the Sn^{IV} signal, a doublet typical for the presence of Sn^{II} was clearly identified whereas the isomer shift was slightly higher than for isolated SnO indicating a strong interaction with the zeolite framework [63,68]. After *in situ* reduction in CO (3 mL min^{-1}) at 723 K for 2 h the average composition of the tin oxide nanoparticles was $SnO_{1.65}$. Reduction with H_2, however, led to the formation of a large amount (about 70%) of metallic Sn^0 (Fig. 6). The Sn^0 cannot be fully re-oxidized and the partial oxidation stops at the oxidation state Sn^{II}. Although Mössbauer spectra indicate the presence of only about 25% of not fully re-oxidized Sn, the reflectance is strongly suppressed, pointing to the presence of SnO par-

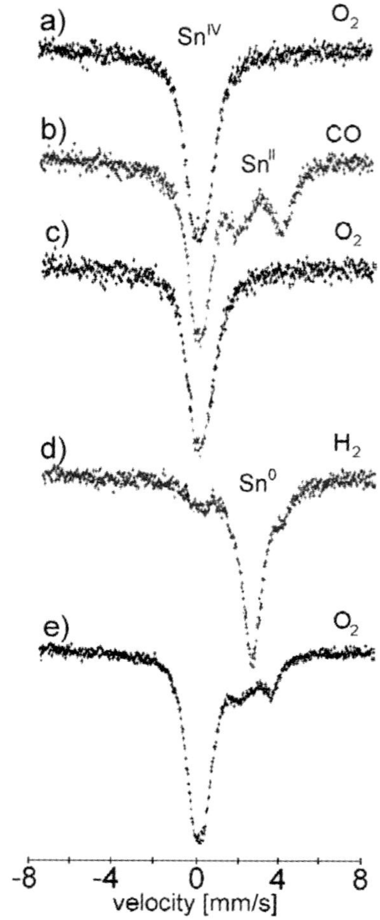

Fig. 6. Mössbauer spectra of SnO_2 nanoparticles in zeolite Y (impregnation, 8 wt.-% Sn) after (a) oxidation with O_2, (b) reduction with CO, (c) re-oxidation with O_2, (d) reduction with H_2, and (e) re-oxidation with O_2, each at 723 K. With permission from [63].

ticles, which also absorb strongly at 620 nm. Possibly a core of SnO in the particles is surrounded by a shell of SnO_2 preventing complete re-oxidation [63].

8.3.2 Tin Oxide Clusters in Mesoporous Materials

Nanosized SnO_2 particles can also be stabilized in SiO_2 matrices directly during a sol-gel process catalyzed with bases or acids using organic precursors such as $Sn(i\text{-}OPr)_4$ or in the pores of Si-MCM-41. For the latter, again, impregnation with

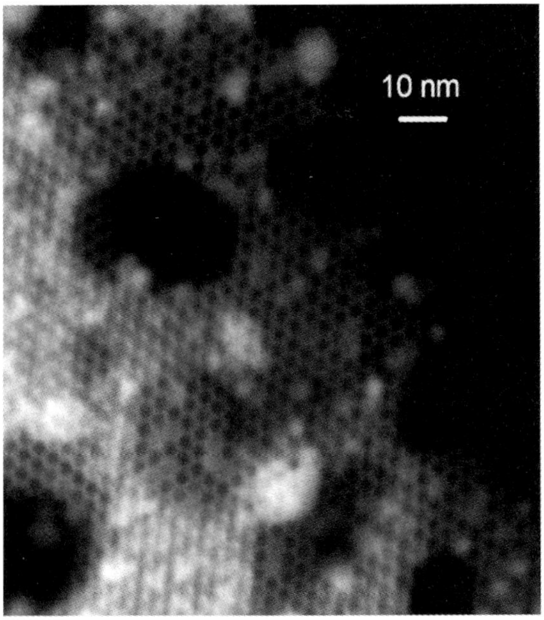

Fig. 7. HR-STEM micrograph of a Si-MCM-41 loaded with SnO_2 by impregnation. The gray pores indicate the presence of 2D SnO_2 nanocarpets inside.

$Sn(C_4H_9)_2(ac)_2$, which can performed up to a loading of 40 wt.-% Sn without degradation of the molecular sieve, is preferable to CVD [69]. The decrease of the mesopore volume of Si-MCM-41 with increasing SnO_2 loading is so small that blockage of the pore mouths can be ruled out. On HR-STEM micrographs no distinct SnO_2 nanoparticles can be observed. However, some of the pores appear gray, indicating an increase of the electron density in the interior and, thus, the presence of Sn oxide species (Fig. 7). Instead of clustering to 3D aggregates, the SnO_2 particles form carpet-like 2D structures on the inner walls of the matrix (Sketch 2) due to strong interaction of the introduced $Sn(C_4H_9)_2(ac)_2$ precursors with the silanol groups of the Si-MCM-41. During the hydrolysis step, the bound Sn^{IV} oxide species are crosslinked with each other without cleaving the bonds to the matrix [69].

In the case of the sol-gel route, a basic catalysis leads to materials with mesopores mainly between 4 and 8 nm wide, hosting SnO_2 particles with a broad size-distribution and mean diameters of 3–5 nm. The acidic route leads to relatively compact samples containing exclusively micropores [70]. In the latter case, TEM shows particles with diameters between 10 and 20 nm, indicating that the SnO_2 particles are part of the SiO_2 network formed in the sol-gel process.

In the case of sol-gel SnO_2/SiO_2 samples, distinct changes of reflectance during reduction/oxidation cycles were only detected for the sample prepared via basic catalysis. Here the SnO_2 nanoparticles are encapsulated in the mesopores of the SiO_2 network where they are highly accessible for the gas molecules. However, if

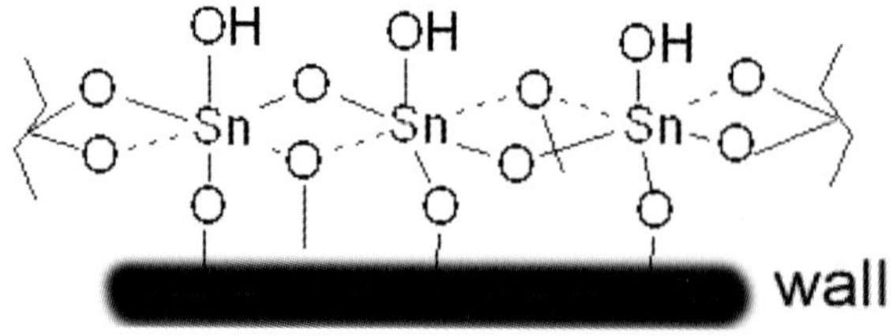

Sketch 2. 2D Sn(IV)-oxide species anchored onto the inner walls of the Si-MCM-41 channels. With permission from [69].

sol-gel SnO_2/SiO_2 and impregnated SnO_2/Si-MCM-41 samples were compared regarding their optical response towards CO, it became obvious that the reversibility, and thus the sensing stability, were much better for the Si-MCM-41 sample [69]. For the sol-gel sample the value of reflectance obtained after a complete cycle decreases continuously, whereas it remains almost constant for the Si-MCM-41 sample. This documents that in the sol-gel sample during the partial reduction and re-oxidation a further agglomeration of the particles takes place. In the Si-MCM-41, however, the carpet-like SnO_2 network is tightly bound and is structurally not altered during reduction and re-oxidation, although Sn^{II} species are formed. To determine the response time for CO under realistic conditions, pulses of different $CO:O_2$ ratios were added to this flow for 60 s and the response in the reflectance of the sample was recorded (Fig. 8). The change in the reflectance was found to increase proportionally with the increasing amount of CO in the mixture and the response time was always less than about 15 s [69]. The detection limit in air was

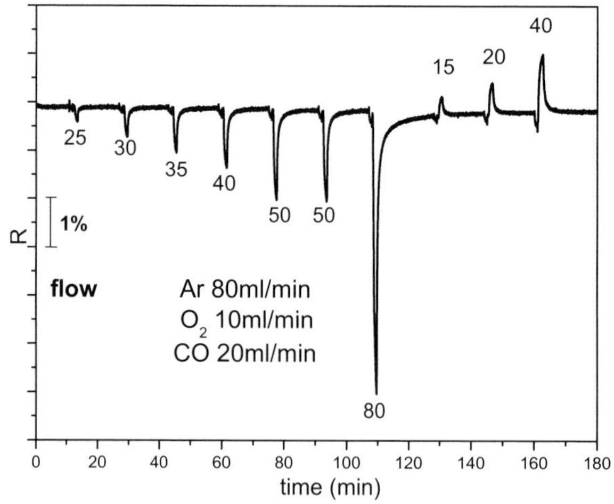

Fig. 8. Monitoring by changes in the reflectance, of pulsed deviations from a stoichiometric mixture of CO: O_2 = 2:1 over a pellet of SnO_2/Si-MCM-41 prepared by impregnation. The numbers give the altered fluxes of CO (downward signals) or O_2 (upward signals).

extremely low at 10 ppm CO; and H_2 and NH_3 could be detected down to concentrations of 5 and 50 ppm, respectively [71]. This demonstrates that SnO_2 particles stabilized as a highly dispersed and stable network in the regular pores of Si-MCM-41 possess a high potential for the development of sensors based on optical detection.

8.4 Vanadium Oxide Clusters

The reduction/oxidation behavior has been studied in detail for vanadium oxide clusters hosted in mesoporous molecular sieves of the M41S type [72]. The hydrothermal synthesis of V-MCM-41 was performed with $VOSO_4 \cdot 3\,H_2O$ dissolved in water as the vanadium source [72]. Multiple impregnation was carried out according to a method described by Van der Voort et al. with vanadyl acetylacetonate ($VO(acac)_2$) [73]. CVD was done with $VOCl_3$ as reactive agent [72]. All the samples were calcined in a dry oxygen stream at 673 K for 12 h (heating rate: 3 K min^{-1}) and rehydrated. Data extracted from N_2 adsorption isotherms and XRD showed that the adsorption and calcination of the vanadyl acetylacetonate complex on the surface of siliceous MCM-41 does not affect its structure. Loading with $VOCl_3$ by CVD for 5 min caused only slight damage, longer periods of treatment (30 min) or several cycles, however, led to substantial destruction. Direct hydrothermal synthesis of V-MCM-41 resulted in a less perfect material, containing about 10% of an amorphous phase [72,74].

For all the hydrated V-MCM-41 samples the absorption starts around 20 000 cm^{-1}, such as for bulk V_2O_5, exhibiting three maxima around 27 500, 40 000, and 47 500 cm^{-1}. The low-energy maximum around 27 500 cm^{-1} is blue-shifted compared to that of bulk V_2O_5 (22 000 cm^{-1}) [75]. Absorption spectra with very similar maxima positions in the high-energy region, around 38 500 and 47 500 cm^{-1}, had been found in Y-zeolites ion-exchanged or impregnated with different vanadium compounds [76]. After heating to 773 K the signal at 27 500 cm^{-1} disappeared almost completely and a corresponding increase of the intensity at wavelengths higher than about 35 000 cm^{-1} was observed. This indicates that a part of the V^V species changed from octahedral coordination, which was established by the adsorption of water, to tetrahedral coordination. In all the V-MCM-41 samples studied, most of the vanadium atoms are mononuclearly dispersed in the pores of the Si-MCM-41 support and tetrahedrally coordinated with oxygen in accordance with assignments given in the literature for $O \rightarrow V$ charge transfer bands at 34 000, 40 000, and 47 500 cm^{-1} [77,78]. From the intensities of a shoulder at 31 000 cm^{-1}, resulting from the absorbance of chain-like oligomeric clusters [79], it was deduced that the content of oligomeric V^V can be assumed to decrease in the order V-MCM-41 (CVD) > V-MCM-41 (impregnated) > V-MCM-41 (hydrothermally synthesized) [72]. Photoelectron spectra of the samples V-MCM-41 (CVD) and V-MCM-41 (impregnated) exhibited very weak signals of V $2p_{3/2}$ electrons with binding energies of 516.1 ± 0.1 eV. In V-ZSM-5 zeolites prepared by CVD a similar binding energy was assigned to very small vanadium oxide clusters in the pores of the zeolite [43].

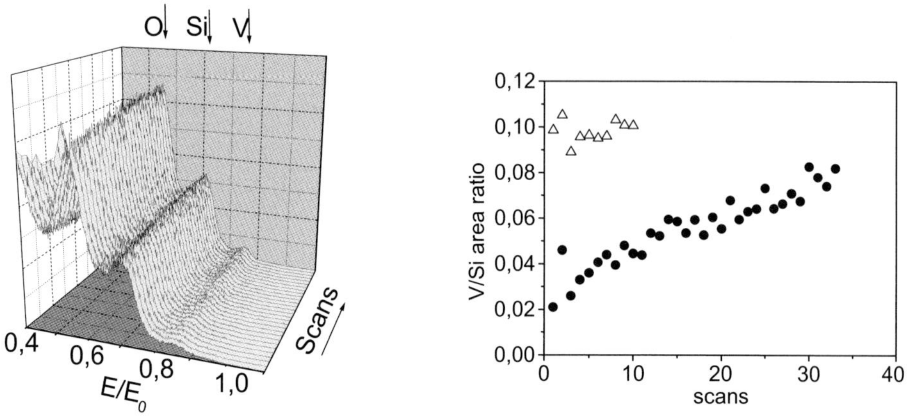

Fig. 9. Left: development of Ion scattering spectroscopy (ISS) signal intensities for the V $2p_{3/2}$, O 1s, and Si 2p electrons during a sputtering experiment with He^+ ions. Right: V/Si area ratios resulting from the ISS experiments with a MCM-41 four-fold impregnated with vanadyl acetylacetonate; ISS of the as-synthesized sample (•) and ISS after prior sputtering with He^+ ions for 2 h (△). With permission from [80].

Since all the V/Si-ratios found by XPS were much lower than those determined by chemical analysis of the bulk, the external surface was found to be depleted of vanadium [72]. This result was further confirmed by sputtering experiments with He^+ ions [80]. Figure 9 shows that with increasing number of sputter scans the intensity of the vanadium signal increases, whereas the signal intensities of Si and O remain almost constant. With every scan approximately 0.1 layers of Si-MCM-41 are removed. The estimated V/Si area ratio increases monotonically and is tripled after about 30 scans (Fig. 9), indicating that the progressing removal of SiO_2 layers exposes the vanadium species located in the pores of the V-MCM-41 to the surface.

8.4.1
Reduction and Re-oxidation

Mononuclear as well as oligomeric vanadium oxide species in the pores of MCM-41 can be reduced by H_2. The reduction results in: a decrease of the absorption at wavenumbers higher than about 30 000 cm^{-1} and the appearance of a distinct, very broad absorption between 12 000 and 20 000 cm^{-1} with a weak maximum around 17 000 cm^{-1}. The evolution of the absorption intensity at a fixed wavenumber (17 000 cm^{-1}) with time during reduction and re-oxidation is depicted in Fig. 10 for a V-MCM-41 prepared by CVD. Changes in the absorption intensity during reduction can be clearly distinguished from the noise signal after 5–10 s, whereas after exposure to oxygen the signal alterations appear within 1 s. The changes in $F(R)$ are reversible following a complete cycle and are proportional to the concentration of reducible vanadium (V^{red}) in the sample, which according to H_2-TPR re-

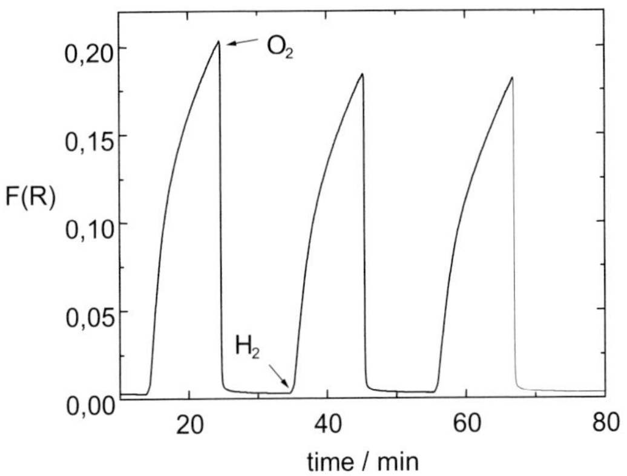

Fig. 10. $F(R)$ values for reduction and oxidation cycles of V-MCM-41(CVD) at 773 K, taken at 17 000 cm^{-1}. The $F(R)$ values are proportional to the concentrations of V^{III} and/or V^{IV}. With permission from [72].

sults could be calculated to be about 95, 70, and 30% in the samples V-MCM-41 (impregnated), V-MCM-41 (CVD), and V-MCM-41 (hydrothermally synthesized), respectively. In agreement with presumptions expressed for V-MCM-41 on the basis of ESR and NMR results [81], V^V species buried in the walls of Si-MCM-41 or in amorphous SiO_2 could not be reduced as long as they were tetrahedrally saturated with O–Si groups [72,82]. Mononuclearly bound VO_x species were reducible with a lower rate than oligomeric clusters. In the samples prepared by CVD the amount of oligomeric clusters decreases with the number of loading, whereas it increases with the number of impregnations [80]. During a second or third CVD loading, some of the initially weakly bound and volatile V-oxide clusters are possibly leached out. The high volatility also explains why during the first cycles of reduction and re-oxidation the number of reducible vanadium species decreases slightly from cycle to cycle (Fig. 10).

8.5 Conclusions

Highly dispersed mononuclear and clustered Ti-, V-, and Sn-oxide species can be stabilized in the pores of FAU-type zeolites as well as of mesoporous molecular sieves of the M41S type by post-synthetic treatment. The structure and reduction behavior of the embedded metal-oxide clusters can be tailored by the experimental methods applied for their encapsulation: hydrothermal synthesis, CVD, ion exchange, or impregnation. A combination of different analytical methods, such as DR-UV/vis, XANES/EXAFS, XPS, Mössbauer spectroscopy, TPR, XRD, TEM physisorption, as well as time-resolved optical reduction/oxidation studies has led to a complete analysis of the structural properties of the encapsulated clusters and the determination of their reactivity. The optical changes can be correlated to the number of oxygen vacancies in the particles.

The observed properties of the molecular sieve supported metal-oxide species in H_2, CO, and O_2 atmospheres at 773 K, that is, (1) their stability as demonstrated by a complete reversibility of the extinction for a large number of redox cycles, (2) their high sensitivities down to concentrations of the reductive gases of less than 10 ppm, (3) their short response times of only several seconds, and (4) the prospect of a more feasible miniaturization compared to most of the established oxygen sensors, render these composites interesting materials for alternative gas sensing using optical detection.

Best sensing results have been found for (1) distorted mononuclear Ti^{IV} oxide units bound in the supercages of zeolite NaY by CVD at elevated temperatures, (2) 3 nm SnO_2 particles hosted in mesopores of a NaY, and (3) carpet-like 2D cluster networks of TiO_2 or SnO_2 in the channels of Si-MCM-41. In the case of V-oxides, however, their high volatility diminishes the stability of the composites.

Acknowledgements

Financial support by the German Science Foundation (DFG, SCHU 426-9, and WA 1116-2) is gratefully acknowledged. We thank Dr. C. Kuebel (FEI Company, Eindhoven, Netherlands) for taking the TEM micrograph of SnO_2/Si-MCM-41.

References

1 F. Quaranta, R. Rella, P. Sicliano, S. Capone, M. Epifani, L. Vasenelli, A. Licciuli, A. Zocco, *Sens. Act. B* **1999**, *58*, 350.
2 K. Wada, M. Egashira, *Sens. Act. B* **2000**, *62*, 211.
3. Y. Amao, K. Miyakawa, I. Okura, *J. Mater. Chem.* **2000**, *10*, 305 and references therein.
4 S.R. Davis, A.V. Chadwick, J.D. Wright, *J. Mater. Chem.* **1998**, *8*, 2065.
5 V. Lantto, in Gas Sensors: Principles, Operations and Developments, G. Sberveglieri (ed.), Kluwer, Dordrecht **1992**, p. 117.
6 D.E. Williams, in Solid State Gas Sensors, P.T. Moseley, B.C. Torfield (eds.), Adam Hilger, Bristol **1987**, p. 71.
7 H. Schaumburg, Sensoren, Teubner, Stuttgart **1992**, p. 437.
8 H. Tang, K. Prasad, F. Levy, *Sens. Act. B* **1995**, *26–27*, 71.
9 W. Göpel, J. Hesse, J.N. Zemel, Sensors, VCH, Weinheim **1989**, p. 182.
10 N. Taguchi, UK Patent, 1280809, **1970**.
11 C. Xu, J. Tamaki, N. Miura, N. Yamazoe, *Sens. Act. B* **1991**, *3*, 147.
12 D.M. Antonelli, J.Y. Ying, *Angew. Chem. Int. Ed. Engl.* **1996**, *35*, 426.
13 Z. Zhong, Y. Yin, B. Gates, Y. Xia, *Adv. Mater.* **2000**, *12*, 206.
14 Y. Matsumoto, Y. Ishikawa, M. Nishida, S. Ii, *J. Phys. Chem. B* **2000**, *104*, 4202.
15 N. Serpone, D. Lawless, R. Khairutdinov, *J. Phys. Chem.* **1995**, *99*, 16 646.
16 S.R. Davis, A. Wilson, J.D. Wright, *Proc. IEE: Circ. Dev. Syst.* **1998**, *145*, 379.
17 V. Demarne, R. Sanjiers, in Gas Sensors: Principles, Operations and Developments, G. Sberveglieri (ed.), Kluwer, Dordrecht **1992**, p. 89.

18 G.B. Barbi, J.P. Santos, P. Serrini, P.N. Gibson, M.C. Horillo, L. Manes, *Sens. Act. B* **1995**, *25*, 559.

19 S.G. Ansari, P. Boroojerdian, S.K. Kulkarni, S.R. Sainar, R.N. Karekar, R.C. Aiyer, *J. Mater. Sci.: Mater. Electron.* **1996**, *7*, 267.

20 Z. Zhong, Y. Yin, B. Gates, Y. Xia, *Adv. Mater.* **2000**, *12*, 206.

21 C. Sanchez, F. Ribot, *New J. Chem.* **1994**, *18*, 1007.

22 C.J. Brinker, G.W. Scherer, *Sol-Gel Science: The Physics and Chemistry of Sol-Gel Processing*, Academic Press, San Diego **1990**, p. 60.

23 J.C. Jansen in Introduction to Zeolite Science and Practice, H. van Bekkum, E. Flanigan, P.A. Jacobs, J.C. Jansen (eds.), Elsevier, Amsterdam **2001**, p. 175.

24 J.Y. Ying, C.P. Mehnert, M.S. Wong, *Angew. Chem. Int. Ed.* **1999**, *38*, 58.

25 M. Wark, H.-J. Schwenn, M. Warnken, N.I. Jaeger, B. Boddenberg, *Stud. Surf. Sci. Catal.* **1995**, *97*, 205.

26 J. Klaas, G. Schulz-Ekloff, N.I. Jaeger, *J. Phys. Chem. B* **1997**, *101*, 1305.

27 G. Li, S. Kawi, *Sens. Act. B* **1999**, *59*, 1.

28 H.B. Lee, H.M. Lim, C.S. Han, *Bull. Korean Chem. Soc.* **1998**, *19*, 1002.

29 K. Inamura, R. Iwamoto, A. Iino, T. Takyu, *J. Catal.* **1993**, *142*, 274.

30 M. Stockenhuber, M.J. Hudson, R.W. Joyner, *J. Phys. Chem. B* **1999**, *103*, 3370.

31 D. Wang, J.H. Lunsford, M.P. Rosynek, *J. Catal.* **1997**, *169*, 347.

32 K. Möller, T. Bein, S. Özkar, G.A. Ozin, *J. Phys. Chem.* **1991**, *95*, 5276.

33 V.S. Komarov, L.P. Shirinskaya, N. Bokhan, *Russ. J. Phys. Chem.* **1976**, *50*, 1478.

34 L. Machese, T. Maschmeyer, E. Gianotti, S. Coluccia, J.M. Thomas, *J. Phys. Chem. B* **1997**, *101*, 8836.

35 K. Ikeue, H. Yamashita, M. Anpo, T. Takewaki, *J. Phys. Chem. B* **2001**, *105*, 8350.

36 K.J. Chao, C.N. Wu, H. He, J. Klinowski, L. Kevan, *J. Phys. Chem. B* **1997**, *101*, 6341.

37 B.I. Whittington, J.R. Anderson, *J. Phys. Chem.* **1993**, *97*, 1032.

38 A. Corma, *Chem. Rev.* **1997**, *97*, 2373.

39 A. Tuel, *Stud. Surf. Sci. Catal.* **1998**, *117*, 159.

40 J. Weitkamp, U. Rymsa, M. Wark, G. Schulz-Ekloff, in Molecular Sieves–Science and Technology, Vol. 3: Modification, J. Weitkamp, H.G. Karge (eds.), Springer, Berlin **2002**, p. 339–414.

41 G. Grubert, M. Wark, N.I. Jaeger, G. Schulz-Ekloff, O.P. Tkachenko, *J. Phys. Chem. B* **1998**, *102*, 1665.

42 T.M. Jaramillo, A. Ivanovskaya, E.W. McFarland, *J. Comb. Chem.* **2002**, *4*, 17.

43 M. Wark, M. Koch, A. Brückner, W. Grünert, *J. Chem. Soc.: Faraday Trans.* **1998**, *94*, 2033.

44 M.A. Uguina, D.P. Serano, G. Ovejero, R. Van Griecken, M. Camacho, *Appl. Catal. A* **1995**, *124*,0 391.

45 G. Grubert, PhD Thesis, University of Bremen, Shaker Verlag, Aachen **1999**.

46 G. Grubert, M. Stockenhuber, O.P. Tkachenko, M. Wark, *Chem. Mater.* **2002**, *14*, 2458.

47 X. Liu, K.K. Iu, J.K. Thomas, *J. Chem. Soc.: Faraday Trans. I* **1993**, *89*, 1861.

48 P. Prochnow, G. Schulz-Ekloff, M. Wark, J. K. Thomas, A. Zukal, J. Rathousky, in *Proc. 13th Int. Zeolite Conf.*, A. Galarneau, F. Di Renzo, F. Fajula; J. Vedrine (eds.), *Stud. Surf. Sci. Catal.* **2001**, *135*, 359 (22-P-06).

49 P. Kubelka, F. Munk, *Z. Tech. Phys.* **1931**, *12*, 593.

50 G. Kortüm, G. Schreyer, *Angew. Chem.* **1955**, *67*, 694.

51 C. K. Jørgensen, *Proc. Inorg. Chem.* **1970**, 12, 101.

52 A.B.P. Lever, Inorganic Electronic Spectroscopy, Elsevier, Amsterdam **1984**.

53 B. Aronson, A.E. Blanford, A. Stein, *Chem. Mater.* **1997**, *9*, 2842.

54 W. Zhang, M. Fröba, J. Wang, P.T. Tanev, J. Wong, T. Pinnavaia, *J. Am. Chem. Soc.* **1996**, *118*, 9164.

55 P. Behrens, J. Felsche, S. Vetter, G. Schulz-Ekloff, N.I. Jaeger, W. Niemann, *J. Chem. Soc.: Chem. Commun.* **1991**, 678.

56 M. Anpo, H. Yamashita, Y. Ichihashi, N. Fujii, M. Honda, *J. Phys. Chem.* **1997**, *101*, 2632.

57 X.S. Zhao, G.Q. Lu, G.J. Millar, H.Y. Zu, *J. Phys. Chem. B* **1997**, *101*, 6526

58 M. Wark, G. Grubert, in *Proc. EUROMAT '99 Conf.*, Vol. 9, Interface Controlled Materials, M. Rühle, H. Gleiter (eds.), Wiley-VCH, Munich **2000**, p. 154.

59 G. B. Raupp, J.A. Dumesic, *J. Phys. Chem.* **1985**, *89*, 5240.

60 W. Göpel, G. Rocker, R. Feierabend, *Phys. Rev.* **1983**, *B 28*, 3427.

61 H. Schaumburg, Sensoranwendungen, Teubner, Stuttgart **1995**, p. 343

62 M. Warnken, M. Wark, G. Grubert, N.I. Jaeger, in Dresdner Berichte zur Sensorforschung, Vol. 5, Chemie- und Biosensoren, J.P. Baselt, G. Gerlach, W. Göpel (eds.), Dresden University Press **1998**, p. 39.

63 M. Warnken, K. Lazar, M. Wark, *Phys. Chem. Chem. Phys.* **2001**, *3*, 1870.

64 G. Schulz-Ekloff, *Stud. Surf. Sci. Catal.* **1991**, *69*, 65.

65 Y. Altindag, Z. Bastl, M. Wark, unpublished results.

66 H.-J. Schwenn, M. Wark, G. Schulz-Ekloff, H. Wiggers, U. Simon, *Colloid Polym. Sci.* **1997**, *275*, 91.

67 M. Warnken, G. Grubert, N.I. Jaeger, M. Wark, *Proc. 12th Int. Zeolite Conf.*, Vol. III, M.M.J. Treacy, B.K. Marcus, M.E. Bisher, J.B. Higgins (eds.), MRS, Pennyslvania **1999**, p. 2249.

68 M.C. Hobson Jr, S.L. Goresh, G.P. Khare, *J. Catal.* **1993**, *142*, 641

69 Y. Altindag, A. Jitianu, M. Wark, Nanoporous Materials III, A. Sayari, M. Jaroniec (eds.), *Stud. Surf. Sci. Catal.* **2002**, *141*, 653.

70 A. Jitianu, Y. Altindag, M. Zaharescu, M. Wark, *J. Sol-Gel Sci. Tech.* **2003**, *26*, 483.

71 Y. Altindag, M. Wark, unpublished results.

72 G. Grubert, J. Rathousky, G. Schulz-Ekloff, M. Wark, A. Zukal, *Micropor. Mesopor. Mater.* **1998**, *22*, 225.

73 P. Van Der Voort, I.V. Babitch, P.J. Grobet, A.A. Verbreckmoes, E.F. Vansant, *J. Chem. Soc.: Faraday Trans.* **1996**, *92*, 3635.

74 S. Gontier, A. Tuel, *Microporous Materials* **1995**, *5*, 161.

75 C.K. Jørgensen, *Mol. Phys.* **1959**, *2*, 309.

76 C.A. Trujillo, U. Navarro Uribe, P.-P. Knops-Gerrits, L.A. Oviedo A., P.A. Jacobs, *J. Catal.* **1997**, *168*, 1.

77 M. Schraml-Marth, A. Wokaun, M. Pohl, H.-L. Krauss, *J. Chem. Soc.: Faraday Trans.* **1991**, *87*, 2635.

78 G. Centi, S. Perathoner, F. Trifiro, A. Aboukais, C.F. Auissi, M. Guelton, *J. Phys. Chem.* **1992**, *96*, 2617.

79 G. Lischke, W. Hanke, H.-G. Jerschkewitz, G. Öhlmann, *J. Catal.* **1985**, *91*, 54.

80 G. Grubert, W. Grünert, J. Rathousky, A. Zukal, G. Schulz-Ekloff, M. Wark; in *Proc. 12th Int. Zeolite Conf.*, Vol. II, M.M.J. Treacy, B.K. Marcus, M.E. Bisher, J.B. Higgins (eds.), MRS, Pennyslvania **1999**, p. 825.

81 Z. Luan, J. Xu, H. He, J. Klinowski, L. Kevan, *J. Phys. Chem.* **1996**, *100*, 19 595.

82 G. Grubert, M. Wark, M. Koch, G. Schulz-Ekloff, *Stud. Surf. Sci. Catal.* **1996**, *105*, 1077.

9
From Stoichiometric Carbonyl Complexes to Stable Zeolite-Supported Subnanometer Platinum Clusters of Defined Size

Martin Beneke, Nils I. Jaeger, and Günter Schulz-Ekloff*

9.1
Introduction

Preparation and stabilization of small metal clusters on various supports as well as characterization of their size-dependent electronic, adsorptive, and catalytic properties are of continued interest [1–5]. Standard techniques used for this purpose usually result in a rather broad size distribution even for particles synthesized within the cages of zeolite hosts. Therefore, the unambiguous characterization of the properties of very small supported metallic clusters is often hampered by a size distribution and desired particle sizes can usually be realized only with considerable effort [6]. Ligand-stabilized metal clusters both of variable and uniform size [2,5] are available as precursors for the preparation of nanoparticle catalysts of uniform cluster size. Provided that agglomeration and sintering can be avoided in the course of the removal of the ligands and the stable anchoring of the clusters to a support can be achieved, highly active and selective nanoparticle catalysts can be designed [7].

The well-defined cage and channel system of molecular sieves has been shown to be a valuable matrix for the preparation and stabilization of metal clusters of variable size up to a few nanometers in diameter and with narrow size distribution. Strong metal–support interactions can be avoided and the clusters remain accessible to suitable probe molecules. The state of the art has been reviewed [4]. As a promising route towards the preparation of well-defined zeolite-supported metal clusters, the "ship-in-the-bottle" synthesis of carbonyl complexes within the zeolite cages has been established in the last ten years for a number of metals, such as Pd [8,9], Ru [10], and Pt [11–13]. For anionic Pt-carbonyl complexes, the influence of the zeolite host (basicity, nature of counter ions) and of the preparation conditions (water content, CO partial pressure, temperature) on the size and stability of the complexes has been studied in considerable detail [13–15].

Based on the results of these studies, the conditions for the preparation of extraordinarily stable subnanometer Pt clusters within the channel and cage structure of zeolite hosts could be determined.

9.2 Chemistry Within Zeolite Cages

To date, two routes for the "ship-in-the-bottle" synthesis of Pt Chini complexes ($[Pt_3(CO)_6]_n{}^{2-}$) in zeolites, predominantly faujasites, have been reported. The first concerns the carbonylation of Pt^{2+} cations in zeolites by CO and traces of water [16–19], which can be expressed by the stoichiometry

$$3n\,Pt^{2+} + (9n+1)CO + (3n+1)H_2O \rightarrow [Pt_3(CO)_6]_n^{2-} + 2(3n+1)H^+ + (3n+1)CO_2 \quad (1)$$

The second route [11,14,15,17,20–23] is the direct carbonylation of the $[Pt(NH_3)_4]^{2+}$ cations by CO in the presence of small amounts of water. The carbonylation is assumed to proceed according to a scheme suggested previously [15]

$$3n[Pt(NH_3)_4]^{2+} + (3n+1)H_2O + (9n+1)CO \rightarrow [Pt_3(CO)_6]_n^{2-} + 2(3n+1)NH_4{}^+ + 2(3n-1)NH_3 + (3n+1)CO_2 \quad (2)$$

Water is assumed to play the dominant role providing H atoms and consequently protons through the WGS reaction [24].

The stoichiometry of platinum carbonyl dianions $[Pt_3(CO)_3(\mu-CO)_3]_n^{2-}$ is given by the number n (nuclearity) of stacked triangular building blocks, consisting of Pt_3 triangles connected by metal Pt–Pt bonds, in which each Pt atom is linked to one linear-bonded and two bridge-bonded CO ligands [25]. Nuclearities ranging from 1 to 10 are reported for dianionic complexes in solution [25], whereas n values of 2, 3, and 5 are reported for the Pt carbonyl complexes in faujasites, of which the clear assignment is impeded by the coexistence of complexes with different nuclearities [15].

9.2.1 Formation of Pt Carbonyls Monitored by FTIR, EXAFS, and UV/vis Spectroscopy

Figure 1 shows the monitoring of the progress of the Pt-carbonyl formation during the treatment of the $[Pt(NH_3)_4]^{2+}$-exchanged (6 wt.-% Pt) NaX with 5×10^4 Pa of CO at 373 K by infrared spectroscopy [26]. The sample was evacuated for 1 min at 298 K, which gives a mild dehydration prior to the carbonylation. The development of the bands for the stretching CO vibrations, assigned to the linear-bonded CO (2020–2050 cm^{-1}) and the bridge-bonded CO (1778–1808 cm^{-1}) that are characteristic of a $[Pt_3(CO)_6]_2^{2-}$ complex [14,15,25], are displayed. The shift of δ_s for the ammine ligands to higher wavenumber is caused by the loss of water ligands [23]. This results in a symmetry distortion of the Pt tetrammine complex. Stronger zeolite dehydration results in the shift of Pt–CO vibrations from 2048 to 2024 cm^{-1} and of bridged CO vibrations from 1778 to 1800 cm^{-1} [27]; owing to mild sample dehydration, the CO bands at 2049 and 1778 cm^{-1} in Fig. 1 are the most intense.

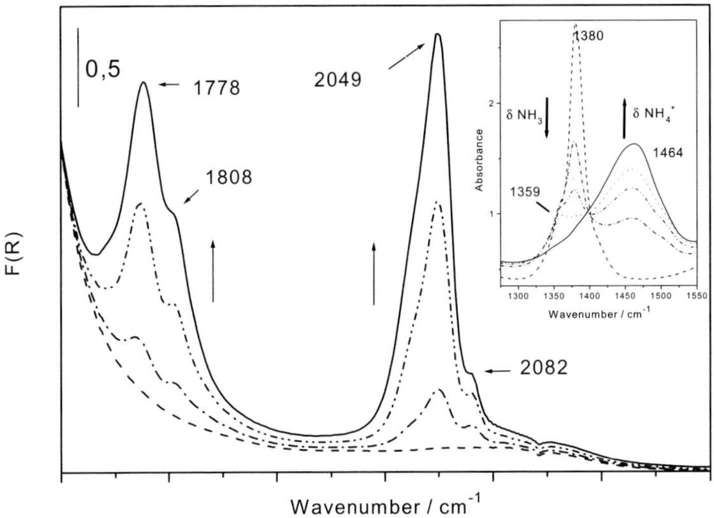

Fig. 1. Formation of $[Pt_3(CO)_6]_2^{2-}$ from $[Pt(NH_3)]_4^{2+}$ (6 wt.-% Pt in NaX) in CO of 5×10^4 Pa at 373 K after mild dehydration (298 K, evacuated 1 min), monitored by FTIR spectra: dashed, 2 min; solid line, 3,6, and 12 h (from bottom to top) solid bold, 43 h. Reprinted with permission from [26].

The isosbestic point at about 1400 cm^{-1} in the N–H vibrations indicates that the NH_3 to NH_4^+ conversion is close to thermodynamic equilibrium.

The platinum in $[Pt(NH_3)_4]^{2+}$-exchanged NaX is almost completely converted into a dianionic $[Pt_3(CO)_6]_2^{2-}$ complex upon reaction with CO. The virtually complete conversion of Pt tetrammine dications follows from the disappearance of the bands of the N–H deformation vibration of NH_3 ligands (1380–1348 cm^{-1}). Formation of the hexaplatinum carbonyl dianion is confirmed by the characteristic FTIR and UV/vis spectra [14,15,25,28].

Figure 2 depicts the normalized absorbance of the white line of the Pt L_3 edge in the course of the formation of the Chini complex. Spectra were taken in 30 min intervals (sampling number). The final measurement (sampling number 18) was done after thermal decomposition of the Chini complex in vacuum at 573 K. The upper curve represents the Pt L_3 edge of the Pt in the Chini complex, the lower curve the reference data of a Pt foil, measured in parallel in the same beam. Bars mark the standard measurement deviation.

The EXAFS results (Fig. 2) as well as results from XP-Spectroscopy [26] clearly indicate that the Pt atoms largely preserve their positive charge in the dianionic complex. The intensity of the white line can be used as a measure for the probability with which electrons can be excited from the $2p_{3/2}$ into the Pt 5d states that are not completely occupied. If the platinum is reduced to the zero valent state these states are more filled and the excitation probability is small.

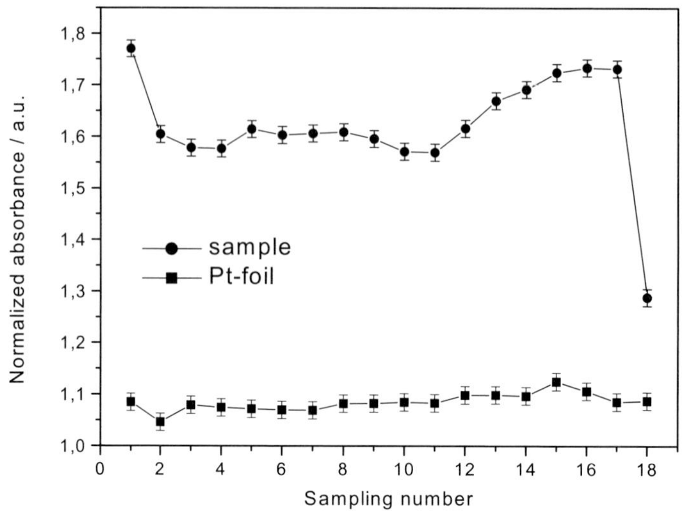

Fig. 2. Normalized absorbance of sample with 10 wt.-% of Pt and Pt-foil during EXAFS measurement. Sampling number: 1, vacuum 298 K; 2, vacuum 363 K; 10^5 Pa CO; (3–7) 363 K; (8–10) 393 K; (11–16) 423 K; 17, vacuum 453 K; 18, vacuum decomposition at 573 K. Reprinted with permission from [26].

The relative change in the intensity of the white line of the platinum can be correlated with the preparation steps of the carbonyl complex. The initial measurement (sampling no. 1 in Fig. 2) represents the state of Pt^{2+} in the original hydrated ion-exchanged specimen. Dehydration in vacuum at 363 K leads to an increase in electron density on Pt^{2+} due to the remaining NH_3 ligands (no. 2). The addition of CO starts a reduction process via the water-gas shift reaction (no. 3). The formation of the carbonyl complex between 363–393 K is a slow process due to an assembling of Pt atoms to the polynuclear complex via migration, and at sampling no. 11 only a small fraction of the Pt^{2+} has been converted. An increase of the reaction temperature to 423 K leads to complete carbonylation within 3.5 h and to the increase of the intensity of the white line (nos. 12–17) to a value even above that obtained for the dehydrated sample (no. 2). In spite of the reductive carbonylation the Pt atoms largely preserve their positive charge in the dianionic complex. The negative charge is predominantly located in the π-electron states of the CO ligands [29]. Consequently, a significant decrease of the intensity of the white line towards the reference value of a Pt foil can be observed only after the complex has been decomposed at 573 K in vacuum (Fig. 2, sampling no. 18), in which strongly reducing conditions are provided.

Figure 3 presents DRIFT spectra of Chini complexes obtained by carbonylation of a calcined Pt/NaX(4 wt.-% Pt) sample at 353 K for 20 h (dashed line) and after additional carbonylation at 388 K for 50 h (solid line). In the region of the fundamental CO bond stretching vibrations three bands at 1829, 2052, and 2080 cm^{-1}

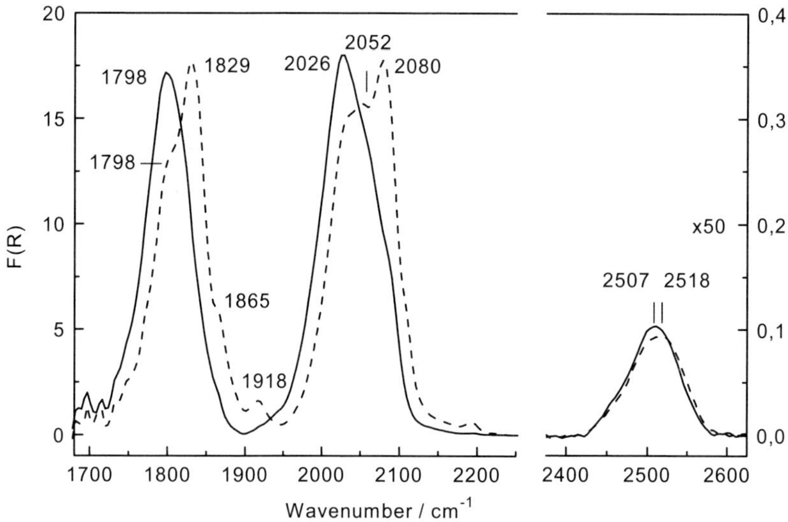

Fig. 3. DRIFT spectra of Pt carbonyl complexes, formed by carbonylation (60 kPa CO) of the Pt/NaX sample at 353 K for 20 h (dashed) and at 388 K for 50 h (solid), representing the bands of the bridge-bonded (1700–1920 cm^{-1}) and the linear-bonded (1900–2100 cm^{-1}) CO as well as of the combination bands for the latter (2400–2600 cm^{-1}). Reprinted with permission from [33].

and a shoulder at 1798 are visible (dashed spectrum). The two low-frequency bands at 1798 and 1829 cm^{-1} correspond to the bridge-bonded CO molecules, while the 2052 and 2080 cm^{-1} bands are attributed to the linearly bonded CO [25, 28]. In the combination mode region the low-intense single absorption band at 2507–2518 cm^{-1} is clearly visible. It belongs to the combination of the fundamental stretching vibrations of CO and Pt–C bonds of the linearly bonded CO molecules [30–32].

The spectrum of the complexes obtained after additional carbonylation at higher temperature (388 K) contains only two strong absorption bands at 1798 and 2026 cm^{-1} corresponding to the bridged and linearly coordinated CO molecules, respectively. In addition, the combination mode is represented by a single band with a maximum at 2507 cm^{-1}.

At lower temperature (353–363 K) the preferential nuclearity $n = 3$ follows from the dominating band positions (Fig. 1) of the linearly (2080 cm^{-1}) and bridged bonded (1830 cm^{-1}) CO [13,26]. At higher temperature (383 K) the nuclearity $n = 2$ is nearly exclusively achieved, gleaned from the maxima of linearly (2026 cm^{-1}) and bridged bonded (1798 cm^{-1}) CO [14,25].

The shift to lower nuclearity can be considered as a reductive conversion

$$2[Pt_3(CO)_6]_3^{2-} + CO + H_2O \leftrightarrow 3[Pt_3(CO)_6]_2^{2-} + 2\,H^+ + CO_2 \qquad (3)$$

owing to the reductive power of hydrogen in *statu nascendi* provided by the low

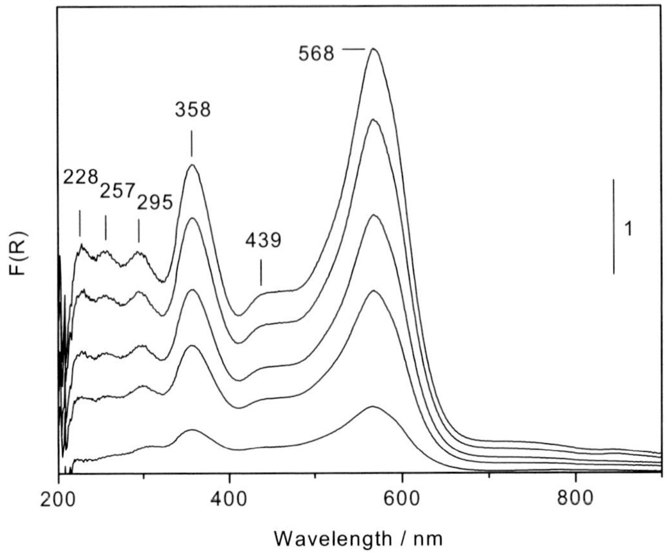

Fig. 4. Development of the UV/vis spectra in time during direct carbonylation of $[Pt(NH_3)]_4^{2+}$-exchanged (7.5 wt.-% Pt) NaEMT at 10^5 Pa CO and 363 K, after previous dehydration in a vacuum at RT for 30 min. From bottom to top: 10, 30, 50, 70, and 90 h. Reprinted with permission from [34].

temperature water gas shift reaction. The formation of the complex with $n = 2$ is possible due to the basicity of NaX [15].

The development of the UV/vis spectra in time during direct carbonylation of the platinum tetraammine complexes in NaEMT in the temperature range 363–423 K and at a CO pressure of about 10^5 Pa is exemplified in Fig. 4. The two dominating bands at 568–569 and 358 nm are assigned to two symmetry-allowed electric dipole transitions, being possible for the assumed D_{3h} symmetry in the x, y plane (high-energy band) and the z direction (low-energy band) [17]. A weak band appears at 431–439 nm. The dominant bands exhibit monotonic and simultaneous growth with temperature and time at constant relative ratio and are, therefore, assigned to the generation of a platinum carbonyl complex of uniform nature. The color of the sample changes from white to deep violet if $T \leq 393$ K and/or $p(CO) \gg 10$ kPa. The formation of the carbonyl complexes at favorable conditions, that is, at 363 K and 0.1 MPa CO pressure, is accompanied by the appearance of a well-structured absorption below 300 nm, exhibiting slight maxima around 228, 257, and 295 nm (Fig. 4). Such absorptions have been found repeatedly, although this range of absorption is frequently affected by various artifacts that were attributed to undesired scattering effects or limited reproducibilities in spectra processing such as subtraction of background spectra.

The formed negatively charged platinum carbonyl complex is stabilized by the mixing of the $2\pi^*$ orbitals of the CO with the valence $6p_z$ orbitals of the platinum atoms [35–37]. The low-energy HOMO–LUMO transition in the z direction is of the ligand–ligand ($A_2'' \rightarrow A_1'$ symmetry) type and the high-energy one in the x, y plane of the metal–ligand charge-transfer ($E' \rightarrow A_1'$ symmetry) character.

The situation is unique for the Pt carbonyl complexes in NaEMT, exhibiting two dominating distinct bands at 568 and 358 nm in the UV/vis spectra (Fig. 4). They characterize a complex of nuclearity $n = 3$, which has frequently been detected in faujasites [14] but never before with such a dominance. The weak band at 439 nm points to the presence of a minor amount of complexes with a nuclearity of $n = 2$ [14]. The reason for the dominance of triplane complexes is found in the peculiar structure of the zeolite EMT [38–40]. The hypercages of the EMT structure enable a highly relaxed accommodation of a complex with $n = 3$ (Fig. 5), in which the nearly three-fold symmetry of the complex fits to the perfect three-fold symmetry axis of the hypercage, being parallel to three lateral elliptic windows (0.74 × 0.65 nm) and normal to two circular windows (0.71 nm). The relaxed accommodation is further favored by the about 10 % larger elongation of the hypercage (about 1.4 nm) along the three-fold axis compared to the faujasite supercage dimensions. Any accommodation of a complex of even $n = 2$ in a hypocage of the EMT structure is not possible due to its small elongation along the three-fold axis (0.69 nm). Although the average distance between Pt triangles of the complex is around 0.3 nm [25], simple molecular geometry demonstrates that complexes with $n > 3$ do not fit into the hypercage. The EMT structure is not able to link complexes in neighboring cages, as is possible for cubic faujasite, since the hypercages of EMT

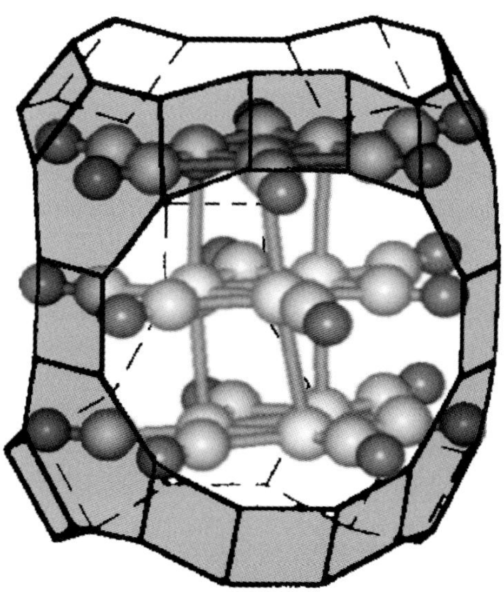

Fig. 5. Geometric modeling of the accommodation of a $[Pt_3(CO)_6]_3^{2-}$ complex in the hypercage of EMT. Reprinted with permission from [34].

are not adjoined through their elliptic windows, but are separated by the hypocages, and are connected through their circular windows under screwing of the symmetry axis by 60°, so that the adjacent complexes cannot be linked to larger "tinker-toy" triangular blocks, for example to form stacked hexaplane complexes of nearly D_{3h} symmetry, in principle.

9.3 Reversible Decomposition of the Complex

9.3.1 Decomposition in Oxygen

The process of oxidative decomposition of the dianionic Pt carbonyl complex was shown to proceed stepwise [26], that is the bridge-bonded CO ligands are removed more easily than the linear-bonded ones. Since the removal proceeds via an oxidation of the CO to CO_2 the preferential removal of the bridge-bonded ligands points to their higher reactivity. The most exciting phenomena were the reappearance of ammonia ligands in the oxidized complex and the rapid rebuilding of the initial biplane complex upon recarbonylation. The latter phenomenon is indicative of a structure of the oxidized complex in which the hexanuclear Pt skeleton is preserved. The stoichiometry of the oxidative decomposition of the dianionic complex could hence be described by Eq. (4)

$$[Pt_3(CO)_6]_2^{2-} + 14\,NH_4{}^+ + 9\,O_2 \\ \rightarrow [Pt_3O_3(NH_3)_3]_2 + H_2 + 12\,H^+ + 8\,NH_3 + 12\,CO_2 \qquad (4)$$

Eq. (4) considers the reappearance of the ammine bands in the FTIR spectra and the preservation of the hexanuclear Pt skeleton. The scheme could be supported by quantitative measurements of the hydrogen amount required for the reduction of the oxidized complex and the amount of ammonia released [26]

$$[Pt_3O_3(NH_3)_3]_2 + 6\,H_2 \rightarrow 6\,H_2O + 6\,NH_3 + 6\,Pt \qquad (5)$$

The recarbonylation led to the original complex according to Eq. (6)

$$[Pt_3O_3(NH_3)_3]_2 + 12\,H^+ + 19\,CO + H_2O \\ \rightarrow [Pt_3(CO)_6]_2^{2-} + 6\,NH_4{}^+ + 8\,H^+ + 7\,CO_2 \qquad (6)$$

Figure 6 depicts successive FTIR spectra obtained during recarbonylation. The top spectrum obtained after 12 h in 5×10^4 Pa CO and 373 K corresponds to the final spectrum in Fig. 1. The removal of oxygen from the complex upon recarbonylation could be monitored by the rapid formation of the original carbonyl complex (Fig. 6, inset).

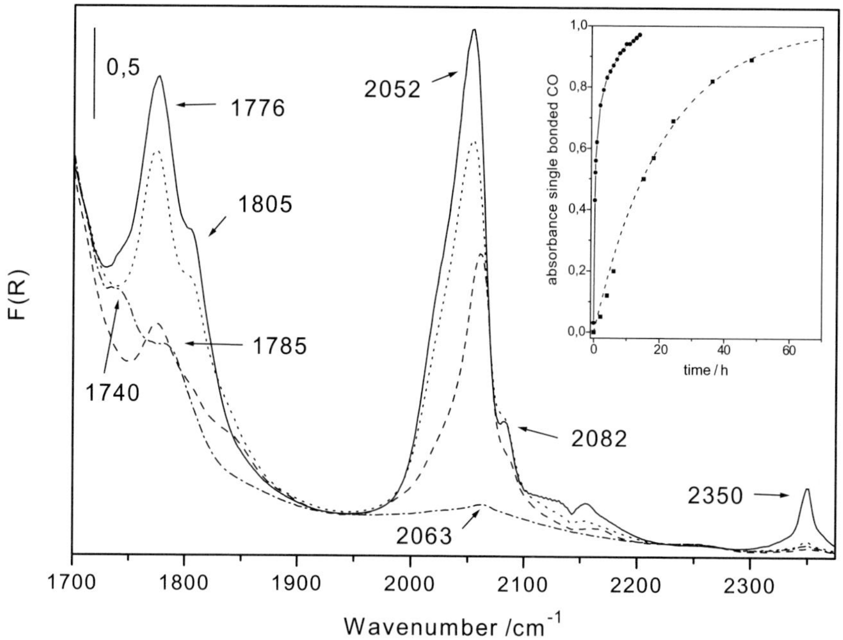

Fig. 6. Successive FTIR spectra obtained during recarbonylation 5×10^4 Pa CO, 373 K following the oxidative removal of CO. Top spectrum recorded after 12 h corresponds to the final spectrum in Fig. 1. Inset: time dependence of the integrated bands of on-top bonded CO, squares = first carbonylation, circles = recarbonylation.

9.3.2
Decomposition in Vacuum

Heating of a carbonylated sample (Fig. 3) in vacuum up to 773 K leads to the disappearance of all absorption bands of coordinated CO molecules both in the fundamental and combination mode regions. This indicates a complete decomposition of the Chini complexes. Oxidation of the Chini complexes by protons provides the driving force for formation of the metal clusters upon thermal decomposition of Pt carbonyls in vacuum

$$[\mathrm{Pt_3(CO)_6}]_n^{2-} + yH^+ \rightarrow [\mathrm{Pt}_{3n}]^{y-2} + \frac{y}{2}\mathrm{H_2} + 6n\,\mathrm{CO} \tag{7}$$

The result of the recarbonylation of the decomposed sample with CO for 100 h at 353 K and after additional heating of the sample in CO atmosphere at 388 K for 100 h is depicted in Fig. 7.

There are several indications that sintering of the platinum clusters can be neglected, and that the nuclearity of the preceding Chini complexes is preserved. First, decomposition temperatures up to 773 K in vacuum do not result in a significant aggregation of the metal clusters. Indeed, the small metal particles are in-

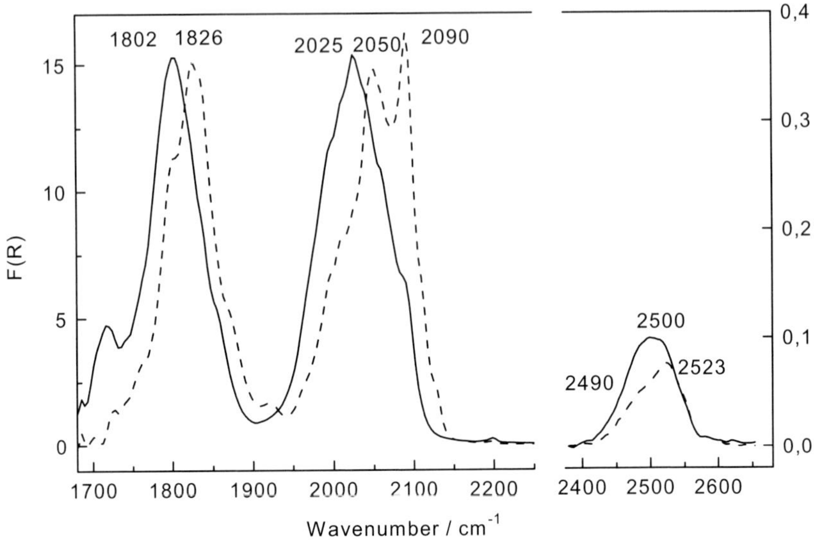

Fig. 7. DRIFT spectra Pt carbonyl complexes formed by recarbonylation (60 kPa CO) of the Pt clusters obtained after decomposition (773 K) of the initial complexes (Fig. 3), using the conditions 353 K for 100 h (dashed) and at 388 K for 100 h (solid). Reprinted with permission from [33].

visible by electron microscopy, while recarbonylation of the metallic Pt clusters to the initial Chini complexes (Fig. 7) preferentially yields the complexes with $n = 3$ at the lower (353 K) and with $n = 2$ at the higher (388 K) temperature. Moreover, XPS results confirm that no aggregation of the Pt has taken place on the outer surface [33]. Such behavior cannot be expected, if significant sintering takes place. In contrast, the formation of Chini complexes is not observed for the samples containing Pt clusters that are readily visible in transmission electron micrographs, that is, for Pt cluster with sizes exceeding 1 nm. The extreme stability of the tiny Pt clusters (<20 Pt atoms) at 773 K, for which easy sintering could be expected, might be attributed to repulsive forces between the Pt clusters, pointing to an excess of negative charge on the metal particles or values $y < 2$ (Eq. 7), respectively.

Second, the carbonylation of the Pt clusters, obtained from the thermal decomposition, requires much less water than the initial reductive carbonylation of Pt^{2+}-loaded NaX. It proceeds with strongly dehydrated samples in the former case, whereas only a slight dehydration is allowed in the latter one. The stoichiometry for carbonylation of Pt clusters, requiring less water than according to Eq. (2), is

$$[\mathrm{Pt}_{3n}]^{y-2} + \left(6n + \frac{y}{2}\right)\mathrm{CO} + \frac{y}{2}\mathrm{H_2O} \leftrightarrow [\mathrm{Pt_3(CO)_6}]_n^{2-} + y\mathrm{H}^+ + \frac{y}{2}\mathrm{CO} \tag{8}$$

This phenomenon also supports the assumed negative charge on the Pt clusters.

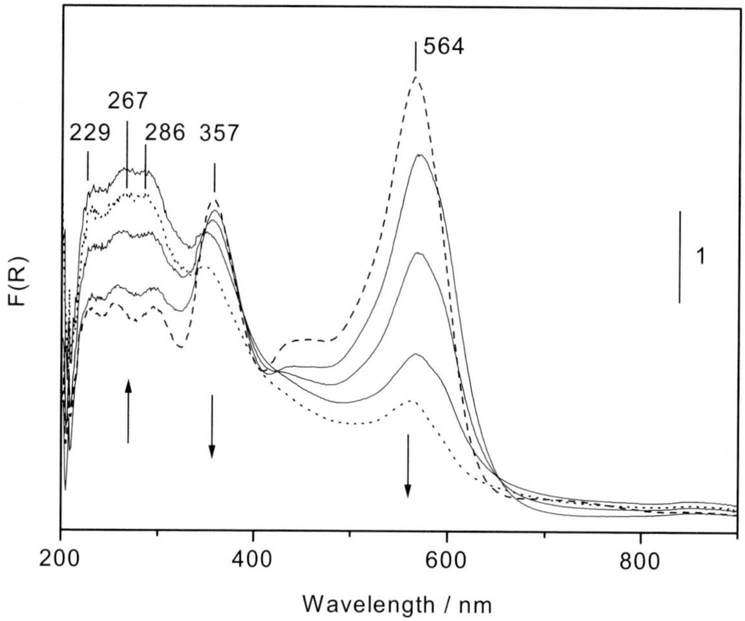

Fig. 8. Development of the UV/vis spectra in time during decomposition of a platinum carbonyl complex, starting from the sample prepared at 363 K (Fig. 4, top) cooling to RT and evacuation (dashed line), increasing the temperature (solid lines from top to bottom: 348, 373, and 423 K) and cooling to RT (bottom, dotted line). Reprinted with permission from [34].

9.4 Stable Subnanometer Platinum Clusters

The evolution of the UV/vis spectra during the decomposition of the carbonyl complex, produced at 363 K (Fig. 4), in a vacuum at 423 K is depicted in Fig. 8. A strong decrease of the characteristic carbonyl bands (357 and 564 nm) and an increase of the absorption below 300 nm is observed.

The appearance of absorptions below 300 nm (Figs. 5 and 8) might be attributed to the generation of smaller zerovalent Pt clusters, for example Pt_9, as side-products under mild conditions. Under less favorable conditions, that is, 423 K or low CO pressure, larger clusters might be formed [34].

The absorptions below 300 nm are tentatively assigned to d–d excitations of the Pt clusters, enabled by a metal–insulator transition. Such a transition can be expected from size-quantization effects that are observed for Pt clusters in faujasites at 80 K, analyzing the temperature dependence of Pt-NMR spectra [41]. Presumably, this metal–insulator transition occurs at elevated temperature for the Pt cluster of the carbonyl complex and the smaller naked clusters generated under the milder conditions. The shell of CO ligands for the carbonyl complex as well as sterical or electrostatic barriers in zeolite cages for the naked clusters might suppress the breathing mode vibrations, which usually cause a broadening of the one-electron energy level so that the effective one-electron energy distribution is continuous and the expected size-quantization effect does not appear.

The relatively high energies in the range 200–300 nm (4–6 eV) for transitions between states close to the Fermi level point to a high fraction of s states contributing to these transitions [42]. Such a high fraction of 6s states in the hybrids with the d atomic states is indeed evaluated for bare platinum clusters using density functional calculations [43]. The larger clusters, which are assumed to be formed at conditions being less favorable for the stabilization of the carbonyl complex, are expected to have higher local densities of state at the Fermi level and, thus, absorb light in the UV as well as visible range. Since this absorption cannot be reflected, due to the small size of the clusters and the limited motion of the electrons, it is manifested in the increase of the base-line in the applied spectral range or the grayish color, respectively. These properties of the larger naked Pt clusters, formed beside the carbonyl complexes, affect the clear detection of the size-quantization effect in the latter species.

Heating of the sample in vacuum at temperatures up to 773 K followed by CO adsorption at 13 Pa results in the FTIR spectrum shown in Fig. 9 (solid line). Similarly to the previous cases it contains the bands at 1788 and 1849 cm^{-1} (bridge-bonded CO), the bands at 2028 and 2050 cm^{-1} (shoulder), the band at 2071 cm^{-1} (linearly bonded CO), and a band of low intensity at 2503 cm^{-1} (combination mode). The band at 2071 cm^{-1} in this spectrum is the strongest. Successive evac-

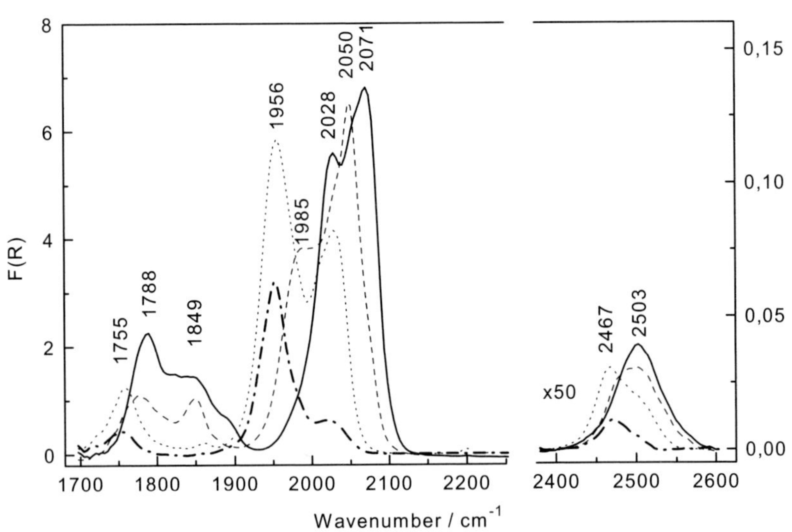

Fig. 9. DRIFT spectra of CO on Pt clusters in NaX, obtained after decomposition (773 K) in vacuum of the initial complexes (Fig. 3) using the conditions: chemisorption of CO (10 Pa) at 300 K (solid) and stepwise removal of the chemisorbed CO in vacuum at 373 K (dashed), 473 K (pointed), and 573 K (dashed-pointed). Reprinted with permission from [33].

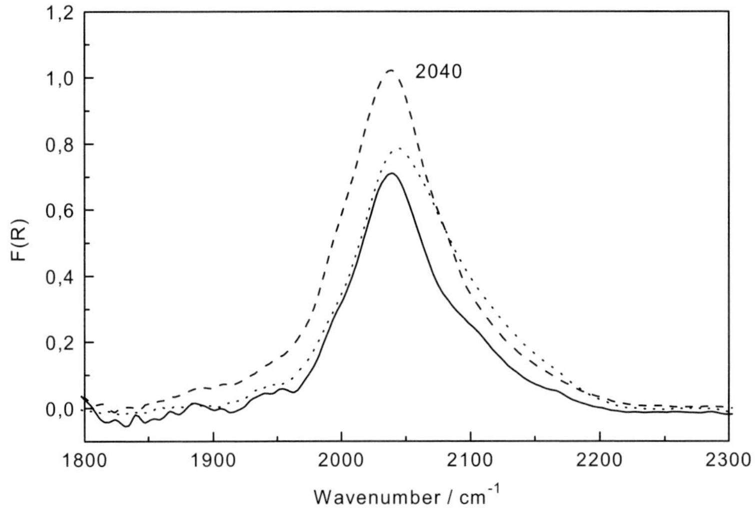

Fig. 10. DRIFT spectra of terminal platinum hydrides formed by dissociative chemisorption of hydrogen at 300 K on Pt clusters in NaX, obtained after decomposition at temperatures up to 773 K in vacuum of the initial complexes formed by carbonylation (60 kPa CO)of the PtNaX sample at 353 K for 20 h. Reprinted with permission from [33].

uation of CO at room temperature or at 373 K (Fig. 9, dashed line) is accompanied by a gradual shift of the band maximum down to 2050 cm^{-1} and by the appearance of a new band at 1985 cm^{-1}. After CO evacuation at 473 K the latter is shifted towards 1956 cm^{-1} and strongly increases in intensity (Fig. 9, dotted line). Further, the high-frequency bands at 2050 and 2071 cm^{-1} disappear, while the spectrum of the bridge-bonded CO transforms into a single band at 1755 cm^{-1}, and a new additional combination band at 2467 cm^{-1} is developed. At higher evacuation temperature of 573 K all the bands are decreased in intensity without noticeable change of the positions of their maxima, and are completely removed in vacuum at 673 K.

Finally, Fig. 10 displays DRIFT spectra of terminal Pt hydrides [44], formed due to the dissociative hydrogen chemisorption on metallic Pt at room temperature, on the samples obtained by the decomposition of the Chini complex in vacuum at 673 and 773 K. Both spectra show single bands of comparable intensities at 2040 cm^{-1}.

9.5 Electron Donor Properties of Pt Clusters Derived from Chini Complexes

The disappearance of all bands in the IR and UV/vis spectra shows that Chini complexes are completely decomposed in vacuum at 773 K. As follows from Fig. 9, readsorption of CO on the resulting metal particles yields spectra of high complexity. The small platinum particles are surprisingly stable against sintering even at 773 K.

The lower Kubelka–Munk values in Fig. 9 compared with those in Figs. 3 and 7 indicate that a smaller number of adsorption sites is available on the clusters formed by decomposition of the Chini complex. The loss could be due to a change in the geometry of the clusters and to an interaction with the support. This would also explain the decrease in the ratio of intensities between bridged and linearly bonded CO. A low ratio is typical for CO adsorption on large metallic Pt clusters. It should be noted, however, that the ratio observed in Fig. 9 is still relatively close to the original Chini complex.

Figure 7 demonstrates the reformation of the original Chini complex following the decomposition of the complex at 773 K (Fig. 9) even including the reduction of the trimer that is formed first at the lower temperature. This can be taken as support for the interpretation of the spectra shown in Fig. 9 as being due to CO adsorption on Pt clusters containing the same number of atoms as the parent Chini complex.

Inspection of the spectra of the fully recarbonylated sample (Fig. 7) shows a loss of less than 10 % for linearly bonded CO compared to Fig. 1 deduced from the integral absorptions. In case the loss can be attributed to the agglomeration of larger Pt clusters it could not be detected in Fig. 7.

Evacuation of the sample at 373 K results in a decrease of the band at 2023–2029 cm^{-1}. Evacuation of adsorbed CO at 473 K and higher temperatures gradually converted this band into a new one centered at 1956 cm^{-1}. There could be two reasons for such transformation. The first one is connected with the dipole–dipole coupling between linearly adsorbed molecules that leads to the intensity redistribution between the low- and high-frequency bands of CO linearly adsorbed on the neighboring metallic sites with different electron donor properties [45,46]. The second reason could be the transformation of the initial type of the linear bonding into another type characterized by a stronger metal–carbon bond. Analysis of the spectra in the combination mode region allows the discrimination between these two effects.

It is established both theoretically and experimentally [30,31] that the frequency of the combination mode of linear Pt–CO complexes is close to the sum of the frequencies of the singleton of the linearly adsorbed CO and of the Pt–C bond vibration. Therefore, the combination mode does not depend on the dipole–dipole interaction between linearly adsorbed CO molecules or on the surface coverage, respectively. Moreover, in the combination mode region there is no intensity redistribution between the low- and high-frequency bands [47]. Such an effect is characteristic only for the bands of fundamental CO bond stretching vibrations of CO molecules linearly adsorbed on neighboring metallic sites [45,46]. Thus, if the spectrum transformation under discussion would be caused by an intensity redistribution effect, one should detect in the combination region only the decrease of intensity of the single band connected with the decrease of the amount of linearly adsorbed CO, but not the appearance of a new combination mode. In our case (Figs. 7 and 9) the transformation of the band at 2023–2029 cm^{-1} into that one at approx. 1956 cm^{-1} is definitely accompanied by the appearance of a new absorption band at 2470 cm^{-1} in the combination mode region. This demonstrates a transformation of one type of linear bonding into another.

Comparison of the band intensities in the fundamental region of the linear Pt–CO complexes with those of the corresponding combination bands enables the assignment of the combination bands at 2470 and 2503 cm^{-1} to linear Pt–CO species with two different frequencies of the fundamental CO bond stretching vibrations at about 1957 and 2023–2039 cm^{-1}, respectively. Thus, the corresponding frequencies of Pt–C bond vibrations, calculated as a frequency difference of combination mode and fundamental CO bond vibration, are about 510 and 480 cm^{-1}. The Pt–C vibration of the higher frequency is then correlated with a thermally more stable form of CO adsorption connected with the band at 1960 cm^{-1} compared with that at 2023–2039 cm^{-1}.

The transformation of Pt–CO complexes with the fundamental CO bond vibration frequency of 2023–2029 cm^{-1} from the linearly adsorbed CO into one with the stretching frequency of 1957 cm^{-1} is a reversible process. After the sample cooling to room temperature, followed by CO adsorption at 13 Pa, the initial spectra are restored. This reversibility indicates that the formation of a new type of linear species cannot be attributed to a carbon deposition on the metallic surface due to CO dissociation (Boudouard reaction) [26,48]. The observation of a reversible recarbonylation of metallic Pt into Chini complexes also favors this conclusion.

The new type of the linear Pt–CO bonding that appears at low metal surface coverage has an extremely low CO stretching frequency of 1957 cm^{-1}. Similar low frequencies of singletons of linearly adsorbed CO were earlier also observed for metallic Pt dispersed on such basic carriers as alkaline and alkaline-earth forms of zeolites [49–52], MgO stabilized by alumina [53], and hydrotalcites [54]. The strong decrease of the singleton frequencies in those cases was explained by a negative charging of the supported particles due to electron density transfer from basic oxygen anions to the metal. This results in an increase of the extent of back-donation of electrons from the metal atoms into antibonding $2\pi^*$ orbitals of coordinated CO molecules. A similar strong red shift of the CO stretching vibration and an increase of the Pt–C bond strength has been shown in a density functional study of the CO adsorption on negatively charged Pt_4 clusters in zeolites [55].

The assumption of a negative charge on the Pt particles formed by decomposition of the Chini complexes is also supported by the spectra of terminal platinum hydrides formed due to dissociative hydrogen adsorption on the metallic Pt surface (Fig. 10). The negative charging follows from a low frequency of the Pt–H bond vibration (2040 cm^{-1}) that is by more than 50 cm^{-1} lower than that characteristic of hydrides on neutral Pt particles [54]. This negative charge could be a result of the electron density redistribution between basic oxygen anions of the zeolite framework and the metallic Pt clusters or it could arise from electrons remaining on the metallic particles formed by thermal decomposition in vacuum of anionic Chini complexes.

The appearance of a new type of linear Pt–CO complex, exhibiting the extremely low frequency around 1957 cm^{-1}, needs further considerations of its origin. This band appears at temperatures of evacuation of preadsorbed CO above 373 K in vacuum (Fig. 9). It is accompanied by a partial removal of bridged bonded CO, whose band intensity decreases to a larger extent than those of the linearly bonded CO. This means, that the fraction of the electron donor capacity of the Pt clusters, needed for the bridged bonded CO molecules, becomes available after their removal, and can now be used additionally for the bonding of the residually adsorbed CO molecules. This increased donor capacity is responsible for the appearance of new bands for the linearly bonded (1957 cm^{-1}) and the bridged bonded (1755–1760 cm^{-1}) CO molecules.

9.6 Conclusions

A route to the formation of stable subnanometer platinum clusters could be established via the decomposition of stoichiometric anionic platinum carbonyl (Chini) complexes synthesized within the cages of supporting zeolites. The Chini complexes were prepared via direct carbonylation of $[Pt(NH_3)_4]^{2+}$ exchanged zeolites. The size of the complexes can be controlled by the steric constraint of the zeolite cages, the chemical properties of the zeolite matrix and by the experimental conditions.

The mechanisms of formation and decomposition as well as the stability of the complexes can be monitored by *in situ* UV/vis and FTIR spectroscopy.

Experimental conditions could be established for the decomposition of the complexes under oxygen and in vacuum to subnanometer platinum clusters corresponding in size to the skeleton of the precursor complex. This could be inferred from the observation of a size quantization effect and from the rapid and almost quantitative recarbonylation of the cluster to the initial carbonyl complex. The vacuum decomposition of Chini complexes formed in NaX zeolite resulted in metal clusters of surprising thermostability. The platinum clusters preserved a low nuclearity; no agglomeration to bigger particles could be observed even at 773 K. The properties of the clusters could be characterized by DRIFT spectroscopy. Analyzing the stretching vibrations of chemisorbed CO and of terminal platinum hydrides revealed a strong electron donor capacity with a red shift of the CO stretching vibration predicted in theoretical studies. Stable platinum clusters of uniform size are now available for further studies of their electronic as well as their catalytic properties.

Acknowledgements

Financial support by the Deutsche Forschungsgemeinschaft is gratefully acknowledged, (Ja 346/15; 436 RUS 113/493; 436 TSE 113/30).

References

1 D. Barthomeuf, *Catal. Rev.* **1996**, *38*, 521.
2 G. Schmid (ed.), Clusters and Colloids, VCH, Weinheim **1994**, pp. 555.
3 W. Ekardt (ed.), Metal Clusters, Wiley-VCH, Weinheim **1999**, pp. 286.
4 G. Schulz-Ekloff, Metal Clusters in Zeolites, in: Comprehensive Supramolecular Chemistry, J.L.

Atwood, J.E.D. Davies, D.D. MacNicol, F. Vögtle (eds.), Vol. 7, Pergamon, Oxford **1996**, p. 549.

5 B.C. Gates, Supported Metal Cluster Catalysts, in: Handbook of Heterogeneous Catalysis, G. Ertl, H. Knözinger, J. Weitkamp (eds.), Vol. 2, Wiley-VCH, Weinheim **1997**, p. 793.

6 G. Ertl, H. Knözinger, J. Weitkamp (eds.), Handbook of Heterogeneous Catalysis, 2.2 Supported Catalysts, Wiley-VCH, Weinheim **1997**, Vol. 1, pp. 191–286.

7 R. Raja, T. Khimyak, J.M. Thomas, S. Hermans, B.F.G. Johnson, *Angew. Chem.* **2001**, *113*, 4774.

8 L.L. Sheu, H. Knözinger, W.M.H. Sachtler, *Catal. Lett.* **1989**, *2*, 129.

9 L. Drozdova, J. Novakova, G. Schulz-Ekloff, N.I. Jaeger, *Micropor. Mesopor. Mater.* **1999**, *28*, 395.

10 A.-M. Liu, T. Shido, M. Ichikawa, *J. Chem. Soc.: Chem. Commun.* **1995**, 507.

11 A. de Mallmann, D. Barthomeuf, *Catal. Lett.* **1990**, *5*, 293.

12 G.J. Li, T. Fujimoto, A. Fukuoka, M. Ichikawa, *Catal. Lett.* **1992**, *12*, 171.

13 G. Schulz-Ekloff, R.J. Lipski, N.I. Jaeger, P. Hülstede, L. Kubel-kova, *Catal. Lett.* **1995**, *30*, 65.

14 L. Kubelková, J. Vylita, L. Brabec, L. Drozdová, T. Bolom, J. Nová-ková, G. Schulz-Ekloff, N.I. Jaeger, *J. Chem. Soc.: Faraday Trans.* **1996**, *92*, 2035.

15 L. Kubelkova, L. Drozdová, L. Brabec, J. Novakova, J. Kotrla, P. Hülstede, N.I. Jaeger, G. Schulz-Ekloff, *J. Phys. Chem.* **1996**, *100*, 15 517.

16 B.C. Gates, *Chem. Rev.* **1995**, *95*, 511.

17 M. Ichikawa, in: Dynamic Processes on Solid Surfaces, K. Tamaru (ed.), Plenum, New York **1993**, p. 149.

18 T. Yamamoto, T. Shido, S. Inagaki, Y. Fukushima, M. Ichikawa, *J. Phys. Chem. B.* **1998**, *102*, 3886.

19 W.M.H. Sachtler, Z. Zhang, *Adv. Catal.* **1993**, *39*, 129.

20 M. Ichikawa, *J. Chem. Soc.: Chem. Commun.* **1976**, 11.

21 M. Ichikawa, *Chem. Lett.* **1976**, 335.

22 Y. Yokomichi, T. Nakyama, O. Okada, Y. Yokoi, L. Takahashi, H. Uchida, H. Ichikawa, R. Yamaguchi, H. Matsui, T. Yamabe, *Catal. Today* **1996**, *29*, 155.

23 L. Drozdová, L. Kubelková, *J. Chem. Soc.: Faraday Trans.* **1997**, *93*, 2597.

24 H. Bischoff, N.I. Jaeger, G. Schulz-Ekloff, L. Kubelková, *J. Mol. Catal.* **1993**, *80*, 95.

25 J.C. Calabrese, F.D. Dahl, P. Chini, G. Longoni, S.J. Martinengo, *J. Am. Chem. Soc.* **1974**, *96*, 2614.

26 Z. Bastl, M. Beneke, L. Brabec, N.I. Jaeger, P. Hülstede, J. Novakova, G. Schulz-Ekloff, *Phys. Chem. Chem. Phys.* **2000**, *2*, 3099.

27 M. Beneke, L. Brabec, N.I. Jaeger, J. Novakova, G. Schulz-Ekloff, *J. Mol. Catal. A* **2000**, *157*, 151.

28 G. Longoni, P. Chini, *J. Am. Chem. Soc.* **1976**, *98*, 7225.

29 R. Harmsen, R.A. van Santen, private communication.

30 N.B.J. Persson, R. Ryberg, *Chem. Phys. Lett.* **1990**, *174*, 443.

31 A.V. Zaitsev, V.Yu. Borovkov, V.B. Kazansky, *Kinet. Catal.* **1992**, *33*, 517.

32 C. De la Cruz, N. Sheppard, *Spectrochim. Acta A* **1994**, *50*, 271.

33 A.I. Serykh, O.P. Tkachenko, V.Y. Borovkov, V.B. Kazansky, M. Beneke, N.I. Jaeger, G. Schulz-Ekloff, *Phys. Chem. Chem. Phys.* **2000**, *2*, 5647.

34 L. Drozdova, L. Brabec, J. Novakova, M. Beneke, N.I. Jaeger, G. Schulz-Ekloff, *Micropor. Mesopor. Mater.* **2000**, *35–36*, 511.

35 K.W. Chang, R.G. Wooley, *J. Phys. C* **1979**, 2745.

36 D.W. Bullet, *Chem. Phys. Lett.* **1985**, *115*, 450.

37 D.J. Underwood, R. Hoffmann, K. Tatsumi, N. Nakamura, Y. Yamamoto, *J. Am Chem. Soc.* **1985**, *107*, 5968.

38 J. Weitkamp, R. Schumacher, U. Weiss, *Chem.-Ing.-Techn.* **1992**, *64*, 1109.

39 F. Delprato, L. Delmotte, J.-L. Guth, L. Huve, *Zeolites* **1990**, *10*, 546.

40 W.M. Meier, D.H. Olson, in: Atlas of Zeolites Structure Types,

Butterworth-Heinemann, London **1992**, p. 88.

41 Y. Tong, D. Laub, G. Schulz-Ekloff, A.J. Renouprez, J.J. van der Klink, *Phys. Rev. B* **1995**, *52*, 8407.

42 J.W. Niemantsverdriet, in: Spectroscopy in Catalysis, VCH, Weinheim **1995**, p. 60.

43 H. Gronbeck, W. Andreoni, *Chem. Phys. Lett.* **1997**, *269*, 385.

44 W.A. Pliskin, R.P. Eischens, *Z. Phys. Chem. N. F.* **1960**, *24*, 11.

45 B.N.J. Persson, R. Ryberg, *Phys. Rev.* **1981**, *B24*, 6954.

46 B.N.J. Persson, F.M. Hoffmann, *J. Electron. Spectr.* **1987**, *45*, 215.

47 V.Yu. Borovkov, V.B. Kazansky, *Catal. Lett.* **1994**, *23*, 107.

48 H. Bischoff, N.I. Jaeger, G. Schulz-Ekloff, *Z. Phys. Chem.* **1990**, *271*, 1093.

49 C. Besoukhanova, J. Guidot, D. Barthomeuf, M. Breyusse, J.R. Bernard, *J. Chem. Soc.: Faraday Trans. I* **1981**, *77*, 1595.

50 V.B. Kazansky, V.Yu. Borovkov, N. Sokolova, N.I. Jaeger, G. Schulz-Ekloff, *Catal. Lett.* **1994**, *23*, 263.

51 G. Larsen, G.L. Haller, *Catal. Lett.* **1989**, *3*, 103.

52 B.L. Mojet, J.T. Miller, C. Koningsberger, *J. Phys. Chem. B* **1999**, *103*, 2724.

53 V.B. Kazansky, V.Yu. Borovkov, E.G. Derouane, *Catal. Lett.* **1993**, *19*, 327.

54 V.B. Kazansky,V.Yu. Borovkov, A.I. Serykh, *Catal. Lett.* **1997**, *49*, 35.

55 A.M. Ferrari, K.M. Neyman, T. Belling, M. Mayer, N. Rösch, *J. Phys. Chem. B* **1999**, *103*, 216.

10
Recent Advances in the Synthesis of Mesostructured Aluminum Phosphates

*Michael Tiemann and Michael Fröba**

10.1
Introduction

10.1.1
Background

Over the last years, the concept of utilizing supramolecular assemblies of surfactants as structure-directing agents for the synthesis of mesoporous materials, introduced in 1992 for silica and aluminosilicate phases [1], has been applied to the preparation of a large number of mesostructured aluminum phosphates. These materials are inorganic/organic composites in which the organic component (the surfactant) is periodically arranged within an inorganic (aluminum phosphate) matrix. If the inorganic part, which is usually amorphous, is thermally stable and interconnected in all three dimensions, the surfactant may be removed from the solid material (e.g., by calcination or solvent extraction), which leads to regularly arranged mesopores.

Taking into account the important role of microporous aluminum phosphates and silico-aluminum phosphates (crystalline "AlPO4-*n*" and "SAPO-*n*" with uniform pore diameters below 1 nm [2,3]) as size-selective heterogeneous catalysts [4], there is an obvious demand for larger pores with an equally regular arrangement. Mesoporous aluminum phosphates exhibit uniform pore diameters between 2 and 4 nm. In the following, the literature on mesostructured aluminum phosphates is briefly summed up; a more detailed review on this topic has recently been published [5].

10.1.2
Nanostructure

Many of the mesostructured aluminum phosphates that have been reported in the literature exhibit lamellar structures, that is they consist of inorganic sheets that are separated from each other by surfactant bilayers [6–20]. Naturally, these phases do not maintain their periodic order upon removal of the surfactant; they do not exhibit any mesopores and are therefore not suitable for catalytic applications. However, they may still be interesting, for mechanistic studies for example.

On the other hand, a number of mesostructured aluminum phosphates with nonlamellar structures have been reported. For most of them it was possible to remove the surfactant by calcination or solvent extraction without collapse of the mesostructure, which led to the generation of mesopores. In these materials the surfactant molecules are arranged in tubular arrays that are either linear and periodically ordered to form hexagonal structures [21–26] (comparable to MCM-41 silica [1]), or worm-like and more or less randomly ordered [10,16,27–33] (as in case of MSU-1 silica [34] or HMS-1 [35] silica). Below we will show an interesting inverse hexagonal structure synthesized in under nonaqueous conditions [36].

10.1.3 Catalytic Potential

In the light of potential applications of mesoporous aluminum phosphates in the field of heterogeneous catalysis, particular interest lies in the incorporation of silicon and/or (transition) metals into the inorganic network. The generation of acidic sites or reactive metal ion valence states in the materials is necessary for many catalytic applications, such as oxidation reactions. Transition metal-substituted microporous aluminum phosphates ("MAPO-*n*") and silico-aluminum phosphates ("MAPSO-*n*") are used as catalysts for various oxidation reactions of alkanes, cycloalkanes, or phenols as well as for methanol conversion [4]. Mesoporous aluminum phosphates containing Si [6,21,29] and/or Co [6], Mn [29], V [28,29], Ga [21], Mg [22,31], or Ti [19,26] have been reported.

10.1.4 Synthesis Conditions

For most of the mesostructured aluminum phosphates reported so far, especially for the stable porous materials, structure-direction was achieved by the utilization of long-chain cationic *n*-alkyl trimethylammonium surfactants [13–15,19,20,22–31]. In other cases neutral *n*-alkyl amines were used, yielding lamellar mesostructures or microporous hexagonal materials [6–10,16,18]. Mesoporous aluminum phosphate-based materials were also obtained from a synthesis using anionic sodium dodecyl sulfate [21]. Finally, alkyl phosphate surfactants have been used for the syntheses of some lamellar phases [11,12,17].

The majority of the syntheses was carried out under aqueous reaction conditions [6,8,10–17,19–21,23–31]. In some cases unbranched primary alcohols and/or (tetra)ethylene glycol were combined with small quantities of water [7,9,18], but nonaqueous syntheses of mesostructured aluminum phosphates have so far not led to porous materials. Below we will summarize the first syntheses of mesoporous aluminum phosphates under basically nonaqueous conditions [32,33]; these are also the first syntheses that utilize primary *n*-alkyl amine surfactants to yield porous phases with large pore diameters.

10.1.5
Short-Range Structural Order

In spite of their periodic order on the nanometer scale, mesoporous (or mesostructured) aluminum phosphates, unlike their microporous analogues, are not crystalline. Their synthesis is difficult to control with respect to the stoichiometric composition of the products, which often quite drastically differs from the "ideal" ratio of Al:P of 1:1. This may be due to an incomplete condensation of the inorganic reactants, resulting in a partial coordination of Al or P with terminal (i.e., nonbridging) groups such as O, OH, or OH_2. In addition, a mesostructured aluminum phosphate may contain more or less significant quantities of aluminum oxide and/or oxyhydroxide species; these may either be part of the mesostructured product or be a second (possibly X-ray amorphous) phase. Apart from some lamellar materials, in which the inorganic parts actually exhibit crystalline structures on the atomic level [6,10,18], no mesostructured aluminum phosphates with exact stoichiometries of Al:P of 1:1 have been reported so far. Below we will show a new, rational synthesis route that overcomes this problem by utilization of a single-source molecular precursor with a pre-defined stoichiometric composition [33].

10.2
Inverse Hexagonal Mesostructured Aluminum Phosphates

Under aqueous reaction conditions the synthesis of mesostructured aluminum phosphates from $Al(O^iPr)_3$ and H_3PO_4 with the utilization of *n*-dodecyl phosphate ($C_{12}H_{25}OPO(OH)_2$) as a structure-director has led to lamellar materials [11,12]. A low-temperature synthesis (10–25 °C) under similar conditions, but with ethanol instead of water as the reaction medium, leads to products that consist of two distinct nanostructured phases, one with a lamellar, and another with a hexagonal structure. These two phases evolve competitively; their relative amounts in the products depend systematically on the synthesis temperature as well as on the reaction time, with the lamellar phase dominating at higher synthesis temperatures and/or longer reaction duration. Powder X-ray diffraction (XRD) patterns are shown in Fig. 1. Thermal analysis shows that at elevated temperatures (about $>35\,°C$) the hexagonal phase irreversibly transforms into the lamellar phase; this was monitored quantitatively by differential scanning calorimetry (DSC). As expected, the enthalpy of this transition with respect to the overall sample weight systematically depends on the relative amount of the hexagonal phase in the respective sample (Fig. 2).

The investigation of the hexagonal phase reveals some significant properties that are not typical of the usual hexagonal inorganic mesostructures prepared by this kind of synthesis. In the latter the surfactant molecules are arranged in rod-like assemblies with the polar head groups facing outwards. Mesostructured inorganic/surfactant composite materials with this kind of hexagonal structure usually exhibit d_{100} values in the range 3–4 nm (for C_{12} surfactants). Contrary to that, the

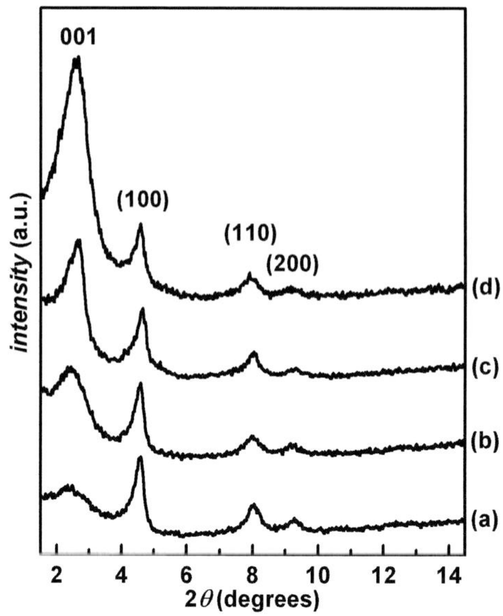

Fig. 1. Powder XRD diagrams of mesostructured aluminum phosphates prepared from $Al(O^iPr)_3$ and H_3PO_4 in ethanol with C_{12}-PO_4 (20 wt.-%) at various temperatures: (a) 10 °C, (b) 25 °C, (c) 40 °C, (d) 60 °C. The relative amount of the lamellar phase (the relative intensity of its 001 reflection) depends on the synthesis temperature. The hexagonal phase is indexed in parentheses.

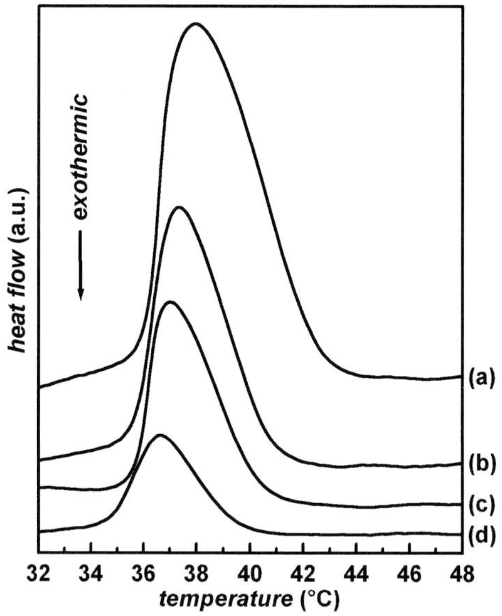

Fig. 2. DSC diagrams showing the transition of the hexagonal phase to the lamellar phase in the samples shown in Fig. 1, prepared at variable temperatures: (a) 10 °C; (b) 25 °C; (c) 40 °C; (d) 60 °C. The enthalpy depends on the relative amount of the hexagonal phase; the curves are normalized with respect to the overall sample weight.

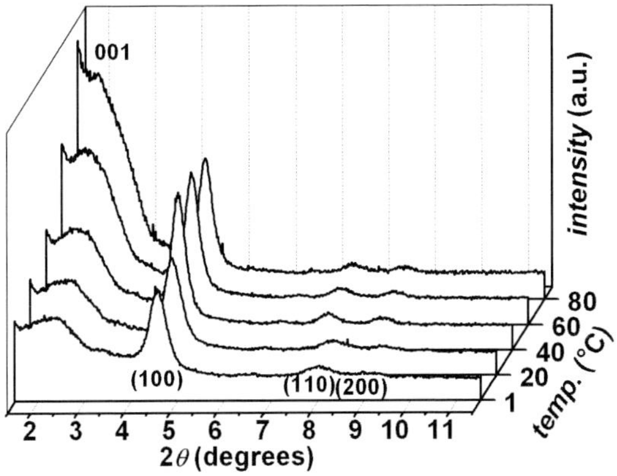

Fig. 3. Schematic representation of the inverse hexagonal mesostructured aluminum phosphate (cross section through the *hk*0 plane); in the rod-like surfactant assemblies the polar head groups are turned inwards, with single inorganic domains (dark areas) in the centers.

hexagonal structure in the system studied here has a d_{100} value that is surprisingly low (1.88 nm). The material is thermally unstable even at rather low temperatures (>35 °C), as mentioned above. Therefore, we suggest a different hexagonal surfactant arrangement in which the surfactant molecules are assembled in an inverted arrangement, so the polar head groups are located inside of the rods and the hydrophobic chains are turned outwards (Fig. 3). In this case the inorganic part is encapsulated in the centers of these assemblies forming individual domains that are not interconnected with each other and extend in one direction only.

n-Dodecyl phosphate is completely soluble in ethanol under the conditions used in the syntheses. The pure surfactant solutions are optically isotropic, that is nonbirefringent under polarized light as was verified by polarized-light optical microscopy (POM); they do not show any small angle X-ray scattering (SAXS) reflections over the entire temperature range (20–90 °C) and concentration region (5–50 wt.-%) that was studied. Hence, the evolution of a mesostructure during the synthesis is obviously induced by the presence of the inorganic reactants; the formation of the mesostructure is a highly co-operative process. The lack of any mesoscopic structure in the absence of the inorganic components is consistent with the fact that the alcohols have a considerably lower polarity than water (i.e., lower dielectric constants); thus, the self-aggregation of the surfactant molecules into micellar assemblies is relatively disfavored in alcohols as compared to water.

The influence of both the temperature and the reaction time on the relative amounts of the two phases was monitored by *in situ* SAXS studies during the reaction.

Figure 4 shows the SAXS patterns of a synthesis mixture in ethanol at variable temperatures; upon heating the 001 reflection of the lamellar phase grows in intensity relative to those of the hexagonal phase. The temporal evolution of the

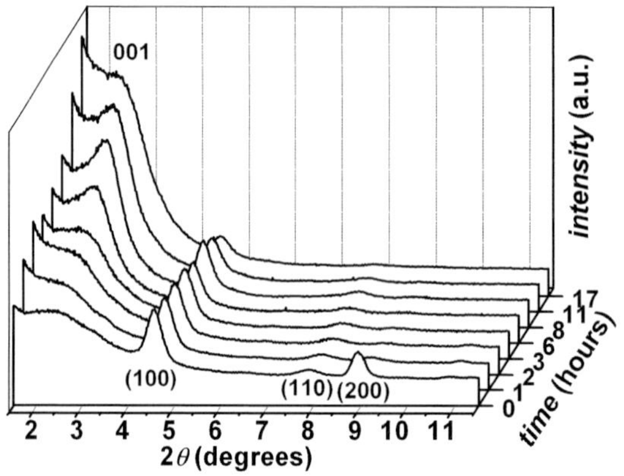

Fig. 4. Thermal evolution of the SAXS diagrams of the alcoholic synthesis mixture: C_{12}-PO_4/ethanol (20/80 w/w), $Al(O^iPr)_3$, H_3PO_4. (Equimolar amounts of C_{12}-PO_4, $Al(O^iPr)_3$, and H_3PO_4.) The lamellar phase and the hexagonal phase (in parentheses) are indexed.

SAXS patterns of a synthesis mixture with the same composition at constantly 25 °C is shown in Fig. 5; the lamellar phase becomes more and more dominant in the course of time.

These results show that the synthesis eventually leads to a single lamellar phase. The inverted hexagonal structure is an intermediate phase that co-exists with the

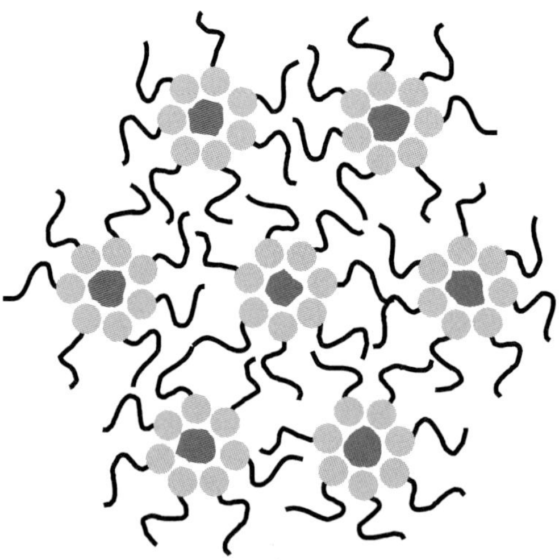

Fig. 5. Temporal evolution of the SAXS diagrams of the alcoholic synthesis mixture: C_{12}-PO_4/ethanol (20/80 w/w), $Al(O^iPr)_3$, H_3PO_4. (Equimolar amounts of C_{12}-PO_4, $Al(O^iPr)_3$, and H_3PO_4.) The lamellar phase and the hexagonal phase (in parentheses) are indexed.

lamellar phase over a certain time interval at the beginning of the reaction, particularly at low temperatures; it finally transforms into the lamellar phase. It is interesting to discuss these findings in the light of mechanistic studies that have been made on the synthesis of hexagonal mesostructured MCM-41 silica materials with cationic ammonium surfactants as structure-directors [37]. The concept of "charge density matching" provides an explanation for a transformation of an intermediate lamellar phase to the final (noninverted) hexagonal silica material: The surface area that each cationic surfactant head group exposes to the anionic silicate oligomers (at the surfactant/silicate interface) depends on the charge density of these silica oligomers. As the condensation of the silicate species proceeds, the charge density is diminished, which leads to an increase of the surfactant head group's interface area; thus, a curvature of the surfactant arrangement is induced, resulting in the formation of columnar micelles (with the head groups turned outwards).

In the system studied here, the transformation of the inverted hexagonal aluminum phosphate mesostructure into the lamellar phase corresponds to a similar increase of each surfactant head group's surface area at the surfactant/inorganic interface. However, the charges are different as compared to the silica synthesis: The surfactant phosphate head group is anionic. For the inorganic species involved during the generation of the final aluminum phosphate network the charge situation is more complex; both cationic (Al^{3+}) and anionic species (${H_xPO_4}^{(3-x)-}$, $x < 3$) are present at the initial stages of the reaction; in aqueous solution they are assumed to form soluble "aggregation oligomers" containing Al–O–P linkages [38], which will carry both positive and negative charges, corresponding to incompletely connected Al and phosphate units at the "loose ends" of the oligomers. A similar situation may be assumed for the ethanolic solution. At any rate, the density of positive charges within the oligomers, which "match" the negatively charged surfactant head groups, will decrease in the course of the reaction. Thus, the observation of an inverted hexagonal intermediate that transforms into a lamellar phase is consistent with the "charge density matching" concept.

10.3 Tubular Mesoporous Aluminum Phosphates

Mesoporous aluminum phosphates with disordered arrangements of tubular pores and specific surface (BET = Brunauer–Emmett–Teller method) areas up to about 600 $m^2 g^{-1}$ can be prepared by utilization of long-chain primary alkyl amines ($C_nH_{2n+1}NH_2$, $n = 12, 14, 16$) as structure-directing agents in ethanol. This is comprehensively described elsewhere [31] for the synthesis from $Al(O^iPr)_3$ and H_3PO_4, the regular precursors for mesostructured aluminum phosphates. This is the first successful synthesis of porous (i.e., nonlamellar) mesostructured aluminum phosphates under predominantly alcoholic conditions, which considerably extends the range of synthetic opportunities. In particular, the alcoholic synthesis route makes it possible to use new kinds of inorganic precursors, which are not suitable for aqueous conditions. The following section describes the first synthesis of mesoporous aluminum phosphates from a single-source precursor.

As mentioned above, a major problem in the synthesis of mesostructured aluminum phosphates is the rational control over the stoichiometric composition and short-range structural order in the products. The concept of using single-source molecular precursors offers a promising way to solve these problems. In order to serve as a potential precursor for a mesostructured aluminum phosphate with an ideal AlPO4 stoichiometry, a molecular unit must exhibit several crucial properties.

1. The building block, that is the core unit of the precursor, must contain equal amounts of alternating Al and P atoms connected with each other via bridging O atoms.
2. During the synthesis of the mesostructured material it must be possible to hydrolytically remove the additional ligands from the building block.
3. The building block itself must be stable enough to maintain its structure during the reaction.
4. The precursor must be soluble in the (polar) medium in which the structure-directed synthesis carried out.

A synthesis for a tetrameric molecular aluminum phosphate complex that meets the above-mentioned requirements was reported in 1975 [39]. The complex consists of an $Al_4P_4O_{12}$ core; each P atom carries one OH group whereas each Al atom is coordinated by three ethanol ligands, resulting in an overall six-fold coordination. A fourth ethanol ligand per Al atom is present in the coordination sphere, but not directly bound to any of the corner atoms of the core. Finally, a total of four chlorine atoms are also part of the tetrameric complex; the empirical formula is $[Al(PO_4)(HCl)(C_2H_5OH)_4]_4$ (Fig. 6). The precursor is soluble and stable in ethanol as well as in methanol. At room temperature the ligands serve as "protective groups" that prevent the tetrameric units from linking together into a long-range $AlPO_4$ structure; at elevated temperature (>50 °C) the ligands are removed from the core, which leads to the precipitation of amorphous aluminum phosphate. The complex is not stable under aqueous conditions; hydrolysis not only of the ligands but also of the Al–O–P linkages in the core is observed when water is added. Owing to these solution properties and hydrolytic behavior the complex is suitable for the structure-directed synthesis of mesostructured aluminum phosphates; such a synthesis must be carried out under strictly nonaqueous conditions.

The alcoholic synthesis of mesoporous aluminum phosphates from the single-source precursor with long-chain primary alkyl amines ($C_nH_{2n+1}NH_2$, $n = 12$, 14, 16) as structure-directing agents leads to disordered tubular mesostructures similar to those obtained from two separate sources of Al and P under otherwise the same conditions. Figure 7 shows the powder XRD patterns of the *as-synthesized* as well as of the porous samples; the single, relatively broad reflection at low diffraction angle is typical of the disordered arrangement of tubular surfactant arrays within the inorganic matrix. The surfactant can be removed quantitatively by solvent extraction as confirmed by IR spectroscopy and elemental analysis. The d spacing depends on

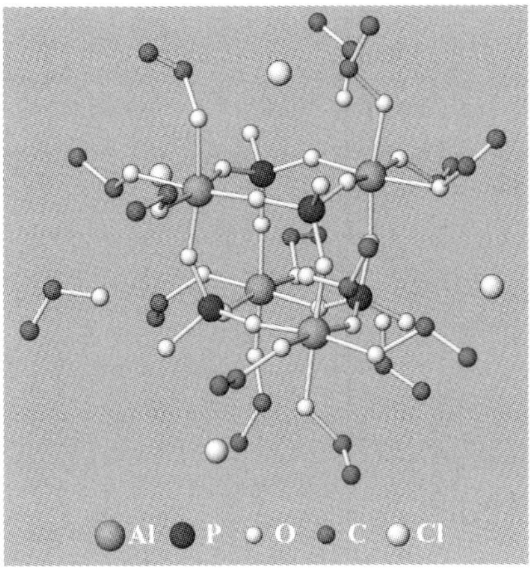

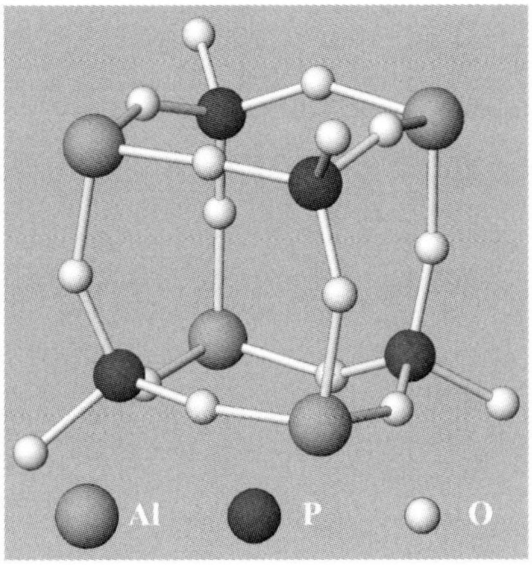

Fig. 6. (Left) Schematic representation of the molecular structure of $[Al(PO_4)(HCl)(C_2H_5OH)_4]_4$. P is coordinated by four O atoms, three of which are bridging towards Al; the fourth is terminal. Al is coordinated by six O atoms, with three of them bridging towards P and the other three belonging to ethanol ligands. Additional ethanol ligands as well as Cl atoms are found in the outer coordination shell. (Right) The same view without the ethanol and Cl ligands clarifies the cube-like structure of the core unit. (Drawings were generated with *Atoms for Windows 3.2*; structural data were taken from reference [39]; H atom positions are not known.)

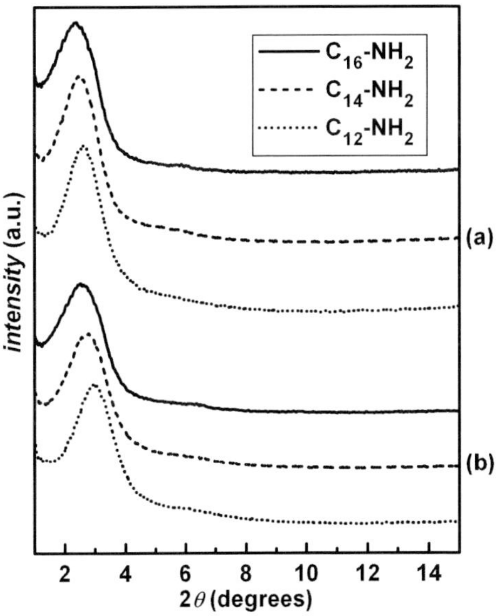

Fig. 7. Powder XRD patterns of mesostructured aluminum phosphates prepared under alcoholic conditions from $[Al(PO_4)(HCl)(C_2H_5OH)_4]_4$ with C_{12}-NH_2, C_{14}-NH_2, and C_{16}-NH_2 (about 15 wt.-%): (a) as synthesized; (b) after extraction of the surfactant.

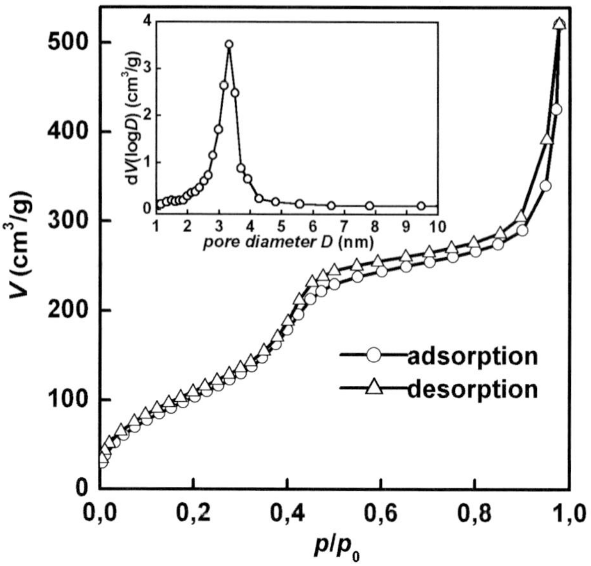

Fig. 8. Nitrogen adsorption/desorption isotherm of a mesoporous aluminum phosphate prepared under alcoholic conditions from $[Al(PO_4)(HCl)(C_2H_5OH)_4]_4$ with C_{16}-NH_2; the specific BET surface area is 410 $m^2\ g^{-1}$. Inset: Pore size distribution (diameter) as calculated by the BJH method from the desorption branch of the isotherm.

the surfactant chain length and slightly shrinks upon surfactant removal. Nitrogen physisorption (Fig. 8) confirms the mesoporous nature of the samples: The adsorption/desorption isotherms are of type IV with a well-defined step at $p/p_0 = 0.35$–0.45, indicating capillary condensation. The pore size (diameter) distribution (calculated by the BJH = Barrett–Joyner–Halender method from the desorption branch) has its maximum at about 3.3 nm; the specific surface area (calculated by the BET equation for $p/p_0 = 0.05$–0.2) is 410 m^2g^{-1}. Hysteresis is attributable to textural porosity [40,41].

Figure 9 shows the ^{27}Al MAS NMR (magic angle spinning nuclear magnetic resonance) spectra of a typical sample both *as-synthesized* and after surfactant removal. Both spectra exhibit a resonance at about 42 ppm, which corresponds to tetrahedral $Al(OP)_4$ groups, and another resonance at about −7 ppm, which is attributable to six-fold coordinated Al with P in the second coordination shell and, presumably, additional H_2O or OH groups, that is $Al(OP)_x(H_2O)_{6-x}$ $(x \leq 4)$. Additionally, there is a weakly resolved resonance around 20 ppm that has been identified [30] as corresponding to five-fold coordinated Al, again with P in the second shell, that is, $Al(OP)_x(H_2O)_{5-x}$. The ^{31}P MAS NMR spectra of the same samples are shown in Fig. 10; they exhibit relatively broad signals between 0 and −30 ppm, which may be attributed to four-fold coordinated P with O–Al (tetrahedral and/or octahedral Al) and various amounts of H_2O or OH groups, that is, $P(OAl)_x(H_2O)_{4-x}$.

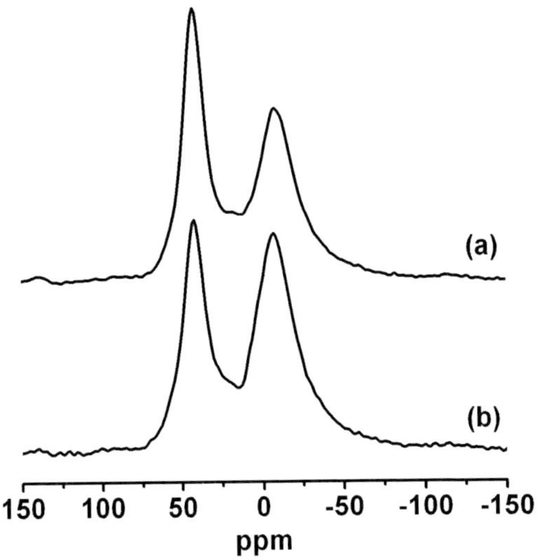

Fig. 9. ^{27}Al MAS NMR spectra of mesostructured aluminum phosphates prepared from $[Al(PO_4)(HCl)(C_2H_5OH)_4]_4$ with C_{16}-NH_2: (a) as synthesized; (b) after surfactant extraction. The two resonances at about 42 and −7 ppm correspond to $Al(OP)_4$ and $Al(OP)_x(H_2O)_{6-x}$, respectively.

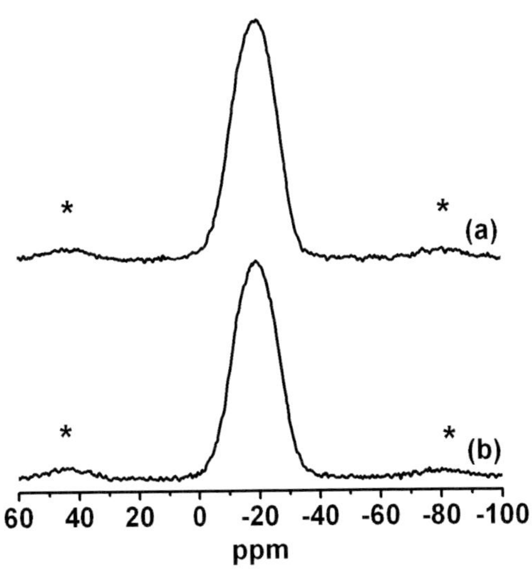

Fig. 10. ^{31}P MAS NMR spectra of mesostructured aluminum phosphates prepared from $[Al(PO_4)(HCl)(C_2H_5OH)_4]_4$ with C_{16}-NH_2: (a) as synthesized; (b) after surfactant extraction. The spectra show broad lines that may consist of several resonances corresponding to $P(OAl)_x(H_2O)_{4-x}$. * = spinning side bands.

Tab. 1. Relative molar ratios of Al, P, C, and N (according to elemental analysis) in some representative mesostructured aluminium phosphate samples as synthesised under alcoholic conditions **(a)** from $[Al(PO_4)(HCl)(C_2H_5OH)_4]_4$ and **(b)** from $Al(O^iPr)_3$ and H_3PO_4.

Synthesis	*Surfactant*	*Elemental analysis (relative molar amounts)*				*Al/P*	*C_n-NH_2/P*
		Al	*P*	*C*	*N*		
(a)	C_{12}-NH_2	0.117	0.118	0.710	0.057	0.99	0.50
(a)	C_{14}-NH_2	0.100	0.101	0.752	0.052	1.00	0.53
(a)	C_{16}-NH_2	0.099	0.099	0.754	0.048	1.00	0.48
(b)	C_{12}-NH_2	0.140	0.125	0.682	0.053	1.12	0.45
(b)	C_{14}-NH_2	0.117	0.102	0.730	0.050	1.14	0.51
(b)	C_{16}-NH_2	0.122	0.108	0.723	0.044	1.14	0.42

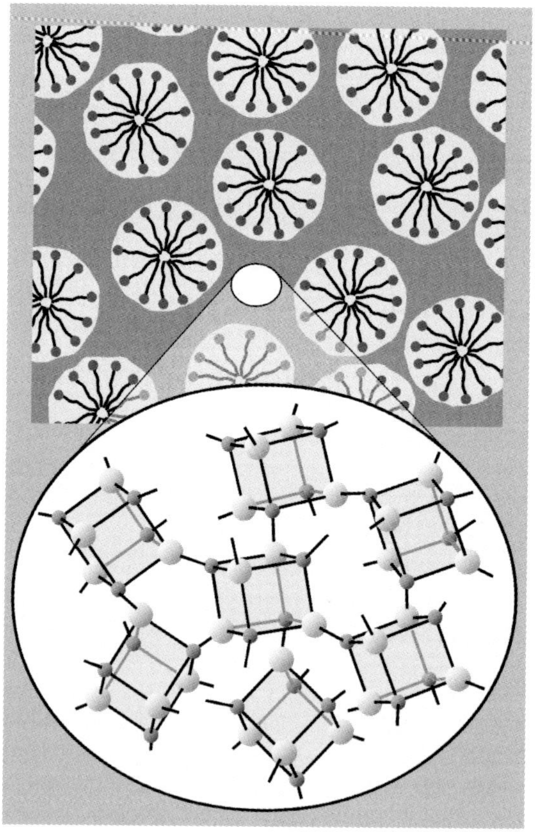

Fig. 11. Schematic representation of a mesostructured aluminum phosphate prepared from the single-source molecular precursor $[Al(PO_4)(HCl)(C_2H_5OH)_4]_4$. The inorganic matrix consists of cube-like $Al_4P_4O_{12}$ core units connected with each other by oxygen bridges; for clarity, oxygen atoms (Fig. 6) are not shown.

The most important property of the products from the single-source synthesis is the stoichiometry. Elemental analysis shows that in the samples prepared from $[Al(PO_4)(HCl)(C_2H_5OH)_4]_4$ the Al/P ratio is always unity whereas an excess (about 12 %) of Al is found in samples prepared from $Al(^iOPr)_3$ and H_3PO_4 under otherwise the same conditions (Table 1). The molar composition suggests that the core unit of the precursor may remain intact during the synthesis, thus serving as a building block for the inorganic network (Fig. 11).

10.4 Conclusions

Nonaqueous synthesis conditions give way to novel mesoporous or mesostructured aluminum phosphates. When *n*-dodecyl phosphate is used as a structure-director under basically ethanolic conditions, an aluminum phosphate/surfactant composite with a previously unobserved inverted hexagonal structure can be obtained. The utilization of primary alkyl amines leads to materials with randomly ordered tubular mesopores. Under these conditions it is possible to use a single-source molecular precursor, which allows for the first rational synthesis of ordered mesoporous aluminum phosphates with strict 1:1 molar ratios of Al and P.

Acknowledgements

SAXS measurements were performed at Hamburger Synchrotronstrahlungslabor (HASYLAB). We thank Sérgio S. Funari and Gert Rapp for their help at the beamline and valuable discussion of the data. We thank Marcus Schulz and Christian Jäger at Friedrich-Schiller University, Jena for recording of the NMR spectra and for helpful discussion of the data. Financial support by the Deutsche Forschungsgemeinschaft (Fr 1372/5-1) and the Fonds der Chemischen Industrie is gratefully acknowledged.

References

1 J.S. Beck, J.C. Vartuli, W.J. Roth, M.E. Leonowicz, C.T. Kresge, K.D. Schmitt, C.T.-W. Chu, D.H. Olson, E.W. Sheppard, S.B. McCullen, J.B. Higgins, J.L. Schlenker, *J. Am. Chem. Soc.* **1992**, *114*, 10 834.

2 S.T. Wilson, B.M. Lok, C.A. Messina, T.R. Cannan, E.M. Flanigen, *J. Am. Chem. Soc.* **1982**, *104*, 1146.

3 B.M. Lok, C.A. Messina, R.L. Patton, R.T. Gajek, T.R. Cannan, E.M. Flanigen, *J. Am. Chem. Soc.* **1984**, *106*, 6092.

4 For review see for example: M. Hartmann, L. Kevan, *Chem. Rev.* **1999**, *99*, 635.

5 M. Tiemann, M. Fröba, *Chem. Mater.* **2001**, *13*, 3211.

6 B. Kraushaar-Czarnetzki, W.H.J. Stork, R.J. Dogterom, *Inorg. Chem.* **1993**, *32*, 5029.

7 G.A. Ozin, S. Oliver, *Adv. Mater.* **1995**, *7*, 943.

8 A. Sayari, I. Moudrakovski, J.S. Reddy, C.I. Ratcliffe, J.A. Ripmeester, K.F. Preston, *Chem. Mater.* **1996**, *8*, 2080.

9 Q. Gao, J. Chen, R. Xu, Y. Yue, *Chem. Mater.* **1997**, *9*, 457.

10 S. Cheng, J.-N. Tzeng, B.-Y. Hsu, *Chem. Mater.* **1997**, *9*, 1788.

11 M. Fröba, M. Tiemann, *Chem. Mater.* **1998**, *10*, 3475.

12 M. Schulz, M. Tiemann, M. Fröba, C. Jäger, *J. Phys. Chem. B* **2000**, *104*, 10 473.

13 Y.Z. Khimyak, J. Klinowski, *Chem. Mater.* **1998**, *19*, 2258.

14 T. Kimura, Y. Sugahara, K. Kuroda, *Chem. Mater.* **1999**, *11*, 508.

15 J.O. Perez O., B.B. Borade, A. Clearfield, *J. Mol. Struct.* **1998**, *470*, 221.

16 M. Eswaramoorthy, S. Neeraj, C.N.R. Rao, *Micropor. Mesopor. Mater.* **1999**, *28*, 205.

17 H. Tanaka, M. Chikazawa, *J. Mater. Chem.* **1999**, *9*, 2923.

18 P. Feng, X. Bu, G. D. Stucky, *Inorg. Chem.* **2000**, *39*, 2.

19 X.S. Zhao, G.Q. Lu, *Micropor. Mesopor. Mater.* **2001**, *44–45*, 185.

20 Z.-Y. Yuan, T.-H. Chen; J.-Z. Wang; H.-X. Li, *Mater. Chem. Phys.* **2001**, *68*, 110.

21 B.T. Holland, P.K. Isbester, C.F. Blanford, E.J. Munson, A. Stein, *J. Am. Chem. Soc.* **1997**, *119*, 6796.

22 Y.Z. Khimyak, J. Klinowski, *Phys. Chem. Chem. Phys.* **2001**, *3*, 1544.

23 T. Kimura, Y. Sugahara, K. Kuroda, *Micropor. Mesopor. Mater.* **1998**, *22*, 115.

24 P. Feng, Y. Xia, J. Feng, X. Bu, G.D. Stucky, *Chem. Commun.* **1997**, 949.

25 B. Chakraborty, A.C. Pulikottil, B. Viswanathan, *Appl. Catal. A* **1998**, *167*, 173.

26 M.P. Kapoor, A. Raj, *Appl. Catal. A* **2000**, *203*, 311.

27 Z. Luan, D. Zhao, H. He, J. Klinowski, L. Kevan, *J. Phys. Chem. B* **1998**, *102*, 1250.

28 Z. Luan, D. Zhao, L. Kevan, *Micropor. Mesopor. Mater.* **1998**, *20*, 93.

29 Z. Luan, D. Zhao, H. He, J. Klinowski, L. Kevan, *Stud. Surf. Sci. Catal.* **1998**, *117*, 103.

30 S. Cabrera, J.E. Haskouri, C. Guillem, A. Beltrán-Porter, D. Beltrán-Porter, S. Mendioroz, M.D. Marcos, P. Amorós, *Chem. Commun.* **1999**, 333.

31 N.C. Masson, H.O. Pastore, *Micropor. Mesopor. Mater.* **2001**, *44–45*, 173.

32 M. Tiemann, M. Schulz, C. Jäger, M. Fröba, *Chem. Mater.* **2001**, *13*, 2885.

33 M. Tiemann, M. Fröba, *Chem. Commun.* **2002**, 406.

34 S.A. Bagshaw, E.Prouzet, T.J. Pinnavaia, *Science* **1995**, *269*, 1242.

35 P.T. Tanev, T.J. Pinnavaia, *Science* **1995**, *267*, 865.

36 M. Tiemann, M. Fröba, G. Rapp, S.S. Funari, *Chem. Mater.*, **2000**, *12*, 1342.

37 A. Monnier, F. Schüth, Q. Huo, D. Kumar, D. Margolese, R.S. Maxwell, G. D. Stucky, M. Krishnamurty, P. Petroff, A. Firouzi, M. Janicke, B.F. Chmelka, *Science* **1993**, *261*, 1299.

38 S. Oliver, A. Kuperman, G.A. Ozin, *Angew. Chem. Int. Ed.* **1998**, *37*, 46.

39 J.E. Cassidy, J.A.J. Jarvis, R.N. Rothon, *J. Chem. Soc., Dalton Trans.* **1975**, 1497.

40 Y. Long, T. Xu, Y. Sun, W. Dong, *Langmuir* **1998**, *14*, 6173.

41 C.G. Sonwane, S.K. Bhatia, *Langmuir* **1999**, *15*, 2809.

11
Organic/Inorganic Functional Materials for Light-Emitting Devices Based on Conjugated Bisphosphonates

*Sabine Stockhause, Peter Neumann, Michael Kant, Ulrich Schülke, and Sigurd Schrader**

11.1
Introduction

The fabrication of thin layers is essential for many technical applications. To guarantee the miniaturization of devices the prepared films need to be thinner and denser than can be produced by spin coating. For many applications the molecules have to show an ordered structure within these thin films to enhance or even cause certain desired properties. These ordered structures are used for many different applications in microelectronics, catalysis, and optics.

11.1.1
Phosphates and Phosphonates: Structure and Intercalation

The best example for a perfectly ordered structure is the single crystal. In this respect, the transition metal phosphates and phosphonates are an interesting class of compounds because they are easy to synthesize and crystallize [1–10]. Besides, they show a layered structure that offers numerous possibilities of intercalation [11–15]. This allows the insertion of molecules with certain properties adjusted to different applications, such as catalysts [16–19], preparative-scale separation of enantiomers [20], proton conductors [21,22]. The intercalation of the zirconium phosphate structure with ionophors or chromophors has already been described [23–26] and was the starting point of this work. These materials are then used as membranes [27], selective electrodes, or NLO (nonlinear optics) materials [28–30]. However, only the insertion of small organic molecules into the layered phosphate structure is possible without destroying the crystalline structure of the transition metal phosphate. If trying to insert larger organic molecules with desired properties between the phosphate layers the crystal is destroyed and therefore looses the pre-orienting effect of the inner surface of the structure.

Like most transition metal phosphates [31,32], in the crystalline state α-zirconium phosphate and its organic equivalents (e.g., α-zirconium phenylphosphonate) have a layered structure consisting of Zr–O– layers that are separated by a gap of

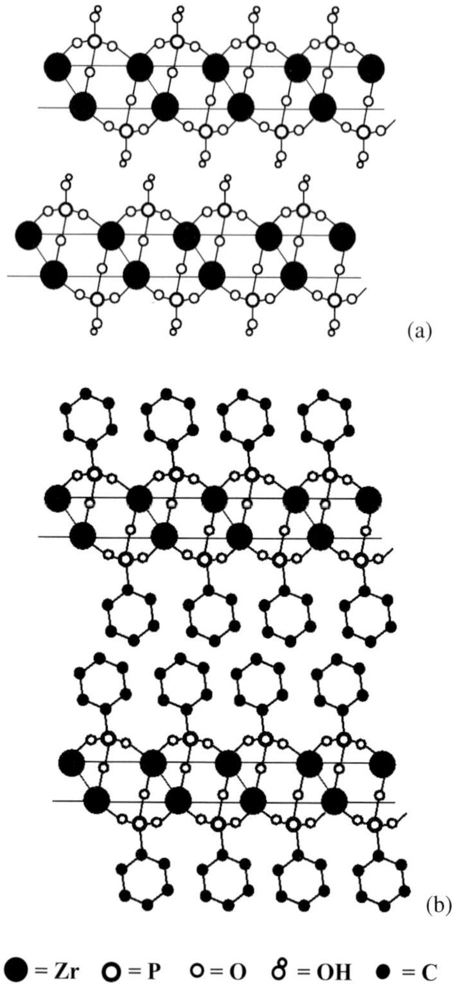

Fig. 1. Layered structure of the α-zirconium phosphate (a) and its organic equivalents α-zirconium phenylphosphonate (b) in the crystalline state, consisting of Zr–O– layers that are separated by a gap of about 7 or 14 Å, respectively.

about 7 or 14 Å, respectively [33,34]. The OH-groups or phenyl moieties penetrate into these gaps (Fig. 1).

As these compounds are hardly soluble it should be possible to build these structures layer by layer [35]. In this respect, the system of the transition metal phosphates is of interest because there exists a strong affinity between the transition metal and the phosphate group that should guarantee a good yield of the reaction building up the next layer. And as the compounds are not reacting with

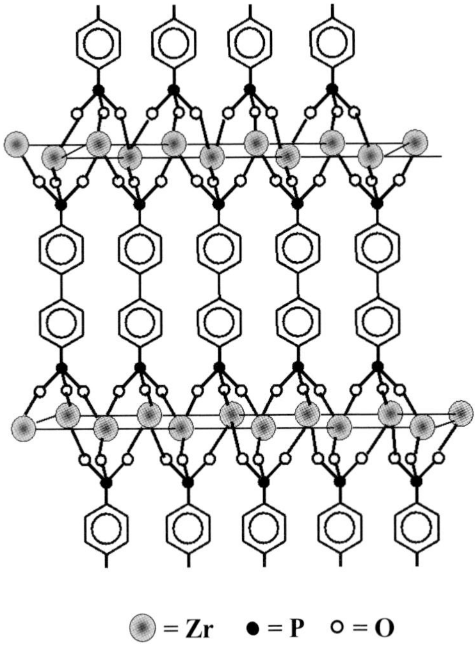

Fig. 2. Idealized structure of Zr-bisphosphonate films prepared by self-assembly. Note the structural analogy to the crystal structures of the zirconium phosphates as given in Fig. 1a.

itself, the creation of more than one monolayer in one step is not possible. In addition, this system offers a great variety of layered crystal structures so that it should be possible to build up many different structures for many different applications, such as asymmetric structures for nonlinear optical devices, conjugated structures for electron transfer, incorporation of dyes for fluorescence.

11.1.2
Self-Assembly Technique

For the layer-by-layer construction of ordered films different techniques are available, such as the Langmuir–Blodgett technique (LB) and self-assembly (SA). The idealized structure of these films is shown in Fig. 2. Comparison with Fig. 1 shows the structural analogy to the crystal structures of the zirconium phosphates. Both structures consist of alternating organic and inorganic layers with similar bond distances.

In the LB technique, a monomolecular ordered layer is formed on a water surface and transferred onto the substrate [36–43]. Drawbacks of this technique are the necessity of planar substrates and the sensitivity towards contamination. For

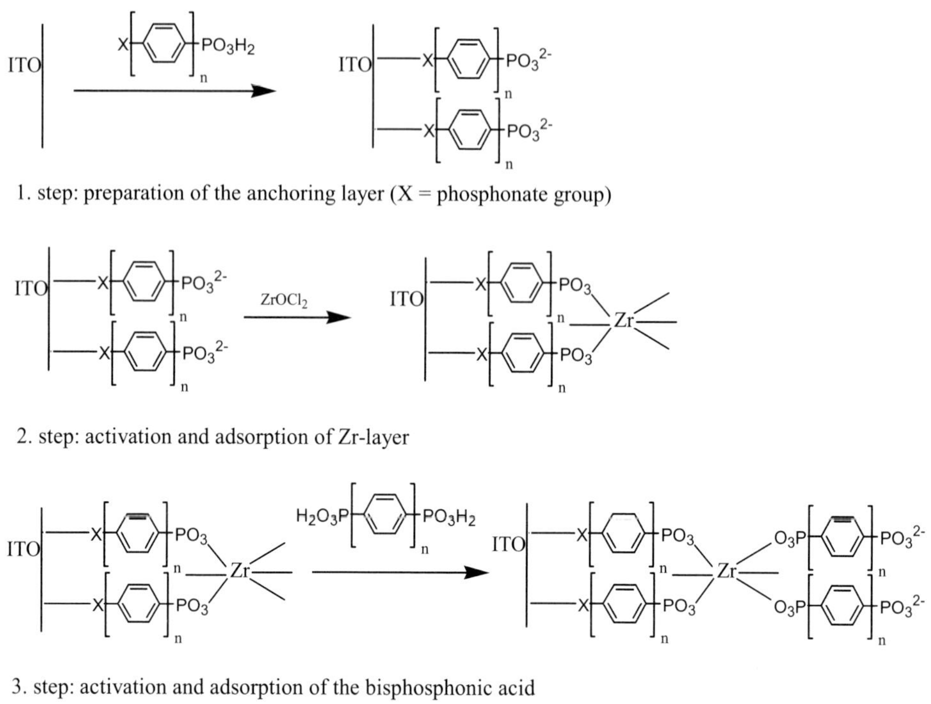

Fig. 3. Scheme of the multilayer preparation as sequence of alternate adsorption of Zr^{4+} and bisphosphonic acid onto the previous layer. The two components can be used as solutions in water or in organic solvents.

the preparation of only one layer a large number of mechanical steps is needed. The most important disadvantage is the creation of only metastable layers as they are only held together by van der Waals interactions and often tend to randomize if stabilizing forces such as H-bonds are missing. Only if functional groups are incorporated into the amphiphiles covalent bonds to neighboring molecules can be formed which then stabilize the given structure.

To overcome these drawbacks the technique of self-assembly was developed [43–47]. This approach was first documented by Sagiv et al. [48,49]. However, the chlorosilanes they used were only moderately suitable for this technique as they showed large structural defects after the assembly of only two or three layers.

The technique of self-assembly is based on the spontaneous adsorption from solution of thermodynamically stable surface layers. It consists of a sequence of alternate activation and adsorption steps, which is illustrated in Fig. 3. Two compounds are alternately deposited as an insoluble monolayer on the previous layer to form thermodynamically stable ordered multilayers. In this irreversible chemisorption process the compound is bound by covalent or ionic interaction to the surface so that the film is quite stable against mechanical and chemical influences. The driving force for the deposition of these organized layers is the thermodynamic stability of the solid state of this class of compounds, which crystallizes spontaneously from a solution containing Zr^{4+} and phosphonic acid. The main difficulty of this technique lies in the deposition of perfect monolayers because each defect will be carried through all the following layers and makes the construction of an ordered multilayer film impossible.

11.1.3 Self-Assembly of Zirconium Phosphonates

Zirconium phosphonates are an ideal choice for this kind of synthesis [30] because both components (Zr^{4+} salt and phosphonates) are of good solubility but form together an insoluble layered structure. The incorporation of various organic moieties is possible as the work in the field of zirconium phosphonate self-assembly layers shows.

In 1988 Mallouk et al. [50] were the first to report the fabrication of ordered zirconium phosphonate multilayers by self-assembly. As organic moiety they used aliphatic bisphosphonic acids and later they investigated the influence of the chain length on the deposition process. Silicon and gold were used as substrates. Already this early work showed that the most important step in this process is the formation of a dense and perfectly ordered anchoring layer to allow the deposition of an ordered multilayer film. Besides, it is necessary to find the most suitable anchoring layer for each substrate that is strongly adsorbed. The growth of the film is monitored by ellipsometry; the film thickness increases by 17 Å for each adsorption cycle, which resembles the interlayer distance in the crystal structure. A stepwise growth with each adsorption cycle was found.

Mallouk et al. [51] discovered later that for silicon the phosphonic acid as well as the Zr^{4+} can be used as anchoring layer. X-ray investigations proofed that the structure of the multilayer resembles that of the crystal.

For various applications (e.g., organic light-emitting diodes (OLEDs), organic field-effect transistors (OFETs), solar cells, integrated circuits, photodiodes) the use of conjugated bisphosphonic acids was investigated [29]. Also, different substrates were tested. Indium–tin oxide (ITO) was a promising substrate, and it is already used as an electrode in various microelectronic devices. However, a suitable anchoring layer needed to be found. By using bisphosphonic acid as an anchoring layer, it was possible to deposit more than 20 layers for the first time, which made the films suitable for applications [52].

In 1990 Katz et al. synthesized multilayers of organic zirconium phosphonates with a polar order [53]. These asymmetric layers were created by the introduction of a third step into the deposition sequence (Fig. 4). They used a monophosphonic acid instead of the bisphosphonic acid so that these organic molecules can be attached with only one end to the zirconium layer. After deposition of the phosphonic acid the other end of the organic molecule is converted into a phosphonic

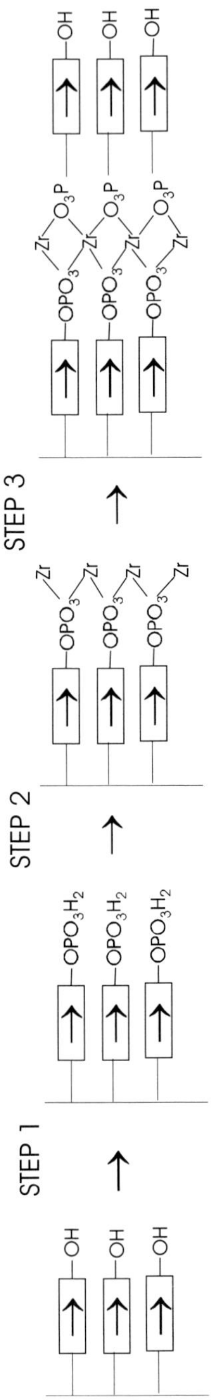

Fig. 4. Scheme of the layer deposition sequence for the formation of layers with polar order [53].

group to which the next zirconium layer can be attached. Silicon and gold served as substrates. These layers can be used in NLO applications and by choosing the most appropriate organic moiety the NLO properties of the multilayer film can be optimized.

Best results were found by using azo dyes as phosphonic acid [28,54,55] such as

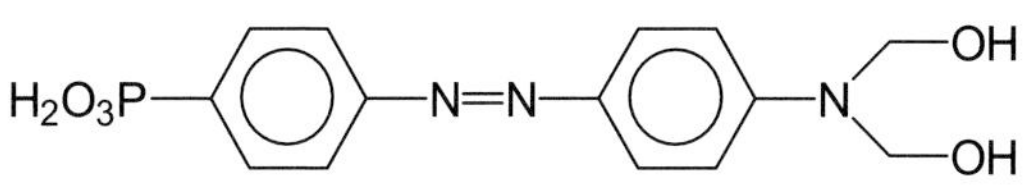

The polar order of these dyes in the layers led to good NLO properties of the film. It was found that the SHG was proportional to the square of the layer thickness. Numerous other dyes were incorporated into the zirconium phosphonate layers, such as thiophenes and quinodimethane [28,29].

The first investigations of the structure of these layers were carried out in 1993 on zirconium alkylphosphonate mono- and multilayers [56]. Thermally stable surface layers of predictable thickness were obtained, which showed an organized structure according to fluorescence measurements. However, infrared (IR) investigations showed that only few of the aliphatic chains were well ordered. Thus no defined long-range order seems to exist in these films, which might be due to a yield lower than 100 % during the deposition of each layer or the later desorption of some molecules from the surface.

Spectroscopic investigations confirmed the theory of the partly disordered component in the film [57]. The disorder was assigned to the conformation of the alkyl chains, which is not uniform throughout the film. This result was explained by the lateral distances of the organic molecules in the film, which is controlled by the inorganic layers. These distances are, therefore, so large that the conformation of the alkyl chain does not matter.

Further investigation of the film growth of self-assembly layers of aliphatic phosphonates followed in 1995 when atomic force microscopy (AFM) was used to monitor the surface structure [58]. This showed that the phosphonate starts growing as islands in the zirconium layer and only after a longer dipping time these islands grow together and cover the whole surface as a uniform film. Only then should the next layer be deposited.

Another interesting approach is the use of paraphenylenevinylene oligomer phosphonates in these systems [59]. The resulting multilayers showed almost the same behavior as polyparaphenylenevinylene (PPV): similar fluorescence spectra and fluorescence decay curves.

To continue the work into the direction of LEDs from self-assembled zirconium phosphonate films, the use of ITO as a substrate was advantageous. With conjugated bisphosphonates in the organic layer light emission should be possible. In this work several conjugated bisphosphonic acids were used to vary the organic layer of these films. Also the use of different transition metals seems promising for optimizing the layer deposition and device performance.

11.2 Chemistry of Bisphosphonates

11.2.1 Material Class, Material Properties

Phosphoric acid (H_3PO_4) as well as its derivatives, phosphonic acid ($R{-}PO_3H_2$) and bisphosphonic acid ($H_2O_3P{-}R{-}PO_3H_2$), are moderately strong acids. If the organic moiety R is small they have a good solubility in water and many polar solvents, and their dissociation is (depending on the nature of the solvent) more or less strongly present. With larger organic moieties the properties of the acid are dominated by the organic part of the molecule.

The conjugated aromatic bisphosphonic acids that were used in the present study for multilayer formation show a particularly strong decrease in the solubility of the free acid in water with increasing number of organic rings. So biphenylbisphosphonic acid (BPBP) is moderately soluble in water, but terphenylbisphosphonic acid (TPBP) and quaterphenylbisphosphonic acid (QPBP) had to be used in form of their ammonium salts, which are more soluble.

11.2.2 Synthesis of Bisphosphonates

The synthesis of the aliphatic bisphosphonic acids is quite straightforward. They can be synthesized in a simple Michaelis–Arbusov reaction [60–63] from alkyl halides R–X (with X = Cl, Br, I) and trialkyl phosphites ($P(OR)_3$) (Eq. 1, Δ = heat)

$$R'{-}X + P(OR)_3 \xrightarrow{\Delta} R'{-}P(O)(OR)_2 + R{-}X \tag{1}$$

Aryl halides, on the contrary, react only in the presence of a catalyst such as copper [64], nickel [65], or palladium compounds [66] with trialkyl phosphites (Eq. 2, cat = catalyst)

$$Ar{-}X + P(OR)_3 \xrightarrow{\Delta + cat} Ar{-}P(O)(OR)_2 + R{-}X \tag{2}$$

Biphenylbisphosphonic acid diethylester was synthesized by conversion of 4,4′-dibrom-biphenyl with triethyl phosphite in the presence of nickel chloride in biphenyl as a solvent. The following hydrolysis of the ester with concentrated hydrobromic acid at boiling temperature leads to the free biphenylbisphosphonic acid (BPBP).

The synthesis of the terphenylbisphosphonic acid and the quaterphenylbisphosphonic acid by hydrolysis of their alkyl esters turned out to be a lot more complicated because of the low solubility of the partially hydrolyzed intermediate products. Therefore, the time for the complete hydrolysis exceeds several days. Only after elaborate separation of the product mixture the free acids could be isolated in only moderate yield.

An alternate route offers the replacement of the triethyl phosphite by the more reactive tris(trimethylsilyl) phosphite ($P(OSiMe_3)_3$), which is easier to hydrolyze.

The tris(trimethylsilyl) phosphite reacts very well with the alkyl bromides and alkyl iodides [67] following the Arbusov mechanism. Besides, it also behaves similar to the trialkyl phosphites [65] in the presence of nickel chloride as catalyst and reacts, therefore, easily with aryl bromides (Eq. 3).

$$\mathrm{Ar{-}X + P(OsiMe_3)_3 \xrightarrow{\Delta+cat} Ar{-}P(O)(OSiMe_3)_2 + Me_3Si{-}X} \tag{3}$$

The resulting arylphosphonic acid bis(trimethylsilyl)esters $\mathrm{Ar{-}P(O)(OSiMe_3)_2}$ can then be converted by addition of water or a water/methanol mixture in a short time and at room temperature into the corresponding acids [68]. For the synthesis of arylbisphosphonic acids with three or more aromatic rings this route offers great advantages compared with the synthesis with trialkyl phosphites.

11.3
Preparation of Zirconium Phosphonate Multilayers by Self-Assembly

11.3.1
General

In the field of organic transition metal phosphonates a broad spectrum of materials can be used for the synthesis of self-assembled layers. In our work we concentrated on the following.

ITO was mainly used as substrate because it is already used as an electrode in an LED, one of the possible application of the prepared multilayers. With the choice of the substrate also the anchoring layer is given as only a few substances can serve as anchoring layer on a given substrate.

For these investigation we have chosen zirconium as transition metal because it shows the best and most regular film growth and the best yield in the reaction with the bisphosphonate and good crystallization properties [1–10].

As the organic part of the layer the shortest aromatic bisphosphonic acids were investigated, as they should be suitable for the use in LEDs. Different chain lengths were chosen to investigate the influence of the length of the aromatic system on film growth.

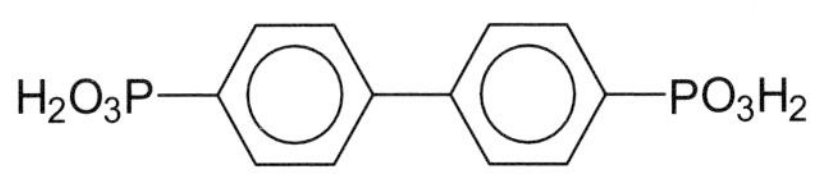

BPBP

TPBP

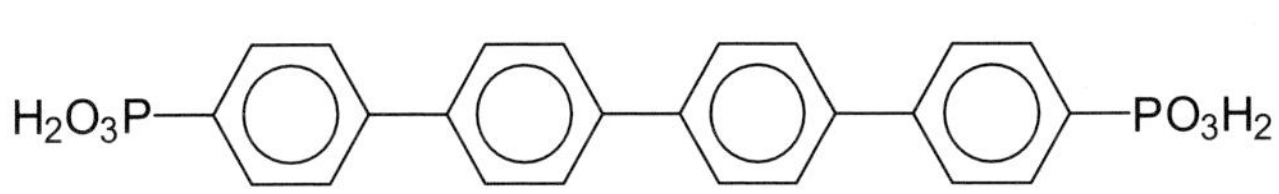

QPBP

If using even longer organic chains the self-assembly of ordered structures is disturbed because these chains then tangle or bind with both ends to the surface.

11.3.2 Substrate Preparation and Anchoring Layer

11.3.2.1 Substrate preparation

The first step in the synthesis is the cleaning and pre-treatment of the substrate. For ITO a special treatment is necessary to reduce the surface roughness of the substrate because otherwise the growth of an ordered film would be inhibited. In order to reduce the roughness of the ITO, several pre-treatment procedures (e.g., O_2 plasma [69–72], UV ozone [69,70,73], acid treatment [74–76]) were used and the surface roughness monitored by AFM. Also, ITO samples from different providers have been compared, and the smoothest have been chosen for further experiments.

As a last pre-treatment step a mixture of concentrated aqueous solution of ammonia and 30 % solution of hydrogen peroxide (3:1) in combination with ultrasound showed to be the most important pre-treatment step for the ITO substrates. The result also depends on the time the substrates stay in this solution. However, even after pre-treatment the roughness is still three times higher than that of silicon wafers, for example. Also, after more time or at higher temperatures the ITO layer is attacked by the solution of NH_3 and H_2O_2. However, ellipsometry showed that under the conditions used the layer thickness of the ITO stayed constant. Because of the surface roughness and the required pre-treatment ITO is not the most suitable substrate for the self-assembly process itself but of great importance for several applications in which it can serve as an electrode.

11.3.2.2 Anchoring layer

The most important step in the self-assembly process is the preparation of a dense and highly ordered anchoring layer, which then allows the deposition of an organized film. Even after finding the most suitable compound for the given substrate the quality of the anchoring depends on many different factors such as concentration of the solution used, temperature, pH, and time of exposure to the solution.

For ITO, bisphosphonic acid was a suitable anchoring layer. The pre-treated ITO substrates were exposed to a solution of 0.5 mM of bisphosphonic acid or their acidic ammonium salts at 50 °C for 3 h [52]. It is difficult to control the deposition of the anchoring layer with ellipsometry because the layer is only a few Ångstrøms thick.

11.3.3 Multilayer Formation

The best demonstration of the suitability of the bisphosphonic acid as the anchoring layer is the possibility of depositing a self-assembled film on the pre-treated

ITO and the anchoring layer. As illustrated in Fig. 3, the multilayer preparation is a sequence of alternate adsorption of Zr^{4+} and bisphosphonic acid onto the previous layer. The two components can be used from solution in water or organic solvents. In this case the crystallization from water was easier than the crystallization from organic solvents, so the compounds were used from aqueous solution for further preparation steps. The layer formation is influenced for each step by the same parameters as discussed for the anchoring layer. An important step is also the cleaning and activation of the surface before depositing the next layer.

For the formation of the multilayers a 2 mM solution of ZrOCl in water was used. The bisphosphonic acid or the acidic ammonium salts were used as a 0.5 mM or 1 mM solution in water with a pH of 6. The samples were exposed to these solutions for times of 20 min to 2 h at temperatures between 20 and 80 °C. After each adsorption step the samples were washed with water and dried. Films of 52 layers were prepared in this way and their stepwise growth was monitored by ellipsometry [52].

It was found that the layer thickness grows linearly with the number of adsorption cycles (Fig. 5). The figure shows also that the absolute thickness depends on the deposition conditions. Even samples prepared under similar conditions can show differences in the layer thickness so that it is difficult to reproduce a certain layer thickness. All samples show a linear growth of the film with the number of adsorption cycles but the thickness per layer varies in dependence on the used bisphosphonic acids and the preparation conditions. For the TPBP single layer a value of 18.4 Å can be calculated from the length of the molecule. Compared with the average experimental value for the thickness of a single layer the theoretical

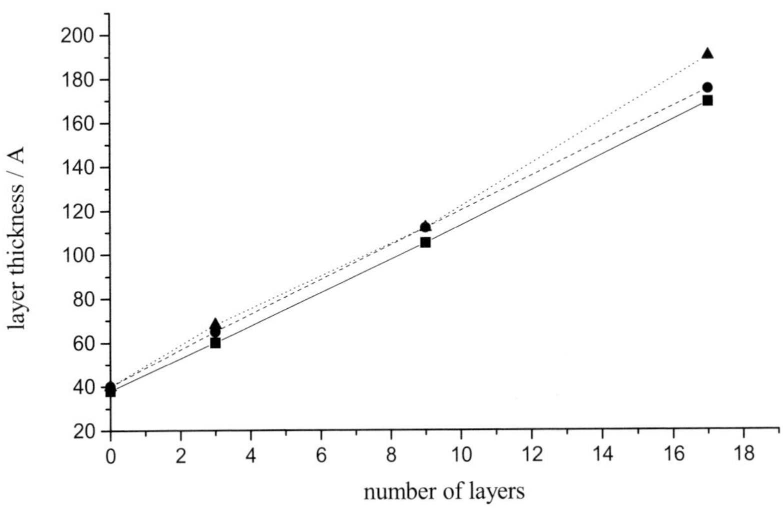

Fig. 5. Thickness of Zr-terphenyl bisphosphonate multilayers determined by ellipsometry (multiangle, at 632.8 nm) as function of the number of deposition cycles for different deposition conditions (■, T = 25 °C, reaction time = 1 h; ●, T = 50 °C, reaction time = 1 h; □, T = 50 °C, reaction time = 1.5 h).

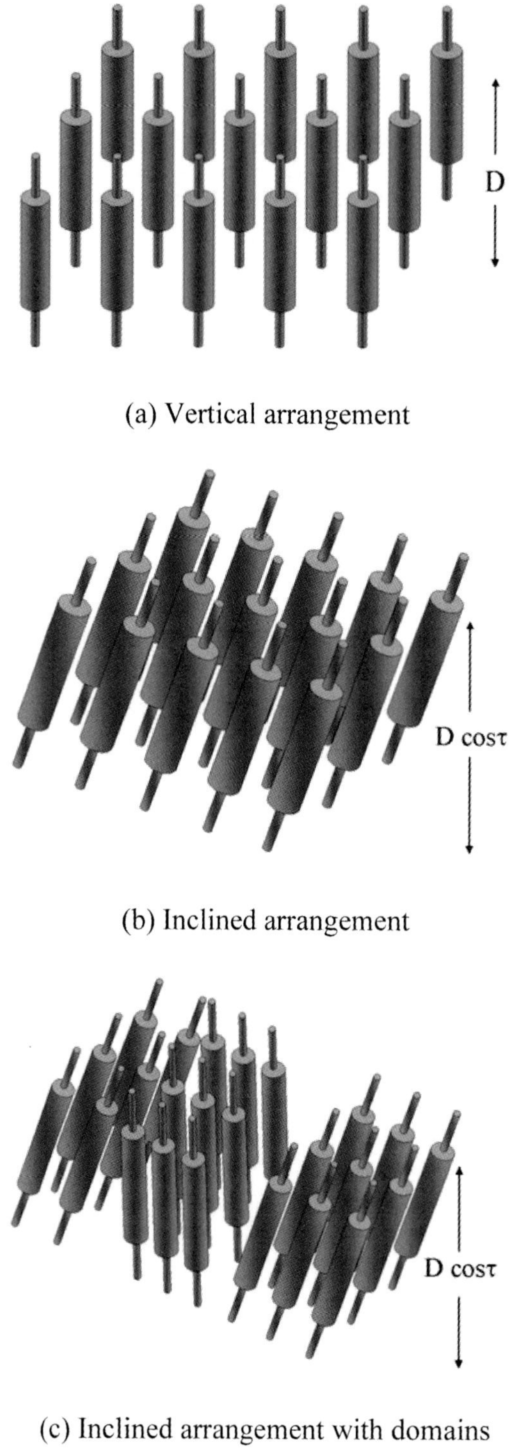

Fig. 6. Schematic presentation of molecular orientation in Zr bisphosphonate multilayers: (a) monodomain with molecular orientation perpendicular to the substrate surface, (b) inclined monodomain, (c) inclined domains with different director orientation; D = molecular length, τ = angle of inclination.

value is higher. This difference between experimental and theoretical layer thickness should be caused by an inclination of the molecules relative to the substrate surface. This explanation can be proofed by structural investigations of the film.

11.3.4 Structural Investigations

To investigate the order of the molecules in these self-assembled films two methods were employed.

11.3.4.1 NEXAFS

As shown in the previous section the results from the ellipsometric measurements gave rise to the assumption that the organic molecules are not standing upright on the surface but are inclined by a certain angle relative to the surface. With the help of NEXAFS measurements (carbon near-edge X-ray adsorption fine structure) it was possible to confirm this assumption [77]. An inclination angle of 45° to 60° was found for the bisphosphonic acids, which is in good agreement with the values calculated from the ellipsometric measurements. However, what cannot be excluded by this investigation is the existence of domains of different inclination angles or different mean director orientation (Fig. 6).

11.3.4.2 X-ray Investigations

X-ray diffraction was used to investigate 20-layer thick films of self-assembled zirconium phosphonates. The spectra (Fig. 7) do not show any Bragg peaks. This might be due to an insufficient crystallinity of the film but it might also confirm the assumption that domains of different inclination angles are present in the film and no long-range order of the organic moieties is, therefore, obtained. These results are in good agreement with the earlier work described in the literature [56].

11.3.5 Automatic Deposition

For a more efficient and reproducible deposition of the self-assembly films, equipment for automatic deposition has been developed. Since the equipment is a closed system it reduces the external influences during the deposition procedure. Figure 8 shows this equipment. The reaction vessel can be heated and the temperature can be varied in a precisely controlled manner. The pumps and valves that control the different reaction and rinsing processes are operated by a computer so that the solutions for deposition as well as for cleaning can be pumped into the vessel at a programmed time and be automatically removed after reaction. With this system two vessels can be operated at the same time and by independent programs. The ultrasound unit improves the cleaning procedure between two deposition cycles. Intensive cleaning is necessary to remove particles from the surface because they

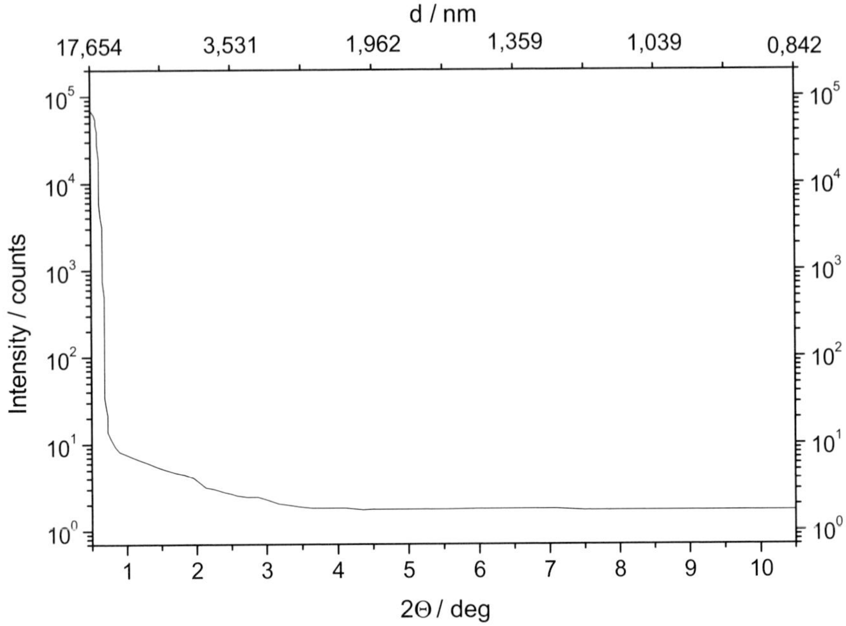

Fig. 7. Low-angle X-ray diffraction pattern of a 20-layer thick film of self-assembled zirconium phosphonate. Note that the spectrum does not show any Bragg peaks [86].

might act as nuclei for the growth of crystals and disturb the deposition of uniform thin layers.

11.4 Applications

Dielectric, luminescent, photocharge generating, and nonlinear optical films have been designed and prepared from transition-metal phosphonate surface multilayers [30]. This class of materials is very promising for such applications because different functions can be produced by variation of the incorporated organic moiety, the chosen transition metal, the obtained molecular packing, and the actual molecular orientation.

In the present study the electroluminescent behavior of the prepared multilayers was investigated. Since the discovery of efficient electroluminescence at low voltages for conjugated low-molecular organic compounds [78] and for conjugated organic polymers [79] in the 1980s and 1990s, organic electroluminescence has become a fast developing field of science and technology in which many fundamental aspects of charge and energy transfer and many aspects of technological

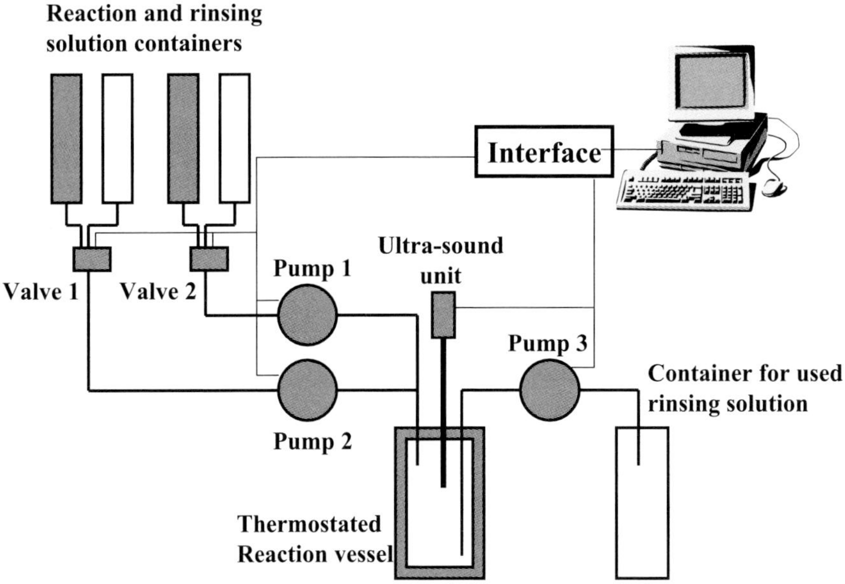

Fig. 8. Schematic presentation of the computer-controlled equipment for automatic deposition of phosphonate multilayers. The reaction vessel can be heated and the temperature can be varied in a precisely controlled manner. Pumps and valves operate the different reaction and rinsing processes. Two vessels (only one is shown) can be operated independently at the same time.

relevance of organic materials have been addressed [80–85]. Meanwhile, some applications based on organic electroluminescence are already in production such as low information content displays in mobile phones, car radios, CD players, or similar applications in the field of consumer electronics.

Transition-metal phosphonates are promising materials for the manufacture of light-emitting devices, which arises from their stability and the large number of different functions that can by realized by their structural variation.

For the investigation of electroluminescence we prepared single-layer devices on a pre-treated ITO/glass substrate. A Zr/TPBP multilayer (about 50 monolayers) was deposited by self-assembly, and an aluminum electrode was evaporated on top of the film. The active area of the LED was 2 mm $\times$ 2 mm. Figure 9 shows the fluorescence emission spectra of such multilayer devices, which are almost identical with the electroluminescence spectrum of this diode. So the Zr/TPBP diode is a blue emitting device, as expected from the active chromophor of these layers.

Figure 10 shows the current–voltage characteristic on a linear scale. A positive sign of the driving voltage means plus at the anode and minus at the cathode side of the diode. We see clearly a rectification behavior of this device but also some

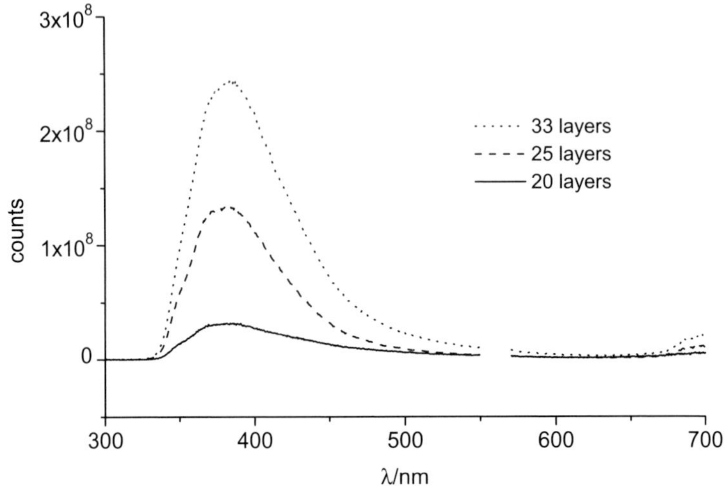

Fig. 9. Fluorescence spectra of zirconium terphenyl bisphophonate (Zr/TPBP) multilayers of different thicknesses [87]

hysteresis in the current–voltage characteristics. This can be explained by charge accumulation in traps prominent in the phosphonate layer. Further investigation of this effect is underway.

Electroluminescence is observed at an onset voltage of 7.5 V in forward direction and less than −11 V in the reverse direction of the diode. For a blue-emitting diode

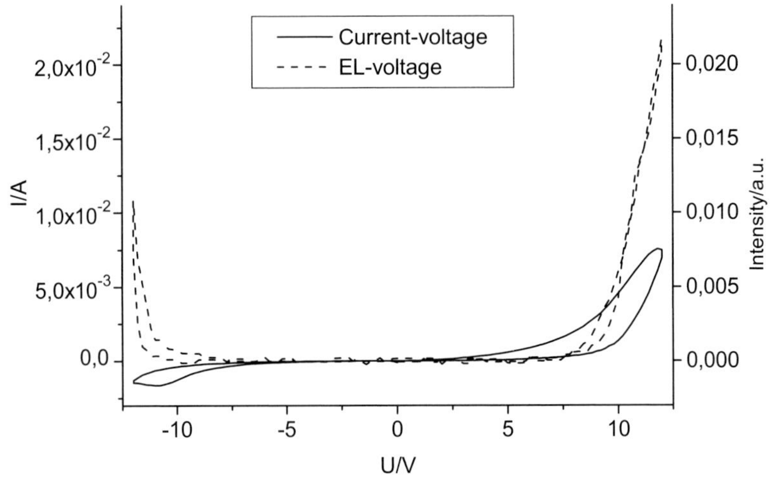

Fig. 10. Current–voltage and electroluminescence–voltage characteristics of an LED with the structure ITO–Zr/TPBP(multilayer)–Al [88]

this value of onset voltage is typical. It can be reduced by further improvement of the injection conditions at the electrodes. These investigations are underway.

A strong increase in electroluminescence is observed above the onset voltage in either direction. The efficiency of these devices is still very low in comparison to other published results on blue light-emitting devices, such as based on polyfluorene. This is not surprising, since we investigated in that initial stage only single-layer devices in which exciton quenching near the electrodes is one possible source of low efficiency. We expect a further increase of device efficiency by improving the device architecture towards a two-layer device. An interesting feature of this single layer device is that it is an ambipolar device because light-emission occurs in both poling directions. This is a hint for injection of carriers of either sign.

11.5 Conclusions

Bisphosphonic acids ($H_2O_3P{-}R{-}PO_3H_2$) are interesting compounds that can be used for formation of functional multilayers by self-assembly. If the organic moiety R is small they show a good solubility in water and many polar solvents. In water they are moderately strong acids. With larger organic moieties the properties of the acid are more and more dominated by the organic part of the molecule. For the formation of organic/inorganic multilayers by self-assembly a chemical reaction between the deposited bisphosphonic acid and a transition metal (e.g., zirconium) is necessary.

The deposition of such conjugated aromatic zirconium bisphosphonates from solution leads to multilayers of sufficient thickness. They are, therefore, an important class of organic-inorganic functional materials, which can be used for applications such as sensing, separation, catalysis, and various applications on optoelectronics and photonics.

For application in electronic devices it is desirable to use ITO as substrate material for the multilayer deposition because it can serve as an electrode in the device. However, special pre-treatment is necessary to ensure good layer-by-layer film deposition. The most important step in this deposition process is the formation of a dense and highly ordered anchoring layer, which implies the right choice of compounds serving as anchoring layer and the best deposition conditions. For ITO substrates the bisphosphonic acid itself can be used as anchoring layer. By the sequential self-assembly process films of zirconium bisphosphonates of up to 58 layers can be deposited. The adsorbed molecules of the phosphonic acid are cross-linked by each zirconium layer. Ellipsometry measurements show a linear growth of the film thickness with the number of adsorption cycles. Within the film the organic moieties are inclined relative to the substrate surface by 45° to 60° and form domains with different directions of inclination so that no long-range order exists. Incorporating these zirconium bisphosphonate films in light emitting device structures with aluminum as top electrode leads to devices emitting in the blue region of the spectrum. Current–voltage and electroluminescence–voltage characteristics show an ambipolar behavior of these devices.

Acknowledgements

The authors would like to thank Prof. Dr. L. Brehmer, Dr. J. Reiche, and P. Imperia, (University of Potsdam, Institute of Physics, Dept. of Condensed Matter Physics, Germany) for their support of this work and many helpful discussions. Many thanks go to Prof. Ch. Wöll and M. Wühn (University of Bochum, Germany) and Dr. W. Braun (Berliner Electron Synchrotron BESSY I) for their support on carrying out NEXAFS measurements. Financial support of German Research Council under project number SCHU-1101/1-3 is gratefully acknowledged.

References

1 A. Clearfield, J. A. Stynes, *J. Inorg. Nucl. Chem.* **1964**, *26*, 117.
2 G. Alberti, E. Torracca, *J. Inorg. Nucl. Chem.* **1968**, *30*, 318.
3 G. Alberti, U. Constantino, R. Guilietti, *J. Inorg. Nucl. Chem.* **1980**, *42*, 1062.
4 M. B. Dines, P. M. DiGiacomo, *Inorg. Chem.* **1981**, 92.
5 G. Alberti, M. Casciola, R. Vivani, R. K. Biswas, *Inorg. Chem.* **1993**, *32*, 4600.
6 G. Alberti, S. Murcia-Mascaros, R. Vivani, *Mater. Chem. Phys.* **1993**, *35*, 187.
7 G. L. Rosenthal, J. Caruso, *J. Solid State Chem.* **1993**, *107*, 497.
8 G. Alberti, F. Marmottini, *Angew. Chem.* **1994**, *106*, 1655.
9 G. Alberti, S. Murcia-Mascaros, R. Vivani, *Mater. Sci. Forum* **1994**, *87*, 152.
10 A. Clearfield, J. D. Wang, Y. Tian, E. Stein, C. Bhardwaj, *J. Solid State Chem.* **1995**, *117*, 275.
11 A. Clearfield, Ch. 1 in Inorganic Ion Exchange Materials, A. Clearfield (ed.), CRC Press, Boca Raton, FL **1982**.
12 G. Alberti, Ch. 2 in Inorganic Ion Exchange Materials, A. Clearfield (ed.), CRC Press, Boca Raton, FL **1982**.
13 G. Alberti, U. Costantino, *Vol. 5*, Oxford University Press, **1991**.
14 C. A. C. Sequeira, M. J. Hudson, Multifunctional Mesoporous Inorganic Solids, Kluwer, Dordrecht **1993**.
15 A. Clearfield, in Progress in Intercalation Research, Kluwer, Dordrecht **1994**, p. 223.
16 U. Costantino, F. Marmottini, M. Curini, O. Rosati, *Catal. Lett.* **1993**, *22*, 333.
17 M. Alcantara-Rodriguez, E. Rodriguez-Castellon, A. Jimenez-Lopez, *Langmuir* **1999**, *15*, 1115.
18 F. J. Perez-Reina, E. Rodriguez-Castellon, A. Jimenez-Lopez, *Langmuir* **1999**, *15*, 2047.
19 M. Jurado-Gonzales, D. L. Ou, B. Ormsby, A. C. Sullivan, J. R. H. Wilson, *Chem. Commun.* **2001**, *1*, 67–68.
20 M. E. Garcia, J. L. Naffin, N. Deng, T. E. Mallouk, *Chem. Mater.* **1995**, *7*, 1968.
21 G. Alberti, M. Casciola, U. Costatino, A. Peraio, E. Montoneri, *Solid State Ionics* **1992**, *50*, 315.
22 G. Alberti, M. Casciola, R. Palombari, A. Peraio, *Solid State Ionics* **1992**, *58*, 339.
23 G. Alberti, R. Vivani, F. Marmottini, P. Zappelli, *J. Porous Mater.* **1998**, *5*, 205.
24 H. E. Katz, G. R. Scheller, T. M. Putvinski, M. L. Schilling, W. L.

Wilson, C. E. D. Chidsey, *Science* **1991**, *254*, 1485.
25 L. A. Vermeulen, M. A. Thompson, *Chem. Mater.* **1994**, *6*, 77.
26 R. Hoppe, G. Alberti, U. Costatino, C. Dionigi, G. Schulz-Ekloff, R. Vivani, *Langmuir* **1997**, *13*, 7252.
27 M. R. El-Sourougy, E. E. Zaki, H. F. Ali, *J. Membr. Sci.* **1997**, *126*, 107.
28 H. E. Katz, M. L. Schilling, S. Ungashe, T. M. Putvinski, G. Scheller, C. E. D. Chidsey, W. L. Wilson, *Proc. SPIE* **1991**, *1560*, 370.
29 H. E. Katz, M. L. Schilling, C. E. D. Chidsey, T. M. Putvinski, R. S. Hutton, *Chem. Mater.* **1991**, *3*, 699.
30 H. E. Katz, *Chem. Mater.* **1994**, *6*, 2227.
31 G. Cao, H. Lee, V. M. Lynch, T. E. Mallouk, *Inorg. Chem.* **1988**, *27*, 2781.
32 G. Cao, V. M. Lynch, J. S. Swinnea, T. E. Mallouk, *Inorg. Chem.* **1990**, *29*, 2112.
33 G. Alberti, *Comprehensive Supramolecular Chemistry, Vol. 7*, **1996**, Pergamon-Elsevier, Oxford.
34 A. Clearfield, *Chem. Mater.* **1998**, *10*, 2801.
35 G. Cao, H.-G. Hong, T. E. Mallouk, *Acc. Chem. Res.* **1992**, *25*, 420.
36 K. B. Blodgett, *J. Am. Chem. Soc.* **1935**, *57*, 1007.
37 K. B. Blodgett, I. Langmuir, *Phys. Rev.* **1937**, *51*, 964.
38 W. A. Barlow, in Thin Films Science and Technology, Vol. 1, Elsevier, Amsterdam **1980**, p. 288.
39 G. Roberts, Plenum Press, New York, **1990**, p. 425.
40 A. Barraud, in *Proc. 5th Int. Conf. Langmuir–Blodgett Films*, Paris, France, 26–30 August **1991**, Elsevier, Amsterdam **1992**, p. 858.
41 R. H. Tredgold, Order in Thin Organic Films, CUP, Cambridge **1994**.
42 M. C. Petty, Langmuir–Blodgett Films: An Introduction, CUP, Cambridge **1996**.
43 A. Ulman, An Introduction to Ultrathin Organic Films: from Langmuir–Blodgett to Self-Assembly, Academic Press, Boston **1991**.
44 G. Decher, J. D. Hong, *Ber. Bunsenges. Phys. Chem.* **1991**, *95/11*, 1430.
45 N. Tillmann, A. Ulman, T. L. Penner, *Langmuir* **1989**, *5*, 101.
46 A. Ulman, *Adv. Mater.* **1990**, *2*, 573.
47 G. M. Whitesides, P. E. Laibinis, *Langmuir* **1990**, *6*, 87.
48 L. Netzer, J. Sagiv, *J. Am. Chem. Soc.* **1983**, *105*, 674.
49 R. Maos, J. Sagiv, *J. Coll. Interface Sci.* **1984**, *100*, 465.
50 H. Lee, L. J. Kepley, H.-G. Hong, S. Akther, T. E. Mallouk, *J. Phys. Chem.* **1988**, *92*, 2597.
51 H. Lee, L. J. Kepley, H.-G. Hong, T. E. Mallouk, *J. Am. Chem. Soc.* **1988**, *110*, 618.
52 S. Stockhause, P. Neumann, S. Schrader, M. Kant, L. Brehmer, *Synthetic Metals* **2002**, *127*, 295.
53 T. M. Putvinski, M. L. Schilling, H. E. Katz, C. E. D. Chidsey, A. M. Mujsce, A. B. Emerson, *Langmuir* **1990**, *6*, 1567.
54 H. E. Katz, W. L. Wilson, G. Scheller, *J. Am. Chem. Soc.* **1994**, *116*, 6636.
55 H. E. Katz, G. Scheller, T. M. Putvinski, M. L. Schilling, W. L. Wilson, C. E. D. Chidsey, *Science* **1991**, *254*, 1485.
56 M. L. Schilling, H. E. Katz, S. M. Stein, S. F. Shane, W. L. Wilson, S. Buratto, S. B. Ungashe, G. N. Taylor, T. M. Putvinski, C. E. D. Chidsey, *Langmuir* **1993**, *9*, 2156.
57 B. L. Frey, D. G. Hanken, R. M. Com, *Langmuir* **1993**, *9*, 1815.
58 H. Byrd, J. L. Snover, M. E. Thompson, *Langmuir* **1995**, *11*, 4449.
59 H. E. Katz, S. F. Bent, W. L. Wilson, M. L. Schilling, S. Ungashe, *J. Am. Chem. Soc.* **1994**, *116*, 6631.
60 A. Michaelis, R. Kaehne, *Berichte der Deutschen Chemischen Gesellschaft* **1898**, *31*, 1048.
61 A. Arbusov, *Journal of the Russian Physico-Chemical Society, Chemical Part (russ.)* **1906**, *38*, 687.
62 A. Arbusov, *Journal of the Russian Physico-Chemical Society, Chemical Part (russ.)* **1910**, *42*, 395.
63 G. M. Kosolapoff, The Synthesis of Phosphonic and Phosphinic Acids, Organic Reactions, Vol. VI, Wiley, New York **1951**.

64 P. TAVS, F. KORTE, *Tetrahedron* **1967**, *23*, 4677.
65 P. TAVS, *Chem. Ber.* **1970**, *103*, 2428.
66 D. J. MARTIN, German Patent DE 2118223 **1970/1971**.
67 L. WOZNIAK, J. CHOJNOWSKI, *Tetrahedron* **1989**, *45*, 2465.
68 M. SEKINE, I. YAMAMOTO, A. HASHIZUME, T. HATA, *Chem. Lett.* **1977**, 485.
69 T. A. BEIERLEIN, W. BRUTTING, H. RIEL, E. I. HASKAL, P. MULLER, W. RIESS, *Synth. Met* **2000**, 0379.
70 X. M. DING, L. M. HUNG, L. F. CHENG, Z. B. DENG, X. Y. HOU, C. S. LEE, S. T. LEE, *Appl. Phys. Lett.* **2000**, *76*, 2704.
71 K. FURUKAWA, Y. TERASAKA, H. UEDA, M. MATSUMURA, *Synthetic Metals* **1997**, *91*, 99.
72 C. C. WU, C. I. WU, J. C. STURM, A. KAHN, *Appl. Phys. Lett.* **1997**, *70*, 1348.
73 Q. WANG, W. GUO, M. C. Y. CHAN, S. LIU, E. H. LI, C. C. SURYA, *Proc. SPIE-Int. Soc. Opt. Eng.* **2000**, *4078*, 289.
74 Q. T. LE, F. NUESCH, E. W. FORSYTHE, L. J. ROTHBERG, Y. GAO, *Proc. SPIE-Int. Soc. Opt. Eng.* **1999**, *3797*, 301.
75 F. NÜESCH, L. J. ROTHBERG, E. W. FORSYTHE, Q. T. LE, Y. GAO, *Appl. Phys. Lett.* **1999**, *74*, 880.
76 A. ANDERSSON, N. JOHANSSON, P. BRÖMS, N. YU, D. LUPO, W. R. SALANECK, *Adv. Mater.* **1998**, *10*, 859.
77 S. SCHRADER, M. WÜHN, C. WÖLL, P. NEUMANN, unpublished results **1997**.
78 C. W. TANG, S. A. VANSLYKE, *Appl. Phys. Lett.* **1987**, *51*, 913.
79 J. H. BURROUGHES, D. D. BRADLEY, A. R. BROWN, R. N. MARKS, K. MACKAY, R. H. FRIEND, P. L. BURNST, A. B. HOLMES, *Nature* **1990**, *347*, 539.
80 P. L. BURN, A. B. HOLMES, A. KRAFT, D. D. C. BRADLEY, A. R. BROWN, R. H. FRIEND, R. W. GYMER, *Nature* **1992**, *356*, 47.
81 J. R. SHEATS, H. ANTONIADIS, M. HUESCHEN, W. LEONARD, J. MILLER, R. MOON, D. ROITMAN, A. STOCKING, *Science* **1996**, *273*, 884.
82 T. TSUTSUI, E. AMINAKA, C. P. LIN, D.-U. KIM, *Phil. Trans. R. Soc. Lond. A* **1997**, *355*, 801.
83 V. BULOVIC, P. E. BURROWS, S. R. FORREST, *Semicond. Semimetals* **2000**, *64*, 255.
84 M. GROSS, D. C. MULLER, H.-G. NOTHOFER, U. SCHERF, D. NEHER, C. BRAUCHLE, K. MERRHOLZ, *Nature* **2000**, *405(6787)*, 661.
85 Y. SHIROTA, *J. Mater. Chem.* **2000**, *10*, 1.
86 J. REICHE, S. SCHRADER, S. STOCKHAUSE, unpublished results **2001**.
87 S. SCHRADER, P. NEUMANN, unpublished results **1999**.
88 P. IMPERIA, S. SCHRADER, unpublished results **2000**.

12
Prussian Blue Derived, Organometallic Coordination Polymers with Nanometer-Sized Cavities

R. Dieter Fischer, Hilka Hanika-Heidl, Min Ling, and Rolf Eckhardt*

12.1
Introduction

Prussian blue (PB) may be considered as one of the archetypes of voluminous three-dimensional (3D) coordination polymers, in which six-coordinated transition metal ions are interlinked by quasi-linear μ-CN spacers [1]. For instance, a xenon atom may tightly be encapsulated in the cuboid voids of the PB-derivative $[Cd^{II}Pt^{IV}(CN)_6]$ [2]. Pauling coined the term *super*-Prussian blue (SPB) for the compound $[Ag^{I}{}_3Co^{III}(CN)_6]$ (**1**) as early as in 1968 [3], since its framework structure involves almost linear Co–CN $\rightarrow$ Ag $\leftarrow$ NC–Co fragments, which are twice as long as the Fe^{II}–CN $\rightarrow$ Fe^{III} linkages of common PB (i.e., about one nanometer). The first crystallographically established organometallic SPB derivative is $[(Me_3Sn)_3Co^{III}(CN)_6] = {}^3_\infty[Co^{III}\{\mu\text{-}CNSn(Me_3)NC\}_3]$ (**2**), in which the silver atoms of **1** are replaced by tin atoms carrying three methyl groups perpendicular to the N–Sn–N axis [4,5]. Formally, a –CN $\rightarrow$ $Sn(Me_3)$ $\leftarrow$ NC– spacer would result if a stanna-isocyanide molecule Me_3SnNC were inserted into one N $\rightarrow$ M bond of a M–CN $\rightarrow$ M′ linkage

$$\underset{\text{Spacer I}}{\text{M–CN} \rightarrow \text{M}'} \xrightarrow{Me_3SnNC} \underset{\text{Spacer II}}{\text{M–CN} \rightarrow \text{Sn(Me}_3) \leftarrow \text{NC–M}'} \quad (1)$$

However, the preparation of corresponding organometallic SPB derivatives is generally based upon metathesis reactions of salts containing a $[M(CN)_n]^{q-}$ anion and likewise water-soluble R_3ECl (E = Sn or Pb). In the presence of water, one H_2O molecule is frequently inserted spontaneously into one N $\rightarrow$ Sn bond of spacer II (Eq. 1), affording frameworks involving partially also the slightly elongated spacer III

$$\text{Spacer II} \xrightarrow{H_2O} \underset{\text{Spacer III}}{\text{–CN} \rightarrow \text{Sn(R}_3) \leftarrow \text{O(H)–H}\cdots\text{NC–}} \quad (2)$$

An early example of a SPB derivative containing the spacers II and III is the host–guest compound $[(Ph_3Sn)_3Fe^{III}(CN)_6 \cdot H_2O \cdot 2MeCN]$ (**3**, Table 3) that is

Tab. 1. Crystallographically established guest-free, *homo*leptic systems $^{m}_{\infty}[M\{\mu\text{-}XE(R_3)X\}_n]$ (m = 2 or 3; n = either 2, 3 or 4) containing either spacer II (X = CN, E = Sn, R = Me) or a closely related spacer.

No.	M	E	X	R	m	n	Ref.
4, 5	Ni(II), Pd(II)	Sn	CN	Me	2	2	9, 10
2, 6	Co(III), Fe(III)[a]	Sn	CN	Me	3	3	4,5
7	Co(III)	Pb	CN	Me	3	3	5
8, 9	Co(III), Fe(III)	Sn	CN	*n*-Bu	3	3	11
10	Rh(III)	Sn	NCS	Me	3	3	12
11, 12	Mo(IV), W(IV)	Sn	CN	Me	3	4	8, 10, 13, 62
13, 14	Mo(VI), Cr(VI)[a]	Sn	O	Me	2	2	14, 15, 61

a: owing to powder X-ray diffractometry

$^{3}_{\infty}[Fe^{III}\{\mu\text{-}CNSn(Ph_3)NC\}_2\{\mu\text{-}CNSn(Ph_3)O(H)H\cdots NC\}\cdot 2MeCN]$ [6]. To better clarify the M···M connectivities in well-understood crystal structures, the latter description will, at least in Tables 1–3, be preferred to the more conventional formulae.

Tin-coordinated H_2O molecules also tend to generate hydrogen bonds with acceptors L different from a terminal cyanide ligand, leading occasionally to consid-

Tab. 2. Crystallographically established guest-free, *hetero*leptic coordination polymers containing, *inter alia*, -CN-Sn(R_3)-NC- spacers.

No.	Compound [a]	nD	NF [b]	Ref.
15	$[Cu^{I}\{\mu\text{-}CNSn(Me_3)NC\}\{\mu\text{-}pyz\}]$	3D	3	19
16	$[Cu^{I}\{\mu\text{-}CNSn(Me_3)NC\}\{\mu\text{-}cpy\}]$	3D	4	10
17	$[Cu^{I}\{\mu\text{-}CNSn(Et_3)NC\}\{\mu\text{-}bpy\}]$	2D	1	20
18	$[Cu^{I}\{\mu\text{-}CNSn(Me_3)NC\}\{\mu\text{-}pym\}]$	2D	1	20
19	$[Cu^{I}_2\{\mu_3\text{-}CNSn(Me_3)NC\}_2\{\mu\text{-}bpy\}]$	3D	2	21
20	$[Cr^{VI}\{\mu\text{-}OSn(Me_3)O\}\{\mu\text{-}OSn(Me_3)O(H)Sn(Me_3)O\}]$[c]	3D	2	22, 15, 61
21–23	$[M^{II}\{\mu\text{-}CNSn(Me_3)NC\}_2\{CNSn(Me_3)OH_2\cdot H_2O\}_2]$ (M = Fe, Ru, Os)	2D	1	23, 10, 8
24, 25	$[M^{II}\{\mu\text{-}CNSn(Me_3)NC\}_2\{CNSn(Me_3)OH_2\cdot cpy\}_2]$ (M = Fe, Ru)	3D	1	10
26, 27	$[M^{II}\{\mu\text{-}CNPb(Me_3)NC\}_2\{CNPb(Me_3)OH_2\}_2]$ (M = Fe, Ru)	2D	1	24
28–30	$[M^{II}\{\mu\text{-}CNSn(Me_3)NC\}_2\{CNSnMe_3\}_2]$ [d] (M = Fe, Ru, Os)	2D	1	20, 25, 27
31	$[Fe^{II}\{\mu\text{-}CNSn(Me_3)NC\}_2\{\mu\text{-}CNSn(Me_3)O(H)H\cdots(diox)\cdots HO(H)Sn(Me_3)NC\}]$ [e]	3D	2	7, 8

[a]: pyz = pyrazine, cpy = 4-cyanopyridine, bpy = 4,4′-bipyridine, pym = pyrimidine, diox = dioxane; [b]: number of equivalent and independent, interpenetrating frameworks; [c]: containing the spacer –O–Sn(Me_3)–O– (IIa); [d] characterized by solid-state NMR and ESCA spectroscopy; [e] containing spacer IV with L = dioxane.

Tab. 3. Crystallographically established host/guest systems with host frameworks containing, inter alia, spacer II (exceptions: **48** and **49**).

No.	Compound [a]	nD	Ref.
35	$[Co^{III}\{\mu\text{-}CNSn(Me_3)NC\}_3 \cdot 0.25\ N_2O_4]$	3D	31
3	$[Fe^{III}\{\mu\text{-}CNSn(Ph_3)NC\}_2\{CNSn(Ph_3)OH_2 \cdots NC\} \cdot 2\ MeCN]$	3D	6
36–39	$[Mo^{IV}\{\mu\text{-}CNSn(Me_3)NC\}_4 \cdot xG]$ ($G = N_2O_4$; THF; pyz; pym) [b]	3D	31, 33, 8, 62
40	$[W^{IV}\{\mu\text{-}CNSn(Me_3)NC\}_4 \cdot xG]$ (G = THF) [b]	3D	33, 8, 62
41	$[Ni^{II}\{\mu\text{-}CNSn(Ph_3)NC\}_2 \cdot Ph_3SnOH \cdot 0.8\ MeCN \cdot 0.2\ H_2O]$	3D	34
42	$[(Co^{III}Cp_2{}^+)Fe^{II}\{\mu\text{-}CNSn(Me_3)NC\}_3{}^-]$ [b]	3D	35
43	$[(MV^{2+})_{0.5}Ru^{II}\{\mu\text{-}CNSn(Me_3)NC\}_3{}^-]$ [b]	3D	36
44	$[(n\text{-}Bu_4N^+)_{0.5}Fe^{II}\{\mu\text{-}CNSn(Me_3)NC\}_{2.5}\{CNSn(Me_3)OH_2\}^-]$	3D	37
45	$[(n\text{-}Bu_4N^+)Cu^I\{\mu\text{-}CNSn(Et_3)NC^-\}_2]$	3D	18
46	$[(n\text{-}Bu_4N^+)_2Ni^{II}{}_2\{\mu\text{-}CNSn(Me_3)O(H)Sn(Me_3)NC\}_2(CN)_4\}^{2-}]$	2D	38
47	$[(n\text{-}Pen_4N^+)_2Ni^{II}{}_2\{\mu\text{-}CNSn(Me_3)O(H)Sn(Me_3)NC\}_2(CN)_4\}_2{}^-]$	2D	43
48	$[(Me_2Sn)_3\{Co^{III}(CN)_6\}_2 \cdot 6H_2O]$ [c]	3D	44
49	$[(R_2Sn)_3\{Co^{III}(CN)_6\}_2 \cdot xG]$ [c,d]	3D	45
50	$[(n\text{-}Bu_4N^+)Cu^I{}_2\{\mu\text{-}CNSn(Me_3)NC\}\{\mu\text{-}CN\}_2{}^-]$	2D	46

a: Each formula comprising a cationic guest G^{n+} starts with the guest's symbol, whereas uncharged guests G are found at the end of the formula; b: THF = tetrahydrofuran, x = *ca.* 1; pyz = pyrazine, x = *ca.* 0.5; pym = pyrimidine, x = *ca.* 1; Cp = η^5-C_5H_5; MV = methylviologen (N,N′-dimethyl-4,4′-bipyridinium); c: each Sn atom carries four cyanide N atoms; d: R = vinyl, G = H_2O/THF.

erably more extended spacers such as

$$-CN \rightarrow Sn(Me_3) \leftarrow O(H)\text{–}H \cdots L \cdots H\text{–}O(H) \rightarrow Sn(Me_3) \leftarrow NC\text{–} \quad \text{spacer IV} \tag{3}$$

L = 1,4-dioxane (diox) [7] or 4,4′-bipyridine (bpy) [8]. Moreover, both hydrogen atoms of a tin-coordinated water molecule may be involved in the fixation of either bridgehead or guest atoms (see below). More recently, even the *trans*-$Me_3Sn(OH_2)_2{}^+$ cation was found to play a role intermediate between that of a guest and part of a spacer [8,62], for example

$$-CN \cdots H\text{–}(H)O \rightarrow Sn(Me_3) \leftarrow O(H)\text{–}H \cdots NC\text{–} \quad \text{spacer V} \tag{4}$$

12.2 Guest-*Free* Homoleptic SPB Derivatives

In the following, all 1D, 2D, and 3D systems involving exclusively spacer II (Eq. 1) will be designated as "homoleptic", irrespective of the absence or presence of guests. Consequently, "heteroleptic" systems contain at least one different spacer (or nonbridging ligand) in addition to spacer II. Although quite a few guest-free SPB derivatives of the type $[(R_3E)_qM(CN)_{2n}] = {}^m_\infty[M\{\mu\text{-}CNE(R_3)NC\}_n\}]$ (for $q = n$) have been described, only a limited number of examples have yet been subjected to single-crystal X-ray studies (Table 1 and Fig. 1), owing to considerable difficulties in arriving at suitable single crystals. The only homoleptic system included in Table 1 (the structure of which has been deduced simply from powder X-ray diffraction (powder-XRD)) is $[(Me_3Sn)_3Fe^{III}(CN)_6]$ (**6**). Also included in Table 1 is the compound $[(Me_3Sn)_3Rh^{III}(SCN)_6]$ (**10**) the notably elongated, although nonlinear, $SCN \rightarrow Sn(Me_3) \leftarrow NCS-$ spacer of which leads to a crystal lattice consisting of two individual and equal, mutually interwoven frameworks. While the methyl groups bonded to the tin atoms of **2** still help avoiding similar self-catenation, (i.e., spontaneous interpenetration) it should be recalled that the lattice of Pauling's SPB **1** already consists of three interpenetrating frameworks [3].

Corresponding 1D coordination polymers, $[(R_3E)M^I(CN)_2] = {}^1_\infty[M^I\{\mu\text{-}CNE(R_3)NC\}]$, with M = Au or Ag could so far not be crystallized appropriately, although the compound $[(Et_3Sn)Au^I(CN)_2]$ was studied by solid-state NMR spectroscopy at an early stage [16]. It would be of some interest to see whether notable intermetallic $(M\cdots M)$ bonding would, as in the case of corresponding $[Tl^IM^I(CN)_2]$ systems (M = Ag or Au) [17], lead to structures of higher dimensionality.

Examples of 3D polymers involving tetrahedral $\{M(CN)_4\}$ fragments are also absent in Table 1. Instead of the generation of diamond-like frameworks that would undergo multiple interpenetration owing to their exceedingly large voids, various "derailment pathways" leading to products of other stoichiometrics than $[(Me_3Sn)_2M^{II}(CN)_4] = {}^3_\infty[M^{II}\{CNSn(Me_3)NC\}_2]$ (M = Zn, Cd) are obviously preferred [18]. Even the homoleptic compound $[(Me_3Sn)_2Mo^{VI}O_4]$ with slightly shorter spacers of the type $-O-Sn(Me_3)-O-$ (spacer IIa), in which each cyanide group is replaced by one oxygen atom, circumvents the formation of a diamond-like framework in favor of a layered structure, in which the Mo atoms are still tetrahedrally surrounded by the oxygen atoms [14]. Guest-free heteroleptic systems with four bidentate bridges have, however, been described (Table 1). The two guest-free compounds **11** and **12** with eight-coordinate transition metal ions (M = Mo, W) display the same framework architecture as their likewise homoleptic host–guest congeners containing for example, about one tetrahydrofuran (THF) molecule per formula unit (Table 2). Strict retention of the architecture of the guest-free host, also after the uptake of a guest, is comparatively rare in the case of supramolecular SPB-derivatives (see below). Hence it is not surprising that, as a result of an increasing space demand of the tin-bonded alkyl group R (e.g., from R = Me to n-Bu in the case of $[(R_3Sn)_3M^{III}(CN)_6]$) systems, the topology of the $\{Sn_3M(CN)_6\}$ skeleton also turned out to change significantly [11]. According to solid-state NMR results, even a transition from R = Me to R = Et seems to be accompanied by structural variations [5].

12.3
Guest-Free *Hetero*leptic systems

In contrast to Table 1, a notable fraction of the compounds listed in Table 2 (**15–20**) comprises tetrahedrally coordinated transition metal ions (mainly Cu(I)). However, only three of them (**15**, **16**, **20**) display slightly distorted diamond-like frameworks (Fig. 2), owing to the presence of two spacers generating different $M\cdots M$ dis-

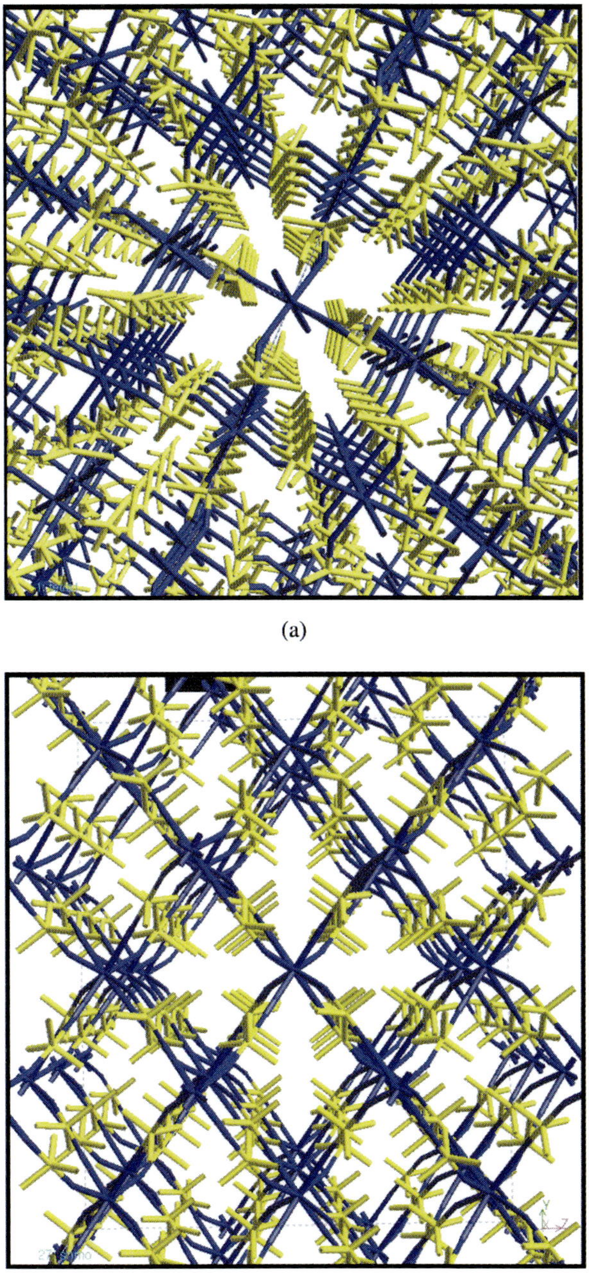

Fig. 1. Perspective view along the main channels of the homoleptic 3D systems **2** (a) and **11** (b). The $\{M(CN)_{2n}\}$ fragments ($n = 3$ or 4) are blue, the $\{Me_3Sn\}$ fragments are yellow.

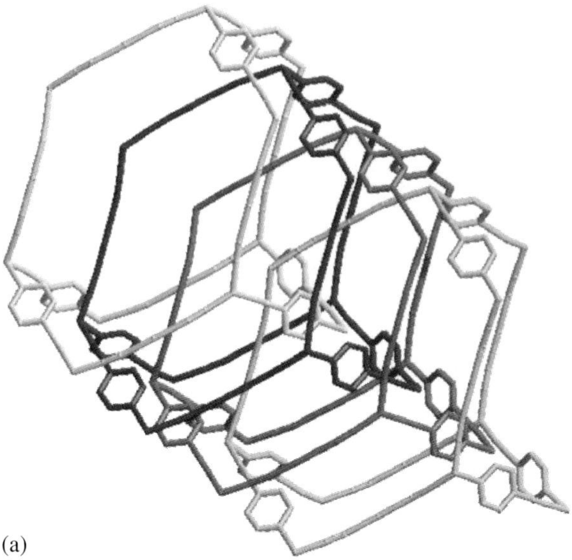

(a)

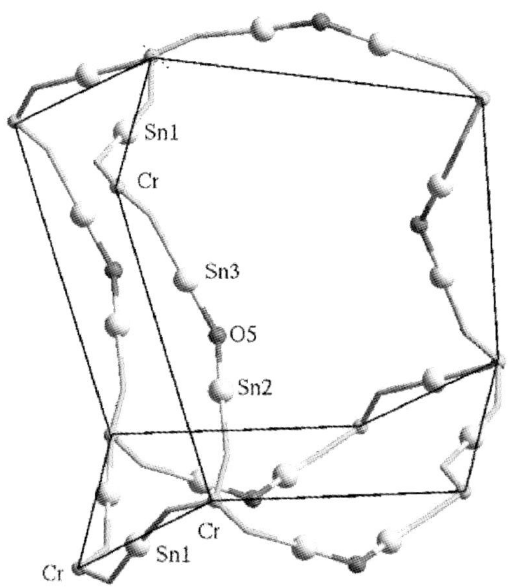

(a)

Fig. 2. Distorted-adamandoid fragments of the four equivalent, interpenetrating frameworks constituting the lattice of **16** (a), and one adamandoid cage from the two interpenetrating frameworks of **20** (b). Tin-bonded CH_3 groups are omitted for clarity. In (a), each $CNSn(Me_3)NC$ spacer is simply represented by one straight line.

tances. The crystal lattices of all three compounds involve interpenetrating frameworks. Strictly speaking, **15**, **16**, and **20** may be considered as remote derivatives of likewise polymeric $M(CN)_2$ ($M = Zn$ or Cd) rather than of PB. Although the crystal structure of **20** was described decades ago [22], its distinct supramolecular architecture has been elucidated only recently, making use of the powerful CERIUS^2 software [15,61]. Interestingly, the two interpenetrating frameworks of **20** are tied together via relatively short $(Sn_2)\mu$-O–H···O=Cr hydrogen bonds. On the other hand, the structures of the two compounds **17** and **18** owe their surprising compactness to the formation of stacked, but notably puckered, sheets. A related structural pattern is also displayed by the R_3E–free host–guest compound $[Cu^I\{\mu\text{-NCNCN}\}\{\mu\text{-bpy}\}_{0.5}MeCN\cdot 0.25\ bpy]$, although here adjacent stacks of infinite ladders just interdigitate [26].

A more exceptional situation is realized by the crystal structure of compound **19**, in which spacer II appears to be tridentate, in that one of its cyanide carbon atoms connects asymmetrically two copper atoms. The copper pairs and spacer II give rise to parallel oriented, infinite layers (Fig. 3), which are tied together vertically by bpy molecules. Two of the open 3D frameworks resulting thereafter undergo optimal interpenetration. While all copper-containing coordination polymers of Table 2 can be obtained by reacting a solution of R_3SnCl and the respective nitrogen base with $K_3[Cu^I(CN)_4]$, their distinct supramolecular architectures are strictly controlled by the nature of the base and/or the alkyl group R. Somewhat unexpectedly, the 18-electron system $[Cu^I(CN)_4]^{3-}$ is readily attacked by various heterocyclic nitrogen bases. A convenient synthesis of the serendipitously found polymer **20** [22] was developed in our laboratory [15,61] (Eq. 5)

$$\underset{\mathbf{14}}{[(Me_3Sn)_2Cr^{VI}O_4]} + Me_3SnOH \rightarrow \underset{\mathbf{20}}{[(Me_3Sn)_3Cr^{VI}O_4(OH)]} \quad (5)$$

Compound **14** is suggested to be isostructural with its molybdenum homologue **12** [14], owing to almost identical X-ray powder diffractograms of the two compounds [15]. The reaction depicted by Eq. (5) is thus reminiscent of that described by Eq. (1), in that now formally a Me_3SnOH molecule is inserted into one of the Sn–O bonds of spacer IIa. The water-insoluble polymer **13** can, however, not be converted into the Mo-containing homologue of **20**.

The compounds **21–30** listed in Table 2 involve hexacoordinated transition metal ions and, in addition to spacer II, either a second, not strictly bridging ligand or (in **25** and **31**) the strongly elongated spacer IV (Eq. 3) with L = 4-cyanopyridine or dioxane. The supramolecular architecture of **31** may best be rationalized in terms of ABAB stacks of truncated $^2_\infty[Fe^{II}\{\mu\text{-spacerII}\}_2]$ layers similar to those of $[(Me_3Sn)_2M^{II}(CN)_4]$ (see above), the Fe atoms of "adjacent" layers being, however, held apart by spacer IV. Two of the open frameworks resulting in this way (of layers designated as A and B) interpenetrate therefore, in that each dioxane molecule of framework A is embedded amidst the almost square-like array provided by each B-layer, and *vice versa*. Figure 4 displays the view upon one particular B-layer containing the notably tilted dioxane molecules of the complementary A-framework. Inter-

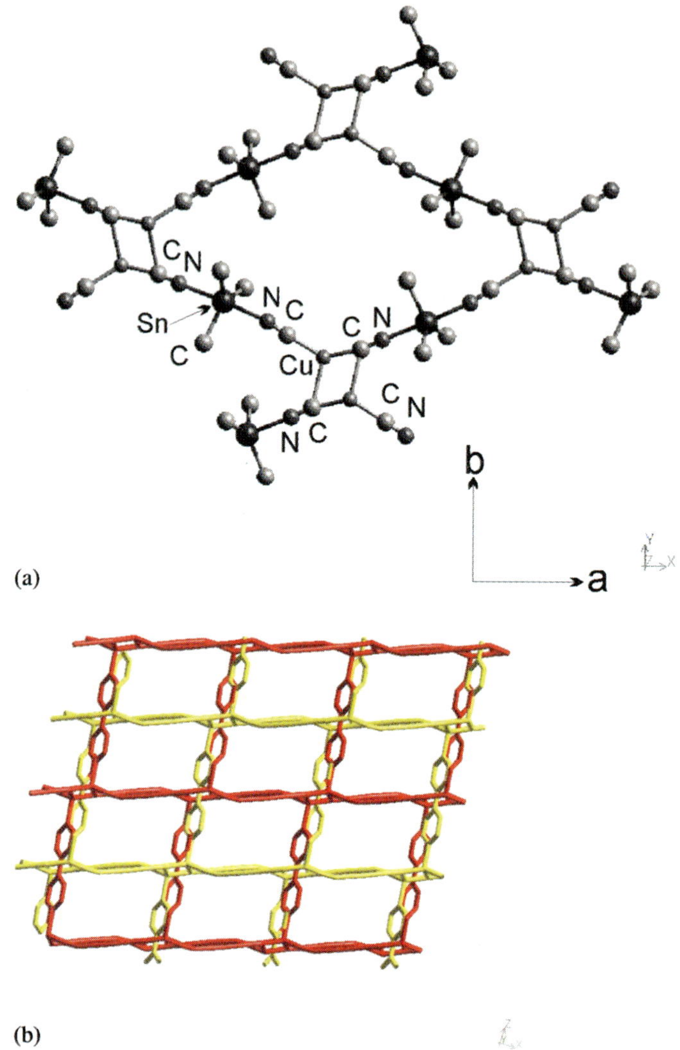

Fig. 3. One single macrocyclic building block constituting the infinite layers of compound **19** (a), and a view along the stacked layers of **19** (b). The two interpenetrating frameworks (yellow/red) contain bpy molecules as vertical spacers. Methyl groups (bonded to tin atoms) have been omitted in (b).

estingly, only when the already published structural analysis of **31** [6] was revisited making use of the CERIUS^2 software, was it realized that the architecture may be understood in terms of two interwoven, strongly distorted, six-connected frameworks [8]. The structure of **25** differs strongly from that of **31** as here the two hydrated Me_3Sn units are *cis*-oriented, one of them being strongly disordered [10,62].

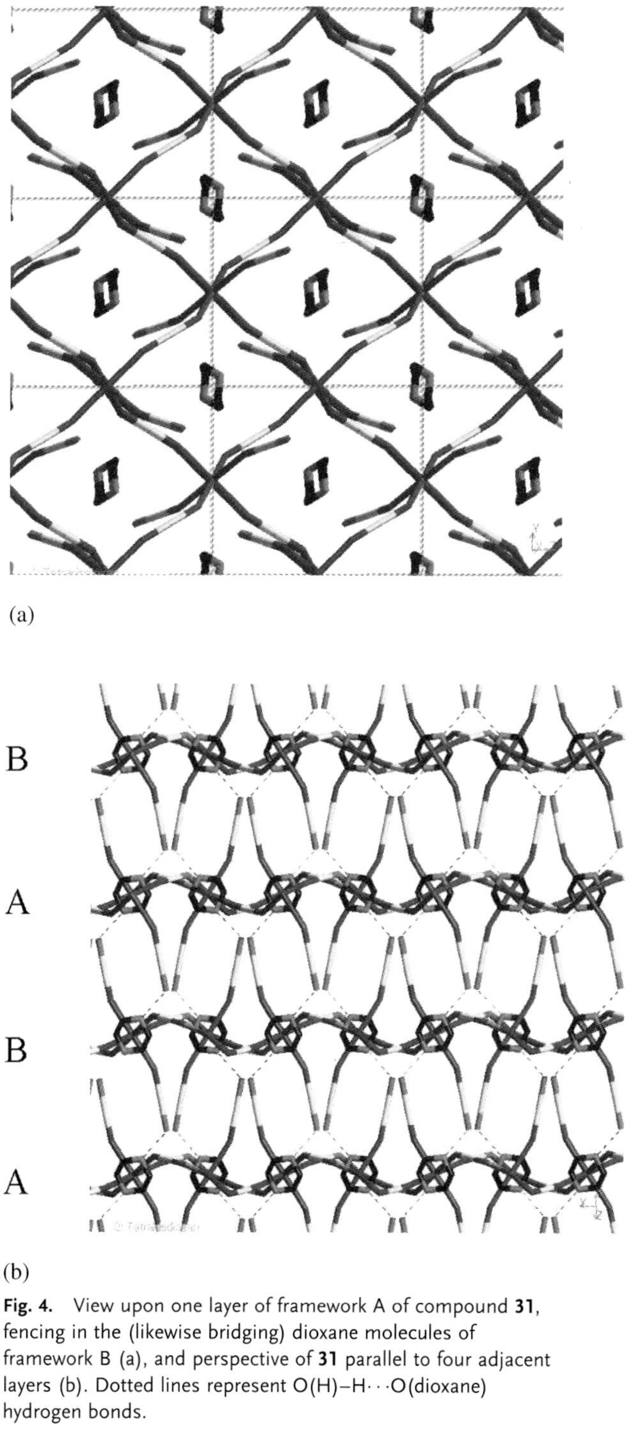

Fig. 4. View upon one layer of framework A of compound **31**, fencing in the (likewise bridging) dioxane molecules of framework B (a), and perspective of **31** parallel to four adjacent layers (b). Dotted lines represent O(H)–H···O(dioxane) hydrogen bonds.

The primarily nonbridging ligands present in the compounds **21**, **23**, **26**, and **27** have in common that they help accomplish strict pentacoordination for all four tin atoms of the basic $[(Me_3Sn)_4M^{II}(CN)_6]$ unit. The three homologues **28–30** seem to involve, according to their characteristic solid-state ^{119}Sn chemical shifts [25], two essentially four-coordinate tin atoms per formula unit. All attempts to obtain single crystals of either **28**, **29**, or **30** have so far resulted in the formation of more readily crystallizing adducts in which all tin atoms are again five-coordinate. A similar situation has been observed for the trimethyllead-containing homologues, although in contrast to **21–23** (E = Sn) with E = Pb the formation of dihydrates is preferred [24]. In the latter systems (with M = Fe, Ru), the corresponding $CNPb(Me_3)OH_2$ ligands have turned out to adopt a quasi-bridging functionality in that one O–H···NC hydrogen bridge connects two adjacent layers, although here the respective cyanide N atom is already engaged in a coordinative N → Sn bond. In the alternative tetrahydrates (with E = Sn), the Sn-bonded H_2O molecules carry, via O–H···O hydrogen bonds, another H_2O molecule, which anchors too loosely to nitrogen atoms of remote $[\text{-CN–Sn–NC–M-}]_n$ chains to justify here the assumption of a distinct "bridge". As single crystals of base-free $[(SnMe_3)_4Fe^{II}(CN)_6]$ have never been accessible, two independent EXAFS study have been carried out, which did, however, not reveal two significantly different Sn–N distances [20,27]. Although by EXAFS the asymmetric unit cannot be obtained either, a detailed solid-state NMR investigation of the reliably anhydrous, isomorphous compounds **28**, **29**, and **30** (involving the nuclei ^{13}C, ^{15}N, and ^{119}Sn) has specified both their asymmetric unit and the local symmetry of the $\{M(CN)_6\}$ fragments quite satisfactorily. These features differ slightly from those of the crystalline dihydrates (with E = Pb) and tetrahydrates (with E = Sn), which result supports further the view that the strictly anhydrous compounds actually represent a special subgroup.

Another interesting class of heteroleptic coordination polymers with Fe(II) as central metal ion involves, in addition to spacer II, a bidentate heterocyclic nitrogen base μ-L such as bpy, pyrazine (pyz), pyrimidine (pym), or *trans*-bipyridylethylene (bpe) and, most probably, also a terminal $CNSnMe_3$ ligand [28]. The analytically and spectroscopically well-characterized, red polymers of the general composition $[(SnMe_3)_6Fe^{II}_2(\mu\text{-L})(CN)_{10}]$ (**32–34**) are obtained in high yields by coprecipitation from aqueous solution according to Eq. (6)

$$Na_6[Fe^{II}_2(\mu - L)(CN)_{10}] + 6Me_3SnCl \xrightarrow{-6\,NaCl} \underset{\mathbf{32\text{–}34} \quad (L = bpy)}{[(Me_3Sn)_6Fe^{II}_2(\mu - L)(CN)_{10}]} \quad (6)$$

The SPB-derivatives **32–34** resemble the members of the subclass $[(Me_3Sn)_4M^{II}(CN)_6]$ in that so far all attempts to generate single crystals suitable for X-ray crystallography have remained unsuccessful. From stacks of initially frozen layers comprising alternantly pure H_2O and solutions of each reactant, Eckhardt has obtained red single crystals built up, however, of another new coordination polymer of $[Fe^{II}(CN)_6]^{4-}$ and no longer of $[Fe^{II}L(CN)_5]^{3-}$ fragments [8,62]. Infrared (IR) spectroscopic results indicate that the exclusively terminal cyanide ligands of the $[Fe^{II}_2(\mu\text{-L})(CN)_{10}]^{6-}$ anions have in fact become bridging units in the bulk polymers **32–34** and that the heterocyclic ligands L are still present.

Moreover, solid-state ^{119}Sn NMR studies strongly suggest that, as in the case of $[(Me_3Sn)_4M(CN)_6]$ (see above), both five- and four-coordinate tin should be present. Actually, according to the formula $[(Me_3Sn)_6Fe^{II}{}_2(\mu\text{-}L)(CN)_{10}]$ all six tin atoms would require twelve cyanide nitrogen atoms to become five-coordinate centers, but only ten terminal cyanide ligands are available in the starting anion $[Fe^{II}{}_2(\mu\text{-}L)(CN)_{10}]^{6-}$. We therefore consider the members of the family $[(SnMe_3)_6Fe^{II}{}_2(\mu\text{-}L)(CN)_{10}]$ as guest-free, 3D coordination polymers with five $\{Me_3Sn\}$ fragments involved in spacers of type II and one nonhydrated, terminal $CNSnMe_3$ (or quasi stanna-isocyanide) ligand.

Finally, a quite singular class of guest-free heteroleptic 3D polymers deserves attention that contain, in addition to spacer II, the closely related spacer $-CN-Sb^{V}(Me_3)-NC-$. Thus, compounds of the net composition $[(Me_3Sn^{IV})_2 \cdot (Me_3Sb^{V})M^{II}(CN)_6]$ [29] have been obtained both by spontaneous co-precipitation from Me_3SnCl, Me_3SbBr_2, and $K_4[M^{II}(CN)_6]$ (M = Fe, Ru) and by exchange of Me_3Sn^+ and G^+ (1:1) by one Me_3Sb^{2+} ion in an already polymeric host–guest system (Eq. 7)

$$[(G)(Me_3Sn)_3M(CN)_6] \xrightarrow{Me_3SbBr_2/H_2O} [(Me_3Sn)_2(Me_3Sb)M(CN)_6] \quad (7)$$

Successfully applied educts have been host–guest systems with G^+ = for example, Et_4N^+, $CoCp_2{}^+$ ($Cp = \eta^5 - C_5H_5$, see below), and even Me_3Sn^+. Single crystals have not been obtained, mainly because the new products are extremely insoluble. The powder X-ray diffractograms of $[(Me_3Sn)_2(Me_3Sb)Fe^{II}(CN)_6]$ and $[(Me_3Sn)_3Co^{III}(CN)_6]$ (**2**) differ notably, and from IR results a virtually statistical distribution of the two similar spacers has been deduced [29].

12.4
Host-Guest Systems with Uncharged or Cationic Guests

From the compounds with homoleptic 3D frameworks listed in Table 1, the SPB-derivatives **2**, **6**, and **7** would appear to be the most promising host systems for comparatively large uncharged guest molecules (involving 10–20 atoms). However, **6** is quite reactive both as a powerful oxidant of potential guests (see below) and owing to light-induced, intramolecular oxidation of part of its cyanide ligands to cyanogen [27,30]. Compound **7** offers, on the other hand, slightly less internal space than **2** and **6**, in view of its significantly smaller (less than 180°) C–N–Pb angles [5] in the infinite [-Pb–NC–M–CN-] chains. Actually, the internal surface of the basic SPB **1** (see above) seems, at least according to BET measurements (N_2 uptake under standardized conditions [31]), to be more than 30 times larger than those of **2** and **11**, although, in contrast to **2** and **11**, three individual 3D frameworks interpenetrate in the lattice of **1**. On the other hand, the tin-coordinated methyl groups of **2** require substantial space within the $^3_\infty[Co^{III}\{\mu\text{-}CNSnNC\}_3]$ skeleton. Moreover, not only do the methyl groups rotate rapidly about their C–Sn axes, but variable-temperature solid-state ^{13}C NMR studies [32] have clearly shown that, down to notably low temperatures, the Me_3Sn groups also rotate rapidly about their N–Sn–N or N–Sn–O axes. This dynamics is likely to reduce the chances of small inert molecules to anchor satisfactorily on the internal surface.

Again quite unexpectedly, **2**, **6**, and **11** were found to adsorb notably higher amounts of NO_2 than **1** (about 0.5, 1.5, 0.9, and 0.2 mol mol^{-1}, respectively) and, moreover, of mesoporous SiO_2 with a BET surface of 1247 m^2g^{-1} (NO_2 uptake: only 0.14 mol mol^{-1}). Interestingly, subsequent adsorption/desorption/adsorption cycles have revealed that the capacity for NO_2 uptake by **2** and **11** even increases slightly, but significantly, during the first 4–5 cycles [31]. This feature is, however, neither reflected by the IR absorptions of the host nor by powder-XRD. According to the diamagnetism (and colorless appearance) of all samples **2**$\cdot xNO_2$ and **10**$\cdot yNO_2$, only diamagnetic, colorless N_2O_4 seems to be incorporated. Both the powder-XRD and the solid-state NMR spectra (^{13}C and ^{119}Sn) of the adducts remain almost identical with corresponding diffractograms and spectra, respectively, of the N_2O_4-free hosts. **2**$\cdot xN_2O_4$ ($x \leq 0.3$) and **11**$\cdot yN_2O_4$ ($y \leq 0.4$) may be used as convenient sources of NO_2 in view of various chemical experiments. For instance, in an atmosphere of SO_2 (with traces of H_2O), sulfuric acid is rapidly emerging on the surface of the polymers as a viscous film, whereas in an atmosphere of NH_3 the expected comproportionation to N_2 (along with H_2O) does not take place [31]. It is quite remarkable that nitrogen dioxide (chemically almost as aggressive as elemental chlorine) does not attack the organometallic constituents of the host SPBs. Interestingly, CO_2, SO_2, NH_3, and NO are not adsorbed by **2** and **11**.

On the other hand, the Fe(II) compound $[(Me_3Sn)_4Fe^{II}(CN)_6]$ (**30**) reacts with NO_2 more readily than both **2** and $K_4[Fe^{II}(CN)_6]$, affording most probably first a 1:1 composite of **6** and Me_3SnNO_2 (or Me_3SnONO). In an $NO_2/O_2/N_2$ atmosphere containing 14.2% of NO_2, even an adduct of the approximate composition **30**$\cdot$4 NO_2 was obtained, suggesting that the framework of the primary composite is capable to trap almost three further NO_2 molecules. After washing the composite with H_2O, the sparingly soluble, orange residue could be identified as **6**, whereas in the aqueous solution $Me_3Sn\cdot aq^+$ cations along with the anions NO_2^- and NO_3^- were found [31]. Various attempts have been made to reduce (e.g., by means of SO_2), all Fe(III) back to Fe(II) in order to recollect the initially used polymer **31** and to simultaneously "condition" the remaining nitrate. During the course of these investigations it has, however, turned out that for example, the PB derivative $[NaDy^{III}Fe^{II}(CN)_6\cdot 2H_2O]$ also incorporates NO_2 (1:1) and can, thereafter, more conveniently be recycled [31].

The Fe(II) system $[(Me_3Sn^{IV})_2(Me_3Sb^V)Fe^{II}(CN)_6]$ (see above) was found to absorb even 7.7 moles of NO_2 under very mild conditions (ambient pressure, 5.6% NO_2, 24 h) [31]. Relevant properties of the resulting adduct still deserve further exploration. In view of the excellent disposal of Fe(III)-containing polymers for NO_2-uptake it is not surprising that neat **6** absorbs about three times more NO_2 than **2** (1.5 vs. 0.5 mol mol^{-1}).

In Table 3, all of the so far crystallographically established 3D and 2D host–guest systems with both uncharged and cationic guests are collected. Assemblies containing additional compounds that may less clearly be described as "heteroleptic guest-free" or as "host–guest" systems will briefly be mentioned in the following section. Generally, ionic guests could be localized crystallographically without notable disorder, whereas uncharged guests remain either undetectable or turn out to be strongly disordered. Usually, the number of uncharged guest molecules per formula unit is not well-defined either. Encapsulation of NO_2 by **2** and **11**, and of THF, pyz, and pym by **11**, seems to belong to the comparatively few cases in which the structural pattern of a guest-free, homoleptic SPB representative remains unchanged after the uptake of the guest. In principle, the infinite channels in the lattices of, for example, **2** and **11** (Fig. 1) would also be attractive for the encapsulation of molecular chains. Interestingly, chains of probably protonated, semiconducting polypyrrole seem to occupy the channels of a corresponding negatively charged $^3_\infty[Fe^{II}\{\mu\text{-}CNSn(Me_3)NC^-\}_3]$ host [39]. Actually, the homoleptic Fe(III) system **6** readily absorbs, and oxidizes, molecules such as pyrrole, aniline, ferrocene, cobaltocene. The guest-free and anhydrous precursor of **3** is, on the other hand, still unknown [6].

The most spectacular uncharged guest was found in the assembly **41** as the cyclic $(Ph_3SnOH)_3$ unit [34]. The space required for these large specimens is provided by a 3D host structurally reminiscent of the NbO lattice [40]. Thus, the Ni atoms adopt formally positions of both the Nb and O atoms. The high space demand of the phenyl groups is likely to discard here the formation of layered $[Ni^{II}\{\mu\text{-}CNSn(Ph_3)NC\}_2]$ homologs of (guest-free) **4**, but appropriately dimensioned guests are then needed to stabilize the large cavities of the alternative 3D framework. Recalling that the NbO structure may formally be derived, like that of PB, from the cubic NaCl lattice [40], but affords cavities with edges twice as long as for SPB systems, **41** could even be regarded as a "super-SPB" system.

Quite spontaneously, both the NbO-related host and its $(Ph_3SnOH)_3$ guest result from Ph_3SnCl and $K_2[Ni^{II}(CN)_4]$ in the presence of H_2O [34], although neat Ph_3SnOH is known [41] to form infinite polymeric chains. On the other hand, $\{(Me_3Sn)_2OH\}^+$ ions result from Me_3SnCl and $K_2[Ni^{II}(CN)_4]$ in the presence of R_4N^+ halides and H_2O, which assemble to the metallacyclic dianion $[Ni^{II}{}_2(CN)_4\{\mu\text{-}CNSn(Me_3)O(H)Sn(Me_3)NC\}_2]^{2-}$ [42]. The R_4N^+ salts may formally be considered as insertion products of R_4NOH into **4**. Depending on the length of the group R in the R_4N^+ counter cation, these 16-membered rings assemble further, via O–H···NC hydrogen bonds, either to infinite layers (R = nBu, **46**) or to infinite ribbons (R = n-Pen, **47**). A related 3D assembly built up of infinite, stapled ribbons interlinked by $SnOH_2\cdots NCCo$ hydrogen bonds with two H_2O guest molecules has the composition $[(Me_2Sn)_3\{Co^{II}(CN)_6\}_2 \cdot 6H_2O]$ (**48** [44]). With larger alkyl ligands, however, 3D host–guest systems $[(R_2Sn)_3\{Co^{III}(CN)_6\}_2 \cdot xG]$ result (**49** [45]).

A heteroleptic host–guest system involving like **46** negatively charged layers between which $n\text{-}Bu_4N^+$ counter-cations are intercalated (compound **50** [46]) was initially obtained during the attempted synthesis of the still unknown, 3D homologue $[(n\text{-}Bu_4N)Cu^I\{\mu\text{-}CNSn(Me_3)NC\}_2]$ of **45** (which contains tin-bonded ethyl groups, see below) [18]. Yet, in **50** three-coordinate Cu(I) ions are interlinked by the spacers I and II (2:1) in a distorted honeycomb-like pattern.

Interestingly, two congeners of **50** have recently been described, one of which contains "naked" type II spacers $-CN-Cu^I-CN-$ instead of the Me_3Sn-containing spacer II [47]. In the other congener (**51**) each of the common type II spacers of **50** has lost one CH_3 group and is interlinked instead with a corresponding $\{Me_2Sn\}$

fragment via a trimethylene bridge. Both detailed powder X-ray and solid-state NMR studies confirm the view that the trimethylene tethers connect adjacent layers of **51** in such a way that **51** may be considered as a heteroleptic 3D derivative of **50**, involving now one two- and one tetradentate spacer and again nBu_4N^+ cations as interlayer guests [46].

Compound **45** is the only crystallographically established homoleptic coordination polymer containing a tetrahedrally coordinated metal ion and spacer II (however, with triethyltin groups). Although both the ethyl and n-butyl groups (of the nBu_4N^+ guest) require considerable space, the large voids of an ideally diamond-like $^3_\infty[Cu^I\{\mu\text{-CNSnNC}\}_2{}^-]$ skeleton cannot even be optimally stuffed, and the actually resulting framework is still found to be strongly distorted [18]. A number of slightly modified congeners of **45** have, in contrast to **50**, the correct stoichiometry of a potential quasi-diamond-like framework, too [13,48], but none of these macrocrystalline products have been subjected to single-crystal X-ray crystallography.

In contrast to the situation discussed for compound **45**, the free space available in the voids of SPB-like $^3_\infty[M^{II}\{\mu\text{-CNSn}(Me_3)NC\}_3{}^-]$ systems (containing in average already nine methyl groups) does not suffice to accommodate one complete nBu_4N^+ ion. Instead, both co-precipitation and ion exchange reactions (the latter starting from **28**) do not afford a host–guest system of the initially expected stoichiometry $[(nBu_4N)(Me_3Sn)_3Fe^{II}(CN)_6]$, but only the more nBu_4N-deficient system $[(nBu_4N)_{0.5}(Me_3Sn)_{3.5}Fe^{II}(CN)_6{\cdot}H_2O]$ (**44**). The crystal structure analysis of **44** has revealed the presence of type II spacers and of a virtually nonbridging $CNSn(Me_3)OH_2$ ligand (Table 3) reminiscent of hydrated trimethylstannaisocyanide [37]. Interestingly, each six-coordinate Fe(II) ion is connected with two adjacent Fe(II) ions by pairs of type II spacers, leading here to the topology of a three-connected net. In contrast, for example, **4**, **2**, and **11** display topologies of four-, six-, and eight-connected nets, respectively. The nBu_4N^+ ions of **44** occupy infinite, straight channels of rectangular cross section, and corresponding to the chirality of six-coordinate complexes with at least two chelating ligands, the single crystal of **44** belonged to the chiral space group $P2_12_12$.

The crystallographically established assemblies **42** and **43** are members of the comparatively large class of "genuine" host–guest systems $[(G^+)(Me_3Sn)_3{\cdot}M^{II}(CN)_6]$ (built up of six-connected nets), in which the guest ion is neither too small nor too large to fit into the negatively charged $^3_\infty[M^{II}\{\mu\text{-CNSn-NC}\}_3{}^-]$ skeleton. Actually, the architecture of the respective $^3_\infty[M^{II}\{\mu\text{-CNSnNC}\}_3{}^-]$ skeleton of for example, **2**, **7**, **8**, **42**, and **43** varies considerably with the shape of the actual guest (and also with the space demand of the tin-bonded group R), indicating that the 3D skeleton adapts quite flexibly to the steric requirements of the guest. This unusual feature is supported by systematic studies of the powder X-ray diffractograms of numerous other host–guest systems of the above type. For instance, the powder-XRD of $[(Et_4N)(Me_3Sn)_3Fe^{II}(CN)_6]$, the guest of which does not contain any "heavy" element (to interfere notably by intense guest-specific reflections), does not closely resemble the diffractograms of **2** and **43**. The structural flexibility results mainly from the fact that in none of the host–guest systems studied the ideal SPB configuration with strictly linear [-M–CN–E–NC-] chains is realized (see above). The "initially" notably distorted, and hence more compact, frameworks have numerous degrees of freedom to adopt stepwise, according to the guest's shape, a variety of more spacious structural patterns.

Both **42** and **43** belong to families comprising specifically homologues with either metallocenium [35,49] or viologen [28] cations as guests. Interestingly, the ferrocenium-containing homologue of **42** and the Fe(II)-containing homologue of **43** are both deep blue in color due to pronounced charge transfer bands (from the central Fe^{2+} ion to either the ferrocenium or viologen guest ion) [28,36]. Several SPB-derivatives may thus display even the same color as their famous Fe(II)/Fe(III)- archetype "PB"! On the other hand, **43** can, as its homologues with M = Fe or Os, be reduced by dithionite anions to another type of host–guest system, which is constantly blue (for $M^{II} = Fe^{II}, Ru^{II}, Os^{II}$) [28] because of the specific light absorption of the now entrapped methylviologen radical cation, MV^{+} (Eq. 8)

$$2[(MV^{2+})_{0.5}(Me_3Sn)_3M^{II}(CN)_6] \xrightarrow{(S_2O_4{}^{2-})} [(MV^{+})(Me_3Sn)_3M^{II}(CN)_6]\downarrow + [M^{II}(CN)_6]^{4-} + 3\,Me_3Sn{\cdot}aq^{+} \qquad (8)$$

The three MV^{2+}-containing homologues with Fe, Ru, and Os are blue, brick red, and violet, respectively, according to the decreasing tendency of the central M^{2+} ion to transfer one electron to the guest. Somewhat unexpectedly, the widest of the parallel running channels of **42** are alternately filled with cobaltocenium ions and with tin-bonded methyl groups only (Fig. 5). A similar feature was found in the crystal structures of two close congeners of **1** of the type $^{3}_{\infty}[(M^{+})M'\{\mu$-

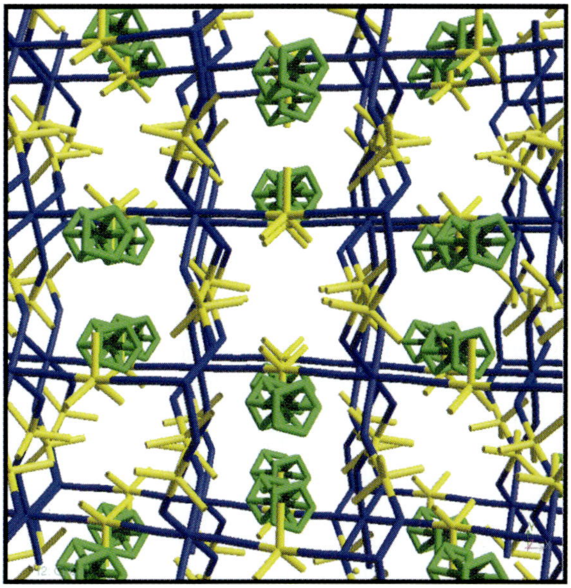

Fig. 5. View along the main channels of **42** (green: cobaltocenium guest cations).

NCM″CN}$_3$$^-$] (with M′ = Co(II) or Cd(II) and M″ = Au(I) or Ag(I)) [50], in which the M^+ guest ions (K^+ or Rb^+) occupy again only every other channel. Systematic studies based on the known crystal structures of numerous host–guest systems suggest that weak, but significant C–H···NC hydrogen bonds between the organic or organometallic guest cation and terminal (or even bridging) cyanide N atoms of the host framework help supporting the nondisordered positioning of the guest molecules in the voids [51,52].

Co-precipitation of the afore-mentioned dinuclear anions $[Fe^{II}{}_2(\mu\text{-L})(CN)_{10}]^{6-}$ (L = pyz, bpy, bpe) with the cations $Me_3Sn{\cdot}aq^+$ and potential guest ions G^+ (G^+ = either Et_4N^+, nBu_4N^+, $CoCp_2{}^+$, or 0.5 MV^{2+}) in H_2O has led to analytically and spectroscopically well-characterized, orange-red host–guest systems of the type $[(G^+)(Me_3Sn)_5Fe^{II}{}_2(\mu\text{-L})(CN)_{10}]$ [28,53]. Attempts to arrive at the same products by ion exchange (Me_3Sn^+ vs. G^+) starting from the guest-free systems $[(Me_3Sn)_6Fe^{II}{}_2(\mu\text{-L})(CN)_{10}]$ (as has been possible with $[(Me_3Sn)_4Fe^{II}(CN)_6]$) were unsuccessful because of the extremely low solubility of the L-bridged systems in water. Interestingly, in contrast to the guest-free compounds that gave rise to two ^{119}Sn solid-state NMR signals almost 100 ppm apart from each other, none of the host–guest systems displayed any low-field signals typical of four-coordinate tin. Actually, here the five $\{Me_3Sn^+\}$ fragments meet just ten cyanide N atoms to form five {*trans*-$Me_3Sn(NC)_2$} units so that no four-coordinate tin will be expected. Somewhat surprisingly, the color of the compound with G^+ = 0.5 MV^{2+} is again orange red (as for its precursor), and not blue as for the Fe(II)-homologue of **43**, suggesting that maybe less favorable steric conditions could hamper the otherwise expected charge transfer, in spite of the quite promising electron releasing power of the $\{Fe^{II}(\mu\text{-L})Fe^{II}\}$ components. Unfortunately, all attempts to arrive at single crystals of the new $\{Fe^{II}(\mu\text{-L})Fe^{II}\}$-derived polymer have so far not been successful.

12.5
Truncated and Expanded SPB Derivatives

Somewhat surprisingly, not only polymers of the types $[(Me_3Sn)_4M^{II}(CN)_6]$ and $[(Me_3Sn)_6Fe^{II}{}_2(\mu\text{-L})(CN)_{10}]$ are capable of exchanging readily one Me_3Sn^+ ion by, among others, one Et_4N^+ guest cation. Also most of the members of the $[(Me_3Sn)_3M^{III}(CN)_6]$ family have been found to release one Me_3Sn^+ ion, and to incorporate one R_4N^+ ion instead, along with a distinct number of water molecules [54–56]. These "truncated" (in view of Me_3Sn) products of the general composition $[(R_4N)(Me_3Sn)_2M^{III}(CN)_6{\cdot}xH_2O]$ are even less water-soluble than their R_4N-free precursors if R = n-propyl, n-butyl, or n-pentyl (but not ethyl!), and the corresponding water content turns out to be x = 2.0, 1.0, and 0.5, respectively. Although part of the initially six coordinative N → Sn bonds per formula unit has been replaced by O → N and $OH_2{\cdots}NC$ interactions, the crystal structures of these truncated systems still reflect the familiar 2D square grid patterns.

Expanded SPB derivatives look, on the other hand, formally more like host –guest systems, for example, [**2**·3H_2O·3/2bpy] (**52**), [**6**·4H_2O·bpy] (**53**), [**11**·2H_2O·bpy] (**54**), [**11**·2H_2O·bpe] (**55**), [**28**·H_2O·3/2bpy] (**56**) [10,62]. Each M atom of **52–55** carries two $CNSnMe_3 \cdot OH_2$ ligands, which are, however, further involved in two strong O–H···N hydrogen bonds to one terminal cyanide ligand and to the nitrogen atom of a bpy (or bpe) molecule, respectively. For instance, the supramolecular assembly **53** is built up primarily of infinite, parallel-oriented "strands", in which Fe(III) ions are connected pairwise by two type III spacers (in that pairs of cyclohexane-like $Fe_2O_2N_2$ chairs share one Fe atom). Suitably positioned strands of **53** are then interlinked to wavy sheets by bpy units via two O–H···N(bpy) hydrogen bonds (Fig. 6). Most interestingly, these sheets display wide pores that accommodate planar $SnMe_3^+$ ions the central Sn atoms of which lie unusually far away from the two closest-lying N atoms of terminal cyanide ligands (Sn···N distances: 2.651(4) and 2.638(4) Å). These extremely long Sn···N distances (as compared with 2.34 Å in common type II and III spacers) leave considerable doubt in significantly strong coordinative N → Sn bonding. Zeolitic H_2O molecules are incorporated between adjacent sheets, without, however, "critically" approaching the planar Me_3Sn^+ ions. The particular framework of **53** thus seems to be held together by a combination of type III spacers, the rather long, new spacer –CN···H–O(Sn)–H···N(bpy)N···H–O(Sn)–H···NC– and by spacers of type IV with L = bpy. **53** might thus be considered as a potential host–guest system with the long-sought [56], "naked" Me_3Sn^+ cation as guest.

12.6 Conclusions

Primarily, the space required by any 3D SPB (or PB) derivative depends just on the dimensions of its "naked" $^3_\infty[M\{\mu\text{-CNENC}\}_3]$ (or $^3_\infty[M\{\mu\text{-CN}\}_3]$) skeleton. In the absence of self-catenation, which situation is successfully realized by substituent-carrying spacers, the infinite [-M–CN–E–NC-] chains generating each skeleton turn out to avoid linearity (i.e., they tend to be shortened) to minimize empty voids, while the actual space demand of groups coordinated to E and of encapsulated guests, respectively, will counteract this tendency. The net result of this interplay is clearly reflected by the notable variation of the corresponding formula volume, V_f (Eq. 9). In Table 4

$$V_f = \text{NF} \cdot M_r/\rho_{calc} = 0.602\ \text{NF} \cdot V/Z \tag{9}$$

the V_f values of selected SPB derivatives with different spacers and guests, respectively, are listed for comparison.

Taking the V_f value of $[CsMn^{II}Co^{III}(CN)_6]$ (175 Å^3) as representative for an unshrinked PB systems, the experimental V_f values of all SPB systems containing exclusively type II spacers (**1, 2, 7–9, 42, 43, 57**, and **59**) turn out to be notably lower than the V_f value of about 1000 Å^3 extrapolated for ideal SPB with strictly linear chains. Yet, the V_f value of one singular framework of **1** exceeds the V_f values of **2** and **7**, probably because it is less compressed than the non-interpenetrating frameworks of **2** and **7**. Owing to the high space demand of nine Sn-bonded tert-butyl groups (per formula unit), the V_f values of **8** and **9**

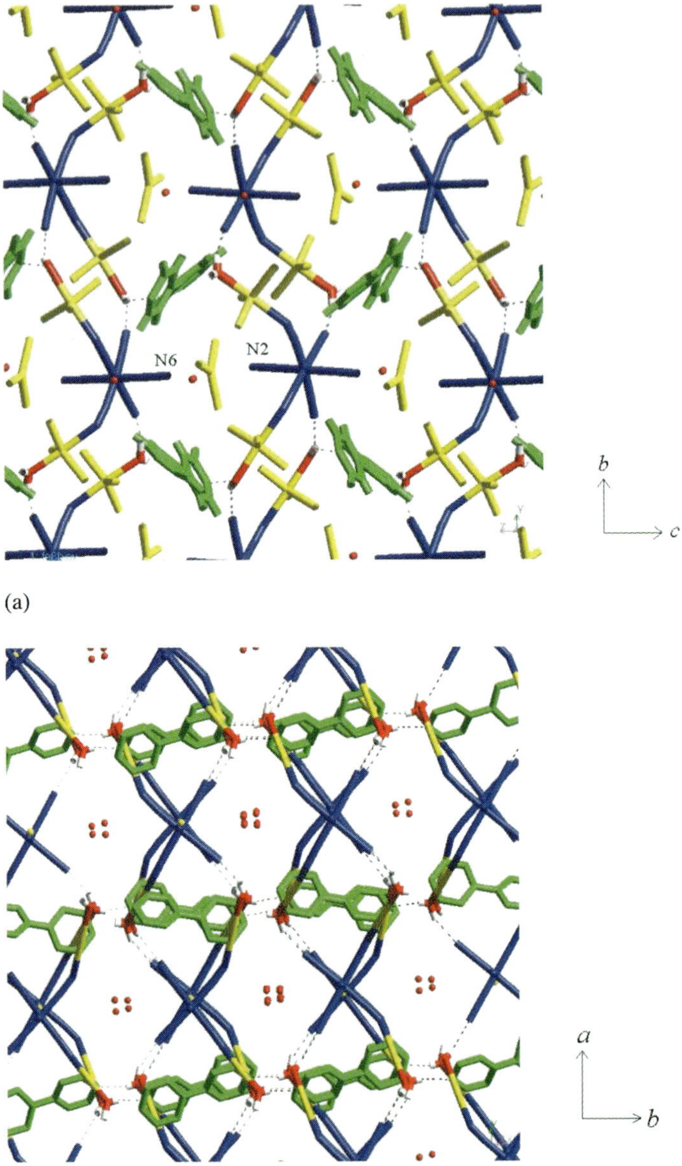

Fig. 6. View along *a* on the structure of **53**. (a): Isolated $\{Me_3Sn\}$ units are supposed to involve three-coordinate Sn. (b): Perspective of **53** along *c* (dotted lines: O–H···N (cyanide) or O–H···N(bpy) hydrogen bonds; red: H_2O).

Tab. 4. Formula volumes, V_f, of selected SPB and related systems (NF: number of interpenetrating frameworks).

No.	*SPB-Derivative* [a]	*NF*	V_f [$Å^3$]
7	$^3_\infty[Co^{III}\{\mu\text{-}PbMe_3(NC)_2\}_3]$	1	435
2	$^3_\infty[Co^{III}\{\mu\text{-}SnMe_3(NC)_2\}_3]$	1	460
43	$^3_\infty[(MV)_{0.5}Ru^{II}\{\mu\text{-}SnMe_3(NC)_2\}_3]$ [b]	1	495
42	$^3_\infty[(Co^{III}Cp_2)Fe^{II}\{\mu\text{-}SnMe_3(NC)_2\}_3]$ [c]	1	522
1	$^3_\infty[Co^{III}\{\mu\text{-}Ag(NC)_2\}_3]$	3	556
57	$^3_\infty[Fe^{II}\{\mu\text{-}Ni^{II}(en)_2(NC)_2\}_3(PF_6)_2]$ [d] [57]	1	585
58	$^3_\infty[Eu^{III}(H_2O)_3\{\mu\text{-}Ag^{I}\ (CN)_2\}_3]$ [e] [58]	3	646
59	$^3_\infty[Cd^{II}\{\mu\text{-}pyz\}\{\mu\text{-}\ Ag^{I}(CN)_2\}\{\mu\text{-}Ag^{I}{}_2(CN)_3\}]$ [f] [59]	3	695
8/9	$^3_\infty[M^{II}\{\mu\text{-}Sn(nBu)_3(NC)_2\}_3]$ (M = Co/Fe)	1	774/785
3	$^3_\infty[Fe^{III}\{\mu\text{-}SnPh_3(NC)_2\}_2\{\mu\text{-}SnPh_3(NC)(OH_2{\cdot}NC\}]$ [g]	1	914
10	$^3_\infty[Rh^{III}\{\mu\text{-}SnMe_3(NCS)_2\}_3]$	2	960
31	$^3_\infty[Fe^{II}\{\mu\text{-}SnMe_3(NC)_2\}_2\{\mu\text{-}(SnMe_3{\cdot}OH_2)_2diox\}]$ [h]	2	1360

[a] groups in round brackets (outside braces) indicate guest ions, while groups in braces are metal connectors (spacers); [b] MV = methylviologen cation (2+); [c] Cp = cyclopentadienyl ligand; [d] en = ethylenediamine ligand; [e] with six cyanide nitrogen atoms at the corners of a trigonal prism; [f] pyz = pyrazole; [g] containing the spacers II and III; [h] diox = dioxane.

come closer to the extrapolated V_f value. Compound **57** involves bis-(ethylenediamine)nickel(II) fragments instead of R_3Sn units and two encapsulated PF_6^- anions as guest anions [57]. Obviously, with the exception of N_2O_4 as guest (see above) the free space available within the frameworks of **2** and **7** is too small to accommodate guests of the size of MV^{2+}, $CoCp^+$, and PF_6^-, and the uptake of these comparatively large guests by negatively charged frameworks (with M(II) instead of M(III)) is usually accompanied by a notable increase in V_f (Table 4). Still higher V_f values result for heteroleptic SPBs with spacers longer than those of type II (**3**, **31**, and **59**) and for the homoleptic system **10**.

The V_f value of the "naked" framework of the super-SPB system **41** (see above) (927 $Å^3$) compares well with the values of **3** and **10**. Obviously, a table similar to Table 4 will be necessary to inspect the variation of V_f of frameworks based upon four-coordinate transition metal ions. Although different insights into the supramolecular architecture have so far been of paramount interest (focusing also on so-called weak interactions), it might be added in concluding that controlled thermolysis of numerous SPB derivatives under oxidative and reductive conditions has turned out to afford various, both amorphous and crystalline, oxidic, or intermetallic phases of promising interest for application for example, as heterogeneous catalysts [27,60].

Most recently, even the voluminous cation $\{KC\ 222\}^+$ with the well-known cryptand 222 = $N(CH_2OCH_2OCH_2)_3N$ could be encapsulated by a SPB framework built up (like compound **3**) of the spacers II and III (2:1) and the $[Ru(CN)_6]^{4-}$ building block [10]. The formula volume V_f (see Table 4) of $[\{KC\ 222\}Ru^{II}\{\mu\text{-}CNSn(Me_3)NC\}_2\ \{\mu\text{-}CNSn(Me_3)O(H)\cdots NC\}]$ amounts to 728 $Å^3$.

References

1 For extensive information on Prussian blue, the following review should be consulted: K.R. Dunbar, R.A. Heintz, *Progr. Inorg. Chem.* **1997**, *45*, 283.

2 M. Kämper, M. Wagner, A. Weiß, *Angew. Chem.* **1979**, *91*, 517; *Angew. Chem. Int. Ed. Engl.* **1979**, *18*, 486.

3 L. Pauling, P. Pauling, *Proc. Nat. Acad. Sci. USA* **1968**, *60*, 362.

4 K. Yünlü, N. Höck, R.D. Fischer, *Angew. Chem.* **1985**, *97*, 863; *Angew. Chem. Int. Ed. Engl.* **1985**, *24*, 879.

5 U. Behrens, A.K. Brimah, T.M. Soliman, R.D. Fischer, D.C. Apperley, N.A. Davies, R.K. Harris, *Organometallics* **1992**, *11*, 1718.

6 J. Liu, W.T.A. Harrison, A.J. Jacobson, *Inorg. Chem.* **1996**, *35*, 4271.

7 M. Adam, A.K. Brimah, R.D. Fischer, X.-F. Li, *Inorg. Chem.* **1990**, *29*, 1595.

8 R. Eckhardt, PhD Thesis, University of Hamburg, Germany, **2002**; www.sub.uni-hamburg.de/emedien.

9 R. Eckhardt, R.D. Fischer, *Inorg. Chem. Commun.* **2000**, *3*, 433.

10 H. Hanika-Heidl, PhD Thesis, University of Hamburg, Germany, **2003**.

11 T. Niu, J. Lu, X. Wang, J.D. Korp, A. J. Jacobson, *Inorg. Chem.* **1998**, *37*, 5324.

12 E. Siebel, R.D. Fischer, *Chem. Eur. J.* **1997**, *3*, **1987**.

13 J.-U. Schütze, R. Eckhardt, R.D. Fischer, D.C. Apperley, N.A. Davies, R.K. Harris, *J. Organomet. Chem.* **1997**, *534*, 187.

14 U. Behrens, A.K. Brimah, K. Yünlü, R.D. Fischer, *Angew. Chem.* **1993**, *105*, 117; *Angew. Chem. Int. Ed. Engl.* **1993**, *32*, 82.

15 E.-M. Poll, PhD Thesis, University of Hamburg, Germany. **2000**; www.sub.uni-hamburg.de/emedien.

16 D.C. Apperley, N.A. Davies, R.K. Harris, A.K. Brimah, S. Eller, R.D. Fischer, *Organometallics* **1990**, *9*, 2672.

17 Z. Assefa, F. DeStefano, M.A. Garepapaghi, J.H. LaCasce, Jr., S. Ouellete, M.R. Corson, J.K. Nagle, H.H. Patterson, *Inorg. Chem.* **1991**, *30*, 2868; see also: M.A. Omary, T.R. Webb, Z. Assefa, G.E. Shankle, H.H. Patterson, *Inorg. Chem.* **1998**, *37*, 1380.

18 A.K. Brimah, E. Siebel, R.D. Fischer, N.A. Davies, D.C. Apperley, R.K. Harris, *J. Organomet. Chem.* **1994**, *475*, 85.

19 E. Siebel, A.M.A. Ibrahim, R.D. Fischer, *Inorg. Chem.* **1999**, *38*, 2530.

20 E. Siebel, PhD Thesis, University of Hamburg, Germany, **1998**.

21 A.M.A. Ibrahim, E. Siebel, R.D. Fischer, *Inorg. Chem.* **1998**, *37*, 3521.

22 A.M. Domingos, G.M. Sheldrick, *J. Chem. Soc. Dalton Trans.* **1974**, 477.

23 U. Behrens, A.K. Brimah, R.D. Fischer, *Organomet. Chem.* **1991**, *411*, 325.

24 A.K. Brimah, P. Schwarz, R.D. Fischer, N.A. Davies, R.K. Harris, *J. Organomet. Chem.* **1998**, *568*, 1.

25 S. Eller, P. Schwarz, A.K. Brimah, R.D. Fischer, D. Apperley, N.A. Davies, R.K. Harris, *Organometallics* **1993**, *12*, 3232.

26 S.R. Batten, A.R. Harris, P. Jensen, K.S. Murray, A. Ziebell, *J. Chem. Soc. Dalton Trans.* **2000**, 3829.

27 M. Rehbein, M. Epple, R.D. Fischer, *Solid State Sci.* **2000**, *2*, 473.

28 S. Eller, PhD Thesis, University of Hamburg, Germany, **1992**.

29 S. Eller, S. Dülsen, R.D. Fischer, *J. Organomet. Chem.* **1990**, *390*, 309.

30 R. Tarhouni, PhD Thesis, University of Hamburg, Germany, **1996**.

31 M. Ling, PhD Thesis, University of Hamburg, Germany, **2001**; www.sub.uni-hamburg.de/emedien.

32 D.C. Apperley, N.A. Davies, R.K. Harris, S. Eller, P. Schwarz, R.D. Fischer, *J. Chem. Soc. Chem. Commun.* **1992**, 740; R.K. Harris, M.M. Sünnetçioğlu, R.D. Fischer, *Spectrochim. Acta* **1994**, *50A*, 2069; *Spectrochim. Acta* **1995**, *51A*, 1389.

33 J. Lu, W.T.A. Harris, A.J. Jacobson, *Angew. Chem.* **1995**, *107*, 2759; *Angew. Chem. Int. Ed. Engl.* **1995**, *34*, 2311.

34 T. Niu, X. Wang, A.J. Jacobson, *Angew. Chem.* **1999**, *111*, 2059, *Angew. Chem. Int. Ed. Engl.* **1999**, *38*, **1934**.

35 P. Schwarz, E. Siebel, R.D. Fischer, D.C. Apperley, N.A. Davis, R.K. Harris, *Angew. Chem.* **1995**, *107*, 1311; *Angew. Chem. Int. Ed. Engl.* **1995**, *34*, 1197.

36 S. Eller, M. Adam, R.D. Fischer, *Angew. Chem.* **1990**, *102*, 1157; *Angew. Chem. Int. Ed. Engl.* **1990**, *29*, 1126.

37 P. Schwarz, S. Eller, E. Siebel, T.M. Soliman, R.D. Fischer, D.C. Apperley, N.A. Davies, R.K. Harris, *Angew. Chem.* **1996**, *108*, 1611; *Angew. Chem. Int. Ed. Engl.* **1996**, *35*, 1525.

38 E. Siebel, R.D. Fischer, J. Kopf, N.A. Davies, D.C. Apperley, R.K. Harris, *Inorg. Chem. Commun.* **1998**, *1*, 346.

39 P. Brandt, R.D. Fischer, E.S. Martinez, R.D. Calleja, *Angew. Chem.* **1989**, *101*, 1275; *Angew. Chem. Int. Ed. Engl.* **1989**, *28*, 1265; P. Brandt, U. Illgen, R.D. Fischer, E.S. Martinez, R.D. Calleja, *Z. Naturforsch.* **1993**, *48b*, 1565; A.M.A. Ibrahim, *J. Mater. Chem.* **1998**, *8*, 841.

40 See: A.F. Wells, Structural Inorganic Chemistry (5th edn.), Clarendon, Oxford **1984**, pp. 87, 241, 538; for a NbO-like supramolecular assembly see: K.N. Power, T.L. Hennigar, M.J. Zaworotko, *J. Chem. Soc. Chem. Commun.* **1998**, 595.

41 C. Glidewell, D.C. Liles, *Acta Crystallogr.* **1978**, *B34*, 129.

42 T.M. Soliman, S.E.H. Etiaw, G. Fendesak, R.D. Fischer, *J. Organomet. Chem.*, **1991**, *415*, C1.

43 E.-M. Poll, R.D. Fischer, *Inorg. Chem. Commun.* **2000**, *3*, 259.

44 E. Siebel, R.D. Fischer, N.A. Davies, D.C. Apperley, R.K. Harris, *J. Organomet. Chem.* **2000**, *604*, 34.

45 T. Niu, A.J. Jacobson, *Inorg. Chem.* **1999**, *38*, 5346.

46 E.-M. Poll, J.-U. Schütze, R.D. Fischer, N.A. Davies, D.C. Apperley, R.K. Harris, *J. Organomet. Chem.* **2001**, *621*, 254.

47 G.A. Bowmaker, H. Hartl, V. Urban, *Inorg. Chem.* **2000**, *39*, 4548.

48 A.M.A. Ibrahim, *J. Organomet. Chem.* **1998**, *556*, 1.

49 U. Nolte, PhD Thesis, University of Hamburg, Germany, **1995**.

50 S.C. Abrahams, J.L. Bernstein, R. Liminga, E.T. Eisenmann, *J. Chem. Phys.* **1980**, *73*, 4585; B. F. Hoskins, R. Robson, N.V.Y. Scarlett, *J. Chem. Soc. Chem. Commun.* **1994**, 2025.

51 E. Siebel, P. Schwarz, R.D. Fischer, *Solid State Ionics* **1997**, *101–103*, 285.

52 P. Schwarz, E. Siebel, R.D. Fischer, N.A. Davies, D.C. Apperley, R.K. Harris, *Chem. Eur. J.* **1998**, *4*, 919.

53 M.S.I. Sayed Ahmed, PhD Thesis, Tanta University, Tanta, Egypt, **1996**.

54 E.-M. Poll, S. Samba, R.D. Fischer, F. Olbrich, N.A. Davies, P. Avalle, D.C. Apperley, R.K. Harris, *J. Solid State Chem.* **2000**, *152*, 286.

55 E.-M. Poll, F. Olbrich, S. Samba, R.D. Fischer, P. Avalle, D.C. Apperley, R.K. Harris, *J. Solid State Chem.* **2001**, *157*, 324.

56 See: J.B. Lambert, S. Zhang, S.M. Ciro, *Organometallics* **1994**, *13*, 2430.

57 N. Fukita, M. Ohba, H. Okawa, K. Matsuda, H. Iwamura, *Inorg. Chem.* **1998**, *37*, 842.

58 Z. Assefa, R.J. Staples, J.P. Facklar, Jr., *Acta Crystallogr. C* **1995**, *51*, 2527.

59 T. Soma, H. Yuge, T. Iwamoto, *Angew. Chem.* **1994**, *106*, 1746; *Angew. Chem. Int. Ed. Engl.* **1994**, *33*, 1665.

60 M. Rehbein, R.D. Fischer, M. Epple, *Thermochim. Acta* **2002**, *382*, 143.

61 E.M. Poll, M. Rehbein, M. Epple, R.D. Fischer, *Supramolec. Chem.*, in press.

62 R. Eckhardt, H. Hanika-Heidl, R.D. Fischer, *Chem. Eur. J.* **2003**, *9*, in press.

Part 2
Structure and Dynamics of Guest–Host Composites Based on Nanoporous Crystals

Structure and Dynamics of Guest–Host Composites Based on Nanoporous Crystals

Ferdi Schüth

Guest–host systems are composites made up of at least two components, the guest and the host, and in the context of this book this is a nanoporous crystal. Each of the components is characterized by its own structure and dynamics. However, when they are combined to form the guest–host system, the complexity increases substantially. First, the new guest–host composite has its composite structure and dynamics: possibly an ordered arrangement of the guest species in the pore system of the host, often with complex dynamic modes. In addition, guest and host can have marked effects upon each other, which can even lead to changes in the zeolitic structural framework, which we otherwise consider to be rigid. Zeolite rho is a prominent example, in which remarkable structural changes are observed upon cation exchange or hydration [1]. To understand the performance of such composites in the applications that are discussed in this book, information on the structure and dynamics of the guest–host systems is mandatory.

The global structural effects are often not very pronounced, and so it is necessary to use sophisticated state-of-the-art techniques such as synchrotron X-ray diffraction or X-ray absorption spectroscopy to elucidate structural details. In addition, theoretical calculations are a tool of increasing importance for supporting the interpretation of experimental data, or for obtaining information that is inaccessible by experiment.

The importance of advanced experimental techniques for analyzing the structure and dynamics of zeolite-based guest–host systems was discussed several years ago by J. Parise [2], who listed four important developments.

1. Bright laboratory or synchrotron radiation sources for analyzing small single crystals of only a few 10 μm in size.
2. Powder diffraction with synchrotron sources having high angular resolution, high brightness, and excellent signal-to-noise ratio, which eases structure solution from powder data.
3. EXAFS for analyzing the immediate environment of a guest species and possible interactions with the host.

4. Neutron diffraction as the technique of choice for the location of hydrogen atoms.

To this list should be added NMR spectroscopy as a more local probe, which can also be used to address other time domains for the analysis of dynamic behavior (as an alternative to other methods mentioned) [3], and quasielastic neutron scattering, also for the analysis of the dynamic behavior of guest species [4]. NMR spectroscopy has been used to analyze the dynamics of aromatic molecules in zeolite pores and of hydrogen bonded complexes, due to the relevance of this problem in zeolite catalysis [5–7]. Koller et al. also give examples of how experiment and theory can assist each other in obtaining deeper understanding of the factors governing the interaction of a guest species with the zeolite host framework [7].

Theory has gone a long way from rationalization of phenomena occurring in guest–host systems to the predictive power it has achieved in some fields now. Some review papers describe the achievements in different areas. Henson and Cheetham gave a short survey of different techniques used in computational approaches towards adsorbed molecules in microporous hosts, together with some examples [8]. Demkov and Sankey discuss the possibility of treating superlattices formed in periodic porous structures, such as zeolites, using electronic structure theory [9]. They also give an overview of theoretical methods available for modeling zeolite-based guest–host systems. Auerbach has recently published a relatively extensive review of theoretical studies of the dynamics of adsorbed molecules in zeolite pores, with an emphasis on aromatics in faujasites [10]. Several aspects of intrazeolite host–guest chemistry are included in a review of theoretical studies for understanding zeolite catalysis [11].

Most of the experimental techniques mentioned above, as well as theoretical methods for molecular mechanics and on the quantum chemistry level, have been used in the case studies that comprise the majority of this Part of the book. These studies exemplify the state of the art in the analysis of zeolite-based guest–host systems and highlight the benefit of a close interaction between experiment and theory, illustrating the symbiotic relationship that theory and experiment have in this field.

Chapter 1 is a survey of theoretical studies and the multitude of different theoretical methods available for understanding guest–host systems based on nanoporous crystals. This should provide a basis for allowing the reader easier access to subsequent chapters and the original literature. The other chapters describe case studies of structure and dynamics of different guest–host systems, typically combining various methods to extract as much information as possible.

Chapter 2 describes the structure and dynamics of guest species in zeolite crystals at the macroscopic level. The authors use the technique of interference microscopy to study the uptake and distribution of guest species in the channel system of large crystals with various zeolite structures. This method has become very useful for analyzing the real structure of large zeolite crystals, which are (in spite of their often remarkable morphological perfection) highly defective and can consist of many domains. This can introduce barriers inside the crystals, which can impair the dynamics of sorption processes and the achievable final loading with guest species. Hence, results obtained from microscopic methods cannot always be extrapolated to the macroscopic level.

The next chapters describe the structure and dynamics of guest species at the atomic scale. In Chapter 3 the bassanite system is the focus of attention. The main guest species investigated is a very simple one, water (some experiments with methanol as guest are also reported), but the system is nevertheless rather complex, since the water molecules are dynamically disordered at room temperature: a rather common phenomenon in guest–host systems based on nanoporous crystals. The interaction of the guest with the host system was found to be rather weak in bassanite with water as well as with methanol, since no substantial distortions of the framework occurred upon removal of the guest species. The presence of disorder and the low degree of interaction observed here are probably strongly coupled phenomena. In a weakly interacting system the mobility of the guest species should be high and thus no strict ordering is expected. On the other hand, strong interaction will typically lead to localized adsorption with a clear minimum energy configuration, which should be easily detectable with structural analysis techniques.

The latter was the predominant adsorption mode of the systems studied in Chapter 4. Different organic guest molecules were localized using both powder XRD with synchrotron radiation and neutron powder diffraction, supported by force-field calculations. Several different organic molecules could clearly be located in the pore system of faujasites, and the strong interaction with the host was seen by strong displacements of sodium cations, which balance the framework charges. If mixtures of guests were adsorbed, the individual molecules had the same location as when they were adsorbed alone, which again demonstrates that the guest–host interaction was dominant over the guest–guest interaction. Force-field calculations could rationalize most of the positions observed experimentally.

Such experimental confirmation of the results obtained by theory is not available for Chapter 5. The authors propose a method for localizing adsorbed molecules in nanoporous crystalline materials and illustrate it for the example of thionine in NaY. The method combines the speed of force-field based molecular dynamics simulations with the higher accuracy of quantum chemical calculations by generating different snapshots from a MD simulation and then calculating the structure of the local minima with quantum-chemical methods. Local minima are then grouped by a clustering algorithm and an averaged energy is calculated. Three low-energy situations, which showed appreciable occupation at room temperature, could be identified by this procedure. These configurations are good starting points for the interpretation of experimental data, for instance by Rietveld refinement, and the method seems to be more generally applicable.

Chapter 6 also deals exclusively with theory, but now on the quantum chemical level using density-functional theory. The authors describe the application of DFT calculations in the analysis of zeolite guest–host systems for various examples such as the location of cations in zeolitic materials and the influence of the zeolite properties, the interaction of organic guest molecules with the zeolite cations, and the structure of metal clusters in the void space of zeolites. The chapter highlights the kinds of problems that can now be addressed at the quantum-chemical level. The authors indicate how complex situations still out of the reach of a quantum-chemical treatment should be dealt with, by combined QM/MM methods in which the important part of the system is accurately treated with quantum-chemical methods and the environment modeled at the molecular mechanical level, an approach advocated by many theoreticians working on zeolites or in other fields [12–14].

This last statement clarifies the approach that is a theme running through Part 2. At present, most insight can be gained from combined approaches at different levels, with experiment and theory at the highest level, and the combination of QM and MM methods or the combination of X-ray and neutron diffraction at sub-levels. We now have many different and often complementary methods available to us for in-depth study and understanding of the structure and dynamics of guest–host systems based on nanoporous crystals, and combining them in an intelligent way allows us to elucidate details of the structure and dynamics of the system that were out of reach ten years ago.

References

1 G.M. Johnson, B.A. Reisner, A. Tripathi, D.R. Corbin, B.H. Toby, J.B. Parise, *Chem. Mater.* **1999**, *11*, 2780, with numerous references.
2 J.B. Parise, *J. Incl. Phenom. Mol. Recog. Chem.* **1995**, *21*, 79.
3 C. Dybowski, *J. Incl. Phenom. Mol. Recog. Chem.* **1995**, *21*, 113.
4 H. Jobic, *J. Phys. IV* **2000**, *10*, 125.
5 B. Geil, O. Isfort, B. Boddenberg, D.E. Favre, B.F. Chmelka, F. Fujara, *J. Chem. Phys.* **2002**, *116*, 2184.
6 D.E. Favre, D.J. Schaefer, S.M. Auerbach, B.F. Chmelka, *Phys. Rev. Lett.* **1998**, *81*, 5852.
7 H. Koller, G. Engelhardt, R.A. van Santen, *Top. Catal.* **1999**, *9*, 163.
8 N.J. Henson, A.K. Cheetham, *J. Incl. Phenom. Mol. Recog. Chem.* **1995**, *21*, 137.
9 A.A. Demkov, O.F. Sankey, *Chem. Mater.* **1996**, *8*, 1793.
10 S.M. Auerbach, *Int. Rev. Phys. Chem.* **2000**, *19*, 155.
11 S.P. Bates, R.A. van Santen, *Adv. Catal.* **1998**, *42*, 1.
12 J. Sauer, M. Sierka, *J. Comput. Chem.* **2000**, *21*, 1470.
13 G. Colombo, G. Carrea, *J. Biotechnol.* **2002**, *96*, 23.
14 M.J. Field, *J. Comput. Chem.* **2002**, *23*, 48.

1
Computational Methods for Host–Guest Interactions

Joachim Sauer

1.1
Introduction

This book includes numerous examples of the successful combination of experimental and computational techniques in studies of host–guest interactions in nanostructured materials. The problems discussed in various chapters are different and so different computational techniques are needed and employed. The challenge of nanostructured host–guest materials for computational quantum chemistry or physics is in the large number of atoms that have to be taken into account and in the differences in chemical nature between guest species and host materials.

The unit cells of nanostructured host materials include hundreds of atoms and, in addition, the guest species are not always ordered into unit cells. This means that very large pseudo-unit cells have to be adopted if periodic boundary conditions (PBC) are applied. The techniques available in computational chemistry and physics are based on quantum mechanics but rely on approximations that are usually different for chemically different systems. This means that the methods that work best for inorganic insulating (host) materials are different from the methods typically used to describe organic, metallic, or semiconducting guest species.

1.2
Computational Problems in Host–Guest Chemistry and Physics

The choice of method depends on the aim of the study. There are three typical cases for host–guest systems.

1. The structure of the material is known and the interest is in optical, electronic, or magnetic properties.
2. The structure of the host material is known, but the distribution and location of the guest species is not. This includes cases in which the structure of the material is in general known (e.g., the framework type of a zeolite) but computational techniques are used to determine the specific position of the host atoms in the presence of guest molecules (relaxation).

3. The presence of guest species induces qualitative changes in the host structure.

This book includes examples of all three situations.

1. Sauer and Windiks (Part 3, Chapter 4) start from the known crystal structures of sodium and potassium electrosodalite and present quantum mechanical (density functional) calculations of the magnetic and electronic properties of these materials. Depending on the specific question, either a local (cluster) model is adopted or PBC are applied.
2. Two different forms of thionine adsorbed in dehydrated NaY zeolite, A and B, have been assumed to explain two different fluorescence maxima that can be converted thermally or optically into each other [1]. Brickmann et al. (Part 2, Chapter 5) start from the known structure of the host material zeolite NaY and determine possible adsorption sites of the organic guest thionine in the zeolite cavity. The cubic unit cell of NaY, the number of atoms and the connectivity of the host atoms are maintained, but the specific positions of the zeolite atoms are allowed to adjust to the presence of the thionine guest in the zeolite cavity. A parameterized force field for the zeolite and for the zeolite–dye interaction is used to describe the energy of the system as a function of the structure of the zeolite lattice and the position and orientation of the dye molecule within the zeolite cavity. Two adsorption structures relevant for explaining the fluorescence behavior have been identified (Part 2, Chapter 5, Fig. 8)
3. The different phases formed in the $CaSO_4/H_2O$ system (Part 2, Chapter 3)
   - gypsum: $CaSO_4{\cdot}2\,H_2O$, $2\,CaSO_4{\cdot}4\,H_2O$
   - bassanite: $2\,CaSO_4{\cdot}H_2O$, $2\,CaSO_4{\cdot}H_2O + 3\,H_2O$
   - γ-$CaSO_4$ (soluble anhydride): $CaSO_4$, $2\,CaSO_4 + 4\,H_2O$
   - β-$CaSO_4$ (insoluble anhydride): $CaSO_4$, $2\,CaSO_4 + 4\,H_2O$

   are examples of qualitative structure changes with increasing amount of guest species. All four structures are different states of the same system with the total composition $2\,CaSO_4{\cdot}4\,H_2O$ if the H_2O vapor phase in equilibrium with the solid phase is also considered. In gypsum the oxygen atoms of the H_2O molecules are part of the coordination polyhedra around the Ca ions. On dehydration of gypsum a new $CaSO_4$ framework structure is obtained consisting of corner-sharing SO_4 tetrahedra and CaO_8 octahedra. In bassanite this framework is partially filled with H_2O molecules, but it is empty in γ-$CaSO_4$. If dehydration of bassanite occurs not at moderate temperature, but higher, the framework is not maintained and β-$CaSO_4$ forms.

1.3 Structure Predictions for Host–Guest Systems using Periodic Boundary Conditions

In principle, for all three types of problems the solutions of the Schrödinger equation for the whole system (all atomic nuclei and electrons) would answer all questions, but in practice a hierarchy of approximations is made. The key mathematical object is the potential energy surface (PES) that describes the (electronic) energy of a system with N atoms as a function of the $3N - 6$ internal coordinates. It is defined by the Born–Oppenheimer approximation and represents the potential energy of the motion of the nuclei. The points of the PES are obtained by (approximately) solving the electronic Schrödinger equation for a given configuration of the atoms of this system. The "system" is defined as a set of atoms (number and type of element), that is, as a certain composition. That implies that all compounds having the same total composition of atoms belong to the same system and are described by one PES. For example, the equilibrium structures of the four different phases mentioned above for the $CaSO_4/H_2O$ system represent four different minima on the $2\,CaSO_4{\cdot}4\,H_2O$ PES.

Stable (or metastable) structures of a system (i.e., the possible isomers) are defined as minima on the PES. Educts and products of a reaction are different minima on the PES, while the transition structure is defined as a saddle point on the PES. The computational determination of stable structures corresponds to finding minima on the potential energy surface. Different vibrational states of a molecules (or solid) are stationary solutions of the nuclear motion problem within a potential well around a minimum.

Statistical thermodynamics takes averages over regions of the PES around minima (or saddle points). For a particular isomer of a system at a finite temperature either the (Boltzmann) distribution over the vibrational states of this isomer is considered or a classical molecular dynamics calculation is made by solving the Newtonian equations for the motion of the nuclei using the PES as the potential. Observed structures (atomic positions) correspond to average positions of the nuclei. The cluster of configurations of thionine in zeolite NaY (Part 2, Chapter 5) corresponds to an ensemble of different configurations around the equilibrium configuration; the two different clusters *a* and *b* represent two different isomers of the NaY–thionine host–guest system.

Different electronic states define different PES for the same system. Optical, electronic, or magnetic properties can be calculated from the solutions of the electronic Schrödinger equation for a given position of the nuclei. In many cases, optical, electronic, or magnetic properties can be calculated by (approximately) solving the Schrödinger equation for a given geometric structure. This can be either the equilibrium position of the atoms obtained by computational techniques or the average position of the atoms obtained by experimental techniques.

The concept of the PES orders the different computational techniques into the following groups.

- Methods used for calculating the PES.
  - density functional theory (DFT)
  - tight-binding DFT
  - parameterized interatomic potential functions (force fields)
  - hybrid DFT/force field
- Methods used for moving on the PES.
  - finding stable structures (minima) or transition structures (saddle points)
  - molecular dynamics

- Methods for calculating properties for given structures (points on the PES).
  - semiempirical quantum chemical methods/model Hamiltonians
  - density functional theory
  - quantum chemical *ab initio* methods

For each of the three different tasks two different models can be adopted.

- Periodic boundary conditions.
- Embedded cluster models.

DFT has become a standard approach in computational chemistry/physics. It should be noted that the results obtained are not of uniform quality, but depend on the specific functional used and the basis set employed. Three classes of functionals may be distinguished [2,3]: local density approximation (LDA), gradient-corrected functionals such as PW91, BP86, or BLYP (generalized gradient approximations, GGA), and hybrid functionals, which contain some exact exchange such as B3LYP, B3PW91, or PBE1PBE (adiabatic connection methods, ACM). Basis sets of different size and of two different types are used: atom-centered Gaussian basis sets and plane-wave basis sets. The former have been taken over from molecular calculations, while the latter are the natural choice for systems with 3D periodicity and require replacement of inner core electrons by pseudopotentials.

Hence DFT calculations are defined by a functional/(pseudopotential) basis set and it is this definition of a quantum mechanical model that determines the quality of the results. In the limiting case of very large basis sets, the results depend only on the functional used. Beyond such models, approximate DFT methods have been suggested, known as tight-binding DFT. They use localized basis sets and make additional approximations to limit the number of interactions included. Sometimes adjustable parameters are introduced to compensate partially for the approximations. Examples are codes such as SCC-DFTB [4], SIESTA [5], or Fireball [6]. These codes are computationally very efficient, but do not approximate results for a given functional within numerical limits. They rather represent different quantum mechanical models whose performance for a class of systems has to be investigated. Typical areas of application are biomolecular systems [4,7], endohedral fullerenes [8], and supralattices [9]. Demkov and Sankey studied Si clusters in silica-sodalite and Na-doped Si clathrates [9].

1.4 Structure Predictions for Host–Guest Systems Using Periodic Boundary Conditions

DFT with PBC is computationally expensive [10] and structure predictions for host–guest systems became feasible only in the last decade, for example for methanol in chabazite [11–13]. The codes available such as CPMD [14] or VASP [15] use plane-wave basis sets and replace core electrons with pseudopotentials [10]. Applications are typically for unit cells with about 30–50 atoms (CH_3CN in zeolite chabazite [16]), but more recently systems with about 100 atoms (benzene in mordenite [17]) to 300 atoms in the unit cell (methanol in MFI [18]) have been studied. DFT calculations using PBC would be well suited for simulating the four different structures of the $2\,CaSO_4 \cdot 4\,H_2O$ system and would probably be able to resolve the controversial location of protons in these structures. Such calculations are in progress for bassanite ($2\,CaSO_4 \cdot H_2O$), but are not yet reported in the open literature (B. Winkler, unpublished, see also Part 2 Chapter 3).

For sodium and electrosodalite (SES) (the host–guest systems discussed in Part 3, Chapter 4) three different structures based on experiments have been reported and employed in calculations of magnetic properties. DFT calculations applying PBC are also suited for theoretical structure predictions that would yield the structure of the paramagnetic $(M_4)^{3+}$ clusters (M = Na, K), their position in the sodalite cage, and the relaxed position of the atoms of the sodalite framework. Such simulations would be highly welcome but are not available. The only structure simulation reported for a sodium-doped zeolite uses a hybrid DFT/interatomic potential method [19]. The DFT plane-wave description is limited to the $(Na_4)^{3+}$ clusters; the $4\,Na^+$ cores are represented by pseudopotentials and only one electron per cluster is explicitly considered. The structure of the sodalite framework is assumed to be rigid and the interaction of the Na^+ cores with the framework ions is described by an ion-pair potential.

Lattice energy minimization of solid materials by parameterized interatomic potential functions has been a standard technique for many years [20]. Various potential functions (force fields) are available for material modeling [21]. One of the widely used codes is GULP (general utility lattice program) [22]. The application of this technology to host–guest systems is straightforward but faces the problem that interatomic functions of different functional form are commonly used for different types of systems or interactions. Inorganic materials are described by rigid-ion or shell-model ion-pair potentials [20], organic molecules and biomolecules by valence force fields [23,24], and noncovalent interactions by atom–atom pair potentials [25,26]. In their study of thionine in zeolite NaY, Brickmann et al. (Part 2, Chapter 5) use Lennard–Jones terms for the thionine–zeolite and thionine–thionine interactions and combine them with a force-field type description for the zeolite framework structure. The thionine structure is kept rigid, therefore no force-field terms are required on thionine. The electrostatic part of the thionine–zeolite interaction is calculated from point charges. The point charges on thionine are potential derived charges found from a DFT calculation of thionine within the host environment (point charges representing the zeolite host are present in the Hamiltonian of the guest species thionine).

The interactions of organic guests with inorganic host materials often are of van der Waals nature with dispersion as the dominating attractive energy contribution. Interatomic potential functions naturally account for dispersion, most simply by the $1/r^6$ term of the Lennard–Jones expression. Examples are given by hydrocarbons adsorbed into all-silica zeolites [25] and template molecular ions in zeolites [27]. However, DFT calculations face limits for such host–guest systems because currently available functionals fail to yield reliable results for dispersion interactions [28]. DFT studies of benzene in mordenite [17] and of m-xylene in faujasite [29] reproduced only a fraction (about 20 %) of the observed heat of adsorption. Different attempts are made to overcome this problem. The dispersion energy is calculated separately and added to the DFT interaction energy. The problem is that in the region of overlapping densities, current density functionals (unlike Hartree–Fock) account for some intersystem correlation. Hence, adjustable parameters are needed to damp the dispersion term for small distances. The expressions used to calculate the dispersion term range from functionals of the densities of both interacting subsystems (and involving a 6D numerical integration) [30] to simple parameterized $1/r^6$ terms [7,17]. Recently, a hybrid DFT/force-field approach has been suggested for studying (meta-)stable carbenium ions in zeolite cavities [29]. The DFT description is limited to the (positively charged) guest species and (the negatively charged) AlO_4 site of the zeolite, which is represented by an $(HO)_2Al(OSi(OH)_3)_2$ cluster model. The remainder of the periodic zeolite framework is described by a shell-model ion-pair potential while the interaction of the zeolite wall with the hydrocarbon is described by Lennard–Jones terms. The charges on the hydrocarbon are potential derived charges, which interact with formal (shell-model) charges on the zeolite framework. This partitioning guarantees that van der Waals interactions between the hydrocarbon species and the zeolite are described by Lennard–Jones terms (which are superior to DFT for this purpose), while bond making/bond breaking steps, which limit the life time of the carbenium ion (e.g., proton transfer to the zeolite), are described by DFT. Comparison with full periodic DFT calculations was made and the hybrid DFT/potential function results were similar for reactive steps, but much improved for adsorption energies.

1.5 Cluster Model Studies for Host–Guest Systems

All the approaches described above use PBC. The cluster approach [31,32] is an alternative way of making quantum mechanical calculations on solids feasible. The microscopically infinite solid is replaced by a finite piece of it. The advantage is that quantum chemical codes designed for molecules such as TURBOMOLE [33,34] can be immediately applied. Care is necessary in designing a cluster and in interpreting the results. Three types of error are made when replacing a periodic solid by a finite piece for the purpose of a quantum mechanical calculation.

1. Interrupted charge transfer between the atoms of the cluster and the surrounding.
2. Missing structure constraints arising from the periodic lattice.
3. Neglected crystal potential.

1. If bonds are cut on cluster definition, charge transfer is interrupted and "dangling" bonds are created. For zeolites (and silica) termination of the dangling bonds by hydrogen atoms compensates for these effects [31] and this explains

that cluster models met with much success in zeolite chemistry in cases in which (2) and (3) are not crucial.
2. The geometric structure of a cluster model can be taken either from experiment or from simulations of the periodic crystal using interatomic potential functions. Quantum mechanical calculations on fixed structures are useful for calculating properties, but may be problematic when aiming at energies or vibrational frequencies. A compromise is to fix the outer atoms of the cluster at "observed" atomic positions and to optimize the positions of the remaining internal atoms (constraint optimization).
3. Early it has been suggested that cluster models be made more realistic by including the long-range potential originating from the periodic zeolite structure [35]. This is done by adding this "external" potential to the one-electron part of the Kohn–Sham operator.

The best solution to problems (2) and (3) are hybrid QM/MM calculation [36–38] that combine the quantum mechanical (QM) treatment of the cluster with a molecular mechanics (MM) treatment of the periodic zeolite lattice using interatomic potential functions.

Part 2, Chapter 6 describes model calculations on (nonembedded) clusters for metal cations, and metal clusters in zeolites. The cluster models adopted limit these studies to local structures at pre-defined extra-framework sites. Predictions of vibrational features of adsorbed test molecules then permit inferences on the occupation of different sites under experimental conditions.

For transition metal ions in zeolite matrices, in particular Cu^{2+} and Cu^+ ions, the limits of cluster models without embedding have been analyzed [39]. From the published cluster model results it was not possible to conclude whether the Cu^+ ion preferably binds to two oxygen atoms of a single AlO_4 site or to several oxygen atoms of four-, five-, or six-membered rings containing the AlO_4 unit. It appeared that linear cluster models consisting of one, three, or five TO_4 units ($HO[\text{-}T(OH)_2\text{-}O\text{-}]_nH$ with $n = 1, 3, 5$; one T = Al, all other T = Si), are strongly biased towards two-fold coordination of Cu^+ ions to oxygen atoms of a single AlO_4 tetrahedron, while cyclic cluster models consisting of four, five, or six TO_4 units ($[\text{-}T(OH)_2\text{-}O\text{-}]_n$ with $n = 4, 5, 6$; one T = Al, all other T = Si) are biased towards structures with higher coordination numbers. In contrast, hybrid QM/MM calculations on embedded clusters allow direct comparison of the relative stability of two different types of Cu^+ sites in different zeolite frameworks, as results for MFI [40] and FER [41] show. In both zeolites, sites were located in which Cu^+ is coordinated by two framework oxygen atoms only (type II) or by three or four oxygen atoms on top of an aluminosilicate ring on the channel wall (type I). Experimental evidence for the existence of two different types of sites comes from photoluminescence studies on CuZSM-5 [42]. The hybrid QM/MM calculations [43] permit assignment to coordination types I and II based on the prediction that $3d^{10}(^1S_O)$ $3d^9 4s^1(^1D_2)$ excitation energies are much higher for Cu sites of type I than of type II, while emission energies are similar. Relaxation of framework atoms outside the QM cluster was crucial for modeling the differences between the two types of sites. Similar results have been obtained for Ag^+ ions in MFI and FER frameworks.

1.6 Electronic and Magnetic Properties of Host–Guest Systems

Quantum mechanical calculations of properties are made for host–guest structures known from experiment or earlier structure determinations. An (approximate) solution of the Schrödinger equation is obtained for the following.

- The whole host–guest system using PBC.
- The guest species only, while the influence of the host is included in an approximate way.

A typical question is how the electronic absorption and emission spectrum changes when an organic dye molecule becomes a guest in a host structure. Conceptually, this question is not different from the problem of solvent effects on electronic spectra of organic molecules. Theoretical methods for guest species range from *ab initio* [44] (MR-CI, LR-CC-D, CAS-PT2, CIS) over TD-DFT [45] to semiempirical methods with spectroscopic parameterization (INDO/S, ZINDO, [46]). The host system is most easily represented as an array of point charges included in the Hamiltonian of the guest species. More advanced schemes take the polarizability of the host system into account. These options are available within hybrid QM/MM methods [47–49]. The *ab initio* methods are limited to small molecules, say up to ten atoms, while DFT is applicable to typical dye molecules such as thionine (Chapter 5). After two different low-energy positions of thionine in NaY have been located, as a further step it would be necessary to calculate the energy differences between the fluorescence state and the ground state at the respective adsorption structures to explain the observed shift in the fluorescence maximum. A force field would not be appropriate for this step. Instead, a quantum chemical method (semiempirical method with spectroscopic parameterization or DFT method) would be needed and the zeolite environment would have to be included, for example by point charges, to reveal the differences between the two different adsorption sites.

Calzaferri et al. studied the electronic structure of zeolite A containing Ag^+ ions ($Ag_xNa_{12-x}A$) and found a change in color from white over yellow to red depending on the hydration state [50,51]. QM calculations on a huge chunk of the material (more than one unit cell, 1296 atoms) revealed the nature of the orbitals from which (HOMO, highest occupied molecular orbital) and into which (LUMO, lowest unoccupied molecular orbital) the excitation is made, specifically the contributions that the framework atoms make to these orbitals. The calculations used the semiempirical extended Hückel method in the special implementation of Calzaferri et al. [52], which, when applied with PBC, also yield band structures.

PBC applied to DFT calculations are the method of choice if the electronic excitations of the host structure itself and their change in the presence of guest species are of interest. In this case the band structure is calculated and band gaps are compared with UV/vis spectra. For four different chalkogenide antimonates (Part 3, Chapter 6) the band structures have been calculated by a full potential extended linear augmented plane-wave method [53]. The gaps are too small compared to gap sizes inferred from photoconductivity measurements and UV/vis spectroscopy, which is due to the LDA applied. Moreover, absorption coefficients calculated from the dielectric function have been compared with the measured transmission spectra [53]. For sodium and potassium electrosodalites band structures have also been calculated [54,55]. The sodalite host itself is an insulator with a wide gap and bands with little dispersion. After introducing the additional electron as part of the $(M_4)^{3+}$ clusters, a small gap semiconductor is obtained. The direct gap is only 0.1 eV, but the corresponding photon absorption transition is symmetry forbidden. The smallest indirect gap for allowed transitions is 0.9 eV, which compares favorably with the lowest edge of the measured absorption spectrum of SES of 0.7 eV [56]. These are spin-unrestricted full potential linear augmented plane wave (FLAPW) [57] results employing the PW91 functional. If the silica-sodalite matrix is removed and a hypothetical periodic array of $(M_4)^{3+}$ clusters created (with a compensating background charge) the band gap disappears and a metallic structure is obtained [54]. This computer experiment highlights the important role of the host material for the host–guest electronic structure.

For host–guest systems with a periodic array of spins, DFT can be used to calculate the total energies of different magnetically ordered states. From these energies the magnetic order below the critical temperature can be predicted. Heisenberg spin-coupling parameters can also be derived, albeit in an indirect way, which provide insight in the magnetic interactions and can be compared with values inferred from experiments. Part 3, Chapter 4 provides a detailed discussion for sodium and potassium electrosodalite.

References

1 M. Ehrl, H.W. Kindervater, F.W. Deeg, C. Bräuchle, R. Hoppe, *J. Phys. Chem.* **1994**, *98*, 11 756.

2 P.M.W. Gill, in Encyclopedia of Computational Chemistry, P. von Ragué Schleyer, N.L. Allinger, T. Clark, J. Gasteiger, P.A. Kollman, H.F. Schaefer III, P.R. Schreiner (eds.), Wiley, Chichester **1998**, p. 678.

3 C.J. Cramer, Essentials of Computational Chemistry, Wiley, Chicester **2002**.

4 M. Elstner, T. Frauenheim, E. Kaxiras, G. Seifert, S. Suhai, *Phys. Stat. Sol. B* **2000**, *217*, 357.

5 J.M. Soler, E. Artacho, J.D. Gale, A. Garcia, J. Junquera, P. Ordejon, D. Sanchez-Portal, *J. Phys.: Condens. Matter* **2002**, *14*, 2745.

6 A.A. Demkov, J. Ortega, O.F. Sankey, M.P. Grumbach, *Phys. Rev. B* **1995**, *52*, 1618.

7 M. Elstner, P. Hobza, T. Frauenheim, S. Suhai, E. Kaxiras, *J. Chem. Phys.* **2001**, *114*, 5149.

8 M. Krause, H. Kuzmany, P. Georgi, L. Dunsch, K. Vietze, G. Seifert, *J. Chem. Phys.* **2001**, *115*, 6596.

9 A.A. Demkov, O.F. Sankey, *Chem. Mater.* **1996**, *8*, 1793.

10 D. Marx, J. Hutter, in Modern Methods and Algorithms of Quantum Chemistry, NIC Series, Vol. 3, J. Grotendorst (ed.), Jülich **2000**, NIC Directors, FZ Jülich p. 301.

11 E. Nusterer, P.E. Blöchl, K. Schwarz, *Angew. Chem., Int. Ed. Engl.* **1996**, *35*, 175.

12 R. Shah, J.D. Gale, M.C. Payne, *J. Phys. Chem.* **1996**, *100*, 11 688.

13 F. Haase, J. Sauer, J. Hutter, *Chem. Phys. Lett.* **1997**, *266*, 397.

14 J. Hutter, A. Alavi, T. Deutsch, M. Bernasconi, S. Goedecker, D. Marx, M. Tuckerman, M. Parrinello, CPMD, 3.4.1., Max-Planck-Institut für Festkörperforschung and IBM Research, Stuttgart, **1995–1999**.

15 G. Kresse, J. Furthmüller, *Comput. Mater. Sci.* **1996**, *6*, 15.

16 B.L. Trout, B.H. Suits, R.J. Gorte, D. White, *J. Phys. Chem. B* **2000**, *104*, 11 734.

17 T. Demuth, L. Benco, J. Hafner, H. Toulhoat, F. Hutschka, *J. Chem. Phys.* **2001**, *114*, 3703.

18 F. Haase, J. Sauer, *Micropor. Mesopor. Mater.* **2000**, *35–36*, 379.

19 C.P. Ursenbach, P.A. Madden, I. Stich, M.C. Payne, *J. Phys. Chem.* **1995**, *99*, 6697.

20 C.R.A. Catlow, W.C. Mackrodt (eds.), Computer Simulations of Solids, Lecture Notes in Physics, Vol. 166, Springer, Berlin **1982**.

21 J.-R. Hill, C.M. Freeman, L. Subramanian, in Reviews in Computational Chemistry, Vol. 16, K.B. Lipkowitz, D.B. Boyd (eds.), VCH, New York **2000**, p. 141.

22 J.D. Gale, *J. Chem. Soc., Faraday Trans.* **1997**, *93* 629; GULP (General Utility Lattice Program), Royal Institution/Imperial College, London **1992–1994**.

23 J.R. Maple, in Encyclopedia of Computational Chemistry, Vol. 2, P. von Ragué Schleyer, N.L. Allinger, P.A. Kollman, T. Clark, H.F. Schaefer III, J. Gasteiger, P.R. Schreiner (eds.), Wiley, Chichester **1998**, p. 1025.

24 J.R. Maple, in Encyclopedia of Computational Chemistry, Vol. 2, P. von Ragué Schleyer, N.L. Allinger, P.A. Kollman, T. Clark, H.F. Schaefer III, J. Gasteiger, P.R. Schreiner (eds.), Wiley, Chichester **1998**, p. 1015.

25 A.H. Fuchs, A.K. Cheetham, *J. Phys. Chem. B* **2001**, *105*, 7375.

26 P. Demontis, G.B. Suffritti, *Chem. Rev.* **1997**, *97*, 2845.

27 M.J. Sabater, G. Sastre, *Chem. Mater.* **2001**, *13*, 4520.

28 T.A. Wesolowski, O. Parisel, Y. Ellinger, J. Weber, *J. Phys. Chem. A* **1997**, *101*, 7818.

29 L.A. Clark, M. Sierka, J. Sauer, *J. Am. Chem. Soc.* **2003**, *125*, 2136.

30 Q. Wu, W. Yang, *J. Chem. Phys.* **2002**, *16*, 515.

31 J. Sauer, *Chem. Rev.* **1989**, *89*, 199.

32 J. Sauer, P. Ugliengo, E. Garrone, V.R. Saunders, *Chem. Rev.* **1994**, *94*, 2095.

33 R. Ahlrichs, M. Bär, M. Häser, H. Horn, C. Kölmel, *Chem. Phys. Lett.* **1989**, *162*, 165.

34 R. Ahlrichs, S.D. Elliott, U. Huniar, *Ber. Bunsenges. Phys. Chem.* **1998**, *102*, 795.

35 J. Sauer, K. Fiedler, W. Schirmer, R. Zahradník, in *Proc. 5th Int. Conf. Zeolites*, Naples/Italy **1980**, L.C.V. Rees (ed.), Heyden, London **1980**, p. 501.

36 T.Z. Mordasini, W. Thiel, *Chimia* **1998**, *52*, 288.

37 J. Sauer, M. Sierka, *J. Comput. Chem.* **2000**, *21*, 1470.

38 P. Sherwood, in Modern Methods and Algorithms of Quantum Chemistry, NIC Series, Vol. 3, J. Grotendorst (ed.), NIC Directors, Jülich **2000**, p. 257.

39 J. Sauer, D. Nachtigallová, P. Nachtigall, in Catalysis by Unique Metal Ion Structures in Solid Matrices. From Science to Application, Nato Science Series, Sub-Series II, Vol. 13, G. Centi, B. Wichterlová, A.T. Bell (eds.), Kluwer, Dordrecht **2001**, p. 221.

40 D. Nachtigallová, P. Nachtigall, M. Sierka, J. Sauer, *Phys. Chem. Chem. Phys.* **1999**, *1*, 2019.

41 P. Nachtigall, M. Davidova, D. Nachtigallova, *J. Phys. Chem. B* **2001**, *105*, 3510.

42 C. Lamberti, S. Bordiga, M. Salvalaggio, G. Spoto, A. Zecchina, F. Geobaldo, G. Vlaic, M. Bellatreccia, *J. Phys. Chem. B* **1997**, *101*, 344.

43 P. Nachtigall, D. Nachtigallova, J. Sauer, *J. Phys. Chem. B* **2000**, *104*, 1738.

44 S.D. Peyerimhoff, in Encyclopedia of Computational Chemistry, Vol. 4, P.

von Ragué Schleyer, N.L. Allinger, P.A. Kollman, T. Clark, H.F. Schaefer III, J. Gasteiger, P.R. Schreiner (eds.), Wiley, Chichester **1998**, p. 2646.

45 R. Bauernschmitt, M. Häser, O. Treutler, R. Ahlrichs, *Chem. Phys. Lett.* **1997**, *264*, 573.

46 P. Du, F.U. Axe, G.H. Loew, S. Canuto, M.C. Zerner, *J. Am. Chem. Soc.* **1991**, *113*, 8614.

47 M.A. Thompson, *J. Phys. Chem.* **1996**, *100*, 14 492.

48 J. Gao, *Acc. Chem. Res.* **1996**, *29*, 298.

49 A. Broo, G. Pearl, M.C. Zerner, *J. Phys. Chem. A* **1997**, *101*, 2478.

50 R. Seifert, A. Kunzmann, G. Calzaferri, *Angew. Chem.* **1998**, *110*, 1603; *Angew. Chem., Int. Ed. Engl.* **1998**, *37*, 1521–1524.

51 G. Calzaferri, C. Leiggener, S. Glaus, D. Schürch, K. Kluge, *Chem. Soc. Rev.* **2003**, *32*, 29.

52 G. Calzaferri, R. Rytz, M. Brändle, ICON-EDiT, Universität Bern, Bern **1997**.

53 F. Starrost, E.E. Krasovskii, W. Schattke, J. Jockel, U. Simon, R. Adelung, L. Kipp, *Phys. Rev. B* **2000**, *61*, 15 697.

54 R. Windiks, J. Sauer, *J. Chem. Phys.* **2000**, *113*, 5466.

55 G.K.H. Madsen, B.B. Iversen, P. Blaha, K. Schwarz, *Phys. Rev. B* **2001**, *64*, 195 102.

56 N.P. Blake, V.I. Srdanov, G.D. Stucky, H. Metiu, *J. Phys. Chem.* **1996**, *104*, 8721.

57 P. Blaha, K. Schwarz, J. Luitz, WIEN97 - A Full Potential Linearized Augmented Plane Wave Package for Calculating Crystal Properties, Technical University of Vienna **1999**.

2
Probing Host Structures by Monitoring Guest Distributions

Jörg Kärger and Sergey Vasenkov*

2.1
Introduction

Studying the propagation rate of the guest molecules in nanoporous materials is a task of really interdisciplinary relevance. During the establishment of zeolite science and technology in the 1960s, measuring zeolitic diffusion seemed to be a simple matter of adopting the relevant solutions of the diffusion equations (as compiled, for example, in Crank's standard textbook [1]) to the situation of a given experiment of molecular uptake or release. Recently, however, with the introduction of pulsed field gradient (PFG) NMR it became clear how wrong this was [2–4]. For numerous systems the NMR diffusivities observed turned out to be by up to five orders of magnitude larger than the previously accepted values. In general, these huge discrepancies could be explained by shortcomings in the experimental procedure of the "conventional" sorption experiments, which were limited by processes different from intracrystalline diffusion. Today, a large number of mechanisms that can give rise to such discrepancies are well understood [4–8] and generally taken into account. Nevertheless, there exists a large series of systems still exhibiting substantial differences (albeit not of five orders of magnitude) between the results of different, well-documented diffusion measurements [7,9].

Considering the sequence of introduction of the various techniques of diffusion measurement into zeolite science and technology [10,11] one may note two conclusions: after PFG NMR in the 1970s many more new measuring techniques were introduced; so far, none of the microscopically operating techniques were able to act under nonequilibrium conditions. It was only with the introduction of interference microscopy that the microscopic (intracrystalline) monitoring of molecular transport was extended to the nonequilibrium situation, so that intracrystalline concentration profiles and their evolution under molecular uptake or release have become directly accessible to observation [12–14]. The present chapter describes the application of this novel technique to tracing the structural and diffusional properties of nanoporous materials. In many cases the information provided will be consolidated by combination with the evidence of other techniques: in particular of PFG NMR and IR microscopy. The chapter is organized as follows.

Section 2.2 is an introduction to the principles of diffusion measurements in zeolites by interference microscopy. Section 2.3 describes the first results on transient intracrystalline concentration profiles during molecular uptake. Section 2.4 deals with the problem of whether the interfaces between the different intracrystalline constituents of MFI-type zeolites serve primarily as short circuits for molecular uptake or as internal transport resistances. Section 2.5 discusses host materials consisting of arrays of parallel channels and their special features with respect to mass transfer. Particular emphasis is given to the question whether mesoporous materials of hexagonal symmetry (Section 2.5.1) or zeolites of type AFI (Sections 2.5.2 and 2.5.3) may in fact be considered as "bunches of macaroni", since researchers in the field of single-file diffusion would be particularly happy to use these for experimental studies. By considering ferrierite, Section 2.5.4 presents the first experimental results of molecular distributions in arrays of mutually intersecting channel systems, which are potential candidates for exploiting the controversially discussed possibility of reactivity enhancement by "molecular traffic control".

2.2 Principles of Interference Microscopy

Figure 1 illustrates the measuring principle of interference microscopy when applied to studying intracrystalline zeolitic diffusion. In this technique, the phases of the light rays passing through a particular zeolite crystallite are compared with those passing through the surrounding atmosphere. As a consequence of the difference between the optical densities of the crystal and the surroundings, the phases of these rays generally differ from each other (Fig. 1, $\Delta\varphi \neq 0$). Depending on the phase difference, by superimposing the light rays passing through the crystal and the gas phase one may attain different degrees of interference, covering

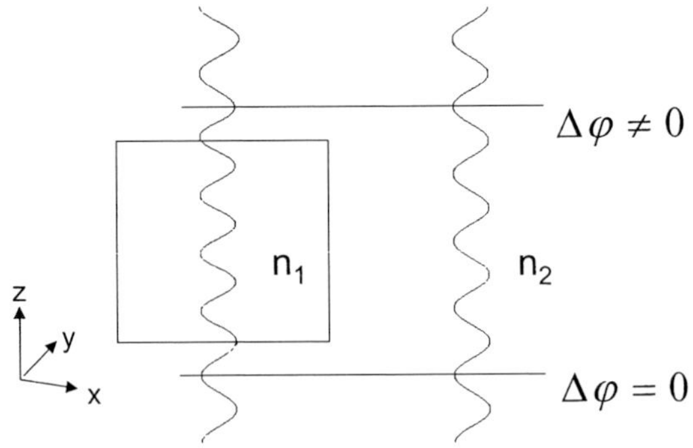

Fig. 1. Measuring principle of interference microscopy.

the total range between complete extinction and maximum amplification. Since the optical density of the crystallites is a function of the concentration of guest molecules, any change in the guest density during adsorption or desorption affects the optical path length through the crystallite and hence its phase and, in particular, the phase difference with the reference ray. Owing to the changes in value of $\Delta\varphi$, the interference pattern is changed, which (in turn) may be used to determine the change in the mean refractive index taken along the light path. Assuming, as a first-order approximation, proportionality between the change in the refractive index and the concentration, the quantity directly accessible in such measurements is the integral

$$\langle \Delta c(x, y, t) \rangle = \frac{1}{L} \int_0^L \Delta c(x, y, z) \mathrm{d}z \tag{1}$$

over the concentration change, where it is assumed that the direction of the light path, coinciding with the direction of integration, coincides with the z axis. The extent of the crystal in the direction of observation is denoted L. The spatial resolution of the measurements, the observation pixel in the xy plane, is typically of the order of 0.45 μm × 0.45 μm.

The experimental arrangement is shown in Fig. 2, which shows how two rays are superimposed upon each other by the shearing mechanism. The phase shifter in the lower light path ensures that the optical path lengths of the primary ray, split by the semitransparent mirror, are kept as close as possible to each other. The zeolite crystallite is contained in a cuvette, where it may be activated prior to the experiment and, subsequently, may be brought into contact with any type of atmosphere or evacuated. A complete description of the experimental arrangement and the data analysis may be found elsewhere [13].

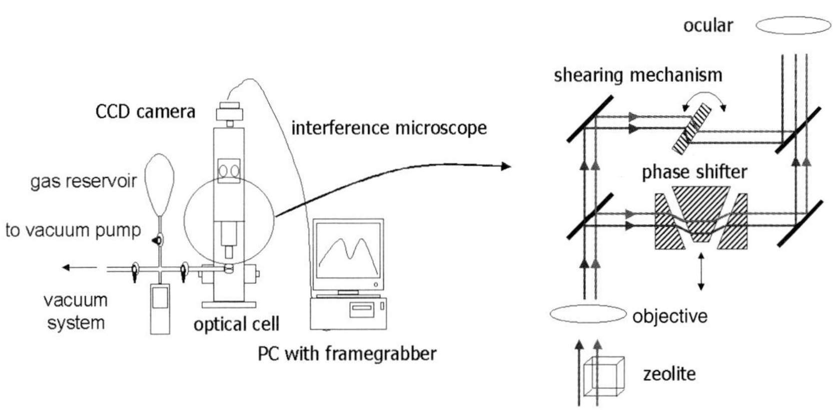

Fig. 2. Experimental set-up: vacuum system (left) and path of light through the interferometer (right).

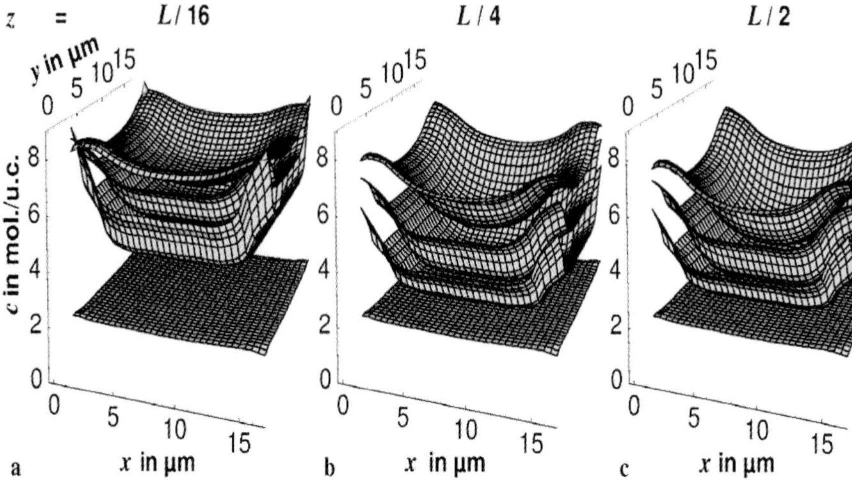

Fig. 3. Evolution of the methanol concentration in zeolite NaCaA at room temperature after increasing the pressure in the surrounding atmosphere from 0.5 to 9.0 kPa. The concentration profiles are shown in x–y planes through the crystal at heights $z = L/16$, $L/4$, and $L/2$ as indicated at the top of the figure. The profiles are monitored (from bottom to top) at times $t = 0$, 10, 80, and 160 s after the pressure increment [13].

2.3
Transient Uptake in Zeolite LTA

Though a microscopic technique, the information directly provided by interference microscopy is the concentration integral in the observation direction rather than the concentration itself. Knowing the crystal symmetry and presupposing an ideal crystalline structure, one may attempt to de-convolute the information contained in the primary experimental data. An ideal model system for such studies is zeolite of type LTA. In fact, this type of zeolite served as a test object in the very first attempts to apply interference microscopy to study diffusion in zeolites [15]. In these early times, however, neither long-distance observation nor data processing had attained the level necessary for this type of experiment.

Figure 3 displays the evolution of the intracrystalline concentration of methanol in zeolite NaCaA during uptake from an initial concentration of about two molecules per large cavity (i.e., per pseudo-unit cell) up to about 8.5 molecules, corresponding to an external pressure of 9 kPa at room temperature. Owing to the cubic symmetry and the cubic shape of the crystallites under study, the intracrystalline concentrations $c(x, y, z)$ could be derived from the primarily accessible information, the concentration integral

$$\int_0^L c(x, y, z)\mathrm{d}z,$$

by a de-convolution procedure assuming that the rate of methanol uptake is exclusively controlled by intracrystalline diffusion, as described previously [13]. Figure 3 shows the evolution of the intracrystalline concentration in three different planes, parallel to one of the six outer crystallite faces. It clearly appears from the representations that the concentration in the volume elements, which are closer to the external surfaces, attain their equilibrium concentrations much more quickly than those in the center. This appears both in the curvature of any individual profile and in the fact that the evolution of the concentrations in the central plane (at $z = L/2$) is behind that for $z = L/4$ and even more behind that for $z = L/12$, in a plane quite close to the outer surface.

In the case of complete diffusion limitation [4,8], the concentration in the volume elements close to the outer surface must be expected immediately to attain the equilibrium values corresponding to the sorbate pressure in the surrounding atmosphere. Such behavior cannot be seen in Fig. 3. However, one must remember that (owing to optical boundary effects) regions close to the crystal faces are excluded from direct observation and do not appear, therefore, in Fig. 3. Moreover, both the influence of (additional) transport resistances on the outer crystal boundary (surface barriers) and the finite rate of heat release during adsorption [4,8] may lead to the concentration close to the crystal boundary only gradually attaining the equilibrium value. As a main virtue of interference microscopy, these limitations do not decisively restrict its potential for measuring intracrystalline zeolitic diffusion as long as one is able to observe nonuniform intracrystalline concentrations and their evolution with time. In this case, the intracrystalline diffusivities follow from a microscopic application of Fick's second law, since one is able directly to deduce the spatial and temporal derivatives of the intracrystalline concentrations. Even if the gradients in the evolving concentration profiles are below the limits of sensitivity, in this way at least a lower limit of the intracrystalline diffusivities may be estimated. In the studies presented earlier [12], the coefficient of intracrystalline transport diffusion of methanol in NaCaA was found to be within the range 8×10^{-14} to 10^{-13} m^2 s^{-1}, which is compatible with corresponding PFG NMR self-diffusion results.

2.4 Evidence of Inner Transport Barriers in Zeolite MFI

With respect to their suitability for diffusion measurement by interference microscopy, LTA-type zeolites are exceptional. Owing to their internal and external symmetry, the directly observable concentration integrals may be transferred to a 3D entity of data points corresponding to the local concentrations within the zeolite crystallite under study. This type of analysis, however, has so far been successfully applied only to single crystals with LTA symmetry, assuming that the rate of uptake is exclusively controlled by the intracrystalline diffusion. For a number of zeolite structure types it is well known that the uptake dynamics may be strongly influenced by the diffusion anisotropy as well as by transport resistances on the crystal surface. In addition, the crystals themselves often cannot at all be considered as ideal single crystals. As an example, Fig. 4 displays two models of MFI

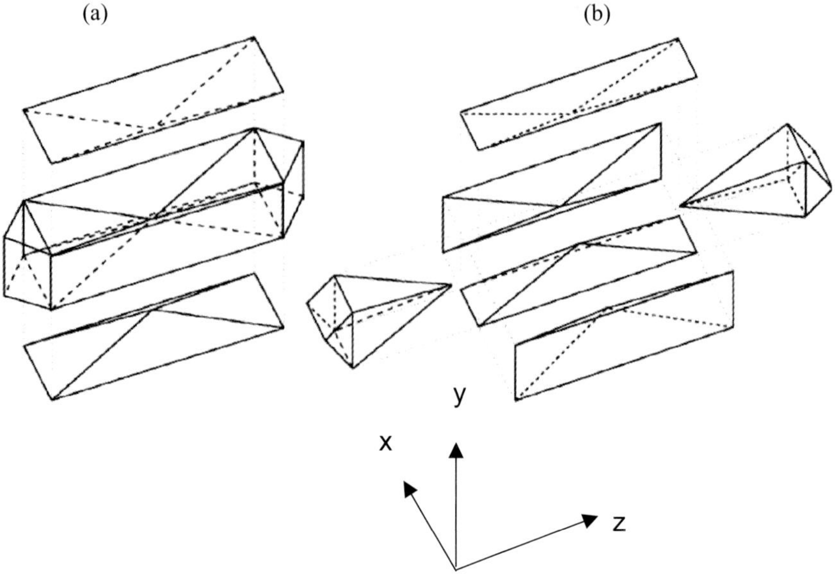

Fig. 4. Schematic representations of the internal structure of silicalite-1 crystals: (a) according to [16–18], (b) according to [19]. From [20].

crystal morphology currently discussed in the literature. According to one of the models [16–18], each crystal is composed of three constituents, that is, the two identical pyramid units and the central component (Fig. 4a). According to the other model [19], the central component of Fig. 4a is not a single section but, in its turn, consists of four components (Fig. 4b), two of which are identical to the pyramid units in Fig. 4a.

Interference microscopy may serve as a valuable tool for detecting regions of different structure, in particular of different sorbate accessibility, and for discriminating the role of their interfaces in molecular transportation. We undertook a detailed study of this type, in which isobutane was used to probe the pore architecture of MFI-type zeolites [20]. Figure 5 shows the microscopic image of a typical silicalite-1 crystal, used in these studies, in two different orientations. The hourglass structure is made visible by using the shearing mechanism of the microscope. The length scale in the x-, y-, and z-directions is shown in micrometers. In Figs. 6 and 7, the evolution of the concentration integrals as determined by interference microscopy (upper representations) is compared with the corresponding results of dynamic Monte Carlo (MC) simulations, which were based on the assumption that the interfaces between the different crystal constituents act either as transport resistances (medium representation) or short-circuits for molecular uptake (lower representation). In the latter case, molecular uptake is assumed to proceed via internal planes, so that sorption might be speculated to proceed "the

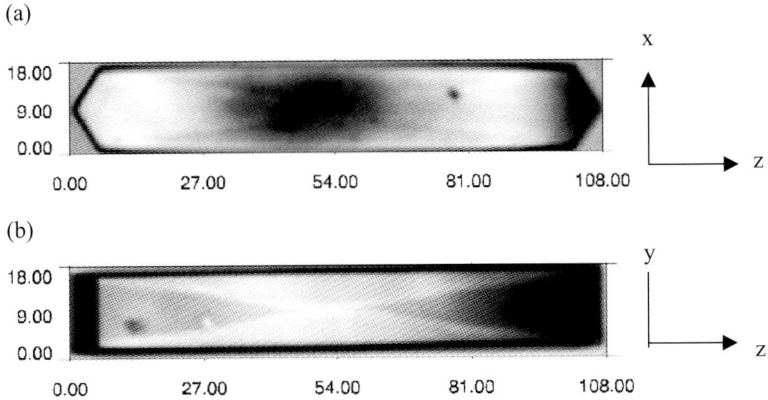

Fig. 5. Microscopic images of a typical silicalite-1 crystal in the two different orientations. The hourglass structure is made visible by using the shearing mechanism of the microscope. The extensions of the crystals in the *x*, *y*, and *z* directions are shown in micrometers [20].

other way round", that is with concentrations increasing faster in central parts of the crystallites than close to the external surface.

The measurements were performed for the crystal orientation shown in Fig. 5b. For this orientation the direction of light propagation, and thus the direction of integration, coincides with the x direction. We will only consider the rectangular part of the crystal (z between 13 and 88 μm in Fig. 5). For this part of the crystal the direction of light is perpendicular to the crystal outer surface exposed to the light. In this case, unambiguous measurements of the concentration profiles are possible. Figure 6 shows the integrated concentration profiles in longitudinal direction, along the middle part of the crystal ($y = 9.8$ μm, left), and near the crystal border ($y = 4.4$ μm, right), while profiles along the "width" of the crystal, through the middle ($z = 51$ μm, left), and closer to the crystal edge ($z = 20$ μm, right), are shown in Fig. 7. The simulations have been carried out by assuming the mutual interdependence of the principal elements of the diffusion tensor, as reported earlier [21], by transition path sampling and the correlation rule of diffusion anisotropy in MFI [4,22]. Further details of the experimental procedure, data analysis, and simulation may be found elsewhere [20]. The results of this study show that fitting of the experimental concentration integrals by the results of the dynamic MC simulations allow us to obtain qualitative and quantitative information on intracrystalline transport even in most complex systems, such as the MFI-type zeolites. Comparison between the experimentally observed integral concentrations and the simulated concentration profiles unanimously yield a satisfactory agreement between the experimental data and the simulations, in which the different constituents of the crystals are assumed to exhibit a modest transport resistance. There is no agreement at all if in the simulations the internal interfaces are assumed to

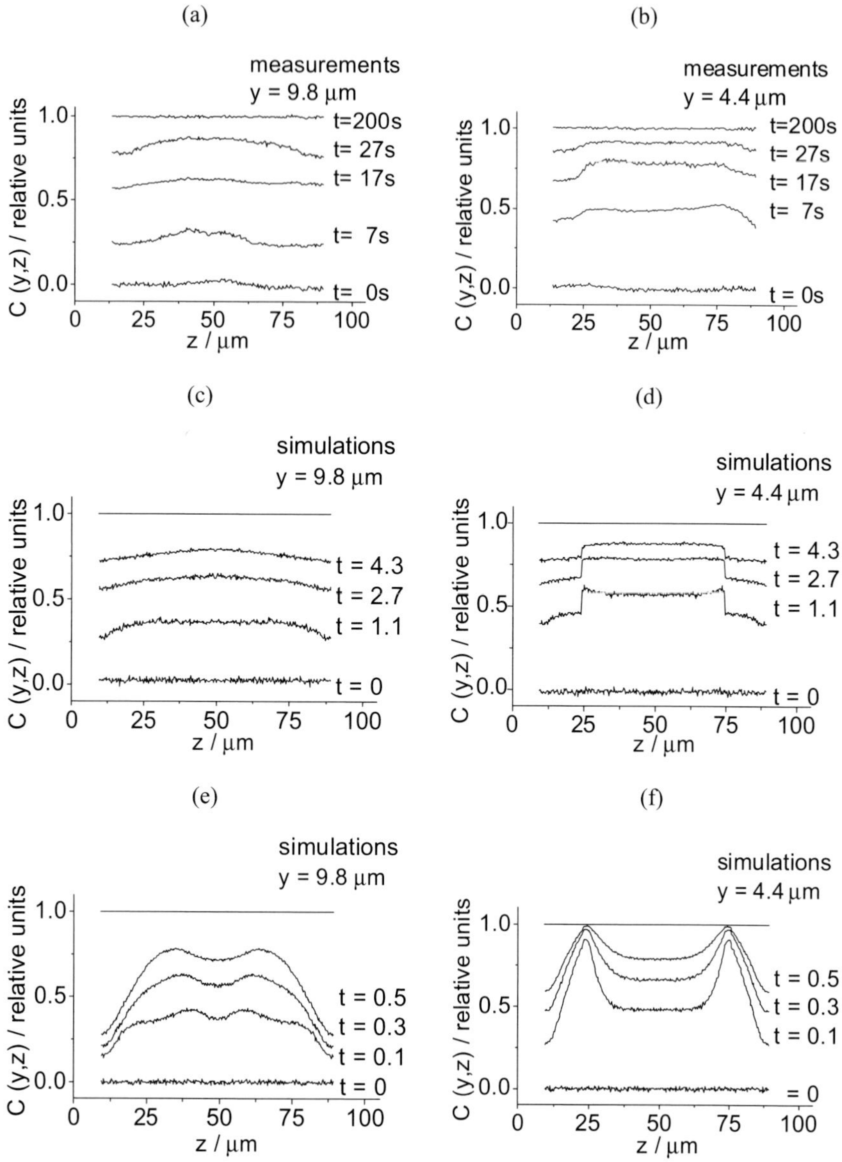

Fig. 6. Intracrystalline concentration profiles of isobutane in a silicalite-1 crystal along z direction during adsorption: (a), (b) profiles measured by interference microscopy; (c), (d) simulated profiles, assuming that the internal interfaces serve only as transport barriers; (e), (f) simulated profiles, assuming that adsorption/desorption may occur through the internal interfaces. For the simulated profiles the time unit is 10^3 elementary diffusion steps. The equilibrium values of $C(y, z)$ after the end of adsorption are equal to 1 [20].

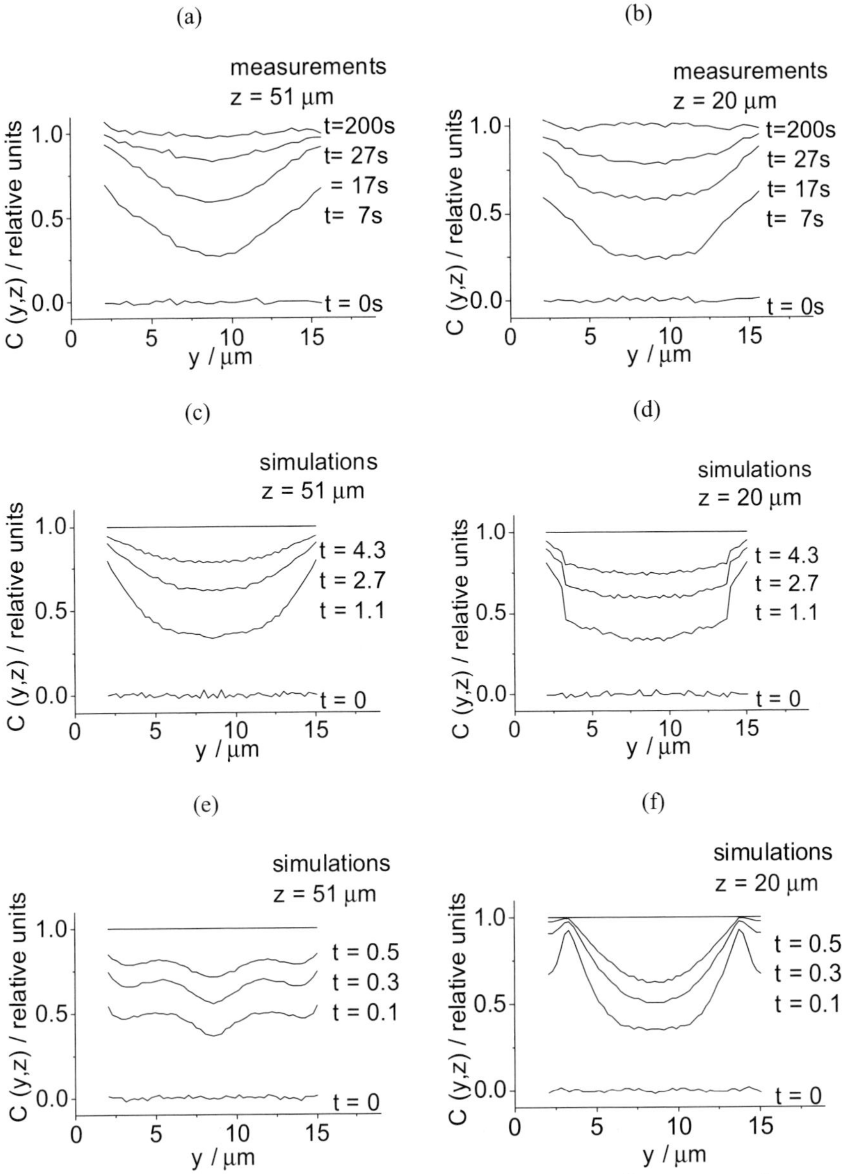

Fig. 7. Intracrystalline concentration profiles of isobutane in a silicalite-1 crystal along the y direction during adsorption: (a), (b) profiles measured by interference microscopy; (c), (d) simulated profiles, assuming that internal interfaces serve only as transport barriers; (e), (f) simulated profiles, assuming that adsorption/desorption may occur through the internal interfaces. For the simulated profiles the time unit is 10^3 elementary diffusion steps. The equilibrium values of $C(y, z)$ after the end of adsorption are equal to 1 [20].

be freely accessible for molecular adsorption. One may exclude, therefore, that molecular uptake may as well proceed via internal interfaces. This conclusion seems to be in contrast with the results obtained previously [18] with the iodine indicator technique, in which iodine was found to be able to distribute over the interfaces. However, in view of the much larger size of the isobutane molecules and the different time scale used in the present studies, such a difference might be quite acceptable.

Recent PFG NMR diffusion studies with methane and n-butane in silicalite-1 and ZSM-5 [23] did reveal that in addition to the transport resistances at the boundaries between the different MFI constituents as revealed by interference microscopy, there have to exist a great number of further internal transport barriers. Their existence follows from a remarkable dependence of the intracrystalline self-diffusivities on the diffusion path length, which can be traced below the micrometer range. Such behavior strongly suggests the existence of transport resistances with mutual spacing in the range of hundreds of nanometers up to micrometers. The formation of MFI-type zeolites is known to proceed via an aggregation of primary particles [24–28]. Some authors even suggested the formation of MFI-type crystals from well-defined so-called nanoblocks [24–26]. Therefore, the formation of the internal barriers might be associated with peculiar structural features in the transition range between these primary particles. One should remember that the intensity of these resistances has to be inferior to the transport resistance at the boundaries between the MFI constituents, since otherwise the latter would remain unobservable by interference microscopy. Moreover, for the probe molecules under study the internal barriers were found to become unimportant above room temperature. This may easily be explained by the fact that with increasing temperature their influence is progressively reduced compared to the diffusional transport resistance of the zeolite bulk phase [29]. Owing to the rather limited influence of such barriers it is probably premature to speculate about whether their existence might contribute to an eventual clarification of the origin of the discrepancy between some results of the different techniques of diffusion measurement.

2.5 Arrays of Parallel Channels

2.5.1 Peculiarities of One-Dimensional Diffusion and Options for its Observation

Particle propagation with the boundary condition that the individual particles are not allowed to mutually exchange their positions is generally referred to as single-file diffusion. Irrespective of the apparently simple premise of this type of transport, an exhausting theoretical treatment of this phenomenon is far from trivial [30,31]. Though being originally introduced as a concept for interpreting matter transfer via ion channels through cell membranes [32,33], it was in particular the adoption of this principle in zeolite science and technology [34,35], which has led to a flood of publications covering both experimental [36–42] and theoretical [43–49] aspects. As the two main prerequisites for an experimental observation of single file diffusion, one has to imply that the channels are infinitely extended and without faults and narrow enough to prevent the guests from passing each other. As a consequence of the high mutual correlation of molecular propagation, under ideal single-file conditions the root mean square displacement of the guest molecules has to increase with the square root of the observation time [30,31,43] rather than with the observation time itself as expected in normal diffusion [50,51]. Practically, the peculiarities of mass transfer may determine the performance of nanoporous materials with single-file properties in both adsorption and separation [52,53], in which the exchange times are found to scale with the third rather than with the second power of the crystal size [45–47], and in catalysis in which it may be observed that the effective activation energy increases rather than remains constant or decreases with increasing temperature [40,41,54,55].

It is one of the challenges in applied material sciences to work out how to unambiguously confirm whether a given nanoporous material obeys the criteria that make it suitable for application under genuine single-file conditions. Within the context of this chapter we would like to illustrate how the various techniques of diffusion measurements may contribute to such a task.

The mesoporous materials of hexagonal symmetry so far considered in our studies did not possess the optical properties that would have made them accessible to diffusion studies by interference microscopy. On the other hand, this type of material has turned out to be an attractive system for tracing the pore architecture by PFG NMR self-diffusion studies. Having channel diameters of several nanometers, typical probe molecules such as simple hydrocarbons clearly cannot be expected to be subject to single-file conditions. Owing to the reduced transverse nuclear magnetic relaxation rates, the possibility of mutual molecular passages, however, significantly improves the measuring conditions of PFG NMR [4,50] compared with the much more narrow channels of, for example, AFI-type zeolites. In this way, it is possible to trace molecular displacements over substantially larger ranges of observation time. As an example, Fig. 8 shows the mean square displacements of water in Si-MCM-41 with a particularly high structural order [56], which may be deduced from the signal attenuation in PFG NMR by assuming molecular propagation described by a diffusion tensor of rotational symmetry [57]. The measurements have been carried out in saturated samples at 263 K, so that water in the intercrystalline space was frozen and hence all displacements observable in the experiments were confined by the shape of the adsorbent particles. Figure 9 displays the Arrhenius plots of the self-diffusivities for an observation time of 10 ms, resulting from the application of the Einstein relation [3,4,50,51]

$$\langle x^2(t) \rangle = 2Dt \tag{2}$$

where x denotes the direction of observation and t the observation time. Figures 8 and 9 reveal three remarkable results.

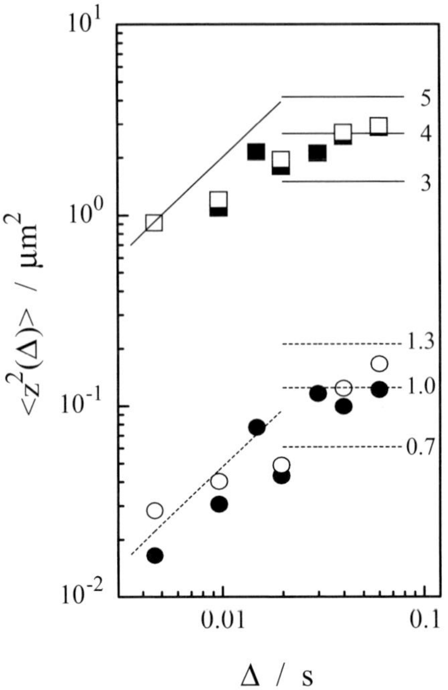

Fig. 8. Dependence of the parallel (■, □) and perpendicular (●, ○) components of the mean square displacement on the observation time for water in two MCM-41 samples at 263 K. The horizontal lines indicate the limiting values for the axial (full lines) and radial (dotted lines) components of the mean square displacements for restricted diffusion in cylindrical rods. The lengths *l* and diameters *d* of the rods are given in micrometers on the lines. The oblique lines (45°), which are plotted for short observation times only, represent the calculated time dependences of the mean square displacements for unrestricted (free) diffusion with $D_{par} = 1.0 \times 10^{-10}$ m^{2}s^{-1} (full line); and $D_{perp} = 2.0 \times 10^{-12}$ m^{2}s^{-1} (dotted line), respectively [57].

1. The limiting values of the observed displacements are in excellent agreement with the behavior (dotted and full lines in Fig. 8) expected on the basis of the particle size, which strongly supports the validity of the data analysis.
2. The diffusivity along the channel axis is smaller by an order of magnitude than in free water, indicating the existence of transport constrictions in the channels with separations smaller than the shortest displacement observed that is notably below 1 μm.
3. There is also the possibility of molecular displacements perpendicular to the main channel axes, though at a rate reduced by a further order of magnitude. This may be caused by a finite permeability of the channel walls, by a finite channel extension, which ensures mutual particle exchange, and/or by channel curvatures, which automatically lead to displacements deviating from a common main direction.

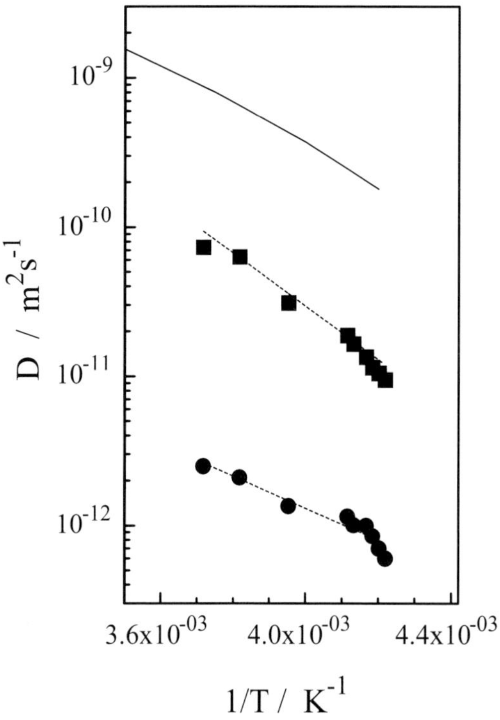

Fig. 9. Dependence of the parallel (■) and perpendicular (•) components of the axisymmetrical self-diffusion tensor on the inverse temperature for water in MCM-41 as measured at 10 ms observation time with PFG NMR. The dotted lines may be used as a guide for the eyes. For comparison, the full line represents the self-diffusion coefficients of super-cooled bulk liquid water [57].

It should be noted that with numerous other mesoporous materials of hexagonal structure, PFG NMR failed to establish similarly clear structure–mobility correlations [58–61] or, at least, could not be confirmed in the way shown in Fig. 8 [62]. Moreover, in the case of a commercial specimen of compacted particles of type MCM-41 [63,64] with benzene as a probe molecule, a remarkable irregularity in the concentration dependence of the intraparticle diffusivity was observed. For medium pore-filling factors the diffusivities pass a pronounced minimum, while being otherwise notably above the diffusivities of the free liquid. Adopting the conception of long-range diffusion [4,65], this behavior might be associated with the observation of the onset of a particularly pronounced adsorption hysteresis around these very concentrations, indicating the formation of a liquid phase in the (macro) pore space, which may notably affect the otherwise unrestricted gas-phase diffusion.

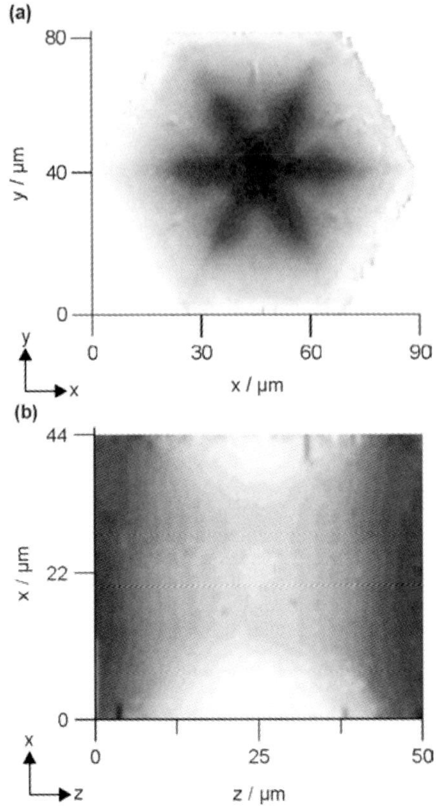

Fig. 10. Equilibrium intracrystalline concentration profile of methanol in a CrAPO-5 crystal. The color intensity is proportional to the integrals of local concentration in the *z* direction (a) and in the *y* direction (b). Darker regions correspond to larger concentration integrals. *x*, *y*, and *z* are the crystallographic directions (the channel direction is *z*). From [68].

2.5.2
Channel Accessibility in AFI-Type Crystals

Owing to their crystallinity, zeolites of type AFI offer much better prospects of structure analysis by interference microscopy than the mesoporous materials considered above. As an example, Fig. 10 displays the equilibrium concentration of methanol in a CrAPO-5 crystal [66,67] in equilibrium with a surrounding methanol atmosphere at 1 mbar, as observed by interference microscopy [68]. The textbook structure of AFI-type zeolites [69] is known to exhibit a hexagonal arrangement of channels with an effective diameter of about 0.7 nm. Therefore, they appear to be an ideal host systems for observing single-file diffusion with sufficiently bulky molecules, such as tetrafluoromethane with a molecular diameter of 0.47 nm [70]. In fact, Fig. 10 shows that the crystal under study is far from an array of parallel equally accessible channels, representing a "bundle of macaroni" of

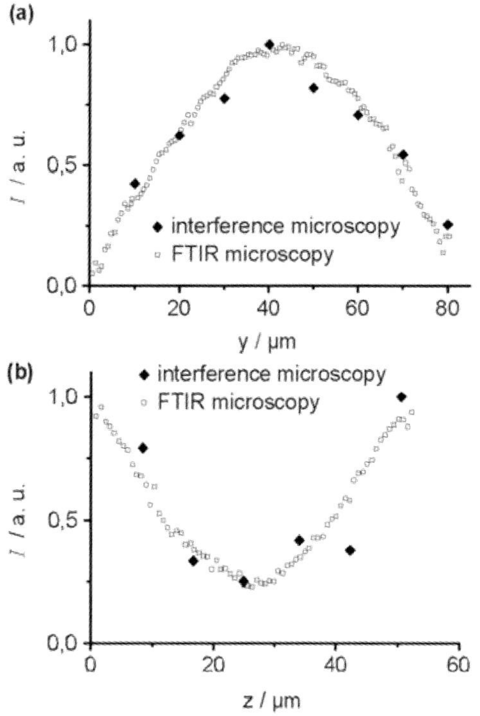

Fig. 11. The mean concentration integrals *I* recorded by FTIR and interference microscopy for a CrAPO-5 crystal: (a) along the *y* direction for *x* values between 35 and 55 µm (Fig. 10); (b) along the *z* direction for *x* values between 12 and 32 µm. *x*, *y*, and *z* are the crystallographic directions (Fig. 10). From [68].

atomistic diameters. The finding of interference microscopy has been confirmed by complementary concentration profiling by IR microscopy. Figure 11 shows the satisfactory agreement between the results of both techniques. It simultaneously illustrates the poorer spatial resolution of IR microscopy compared to interference microscopy. On the other hand, owing to the much higher sensitivity and its sensitivity to particular molecular species [71–73] and even to particular adsorption sites, IR microscopy may be much more than a valuable complement of interference microscopy. The concentration profiles shown in Fig. 10, and confirmed in Fig. 11 by comparing the interference and IR microscopy data, lead to the structure model shown in Fig. 12. It turns out that there is an internal core within the crystal structure, which is predominantly occupied by the methanol molecules, while the remaining part remains essentially empty. This core has the shape of a double-semi-pyramid with a star-like cross section.

The origin of the intergrowth effects in CrAPO-5 may be found in the crystallization history. It was shown previously [74,75] that the dumbbell shape is characteristic of some AFI-type crystals in the intermediate stage of growth. Further

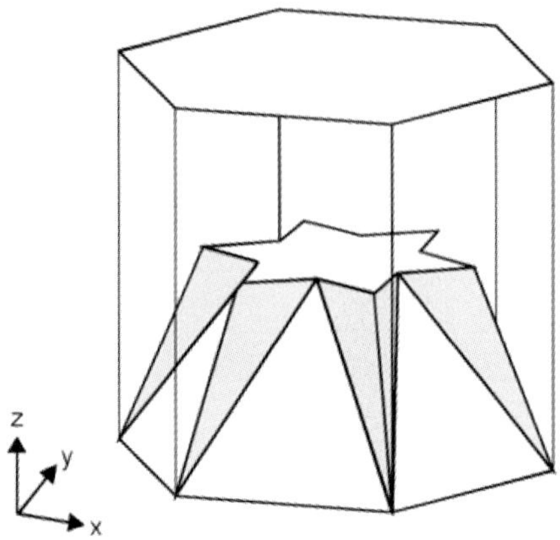

Fig. 12. Suggested internal structure of CrAPO-5 crystals (shown only for the lower part of the crystal). *x*, *y*, and *z* are the crystallographic directions. From [68].

growth leads to filling of the gap in the central part of the crystals. Obviously, though this process leads to the formation of perfectly hexagonally shaped crystals, their internal structure is far from perfect.

It is interesting to note (Fig. 13a) that a similar, although not as pronounced, distribution pattern is observed with water at a vapor pressure 1 mbar [76]. One may conclude, therefore, that the range of essentially excluded accessibility of methanol as reflected by the structure model shown in Fig. 12, likewise represents a region of reduced water concentration under the given conditions. With increasing water pressure (Fig. 13c), however, in parallel with a general concentration enhancement, the differences over the different regions of the intracrystalline space essentially disappear.

To understand the transition from the nonhomogeneous to the homogeneous profile it is helpful to consider the adsorption isotherm of water in CrAPO-5 zeolite. It belongs to the type IV isotherms [77]. Such isotherms exhibit relatively slow adsorption with increasing adsorbate pressure followed by a sudden increase of the amount adsorbed at a certain "critical" pressure. The initial, slow adsorption is usually associated with the adsorption of single molecules on different functional groups of the pore surface. The sudden increase of the concentration in the pores with increasing pressure is explained by the formation of a liquid-like adsorbate phase. It is reasonable to assume that the structure of the different intergrowth components of CrAPO-5 crystals may be slightly different. One of the candidates for the structural factor, which strongly influences intracrystalline water concentration at low loadings and, at the same time, may give rise to some sort of heterogeneity in CrAPO-5, is the content of Cr atoms. Indeed, different Cr contents in the different crystal components may be responsible for the nonhomogeneous intracrystalline concentration profile observed at low water pressure. Clearly, a nonhomogeneous distribution of any other adsorption sites, such as defect sites

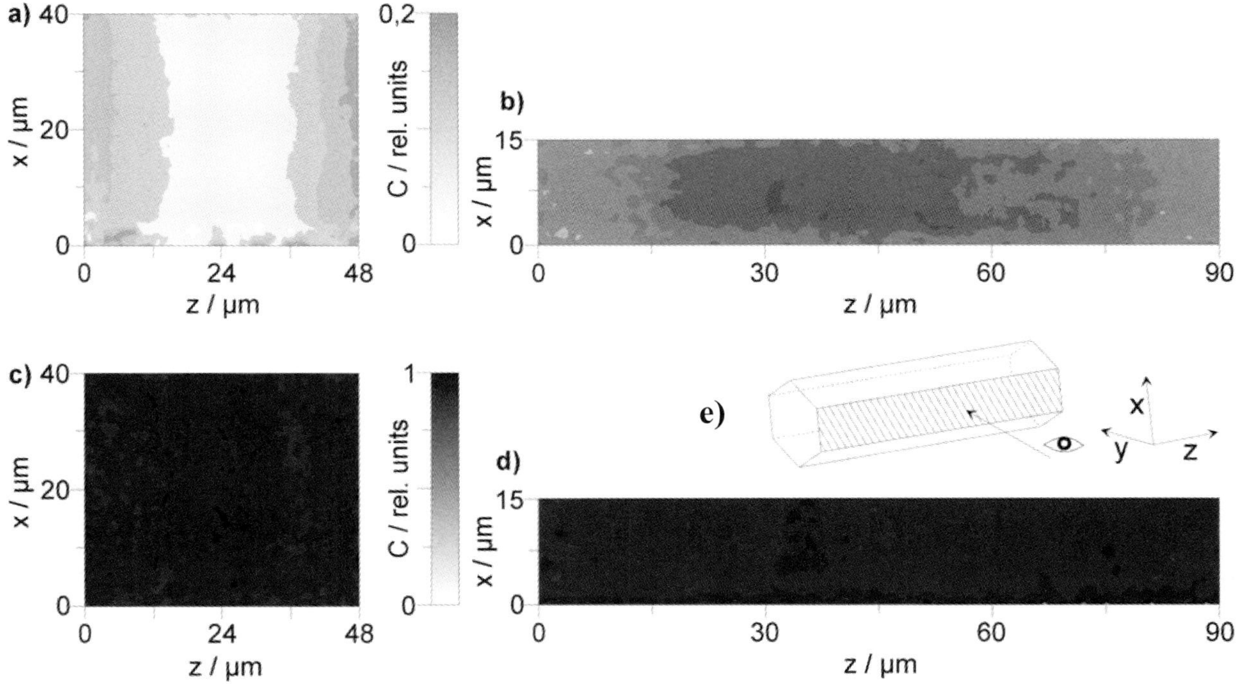

Fig. 13. Intracrystalline concentration profiles of water in the CrAPO-5 (a, c) and SAPO-5 (b, d) crystals integrated along the *y* direction under equilibrium with water vapor at 1 (a, b) and 20 mbar (c, d). The profiles are shown only for the crystal surface marked on the image (e). The channels run along the *z* axis. Darker regions correspond to higher concentration integrals [76].

over the crystal components, may lead to the same result. The concentration and the strength of adsorption sites does not, however, influence the sorption capacity of CrAPO-5 for water. This explains the homogeneous profiles observed at high water pressure, when the total available pore volume is expected to be filled with the liquid-like water [76,77].

The concentration profiles of water observed in SAPO-5 (Figs. 13b and d) can be explained in a qualitatively similar way. These profiles, like those recorded for the CrAPO-5 crystals, show a nonhomogeneous water distribution at low water pressures, which transforms to the homogeneous distribution on pressure increase. The heterogeneities in the structure of SAPO-5 crystals, which were observed earlier by electron microprobe analysis [78], may be responsible for the nonhomogeneous profile of water. The microprobing has revealed that the silicon concentration in the central part of the SAPO-5 crystals was lower by a factor of two to three than that on the crystal margins. Studies of the growth of the SAPO-5 crystals helped to clarify the reasons for nonhomogeneous silicon distribution [78]. It was found that initially "pencil-like" crystals are formed. At the later stages of the crystal growth, the tips of the "pencils" flatten out, which in some cases gives rise to the ideally shaped hexagonal crystals. These later stages proceed under consumption of much higher amounts of silicon than the initial stage. Hence, in agreement with the results of the microprobe sampling, the "pencil-like" core of the crystals may be expected to have depleted silicon content. This finding was correlated with the nonhomogeneous intracrystalline distribution of pyridine species in SAPO-5 at 373 K [73]. The concentration of pyridine species was found to be lower at the crystal edges than in the central part of the crystal. This is in qualitative agreement with the intracrystalline water distribution in SAPO-5 recorded by the interference microscopy method under equilibrium with 1 mbar of water (Fig. 13b). At high water pressures, condensation leading to pore volume saturation occurs. In this case, the water concentration is expected to depend primarily on the accessible pore volume. This explains the homogeneous concentration profile observed in SAPO-5 under equilibrium with 20 mbar of water (Fig. 13d).

2.5.3
Transient Concentration Profiles in AFI-Type Zeolites

In contrast to zeolites with 3D channel networks considered above, zeolites of ideal AFI structure offer the unique chance of allowing the direct determination of local concentrations. This is due to the fact that observation perpendicular to the channel direction yields integral concentrations, which are proportional to the crystal thickness in the observation direction multiplied by the local channel concentration, provided that all channels (being subject to identical boundary conditions) are also structurally identical. With the findings leading to the model of intergrowth structure in Fig. 12, however, for the so far considered specimens such a simple possibility has to be abandoned. It has to be admitted that for a substantially larger amount of investigated AFI-type zeolite crystals, interference microscopy revealed even much more pronounced deviations from the ideal textbook structure of par-

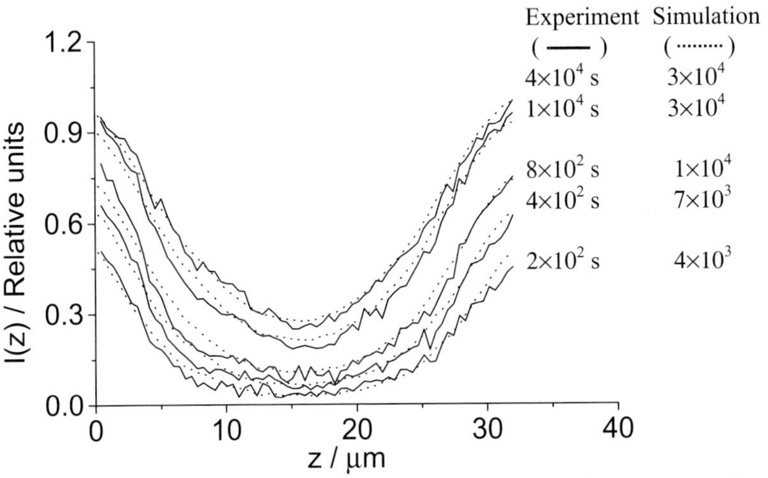

Fig. 14. Intracrystalline concentration of methanol integrated along the y crystallographic direction in CrAPO-5 at different times after the start of the methanol adsorption. The concentration integrals were measured by interference microscopy (solid lines) and were also obtained by dynamic MC simulations (dotted lines). The profile measured after 40 000 s represents the equilibrium concentration profile. For the calculated profiles the time unit is the time of one elementary diffusion step [79].

allel, equally extended channels expected for single crystals. There is no doubt, therefore, that irrespective of the numerous attempts of the last few years [66,67], the synthesis of nanoporous large single crystals with identical, unrestrictedly accessible channels remains a challenging task for the future.

In view of the lack of host systems of AFI structure type, which would allow the immediate observation of transient local concentrations and a direct estimation of the associated transport diffusivities, consistency checks between the observed integrated concentration profiles and the profiles obtained by the MC simulations on the basis of the structure models are particularly important. As an example, Fig. 14 compares the profiles of concentration, integrated perpendicular to the channel direction, for methanol in CrAPO-5, as resulting from interference microscopy [79] with simulation results. The simulations have been carried out by assuming that the molecules perform jumps between adjacent adsorption sites, where both the jump length and the frequency of jump attempts is independent of the local concentration. The channels are assumed to run through the total crystal only in the core with a cross section given by the star shown in Fig. 12. The channels beyond this central range are only accessible on one side, the respective outer face, and end at the interface displayed in Fig. 12. As assumed for establishing this structure model, the space outside of the two opposed semi-pyramids remains inaccessible for methanol.

Figure 14 shows the best fits between the experimentally determined transient integral concentration profiles and the simulation results. The established concentration profiles are in fact found to be nicely reproduced by the model considerations based on the simple model described above, with the inclusion of an additional transport resistance on the crystal surface. On comparing the different time scales, however, one has to admit that this agreement is satisfactory only for short times. While the time intervals between onset of uptake and profile measurement start to increase in roughly the same way, ($4 \times 10^2/2 \times 10^2 = 2$ in the experiments and $7 \times 10^3/4 \times 10^3 = 1.75$ in the simulations for the two first profiles), the ratios between the sorption times for the first and fourth profile, respectively, are, for example, 50 in the experiments and only 7.5 in the simulations. This behavior can be attributed to the well-known fact that the intracrystalline diffusivity may strongly depend on the concentration of guest molecules. Such dependence was neglected in the present simulations. The fit of the experimental profiles by the results of the simulations (Fig. 14) allows us to obtain the methanol diffusivity (0.4×10^{-12} m^2 s^{-1}) and the permeability of surface resistance (0.35×10^{-7} m s^{-1}) in AFI crystals in the limit of small methanol concentrations.

2.5.4
Guest Distribution in Ferrierite

Ferrierite consists of a network of mutually intersecting channels of elliptical cross sections with diameters of 0.42 nm and 0.54 nm ("10-ring" channels), and 0.35 nm and 0.48 nm ("8-ring" channels). Offering in this way two different types of diffusion paths, ferrierite may serve as a model system for experimentally tracing conditions of "molecular traffic control" [80–83]. The concept of molecular traffic control has been introduced as a possibility of reactivity enhancement in heterogeneous catalysis [80,81]. It is based on the assumption that the reactant and product molecules are preferentially accommodated in different parts of the pore system. As a consequence, the mutual hindrance of reactant and product molecules on their diffusion paths to and from the reactive sites is reduced, leading to enhanced exchange rates and hence (under the conditions of diffusion control) to enhanced reactivities. It has been demonstrated by MD simulations that under conditions of multicomponent adsorption different parts of the pore system may in fact accommodate the individual constituents with different probabilities [82,84]. In recent dynamic MD simulations the phenomenon of reactivity enhancement in networks of intersecting channels with mutually excluded accessibility for the reactant and product molecules could in fact be quantified [85,86].

Monitoring the evolution of intracrystalline concentration profiles might be one way of experimentally determining the preferred diffusion paths. As an example, Fig. 15 displays the evolution of the distribution of methanol in a ferrierite crystal during adsorption [87]. The crystal under study is a platelet with an average thickness of 9 μm and an extent of 40 μm × 210 μm, in which the channel network is extended in the plane of the platelet. The concentration profiles recorded during molecular uptake obviously do not comply with the patterns to be expected for uptake limitation by 2D diffusion in the plane of the platelet. As a first remarkable feature, like observed already with the AFI-type zeolites, the apparent equilibrium concentration is found to be far from uniform over the crystal. It turns out that in

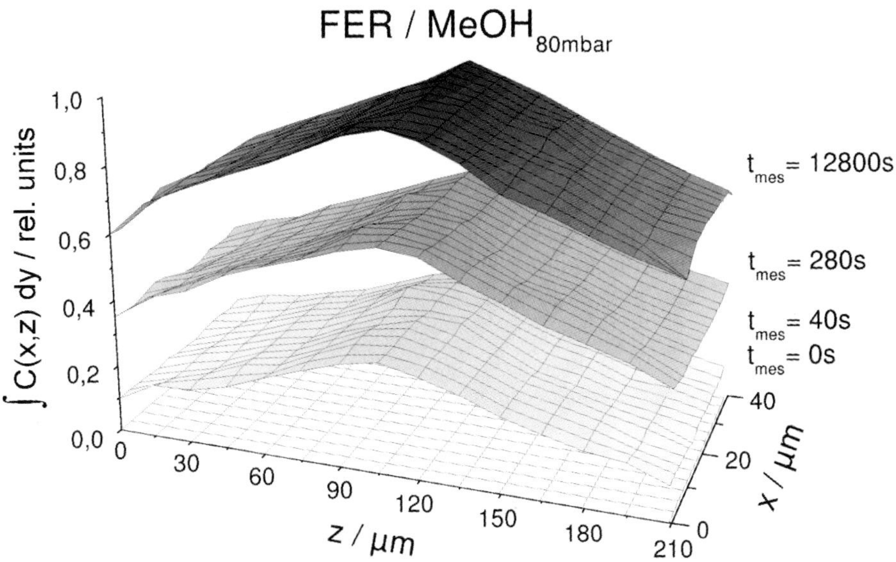

Fig. 15. Profiles of the integrals of intracrystalline concentration of methanol in ferrierite measured by interference microscopy at different times after the start of the methanol adsorption. The profile measured after 12 800 s represents the equilibrium concentration profile. The pressure of methanol vapor in the gas phase surrounding the crystals was kept at 80 mbar [87].

the x direction (the direction of the shorter plate extension) the concentration is essentially uniform, while in the z crystallographic direction (the direction of crystal length) the concentration linearly increases from both ends, forming a roof-like overall concentration profile. This finding is attributed to the experimental observation that the thickness of the platelet does not remain constant in the longitudinal direction. Instead, it increases gradually towards the middle of the crystal, mimicking the shape of the concentration profiles. It is interesting to note that the final shape of the concentration profiles is attained essentially immediately after the onset of adsorption. One has to conclude, therefore, that the overall process is limited by external processes (probably the penetration of a surface resistance), rather than by intracrystalline diffusion. On the basis of these experimental data it is impossible to decide, therefore, whether the ferrierite platelets under study may exhibit the properties desired for host systems under "molecular traffic control".

2.6 Conclusions

With the introduction of interference microscopy in zeolite science and technology, observation of the evolution of intracrystalline concentrations of guest molecules in nanoporous materials has become possible for the first time. From the experimental evidence provided by this technique, the real structure of crystals, which reveal perfect shapes characteristic of single crystals, has been found to deviate decisively from the textbook structure of the given type of zeolites. The existence of such deviations has been confirmed by complementary diffusion studies using pulsed field gradient NMR and interference microscopy. Though it is still premature to speculate about a possible correlation between the observed structural deviations and the discrepancies in the findings of the different experimental techniques on intracrystalline diffusion, the present studies emphasize the need for ideal nanoporous crystallites for a systematic study of the fundamentals of guest arrangement and guest diffusion in nanoporous materials.

Acknowledgements

The samples considered in this chapter have been synthesized in the groups of J. Caro (Hannover), J. Kornatowski (Munich), D. Michel and H. Papp (Leipzig), W. Schwieger (Erlangen), F. Schüth (Mülheim), and J. Weitkamp (Stuttgart). We thank these colleagues for supplying us with their specimens and for numerous stimulating discussions. We are particularly obliged to our coworkers and colleagues, in particular to C. Chmelik, D. Freude, P. Galvosas, O. Geier, B. Knorr, C. Krause, E. Lehmann and F. Stallmach, who have carried out most of the work presented and for their interest in the whole project.

References

1 J. Crank, Mathematics of Diffusion, Oxford **1956**, 347 pp.

2 J. Kärger, J. Caro, *J. Chem. Soc. Faraday Trans. I* **1977**, *73*, 1363.

3 D.M. Ruthven, Principles of Adsorption and Adsorption Processes, Wiley, New York **1984**, 433 pp.

4 J. Kärger, D.M. Ruthven, Diffusion in Zeolites and Other Microporous Solids, Wiley, New York **1992**.

5 N.K. Kanellopoulos, Recent Advances in Gas Separation by Microporous Ceramic Membranes, Elsevier, Amsterdam **2000**, 510 pp.

6 M. Bülow, A. Micke, *Z. Phys. Chem.* **1995**, *189*, 195.

7 J. Kärger, in *Proc. 12th Int. Zeolite Conf.*, M.M.J. Treacy, B.K. Marcus, M.E. Bisher, J.B. Higgins (eds.), Vol. 1, MRS, Baltimore **1999**, p. 35.

8 J. Kärger, D.M. Ruthven, in Handbook of Porous Solids, F. Schüth, K.S.W. Sing, J. Weitkamp (eds.), Wiley-VCH, Weinheim **2002**, p. 2089.

9 H. Jobic, in Recent Advances in Gas Separation by Microporous Ceramic Membranes, N.K. Kanellopoulos (ed.), Elsevier, Amsterdam **2000**, p. 109.

10 J. Kärger, D. Freude, *Chem. Ing. Tech.* **2001**, *73*, 1517.

11 J. Kärger, *Ind. Eng. Chem. Res.* **2002**, *41*, 3335.

12 U. Schemmert, J. Kärger, C. Krause, R.A. Rakoczy, J. Weitkamp, *Europhys. Lett.* **1999**, *46*, 204.

13 U. Schemmert, J. Kärger, J. Weitkamp, *Micropor. Mesopor. Mater.* **1999**, *32*, 101.

14 S. Vasenkov, O. Geier, U. Schemmert, J. Kärger, R.A. Rakoczy, J. Weitkamp, in Fundamentals of Adsorption 7, K.

Kaneko, H. Kanoh, Y. Hanzawa (eds.), International Adsorption Society, Shinjuku, Japan **2002**, p. 53.
15 J. Kärger, R. Danz, J. Caro, *Feingerätetechnik* **1978**, *27*, 539.
16 J. Caro, M. Noack, J. Richter-Mendau, F. Marlow, D. Peterson, M. Griepenstrog, J.J. Kornatowksi, *J. Phys. Chem.* **1993**, *97*, 13 685.
17 C. Weidenthaler, R.X. Fischer, R.D. Shannon, O. Medenbach, *J. Phys. Chem.* **1994**, *98*, 12 687.
18 M. Kocirik, J.J. Kornatowksi, V. Masarik, P. Novak, A. Zikanova, J. Maixner, *Micropor. Mesopor. Mater.* **1998**, *23*, 295.
19 E.R. Geus, J.C. Jansen, H. van Bekkum, *Zeolites* **1994**, *14*, 82.
20 O. Geier, S. Vasenkov, E. Lehmann, J. Kärger, U. Schemmert, R.A. Rakoczy, J. Weitkamp, *J. Phys. Chem. B* **2001**, *105*, 10 217.
21 T.J.H. Vlugt, C. Dellago, B. Smit, *J. Chem. Phys.* **2000**, *113*, 8791.
22 J. Kärger, *J. Phys. Chem.* **1991**, *95*, 5558.
23 S. Vasenkov, W. Böhlmann, P. Galvosas, O. Geier, H. Liu, J. Kärger, *J. Phys. Chem. B.* **2001**, *105*, 5922.
24 R. Ravishankar, C.E.A. Kirschhock, P.-P. Knops-Gerrits, E.J.P. Feijen, P.J. Grobet, P. Vanoppen, F.C. De Schryver, G. Miehe, H. Fuess, B.J. Schoemann, P.A. Jacobs, J. Martens, *J. Phys. Chem. B* **1999**, *103*, 4960.
25 C.E.A. Kirschhock, R. Ravishankar, F. Verspeurt, P.J. Grobet, P.A. Jacobs, J. Martens, *J. Phys. Chem. B* **1999**, *103*, 4965.
26 C.E.A. Kirschhock, R. Ravishankar, L. Van Looveren, P.A. Jacobs, J. Martens, *J. Phys. Chem. B* **1999**, *103*, 4972.
27 P.-P.E.A. De Moor, T.P.M. Beelen, B.U. Komanschek, O. Diat, R.A. Van Santen, *J. Phys. Chem. B* **1997**, *101*, 11 077.
28 P.-P.E.A. De Moor, T.P.M. Beelen, B.U. Komanschek, R.A. Van Santen, *Micropor. Mat.* **1998**, *21*, 263.
29 S. Vasenkov, J. Kärger, *Micropor. Mesopor. Mater.* **2002**, *55*, 139.
30 K. Binder, Application of the Monte Carlo Method in Statistical Physics, Berlin, **1987**, 341.
31 K.W. Kehr, K. Mussawisade, G.M. Schütz, T. Wichmann, in Diffusion in Condensed Matter, P. Heitjans, J. Kärger (eds.), Springer, Heidelberg 2003, in press.
32 E. Neher, *Science* **1992**, *256*, 498.
33 B. Sakmann, *Science* **1992**, *256*, 503.
34 L. Riekert, *Adv. Catal.* **1970**, *21*, 281.
35 J. Kärger, M. Petzold, H. Pfeifer, S. Ernst, J. Weitkamp, *J. Catal.* **1992**, *136*, 283.
36 V. Gupta, S.S. Nivarthi, A.V. McCormick, H.T. Davis, *Chem. Phys. Lett.* **1995**, *247*, 596.
37 V. Kukla, J. Kornatowski, D. Demuth, I. Girnus, H. Pfeifer, L.V.C. Rees, S. Schunk, K.K. Unger, J. Kärger, *Science.* **1996**, *272*, 702.
38 K. Hahn, J. Kärger, V. Kukla, *Phys. Rev. Lett.* **1996**, *76*, 2762.
39 H. Jobic, K. Hahn, J. Kärger, M. Bée, M. Noack, I. Girnus, A. Tuel, G.J. Kearley, *J. Phys. Chem.* **1997**, *101*, 5834.
40 G.D. Lei, B.T. Carvill, W.M.H. Sachtler, *Appl. Catal. A* **1996**, *142*, 347.
41 Z. Kapinski, S.N. Ghandi, W.M.H. Sachtler, *J. Catal.* **1993**, *141*, 337.
42 F.J.M.M. de Gauw, J. van Grondelle, R.A. van Santen, *J. Catal.* **2001**, *204*, 53.
43 J. Kärger, *Phys. Rev. E* **1993**, *47*, 1427.
44 C. Rödenbeck, J. Kärger, H. Schmidt, T. Rother, M. Rödenbeck, *Phys. Rev. E* **1999**, *60*, 2737.
45 C. Rödenbeck, J. Kärger, *J. Chem. Phys.* **1999**, *110*, 3970.
46 P.H. Nelson, S.M. Auerbach, *J. Chem. Phys.* **1999**, *110*, 9235.
47 P.H. Nelson, S.M. Auerbach, *Chem. Eng. J.* **1999**, *74*, 43.
48 H.L. Tepper, J.P. Hoogenboom, N.F.A. van der Vegt, W.J. Briels, *J. Chem. Phys.* **1999**, *110*, 11 511.
49 S. Vasenkov, J. Kärger, *Phys. Rev. E*, **2002**, *66*, 52 601.
50 J. Kärger, F. Stallmach, in Diffusion in Condensed Matter, P. Heitjans, J. Kärger (eds.), Springer, Heidelberg **2003**, in press.

51 J. Kärger, S. Vasenkov, S.M. Auerbach, in Handbook of Zeolite Catalysts and Microporous Materials, S.M. Auerbach, K.A. Carrado, P.K. Dutta (eds.), Marcel Dekker Inc., New York **2003**, in press.

52 J. Caro, M. Noack, P. Kölsch, R. Schäfer, *Micropor. Mesopor. Mater.* **2000**, *38*, 3.

53 J. van de Graaf, F. Kapteijn, J.A. Moulijn, *Micropor. Mesopor. Mater.* **2000**, *36*, 267.

54 C. Rödenbeck, J. Kärger, K. Hahn, *J. Catal.* **1998**, *176*, 513.

55 C. Rödenbeck, J. Kärger, K. Hahn, W. Sachtler, *J. Catal.* **1999**, *183*, 409.

56 U. Oberhagemann, M. Jeschke, H. Papp, *Micropor. Mesopor. Mater.* **1999**, *33*, 165.

57 F. Stallmach, J. Kärger, C. Krause, M. Jeschke, U. Oberhagemann, *J. Am. Chem. Soc.* **2000**, *122*, 9237.

58 E.W. Hansen, F. Courivaud, A. Karlsson, S. Kolboe, M. Stöcker, *Micropor. Mesopor. Mater.* **1998**, *22*, 309.

59 F.P. Matthae, W.D. Basler, H. Lechert, in Studies in Surface Science and Catalysis, L. Bounneviot, F. Béland, C. Danumah, S. Giasson, S. Kaliaguine (eds.), Vol. 117, Elsevier, Amsterdam **1998**, p. 301.

60 F. Courivaud, E.W. Hansen, A. Karlsson, S. Kolboe, M. Stöcker, *Micropor. Mesopor. Mater.* **2000**, *6*, 327.

61 F. Stallmach, A. Graser, J. Kärger, C. Krause, M. Jeschke, U. Oberhagemann, S. Spange, *Micropor. Mesopor. Mater.* **2001**, *44*, 745.

62 L. Frunza, C. Krause, F. Stallmach, J. Kärger, I. Pitsch, H. Kosslick, M.M. Pohl, A. Schönhals, *J. Phys. Chem. B* **2003**, submitted.

63 S. Spange, Y. Zimmerman, A. Gräser, *Chem. Mater.* **1999**, *11*, 3245.

64 C. Krause, F. Stallmach, D. Hönicke, S. Spange, J. Kärger, *Adsorption*, in press.

65 O. Geier, S. Vasenkov, J. Kärger, *J. Chem. Phys.* **2002**, *117*, **1935**.

66 B.V. Padlyak, J. Kornatowski, G. Zadrozna, M. Rozwadowski, A. Gutsze, *J. Phys. Chem. A* **2000**, *51*, 11 837.

67 J.J. Kornatowksi, G. Zadrozna, M. Rozwadowski, B. Zibrowius, F. Marlow, J.A. Lercher, *Chem. Mater.* **2001**, *13*, 4447.

68 E. Lehmann, C. Chmelik, H. Scheidt, S. Vasenkov, B. Staudte, J. Kärger, F. Kremer, G. Zadrozna, J. Kornatowski, *J. Am. Chem. Soc.* **2002**, *124*, 8690.

69 C. Baerlocher, W.M. Meier, D.H. Olson, Atlas of Zeolite Framework Types, Elsevier, Amsterdam **2001**, 302 pp.

70 D.W. Breck, Zeolite Molecular Sieves, Wiley, New York **1974**, p. 636.

71 H.G. Karge, W. Niessen, *Catal. Today* **1991**, *8*, 451.

72 M. Hermann, W. Niessen, H.G. Karge, in Catalysis by Microporous Materials, H.K. Beyer, H.G. Karge, I. Kiricsi, J.B. Nagy (eds.), Elsevier, Amsterdam **1995**, p. 131.

73 W.P.J.H. Jacobs, D.G. Demuth, S.A. Schunk, F. Schüth, *Micropor. Mater.* **1997**, *10*, 95.

74 G.J. Klap, M. Wübbenhorst, J.C. Jansen, H. van Koningsveld, H. van Bekkum, J. van Turnhout, *J. Chem. Mater.* **1999**, *11*, 3497.

75 G.J. Klap, H. van Koningsveld, H. Graafsma, A.M.M. Schreurs, *Micropor. Mesopor. Mater.* **2000**, *38*, 403.

76 E. Lehmann, S. Vasenkov, J. Kärger, G. Zadrozna, J.J. Kornatowksi, Ö. Weiss, F. Schüth, *J. Phys. Chem. B.* **2003**, *107*, **4685**.

77 J.J. Kornatowksi, G. Zadrozna, J. Wloch, M. Rozwadowski, *Langmuir.* **1999**, *15*, 5863.

78 S.A. Schunk, D. Demuth, B. Schulz-Dobrick, K.K. Unger, F. Schüth, *Micropor. Mater.* **1996**, *6*, 273.

79 E. Lehmann, S. Vasenkov, J. Kärger, G. Zadrozna, J.J. Kornatowksi, *J. Chem. Phys.*, submitted.

80 E.G. Derouane, Z. Gabelica, *J. Catal.* **1980**, *65*, 486.

81 J. Weitkamp, L. Puppe, Catalysis and Zeolites, Springer, Berlin **1999**, p. 346.

82 L.A. Clark, G.T. Ye, R.Q. Snurr, *Phys. Rev. Lett.* **2000**, *84*, 2893.

83 P. Bräuer, J. Kärger, N. Neugebauer, *Europhys. Lett.* **2001**, *53*, 8.

84 R.Q. Snurr, J. Kärger, *J. Phys. Chem. B* **1997**, *101*, 6469.

85 N. Neugebauer, P. Bräuer, J. Kärger, *J. Catal.* **2000**, *194*, 1.

86 P. Bräuer, A. Brzank, J. Kärger, *J. Chem. Phys.*, in press.

87 E. Lehmann, S. Vasenkov, J. Kärger, J. Weitkamp, Y. Traa, *J. Catal.*, in preparation.

3
Host–Guest Interactions in Bassanite, $CaSO_4 \cdot 0.5\,H_2O$

Henning Voigtländer, Björn Winkler, Wulf Depmeier, Karsten Knorr, and Lars Ehm*

3.1
Introduction

The interactions between the sulfate "host" lattice and the H_2O "guest" molecule in bassanite, $CaSO_4 \cdot 0.5\,H_2O$, are of interest, as they govern the dehydration and rehydration behavior. This, in turn, is of technological relevance, as bassanite is used for plaster production in the construction industry and for medical purposes, such as in dentistry. Bassanite is metastable and can be obtained from gypsum, $CaSO_4 \cdot 2\,H_2O$, by dehydration. On further dehydration at moderate temperatures, γ-$CaSO_4$ (soluble anhydrite III or AIII) is formed. Dehydration at higher temperatures leads to β-$CaSO_4$ (insoluble anhydrite). The rehydration of γ-$CaSO_4$ to bassanite, and of bassanite to gypsum, is rapid and exothermic.

The crystal structures of bassanite and γ-$CaSO_4$ are closely related. They consist of corner-sharing SO_4- and CaO_8-polyhedra, which form a three-dimensional framework with continuous channels parallel to [001]. These channels have a diameter of about 4.5 Å and are empty in γ-$CaSO_4$, but partially filled with H_2O molecules in bassanite (Fig. 1). The latter is thus a host–guest system and the present study aims at furthering the understanding of the interactions between the guest molecules and the host framework.

The positions of the framework atoms in bassanite have been well established by a number of diffraction studies [1–5] and the location of the oxygen atoms of the water molecules has also been determined [4,5]. A controversy remains with respect to the positions of the hydrogen atoms. Furthermore, different space groups have been proposed for the bassanite structure. Abriel and Nesper [4] experimentally found trigonal symmetry for bassanite (space group $P3_121$) with a statistical distribution of water over three different sites. However, assuming a monoclinic superstructure, the water can be described ordered on one crystallographic site. Only the position of the oxygen atom, OW, has been determined. Bezou et al. [5], on the other hand, found monoclinic symmetry (space group $I121$) for bassanite with two independent water molecules, one with the oxygen atom OW1 on a special position (Wyckoff position $2a$) and the other with the oxygen atom OW2 in a general position (position $4c$). In addition, they also give fractional coordinates for

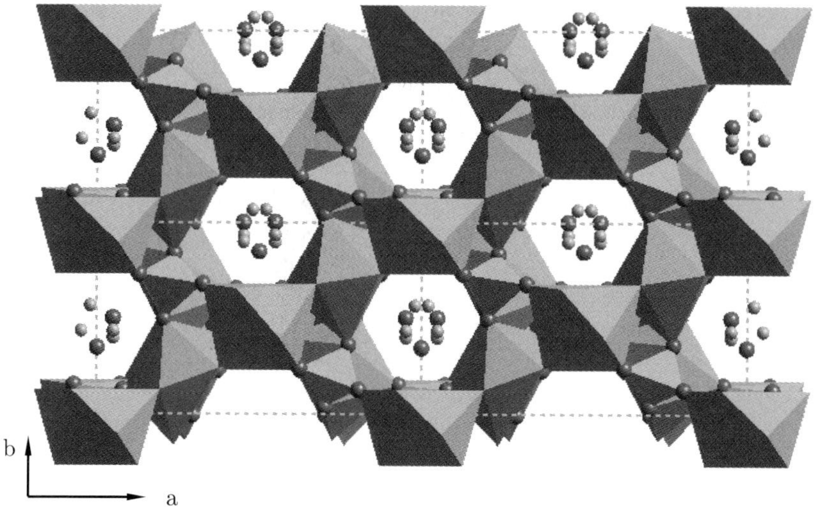

Fig. 1. Crystal structure of bassanite viewed along [001] according to the structural model given by Bezou et al. [5]. Part of the CaO_8 polyhedra are linked by corner-sharing SO_4 tetrahedra. The dotted line indicates the unit cell edges. H_2O occupies the channels; the water oxygen atoms are concentrically arranged around the central axis of the channel. The LL-sheets (see text) are parallel *c* along [010], [110], and [11<Ü->0].

the hydrogen atoms from a combination of neutron and synchrotron diffraction measurements performed at room temperature [5]. All protons occupy general sites (Wyckoff position 4*c*). Very recently, Ballirano et al. [6] published a structural model for bassanite, which is identical to the one presented by Bezou et al. [5], except that no proton positions are given. There is no reference in the publication of Ballirano et al. [6] to the study of Bezou et al. [5].

The number of water molecules that can be incorporated into the channels of bassanite is also discussed controversially. Several authors claim the existence of calcium sulfate subhydrates with water contents greater than $x = 0.5$ per formula unit $CaSO_4$. Abriel [7] reports a subhydrate with $x = 0.81$, based on the refinement of the site occupation factor for the oxygen atom of the crystal water. Bezou et al. [5] describe $CaSO_4{\cdot}0.6\,H_2O$, Kuzel and Hauner [3] report a water content of $x = 0.53$ and 0.66, depending on the partial pressure p_{H2O}, in which the water content has been determined by gravimetric measurements. Lager et al. [2] suggest that water content in excess of $x = 0.5$ would result in a too close proximity of the water molecules, resulting in energetically unfavorable repulsive forces. They note that water contents obtained from gravimetric measurements may be biased by the presence of molecules adsorbed on the grain surfaces. As the dehydration takes place at low temperatures ($T < 360$ K), the distinction between molecules in the channels and adsorbed water is experimentally demanding. Also, Lager et al. [2] point out that a refinement of the water content from X-ray diffraction data is generally problematic as there is a very strong correlation between site occupation factors and the atomic displacement parameters, both of which need to be determined [2].

In addition to experimental difficulties encountered in the diffraction experiments, the characterization of structural aspects is complicated by dynamical disorder of water at ambient conditions. This has been inferred from neutron spectroscopic investigations [8], in which a splitting in the low frequency spectrum between 150 and 3 K (the two temperatures at which experiments have been performed) has been interpreted as a "freezing out" of dynamic disorder.

In contrast to the incomplete understanding of the structure of bassanite, the structure of gypsum is very well understood. In gypsum, the water oxygen is part of the coordination polyhedron surrounding the Ca atoms and the hydrogen positions have been determined reliably by solid state NMR [9] and neutron diffraction [10]. The OH-stretching frequencies in gypsum are 3407 and 3550 cm^{-1} [11], whereas in hemihydrate they are 3560 and 3615 cm^{-1} [12]. While the dehydration of gypsum leads to the very different crystal structure of bassanite, a further dehydration to γ-$CaSO_4$ leads to a very small distortion of the sulfate framework only. These findings imply a fundamental difference between the interaction of the H_2O molecule and the remaining structure in gypsum and bassanite. Owing to the chemical similarity, however, gypsum is well suited as a reference system and this has been exploited here.

As mentioned above, the aim of the present study was to provide a better understanding of host–guest interactions in bassanite. To achieve this, structural studies using X-ray and neutron diffraction were extended to low temperatures, in order to characterize possible structural changes. High-pressure diffraction was also performed, as in such experiments the host–guest interaction was thought to be increased due to the compression of the framework. The experiments were accompanied by quantum mechanical model calculations at the density functional theory level. While successful structural studies are a prerequisite for the understanding of the host–guest interactions, the energetics of such interactions can only be obtained by spectroscopic studies. Solid-state NMR spectroscopy as a well-established tool has been employed here [13]. In addition, we explored the usefulness of deep inelastic neutron scattering (neutron Compton scattering), which in principle can provide direct information about the potential experienced by the hydrogen atoms. However, this is a method still requiring development and has been only rarely applied to chemically complex systems until now. Another approach to better understand host–guest interactions is by changing the guest. It had been shown previously that methanol can be incorporated into the channels of γ-$CaSO_4$ [14]. Here we attempted to extend the range of guests by absorption studies.

3.2 Investigation of the Bassanite Host Lattice

3.2.1 High Resolution Synchrotron Radiation Powder Diffractometry

At HASYLAB (beamline B2), high-resolution powder experiments were performed at 10, 20, 40, and 298 K in order to check whether the proposed change in the dy-

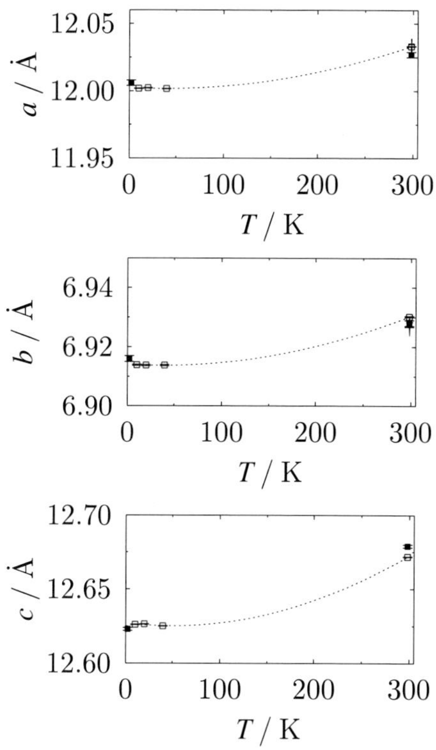

Fig. 2. Evolution of the lattice parameters of bassanite with temperature determined from synchrotron (open symbols) and neutron diffraction (filled symbols), compared with ambient temperature data by Bezou et al. [5] (daggers). All lines are guides to the eyes.

namics of the H_2O molecule [8] leads to structural changes. Rietveld refinements applying GSAS [15] were performed using the structural model given by Bezou at al. [5]. The positions of the water oxygen atoms were determined by difference Fourier analysis.

The powder diffractograms recorded show no significant changes in the intensity distribution as a function of temperature. The temperature dependence of the lattice parameters is plotted in Fig. 2; the corresponding values are given in Table 1. At room temperature the values obtained are in excellent agreement with those reported in the literature [5,6] and consequently support the monoclinic symmetry assignment by Bezou et al. [5]. Between 298 and 10 K, the relative decrease in length is similar for all unit cell axes and equals 0.26% for a, 0.24% for b, and 0.36% for c, thus reducing the unit cell volume by 0.86%. The monoclinic angle β is constant within the error of the cell parameters calculated.

No significant changes in the atomic coordinates of the framework atoms were found at low temperatures. The OW positions do also not change significantly be-

Tab. 1. Lattice parameters for bassanite from high-resolution synchrotron diffraction experiments

T [K]	*a [Å]*	*b [Å]*	*c [Å]*	*ß [°]*	*V [Å³]*
298 [5]	12.0317(4)	6.9269(2)	12.6712(3)	90.27(1)	1056.0(1)
298	12.0332(2)	6.9303(2)	12.6718(2)	90.26(1)	1056.7(1)
40	12.0017(2)	6.9138(2)	12.6260(2)	90.25(1)	1047.6(1)
20	12.0023(2)	6.9138(2)	12.6264(2)	90.25(1)	1047.7(1)
10	12.0019(2)	6.9139(2)	12.6261(2)	90.25(1)	1047.7(1)

tween 298 and 10 K and remain concentrically arranged around the central axis of the channels. The positions of the hydrogen atoms could not be located in the difference Fourier maps.

3.2.2
Neutron Powder Diffraction

Neutron diffraction is a well-established technique for locating proton positions in a crystal structure, since hydrogen has a very large scattering cross section compared to heavier elements [16]. Therefore, neutron powder diffraction experiments were performed on the instrument E9 at the Hahn–Meitner Institute, Berlin. Our aim was to determine the proton positions at several temperatures in order to determine the geometry of the assumed orientational disorder and possible changes with temperature. Since ^1H has a very large incoherent scattering length causing high background and hence a low peak/noise ratio, deuterated bassanite was synthesized by dehydration of bassanite at 393 K for 6 h and subsequent rehydration over a saturated solution of $CaCl_2$ in D_2O. A standard vanadium container of 15 mm diameter was employed and data were collected at 298, 40, 20, and 2 K. The wavelength and peak shape parameters were determined using α-Al_2O_3 as an external standard prior to the measurements.

The diffraction patterns recorded at 20 and 2 K are shown in Fig. 3. All diffractograms measured exhibit a high background with a peak/background ratio of about 4:1. This is indicative of either an incomplete deuteration of the sample or a previously undetected failure in the containment of the sample in an inert-gas atmosphere and hence an exchange of D by H.

The data were analyzed by Rietveld refinements with FULLPROF [17] and commenced from the structural model given by Bezou at al. [5]. The positions of the water atoms were searched for by difference Fourier analysis.

The temperature-dependence of the lattice parameters of deuterated bassanite (HD) is included in Fig. 2; Table 2 contains the lattice parameters for HD at various temperatures.

While the data sets obtained between 20 and 298 K differ only due to thermal expansion, the powder pattern measured at 2 K exhibits a considerably different intensity distribution (Fig. 3b).

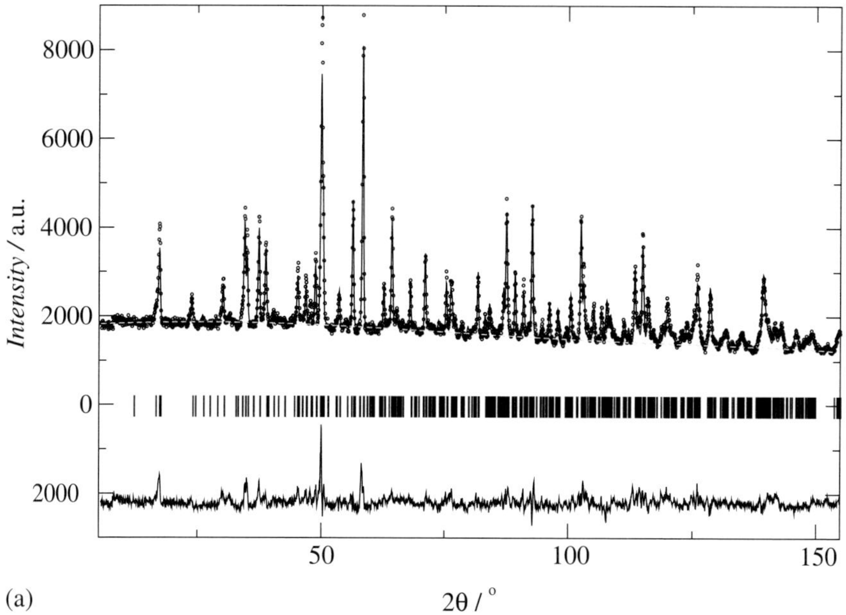

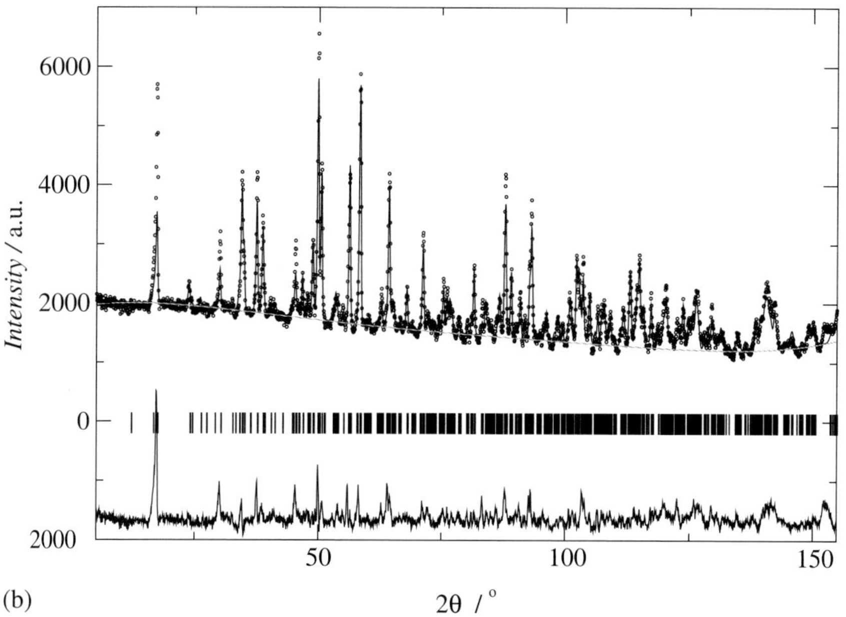

Fig. 3. Measured (dots) and calculated (line) neutron powder diffractogram of deuterated bassanite at 20 K (a) and 2 K (b), based on the structural model given by Bezou et al. [5] without hydrogen. Below: residual curve. Markers denote the peak positions.

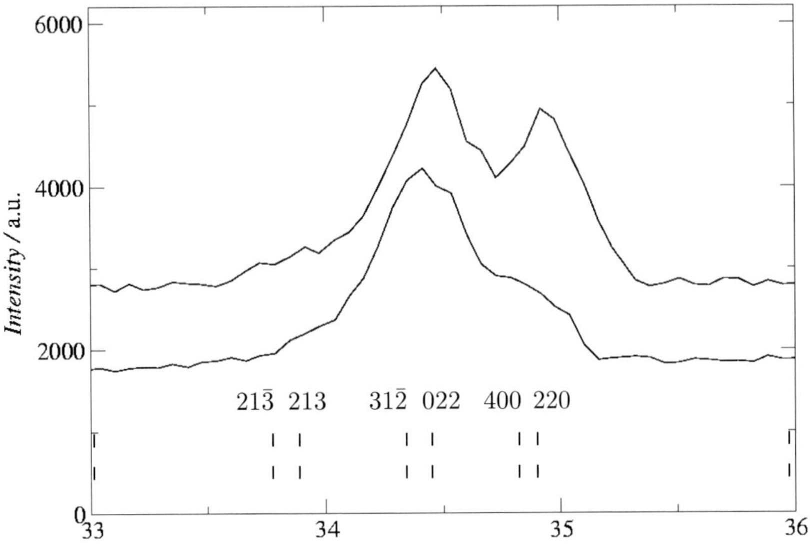

Fig. 3. (c) Details of the diffractograms recorded at 20 K (above) and 2 K (below). Markers denote peak positions for the measurements at 20 K (above) and 2 K (below).

The cell parameters obtained for the measurement at room temperature are in good agreement with the data of Bezou et al. [5]. The Rietveld refinements did not allow the identification of the positions of the hydrogen atoms. This can partially be explained by the insufficient instrumental resolution, which leads to a strong peak overlap at diffraction angles above 75° 2θ. A further detailed analysis strongly indicated an incomplete deuteration. This prevented any further refinement of the structural model and hence an understanding of the intriguing change in the intensity distribution at low temperatures would require additional measurements based on improved sample preparation and handling.

Tab. 2. Lattice parameters for deuterated bassanite at various temperatures from neutron diffraction

T [K]	*a [Å]*	*b [Å]*	*c [Å]*	*ß [°]*	*V [Å³]*
298 [5]	12.0317(4)	6.9269(2)	12.6712(3)	90.27(1)	1056.0(1)
298	12.029(3)	6.929(2)	12.678(2)	90.27(2)	1056.4(2)
40	12.006(2)	6.913(1)	12.654(2)	90.23(4)	1050.2(5)
20	12.007(2)	6.914(1)	12.647(3)	90.24(1)	1049.9(9)
2	12.006(2)	6.916(1)	12.623(2)	90.25(1)	1048.1(4)

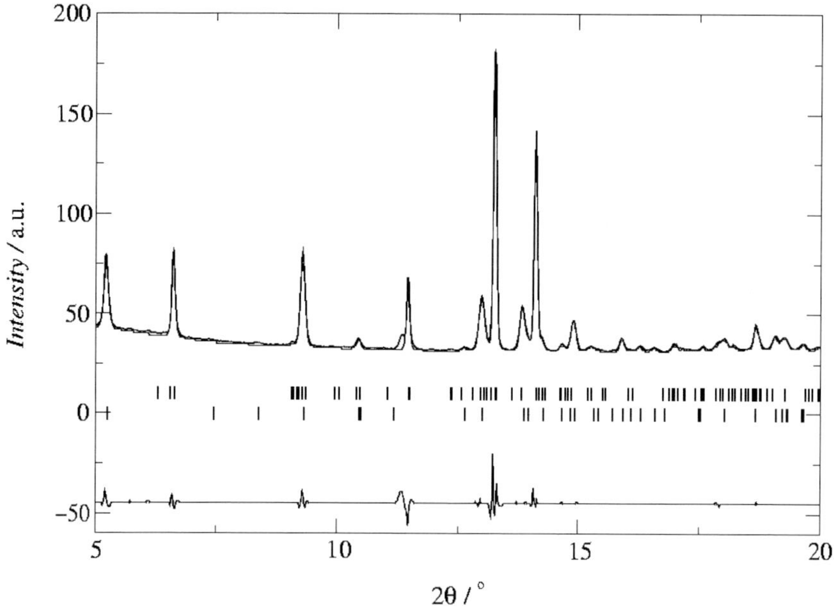

Fig. 4. Example of a high-pressure dataset measured at 1.68(2) GPa together with calculated and difference pattern. Markers denote peak positions of bassanite (upper row) and gypsum (lower row).

3.2.3
High-Pressure Behavior

High-pressure powder diffraction experiments at room temperature were performed at the diffraction beam-line of the ELETTRA Synchrotron in Trieste, Italy. Pressure up to 6.3 GPa was applied using a Merrill–Bassett type diamond-anvil cell [18] with a 4:1 methanol–ethanol mixture as pressure-transmitting medium. The pressure was determined via the ruby fluorescence method following the Piermarini scale [19]. 2D diffraction data were collected with an image plate detector, transformed to conventional powder patterns [20] (Fig. 4) and analyzed by the full pattern-matching method with FULLPROF [17].

The evolution of the lattice parameters as a function of pressure is shown in Fig. 5a. Below 3 GPa, the compression of all crystallographic axes is isotropic, while above 3 GPa the compression of the structure along a is significantly stronger than along b and c. The monoclinic angle β increases by about 1°. The change in the slope of the unit cell volume data (Fig. 5b) as a function of pressure at about 3 GPa indicates a change of the compression mechanism. Consequently, two second-order Birch–Murnaghan Equations of State (EOS) [21] were applied for pressures of 0–3 GPa and 3–6.33 GPa. This resulted in an appropriate description of the data (Fig. 5b) and yields bulk moduli of 67(1) GPa below 3 GPa and 101(1)

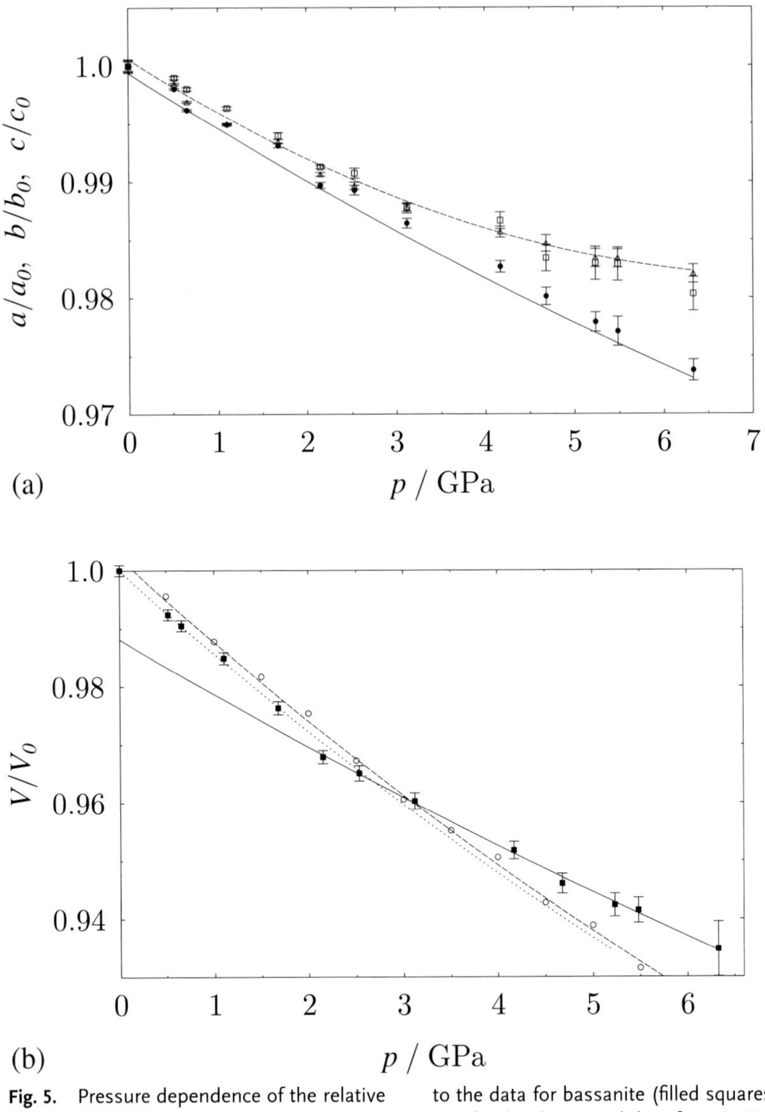

Fig. 5. Pressure dependence of the relative unit cell parameters (a) (circles: a/a_0, squares: b/b_0, triangles: c/c_0, all lines are guides to the eye) and (b) the unit cell volume in bassanite. Below 3 GPa all lattice parameters show an isotropic compression behavior, beyond 3 GPa it becomes anisotropic with compressibilities $K \parallel a > K \parallel b, c$. (b) Experimental and calculated pressure dependence for the unit cell volume of bassanite. Two second-order EOS were fitted to the data for bassanite (filled squares) and one for the theoretical data for γ-$CaSO_4$ (circles). Below 3 GPa the experimentally observed pressure behavior of bassanite (dashed curve) corresponds to that of the theoretical data for γ-$CaSO_4$ (dotted curve). Above 3 GPa bassanite is significantly less compressible than γ-$CaSO_4$. The experimental error of the pressure determination is about 0.03 GPa.

GPa above 3 GPa. The change of the compressibility is attributed to a change in the compression mechanism, as hardening due to nonhydrostatic conditions at this relatively low pressure is unlikely [22].

The structural origin of this behavior may be due to the existence of relatively tightly packed polyhedra "sheets" (called LückenLos- or LL-sheets) lying in planes containing the [001] and one of [010], [110], or [1<Ü->10] (Fig. 1). These sheets do not correspond to those used in the conventional description of layered structures such as micas, but nevertheless can be clearly identified (Fig. 1). Since the compressibilities of coordination polyhedra are generally small compared to volume changes caused by the variation of inter-tetrahedral angles [23], it can be assumed that the reduction of the cell volume is mainly caused by the latter mechanism. This would imply that the crystal structure is less compressible along directions parallel to the sheets than perpendicular to them, which is consistent with the observed orientation of the LL-sheets described above. It is unlikely that the increasing anisotropy of the compressibility with increasing pressure is due to a pressure-induced filling of the channels with molecules from the pressure-transmitting medium, as this should lead to a similar behavior in *a* and *b*.

In Fig. 5b volume data obtained from quantum mechanical calculations for γ-$CaSO_4$ [24] are also shown. For these data the pressure dependence of the volume behavior was fitted with a second-order EOS, resulting in a bulk modulus of 66(2) GPa. It is therefore in very good agreement with the experimentally determined pressure behavior of bassanite up to 3 GPa, and the resulting EOS practically coincide. Above 3 GPa, bassanite becomes much less compressible than γ-$CaSO_4$, as is shown by the equations of state for the two compounds. A detailed analysis of the theoretical data is in progress, but in the absence of any other obvious change the pressure-induced change in the compressibility of bassanite is currently thought to be due to a strong increase in the guest–host interaction.

3.3 Dynamics of H_2O as a Guest Molecule in Bassanite

3.3.1 Nuclear Magnetic Resonance Measurements

^1H Magic Angle Spinning-NMR spectroscopy was employed to determine the T_1 relaxation times for the hydrogen atoms in bassanite, to calculate the activation energy associated with the molecular motion of the water molecules, and hence to further examine the dynamics of the crystal water in bassanite. Measurements were performed at the Institute of Mineralogy at Ruhr University, Bochum, using an FT-NMR spectrometer with a ZrO_2 rotor. Spectra were recorded at 10 K intervals between 150 and 390 K. The spinning speed was 10 kHz for measurements at 295–390 K and 8 kHz for measurements at 290–150 K. The "saturation recovery method" [25] was employed because of the long relaxation time. For the determination of the T_1 relaxation times a pulse sequence of 16 pulses between 1 ms and

15 s was used. Tetramethylsilane was used as a standard for all measurements. For the evaluation of the intensities of the central signals and of the spinning sidebands the program GNUPLOT was employed [26]. Depending on the number of sidebands, up to 22 Lorentzians were fitted to the data. The background was described separately with a combination of a Lorentzian and third-order polynomial function.

The ^1H MAS-NMR spectrum of bassanite at room temperature (Fig. 6a) contains two central peaks at chemical shifts of 0.57(5) and 4.52(1) ppm. The position of the weaker peak is consistent with the chemical shift for ^1H in water-bearing zeolites, which is about 5 ppm [27]. The stronger peak, however, exhibits an extraordinarily small chemical shift, similar to that for ^1H in tremolite, $Ca_2Mg_5Si_8O_{22}(OH)_2$, which is 0.7 ppm [28]. Spinning sidebands due to strong homonuclear dipolar interactions among the protons [28] are present in all measured spectra. The number of spinning sidebands is primarily dependent on the spinning speed and decreases if the sample is spun at higher speeds. As thermal motion of the atoms decreases with temperature, the homonuclear dipolar interactions among the protons increase. This leads to increasing numbers of spinning sidebands with decreasing temperature at constant spinning speed. Consequently, the spectrum recorded at 390 K shows only one pair of spinning sidebands, whereas at 150 K, there are five pairs.

All spectra show a high background, which increases with decreasing temperature. This complicates the determination of the intensities of the weak spinning sidebands and thus limits the precision with which the ratio of the intensities of the central signals can be determined.

The structural model proposed by Bezou et al. [5] implies three different environments for the hydrogen atoms in bassanite. For $H_2O(1)$ (with OW1), both protons are symmetrically equivalent (H1). Consequently, their environment is identical and thus they are indistinguishable by NMR spectroscopy. The OW2 water molecule has two symmetrically independent protons (H2 and H3), which according to the structural model used [5], have slightly different environments. As only two peaks have been observed, we conclude that these small differences cannot be resolved. The total number of H2 and H3 is twice that of H1. Therefore, an intensity ratio of 2:1 for (H2 + H3): H1 would be a confirmation of the structural model [5]. Independent of temperature we observe intensity ratios of about 2.5(1):1. The linewidths of both signals seem to be very similar, but due to the problematic description of the background, a more quantitative assessment is currently not possible. The relative change in linewidth with decreasing temperature is comparable for both signals. Although the agreement between the observed intensities and the intensities implied by the structural model is not perfect, it certainly confirms the presence of at least two symmetrically independent hydrogen sites.

By plotting the summed intensities of the signals versus the dwell time and fitting an exponential function the T_1 relaxation times for the protons in bassanite could be calculated. The values are equal within the error limits for each temperature investigated and vary between 3.3(3) s at 390 K and 13(2) s at 160 K.

The activation energy for the ^1H dynamics was calculated from the relaxation

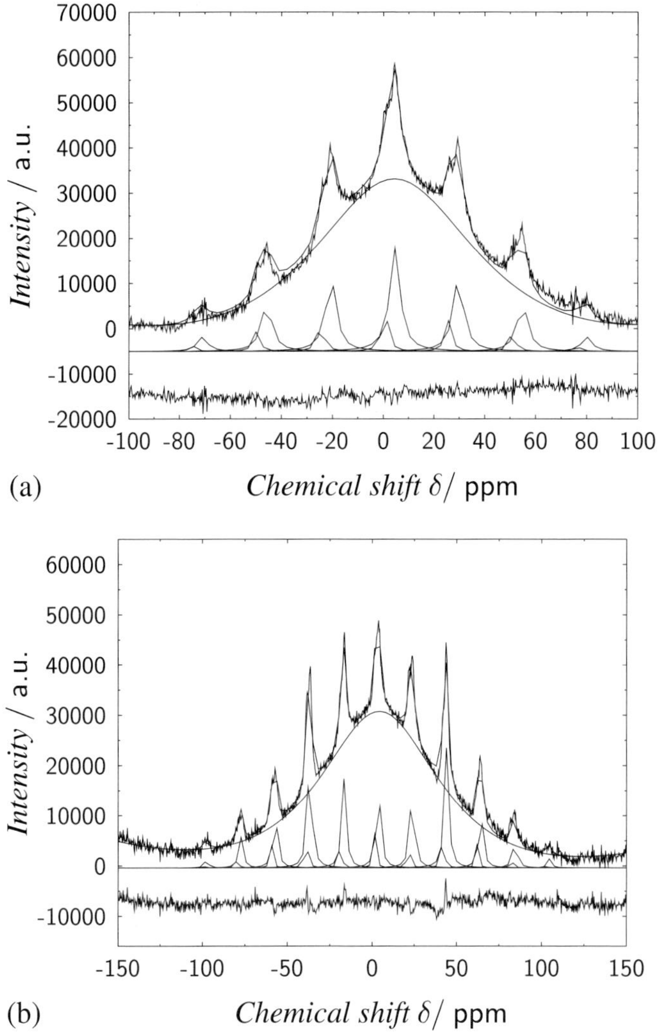

Fig. 6. ^{1}H-MAS-NMR spectra of bassanite at 295 K (a) and 150 K (b) with fitted model. The intensity ratio of the central peaks is 2.4:1 and 2.6:1 at 295 K and 150 K, respectively.

times using a BPP (Bloembergen, Purcell, and Pound) equation [29] and equals 0.032(2) eV for both water molecules.

The minimum of the BPP curve (Fig. 7) is not reached within the temperature interval investigated. Thus, only the low-temperature branch of the BPP equation, that is, the linear region can be used for determining the activation energy. Consequently, no correlation times τ_C can be calculated, but only the τ<Ü=>$_C/C$ ratio, where τ<Ü=>$_C$ is the limit for the correlation time at very high temperatures and C

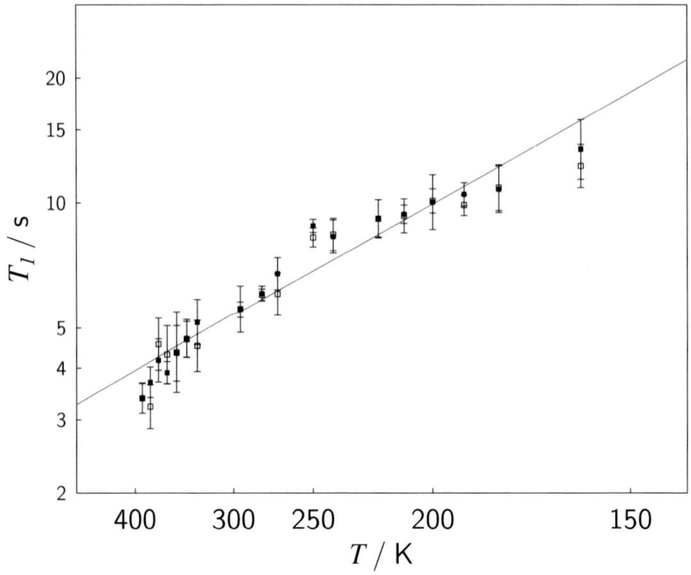

Fig. 7. T_1 relaxation times for ^1H in bassanite with fitted BPP curve.

is a measure for the strength of the dipole interaction of the water molecules. The correlation time describes the time scale on which a molecule moves. It is not equivalent with the relaxation time T_1, which describes the time the spins need to re-orientate in the applied magnetic field. The experimentally determined τ<Ü=>$_C$/C ratio is $2 \times 10^{-17}\ s^{-1}$ for both water molecules in bassanite. Further experiments at temperatures above 400 K are required to reach the minimum of the BPP curve and thus access the correlation times. This temperature range, however, is beyond the stability of bassanite and thus the experiments are probably not feasible.

3.3.2 Deep Inelastic Neutron Scattering

Deep inelastic neutron scattering (DINS) allows measurements of atomic momentum distributions and kinetic energies [30,31]. Specifically, they can be used to investigate the potential experienced by a hydrogen atom in a hydrogen bond [32]. While this method is still in its early development stages, we wanted to assess whether one could distinguish the influence of the different crystal chemical environments of the hydrogen atoms in bassanite and gypsum. We also hoped that a temperature dependence of the spectra would provide additional data on the dynamics of the H_2O molecule in bassanite.

Experiments were therefore performed on synthetic bassanite powder as well as on synthetic gypsum powder and on a natural gypsum single-crystal. The "electron Volt spectrometer" (eVS) at ISIS, Rutherford Appleton Laboratory, UK was employed. The eVS is a filter difference time-of-flight spectrometer [31] with inverse geometry. Application of the filter difference technique defines the energy of the scattered neutrons. Thin filter foils of either Au or U have strong neutron absorption resonance in the eV energy range. They are switched in and out of the scattered neutron beam at five minutes intervals. Foil in and foil out data are stored separately and subtracted afterwards to achieve the difference spectrum. It represents the spectrum of neutrons scattered with a final energy equal to that of the neutron absorption resonance of the filter foils. Because of the fixed energy of the scattered neutrons, the momentum transfer *q*<Üp> and the energy transfer ω can be calculated. This enables calculation of *y*, which is the component of the atomic momentum *p*<Üp> along the direction of *q*<Üp>.

All samples used were 7 cm × 7 cm × 0.2 cm in size, resulting in approximately 5% scattering of the incident radiation, thereby reducing the occurrence of multiple scattering. Data on bassanite powder samples were recorded at 160, 45, and 5 K. Measurements on a gypsum single-crystal and powder gypsum samples were performed at 5 K. Figure 8 shows the momentum distributions for gypsum powder at 4 K and for bassanite powder at 160, 45, and 5 K.

The powder measurements gave kinetic energies of 122(4) meV for the proton in gypsum, with a linewidth of 4.43(7) $Å^{-1}$ (Fig. 8).

Figure 9 shows a contour plot of the proton potential in gypsum derived from the single-crystal measurements at 4 K. The sample was oriented so that the (010) face was approximately perpendicular to the incident beam. For this orientation the scattering is isotropic, that is, there is no observable anisotropy due to the influence of the O–H or the hydrogen bond.

The measurements on bassanite showed no temperature dependence within the experimental errors. The kinetic energies are about 160 meV and thus about 30% larger than in gypsum The linewidth is 5 $Å^{-1}$ (Fig. 8), which is more than 10% wider than in gypsum. While the experiments therefore confirmed that hydrogen atoms in bassanite are not as well localized as those in gypsum, a more quantitative evaluation would require significantly better counting statistics, which will only become possible with more advanced neutron sources than currently available.

3.4 Incorporation of Other Guest Molecules into γ-$CaSO_4$

3.4.1 Experiments Using a Normal-Pressure Flow Device

Experiments were performed at the Institute of Chemical Technology at the University of Stuttgart to find out whether it is possible to incorporate other guest molecules into the polyhedral framework of γ-$CaSO_4$. A normal-pressure flow device [33] was used. Prior to the attempts to incorporate guest molecules, bassanite was dehydrated inside the apparatus at 383 K for 1 h to give γ-$CaSO_4$, which served as the absorber for the potential guest molecules. Then, vaporized methanol

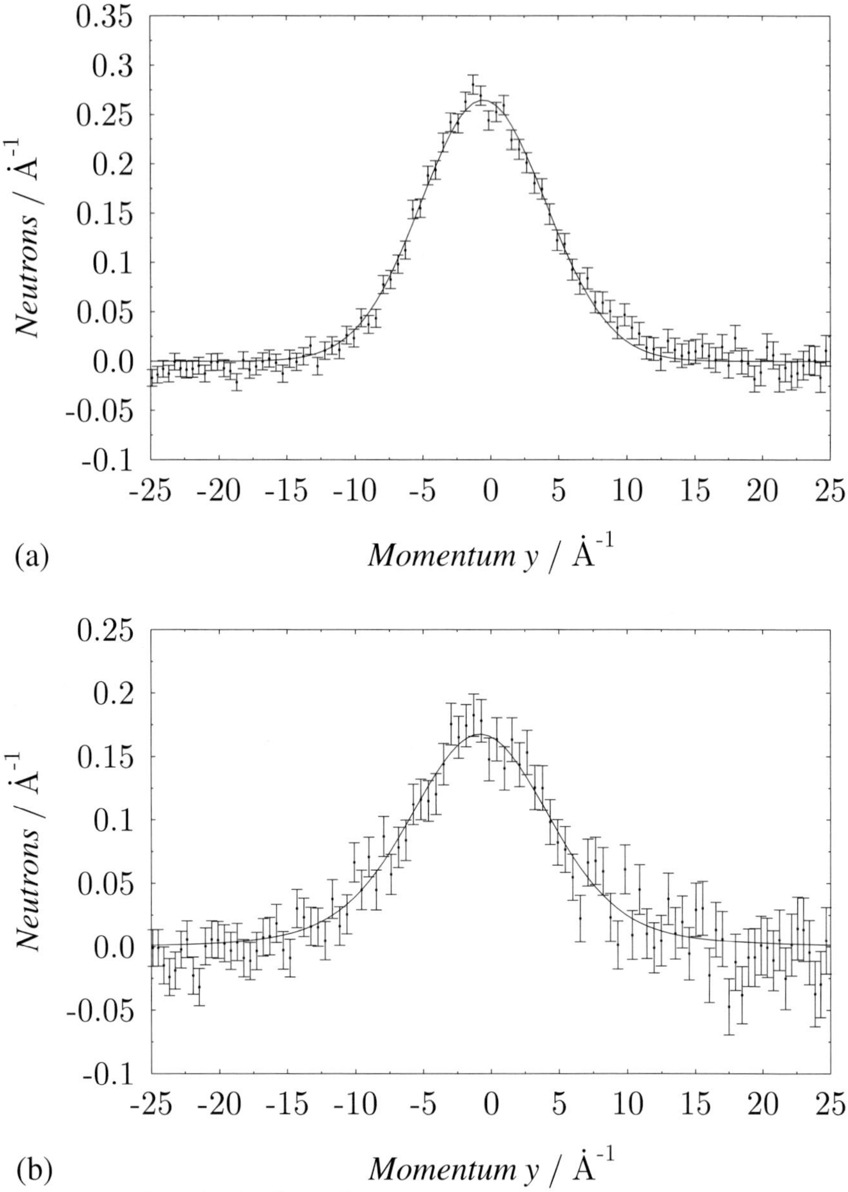

Fig. 8. Summed impulse distribution for gypsum (a) at 4 K and bassanite at 160 K (b), 45 K (c), and 5 K (d). For bassanite, no temperature dependence of the impulse distribution is observable within the calculated error.

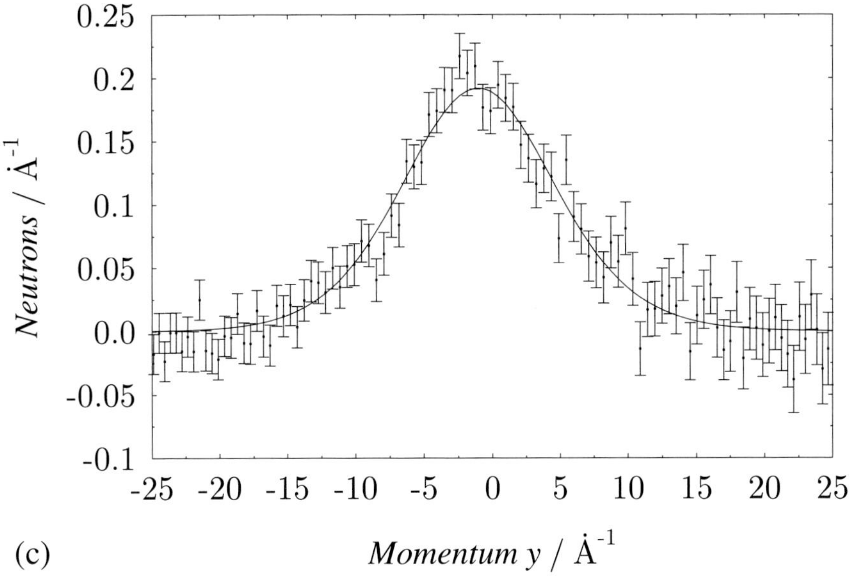

(c)

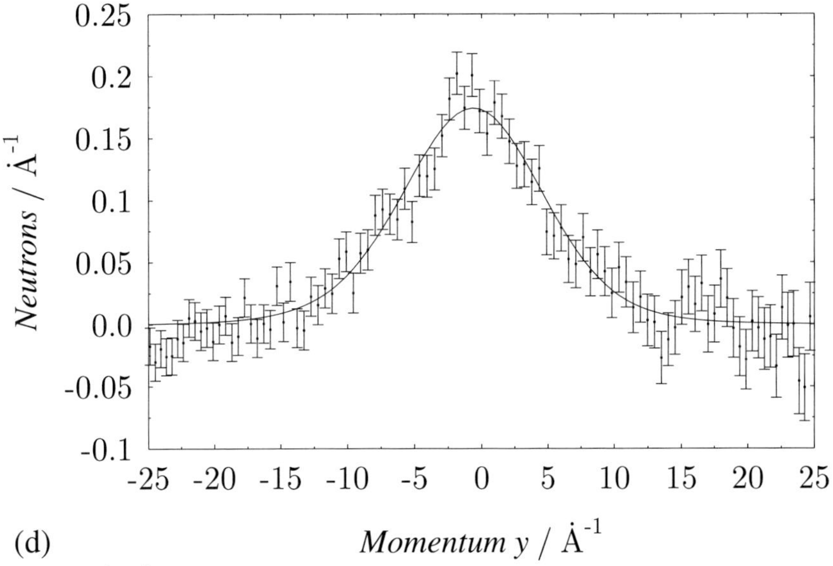

(d)

Fig. 8. *(continued)*

(CH_3OH), ethanol (C_2H_5OH), propanol (C_3H_7OH), acetonitrile (CH_3CN), and water were transported to the sample using N_2 as a carrier gas. The partial pressure of the potential guest molecules was maintained at 2 kPa, the γ-$CaSO_4$ was held at 303 K.

The carrier gas flows through the absorber and both the educt stream and product stream are periodically analyzed in a gas chromatograph. The experiment is

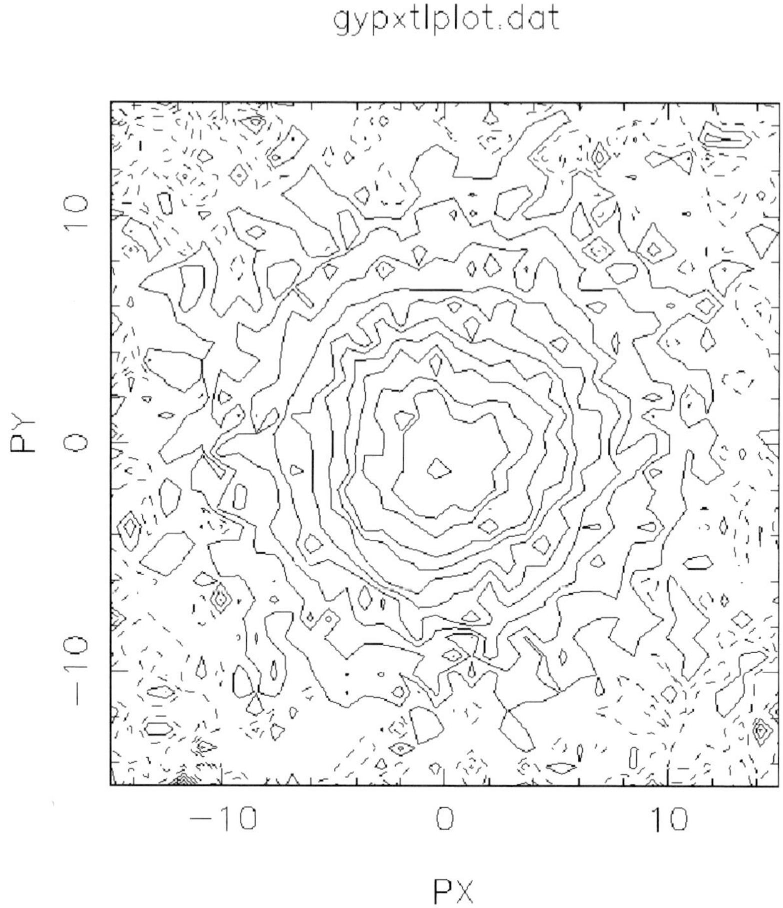

Fig. 9. Contour plot of the proton potential in gypsum from a single-crystal measurement at 4 K. Both axes have the dimension $Å^{-1}$.

stopped when the partial pressure of the potential guest molecule in the product flow equals the partial pressure of the potential guest molecule in the educt flow. For methanol, ethanol, and acetonitrile, this was the case after less than 15 min, whereas for water the equilibration required more than 9 h (Fig. 10).

From the evolution of the partial pressure of the potential guest molecule in the product flow, the amount of the guest molecule absorbed in γ-$CaSO_4$ can be determined. According to the measurements, only water could be absorbed into γ-$CaSO_4$, the other compounds are most probably only adsorbed at the surface of γ-$CaSO_4$. This is also the case for methanol, which can be incorporated during sample synthesis (see below). Hence, there is a fundamental difference between

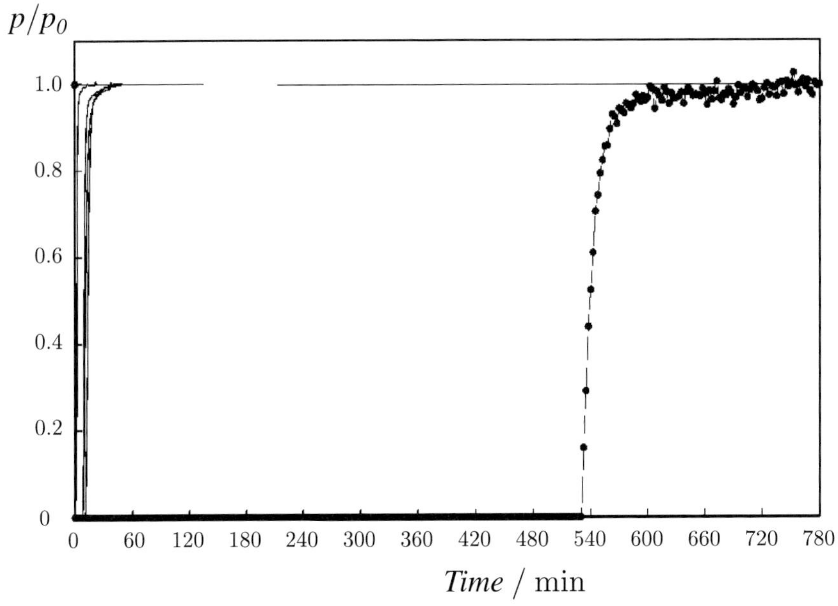

Fig. 10. Absorption curves for different guest molecules in dehydrated bassanite. Absorption decreases with increasing p/p_0 ratio, in which p is the partial pressure of the potential guest molecule in the product stream and p_0 its partial pressure in the educt stream. No other species except water has been incorporated. The amount of water being incorporated corresponds to a weight gain of 4.3 wt.-%.

H_2O and the other molecules investigated here, as only H_2O diffuses into the channels. This may either be due to kinetic effects (i.e., the molecules are too big to easily diffuse into the channels) or it may be due to unfavorable reaction enthalpies. The latter aspect was checked by quantum mechanical calculations of the incorporation of methanol, H_2O, and acetonitrile into γ-$CaSO_4$ by Winkler [34], in which it was found that indeed the reaction energies favor incorporation only in the case of H_2O.

The incorporation of water leads to a weight gain of 4.3 wt.-% while a composition of $CaSO_4{\cdot}0.5\,H_2O$ corresponds to a water content of 6.2 wt.-%. This may be due to the low partial pressure for water in the experimental setup (2 kPa), compared to the partial pressure of H_2O in air of 4.2 kPa [35] at 303 K.

3.4.2
Incorporation of Methanol into the Framework of γ-$CaSO_4$

For a better understanding on how host–guest interactions depend on the incorporated guest molecules, calcium sulfate hemimethanolate, $CaSO_4{\cdot}0.5\,CH_3OH$ (called HM) has been synthesized, following the method given by Reisdorf and

Abriel [14]. No large single-crystals could be obtained; the maximal crystal dimensions of the product being about 5 μm (determined via Scanning Electron Microscopy). Fig. 11 shows a SEM micrograph of the nonidiomorphic reaction product.

HM turned out to be unstable at air, in which it decays within some minutes and transforms in bassanite. Hence, it had to be stored under dry ether, absolute methanol or ethanol. But even under these conditions, HM can be held for a maximum of two or three weeks only. The HM was characterized using a Siemens powder diffractometer, model D500 and turned out to contain small amounts of β-$CaSO_4$, but apart from that was identical within experimental error to that given by Reisdorf and Abriel [14].

3.5 Investigations on Hemimethanolate

3.5.1 High Resolution Synchrotron Radiation Powder Diffractometry

Further experiments on HM were performed at 298, 40, and 10 K at beamline B2, HASYLAB. These experiments aimed at investigating possible temperature-induced changes of the polyhedral framework, and to ascertain the results obtained by the low-resolution results of Reisdorf and Abriel [14] with regard to the space group symmetry assignment (space group $P3_121$). The sample investigated contained a small amount of β-$CaSO_4$. The calculated lattice parameters are in good agreement with those by Reisdorf and Abriel [14] (Table 3). In order to simplify the comparison between the unit cell dimensions of bassanite and HM, the lattice parameters are given for an orthohexagonal setting of the HM unit cell. As with bassanite, the unit cell parameters do not change very much with decreasing temperature. The relative decrease in length of the unit cell axes is 0.22% for a, 0.23% for b, and 0.41% for c, thus reducing the unit cell volume by 0.86%, which is equal to the value obtained for bassanite. In HM, however, the decrease in length of the unit cell axes shows a more distinctive anisotropy than in bassanite.

3.5.2 Nuclear Magnetic Resonance Measurements

For HM ^1H MAS-NMR experiments were performed as described above for bassanite. Owing to the limited stability of HM, the temperature range for these experiments extended only from 150 to 300 K. In addition, a ^{13}C-CP-MAS-NMR spectrum was recorded at 298 K to ascertain that methanol had in fact been incorporated into the polyhedral framework.

The ^{13}C spectrum of HM at room temperature shows two peaks at chemical shifts of 51.2 and 52.6 ppm, (Fig. 13). The chemical shift for methanol in solution is 49.0 ppm [36]. The change of the chemical shift is significant and strongly im-

(a)

(b)

Fig. 11. Scanning electron micrograph of (a) calcium sulfate hemimethanolate (HM) and (b) bassanite specimen. The HM crystal aggregates show no idiomorphic forms, whereas the bassanite specimen is well-crystallized.

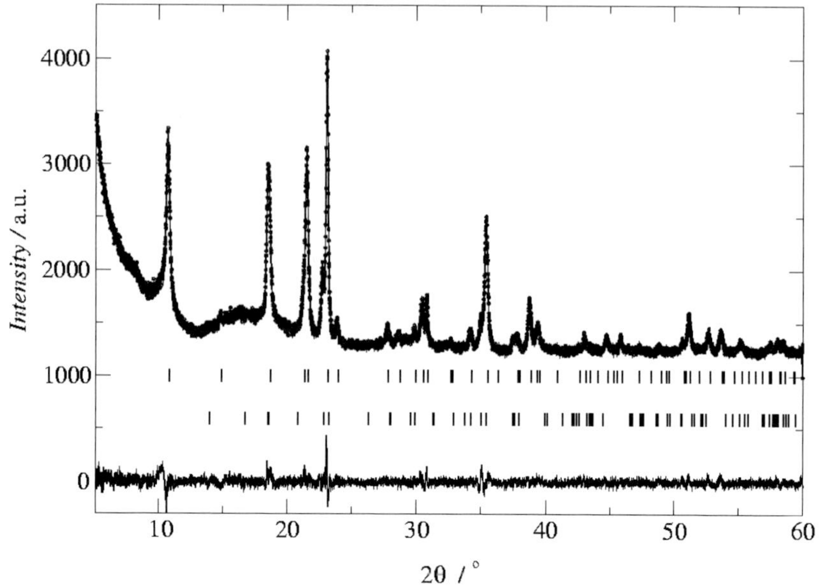

Fig. 12. Calculated and measured powder diffractogram of calcium sulfate hemimethanolate (HM) at 295 K with residual curve (bottom). Markers denote the peak positions for HM (above) and for β-$CaSO_4$ (below).

plies that methanol has in fact been incorporated into the polyhedral framework of γ-$CaSO_4$, and was not adsorbed on the surface. Since methanol contains only one C atom, the existence of two peaks in the ^{13}C spectrum points at two crystallographically different positions for methanol in HM. The intensities of the signals are 1.5:1, hence one can conclude that the sites have different multiplicities or occupancies.

The ^{1}H measurements at room temperature show two peaks at 3.25 ppm and 5.44 ppm (Fig. 14). Methanol in solution has a chemical shift of 3.30 ppm and 4.80 ppm for ^{1}H [36], which confirms the incorporation of methanol into the polyhedral framework of γ-$CaSO_4$.

Tab. 3. Lattice parameters for calcium sulfate hemimethanolate from high-resolution synchrotron diffraction experiments. Note that the lattice parameters are given for an orthohexagonal setting of the unit cell.

T [K]	*a [Å]*	*b [Å]*	*c [Å]*	*ß [°]*	*V [Å³]*
298 [14]	12.024(2)	6.942(1)	12.708(2)	90.000	1060.7(5)
298	12.017(6)	6.938(3)	12.690(6)	90.000	1058.0(9)
40	11.990(6)	6.923(3)	12.640(6)	90.000	1049.3(8)
10	11.990(6)	6.922(3)	12.638(6)	90.000	1049(1)

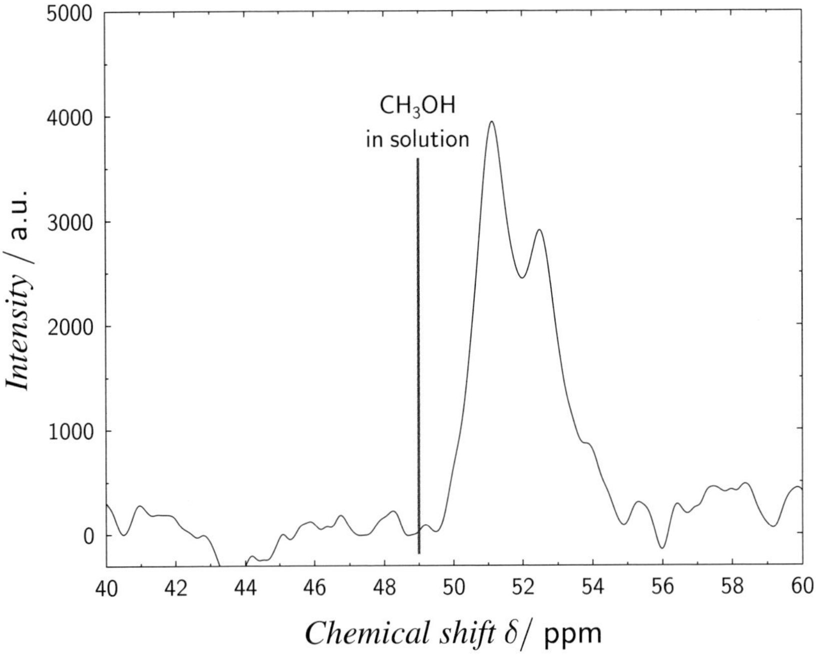

Fig. 13. ^{13}C-MAS-NMR spectrum of calcium sulfate hemimethanolate (HM) at 295 K. The intensities of the peaks are 1.5:1, indicating two different sites with different multiplicities for methanol in HM.

The intensity ratio of the ^{1}H peaks is 2.9:1. The stronger central peak represents the methyl protons, the weaker one represents the hydroxyl protons (Fig. 14). Very weak spinning sidebands, with intensities of about 2–2.5% of the intensity of the central peak can be neglected in a first approximation. Surprisingly, it is impossible to distinguish between the two different methanol molecules in HM. This is probably caused by dynamic disorder of the protons in methanol, which would lead to similar crystal chemical environments for the protons. Moreover, the resolution of the spectrometer may be insufficient to distinguish between the peaks. Table 4 contains selected refined parameters for the spectra.

The T_1 relaxation times for ^{1}H in HM (Fig. 15) were calculated using the program dm2000nt [37]. The calculated values vary between 0.34 and 0.54 s and, hence, are at least ten times shorter than for the protons in bassanite. The corresponding activation energies are equal within the error limits for both the methyl and the hydroxyl protons with values of 0.046(2) and 0.049(2) eV, respectively. The limit for the correlation time at very high temperatures, τ<Ü=>$_{C}$, is $1.3(1)\cdot 10^{-10}$ s for both the methyl and the hydroxyl protons, corresponding to a correlation time τ_C of $7.6(6)\cdot 10^{-10}$ s at 300 K and $C = 8.3(1) \times 10^{8}$ s^{-2}. As it is not possible to distinguish between the two methanol molecules, it is assumed that the correlation time is the same for both molecules.

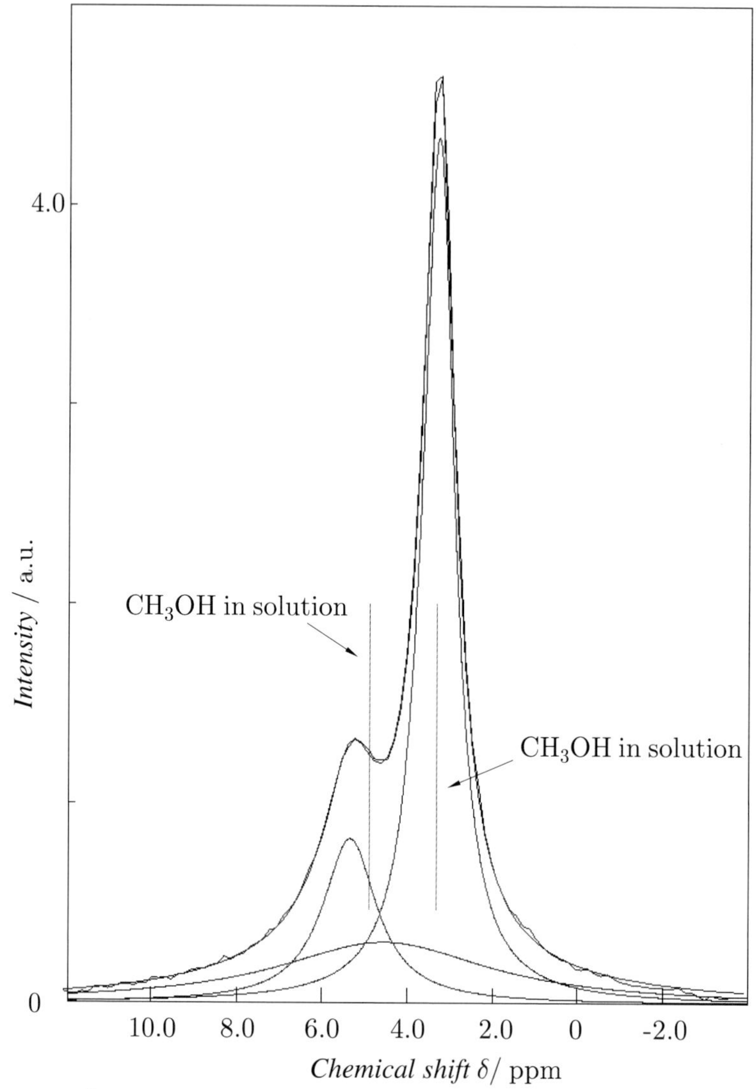

Fig. 14. ^{1}H-MAS-NMR spectrum of calcium sulfate hemimethanolate (HM) at 295 K. Two peaks indicate the incorporation of methanol into HM, with an intensity ratio of 2.9:1. Decreasing temperature induces lowering of the chemical shift. Only very weak spinning sidebands occur, with intensities of about 2–2.5 % of the intensity of the central peaks.

Tab. 4. Selected refined parameters for calcium sulfate hemimethanolate

T [K]	*Linewidths [ppm]* CH_3 // *OH*	*Intensities [au]*
150	0.75 // 0.97	3.03:1
295	0.91 // 1.54	2.91:1

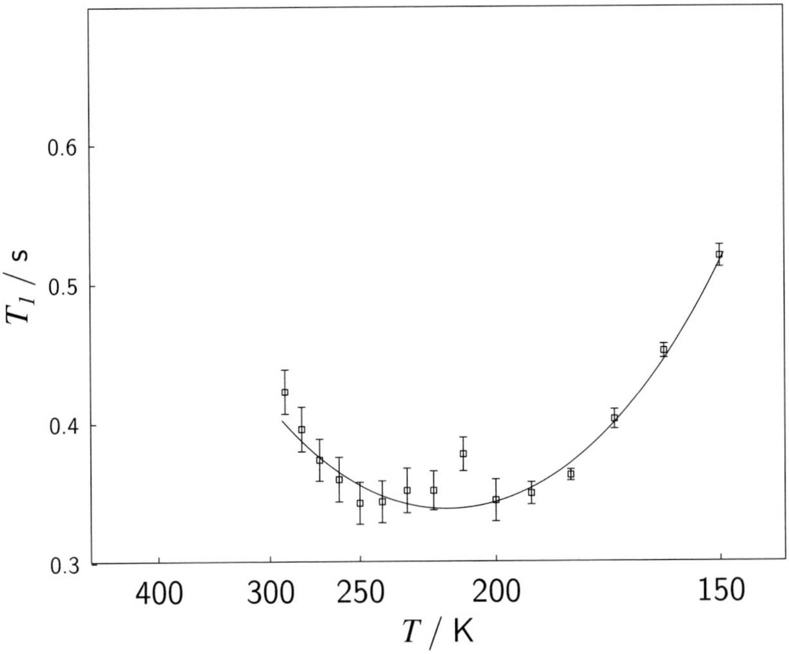

Fig. 15. T_1 relaxation times for ^{1}H of the methyl group in calcium sulfate hemimethanolate (HM) with fitted BPP curve. The T_1 relaxation times for the hydroxyl protons in HM are equal within limits of error.

3.6 Conclusions

With our experimental results reported here, we hope to have extended the information base on the behavior of H_2O molecules in bassanite. Specifically, the NMR measurements confirmed the conclusion of Winkler and Hennion [8] that there is a significant dynamical disorder at ambient conditions. While the correlation time could not yet be determined, the ratio of τ<Ü=>$_{C}/C$ is now known. Also, the NMR experiments provide an independent confirmation of the existence of at least two symmetrically independent hydrogen positions. The very small chemical shift observed for one resonance is unusual. It is, however, similar to that observed in tremolite, which also exhibits a very low chemical shift for the protons. The O–H stretching frequency in tremolite is 3668 cm^{-1} [38] and indicates little or no hydrogen bonding [39–41]. The high-frequency O–H stretching frequency in bassanite (3615 cm^{-1} [12]) also implies the absence of any significant hydrogen bonding. It seems therefore possible to understand the structural origin of the small chemical shift, and this will be attempted in the near future.

The host–guest interactions in bassanite are weak. This can be concluded from earlier work on the thermodynamics of the de- and rehydration [42] and from IR-spectroscopic results [12,43]. Consequently, the dehydration of bassanite to γ-$CaSO_4$ induces only very small distortions of the polyhedral framework. The high-pressure experiments may point at a pressure-induced strengthening of the host–guest interaction. This needs to be confirmed by high-pressure infrared spectroscopy, which is very sensitive to pressure-induced changes in hydrogen bonds [44–46].

Investigations of a possible correlation between the host–guest interaction and the guest were limited to methanol in HM. The only way of incorporating methanol is by using it as a template for the crystallization, as until now we have not found further guests, which can be incorporated. Incorporation of methanol leads to trigonal symmetry with a cell volume slightly larger than that of bassanite. The NMR results for HM show that the host–guest interactions in HM are also weak, because the activation energies correspond to those for H_2O in bassanite.

Acknowledgements

We would like to thank H. Ehrenberg, M. Knapp, and C. Baehtz at HASYLAB, D. Többens at HMI, J. Mayers at ISIS, C.E. Bussetto at ELETTRA, and M. Fechtelkord, I. Wolf, and M. Borowski (NMR spectroscopy) for their help with the experiments and the evaluation of the data. H.V. would also like to thank R. Gläser for his help with the incorporation experiments, and M. Chall and C. Griewatsch for numerous helpful hints. Last, but not least, we thank the German Science Foundation for financial support (De 412/23-1).

References

1 O.W. Flörke, *N. Jb. Mineral., Abh.* **1952**, *84*, 189.
2 G.A. Lager, Th. Armbruster, F.J. Rotella, J.D. Jorgensen, D.G. Hinks, *Am. Mineral.* **1984**, *69*, 910.
3 H.-J. Kuzel, M. Hauner, *Zem.-Kalk-Gips*, **1987**, *40*, 12, 628.
4 W. Abriel, R. Nesper, *Z. Kristallogr.* **1993**, *205*, 99.
5 C. Bezou, A. Nonat, J.-C. Mutin, N. Christensen, M.S. Lehmann, *J. Solid. State Chem.* **1995**, *117*, 165.
6 P. Ballirano, A. Maras, S. Meloni, R. Caminiti, *Eur. J. Mineral.* **2001**, *13*, 985.
7 W. Abriel, *Acta Cryst.* **1983**, *C39*, 956.
8 B. Winkler, B. Hennion, *Phys. Chem. Minerals* **1994**, *21*, 539.
9 G. Hutton, B. Pedersen, *J. Phys. Chem. Sol.* **1969**, *30*, 235.
10 B.F. Pedersen, D. Semmingsen, *Acta Cryst.* **1982**, *B38*, 1074.
11 V. Seidl, O. Knop, M. Falk, *Can. J. Chem.* **1969**, *47*, 1361.
12 J. Bensted, S. Prakash, *Nature* **1968**, *219*, 60.
13 G. Engelhardt, D. Michel, High-Resolution Solid-State NMR of Silicates and Zeolites, Wiley, New York **1987**, p. 485.
14 K. Reisdorf, W. Abriel, *Zem.-Kalk-Gips* **1988**, *41*, 356.
15 A.C. Larson, R.B. von Dreele, GSAS: LANL document code 86-748, **1994**, Los Alamos, USA.
16 G.E. Bacon, in The Determination of Crystal Structures by Neutron-diffraction Measurements, R. Brill (ed.), Advances in Structure Research by Diffraction Methods,

Vol. 1, Vieweg, Braunschweig **1964**, p. 1.

17 J. Rodriguez-Carvajal, in *Abstr. Satellite Meeting of the XV Congr. IUCr*, Toulouse, **1990**, p. 127.

18 L. Merrill, W. Bassett, *Rev. Sci. Instr.* **1974**, *45*, 290.

19 G.J. Piermarini, S. Block, J.D. Barnett, R.A. Fonnan, *J. Appl. Phys.* **1975**, *46*, 2774.

20 S. Vogel, L. Ehm, K. Knorr, G. Braun, *Adv. X-ray Anal.* **2002**, in press.

21 F. Birch, *J. Geophys. Res.* **1983**, *83*, B3, 1257.

22 K. Takemura, *J. Appl. Phys.* **2001**, *89*, 662.

23 R.M. Hazen, L.W. Finger, Comparative Crystal Chemistry, Wiley, New York **1982**, p. 231.

24 B. Winkler (unpublished).

25 W. Kemp, NMR in Chemistry, A Multinuclear Introduction, 1st ed., Macmillan, London **1986**, p. 134.

26 GNUPLOT: *www.cs.dartmouth.edu/gnuplot_info.html*, function file written by M. Chall (unpublished).

27 K.F.M.G.J. Scholle, W.S. Veeman, J.G. Post, J.H.C. van Hooff, *Zeolites* **1983**, *3*, 214.

28 J.P. Yesinowski, H. Eckert, G.R. Rossman, *J. Am. Chem. Soc.* **1988**, *110*, 1367.

29 N. Bloembergen, E.M. Purcell, R.V. Pound, *Phys. Rev.* **1948**, *73(7)*, 679.

30 L. Kosidowski, A.V. Powell, J. Mayers, *Physica B*, **1998**, *241-243*, 335.

31 A.L. Fielding, J. Mayers, Technical Report RAL-TR-2000-013, **2000**.

32 P. Postorino, F. Fillaux, J. Mayers, J. Tomkinson, R.S. Holt, *J. Chem. Phys.* **1991**, *94(6)*, 4411.

33 R. Gläser, PhD Dissertation, Universität Stuttgart, **1997**.

34 B. Winkler (unpublished).

35 Holleman-Wiberg, Lehrbuch der Anorganischen Chemie, 91st–100th edn., de Gruyter, Berlin **1994**, p. 37.

36 Spectral Data Services, NMR Data Acquisition Services, *www.sdsnmr.com/cs_table.html*.

37 D. Massiot, dm2000nt, http://crmht-europe.cnrs-orleans.fr/dmfit/download/download.asp.

38 C. Karr, Jr., Infrared and Raman Spectroscopy of Lunar and Terrestrial Minerals, 1st edn., Academic Press, New York **1975**, p. 218.

39 F.C. Hawthorne, H.D. Grundy, *Can. Mineral.* **1976**, *14*, 334.

40 K. Nakamoto, M. Margoshes, R.E. Rundle, *J. Am. Chem. Soc.* **1955**, *77*, 6488.

41 E. Libowitzky, *Mh. Chem.* **1999**, *130*, 1047.

42 M.C. Ball, R.G. Urie, *J. Chem. Soc. A* **1970**, 528.

43 A. Putnis, B. Winkler, L. Fernandez-Diaz, *Mineral. Mag.* **1990**, *54*, 123.

44 B. Winkler, K. Langer, P.G. Johannsen, *Phys. Chem. Minerals* **1989**, *16*, 668.

45 G.A. Lager, P. Ulmer, R. Miletich, W.G. Marshall, *Am. Mineral.* **2001**, *86*, 176.

46 B. Velde, G. Martinez, *Am. Mineral.* **1981**, *66*, 196.

4
Organic Guest Molecules in Zeolites

Carsten Baehtz and Hartmut Fuess*

4.1
Introduction

The last few years have seen an increasing interest in host–guest systems composed of organic guest molecules and zeolite frameworks as host. Both chemical and physical properties of the guest and these systems change by incorporation. Thus a broad field of various new applications has been opened [1]. Zeolites are well known as catalytic materials but other applications are conceivable in the future. Laser activity of incorporated laser dyes is observed [2,3]. Another promising achievement is the construction of membranes based on zeolites [4,5]. Gas sensitive compounds in zeolites can be used as sensors [6]. The base of the understanding of the changes of the chemical and physical behavior due to the influence of the framework on the guest molecules is a knowledge of the adsorption site of the guest in the framework. This could also be very useful to explain effects such as shape selectivity in heterogeneous catalysis [7,8]

Other investigations concentrated more on the adsorption position of special molecules and combined this knowledge with open force field simulations [9] or the determination of the molecular dynamics [10]. To investigate host–guest interactions of isolated guests more accurately and to neglect guest–guest interaction, we have examined dried zeolites [11] with a content of about one molecule or complex per cage. Our specific interest is focused on tetrathiafulvalene (TTF) and 7,7,8,8-tetracyanoquinodimethane (TCNQ) both separately and together as a charge transfer complex [12] in zeolite faujasite called NaY with the stoichiometry $Na_{52}Al_{52}Si_{140}O_{384}$ (structure type FAU, space group number 227, $a = 24.70$ Å). The charge-transfer activity offers another observable to study such influences by UV/vis spectroscopy [13].

In addition molecules such as naphthalene, anthracene, and 2,3-benzanthracene are incorporated in this host to elucidate the influence of molecule size onto the adsorption position. Additionally chloranil as an electron acceptor is combined with different electron donors and loaded into the zeolite lattice.

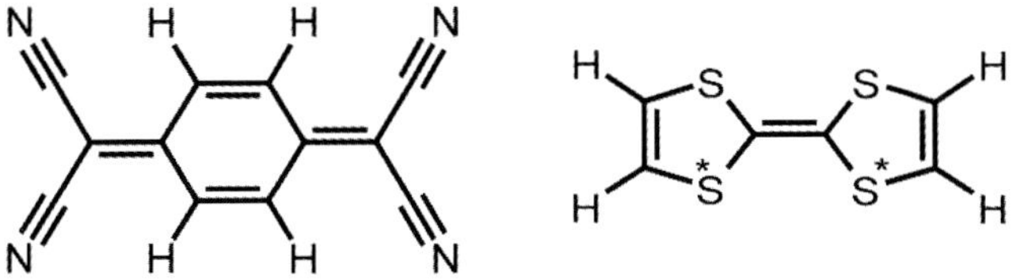

Fig. 1. Stick diagram of TCNQ and TTF.

4.2 Experimental

4.2.1 Localization of Guest Molecules by Powder Diffraction

The localization of guest molecules can be performed by powder diffraction experiments, subsequent Fourier analysis, and Rietveld refinement [14]. If a guest molecule is incorporated, the crystal structure of the host itself remains unchanged. But the diffraction pattern of the loaded host shows changes in intensities of some reflections caused by the guest. These changes therefore reveal information about the adsorption site.

Figure 2 displays as an example the diffraction pattern of TCNQ in NaY measured at beamline B2, HASYLAB Hamburg. In general such experiments are performed at low temperatures, in this particular example at 10 K, to suppress molecular motions of the guest. The residual scattering density (nonframework scattering density) is calculated by Fourier analysis based on these intensity changes. Residual scattering density is displayed in Fig. 3. High intensities of residual scattering density are observed in the center of the 12-ring windows, which connect two supercages.

The 2D Fourier map in Fig. 4 shows the position of TCNQ in these windows more accurately. To reveal the adsorption site of the guest by Rietveld refinement, the so-called rigid body method is used [15]. The conformation of the investigated molecules is fixed, there are no free rotating functional groups due to the existing double and triple bonds. Bond lengths and angles of the molecules are known and under the assumption of no changes by adsorption, they can be fixed during the refinement and the molecule is regarded as one structural unit.

So instead of the fractional atomic parameters of each atom of the molecule only three parameters for position and three parameters for orientation of the rigid body in the lattice of the host are refined. Therefore the number of additional parameters rapidly decreases and is fixed. Furthermore they are independent of the number of atoms of the molecule. The localization of big molecules is therefore in principle easier, because the changes of intensities in the diffraction pattern are more significant, but additionally only the rigid body parameters are needed. The occupation factor, in fact the loading of the host, is in the order of the content known from the synthesis of the samples. The isotropic thermal motion factor of the guest was estimated and fixed. By this method the adsorption site of guest

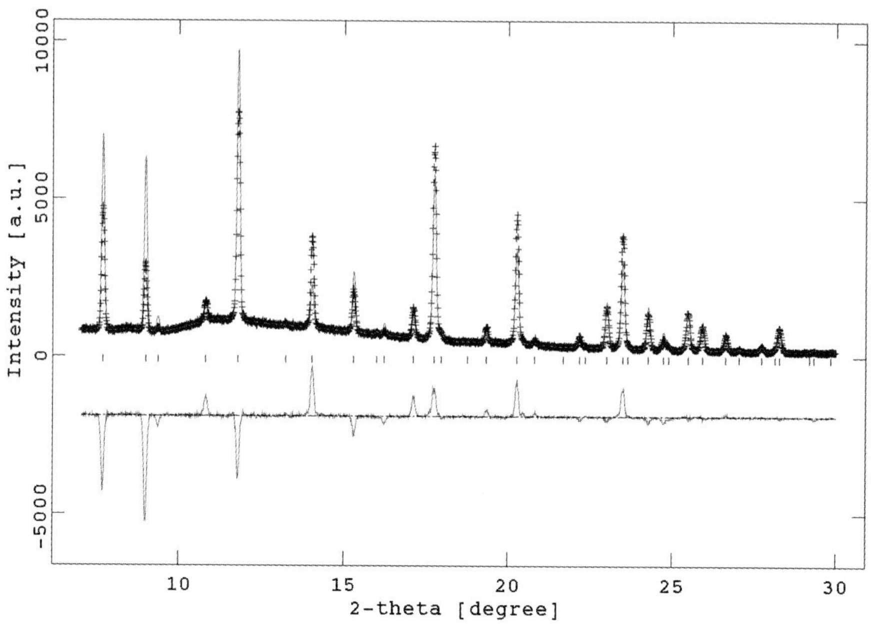

Fig. 2. Diffraction pattern of TCNQ in NaY (crosses) compared to unloaded, dried NaY (solid line), the difference curve displays the changes in intensities due to loading of TCNQ.

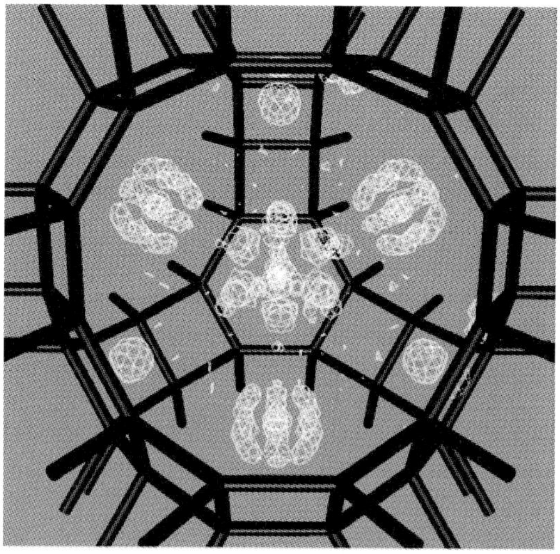

Fig. 3. 3D contour plot of the residual scattering density of TCNQ in NaY.

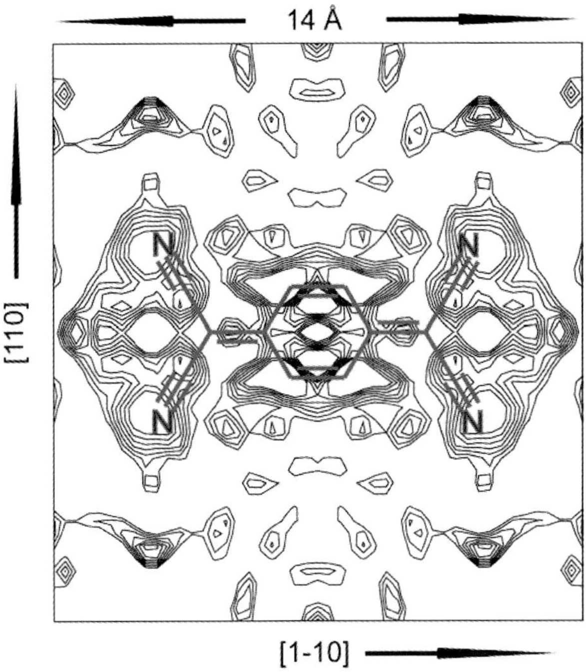

Fig. 4. 2D Fourier map of TCNQ in NaY.

molecules can be localized by Rietveld refinement, the position found in the Fourier map is used as a starting model.

For this purpose high quality powder diffraction data of both the unloaded host as reference to determine its crystal structure very precisely and the loaded zeolite are needed. The use of synchrotron radiation provides data with a very high resolution (Fig. 5). Neutron scattering is very sensitive to organic molecules. Additionally it is possible to observe a large 2θ range, due to the $\sin\theta/\theta$ independence of the neutron scattering lengths (Fig. 14).

The (111) reflection of zeolite faujasite has an asymmetric shape due to the umbrella effect and beam divergence. But this reflection shows a very pronounced change in intensity by loading the zeolite. After some preliminary refinements the reflection has been taken into account for the refinement by application of the asymmetry function of Finger [16].

4.3 Results

4.3.1 TTF and TCNQ in Zeolite Faujasite NaY

The guest molecules are incorporated in the dried host by adsorption from the liquid phase. The loaded host is dried under high vacuum again and sample purity

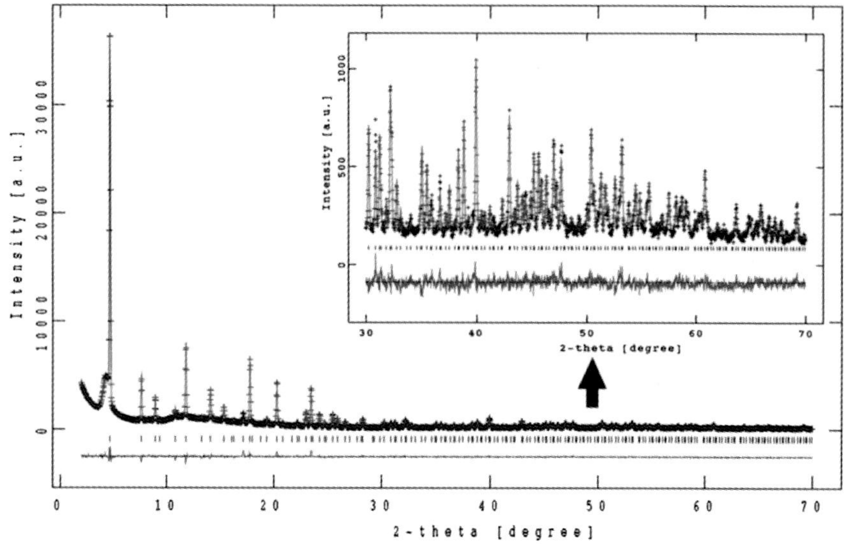

Fig. 5. Observed and calculated diffraction patterns together with their difference curve of TCNQ in NaY, measurement performed at 10 K on beamline B2, HASYLAB, Hamburg, Germany.

is checked with mass spectrometry. The data evaluation is performed as described above and the result is shown in Fig. 5.

TCNQ was located by this method in the center of the 12-ring windows of the zeolite host, while four CN groups point into the two neighboring supercages (Fig. 6). Each group is coordinated to one sodium cation of the host lattice at position II in front of the six-ring window of the sodalite cage. This sodium cation is attracted along the crystallographic [111] axis towards the guest molecule. The sodium–nitrogen distance is 2.7 Å.

This is a position with extreme high local symmetry (3m). In one 12-ring window three crystallographic equivalent positions for the TCNQ-molecule exist, but of course only one can be occupied. TCNQ is also a highly symmetric molecule (point group mmm) and it has got the perfect shape and the right configuration especially of its CN group to fit in this position. Both guest and lattice have the symmetry element 2/m in common. This is the first four-fold coordination of an organic guest molecule known in literature [17]. Open force field simulations using the Burchart Universal force-field [18] and the software package Cerius^2 confirm this position.

In contrast to this is the position of TTF in NaY. Several simultaneous structure refinements of diffraction patterns using neutron and synchrotron radiation give slightly different adsorption positions with some characteristics in common. TTF is in the supercage of NaY, two sulfur-atoms labeled * in Fig. 1 point to the sodium cation II. The distance between the center of mass of TTF and this cation is 3.5–

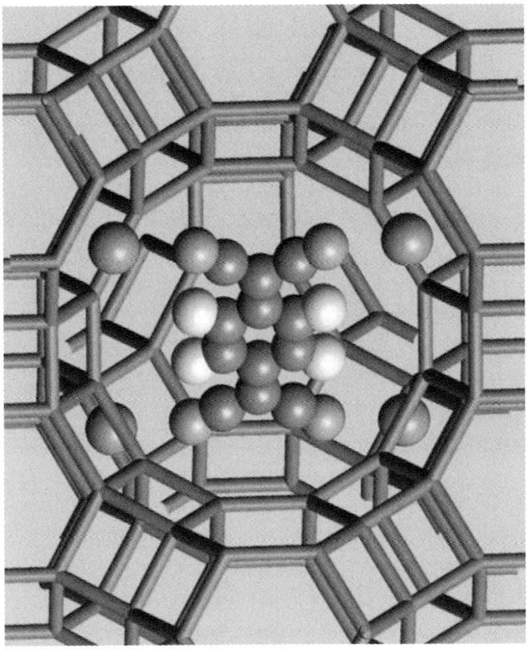

Fig. 6. TCNQ in NaY, only sodium cation at position II, to which TCNQ coordinates, are drawn.

4.4 Å. But the position of the center of mass and the rotation angles are not well defined. Figure 7 displays such adsorption sites.

Open force field simulations give a second position shown in Fig. 8. Here all four sulfur atoms have similar distances to the sodium cation at position II. This position is very close to the first one discussed above. The Rietveld refinement can not confirm its existence, but also does not rule it out.

The structure refinement of the mixed loading of TCNQ and TTF into NaY as a charge transfer complex show that both molecules are at identical positions to those described above.

The influence of adsorption onto the guest molecules can be observed by spectroscopic methods [19,20].

DRIFT measurements show the influence of adsorption onto the absorption band of the CN valence vibration. This band is shifted to higher wavenumbers that is, energies, if TCNQ is incorporated, compared to the pure substance. This shows a polarization of this triple bond, which increases the ionic character and strengthens this bonding [21].

The UV/vis spectra of TTF [22] in NaY displays a bathochromic shift of the whole spectra of 59 nm, which may be compared with the shift of absorption bands to lower energies in UV/vis spectra measured in solutions with increasing solvent polarity. The solvent stabilizes the excited state of the molecule, which are neutral

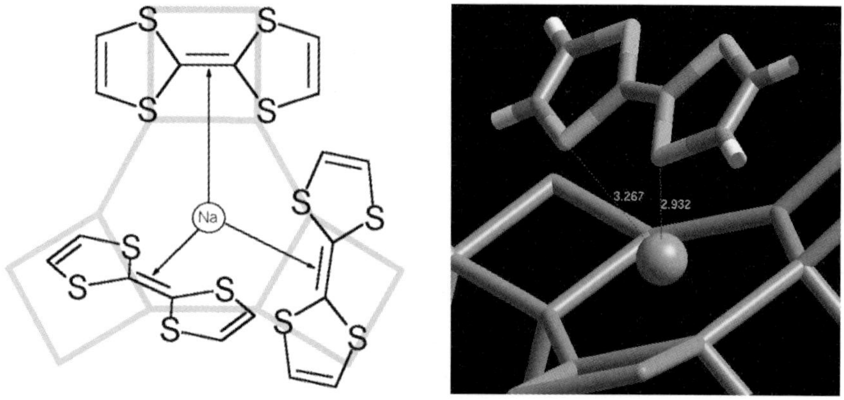

Fig. 7. (a) Representative arrangements of TTF in NaY, (b) position found by OFF-simulations, all positions have the described characteristics in common.

in ground state [23]. In contrast, the bands of the radical ions of TCNQ and TTF as a charge transfer complex shift to higher energies. Such an anomalous effect is known for strong charge-transfer complexes solved in polar or protic solvents [23].

4.3.2
TTF and TCNQ in Zeolite Faujasite HY

TCNQ and TTF are also included in the pores of zeolite faujasite with the stoichiometry of $H_{22}Na_{30}Al_{52}Si_{140}O_{384}$. In this zeolite the sodium cation position II is only occupied by 40% instead of 96% in the case of NaY ($Na_{52}Al_{52}Si_{140}O_{384}$), both

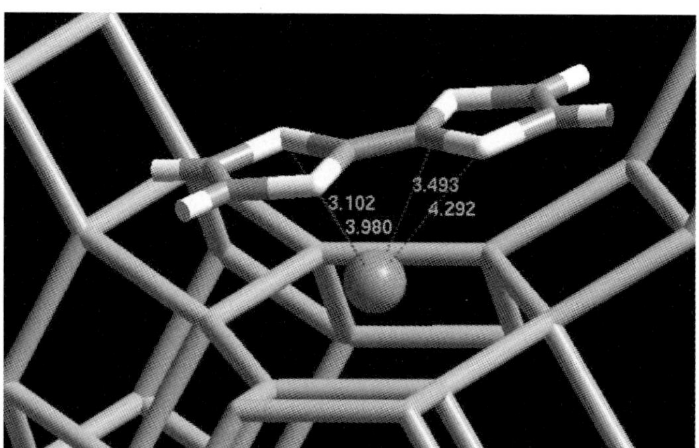

Fig. 8. Second arrangement of TTF in NaY found by OFF-simulations

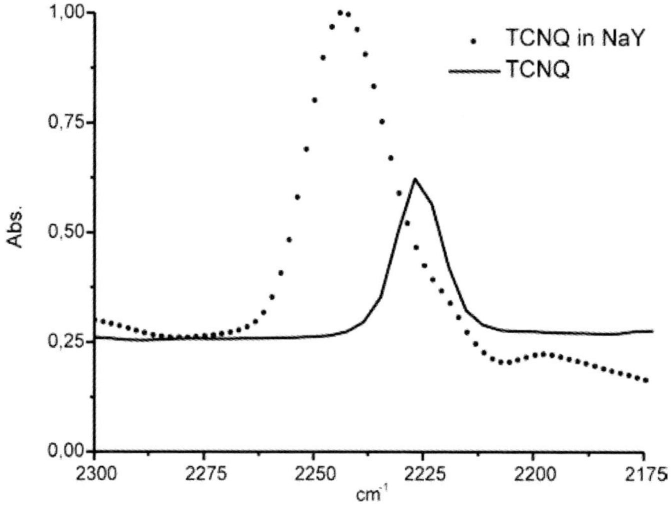

Fig. 9. DRIFT spectra of TCNQ.

found by Rietveld refinement and confirmed by X-ray fluorescence analyses. When the guest molecules are separately incorporated, the same characteristic colors as for NaY as host are obtained (Table 1). The guest molecules are at the same adsorption positions as described above and the spectroscopic results are identical. But in contrast to NaY the attempt to include the charge transfer complex, consisting of TCNQ and TTF, failed. The sample has the brown color of TTF in NaY instead of blue. The UV/vis-spectra confirms that only TTF is included into the cavities of this zeolite. The diffraction pattern shows the crystalline phase of pure

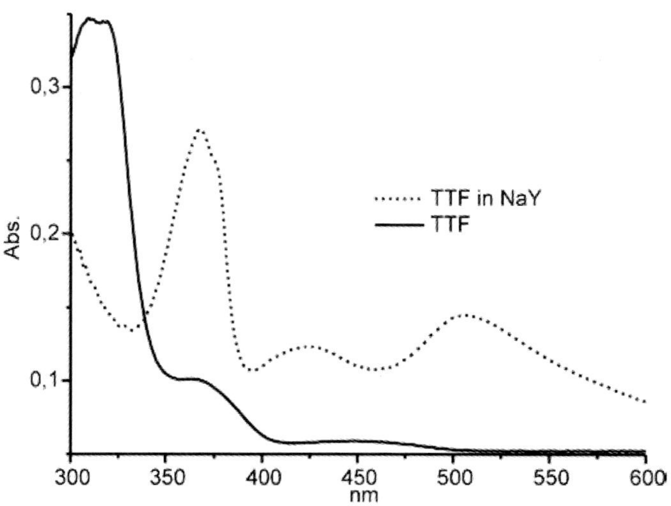

Fig. 10. UV/vis spectra of TTF.

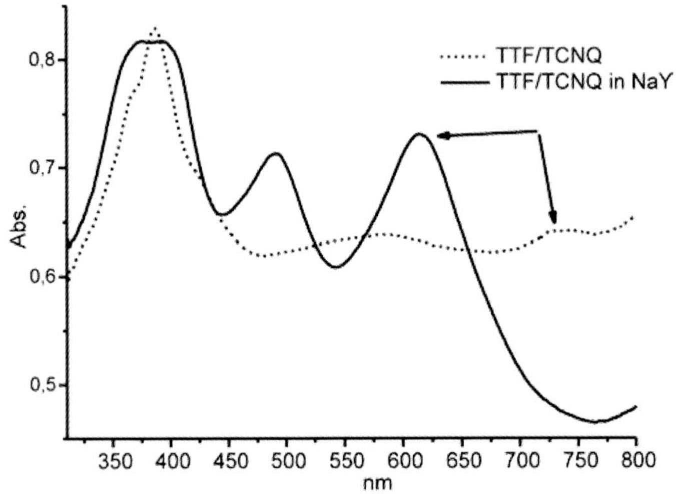

Fig. 11. UV/vis spectra of the charge-transfer complex of TCNQ and TTF.

TCNQ. This could be due to a kinetic effect, TTF penetrates faster into the pores, coordinating itself onto the sodium cation. Owing to the lower concentration of sodium ions at position II no more coordination sites for TCNQ are left and this guest is not adsorbed.

4.3.3 Naphthalene, Anthracene, 2,3-Benzanthracene, and Pentacene in NaY

In this section the effect of shape, size, and chemical properties of the guest molecules naphthalene, anthracene, 2,3-benzanthracene, and pentacene onto the adsorption sites in the zeolite faujasite are investigated.

These molecules are aromatic hydrocarbons with similar chemical properties but different sizes. The electron acceptor abilities correspond to the increase in molecular size. The sodium cation at position II plays again a decisive role in these investigated systems. Fitch et al. [24] investigated the adsorption of benzene in faujasite. They located this molecule in the supercage in front of the sodium ion at position II. Benzene is parallel to this window and therefore perpendicular to [111].

Tab. 1. Colours of the loaded hosts

	TCNQ	*TTF*	*TCNQ/TTF*
Pure substance	yellow	red	dark violet/blue
Incorporated in NaY	yellow	brown	blue
Incorporated in $H_{22}Na_{30}Al_{52}Si_{140}O_{384}$	yellow	brown	brown

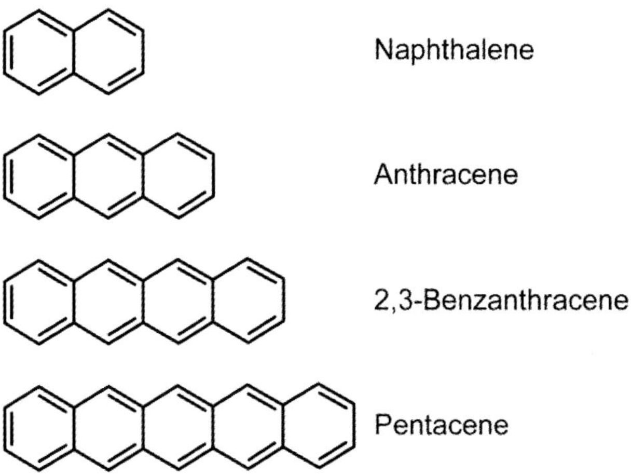

Fig. 12. Stick diagram of the guest molecules.

The localization is performed as described above. Figure 13 displays the view through a 12-ring window in the supercage onto the 6-ring window of the sodalite cage and the three neighboring 4-ring windows of the hexagonal prism. Above the window the white contour plot represents residual scattering density of naphthalene in NaY. The structure refinement confirms this position (Fig. 14). Naphthalene is in the supercage in front of the sodium cation at position II. Anthracene and 2,3-benzanthracene were also found at these positions (Figs. 15–17).

At first glance, all these adsorption positions are similar. One end of the molecule is coordinated to the sodium ion by its π-electron system while the other points into one of the neighboring three 12-ring windows. But a closer look reveals some differences.

Firstly, a shift of the sodium cation at position II compared with the dried and unloaded NaY towards the guest molecule along the [111] axis is observed, which increases in the order of naphthalene $<$ anthracene $<$ 2,3-benzanthracene from 0 up to 0.4 Å (Table 2) with electron donation abilities.

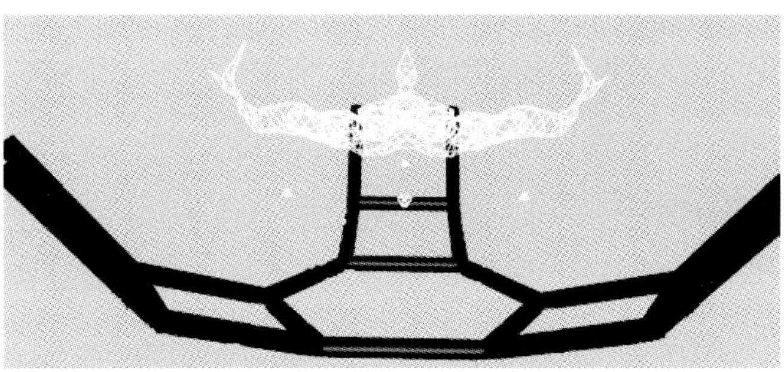

Fig. 13. Residual scattering density of naphthalene in NaY.

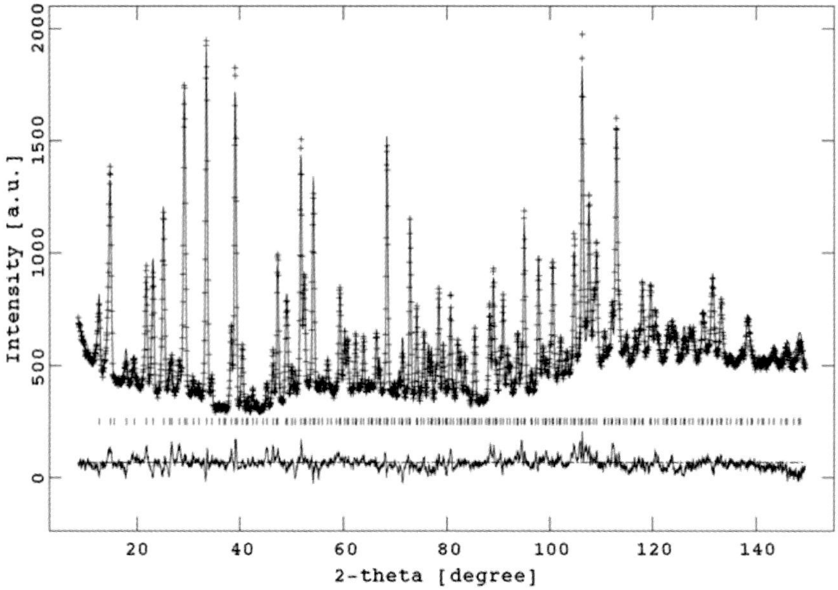

Fig. 14. Observed and calculated neutron diffraction patterns together with their difference curve of naphthalene in NaY, measurement performed at 5 K on D2B, Institute Laue–Langevin, Grenoble, France.

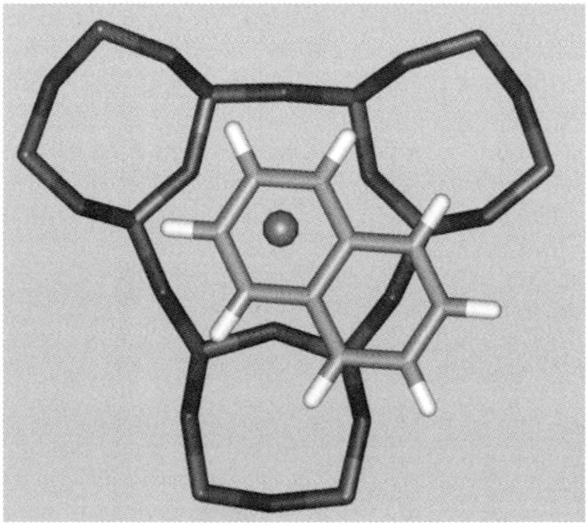

Fig. 15. Naphthalene in NaY.

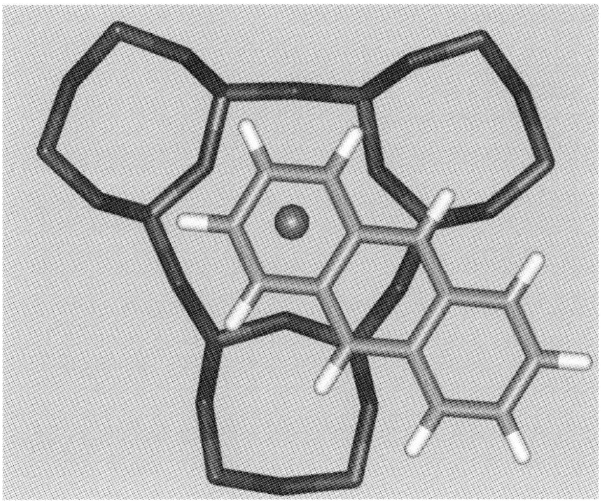

Fig. 16. Anthracene in NaY.

This shows the strong host–guest interactions between these two partners, which define the adsorption position of the molecules. Owing to the loading of one guest molecule per supercage and the fact, that the sodium position II is nearly fully occupied (four atoms per supercage), only about 25% of these cations are linked to one molecule. So the observed shift is an average of the sodium cations. It is not possible to distinguish these two sodium positions, due to small differences and the increasing numbers of parameters in the structure refinement.

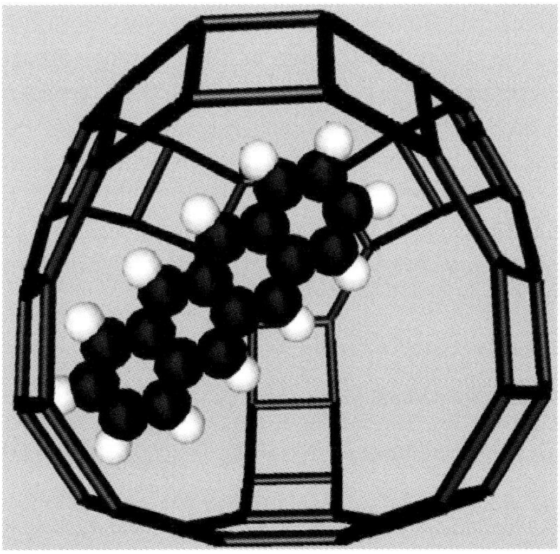

Fig. 17. 2,3-benzanthracene in NaY.

Tab. 2. Shift of the sodium cation at position II in loaded NaY.

	Sodium shift [Å]
Naphthalene	0.0
Anthracene	0.3
2,3-Benzanthracene	0.4

The distance between sodium and the guest molecule, especially of the six carbon atoms of the guest that are closest to the sodium cation at position II, is also of interest. These are the six carbon atoms of the outer six-ring of naphthalene and anthracene.

2,3-benzanthracene is shifted in such a way that the two bridge-building carbon atoms of the first and second fused six-rings of the molecule and the four neighboring carbon-atoms are coordinated to the sodium ion (Fig. 17). The average distance in all cases is about 3.2 Å, but the deviation increases with the size of the molecules as given in Table 3.

Thus the adsorption position differs with size from those in benzene or naphthalene, which might be seen as the ideal one. 2,3-benzanthracene is slightly shifted out of this position. The electron density is higher at the two inner six-rings of the molecule, the coordination with these is favored. The molecule would be too close however to the four-rings of the host lattice and repulsive forces may dominate.

The characteristics of the adsorption position of 2,3-benzanthracene supports the notion that pentacene is not incorporated into NaY because of its size.

The increasing shift of the sodium cation and the constant average distance of sodium to the carbon atoms of the guest leads to increasing distance between the guest molecule and the zeolite lattice with increasing size of the guest.

Two opposing trends determine the adsorption position in these systems. Firstly there are attractive host–guest interactions, especially the interactions between the π-electrons of the guest and the sodium cation, which explains the observed sodium shift. On the other hand are the repulsive interactions with the host lattice, which can be inferred from the differences of the adsorption position of 2,3-benzanthracene compared with naphthalene. 2,3-benzanthracene is, in contrast to pentacene, small enough to penetrate into the host pore system but too big to fit into the ideal adsorption position of benzene or naphthalene [25].

Tab. 3. Distance from the guest molecule to the sodium cation at position II in loaded NaY.

	Average distance [Å]	*Deviation [Å]*
Naphthalene	3.2	± 0.0
Anthracene	3.2	± 0.4
2,3-Benzanthracene	3.1	± 0.7

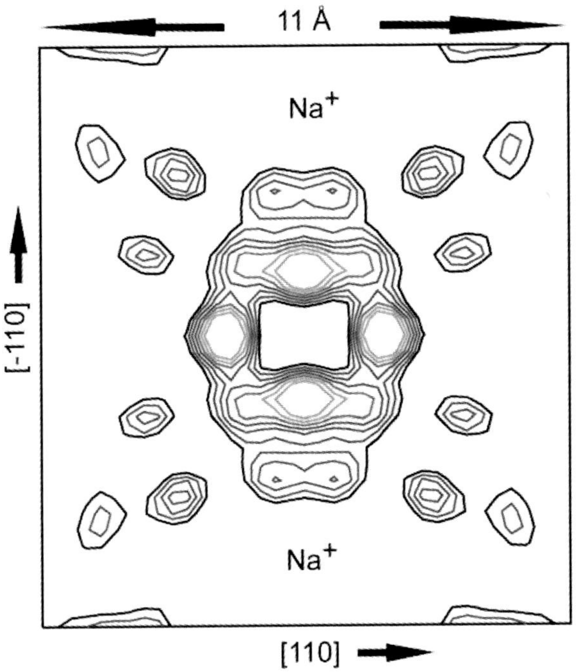

Fig. 18. Residual scattering density of chloranil in NaY.

4.3.4
Chloranil in NaY

Choranil (2,3,5,6-tetra-chloro-1,4-benzo-quinone) as an electron acceptor is incorporated into zeolite NaY with different molecules with electron donation abilities, which have been discussed above. The loaded host has a yellow color, when chloranil is incorporated. The measurements for the location of chloranil in NaY was performed at 5 K on D2B, Institute Laue–Langevin, Grenoble, France. The data evaluation shows a high concentration of residual scattering density among (0.375 0.375 0.250) between two cations at position II, displayed in Fig. 18. The imaginary line linking the cations Na^+ is also a mirror plane of the zeolite lattice.

The result of the crystal structure refinement is shown in Fig. 19. Chloranil is centered at (0.375 0.375 0.250), while the two keto-groups are coordinating to the sodium cations. The molecule is not exactly aligned in between the two cations, it is tilted slightly out of this position. By this the distance between oxygen and sodium increases up to 3 Å. Again the cation is shifted by 0.3 Å along the [111] axis, and not directly towards the guest molecule. Open force field simulations gave another, but very similar position (Fig. 20). Here chloranil is not between these cations but displaced to the side and aligned to the presumed line linking these cations. The distance between oxygen and sodium is 2.9 Å. So the open force field simulations failed in the way that the coordination of chloranil is a similar but a different adsorption site was found.

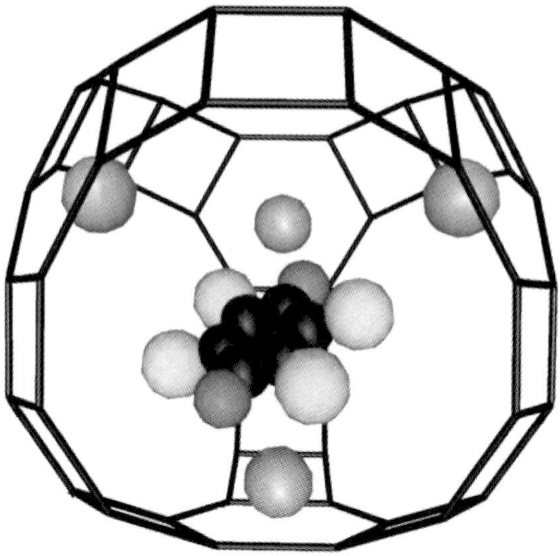

Fig. 19. Arrangement of chloranil found by crystal structure refinement.

Chloranil is combined with naphthalene, a weak electron donor, or TTF, a strong electron donator, and incorporated in NaY. The corresponding loaded zeolites have a pink color and dark brownish color respectively, which is a hint for charge-transfer. The UV/vis spectra show strong absorption bands in both cases of the loaded NaY. Spectra of a suspension of the pure solid charge-transfer complex in

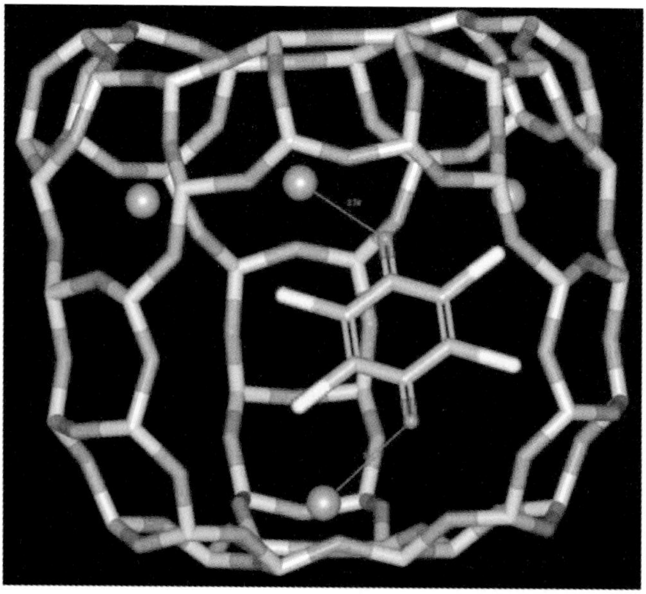

Fig. 20. Arrangement of chloranil found by open-force field simulations.

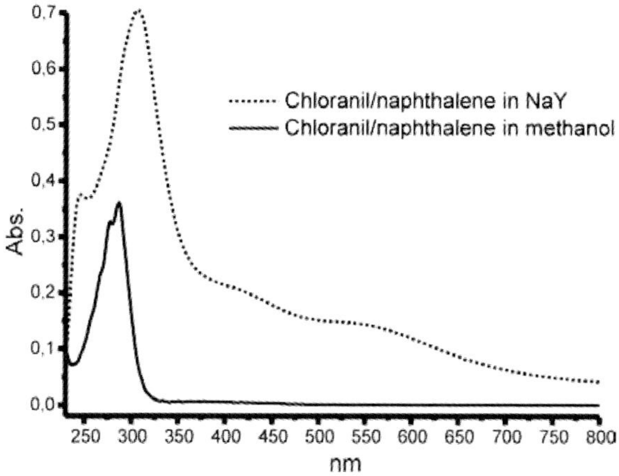

Fig. 21. UV/vis spectra of chloranil and naphthalene.

methanol are used as reference. As pure substances chloranil and naphthalene do not form such a complex, only at higher temperatures, when naphthalene is melted, a charged transfer band is observed. Incorporation in the pore system thus induces the formation of this complex, as shown in Fig. 21. It displays an absorption band at 550 nm. The spectra of chloranil and TTF in NaY show a strong broadening and bathochromic shift of the signal. This is the same effect as in the case of TTF in NaY, described above (Fig. 10).

4.4 Summary

It has been demonstrated, that guest molecules in zeolitic host frameworks can be localized by powder diffraction, Fourier analysis and Rietveld refinement. The

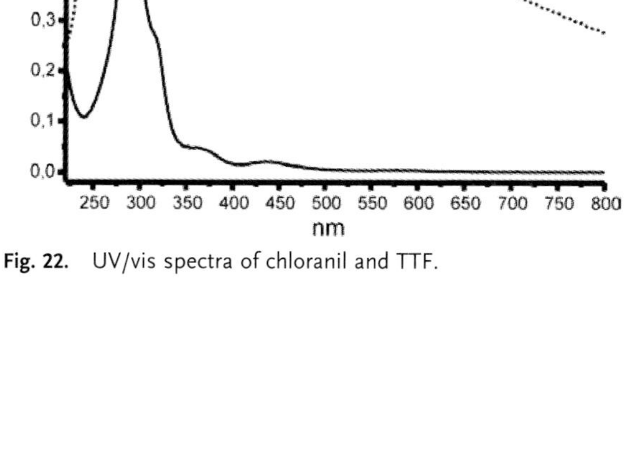

Fig. 22. UV/vis spectra of chloranil and TTF.

main parameters for adsorption are shape and size of an organic molecule. In addition to this, the charge distribution, the functional group and their conformation are important. These aspects are associated with the structure of the pore system of the host and the cation content and distribution therein. Molecules with more than one functional group can be anchored in the host lattice such as TCNQ or chloranil. All investigated and incorporated charge-transfer complexes show the same effect, the molecules are at identical positions as incorporated individually. So host–guest interactions are stronger than guest–guest interactions, but the UV/vis spectra give evidence for charge-transfer activity. So guest–guest interactions are still present. The observed shifts of sodium cations show that the host lattice is also influenced by the adsorption of guest molecules. These shifts give also hints on the strength of host–guest interactions. No super structure reflections are observed, so the complexes or guest molecules are not long-range ordered. The results of the UV/vis measurements can be compared with solution of the substances in a polar or protic solvent. The observed shift by stabilizing excited states of molecules can thus be explained, but the differences have to be mentioned too. A solvent forms one, two or more flexible coordination spheres with interchanging solvent molecules around the dissolved molecule. In contrast to that, the host lattice is rigid, the incorporated molecule is bond to the lattice at its adsorption site. Investigations of molecular dynamics show that such incorporated molecules can be very immobile compared with solutions [26]. Investigations of host–guest interactions will help to understand such complex and diverse systems and provide a basis for molecular simulations and for the development of new catalytic or novel functional materials.

Acknowledgements

We want to thank H. Vogel, Darmstadt University of Technology, Germany, for the support during the DRIFT measurements, H. Spange, University of Technology Chemnitz, Germany, for supporting and performing the UV/vis measurements, and H. Hewat, Institute Laue–Langevin, France, for his assistance during the neutron powder diffraction experiments. Support of this work by the Deutsche Forschungsgemeinschaft and the Fonds der Chemischen Industrie is gratefully acknowledged.

References

1 D. Wöhrle, G. Schulz-Eckloff, *Adv. Mater.* **1994**, *6*, 875.

2 U. Vietze, O. Krauß, F. Laeri, G. Ihlein, F.Schüth, B. Limburg, M. Abraham, *Phys. Rev. Lett.* **1998**, *81*, 4628.

3 M. Bockstette, D. Wöhrle, I. Braun, G. Schulz-Ekloff, *Zeolites* **1997**, *19*, 28.

4 J. Caro, M. Noack, P. Kölsch, R. Schäfer, *Micropor. Mesopor. Mater.* **2000**, *38*, 3–24.

5 W.J.W. Bakker, J. Poppe, F.Kapteijn, J.A. Moulijn, *J. Membr. Sci.* **1996**, *117*, 57.

6 J.L. Meinershagen, T. Bein, *J. Am. Chem. Soc.* **1999**, *121*, 448–449.

7 N.Y. Chen, T.F. Degnan, C.M. Smith, Molecular Transport and Reactions in Zeolites, VCH, New York **1994**.

8 J. Weitkamp, S. Ernst, H. Danns, E. Gallei, *Chem.-Ing.-Tech.* **1986**, *58*, 623.

9 H. Klein, C. Kirschhock, H. Fuess, *J. Phys. Chem.* **1994**, *98*, 12 345.

10 M. Czjzek, H. Fuess, T. Vogt, *J. Phys. Chem.* **1991**, *95*, 5255.

11 K.B. Yoon, T.J. Huh, J.K. Kochi, *J. Phys. Chem.* **1995**, *99*, 7042.

12 H. Basita, D.A. Bonn, T. Timusk, *Phys. Rev. B* **1980**, *42*, 4088.

13 K.B. Yoon, T.J. Huh, D.R. Corbin, J.K. Kochi, *J. Phys. Chem.* **1993**, *97*, 6492.

14 R.A. Young, The Rietveld Method, OUP, Oxford **1993**.

15 R.B. Von Dreele, A.C. Larson, GSAS – General Structure Analysis System, Los Alamos National Laboratory, **1994**, 167–169.

16 L.W. Finger, D.E. Cox, A.P. Jephcoat, *J. Appl. Cryst.* **1994**, *27*, 892.

17 C. Baehtz, H. Ehrenberg, H. Fuess, *Phys. Chem. Chem. Phys.* **2000**, *2*, 5764.

18 E. de Vos Burchart, PhD Thesis, Technical University, Delft **1992**.

19 H. Förster, H. Fuess, E. Geidel, B. Hunger, H. Jobic, C. Kirschhock, O. Klepel K. Krause, *Phys. Chem. Chem. Phys.* **1999**, *1*, 593.

20 K.B. Yoon, *Chem. Rev.* **1993**, *93*, 321.

21 C.R.A. Catlow, Modelling of Structure and Reactivity in Zeolites, Academic Press, New York **1992**, p. 217.

22 D.L. Coffen, J. Q. Chambers, D.R. Williams, P.E. Garrett, N.D. Canfield, *J. Am. Chem. Soc.* **1971**, *93*, 2258.

23 K.M.C. Davis, Molecular Association, R. Foster (ed.), Academic Press, London **1975**, p. 151.

24 A.N. Fitch, H. Jobic, A. Renouprez, *J. Phys. Chem.* **1986**, *90*, 1311.

25 C. Baehtz, H. Fuess, submitted.

26 H. Jobic, A.N. Fitch, J. Combet, *J. Phys. Chem. B*, **2000**, *104*, 8491.

5
Thionine in Zeolite NaY: Potential Energy Surface Analysis and the Identification of Adsorption Sites

*Marco Müller, Stefan M. Kast, Hans-Jürgen Bär, and Jürgen Brickmann**

5.1
Introduction

Phenothiazine and its derivatives play an important role as narcotics, depressants, insecticides, and dyes in medicine, biology, and chemistry [1,2]. Thionine (Fig. 1), the simplest derivative of phenothiazine, is in its cationic form an important dye molecule for the dyeing of paper, office furniture, and lacquers.

The search for new optical and electronic properties of guest–host systems inspired researchers to incorporate thionine molecules into certain zeolites through cation exchange [3–7]. Zeolites consist of a regular framework of corner-sharing SiO_4 and AlO_4 tetrahedra, which form well-defined pores and cavities. The exchange of silicon and aluminum atoms results in a negative net charge of the silicate framework, which is compensated by extraframework cations, which are located at specific positions.

Because the thionine molecule has van der Waals radii of about 7.2×15 Å, confinement effects in hosts such as zeolite L or faujasite-type zeolites are of importance in the study of these systems.

In 1994, Ehrl et al. [5] presented results of optical spectroscopic experiments of the system thionine in dehydrated zeolite NaY, which they compared to the slightly larger oxazine-4 and methylene blue guests in faujasite zeolites. In their study, they found that thionine in NaY was the only system to exhibit spectral hole burning in dehydrated zeolite, while oxazine-4 and methylene blue needed coadsorbed water.

Also, on gradually cooling the system from 200 to 2 K, a strong decrease in the integrated fluorescence intensity was observed, with the maximum of the spectrum shifting from 627 to 640 nm [5]. This spectral shift was found to be reversible and has been associated with two different forms (A and B) of thionine in dehydrated NaY that can convert thermally and optically into each other at low temperatures [5]. The ground state energies of these two forms were estimated to differ by $\Delta E = 2.0$ kJ mol^{-1}. The separating potential barrier was found to be even smaller, $\Delta V = 1.4$ kJ mol^{-1}. Ehrl et al. [5] conclude their paper with an optical term scheme that reflects their experimental findings and which is shown in Fig. 2.

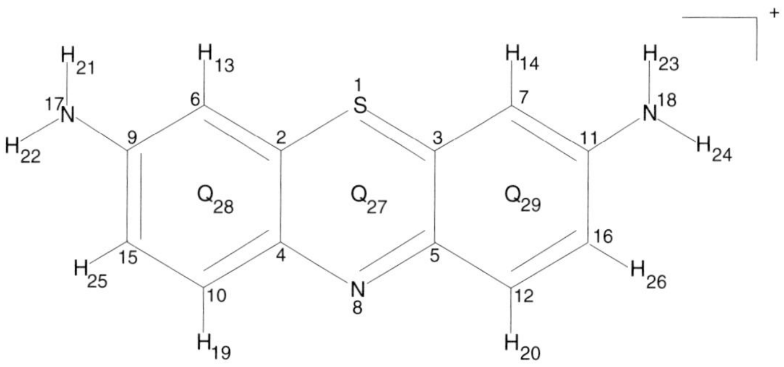

Fig. 1. The thionine cation with atom numbers as used in this text and additional point charges (ghost atoms) in the ring centers.

A detailed microscopic explanation in terms of structures of the interconvertible two forms A, B of thionine in NaY at low temperatures is still lacking. Such an explanation should be possible from knowledge of the adsorption sites of the guest molecule in the host lattice, as the adsorption sites are regions with the highest probability of occupation.

Experimentally, adsorption sites can be determined through X-ray diffraction or neutron scattering [8]. Unfortunately, such experiments have not yet been published for thionine in NaY to the best of our knowledge.

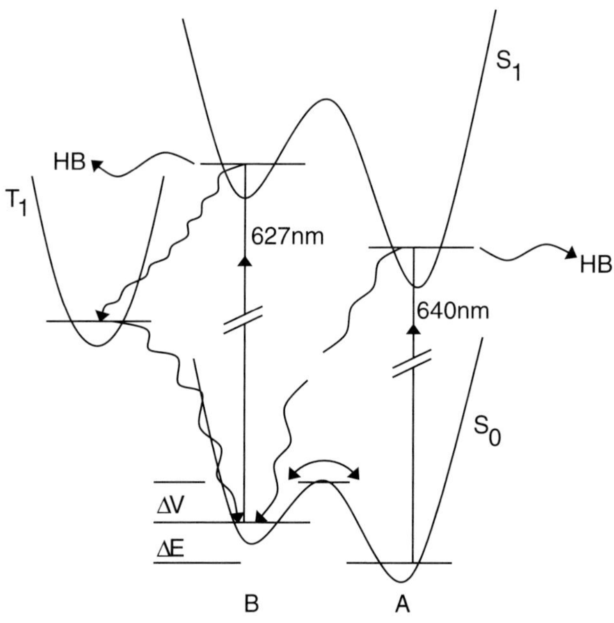

Fig. 2. Optical term scheme of thionine in NaY (after Ehrl et al. [5]).

A different approach to determine adsorption sites is to use computer simulation methods, such as molecular dynamics (MD) simulations based on empirical force fields or Car–Parrinello (CP) simulations. Each of these simulation methods has its own limitations. The results of MD simulations are limited by the potentially inadequate description of polarization effects by standard force fields. Typically, long-range electrostatic forces are the most important contribution to guest–host interactions [9], particularly if the host exhibits strong internal electrostatic fields. Although standard MD simulations with point charge models have been successfully applied in reproducing energetic and structural properties as well as relative free energies [10–12], the determination of properties such as local minima of charged molecules in a strongly polarizing environment cannot easily be treated within this framework. Diffusion coefficients and reaction pathways depend crucially on polarization effects. CP simulations on the other hand, are free of empirical parameters, but the time scale that may be investigated is still too short for medium to large guest–host systems to allow a reliable determination of system properties, although some progress has been made recently [13].

To overcome the limitations of standard force fields, we developed a methodology that is particularly suited for locating adsorption sites of charged guest molecules in a polarizing host environment [14]. The idea behind this technique is to refine an empirical guest–host potential energy surface (PES) to an *ab initio* PES at a number of local minima. These local minima are then analyzed in terms of similarity and occupancy to obtain the adsorption sites.

An important feature of this approach is the availability of the generated set of local minima as a basis for the calculation of thermodynamic properties of a system. To this end the partition function is computed within a discrete state approximation (DSA), by summation of the discrete Boltzmann factors of the minima. This approach has been used earlier in polymer physics where it is known as the rotational isomeric states (RIS) approximation [15].

In this chapter we present an overview of our approach for the localization of adsorption sites in a guest–host system and the application to the system thionine in zeolite NaY. For details of the strategy, the reader is referred to earlier work [14].

5.2 Methods

5.2.1 Determination of Local Minima

The algorithm for the determination of local minima is depicted in the scheme in Fig. 3. At the beginning of our algorithm a vacuum geometry optimization of the guest molecule thionine at the B3LYP/6-31 + G* level [16–23] was carried out (Table 1). A combined thionine/zeolite force field was constructed on the basis of a flexible zeolite model [24,10] and intermolecular parameters for the thionine molecule, which are tabulated in Table 2. The Lennard–Jones parameters were taken

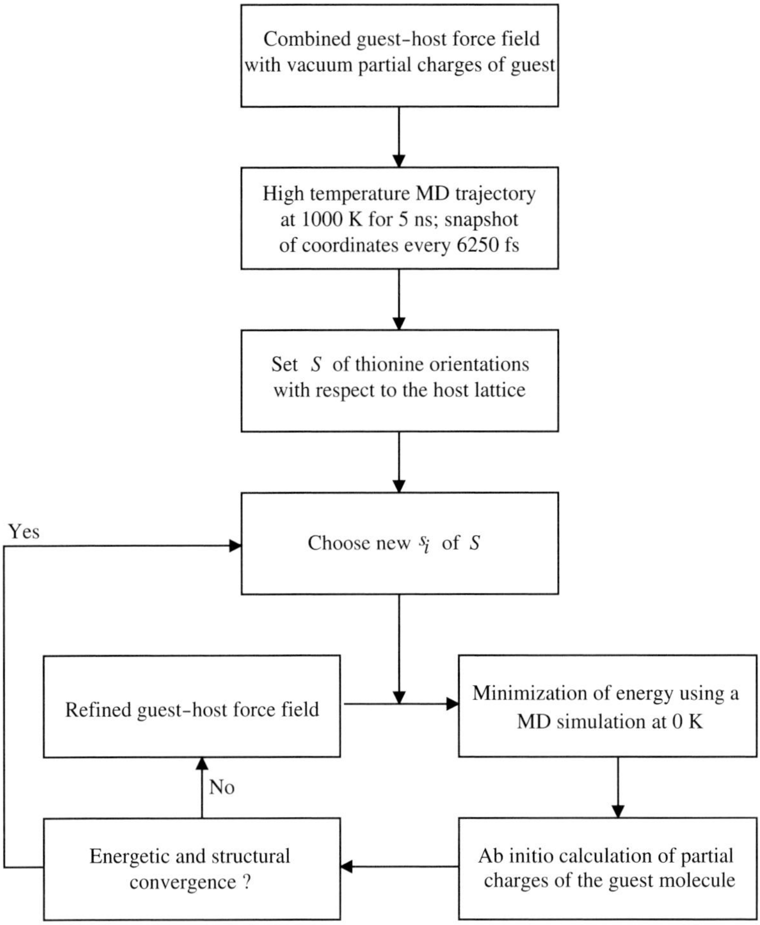

Fig. 3. Schematic presentation of our optimization procedure. Details can be found elsewhere [14].

from the literature [25,26], while the potential derived charges [27] were calculated for the thionine molecule in vacuum.

Throughout the simulations the thionine molecule was kept rigid [28,29]. This approximation is not necessary for the application of the algorithm but seems chemically justified based on the aromaticity of the thionine molecule. A high temperature MD run at 1000 K for 5 ns was used to take 800 snapshots of the spatial coordinates of the system at regular intervals of 6.25 ps. Each of these snapshots was quenched to 0 K before it entered an iterative cycle of *ab initio* calculations to adapt the partial charges of the thionine molecule to the zeolite surrounding and subsequent MD cooling. As a result, 800 local minima of the guest–host system with *ab initio* quality were obtained. For details of the procedure, the reader is referred to our earlier work [14].

Tab. 1. Cartesian coordinates of the vacuum geometry optimized thionine cation at the B3LYP/6-31 + G* level. Atom numbers are given in Fig. 1. The atoms denoted as Q_{27}, Q_{28} and Q_{29} are massless, so called ghost atoms that were added to the ring centers of the thionine molecule in order to allow for more flexibility during the ab initio charge fitting process.

Atom	*x [Å]*	*y [Å]*	*z [Å]*
S_1	0.000	0.000	0.000
C_2	0.000	0.000	1.747
C_3	1.700	0.000	−0.402
C_4	1.240	0.000	2.485
C_5	2.703	0.000	0.636
C_6	−1.220	0.000	2.404
C_7	2.059	0.000	−1.740
N_8	2.463	0.000	1.949
C_9	−1.269	0.000	3.816
C_{10}	1.154	0.000	3.916
C_{11}	3.422	0.000	−2.113
C_{12}	4.076	0.000	0.223
H_{13}	−2.149	0.000	1.838
H_{14}	1.294	0.000	−2.513
C_{15}	−0.047	0.000	4.563
C_{16}	4.430	0.000	−1.095
N_{17}	−2.452	0.000	4.466
N_{18}	3.782	0.000	−3.413
H_{19}	2.091	0.000	4.464
H_{20}	4.825	0.000	1.009
H_{21}	−3.333	0.000	3.970
H_{22}	−2.498	0.000	5.476
H_{23}	3.097	0.000	−4.157
H_{24}	4.755	0.000	−3.690
H_{25}	−0.089	0.000	5.649
H_{26}	5.477	0.000	−1.385
Q_{27}	1.351	0.000	1.069
Q_{28}	3.065	0.000	−0.748
Q_{29}	−0.024	0.000	3.155

5.2.2 Classification of Minima

The analysis of the local minima is based on a clustering process that groups similar local minima together. The comparison of local minima is made possible through the definition of a coordinate system using six geometric distances of thionine coordinates to fixed points in the zeolite supercage (Fig. 4). The chosen zeolite host reference points are the crystallographic supercage center and the two 12-T-window centers closest to the thionine center. The simple geometric distances shown in Fig. 4 allow an easy comparison of thionine orientations with respect to the zeolite host and a correct treatment of the tetrahedral symmetry of the zeolite supercage.

Tab. 2. Nonbonding force field parameters of thionine and the counterion chloride, which is used throughout the simulations to ensure electric neutrality. The charges given here are the vacuum potential derived charges used for the initial high temperature MD run

Atom types	q_i *[e]*	σ *[Å]*	ε *[kJ mol^{-1}]*
S_1	0.041	3.55	1.047
C_2, C_3	−0.119	3.55	0.293
C_4, C_5	0.266	3.55	0.293
C_6, C_7	−0.290	3.55	0.293
N_8	−0.477	3.25	0.711
C_9, C_{11}	0.406	3.55	0.293
C_{10}, C_{12}	−0.134	3.55	0.293
H_{13}, H_{14}	0.210	2.42	0.126
C_{15}, C_{16}	−0.263	3.55	0.293
N_{17}, N_{18}	−0.761	3.25	0.711
H_{19}, H_{20}	0.156	2.42	0.126
H_{21}, H_{23}	0.388	0.00	0.000
H_{22}, H_{24}	0.403	0.00	0.000
H_{25}, H_{26}	0.177	2.42	0.126
Q_{27}	0.254	0.00	0.000
Q_{28}, Q_{29}	0.153	0.00	0.000
Cl	−1.000	4.47	0.494

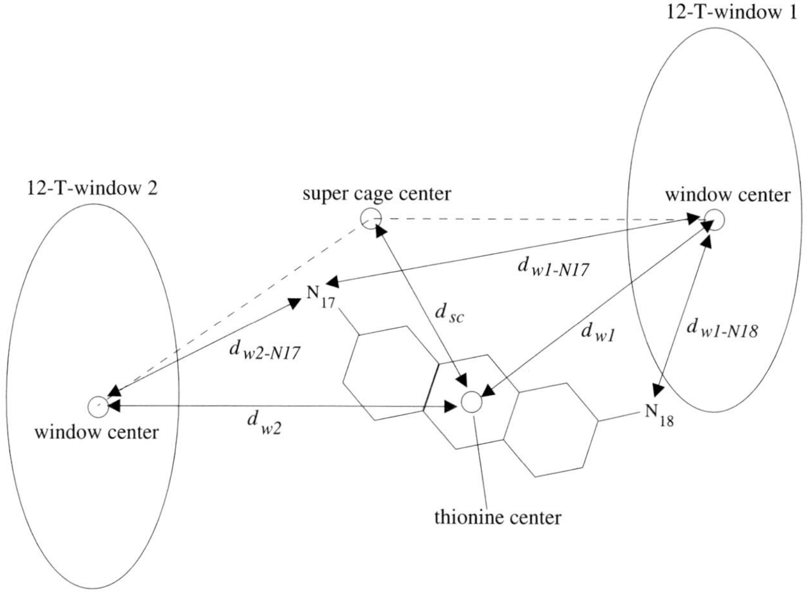

Fig. 4. Schematic representation of the six defined distances: d_{sc}, d_{w1}, d_{w2}, d_{w1-N17}, d_{w1-N18}, d_{w2-N17}.

An agglomerative clustering method was employed to group local minima together according to similarity in the chosen set of coordinates. For each pair of local minima a measure a_{ij} of similarity was calculated as an euclidean distance by

$$a_{ij} = \sqrt{\sum_{k=1}^{6} (\rho_k x_{ijk})^2} \tag{1}$$

where i, j are the index numbers of the local minima under consideration, the sum over k is the sum over the coordinates chosen to describe the local minima, ρ_k are weight functions and the x_{ijk} are given by $x_{ijk} = d_{ik} - d_{jk}$, where d_{ik} and d_{jk} are the coordinate values. The purpose of the weight function is to ensure that the measure of similarity is dimensionless. In our case the ρ_k are all set to 1/Å. Because we are interested in an agglomerative rather than a hierarchical clustering, it is necessary to choose empirically a threshold B that remains constant throughout the clustering process and that is used as a similarity criterion. If the distance of two local minima as defined in Eq. (1) is less than the threshold, these local minima are aggregated to one cluster. The distance between the newly formed cluster and one of the remaining local minima is taken as the maximum distance between the local minimum and the elements of the cluster (complete linkage). In this way clustering proceeds until no distance smaller than the threshold is found.

After the clustering is completed, the cluster energy is calculated by averaging over all elements of the cluster. Note that here and in the following discussion we are focusing on the total energy (the sum of potential and kinetic energy of the system) of a local minimum as determined by the last MD cooling with adapted thionine partial charges [14]. Consequently, the cluster energy is the averaged total energy of the local minimum under consideration.

5.2.3
Discrete State Approximation

The adsorption sites are determined by evaluating the dependence of occupancy of the clustered local minima on the temperature by calculating the Boltzmann factors according to

$$\frac{N_i}{N} = \frac{g_i e^{-E_i/RT}}{Z} \tag{2}$$

where the partition function Z is

$$Z = \sum_i g_i e^{-E_i/RT}, \tag{3}$$

g_i is the number of states found for a local minimum i, which for the current treatment equals the number of elements of cluster i, E_i is the corresponding average energy, T is the absolute temperature in Kelvin, R is the gas constant, and N_i/N gives the occupancy of each energy level i.

The DSA allows the computation of the partition function by making the assumption that knowledge about the local minima of a system (discrete states) of the potential energy surface and their population is sufficient for the treatment of its thermodynamics [15]. The DSA can be successfully applied in the case of near acicular local minima that are separated by potential energy barriers greater than RT. Alternatively, one may view the DSA in the sense of neglecting the entropic contributions of motion within a single minimum well compared to the entropy of flipping between accessible minima. For systems with highly localized structure, the approximation should work well.

From the partition function other thermodynamic functions are analytically accessible. In this chapter we present the heat capacity C_V

$$C_V = \frac{\sum_i \left(\frac{E_i^2}{R} g_i e^{-E_i/RT} \right)}{T^2 Z} - \frac{U^2}{RT^2} \tag{4}$$

where U is the internal energy

$$U = \frac{\sum_i (g_i E_i e^{-E_i/RT})}{Z}. \tag{5}$$

C_V is plotted as a function of temperature and the number of clusters entering the expressions.

Other thermodynamic functions such as the Helmholtz free energy A, the internal energy U, and the partition function Z are discussed in detail elsewhere [30].

5.3 Results and Discussion

5.3.1 Structural Properties

Using a threshold value of $B = 0.7$ the inital set of 800 minima was clustered into 93 groups of almost identical minima. The coordinates and energies of the twelve lowest energetic clusters together with the number of elements are listed in Table 3.

Of the 93 different local minima (clusters), only the energetically lowest three show significant occupancy even at as high a temperature as 750 K, as can be seen from the Boltzmann factors plotted in Fig. 5. Under typical experimental conditions, these three local minima (cluster 1, 2, and 3 in Table 3) can be identified as the adsorption sites, which we call in order of increasing energy α, β, and γ. In the following we describe the adsorption sites α and β in detail and present a comparative view of all three sites together. Coordinates for adsorption site γ and the twelve lowest energetic local minima are listed in Table 3.

Tab. 3. Lowest energetic clusters with their average total energies, rms values and the coordinates (see Fig. 4) that were used for clustering.

Nr.	Size of cluster	E_{total} [kJ mol^{-1}]	RMS [kJ mol^{-1}]	d_{sc} [Å]	d_{w1} [Å]	d_{w2} [Å]	$d_{w1\text{-}N17}$ [Å]	$d_{w1\text{-}N18}$ [Å]	$d_{w2\text{-}N17}$ [Å]
1	13	1.007	0.906	3.35	3.21	5.76	7.76	4.42	2.59
2	14	3.515	1.493	3.56	4.53	4.84	9.51	2.89	2.25
3	14	5.393	1.027	3.98	4.27	4.91	9.18	3.79	1.14
4	14	15.734	0.472	3.65	3.86	5.04	8.89	3.38	2.44
5	3	21.092	0.166	3.35	3.48	5.40	8.55	2.72	2.49
6	13	35.130	0.707	3.88	4.50	4.73	1.36	7.31	9.58
7	1	36.800	0.000	2.95	4.27	4.53	2.59	6.70	9.30
8	1	37.941	0.000	2.89	4.87	4.98	9.83	3.22	4.84
9	13	46.020	0.046	3.02	4.50	5.09	4.91	5.11	9.91
10	29	48.085	2.828	2.88	4.92	4.97	9.88	3.25	4.96
11	3	48.555	0.003	3.67	3.12	6.52	7.96	2.69	3.05
12	1	48.640	0.000	3.22	3.60	5.43	2.54	6.95	10.36

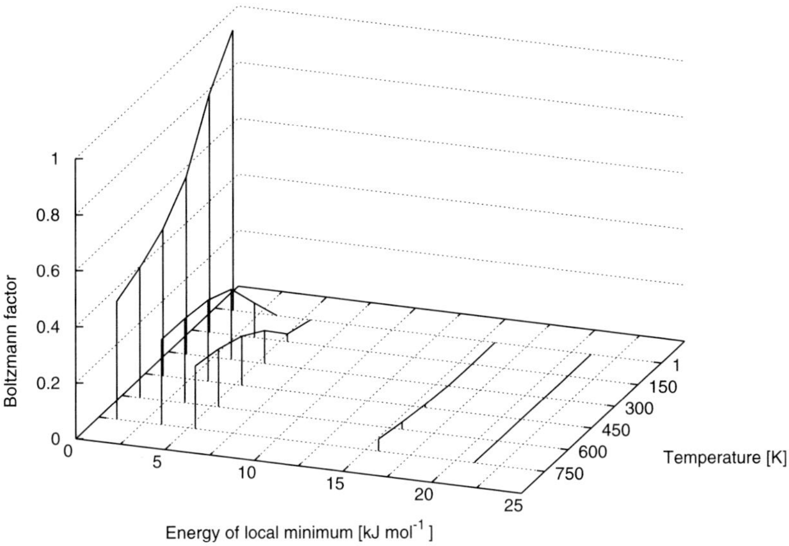

Fig. 5. Boltzmann factors for the energetically lowest five clustered local minima (Table 3). Minima with higher energy have Boltzmann factors close to zero for the given temperature range 0–750 K.

The energetically lowest adsorption site α features the smallest average distance of the thionine atoms to the supercage wall. The average distance of the thionine atoms to the lattice atoms increases from adsorption site α to γ. In Figs. 6 and 7 we show some of the distances of thionine atoms to neighboring atoms that are smaller than 3 Å as dashed lines. Note that during the initial high temperature MD simulation [14] the chloride counterion moved into the vicinity of the thionine molecule and remained there for all generated local minima. Occupying adsorption site α, the positively polarized hydrogens of the thionine molecule, notably H_{22}, H_{23}, H_{25}, can interact favorably with lattice oxygen atoms (distances of 1.8, 1.7, and 2.3 Å, respectively). For adsorption site β we find favorable interactions with lattice oxygens for thionine hydrogens H_{21} and H_{23} with distances of 1.9 and 2.3 Å respectively. The fundamental difference between adsorption sites α and β is the proximity of the thionine ring nitrogen (N_8) atom to the S_{II} sodium atom of the supercage (2.9 Å) in the case of adsorption site α, while for site β the sulfur atom (S_1) of the central thionine ring is the closest heteroatom to the sodium at the S_{II} site (2.4 Å).

The third adsorption site, γ, is very similar to site α with respect to the proximity of the central ring hetero atoms. The closest distance in this case is 2.5 Å for the distance N_8 to the S_{II} sodium atom. This change of coulomb interaction between the lattice sodium atom and the thionine molecule reflects the necessity of considering polarization, as our approach does implicitly.

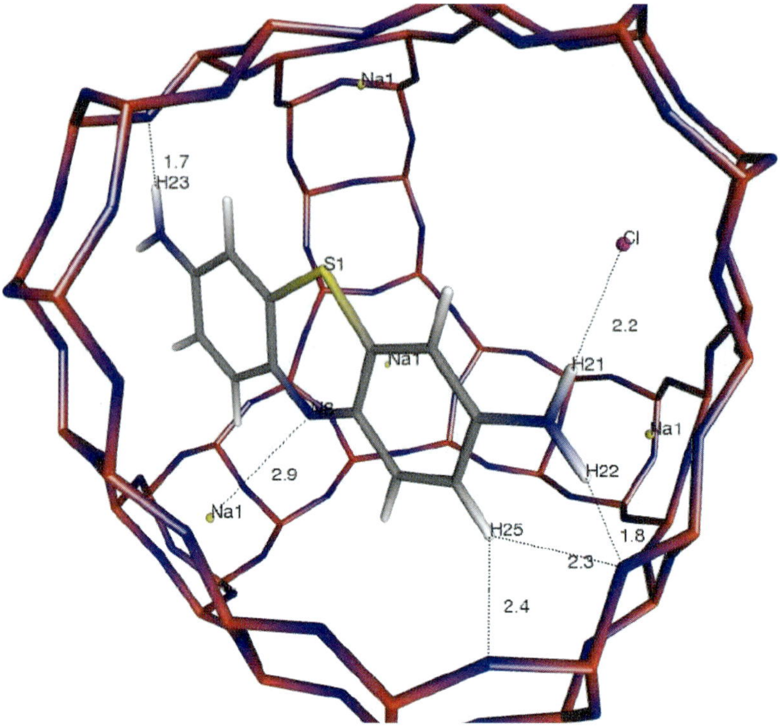

Fig. 6. Adsorption site α.

In Fig. 8 we present a color-coded picture of all three adsorption sites together to visualize the differences described above. Green is used for adsorption site α, yellow for β, and red for site γ.

5.3.2 Energetics

The energy values for the adsorption sites together with their rms values and occupancies at the experimentally relevant temperature of 175 K are summarized in Table 4. The energy differences between adsorption site α and β, and β and γ of about 2.5 kJ mol^{-1} and 1.9 kJ mol^{-1} are very similar to the experimentally found value of 2 kJ mol^{-1} for two different interconvertible forms of thionine in NaY at low temperatures [5]. Because of the higher occupancy values for α and β it is plausible to think of a thionine movement from adsorption site α to site β and back as the experimentally occurring process. This process would involve a rotation-like motion of the thionine and a polarization change of the thionine atoms such that the nearest interaction center to the sodium atom in S_{II} position is not longer the N_8 but the S_1 atom.

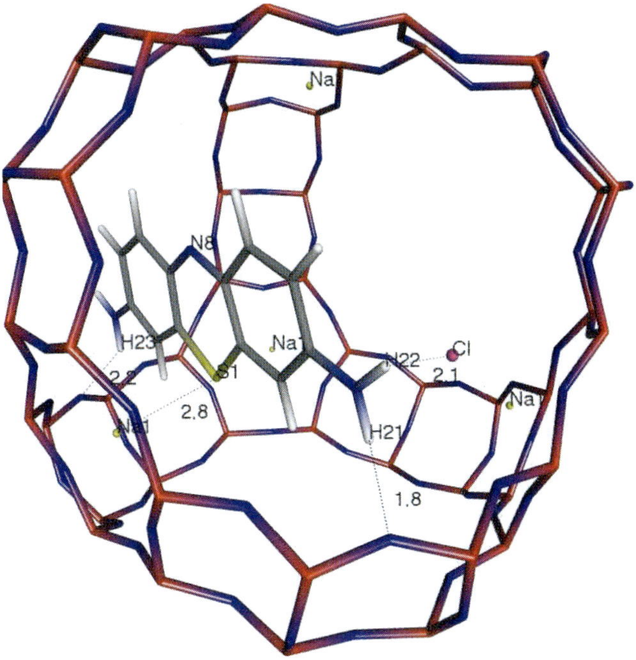

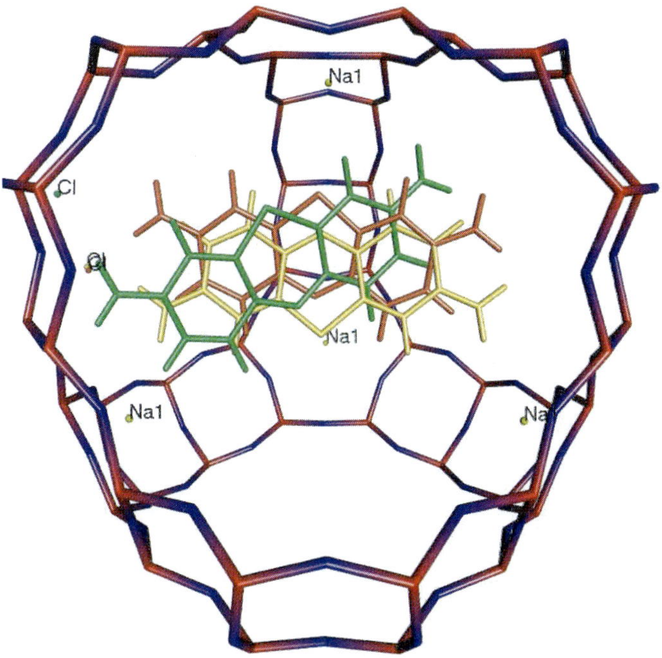

Fig. 8. Color-coded representation of the three adsorption sites (green = α, yellow = β, red = γ).

It is not easily possible to estimate the potential energy barrier for such a transition on the basis of our simulations. However, the required redistribution of partial charges in our thionine rotation model could provide an explanation for the experimentally found shift in the optical absorption spectra of the two thionine forms (from 627 to 640 nm upon cooling). Further clarification could be obtained by the computation of optical spectra on the basis of adsorption geometries here defined. This is the subject of ongoing work.

5.3.3 Thermodynamics

The question whether only three minima are sufficient for characterizing the thermodynamics of the guest–host system can be answered by analyzing the temperature-dependence of thermodynamic quantities as a function of the number of the minima taken into account.

Figure 9 shows the heat capacity C_V versus temperature T. The plot was calculated using the DSA according to Eq. 4. Figure 9 contains two curves that are the result of evaluating the partition function Z (Eq. 3) with a different number of local minima. The solid line is the result of using all 93 clusters for the calculation of Z. The dotted curve results when only local minima with an energy less than 10 kJ mol^{-1} or just the three adsorption sites were employed for the sum in Eq. 3. As can be seen from the plot, the use of only the three adsorption sites (local minima) for the construction of the partition function is at least for temperatures up to 200 K sufficient to reflect the behavior of the thermodynamic functions when all existing information about the PES (all local minima) is used. Therefore, the adsorption sites that we identified in this work are sufficient to represent relevant features of the PES in the low temperature (<200 K) range.

The heat capacity (Fig. 9) features a maximum at about 175–200 K, which can be explained as a typical saturation/frustration effect: relatively larger amounts of energy are needed to populate higher energetic states.

5.4 Summary and Conclusions

In this work we described an algorithm for locating adsorption sites in a guest–host system together with the results of its application to the system thionine in zeolite NaY.

Tab. 4. The energies of the adsorption sites, their corresponding rms values and occupancies at the experimentally relevant temperature of 175 K.

Adsorption site	*E [kJ mol^{-1}]*	*RMS [kJ mol^{-1}]*	*N_i/N (175 K) [%]*
α	1.0	0.9	80.3
β	3.5	1.5	15.4
γ	5.4	1.0	4.3

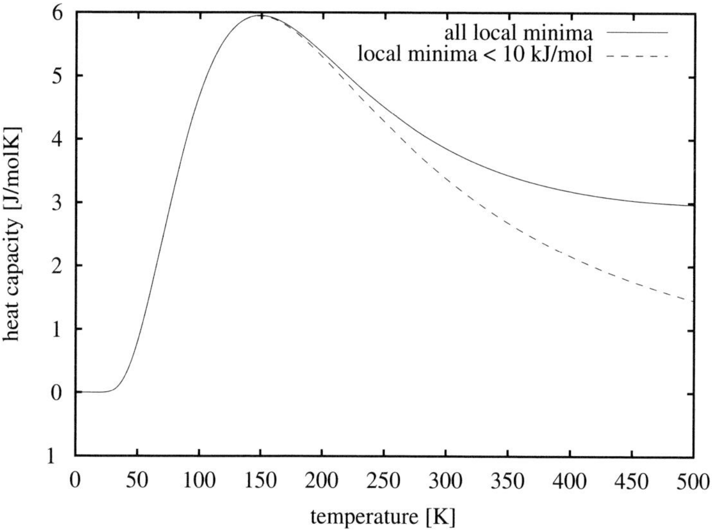

Fig. 9. Heat capacity C_V (Eq. 4) versus temperature T.

The primary result of this methodology is a set of 800 local minima of the *ab initio* PES. An agglomerative clustering approach was used to categorize these minima into 93 groups of similar or identical minima. The computation of Boltzmann factors for these 93 groups of local minima gave the occupancy of local minima at different temperatures, which in turn determine the adsorption sites. In the case of thionine in zeolite NaY, three dominant adsorption sites were identified. The energy difference together with the occupancy numbers of the energetically lowest two adsorption sites make these two plausible candidates for the experimentally observed two ground state conformations of thionine in dehydrated zeolite NaY at low temperatures.

The partition function was calculated on the basis of the set of generated local minima as an approximate picture of the PES (discrete state approximation). Two different sizes of the underlying set of local minima were used to emphasize the relevance of the identified adsorption sites in the low temperature range. From the partition function the heat capacity was obtained, which features a maximum at an experimentally relevant temperature. The results presented here provide the foundation for further work on the detailed microscopic insight into experimental features.

Acknowledgements

We would like to thank the Deutsche Forschungsgemeinschaft Bonn and the Fonds der Chemischen Industrie for financial support.

References

1 C.O. Okafor. *Dyes Pigm.* **1986**, *7*, 249.

2 R.R.E. Gupta. Phenothiazines and 1,4-Benzothiazines Chemical and Biomedical Aspects. Elsevier **1988**.

3 G. Calzaferri, N. Gfeller. *J. Phys. Chem.* **1992**, *96*, 3428.

4 V. Ramamurthy, D.R. Sanderson, D.F. Eaton. *J. Am. Chem. Soc.* **1993**, *115*, 10 438.

5 M. Ehrl, H.W. Kindervater, F.W. Deeg, C. Bräuchle, R. Hoppe. *J. Phys. Chem.* **1994**, *98*, 11 756.

6 X. Li, V. Ramamurthy. *J. Am. Chem. Soc.* **1996**, *118*, 10 666.

7 R.J. Robbins, V. Ramamurthy. *Chem. Commun.* **1997**, *1071–1072.*

8 R. Hoppe, G. Schulz-Ekloff, D. Wöhrle, C. Kirschhock, H. Fuess, L. Uytterhoeven, R. Schoonheydt. *Adv. Mater.* **1995**, *7*, 61.

9 B. Honig, A. Nicholls. *Science* **1995**, *268*, 1144.

10 G. Schrimpf, J. Brickmann. *J. Comput.-Aided Mater. Des.* **1995**, *2*, 49.

11 M.P. Allen, D.J. Tildesley. Computer Simulation of Liquids. Clarendon Press, Oxford **1987**.

12 R. Haberlandt, S. Fritzsche, G. Peinel, K. Heinzinger. Molekulardynamik. Vieweg, Braunschweig **1995**.

13 D. Marx. *Nachr. Chem. Tech. Lab.* **1999**, *47*, 187.

14 M. Müller, H.-J. Bär, S.M. Kast, J. Brickmann. *Chem. Phys. Lett.* **1999**, *311*, 485.

15 P.J. Flory. *Macromolecules* **1974**, *7*, 381.

16 A.D. Becke. *J. Chem. Phys.* **1993**, *98*, 5648.

17 C. Lee, W. Yang, R.G. Parr. *Phys. Rev. B* **1988**, *37*, 785.

18 B. Miehlich, A. Savin, H. Stoll, H. Preuss. *Chem. Phys. Lett.* **1989**, *157*, 200.

19 R. Ditchfield, W.J. Hehre, J.A. Pople. *J. Chem. Phys.* **1971**, *54*, 724.

20 W.J. Hehre, R. Ditchfield, J.A. Pople. *J. Chem. Phys.* **1972**, *56*, 2257.

21 P.C. Hariharan, J.A. Pople. *Mol. Phys.* **1974**, *27*, 209.

22 M.S. Gordon. *Chem. Phys. Lett.* **1980**, *76*, 163.

23 P.C. Hariharan, J.A. Pople. *Theo. Chim. Acta.* **1973**, *28*, 213.

24 G. Schrimpf, M. Schlenkrich, J. Brickmann. *J. Phys. Chem.* **1992**, *96*, 7404.

25 W.L. Jorgensen, T.B. Nguyen. *J. Comp. Chem.* **1993**, *14*, 195.

26 W.L. Jorgensen, E.R. Laird, T.B. Nguyen, J. Tirado-Rives. *J. Comp. Chem.* **1993**, *14*, 206.

27 C.M. Breneman, K.B. Wiberg. *J. Comp. Chem.* **1990**, *11*, 361.

28 G. Ciccotti, M. Ferrario, J.-P. Ryckaert. *Mol. Phys.* **1982**, *47*, 1253.

29 J.-P. Ryckaert. *Mol. Phys.* **1985**, *55*, 556.

30 M. Müller, S.M. Kast, H.-J. Bär, J. Brickmann. *PCCP* **2002**, *4*, 4212.

6
Density Functional Model Cluster Studies of Metal Cations, Atoms, Complexes, and Clusters in Zeolites

Notker Rösch, Georgi N. Vayssilov, and Konstantin M. Neyman*

6.1
Introduction

Subnano- and nano-sized pores and cavities of zeolites are not only of interest for applications as catalysts and sorbents but they can also be used profitably as hosts for metal complexes and small clusters. The properties of such encapsulated metal species commonly depend on their position in the zeolite cavities as well as on the structure and the aluminum content of the zeolite. The last decade has witnessed growing interest in the preparation and characterization of zeolite-hosted metal clusters and ionic complexes [1–3]. The first step in these studies was to clarify the location and properties of single metal cations with the help of X-ray diffraction, nuclear magnetic resonance (NMR), thermochemistry, and infrared (IR) spectroscopy of probe molecules [3–5]. Applications of the latter two methods are also connected with the structure determination of complexes that are formed by extra-framework metal cations with gas-phase molecules [4,5]. Effects of an isomorphic substitution of framework Al^{3+} cations by isovalent metal cations T^{3+} [6] have also been addressed [7]. A quantitative characterization of transition metal clusters of uniform size supported in zeolite pores is experimentally most challenging [1–3,8,9].

Here we review the computational modeling of metal moieties carried out by our group [10–24]; these theoretical efforts roughly followed the experimental studies. In Section 6.2 we describe simulations of the location of metal cations and related electron-deficient atomic species in zeolites [10–13] as well as their influence on properties such as the basicity of zeolite oxygen centers [14,15] or the acidity of neighboring hydroxyl groups [12,13]. We also report on adsorption complexes of these atomic metal moieties (cations) with probe or reagent molecules [10,11,16–20], such as carbon monoxide, dinitrogen, methane, and methanol. In Section 6.3 we discuss model complexes of small Pd, Pt, and Ir clusters with zeolite supports [22–24], focusing on the determination of their structure and charge.

The computational results outlined here have been obtained with the model cluster approach [25] using the method called linear combination of Gaussian-type orbitals fitting-functions density functional (LCGTO-FF-DF) [26]. Most of the calculations were carried out with the powerful parallel computer code ParaGauss [27,28], recently developed and implemented in our group.

6.2 Metal Cations in Zeolites

Since it is not always possible to identify the location of the cations and the structure of their complexes with probe molecules by experimental means alone, theoretical modeling can help substantially in clarifying these two issues as well as the number and types of the oxide support centers interacting with the cations [19,29]. Computational simulations make it possible to distinguish the most relevant structures among the variety suggested by spectral information.

6.2.1 Location of Cations

Zeolite frameworks consist of rings with different number of T-atoms (Si or Al atoms in tetrahedral coordination) connected by oxygen atoms. Owing to the presence of trivalent Al in tetrahedral positions instead of tetravalent silicon, the framework of a zeolite, crystalline alumosilicate, is negatively charged. This charge has to be compensated, by different cations located outside the framework for example, hence easily exchangeable; this allows a wide variation of zeolite properties. In H-forms of zeolites, the charge-compensating cations are protons that are attached to framework oxygen centers, thus forming bridging hydroxyl groups (Al–OH–Si). In alkali forms of zeolites, as well as zeolites exchanged with other metal cations, one expects to find the metal ions near Al positions of the lattice because of the local negative charge there. For this reason our models consist of zeolite fragments that contain one or more Al atoms. The first zeolite model of cationic sites that we applied was two negatively charged clusters of C_{3v} symmetry (Fig. 1) which consist of one Al atom, three OH or $OSiH_3$ groups, and one saturating H atom [10]. The alkali cations interact with the three oxygen centers of the side groups at distances defined by their ionic radii.

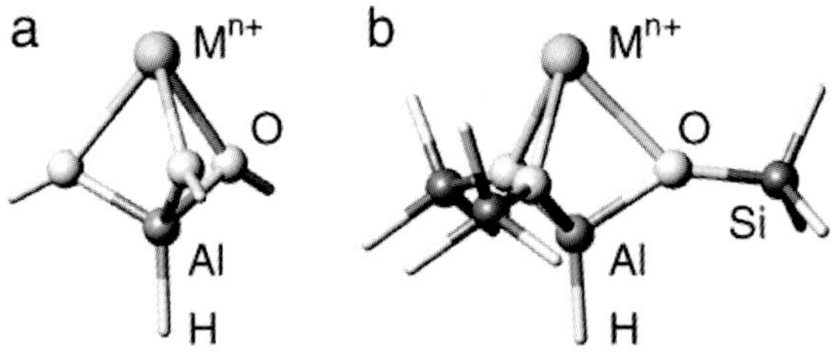

Fig. 1. Cluster models of zeolite sites with alkali ($n = 1$) or alkaline-earth ($n = 2$) cations M^{n+}: (a) standard $[AlH(OH)_3^-][M^{n+}]$, (b) extended $[AlH(OSiH_3)_3^-][M^{n+}]$.

6.2.1.1 Alkali Cations

In a typical zeolite structure, faujasite (FAU), the framework is constructed by four- and six-membered rings with the T-atoms of a ring lying in one plane each. Crystallographic studies suggest five different types of cationic sites in a FAU structure [3,5]. Sites SI, SI′, SII, and SII′ are situated at six-rings while in site SIII the cation is coordinated to a four-ring. However, only site SII from the first group of the cationic positions is accessible to guest molecules because it is located in the supercage, while the other three sites are blocked by the zeolite framework. Site SIII is also in the supercage, but it is occupied by cations only in zeolites with high Al content, that is, with a large concentration of alkali cations, as in X zeolites [30]. The local interaction of the cation with the four- and six-rings is similar in EMT zeolites [31].

We modeled the interaction of a Na^+ cation with six-rings of faujasites containing one (Al-1), two (Al-2) or three (Al-3) aluminum atoms [11]. In cluster Al-2, both possible relative positions of the two Al atoms in the ring as allowed by the Loewenstein rule [32] were considered: *para* (Al-2p) and *meta* (Al-2m). The optimized structures show that Na^+ prefers positions close to oxygen centers bonded to Al atoms rather than those of Si–O–Si bridges. A Na^+ cation interacts more strongly with the oxygen centers oriented inside the ring. The Na^+ cation in the cluster with one Al atom, Na-Al-1, is located almost in the plane of the ring, near the oxygen atoms of Al–O–Si bridges. For the other six-rings, Na-Al-2m, Na-Al-2p, and Na-Al-3, two possible positions of the Na^+ ion were found, one at each side of the ring. The positions are denoted by reference to O centers oriented inside the zeolite ring: *anti* or *syn* (Fig. 2a,b). The potential energy barrier in cluster Na-Al-2p for the

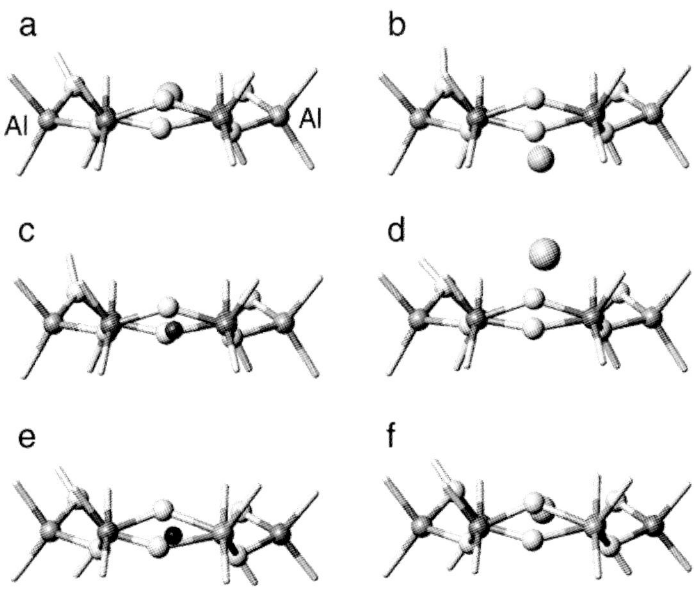

Fig. 2. Location of alkali and alkaline-earth cations at zeolite six-rings: (a) Na(*anti*)-Al-2p, (b) Na(*syn*)-Al-2p, (c) Li-Al-2p, (d) K-Al-2p, (e) Mg-Al-3, (f) Ca-Al-3.

transfer of a Na^+ cation between the two local minima, *anti* and *syn*, is only about 10 kJ mol^{-1}.

After comparison of calculated distances with crystallographic data and calculated vibrational frequencies of CO probe molecules with IR spectra of CO on NaY zeolite, we proposed that the crystallographic SII sites of faujasites actually comprise two cation positions with preference to *anti* position of Na^+ [11].

The positions of Li^+ at the Al-2p six-ring is close to the plane of T-atoms [12] with shortest Li–O distances of 190 pm. The specific location of Li^+ inside the zeolite ring allows the approach of one probe molecule only; the cation cannot be displaced enough so that it might be accessible to a second probe, as in the case of larger cations. This peculiarity of lithium, different from other alkali cations, was experimentally observed for CO and N_2 molecules [33,34].

Two local minima were also found for K^+ at the six-ring Al-2p [12], similar to the *syn* and *anti* positions of a Na^+ cation, but farther away from the plane of T-atoms, compared to Na^+. At variance with Na^+, both positions of K^+ were also observed at the six-ring Al-1 because the potassium cation is too large to move to the center of the ring. The shortest K–O distance of the *anti* position at the six-ring Al-2p is 270 pm.

6.2.1.2 Alkaline-Earth Cations

Two alkaline-earth cations, Mg^{2+} and Ca^{2+}, were studied at the Al-3 ring [12] (Fig. 2e,f). The alkaline-earth cations show trends similar to those of alkali cations of comparable ionic radius: Mg^{2+} is located similarly to Li^+, and Ca^{2+} similarly to Na^+. A magnesium cation is situated close to the plane of T-atoms, with a minimum Mg–O distance of 203 pm. The position of Ca^{2+} at the six-ring Al-3 is at 30 pm above the plane of T-atoms with the shortest Ca–O distance at 229 pm. Such locations of the cations allow simultaneous coordination of more than one probe molecule (CO or N_2). IR spectra suggest that even three CO molecules can be coordinated to a single calcium cation [35], probably because of its stronger electrostatic field compared to Na^+.

6.2.1.3 Rhodium Cation

Coordination of Rh(I) in faujasites was modeled at three types of cationic sites that are typical of these zeolites [19]: a four-ring (T4), a six-ring (T6), or a threefold-hollow position (T5) close to the Al center, modeled as a four-ring with an additional $OSiH_3$ group attached at the Al atom. The calculations show that at four- and six-rings, Rh(I) is bonded to the two oxygen centers of the ring connected to the Al center, as a consequence of their high basicity. At the four-ring, Rh(I) interacts with two of the oxygen centers that are on the same side of the ring, while at the six-ring the cation is exactly in the plane of the T-atoms since this is the only position that allows interaction with the oxygen centers of both Al–O–Si bridges. At the cluster T5, Rh(I) is located in a threefold-hollow position, bound to the oxygen center of the Al–O–Si bridge and the two closest oxygen centers of the four-ring [19].

6.2.2
Influence of Metal Cations on the Properties of Zeolites

The presence of metal cations instead of protons for charge compensation of the zeolite changes also the local properties of the framework. We investigated the influence of the cations on two of the important characteristics of the zeolites: the basicity of framework oxygen centers close to cations [14,15] and the Brønsted acidity of nearby bridging OH groups [12]. In passing, we mention another way of altering the Brønsted acid strength of zeolites, namely isomorphic substitution of *framework* Al^{3+} cations by isovalent ions such as Ga^{3+} and Fe^{3+}; such systems have also been investigated by us computationally [13].

6.2.2.1 Basicity

A direct absolute measure of zeolite basicity is the proton affinity (PA), that is, the energy gained via attachment of a proton to the basic oxygen site under study [14]. Our modeling had two main goals: to clarify the local influence of the Al content and the charge-compensating cations on the PA of certain framework oxygen centers; and to establish a connection between the PA and experimentally measurable quantities in order to construct a general scale of basicity of zeolite oxygen centers.

For the first goal, we calculated the PA of the oxygen centers in the Na^+ containing zeolite clusters described in Section 6.2.1.1. For the *anti* position of Na^+, the most basic oxygen atoms are those situated between two Al atoms in *meta* position and far from the charge-compensating cations of the six-rings Al-2m and Al-3 (Fig. 3); they have PA values of 903 and 886 kJ mol^{-1}, respectively. This location corresponds to oxygen centers between two Al atoms at next-nearest-neighbor T-atom positions of the zeolite. The PA values of the most basic oxygen centers of the other two Na^+ containing clusters, Na-Al-1 and Na-Al-2p, are lower by about 40–50 kJ mol^{-1}. The least basic oxygen centers are those of Si–O–Si bridges in the cluster Na-Al-1, with PA = 698 kJ mol^{-1}. The PA values of oxygen atoms of the six-ring Si-6, which contains only silicon T-atoms, is 771 kJ mol^{-1}; this value is close to the PA of methanol: 775 kJ mol^{-1}. Thus, the oxygen centers of Si-6 are sig-

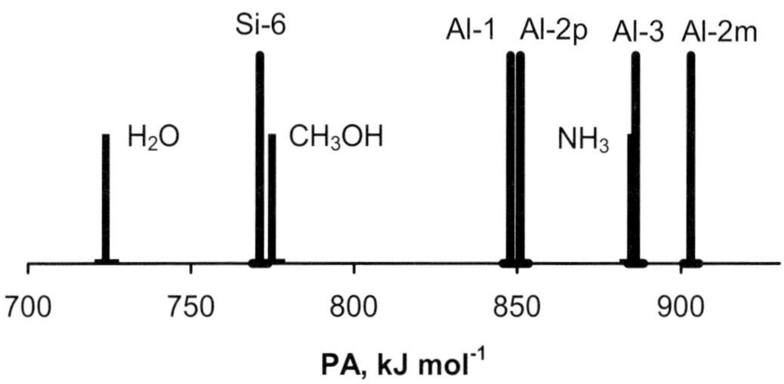

Fig. 3. Proton affinity (kJ mol^{-1}) of the most basic oxygen centers of zeolite model six-rings and of some reference molecules.

nificantly less basic than those of rings containing Al centers. The PA of ammonia, 885 kJ mol^{-1}, is somewhat smaller than the PA values of the strongest basic sites of zeolite rings, while the PA of water is calculated at 724 kJ mol^{-1}, close to that of the less basic zeolite oxygen atoms (Fig. 3). The proton affinities of oxygen atoms of the K^+ cluster model are 23–29 kJ mol^{-1} higher than those of the corresponding Na^+ model. This agrees with the experimentally observed higher basicity of K^+ exchanged zeolites compared with their Na^+ forms.

To correlate the basicity of zeolite oxygen centers with other measurable quantities we investigated the shifts $\Delta E_b(O1s)$ of the 1s core level of these oxygen centers (relative to a reference value as measured by XPS) [14] and the shift of the IR frequency of the OH group of a probe molecule that forms a hydrogen bond with the basic site [15].

The PA values of oxygen atoms in the studied six-rings and their calculated core level shifts $\Delta E_b(O1s)$ correlate very well [14]. From the slope of the high quality correlation line we conclude that a negative shift of the O1s binding energy by 1 eV (toward a less stable O1s level) implies an increase of the PA by 82 kJ mol^{-1}. Experimentally such a linear correlation has been observed for PA and $E_b(1s)$ values of a series of oxygen and nitrogen containing molecules in the gas phase [36]. The correlation of PA and ΔE_b values underlines the close connection between the PA and the energies of orbitals located at the oxygen center, in particular the HOMO (lone pair) of that center.

In addition, we demonstrated that the shifts of the stretching and the deformation O–H vibrational frequencies of methanol adsorbed on alkali-exchanged zeolites correlate with their calculated PA [17]; these shifts are particularly sensitive to the basicity of zeolite oxygen centers. To render this correlation directly applicable to experiment without the detour to calculations, we calibrated the calculated PA using experimental gas-phase PA values [36] of a series of small oxygen-containing organic molecules and water. Via these correlations, *experimental* PA values of zeolite oxygen centers can be derived from *measured* IR frequency shifts [37] of the O–H stretching and deformation bands of methanol adsorbed on alkali-exchanged zeolites [15]:

$$\mathrm{PA}(\nu)/\mathrm{kJ\ mol^{-1}} = 776 - 0.282\ \Delta\nu(\mathrm{OH})/\mathrm{cm^{-1}} \tag{1}$$

$$\mathrm{PA}(\delta)/\mathrm{kJ\ mol^{-1}} = 767 - 1.52\ \Delta\delta(\mathrm{OH})/\mathrm{cm^{-1}} \tag{2}$$

The frequency shifts vanish at PA $= 772 \pm 5$ kJ mol^{-1}. As a specific advantage, this approach allows a direct "experimental" determination of the PA value of a particular type of oxygen centers using two different criteria, Eqs. (1) and (2) thus offer an error check of the measured frequencies.

6.2.2.2 **Brønsted Acidity**

The other type of active centers in zeolites are the bridging OH groups; they act as Brønsted acidic centers in catalytic transformations. An interesting problem is how extra-framework metal cations located in the vicinity of such OH groups affect their properties. As a measure of the Brønsted acidity of bridging OH groups, we calculated their deprotonation energies (DE), namely the energy difference between the initial neutral cluster and the deprotonated cluster. The lower the DE of an OH group, the higher is its Brønsted acidity. The DE value of the cluster H-Al-2p may be taken as reference because it is the lowest among the clusters whose negative charges are compensated by protons only. The presence of an alkali or alkaline earth cation decreases the DE of a bridging OH group; thus, mixed clusters containing both protons and metal cations for charge compensation are more acidic than the purely protonic forms [12]. The lowest DE value was calculated for the bridging OH group of the cluster Ca-Al-3. It is 154 kJ mol^{-1} lower than the DE of the corresponding protonic cluster H-Al-3; the difference ΔDE relative to the reference cluster H-Al-2p is -82 kJ mol^{-1}. This suggests that zeolites containing Ca^{2+} and H^+ close to each other, such as HCaX, are the most acidic among the systems studied. In the order of decreasing acidity, they are followed by the zeolites HNaY and HNaX, represented by the model clusters Na-Al-2p and Na-Al-2m with ΔDE values from −50 to −56 kJ mol^{-1}.

The DE depends both on the effect of the charge-compensating cation on the initial state of the OH group and on the stabilization of the final anionic form of the cluster. The trend that longer OH bonds and lower OH frequencies correspond to lower DE values is rather poor. The DE values of bridging OH groups of the alkali-compensated clusters go through a minimum in the order Li, Na, K; ΔDE of Na-Al–2p is −56 kJ mol^{-1}, while for the other two alkali cations ΔDE is about −40 kJ mol^{-1}. Since the initial-state characteristics R(OH) and ν(OH) of the bridging OH group change with the ionic radius of the cation, the broken trend in ΔDE can be rationalized with the influence of the stabilization of the deprotonated form of these clusters. A stronger stabilization is expected for the deprotonated Na^+ cluster since the ionic radius of Na^+ allows it to interact stronger than the other two alkali cations with all oxygen centers of the zeolite cluster.

The lack of a high-quality correlation between DE on the one hand and the bond lengths R(OH) and the O–H vibrational frequencies ν(OH) on the other hand, suggests that these initial-state characteristics are not representative for the acidity of these hydroxyl groups in zeolites as evaluated by their DE. Similar indications have already been reported in other theoretical studies of zeolite acidity [38,39].

The Brønsted acidity can also be tuned via isomorphic substitution of a *framework* Al with another trivalent atom [6]. The acidic strength and related characteristics of bridging hydroxyl groups of Al-, Ga-, and Fe-framework-substituted zeolites have been studied [13] employing small cluster models $H_3Si(OH)TH_3$. The acidity has been quantified based on a series of pertinent calculated observables: the adsorption energy of a CO probe molecule as well as shifts of the vibrational frequencies and absolute IR intensities of the O–H and C–O modes induced by CO adsorption. The Brønsted acidity was found to decrease in the order Al(OH)-Si > Ga(OH)Si > Fe(OH)Si in agreement with experimental results [40]. It has been demonstrated by us that the O–H frequency shift as a result of CO probe adsorption represents a sensitive observable to differentiate reliably and conveniently between the strength of Brønsted acid sites of zeolites with various framework trivalent metal atoms. On the other hand, the vibrational frequency of

O–H groups without a probe is significantly affected structurally by the angle Si–O–T of the Si(OH)T units and thus cannot be used as indicator of the acidity strength [38].

6.2.3 Interaction of Guest Molecules with Cations

IR spectroscopy of adsorbed probe molecules is a powerful technique for elucidating the location and properties of metal cations in oxides and zeolites. Usually weakly interacting diatomic molecules are employed as probes: carbon monoxide, molecular nitrogen, or hydrogen [4,5,33–35,41]. In an early combined computational and FTIR study, we demonstrated that N_2 molecules can be applied as a probe of Brønsted acidic centers [21]. Homonuclear molecules as probe (e.g., N_2, H_2) are IR active only when adsorbed; this is a distinct advantage over the most common probe CO since the spectra do not need to be corrected for the absorbance of gas phase molecules.

6.2.3.1 Carbon Monoxide

Earlier calculations of a CO probe interacting with *bare* alkali cations qualitatively reproduced the experimentally observed blue shifts of the C–O vibrational frequency in a series of metal cations in zeolites, but notably overestimated the measured values [42]. To improve the accuracy, we designed a more sophisticated, yet simple model of cationic zeolite sites, $AlH(OH)_3^-M^+$ (Fig. 1), containing one Al atom for embedding the cations [10]. Our DF calculations for this type of cluster embedding yielded good agreement with the experimental frequency shifts of CO adsorbed on alkali cations (Li^+ to Cs^+) in zeolites. The computed C–O frequency shift varies from 45 to 21 cm^{-1}, respectively [10], close to the interval of 45 to 12 cm^{-1} measured for alkali exchanged MFI (i.e., ZSM-5) and MOR (mordenite) zeolites at low cation loading [43,44]. We also made an attempt to assign weak puzzling features in the red-shifted area of the IR spectra of CO adsorbed on alkali-exchanged zeolites (only very faint for Li^+ and Na^+ samples) [10]. Exploring an earlier hypothesis, we indeed obtained negative C–O frequency shifts, −17 to −23 cm^{-1}, for CO bound via its O atom to the alkali cations M^+ of the model cluster; these shifts are close to the experimental red shifts of −19 to −23 cm^{-1} [43,44]. Very small free enthalpy changes computed for the reaction $AlH(OH)_3^-M^+/CO \leftrightarrow AlH(OH)_3^-M^+/OC$ at 77 K ($\Delta G^o(77\ K) \leq 2$ kJ mol^{-1} for M = K, Rb, and Cs) are indicative of a small probability (at least 4–9 %) of finding O-bound CO species in zeolites. In combination with other arguments (e.g., additional heating via IR irradiation, notably higher intrinsic IR intensity of the C–O mode for O-bound species than for C-ones) this finding strongly suggests that a sufficient amount of *nonclassically* O-bound CO molecules can be formed under these experimental conditions to allow the manifestation of these species in the IR spectra. The computational result on the increase of the relative number of the O-bound CO molecules in the equilibrium with the radius of the cation nicely fits the experimental IR findings.

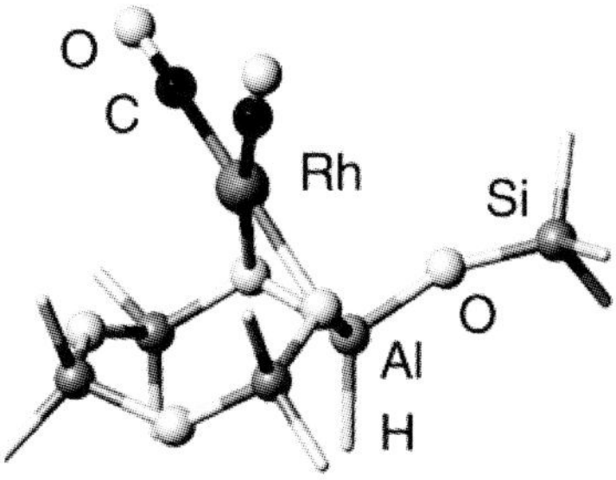

Fig. 4. Structure of the rhodium dicarbonyl complex at a zeolite model cluster.

Considerably stronger adsorption of CO occurs on Rh(I) cations in zeolites: the binding energy (BE) per CO in mono- and dicarbonyls is 215 and 211 kJ mol^{-1}, respectively [19]. Experimental studies of supported rhodium complexes clearly show that the Rh(I)$(CO)_2$ dicarbonyl complex is a very stable surface species [19,45,46] while monocarbonyls can only be formed after special treatment of dicarbonyls [46]. For this reason, we focused on dicarbonyl complexes. Stable structures of the complex Rh(I)$(CO)_2$ were found at the four-ring model clusters T4 and T5 (Section 6.2.1.3), but not at the six-ring Al-1 [19]. In the optimized structure of the complex Rh(I)$(CO)_2$ at the four-rings T4 and T5, both CO groups are oriented in continuation of the O(z)-Rh bonds (Fig. 4). The whole complex including the Rh(I) cation, both zeolite oxygen centers bound to it and ligand CO molecules, exhibits an almost planar, four-ligand pseudo-square coordination, similar to the classical inorganic or organometallic complexes of Rh(I). Calculated vibrational frequencies of the C–O stretching mode of the Rh(I)$(CO)_2$ complex adsorbed on the model four-ring differ by only 1–11 cm^{-1} from the reported IR features [19,45,46]. Our results allowed us to identify the proper structure of the Rh(I)$(CO)_2$ complex supported in DAY zeolite [19] among three possible structural models that emerged from the extended X-ray absorption fine structure (EXAFS) analysis. Close agreement between experimental parameters (IR and EXAFS) and the corresponding quantities calculated for the optimized structure of the complex Rh(I)$(CO)_2$ on a zeolite four-ring corroborated the structure assigned to the system Rh(I)$(CO)_2$/DAY.

Zeolites provide excellent opportunity for producing metal moieties of very small size [1,3]. From this viewpoint, atomic metal species can be considered as a limiting case. In general, metal atoms should not necessarily play a role as charge-compensating cations. Rather, they can be almost neutral, electron-deficient or even electron-enriched, such as small transition metal clusters [1,3,9], depending on their location and interactions in zeolite cavities. The formation of stable monoatomic Pt species in Pt/HMOR has been substantiated by results of an FTIR study of adsorbed CO [47]. An IR adsorption band at rather high wavenumbers, 2123 cm^{-1}, was found not to shift with CO coverage change and to disappear when zeolitic protons are neutralized. It was interpreted as manifestation of CO adsorbed on electron-deficient Pt atoms anchored to zeolitic protons. In our DF study [20] a

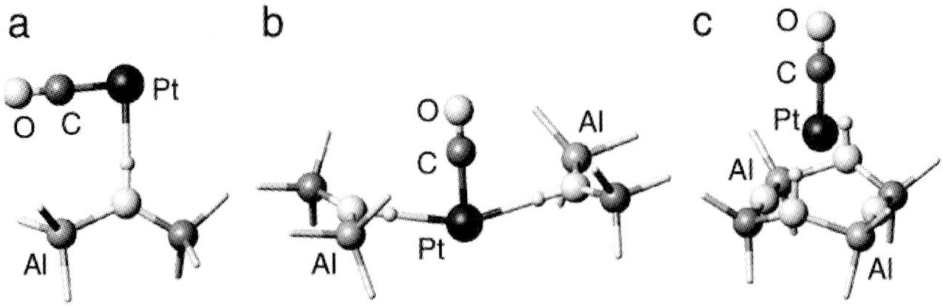

Fig. 5. Cluster models of a PtCO moiety interacting with a zeolite framework. Zeolite model clusters with (a) one, (b) two separated, and (c) two adjacent acidic bridging hydroxyl groups of a four-ring.

series of models of PtCO species anchored to one or two bridging acidic OH groups (Fig. 5) has been considered to rationalize these experimental findings [47] and to better understand interactions of highly dispersed transition metal species with Brønsted acidic sites of zeolites. It was found that anchoring a PtCO moiety by thc protons of acidic hydroxyl groups increases the C O frequency compared to that of free PtCO, but leaves it still smaller than for a CO molecule, in line with experiment. The results for various molecular model complexes support the hypothesis [47] that the atomic Pt species in MOR are electron deficient. An alternative model comprising bare protons attached to PtCO can be ruled out since the calculated CO vibrational frequency is too large. These model results also permitted a discussion of how the C–O vibrational frequency depends on the acidity of the Brønsted groups and on the electronic charge of the Pt species [20].

6.2.3.2 **Nitrogen Molecule**

A computational study of N_2 adsorption on model zeolite clusters supported the proposed observation of the simultaneous adsorption of two N_2 molecules at one alkali cation [34]. We studied the adsorption of one or two probe molecules both on bare alkali cations and on cations embedded in a zeolite fragment [18]. In all cases a linear adsorption of the probe N_2 was found to be the most stable; no stable configuration was obtained for probe molecules adsorbed in side-on fashion to cations at the zeolite models. In linear complexes, the stretching mode ν(N–N) is blue-shifted by 10–25 cm^{-1} compared to the value calculated for N_2 in the gas phase. Frequency shifts $\Delta\nu$(N–N) and intensities I(N–N) of the stretching mode were found to correlate in linear fashion along the series of alkali cations. The blue-shift of the N–N stretching band for monomolecular adsorption is about 21 and 15 cm^{-1} for the clusters Na-Al-2p and K-Al-2p, respectively. The mean blue-shift of the two N–N bands for bimolecular adsorption at the Na^+ cluster, 17 cm^{-1}, is about 4 cm^{-1} smaller than the shift for monomolecular adsorption, while for the K^+ cluster the average shift of bimolecular adsorption is essentially the same as for monomolecular adsorption. This is in line with experimental observations in which the N–N frequency decreases by 2.0–3.5 cm^{-1} upon adsorption of a second probe molecule at a Na^+ center in zeolites NaY, NaEMT, and NaETS, while the frequency of a bis-dinitrogen complex on K^+ containing zeolites is only 0.5 cm^{-1} lower than for the corresponding monomolecular complex [33,34,48].

6.2.3.3 Methane

Methane has been proposed as an alternative probe for cationic forms of zeolites [49,50]; its low adsorption affinity and similarity to other hydrocarbon molecules involved in catalytic reactions are considered as special advantages. We have studied adsorption complexes of CH_4 with cationic sites of alkali (Li, Na, K, Rb, Cs) and alkaline-earth (Mg, Ca, Sr, Ba) forms of zeolites [16]. The goal of the investigation was to rationalize structure and bonding of methane complexes with metal forms of high-silica zeolites as well as their measured vibrational features. We aimed also at clarifying the origin of opposite trends observed in the adsorption-induced alteration of the vibrational frequency and the IR intensity of the symmetric stretching mode ν_1 of CH_4 for cation-exchanged zeolites with different metal loading: for ion-exchanged high-silica zeolites (MOR, MFI = ZSM-5) the radii of alkali [49] and alkaline-earth [50] metal cations (hence their polarizing power) were found to correlate with the vibrational parameters just mentioned while in Na-Y and Cs-Y zeolites with higher metal loading the frequency shift and IR intensity of adsorbed CH_4 were found to be independent of the polarizing power of the metal cation [51].

Our calculations [16] suggest that both on alkali- and alkaline-earth cations in zeolites CH_4 is bound in a three-fold configuration M-H_3CH. Other conceivable structures, two-fold M-H_2CH_2 and one-fold M-HCH_3, are only slightly destabilized transition states. As most important adsorption-induced effects, one notes the activation of the ν_1 vibrational mode, IR forbidden in free CH_4, and a red-shift of ν_1 and other bands that decreases from Li^+ to Cs^+ and from Mg^{2+} to Ba^{2+}. These trends, consistent with experiments for CH_4 in metal-exchanged high-silica zeolites [49,50], have been rationalized in terms of the electrostatic perturbation of the adsorbate as measured by the size and the charge changes of the cations. The calculated trends remain unchanged when the cluster models are extended to include part of zeolite framework. Thus, the observations that both the frequency red-shift and the IR intensity of the ν_1 band of adsorbed methane are larger in Cs-Y than in Na-Y zeolites [51] (one expects opposite trends) can hardly be rationalized by models in which methane interacts exclusively with cationic sites. In response to this, we invoked a simple bifunctional adsorption model $H_2O/Cs^+/CH_4$ [16] in which CH_4 is able to interact *simultaneously* with a cation and an O center representing a framework oxygen of a zeolite. This model yields notably enhanced values for the frequency shift $\Delta\nu_1$ and the IR intensity compared to Cs^+/CH_4, improving the agreement with experiment [51]. The result suggests that the experimental findings for methane in M-Y zeolites can be explained by invoking bifunctional coordination of CH_4 to both a Lewis acidic site M^+ and a basic O site of the zeolite framework.

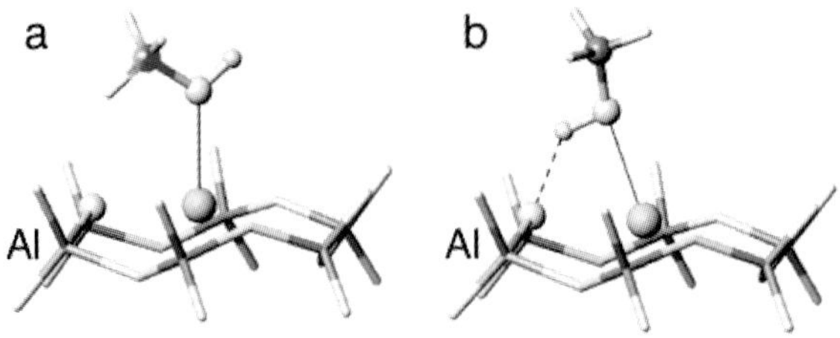

Fig. 6. Interaction of methanol with the zeolite model cluster Na-Al-1: (a) O-bound species, (b) O,H-bound species.

6.2.3.4 Methanol

Methanol is a convenient probe molecule for materials containing both acidic and basic sites in close proximity since it can form a coordination bond with the oxygen atom of its OH group and an acidic surface center; and a hydrogen bond of the proton from the OH group to a basic surface site. Owing to the complexity of the interaction of methanol with cation exchanged zeolites, IR spectra of adsorbed methanol change with the nature of the alkali cation and the Al content of the zeolite (i.e., with the basicity of the material) [37]. To clarify the adsorption modes of methanol on alkali zeolites we modeled its adsorption on zeolite clusters with O atoms of different basicity [17].

Two types of methanol adsorption complexes on zeolite cluster models were considered: an O-bound molecule coordinating only via the methanol oxygen center to an alkali cation (Fig. 6a); and O,H-bound species forming in addition a hydrogen bond between the methanol OH group and a zeolite oxygen center (Fig. 6b). The coordination bond between the methanol oxygen atom and the metal cation furnishes the main contribution to the interaction with Na^+ or K^+ exchanged zeolites. For species bound by this coordination bond only, the BE is calculated at 41–56 kJ mol^{-1}, while the hydrogen bond contributes another 12–18 kJ mol^{-1}. The BE increases with the Lewis acidity of the cation. On the other hand, the contribution of the hydrogen bond to the BE of the molecule increases with the PA of the basic oxygen center of the zeolite cluster. Following these observations, one would expect that methanol will adsorb at available Na^+ or K^+ cations in the zeolites and that it will form a hydrogen bond with a nearby basic zeolite oxygen center. The dominant role of the coordination bond to the cation influences also the reactivity of adsorbed methanol species on Na^+ exchanged zeolites. This bond activates methanol by weakening the CO bond, but it stabilizes the CH bonds of the methyl group. Thus, Na^+ exchanged zeolites act similarly to weak acidic catalysts rather than basic ones. The decrease of the Lewis acidity of the cation and the simultaneous increase of the basicity of the zeolite oxygen centers when going from Li^+ to Cs^+ exchanged zeolites changes the relative contributions of the coordination bond and the hydrogen bond to the BE of methanol on such zeolites. In this way, the catalytic activity of zeolites exchanged with alkali cations varies with the atomic number of the cation.

Although the formation of a hydrogen bond between the methanol hydroxyl group and a zeolite oxygen center yields only a moderate contribution to the BE, it significantly influences the IR spectra of adsorbed methanol via the OH stretching frequency. The red shift of both calculated and experimental frequency of the methanol stretching OH mode after formation of a hydrogen bond depends strongly on the PA value of the zeolite oxygen center [15]. This can be used to estimate the PA of zeolite oxygen atoms (Section 6.2.2.1). Analysis of the OH deformation vibration of bifunctionally adsorbed methanol shows that the corresponding frequency shift $\Delta\delta(OH)$ also depends on the PA of the zeolite oxygen centers to which the guest molecule coordinates. This frequency shift $\Delta\delta(OH)$ was found to correlate with the red shift of the corresponding OH stretching frequency $\Delta\nu(OH)$.

6.3 Transition Metal Clusters in Zeolites

In Section 6.2.3.1 we have already considered the smallest possible transition metal moieties in zeolite pores, *atomic* electron-deficient Rh and Pt species. This section is devoted to reviewing our DF studies of just slightly larger transition metal moieties, namely encapsulated Pd, Pt and Ir particles consisting of four to six atoms. Questions concerning the structure, stoichiometry, charge, and adsorption properties of these small metal clusters in zeolites will be discussed now.

6.3.1 Charge and Adsorption Properties of Small Metal Clusters

6.3.1.1 Electron-Deficient Palladium Clusters

An electron-deficient state of a metal cluster formed in a zeolite cavity can result from its interaction with zeolitic protons [52]. We investigated computationally free and protonated clusters Pd_4 and Pd_6 to understand better the protonation of Pd_n clusters in zeolites and how protonation affects the interaction of these clusters with CO probe molecules [22]. DF calculations showed, in line with the experimental data [53], that protonation reduces the CO adsorption energy and increases the vibrational frequency of adsorbed CO. The proton affinities of the clusters Pd_4 and Pd_6 were calculated at approximately 900 and 950 kJ mol^{-1}, respectively. These values are as large as the proton affinity of such a strong base as ammonia (calculated 885 kJ mol^{-1} [14,22]; experimental 845 kJ mol^{-1}). The calculated data, in particular the large protonation energies as well as the bonding and vibrational properties of CO probes adsorbed on palladium clusters, and their agreement with available experimental results support the proposal that electron-deficient $[Pd_nH]^+$ species can be formed in faujasites (e.g., Pd/NaY) as a result of the interaction of the guest metal particles with zeolitic protons.

6.3.1.2 Pt_4 clusters

Since platinum is especially important in catalysis, much experimental effort has been spent to characterize small Pt particles entrapped in zeolite cages, using the whole arsenal of spectroscopic, structural and chemical methods (e.g., [1,3,9] and references therein). However, the interpretation of experimental data was often ambiguous because the measured values usually result from more than one effect: as a rule, when the electronic state of an encapsulated metal species changes owing to interactions with the zeolite host, so do its size and shape. Quantum chemical model investigations allow to separate these different effects.

Our scalar relativistic DF study [23] aimed at analyzing adsorption properties (binding, geometric and electronic structure, vibrations) of CO probes bound to small platinum clusters; we emphasized the dependence of these properties on the cluster charge and the adsorption site. To mimic encapsulated metal species that become (partially) charged owing to interactions with the zeolite framework, we employed monocarbonyl adsorption complexes of the bare neutral cluster Pt_4 as well as of singly charged Pt_4^+ and Pt_4^- moieties. We have also investigated the systems XPt_4CO ($X = Na, Na^+$) to validate the electronically modified clusters as models of moieties interacting with electron donor or acceptor centers in zeolites; Na was taken as model electron donor and Na^+ as model electron acceptor to induce chemical effects on the electronic structure and the adsorption properties of the cluster Pt_4.

Properties of adsorbed CO molecules were found by us [23] to be rather sensitive to the electronic state and the adsorption site of the Pt_4 particles, in line with experiment. A linear correlation between the effective charge of the metal cluster and the adsorption-induced vibrational frequency shift $\Delta\nu(CO)$ was established for CO at on-top position, the energetically preferred adsorption geometry. With this correlation we estimated the effective charge of metal particles as induced by interaction with their surrounding. The properties of CO adsorbed on Na/Pt_4 and on Na^+/Pt_4 were computed to be very close to those for CO bound to the appropriately charged clusters $Pt_4{}^q$ ($q = -0.35e, +0.68e$), where the charge had been determined from the calculated correlation. These electronically modified clusters, indeed, constitute useful models for describing electron-enriched or electron-deficient states of small platinum particles as induced by metal-support interaction. CO molecules were found to probe the charge of the metal clusters by means of the frequency $\nu(CO)$, irrespective of how this particular state has been generated: either by directly modifying the (partial) charge of the cluster or by charge exchange resulting from metal–ligand or metal–support interaction.

Recently, a band with very low C–O frequency, $\Delta\nu(CO) = -186\ cm^{-1}$, has been measured for terminal CO molecules on Pt_n species formed by decomposition of the Chini complex $[Pt_3(CO)_3(\mu\text{-}CO)_3]_3{}^{2-}$ in NaX zeolite [9]. That frequency shift has been attributed to a negative charge of the encapsulated platinum cluster, in good agreement with our correlation [23].

6.3.2
Structure of Metal Clusters in Zeolite Cages: Case Study of Ir_4

EXAFS spectroscopy is particularly successful in a structural characterization of often very small supported clusters [2,8]. Since a large fraction of the metal species is in contact with the support, a notable part of the EXAFS signal is owing to metal-support contributions, thus providing information about the metal–support interface. Conventional preparation techniques result in nonuniform supported metal species that are difficult to characterize in detail. Thus, a strategy has been developed to prepare nearly uniform encaged metal clusters by synthesis of zeolite-supported metal carbonyl cluster precursors and subsequent removal of the carbonyl ligands with minimal disruption of the metal frame [2,54]. However, decarbonylation commonly leads to fragmentation and/or aggregation of metal clusters. Supported iridium clusters on metal oxides and zeolites (KLTL, NaX, and NaY) have been investigated in great detail [54] because these clusters feature rather stable metal frames such as Ir_4. Notwithstanding the substantial work on such samples, several questions remained, associated with limitations of EXAFS spectroscopy and the lack of other suitable experimental techniques for investigating metal species dispersed in porous solids. To shed light on these open questions, and in particular to clarify how the interaction with a zeolite framework can modify the structural and electronic properties of small supported metal particles, we have carried out a DF investigation of the interaction between Ir_4 clusters and a zeolite fragment, using a cluster model of a six-ring of a faujasite framework (Fig. 7) [24].

Two models were used to describe the zeolite T6 ring: the charged model Z, $Al_3Si_3O_6H_{12}{}^{3-}$, and the neutral model Z-Na, $Al_3Si_3O_6H_{12}Na_3$, in which the negative charge induced by the presence of the three Al centers is compensated by three Na^+ cations, just as in alkali-exchanged zeolites. As to structure and energetics, the two models yield similar descriptions of the main features of the metal–support interaction. The resulting most stable structures, Ir_4(A)/Z and Ir_4(A)/Z-Na (Fig. 7A), are characterized by quite strong GGA binding energies (about 355 and 260 kJ mol^{-1}, respectively). The shortest metal–oxygen distance $d(\text{Ir–O}_1)$ is computed at about 220 pm. This value, in good agreement with reported EXAFS structure results [24], indicates a close approach of the metal to the zeolite fragment. The computed Ir–Ir inter-metal distances are slightly elongated, up to 60 pm, in comparison with the value R(Ir–Ir) = 244 pm calculated for the free metal cluster. This result is at variance with EXAFS simulations which indicate an inter-metal distance R(Ir–Ir) of about 270 pm, very close to the corresponding values for $Ir_4(CO)_{12}$ (269 pm) and bulk iridium, 271.5 pm. The unusually large discrepancy

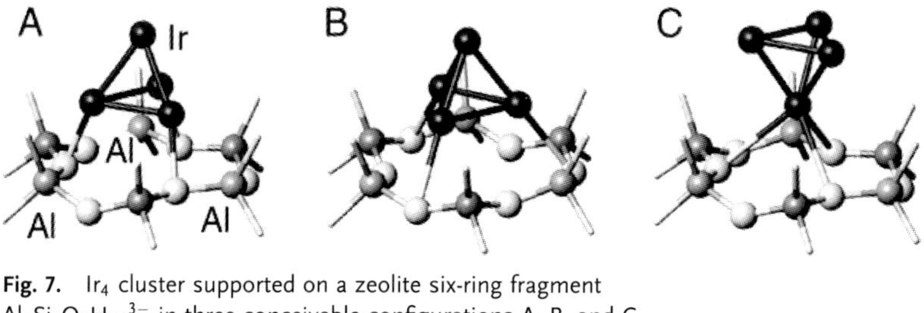

Fig. 7. Ir_4 cluster supported on a zeolite six-ring fragment $Al_3Si_3O_6H_{12}{}^{3-}$ in three conceivable configurations A, B, and C.

between the calculated inter-metal distances of the bare Ir_4 cluster and the EXAFS result (about 20 pm) falls far outside the accuracy of relativistic all-electron DF calculations, in particular for metal–metal distances [55]. We analyzed possible reasons for this discrepancy [24]. For instance, a residuum at a supported cluster is conceivable as a result of the decarbonylation treatment during cluster preparation. Indeed, we found that a carbon atom as extra ligand, located at a two-fold bridge or a three-fold hollow site of the Ir_4 cluster, induces an elongation of the Ir–Ir inter-metal distance comparable to 20 pm. Therefore, further experimental investigation of possible undetected ligands of zeolite-supported clusters is worthwhile.

Furthermore, we employed a CO molecule chemisorbed at the on-top site of the metal cluster to probe changes in the electronic structure of the metal particles induced by the zeolite support. When the zeolite substrate is described by the realistic Z-Na model, the interaction is accompanied by a small electron donation to the cluster, notable through a somewhat enhanced red shift of the C–O stretching frequency.

The structural interpretation based on EXAFS data could be extended and improved by invoking our theoretical results [24] as follows.

Zeolite-supported Ir_4 clusters bear only a small negative charge, if any. This statement may hold in general for noble metal clusters supported on metal oxides and zeolites. This suggestion is in line with the hypothesis of small, electron-rich Pt clusters formed in basic zeolites as a result of electron transfer from highly negatively charged framework oxygen atoms [3].

Zeolite-supported Ir_4 clusters are characterized by a moderate charge rearrangement in the bonding region between the Ir and O atoms, accompanied by a polarization of the electron density toward the top site of the cluster. Such polarization might be responsible for some of the support effects observed in catalysis by supported metal clusters.

The metal–metal distances observed by EXAFS spectroscopy indicate metal clusters that are probably not entirely ligand-free, even after decarbonylation and evacuation under mild conditions. EXAFS data are not sensitive enough to provide evidence of small numbers of ligands (in particular of light atoms) present in addition to the oxygen of the support. Computational model investigations evidently provide a very convincing method for establishing their likely presence, possibly in combination with very accurate IR spectra. Such experiments are expected to be challenging because the metal loading on the supports needs to be low and a treatment under only moderately severe conditions to remove possible ligands (such as C that has been calculated to bind strongly to Ir_4) readily causes restructuring of the metal frame.

Thus, the present results demonstrate how theory, used in concert with experiment, is helping to define opportunities for progress in research with supported metal clusters. We anticipate that powerful computational tools will soon be able to help identify reaction intermediates on supported heavy metal catalysts. Because reaction intermediates are inherently unstable, they seem to be beyond the reach of current experimental methods such as *in situ* EXAFS spectroscopy, which is likely to give evidence only of stable species.

6.4 Future Trends

The results presented in this review illustrate the growing power of computational modeling of structure, spectral features and chemical properties of different types of supported metal cations, complexes or clusters. Simulations of good quality were able to assist in various cases to clarify or rationalize measurements or even to revise some hypotheses derived from experimental information only. Luckily, the power of computers increases continuously and thus in the future will allow to perform "computer experiments" on more realistic model systems with more accurate methods. Especially in the case of zeolite supported species, it is important to account for the influence of the zeolite framework on bonding, location, structure and other properties of guest species. A very promising strategy is to apply a combined method that describes the active site of the zeolite with a quantum mechanical method and treats the rest of the framework at the molecular mechanical level [56]. We recently developed the EPE approach (elastic polarizable environment) for ionic systems [57], which takes both long-range electrostatic and short-range interactions between the surrounding framework and the active site into account; the formalism treats the cluster-environment interaction in a variationally stable and self-consistent fashion. Application of the EPE method, suitably adapted to zeolite-embedded systems, will lead to more realistic representation of experimental situations and one can expect an improved correspondence between calculated and experimental quantities.

Acknowledgements

We thank A.M. Ferrari, B.C. Gates, H. Knözinger, J.A. Lercher, and G.M. Zhidomirov for their cooperation and many stimulating discussions. We also acknowledge the strong effort of the team that developed the program suite ParaGauss [28]; many studies reviewed here profited considerably from using this program. G.N.V. is grateful for a research fellowship of the Alexander von Humboldt Foundation. This work has been supported by the Deutsche Forschungsgemeinschaft (Priority Program 467 "Nanostrukturierte Wirt/Gast-Systeme") and the Fonds der Chemischen Industrie.

References

1 W.M.H. Sachtler, *Acc. Chem. Res.* **1993**, *26*, 383.

2 O. Alexeev, B.C. Gates, *Top. Catal.* **2000**, *10*, 273.

3 D. Barthomeuf, *Catal. Rev.* **1996**, *38*, 521.

4 H. Knözinger, S. Huber, *J. Chem. Soc., Faraday Trans.* **1998**, *94*, 2047.

5 A. Zecchina, C. Otero Areán, *Chem. Soc. Rev.* **1996**, *25*, 187.

6 R.M. Barrer, Hydrothermal Chemistry of Zeolites, Academic, London **1982**, p. 251.

7 I. Mirsojew, S. Ernst, J. Weitkamp, H. Knözinger, *Catal. Lett.* **1994**, *24*, 235.

8 S. Kawi, B.C. Gates, in Clusters and Colloids, G. Schmid (ed.), VCH, Weinheim **1994**, p. 299.

9 A.I. Serykh, O.P. Tkachenko, V.Yu. Borovkov, V.B. Kazansky, M. Beneke, N.I. Jaeger, G. Schulz-Ekloff, *Phys. Chem. Chem. Phys.* **2000**, *2*, 5647.

10 A.M. Ferrari, K.M. Neyman, N. Rösch, *J. Phys. Chem. B* **1997**, *101*, 9292.

11 G.N. Vayssilov, M. Staufer, T. Belling, K.M. Neyman, H. Knözinger, N. Rösch, *J. Phys. Chem. B* **1999**, *103*, 7920.

12 G.N. Vayssilov, N. Rösch, *J. Phys. Chem. B* **2001**, *105*, 4277.

13 P. Strodel, K.M. Neyman, H. Knözinger, N. Rösch, *Chem. Phys. Lett.* **1995**, *240*, 547.

14 G.N. Vayssilov, N. Rösch. *J. Catal.* **1999**, *186*, 423.

15 G.N. Vayssilov, N. Rösch, *Phys. Chem. Chem. Phys.* **2002**, *4*, 146.

16 A.M. Ferrari, K.M. Neyman, S. Huber, H. Knözinger, N. Rösch, *Langmuir* **1998**, *14*, 5559.

17 G.N. Vayssilov, J.A. Lercher, N. Rösch, *J. Phys. Chem. B* **2000**, *104*, 8614.

18 G.N. Vayssilov, A. Hu, U. Birkenheuer, N. Rösch, *J. Mol. Catal. A* **2000**, *162*, 135.

19 J.F. Goellner, B.C. Gates, G.N. Vayssilov, N. Rösch, *J. Am. Chem. Soc.* **2000**, *122*, 8056.

20 A.L. Yakovlev, K.M. Neyman, G.M. Zhidomirov, N. Rösch, *J. Phys. Chem.* **1996**, *100*, 3482.

21 K.M. Neyman, P. Strodel, S.P. Ruzankin, N. Schlensog, H. Knözinger, N. Rösch, *Catal. Lett.* **1995**, *31*, 273.

22 A.L. Yakovlev, G.M. Zhidomirov, K.M. Neyman, V.A. Nasluzov, N. Rösch, *Ber. Bunsenges. Phys. Chem.* **1996**, *100*, 413.

23 A.M. Ferrari, K.M. Neyman, T. Belling, M. Mayer, N. Rösch, *J. Phys. Chem. B* **1999**, *103*, 216.

24 A.M. Ferrari, K.M. Neyman, M. Mayer, M. Staufer, B.C. Gates, N. Rösch, *J. Phys. Chem. B* **1999**, *103*, 5311.

25 N. Rösch, in Cluster Models for Surface and Bulk Phenomena, G. Pacchioni. P.S. Bagus, F. Parmigiani (eds.), *NATO ASI Series B*, Vol. 283, Plenum, New York **1992**, p. 251.

26 B.I. Dunlap, N. Rösch, *Adv. Quantum Chem.* **1990**, *21*, 317.

27 T. Belling, T. Grauschopf, S. Krüger, M. Mayer, F. Nörtemann, M. Staufer, C. Zenger, N. Rösch, in High Performance Scientific and Engineering Computing, H.-J. Bungartz, F. Durst, C. Zenger (eds.), Lecture Notes in Computational Science and Engineering, Vol. 8, Springer, Heidelberg **1999**, p. 439.

28 T. Belling, T. Grauschopf, S. Krüger, F. Nörtemann, M. Staufer, M. Mayer, V.A. Nasluzov, U. Birkenheuer, A. Hu, A.V. Matveev, M.S.K. Fuchs-Rohr, A.M. Shor, K.M. Neyman, D.I. Ganyushin, N. Rösch, ParaGauss 2.2, Technical University, Munich **2001**.

29 A. Hu, K. M. Neyman, M. Staufer, T. Belling, B.C. Gates, N. Rösch, *J. Am. Chem. Soc.* **1999**, *121*, 4522.

30 D.H. Olson, *Zeolites*, **1995**, *15*, 439.

31 J.L. Lievens, J.P. Verduijn, A.J. Bons, W.J. Mortier, *Zeolites* **1992**, *12*, 698.

32 W. Loewenstein, *Am. Mineral.* **1954**, *39*, 92.

33 K. Hadjiivanov, H. Knözinger, *Catal. Lett.* **1999**, *58*, 21.

34 K. Hadjiivanov, P. Massiani, H. Knözinger, *Phys. Chem. Chem. Phys.* **1999**, *1*, 3831.

35 K. Hadjiivanov, H. Knözinger, *J. Phys. Chem. B* **2001**, *105*, 4531.

36 B.E. Mills, R.L. Martin, D.A. Shirley, *J. Am. Chem. Soc.* **1976**, *98*, 2380.

37 M. Rep, A.E. Palomares, G. Eder-Mirth, J.G. van Ommen, N. Rösch, J.A. Lercher, *J. Phys. Chem. B* **2000**, *104*, 8624.

38 A.G. Pelmenshchikov, E.A. Paukshtis, V.G. Stepanov, V.I. Pavlov, E.N. Yurchenko, K.G. Ione, G.M. Zhidomirov, *J. Phys. Chem.* **1989**, *93*, 6725.

39 M. Sierka, U. Eichler, J. Datka, J. Sauer, *J. Phys. Chem. B* **1998**, *102*, 6397.

40 C.T.-W. Chu, C.D. Chang, *J. Phys. Chem.* **1985**, *89*, 1569.

41 T. Yamazaki, I. Watanuki, S. Ozawa, Y. Ogino, *Bull. Chem. Soc. Jpn.* **1988**, *61*, 1039.

42 A.M. Ferrari, P. Ugliengo, E. Garrone, *J. Phys. Chem.* **1996**, *105*, 4129 and references therein.

43 A. Zecchina, S. Bordiga, C. Lamberti, G. Spoto, L. Carnelli, C. Otero Areán, *J. Phys. Chem.* **1994**, *98*, 9577.

44 S. Bordiga, C. Lamberti, F. Geobaldo, A. Zecchina, G. Turnes Palomino, C. Otero Areán, *Langmuir* **1995**, *11*, 527.

45 H. Miessner, I. Burkhardt, D. Gutschick, A. Zecchina, C. Morterra, G. Spoto, *J. Chem. Soc., Faraday Trans. 1* **1989**, *85*, 2113.

46 H. Miessner, *J. Am. Chem. Soc.* **1994**, *116*, 11522.

47 V.L. Zholobenko, G.-D. Lei, B.T. Carvill, B.A. Lerner, W.M.H. Sachtler, *J. Chem. Soc., Faraday Trans.* **1994**, *90*, 233.

48 A. Zecchina, C. Otero Areán, G. Turnes Palomino, F. Geobaldo, C. Lamberti, G. Spoto, S. Bordiga, *Phys. Chem. Chem. Phys.* **1999**, *1*, 1649.

49 T. Yamazaki, I. Watanuki, S. Ozawa, Y. Ogino, *Langmuir* **1988**, *4*, 433.

50 A.Y. Khodakov, L.M. Kustov, V.B. Kazansky, C. Williams, *J. Chem. Soc., Faraday Trans.* **1993**, *89*, 1393.

51 S. Huber, H. Knözinger, *Chem. Phys. Lett.* **1995**, *244*, 111.

52 W.M.H. Sachtler, Z. Zhang, *Adv. Catal.* **1992**, *39*, 129.

53 L.L. Sheu, H. Knözinger, W.M.H. Sachtler, *J. Am. Chem. Soc.* **1989**, *111*, 8125.

54 B.C. Gates, in Catalysis by Di- and Polynuclear Metal Cluster Complexes, R.D. Adams, F.A. Cotton (eds.), VCH, Weinheim **1998**, p. 509.

55 J.F. Goellner, K.M. Neyman, M. Mayer, F. Nörtemann, B.C. Gates, N. Rösch, *Langmuir* **2000**, *16*, 2736.

56 M. Sierka, J. Sauer, *Faraday Discuss.* **1997**, *106*, 41.

57 V.A. Nasluzov, V.V. Rivanenkov, A.B. Gordienko, K.M. Neyman, U. Birkenheuer, N. Rösch, *J. Chem. Phys.* **2001**, *115*, 8157.

Part 3
Electrical Properties and Electronic Structure

Introduction to Part 3

Ulrich Simon

During the last twenty years the chemistry of nanoscaled materials has evolved to a field of research of broad and multidisciplinary interest. These nanomaterials represent, on the one hand, objects whose lateral extension lies in the size range of a few nanometers. These so called nanoparticles, nanocrystals, or clusters (in the current literature there is no common nomenclature) are much smaller than the characteristic length scales, namely, the de Broglie wavelength, the mean free path, and the phase relaxation length, and express themselves in the so-called size quantization effects or quantum size effects (QSEs) and are therefore described as "quantum dots". Although these length scales vary widely from one material to another, in this size range the dimensionality determines the material properties [1]. On the other hand, nanoporous crystalls must also be included in this classification. The most prominent compounds are zeolites and related micro- and mesoporous oxides, which are strongly bounded open-framework structures with pores and channels of nanometer dimensions. The accessibility of these pores for various guest molecules makes these materials important for many applications, e.g., in catalysis, chemical sensing, water treatment, and separation processes [2]. Due to their chemical composition these nanoporous oxides have wide electronic band gaps, and hence these materials are optically transparent as well as electrically insulating at moderate temperature. Furthemore, most of these compounds exhibit ionic conductivity due to mobile cations, which compensate the negative lattice charge and which are located and electrostatically bound inside the channel structures.

Due to structural complementarity zeolites have been recognized as suitable host compounds for stabilizing nanoscaled objects like the above-mentioned metal and semiconductor clusters [3]. By inclusion of nanoscaled guests inside the defined void spaces of the matrix, new nanocomposites can be obtained with tunable optical, magnetic, or electrical properties. These materials have been dicussed as potential candidates for the future nanoelectronics and nanoscaled system integration. In addition, the void spaces provide a restricted volume that allows size effects on the orientational dynamics of polar molecules to be studied. Furthermore, within the last few years, attempts have been made to synthesize nanoporous semiconductors. These new materials are of enormous interest from the scientific and technical points of view, since many new physical properties may be expected from these "exosemiconductors" [4].

This part reports on the electrical properties and the electronic structure of nanoporous solids and nanoscaled host–guest compounds. By means of selected examples it is intended to acquaint the reader with charge transport phenomena in nanoporous solids and how charge transport and relaxation can be affected by host–guest interaction. Furthermore, an insight into the electronic structure of these solids is given that takes into account both the band structure of the host lattice and the electronic structure of the encapsulated guests (ions, clusters, molecules). The chapters in this part highlight recent developments in this field of research by dealing with fundamental questions and technical applications.

The first chapter concentrates on the electrical properties of zeolites, which is based on the mobility of the charge-compensating cations. It takes us back to the first dielectric measurements on zeolites, which already have been performed almost as early as the middle of the last century. Since then, improvements in impedance measuring techniques and their extension over broad frequency and temperature ranges led to models for the local and translational motion of the metal cations that take into account the distribution of the cations over different sites, cation–cation interactions, and Sanderson electronegativity. Most studies have focused on the ionic conductivity in H-form zeolites. Besides the translational motion, protons can perform local on-site jumps in the first coordination sphere of Brønsted sites. While the latter is typically studied experimentally by ^{1}H MAS-NMR spectroscopy, the translational (inter-site) motion of protons has been analyzed by impedance measurements in conjunction with a combined quantum mechanics/interatomic potential function approach. The proton mobility can selectively be affected by H_2O and NH_3, and this led to the development of an NH_3 sensor with an extremely low cross-sensitivity towards CO and NO for selective catalytic reduction (SCR) applications, e.g., in automotive exhaust gas control.

The second chapter is focused on the dynamics of polar guest molecules [ethylene glycol, propylene glycol, poly(isobutyl vinyl ether)] confined in the pore and channel systems of zeolites and mesoporous MCMs. Here it is pointed out that the molecular dynamics, studied by dielectric spectroscopy, are determined by the interplay between confinement and surface effects.

In the third chapter Petkov and Bein give an overview on the subject of conductive guest materials in periodic mesoporous hosts with different pore diameters, topologies, and symmetries. They focus on the synthetic methods for the encapsulation of nanoscale conductors, like metal nanowires or semiconductor quantum dots, taking into account different material morphologies, such as thin films.

In Chapter 4 Sauer and Windiks report on density functional studies on paramagnetic Na_4^{3+} clusters in sodalite. In these clusters a single electron is shared by all four sodium cations, like in an F center (e.g., in Na-doped NaCl), whereas the unpaired electrons form a regular bcc lattice with nearest neighbor electron–

electron distances of 0.767–0.769 nm. The computational approach provides explanations for the electronic and magnetic properties of this host–guest system. It allows the comparison between the experimentally accessible sodium electro sodalite (SES) and hypothetical superstructures such as Na_4^{3+} clusters fixed in free space at the same position as in SES, thus revealing the effect of host–guest interaction on the electronic structure of the guest system.

The superposition of the electronic structures of the zeolite framework, the charge-compensating cations, the solvent molecules, and the guest species is the main emphasis of Calzaferri et al. in Chapter 5. They show that the charge-compensating cations, as well as encapsulated quantum-sized silver sulfide clusters, in zeolite A lead to new electronic states, which lie within the band gap region of the zeolite lattice. Solvent molecules influence the interaction between the metal ions and the zeolite lattice, which again affects the electronic structure. Thus, reversible hydration and dehydration of zeolites leads to a controlled perturbation of the electronic structure, which, e.g., in Ag^+ zeolites is reflected directly in a change in the optical properties.

In the final chapter, cetineites, a new class of nanoporous semiconductors with zeolite-like channel structure, are introduced. Besides synthesis and the structure, the optical and the (photo)conduction properties are described and compared to the electronic structure, which has been obtained from calculations using the AFC ELAPW $\mathbf{k} \cdot \mathbf{p}$ method. These compounds have features which are so far almost unique in nanoporous solids, such as photoconduction that can be tuned via the chemical composition, as well as a pronounced conduction anisotropy.

The following reports will illustrate that research on nanoporous solids is a challenging area of solid-state chemistry and physics. The materials described in the following are excellent objects for studying fundamental questions about size and confinement effects, about localized and collective properties in molecular and nanofabricated superstructures, and about host–guest interactions on the molecular length scale. Especially the electrical properties are of central interest in terms of new technical applications based on chemical sensing and low-dimensional conductivity.

A sketch which visualizes the topics treated in this part is given in Fig. 1. It represents a section through a solid with a hexagonally ordered one-dimensional channel system. The different aspects of host–guest interaction and intrinsic local and integral phenomena determining the electrical properties and the electronic structure are illustrated. These are:

- Thermally activated ion motion (metal cations, protons) along a periodic potential landscape of the polyanionic zeolite lattice.
- Proton mobility supported by guest molecules (NH_3, H_2O).
- Orientational dynamics of confined polar molecules.
- Synthesis of conducting nanoscaled objects (quantum dots, metal nanowires).
- Electronic properties and host–guest interaction of molecular and quantum-sized clusters.
- Electronic structure and photoconductivity of nanoporous solids.

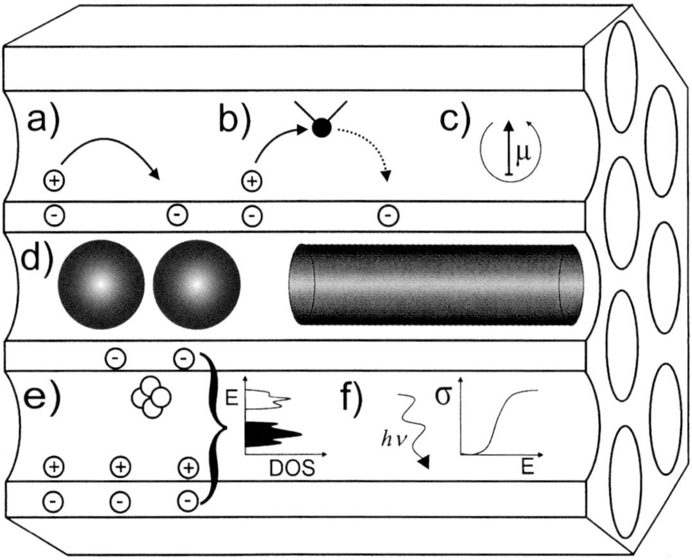

Fig. 1. A hexagonally ordered one-dimensional channel system showing the different aspects of host–guest interaction and intrinsic local and integral phenomena (see text for details).

References

1 See for example: E. Corcoran, *Sci. Am.* **1990**, 122; *Engineering a Small World: From Atomic Manipulation to Microfabrication*, *Science* **1991**, *254*, 1300; P. Avouris, I.W. Lyo, *Science* **1990**, *253*, 173; M.A. Reed, *Sci. Am.* **1993**, 118; M.C. Steigerwald, L.E. Brus, *Acc. Chem. Res.* **1990**, *23*, 183; Y. Wang, *Acc. Chem. Res.* **1991**, *24*, 133; U. Simon, H. Schön in H. S. Nalwa (Ed.) *Handbook of Nanostructured Materials and Nanotechnology*, Academic Press, 1999, Vol. 3, pp. 131–175, U. Simon, in Braunstein, Oro, Raithby (Eds.) *Metal Clusters in Chemistry*, Wiley-VCH, Weinheim, 1999, Vol. 3, pp. 1342–1359; U. Simon, *Adv. Mater.* **1998**, *10*, 1487.

2 H. von Bekkum, E.M. Flamigen, P.A. Jacobs, J.C. Jansen (Eds.), *Introduction to Zeolite Science and Practice, Studies in Surface Science and Catalysis*, Vol. 137, Elsevier, Amsterdam, 2001; D.W. Breck, *Zeolite Molecular Sieves, Structure, Chemistry and Use*, Wiley & Sons, New York, 1974; U. Simon, M.E. Franke, *Microporous Mesoporous Mater.* **2000**, *41*, 1.

3 G.A. Ozin, *Adv. Mater.* 1992, *4*, 612; P. Behrens, G. Stucky in G. Alberte, T. Bein (Eds.), *Comprehensive Supramolecular Chemistry*, Vol. 7, Pergamon, 1996

4 G.A. Ozin in L.V. Interrante, L.A. Caspar, A.B. Ellis (Eds.), *Material Chemistry: An Emerging Discipline, Advances in Chemistry Series* **1995**, 245.

1
Ionic Conductivity of Zeolites: From Fundamentals to Applications

Ulrich Simon and Marion E. Franke*

1.1
Introduction: Historical Survey of Metal Cation Conduction in Dehydrated Zeolites

The electrical properties of zeolites have been the subject of intense studies for more than three decades [15]. The experiments, most of which were performed by Freeman and Stamires and by Schoonheydt and Uytterhoeven [1,2,6–13], focused on the mobility of the exchangeable cations in technically relevant zeolites X, Y, and A by means of dielectric spectroscopy. The authors attribute the ionic conductivity in dehydrated faujasite to the migration of the exchangeable cations. Additionally, a local ionic relaxation process was detected, which is caused by the restricted local motion of the cations inside the supercages. According to this, the charge compensating cations move through the zeolite lattice by overcoming potential barriers of different height. The barrier height is determined by the electrostatic interactions between the cations and the polyanionic lattice, and it can be approximated by a simple Coulomb energy term (Eq. 1)[14]

$$E_c = \frac{1}{4\pi\varepsilon_0}\sum_i \frac{z_i z_c e^2}{r_{ic}}, \qquad (1)$$

where $z_i e$ is the lattice charge, $z_c e$ the cation charge, r_{ic} the cation–lattice distance, and E_c the activation energy of the charge transport. Because of the first-power dependence in $1/r_{ic}$ this is a long-range interaction. Thus, the number and distribution of exchangeable cations over neighboring sites and cavities influence the energetics and mobility [14].

This model, which initially regards just the attractive electrostatic interactions between cation and polyanionic lattice, has been deduced from impedance measurements on the Na^+-form of zeolites X and Y. However, it turned out to fail for zeolites which have been exchanged with different types of cations. Therefore, Jansen and Schoonheydt [8] added a term for repulsive cation–cation interactions, whereas Lortz and Schön [15,16] assumed that an additional influence from steric hindrance, especially for bulky cations like Rb^+, must be taken into account. Consequently, the motion of the exchangeable cations in the zeolite lattice is a highly-cooperative diffusion process over a heterogeneous system of potential barriers, which depends on several parameters such as structure and composition of the zeolite and type and charge of the cation.

To give an indication of this complexity we describe exemplarily in the present section the results of impedance measurements on faujasite exchanged with different types of mono- and divalent cations, reported by Schooheydt et al., which lead to a refined mechanistic picture. Our description is focused on dehydrated samples, since water present in the zeolite lattice significantly influences the mobility with its characteristic energies of the metal ions [17]. These measurements identify a local motion and a translational motion (conductivity) of the cations localized on the cation sites SII and SIII. Because of the different temperature ranges in which these two processes have been observed, they were assigned to a low-temperature (dipolar relaxation, <623 K) and a high-temperature process (diffusion/conductivity, >623 K). This assignment originates from the fact that both processes are thermally activated and thus observed in different temperature ranges, since a limited frequency range of 200 Hz to 20 kHz was available in the experiments [2,14,18]. By means of low-frequency impedance measurements on dehydrated zeolite X and Y over a broad temperature and frequency range, Simon and Flesch analyzed two distinct relaxation modes, which could be observed in modulus plots [19]. The high-frequency (HF) and the low-frequency (LF) signals correspond to the above high- and low-temperature process, respectively. The activation barriers, which are compared in Tab. 1, are higher for both LF and HF process in NaY than in NaX.

In earlier works the difference in activation energies for Na-X and Na-Y was only discussed for the conductivity process, which corresponds to the LF process discussed here. In Ref. [14] the lower activation energy in Na-X is attributed to the lower effective charge of the lattice due to the smaller SiO_2/Al_2O_3 ratio. The authors use the phenomenological model of the Sanderson electronegativity [20,21], which describes the effective electronegativity of the zeolite lattice by taking into account the electronegativity of the constituent elements. Nevertheless, the results presented in Tab. 1 show different values for ΔE_A, although the effect of the Sanderson electronegativity should be the same for the two processes. Therefore, an additional repulsive term has been discussed [19]. Due to the spatial restriction, the latter is assumed to be more pronounced for the local motion of the cations within the α-cage (HF process). Consequently, this process has been attributed to thermally activated hopping of the Na^+ ions between SII sites across the square

Tab. 1. Activation energies (in kJ mol^{-1}) in Na-X and Na-Y and difference between them (ΔE_A).

Zeolite	*LF process*	*HF process*
Na-Y	80	72
Na-X	68	41
ΔE_A	12	31

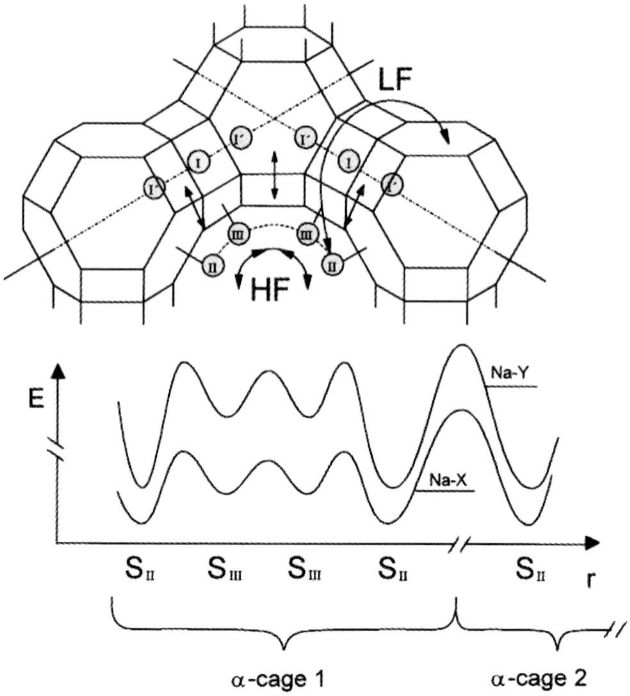

Fig. 1. Illustration of the conductivity processes and the corresponding potential profile in faujasite indicating the different activation energies of the high-frequency (HF) and low-frequency (LF) processes [19] (with kind permission from Kluwer Academic Publishers).

elements of SIII. The rate-limiting step in this process is overcoming the highest potential barrier, i.e., that between SII and SIII (Fig. 1) [22].

1.2 Proton Conduction

In H-form zeolites, in which protons are the charge compensating cations, bridged hydroxyl groups on each AlO_4^- site, so-called Brønsted acid sites, appear. Hence, protons can perform either local on-site jumps in the first coordination sphere of an aluminum site or translational inter-site motion between neighboring aluminum sites. Therefore, the question arose whether the model of charge transport developed for zeolites in their alkali metal forms, as discussed above, can be transferred to their proton forms. This question is of great interest, since the activation energy for creating mobile protons in zeolites might be related to the Brønsted acidity and the overall activity of these technically important heterogenous catalysts [23].

In general, no unifying concept has been developed up to now to relate the acidity of a zeolite strictly to its structure and composition. Based on the definition of the acidity as the enthalpy of deprotonation [24], several attempts have been made to characterize the strength of the O–H bond at the acid site. To this end, IR spectroscopy [25–28] and measurement of the heat of adsorption of NH_3 by microcalorimetry or temperature-programmed desorption [23,29–31] were predominantly used. Additionally, deprotonation energies have been determined directly by means of quantum chemical calculations [32,33]. Since, as described above, characterization of static properties apparently cannot fully characterize the acidity of zeolites (particularly with regard to catalytic activity), the analysis of dynamic properties and their characteristic energies is assumed to be a promising strategy.

Variable-temperature ^1H MAS-NMR spectroscopy has been established to determine directly the proton mobility [34–38]. However, the observed activation energies are in the range of 17–78 kJ mol^{-1} for H-faujasite, H-mordenite and H-ZSM-5 and show no correlation with the zeolite structure and composition. All authors draw the conclusion that the observed proton mobility results from local on-site jumps, since translational inter-site motion would not occur in the temperature range of the experiments because of the high activation energy. Due to the obvious uncertainty of the experimental results, the activation barrier for on-site proton jumps has been determined by quantum chemical calculations. Recent calculations using the combined QM-Pot method determine activation barriers for on-site proton jumps of 70–102, 68–106, and 52–98 kJ mol^{-1} for H-chabasite, H-faujasite, and H-ZSM-5, respectively [39]. The calculated energy barriers are tentatively higher than those resulting from ^1H NMR techniques.

Nevertheless, the question remained open whether protons can perform translational motion through the porous zeolite structure. The deprotonation energy of about 1300 kJ mol^{-1} [40–42] implies that long-range motion of protons in zeolites should not appear at moderate temperatures. This is, however, in contrast to the experimental evidence of translational proton motion in various dehydrated H-forms of zeolites, observed in the range below 673 K [43–45]. The following sections deal with this apparent contradiction.

1.2.1 Impedance Measurements on Dehydrated H-ZSM-5

The electrical properties of dehydrated H-ZSM-5 with a wide range of SiO_2/Al_2O_3 ratios (30–1000) were studied by means of impedance spectroscopy [17,46,47]. The activation energy for proton relaxation strongly depends on the SiO_2/Al_2O_3 ratio, i.e., on the average spatial distance between the Brønsted acid sites (see Fig. 2). This dependence has been analyzed by means of the classical hopping transport theory and the Debye–Hückel theory, as described in the following.

The classical hopping transport theory is accepted to be reasonable as long as the characteristic temperature is small compared to the height of the barriers [48], which is the case in the experiments discussed here. Rice and Roth [49] extended the expression for the temperature-dependent conductivity (in a cubic system) to a

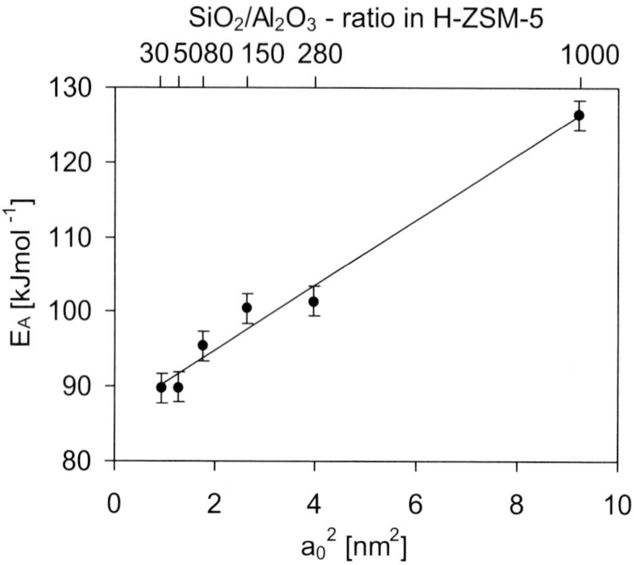

Fig. 2. Plot of the activation energy vs. $a_0{}^2$ (SiO_2/Al_2O_3 ratio), including the regression line [46].

transport theory for quasifree ions, which involves two phenomenological parameters, E_A and τ_0, where E_A is the migration energy, i.e., the energy for creating a quasifree ion, and τ_0 the lifetime of the excited ionic state. They postulate that the total energy E_T of a free ion ($\cong E_A$) is totally transferred into kinetic energy E_x of the ion, given by $E_x = \frac{1}{2} m_{ion} v_x{}^2$ (m_{ion} = ion mass, v = velocity of the ion). For hopping transport between neighboring aluminum sites the mean free path l_0 is given by the hopping length a_0, so that for ions with $E_T \geq E_A$ Eq. (2) holds

$$E_A = a_0^2 \cdot \frac{m_{ion}}{2\tau_0^2}. \tag{2}$$

Therefore, a_0 corresponds to the average spatial separation of the aluminum sites, which is calculated from the volume and aluminum content of the unit cell [46]. Note that this simplification does not take into account the real structure of H-ZSM-5, which may provide preferred conduction paths. With sufficient accuracy a linear dependence of E_A on $a_0{}^2$ is found, which leads to the conclusion that the increase of E_A with increasing SiO_2/Al_2O_3 ratio in H-ZSM-5 reflects proton hopping between neighboring Brønsted sites (cf. Fig. 2).

Another approach to verify this assumption of inter-site proton motion is the Debye–Hückel theory (DHT) for ionic solutions. To apply this theory it was assumed that the zeolite behaves like an isotropic solid electrolyte. Then the Coulomb potential $\varphi(r)$ of a particle, which decays with $1/r$, extends over more than two nanometers. This leads to a distance-dependent overlap of the Coulomb potentials of neighboring aluminum sites, which decreases the activation energy between interacting sites and results in preferred conduction paths for protons.

Both models lead to the conclusion that the increase of E_A with increasing SiO_2/Al_2O_3 ratio in H-ZSM-5 reflects proton hopping between neighboring sites, despite the high deprotonation energy of about 1300 kJ mol^{-1}. This indicates that the translational motion of protons does not require full deprotonation, and this can be understood in terms of proton interaction with Si–O–Si groups bridging neighboring AlO_4^- sites. This picture is consistent with the mechanism of proton conductivity described for compact, non-porous, solid oxide proton conductors, in which the moving protons are assumed to be embedded in the electron density of the lattice oxygen atoms [50].

1.2.2
Quantum Chemical Description of Translational Proton Motion in H-ZSM-5

Assistance from computational chemistry was sought to deepen the above-mentioned model for proton transport in dehydrated zeolites. Therefore, activation barriers for translational proton motion in zeolite H-ZSM-5 were calculated by a combined quantum mechanics/interatomic potential function approach (QM-Pot). The QM-Pot method combines a quantum mechanical description of the reaction site with an interatomic potential function describing the periodic zeolite lattice [32,51].

First, the potential energies of the stable intermediate proton positions and transition structures relevant for the translational proton motion were calculated for H-ZSM-5 with one aluminum atom per unit cell, which is equivalent to a formal SiO_2/Al_2O_3 ratio of 190. The proton starts at the crystallographic Al7-O17H-Si4 position, which proved to be one of the most stable in the orthorhombic modification containing one aluminum atom [52]. The proton moves along a T-O-T chain (T = Si, Al), which connects two neighboring Brønsted sites which are spatially separated by about 1.4 nm. Figure 3 shows the chain of T-O-T sites.

The potential energies of the six stable intermediate proton positions and the five transition structures, taking into account zero-point vibrational energy corrections, are shown in Fig. 4 as a function of their position. Taking the highest energy between initial and final state as the energy to create mobile charge carriers, the translational motion of the proton may appear with an activation energy of about 210 kJ mol^{-1} for a SiO_2/Al_2O_3 ratio of 190.

The energies of equilibrium proton positions connecting the two neighboring Brønsted sites show an almost symmetrical course that reaches its maximum for a proton position in the middle between these sites. This trend has been explained by the Coulomb potential of the two aluminum sites, which decays with $1/r$ [53]. The energies of the transition structures show a similar trend, whereas their absolute value depends on the local structure [54]. Nevertheless, this trend of potential energies suggests that the activation barrier for inter-site proton motion decreases as soon as the Coulomb potential of neighboring aluminum sites starts to overlap, as predicted from the findings of impedance measurements (cf. Section 1.2.1). To

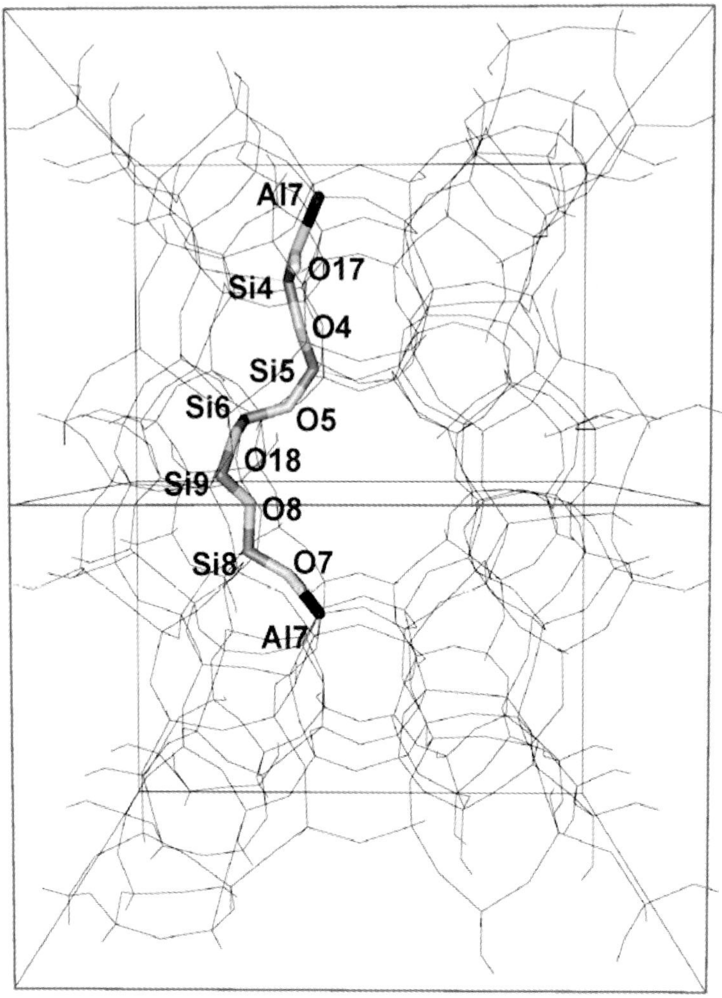

Fig. 3. Chain of T-O-T sites along which the proton moves translationally between two crystallographic identical Brønsted sites with a spatial separation of about 1.4 nm. The proton is located in its initial position.

further verify this assumption, the activation barrier for the initial step of the translational motion leaving AlO_4^- (O17 → O4) was calculated for two additional ZSM-5 models, in which a second aluminum atom per unit cell is added at T5 and T6 sites, respectively. For the resulting Al–Al distances of 1.4, 0.8, and 0.6 nm, the values of E_A are 127, 119, and 83 kJ mol^{-1}, respectively. Thus, the barriers apparently decrease with decreasing spatial Al–Al distance.

In conclusion, theory and experiment agree on two main aspects: the activation barrier to create a mobile proton is much lower than the deprotonation energy of

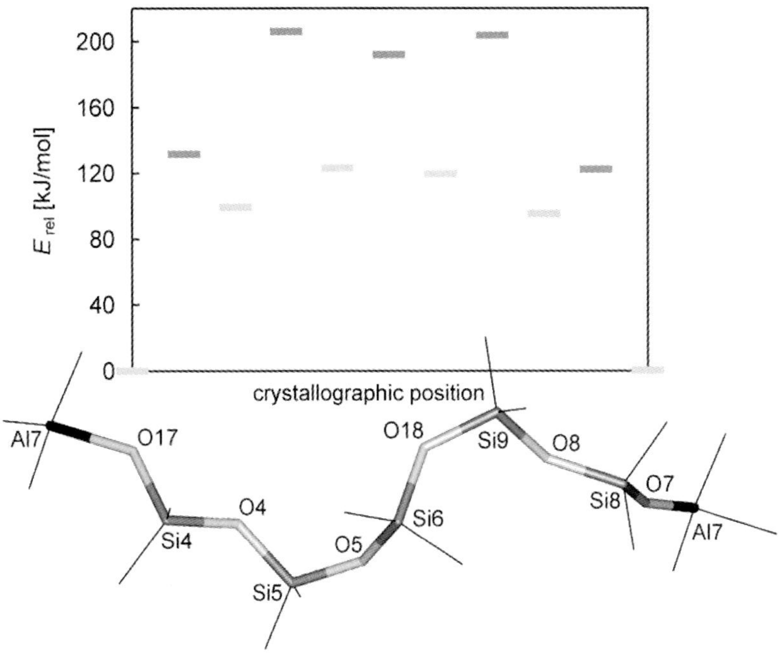

Fig. 4. QM-Pot energies of minimum and transition structures for the proton moving along the indicated T-O-T chain

about 1300 kJ mol^{-1} for an Al-OH-Si site, and is obviously mainly compensated by the energy gained by binding the proton to the bridging Si-O-Si groups. Furthermore, it was found that the activation barrier for inter-site jumps decreases with decreasing SiO_2/Al_2O_3 ratio due to increased Coulomb interaction between neighboring aluminum sites.

1.2.3
Effect of Guest Molecules on Proton Mobility

After having developed a mechanistic picture of charge-carrier energetics in dehydrated zeolites by means of experimental and theoretical methods, in this section the interaction of protons with different guest molecules is discussed. Much theoretical work has been done to study interactions, especially of H_2O and NH_3, with the H-forms of zeolites [24,55–62], but these calculations are restricted to the local proton on-site motion and the determination of the adsorption energy.

From impedance measurements on zeolite H-beta it is known that NH_3 influences the effective proton mobility, which can be directly measured by an increase in conductivity in the presence of NH_3 [45]. Measurements on the Na^+-form of zeolite beta, in which the conductivity is not influenced in the presence of the guest molecule, lead to the conclusion that NH_3 predominantly influences the

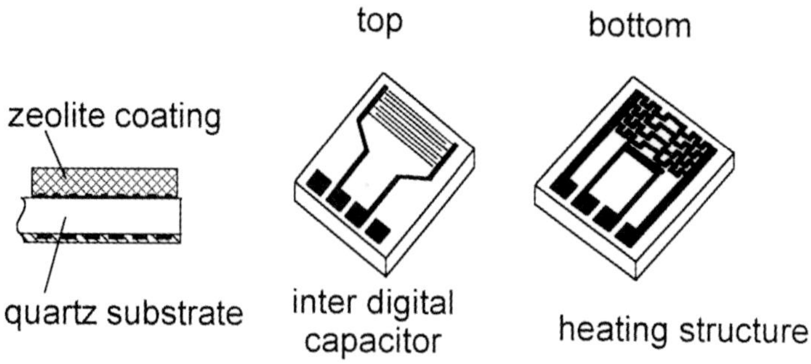

Fig. 5. An idc structure and cross section of a coated zeolite sensor [45]

mobility of the Brønsted acidic protons. For these measurements a special sample geometry is applied, i.e., an inter digital capacitor (idc) with noble metal electrodes on an alumina substrate (Fig. 5). The electrodes are covered with the zeolite film, which allows quick gas diffusion into the sample [17,45]. The bottom side of the substrate consists of a covered platinum heater.

Here we report on the adsorption of the small polar guest molecules H_2O and NH_3 in H-ZSM-5 as a function of the SiO_2/Al_2O_3 ratio. In general, samples with high aluminum content ($SiO_2/Al_2O_3 = 30$, 50, 80) show significant increase in conductance when guest molecules are present. These characteristics are reflected in the Arrhenius plot, shown exemplarily for H-ZSM-5 ($SiO_2/Al_2O_3 = 50$) in Fig. 6. The change in mechanism for different temperature ranges can be described according to the model of Grotthus-like transport and vehicle transport [63]. Accordingly, the proton transport mechanism changes from Grotthus-like reorientation at low temperatures to vehicle transport mechanism at higher temperatures [50], whereby NH_4^+ and H_3O^+ are the mobile species (cf. Fig. 6). The characteristic desorption temperature is 633 K for H_2O, and that of NH_3 is higher (733 K) due to its stronger interaction with the zeolite lattice.

To test the gas-sensing properties of H-ZSM-5, concentration dependent measurements were performed in the range of 0–100 ppm NH_3 at 373 and 673 K. Figure 7 shows the dependence of the relative change of conductance $\Delta Y'_{rel}$ of H-ZSM-5 ($SiO_2/Al_2O_3 = 50$) in the presence of NH_3 in adsorption and desorption cycles. The supporting effect of NH_3 on the conductivity is reversible and more pronounced at 373 than at 673 K. Since the average response time at 373 K is typically 5 s, this sensor is applicable for exhaust gas control in automotive applications and other applications at elevated temperature.

1.3
Application of H-ZSM-5 as NH_3 Sensor for SCR Applications

A prominent example for the need of NH_3 sensors in exhaust gas control is their use in selective catalytic reduction (SCR) systems for minimizing nitrogen oxide (NO_x) emissions from commercial vehicles. In the SCR converter ammonia

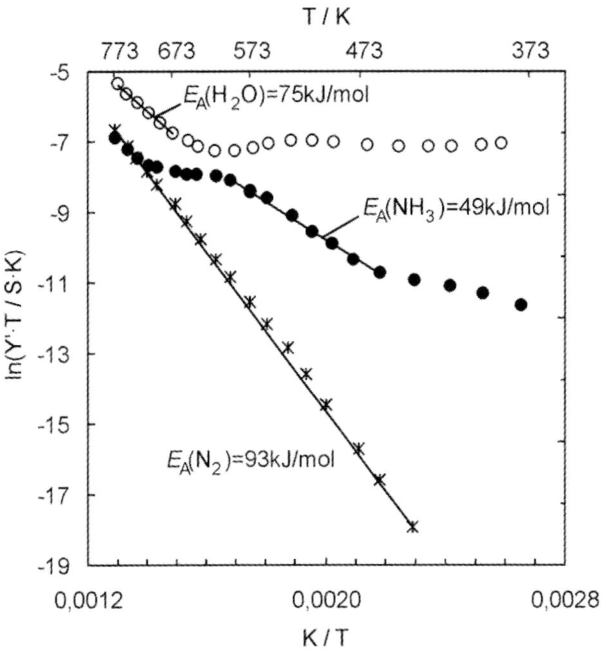

Fig. 6. Arrhenius plot of H-ZSM-5 (SiO_2/Al_2O_3 = 50) in the presence of 100 ppm NH_3 and 3 vol% H_2O.

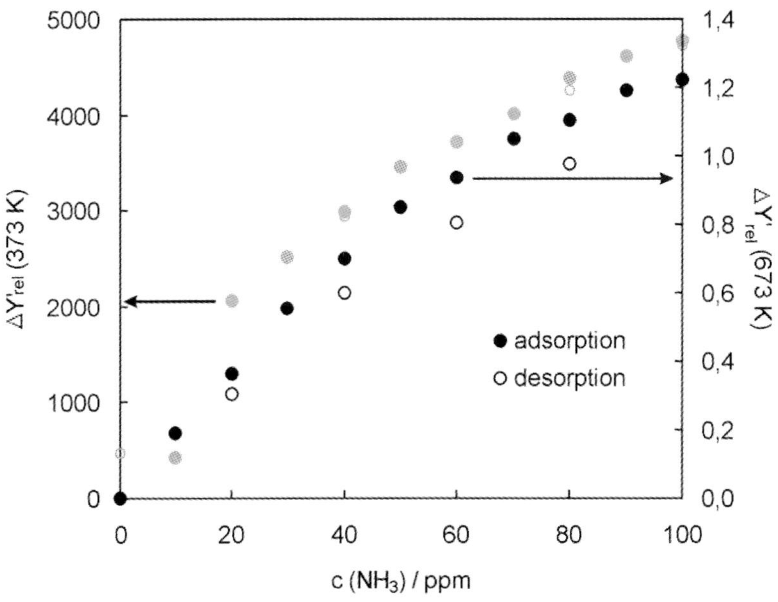

Fig. 7. Dependence of relative conductivity $\Delta Y'_{rel}$ of H-ZSM-5 (SiO_2/Al_2O_3 = 50) in the presence of NH_3 in adsorption and desorption cycles of 0–100 ppm

selectively reduces NO_x. In urea SCR systems an aqueous solution of urea is injected into the exhaust pipe in front of the SCR converter, where it decomposes into ammonia that serves as a reducing agent for NO_x. Exhaust-gas ammonia sensors are required to optimize the injected amount of urea and to assure that no ammonia emissions occur.

Moos et al. report on the development of a selective ammonia gas sensor based on H-ZSM-5 as gas-sensing material [64]. Zeolites are assumed to withstand the harsh conditions of internal combustion engine exhaust gas. As described in Section 1.2.3 the sensor principle is based on the selective and sensitive dependence of the electrical impedance of zeolites on the ammonia concentration of the ambient gas. As the sensor setup an idc structure is used (cf. Fig. 5). It was shown that the structure and composition of the zeolite samples and the operating temperature determine the working frequency, sensitivity, and selectivity of the sensor. At a fixed and setup-specific working frequency, the resistance R depends on the ammonia concentration. From the most promising zeolite material – zeolite H-ZSM-5 with a SiO_2/Al_2O_3 ratio of 140 – a sensor was manufactured and tested on engine test benches. The zeolite film was kept at 693 K. The gas atmosphere contained 10 % oxygen and 5 vol % water in nitrogen. The ammonia concentration was varied stepwise between 0 and 100 ppm. Figure 8 shows a typical test run for the determination of sensitivity and response time. The sensor resistance, measured at 1 MHz, responds immediately and reversibly to changes in ammonia concentration.

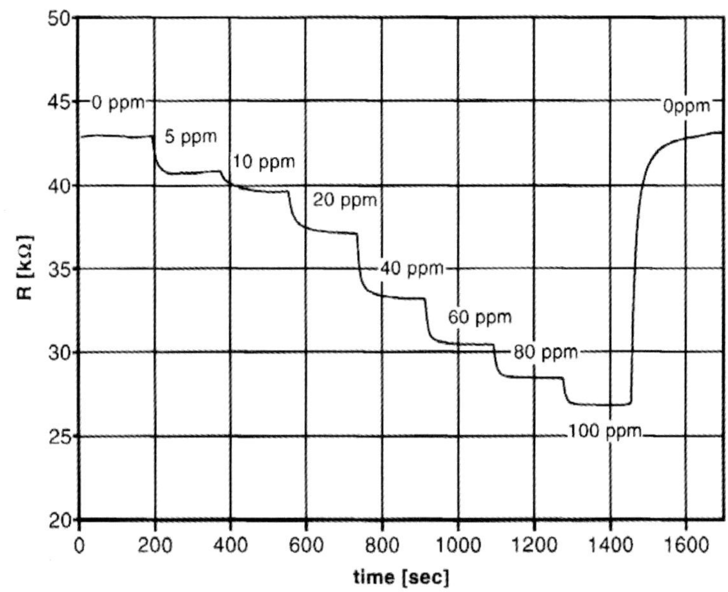

Fig. 8. Resistance of a representative sensor when exposed to different ammonia concentrations [64] (with kind permission from Elsevier).

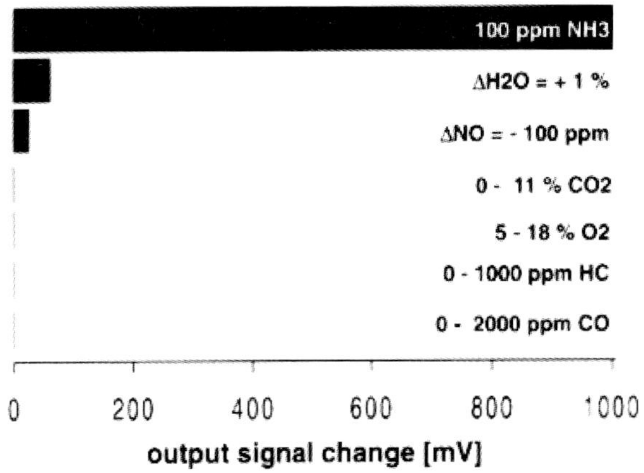

Fig. 9. Comparison of the output signal for 100 ppm ammonia with changes in the concentration of other relevant gaseous compounds. For example, $\Delta NO = -100$ ppm means a decrease in nitrogen oxide concentration of 100 ppm leads to an increase in the sensor output signal by 25 mV. Sensor temperature 693 K [64] (with kind permission from Elsevier).

To describe quantitatively the cross-sensitivity of the sensor, in the range of 0–60 ppm NH_3 the influence of other possible compounds in the engine exhaust gas on the sensor output signal was investigated (Fig. 9). Bars denote the change in output signal of a specially designed microelectronic device for sensor control when the concentration of the specific gas component indicated in or beneath the bar is varied. The sensor shows no cross-sensitivity for CO (0–2000 ppm), CO_2 (0–11 vol %), hydrocarbons (0–1000 ppm), and oxygen (5–18 vol %). Cross-sensitivity to water and nitrogen oxide are very small at the test temperature.

For the desired application the idc with it input/output wires needs to be packed in an exhaust-gas-stable housing (Fig. 10). This provides mechanical protection and long-term stability by reducing interaction with soot and other critical components of the exhaust gas.

1.4 Summary

We have shown that impedance measurements on metal- and proton-exchanged dehydrated zeolites lead to a detailed model for the energetics of the transport processes in microporous ionic conductors. The complex interplay of zeolite structure and composition, as well as the type of ionic species and energetics of ionic relaxation processes has been described for faujasites. Further, this model has been extended to proton-exchanged zeolites, in which, due to the formation of bridged Si-OH-Al groups, local on-site and translational inter-site motion of the proton must be distinguished. In contrast to the conventional picture, it was

Fig. 10. Photograph of a housed sensor. The sensor chip has a width of 0.25 inches (6.35 mm) and a thickness of 635 μm. The thread of the housing is compatible with a common λ probe [64] (with kind permission from Elsevier).

shown that protons undergo translational motion at moderate temperature, despite the high deprotonation energy of about 1300 kJ mol^{-1}. The results of QM-Pot calculations support the view that the proton conductivity measured by impedance spectroscopy results from proton hopping between Brønsted sites. According to this, the energy required to "deprotonate" the $(AlO_4)H$ site is apparently largely compensated by the energy gained by adding the proton to a Si-O-Si bridge. The complementary use of computational methods refined the picture of proton transport in zeolites and therefore turned out to be essential for the development of gas sensors, explicitly for the development of NH_3 sensors as a key element for SCR technology in automotive applications.

References

1 D.C. Freeman Jr., D.C. Stamires, *J. Chem. Phys.* **1961**, *35*, 799.

2 R.A. Schoonheydt, J.B. Uytterhoeven, *Molecular Sieve Zeolites-1, Advances in Chemistry Series 101*, American Chemical Society, **1971**, p. 456.

3 D.C. Stamires, *Clays Clay Miner.* **1973**, *21*, 379.

4 T. Ohgushi, Y. Kawanabe, *Zeolites* **1994**, *14*, 356.

5 U. Simon, C. Möhrke, G. Schön, *Chem. Ing. Tech.* **1995**, *67*, 583.

6 R.A. Schoonheydt, W. De Wilde, *J. Chem. Phys.* **1976**, *80*, 511.
7 R.A. Schoonheydt, W. De Wilde, *J. Chem. Soc. Faraday Trans. 1* **1974**, *70*, 2132.
8 F.J. Jansen, R.A. Schoonheydt, *J. Chem. Soc. Faraday Trans. 1* **1973**, 1338.
9 R.A. Schoonheydt in *Proceedings of the Fifth International Conference on Zeolites*, L.V.C. Rees (Ed.), Heyden, London, **1980**, p. 242.
10 R.A. Schoonheydt, J.B. Uytterhoeven, *Clay Clay Minerals* **1969**, *8*, 71.
11 R. A. Schoonheydt, *J. Phys. Chem. C Suppl.* **1980**, *41*, C6–261.
12 D.C. Stamires, *Clays Clay Minerals* **1973**, *21*, 379.
13 I.R. Beattie, *Trans. Faraday Soc.* **1954**, *50*, 581.
14 W.J. Mortier, R.A. Schoonheydt, *Prog. Solid State Chem.* **1985**, *16*.
15 W. Lortz, PhD Thesis, University of Essen, **1987**.
16 G. Kelemen, W. Lortz, G. Schön, *J. Mater. Sci.* **1989** *24*, 333.
17 U. Simon, M.E. Franke, *Microporous Mesoporous Mater.* **2000**, *41*, 1.
18 D. W. Breck, *Zeolite Molecular Sieves: Structure, Chemistry and Use*, Wiley, New York, **1974**, p. 99.
19 U. Simon, U. Flesch, *J. Porous Mater.* **1999**, *6*, 33.
20 D. H. Olson, E. Dempsey, *J. Catal.* **1969**, *13*, 221.
21 R. T. Sanderson, *Chemical Band and Bond Energy*, Academic Press, New York, **1976**; *Polar Covalenz*, Academic Press, New York, **1983**.
22 S. Beran, J. Dubsky, *Chem. Phys. Lett.* **1980**, *71*, 300
23 W.E. Farneth, R.J. Gorte, *Chem. Rev.* **1995**, *95*, 615.
24 M. Brändle, J. Sauer, *J. Am. Chem. Soc.* **1998**, *120*, 1556.
25 J. Datka, M. Boczar, B. Gil, *Langmuir* **1993**, *9*, 2496–2498.
26 B. Gil, E. Broclawik, J. Datka, J. Klinowski, *J. Phys. Chem.* **1994**, *98*, 930–933.
27 J.A. Lercher, C. Gründling, G. Eder-Mirth, *Catal. Today* **1996**, *27*, 353–376.
28 R.J. Gorte, *Catal. Lett.*, **1999**, *62*, 1, and references therein.
29 A. Auroux, P.C. Gravelle, J.C. Vedrine in *Proceedings of the 5th International Zeolite Conference*, L.V.C. Rees (Ed.), Heyden, London, **1980**, p. 433.
30 N. Katada, H. Igi, J.-H. Kim, M. Niwa, *J. Phys. Chem. B* **1997**, *101*, 5969.
31 Y. Miyamoto, N. Katada, M. Niwa, *Microporous Mesoporous Mater.* **2000**, *40*, 271.
32 U. Eichler, M. Brändle, J. Sauer, *J. Phys. Chem. B* **1997**, *101*, 10 035.
33 M. Sierka, U. Eichler, J. Datka, J. Sauer, *J. Phys. Chem. B* **1997**, *102*, 6397
34 D. Freude, W. Oehme, H. Schmiedel, B. Staudte, *J. Catal.* **1974**, *32*, 137.
35 H. Ernst, D. Freude, T. Mildner, H. Pfeifer, in *Proceedings of the 12th International Zeolite Conference*, M.M.J. Treacy, B.K. Marcus, M.E. Bisher, J.B. Higgins (Eds.), Vol. 4, Materials Research Society, Warrendale, **1999**, p. 2955.
36 D. Sarv, T. Tuherm, E. Lippma, K. Keskinen, A. Root, *J. Phys. Chem.* **1995**, 99 13 763.
37 T. Baba, N. Komatsu, Y. Ono, H. Sugisawa, *J. Phys. Chem. B* **1998**, *102*, 804.
38 T. Baba, Y. Inoue, H. Shoji, T. Uematsu, Y. Ono, *Microporous Mater.* **1995**, *3*, 647.
39 M. Sierka, J. Sauer, *J. Phys. Chem. B* **2001**, *105*, 1603.
40 G.J. Kramer, R.A. van Santen, C.A. Emels, A.K. Nowak, *Nature* **1993**, *363*, 529.
41 G.J. Kramer, R.A. van Santen, *J. Am. Chem. Soc.* **1993**, *115*, 2887.
42 A. Redono, P.J. Hay, *J. Phys. Chem.* **1993**, *97*, 11 754.
43 T. Hibino, T. Akimoto, H. Iwahara, *Solid State Ionics* **1993**, *67*, 71.
44 M.B. Sayed, *Microporous Mater.* **1996**, *6*, 181.
45 U. Simon, U. Flesch, W. Maunz, R. Müller, C. Plog, *Microporous Mesoporous Mater.* **1998**, *21*, 111.
46 M.E. Franke, U. Simon, *Solid State Ionics* **1999**, *118*, 311.

47 M.E. Franke, U. Simon, *Phys. Stat. Solidi (b)* **2000**, *218(1)*, 287.
48 H. Böttger, V.V. Bryksin, *Hopping Conduction in Solids*, Wiley-VCH, Weinheim, **1985**.
49 M.J. Rice, W.L. Roth, *J. Solid State Chem.* **1972**, *4*, 294.
50 K.D. Kreuer, *Chem. Mater.* **1996**, *8*, 610.
51 M. Sierka, J, Sauer, *J. Chem. Phys.* **2000**, *112*, 6983.
52 K.-P. Schröder, J. Sauer, M. Leslie, C.R.A. Catlow, *Zeolites* **1992**, *12*, 23.
53 M.E. Franke, M. Sierka, J. Sauer, U. Simon, *Mat. Res. Soc. Symp. Proc.* **2001**, *658*, GG7.4.1.
54 M.E. Franke, M. Sierka, U. Simon, J. Sauer, *Phys. Chem. Chem. Phys.*, **2002**, *4(20)*, 5207.
55 M. Krossner, J. Sauer, *J. Phys. Chem.* **1996**, *100*, 6199.
56 H. Jobic, A. Tuel, M. Krossner, J. Sauer, *J. Phys. Chem.* **1996**, *100*, **19 545.**
57 M. Brändle, J. Sauer, *J. Molec. Catal. A* **1997**, *119*, 19.
58 M. Brändle, J. Sauer, R. Dovesi, N.M. Harrison, *J. Chem. Phys.* **1998**, *109*, 10 379.
59 V. Termath, F. Haase, J. Sauer, J. Hutter, M. Parrinello, *J. Am. Chem. Soc.* **1998**, *120*, 8512.
60 K. Schwarz, E. Nusterer, P.E. Blöchl, *Catal. Today* **1999**, *50*, 501.
61 L. Benco, T. Demuth, J. Hafner, F. Hutschka, *Chem. Phys. Lett.*, **2000**, *324*, 373.
62 L. Benco, T. Demuth, J. Hafner, F. Hutschka, *Chem. Phys. Lett.*, **2000**, *330*, 457.
63 P. Colomban (Ed.), *Chemistry of Solid State Materials: Proton Conductors*, Cambridge University Press, **1992**.
64 R. Moos, R. Müller, C. Plog, A. Knezevic, H. Leye, E. Irion, T. Braun, K.-J. Marquardt, K. Binder, *Sensors Actuators B*, **2002**, *83*, 181

2
Molecular Dynamics in Confined Space

Friedrich Kremer, Andreas Huwe, Annett Gräser, Stefan Spange, and Peter Behrens*

2.1
Introduction

The molecular and collective dynamics in confined space is determined by the counterbalance between surface and confinement effects [1]. The former result from interactions of a host system with guest molecules at the interface between them, and the latter originate from the inherent length scale on which the underlying molecular fluctuations take place. Surface effects cause a decrease and confinement effects an increase in the molecular dynamics with decreasing spatial dimensions of the confining space (Fig. 1). Hence, in glass-forming systems [2–11] increases and decreases, respectively, are observed in the calorimetric glass transition temperature. Evidently this counterbalance must depend sensitively on the type of confined molecules (glass-forming liquids, polymers, liquid crystals), on the properties of the (inner) surfaces (wettable, nonwettable), and on the architecture of the molecules with respect to the walls (grafted, layered, or amorphous systems). This article exemplifies the above described interplay between surface and confinement effects for the following systems: (1) ethylene glycol (EG) in zeolites, (2) propylene glycol (PG) in native and silanized mesoporous MCM, and (3) poly(isobutyl vinyl ether) (PIBVE) in mesoporous MCM materials.

2.2
Ethylene Glycol in Zeolites

Zeolites [12,13] offer the unique possibility to vary the dimension and the topology of spatial confinement on a subnanometer scale in a controlled manner. Silica sodalite consists of identical β-cages with an inner diameter of 0.6 nm. Ethylene glycol (EG) is one of the structure-directing agents which control the formation of silica sodalite [14,15]. Exactly one EG molecule becomes occluded in each sodalite cage during synthesis and cannot escape from it unless the cage is thermally decomposed [15]. Silicalite-1, zeolite beta, and AlPO4-5 have channel-like pore sys-

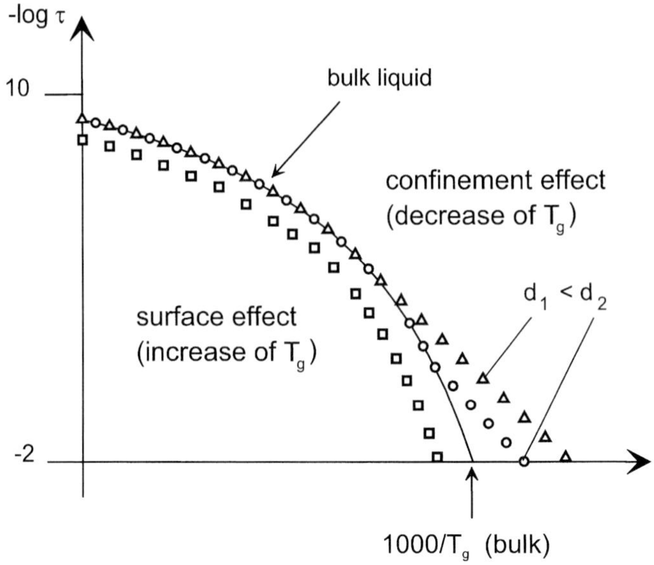

Fig. 1. Scheme of the molecular dynamics in confined space as a counterbalance between surface and confinement effects.

tems (see Fig. 2). In silicalite-1, which consists of pure SiO_2, rings of 10 Si and 10 O atoms form a three-dimensional pore system with two types of elliptical channels having cross sections of 0.56 × 0.53 nm and 0.55 × 0.51 nm [24]. In zeolite beta (12-ring system) the channels in the [100] and [010] directions have diameters of 0.76 × 0.64 nm, whereas the channels in the [001] direction have smaller pores (0.55 × 0.55 nm) [16]. The Si/Al ratio of the sample was 40, to reduce the number of counterions in the channels. $AlPO_4$-5 has a one-dimensional pore system. In this aluminophosphate, the channels with diameters of 0.73 nm are arranged in a hexagonal array.

Apart from sodalite, which is already loaded with EG after synthesis, all nanoporous hosts were heated to 600 K with a temperature increase of 20 K h^{-1} and evacuated at mbar 10^{-5} for 36 h to remove water and other volatile impurities. Then the zeolitic host systems were filled with EG from the vapor phase in a closed vacuum chamber at 448 K. The samples were cooled to room temperature and remained in the vacuum chamber for 24 h before the dielectric measurements were carried out.

Isothermal data (Fig. 3) of the dielectric loss ε'' were fitted by a superposition of a relaxation function given by Havriliak and Negami (HN) and a conductivity contribution (Eq. 1) [17,18].

$$\varepsilon'' = \frac{\sigma_0}{\varepsilon_0}\frac{a}{\omega^s} + \mathrm{Im}\left[\frac{\Delta\varepsilon}{(1+(i\omega\tau)^\alpha)^\gamma}\right] \tag{1}$$

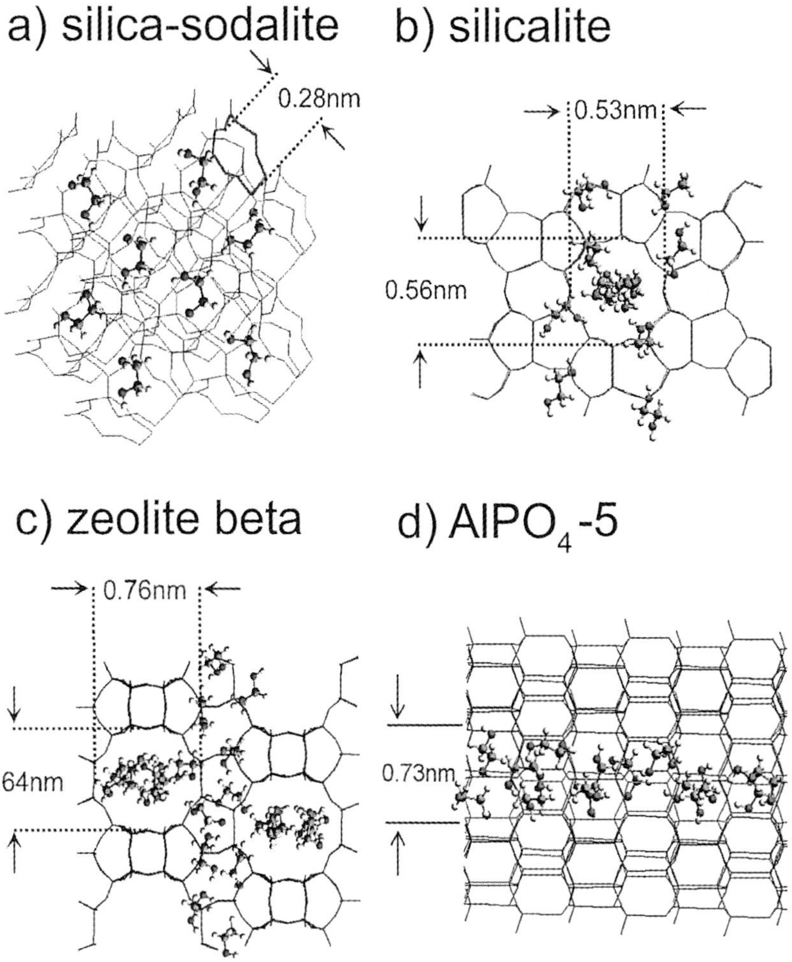

Fig. 2. Scheme of the zeolitic host systems in which the guest molecule ethylene glycol was confined. a) Silica sodalite (SiO_2) has cubic cages with a lattice constant of 0.89 nm. The cages are connected by channels with a diameter of 0.28 nm. Only one molecule is confined to each cage. b) Silicalite consists of pure SiO_2 and has a three-dimensional pore system with two different types of elliptical channels having cross sections of 0.56 × 0.53 nm and 0.55 × 0.51 nm. c) Zeolite beta is an aluminosilicate with a three-dimensional pore system having pore diameters of 0.76 × 0.64 nm and 0.55 nm. d) $AlPO_4$-5 is an aluminophosphate with one-dimensional channels (diameter 0.73 nm) arranged in hexagonal array. Taken from Ref. [1] with permission.

In this notation, ε_0 is the permittivity of vacuum, σ_0 the DC conductivity, $\Delta\varepsilon$ the dielectric strength, and α and γ describe the symmetric and asymmetric broadening of the relaxation peak. The exponent $s = 1$ holds for pure electronic conduction; deviations ($s < 1$) are caused by electrode polarization or Maxwell–Wagner polarization effects. The factor a has the dimension s^{1-s}. The uncertainty in the

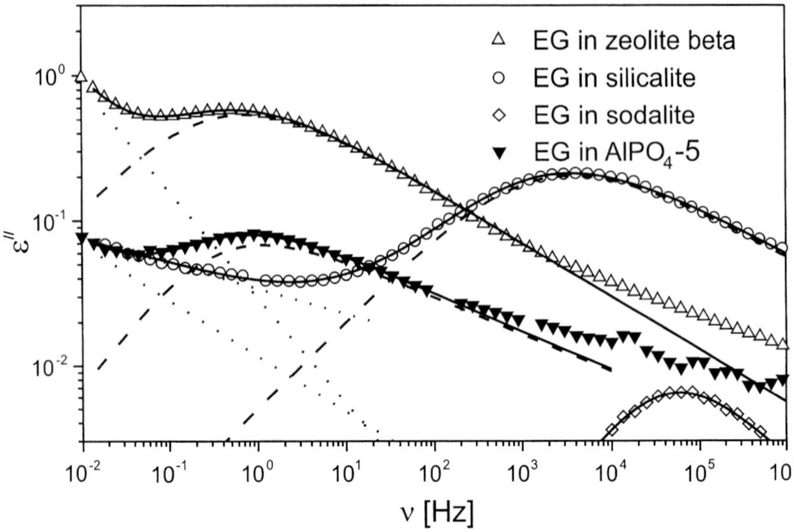

Fig. 3. Dielectric loss ε'' versus frequency for ethylene glycol (EG) confined in zeolitic host systems. The solid lines are a superposition of a Havriliak-Negami-relaxation (dashed line), and a conductivity contribution (dotted line). Taken from Ref. [1] with permission.

determination of log τ is ≤ 0.1 decades, and it is less than 5 % for $\Delta\varepsilon$. Due to the fact that ε' and ε'' are connected by the Kramers–Kronig relations, a fit in ε' does not improve the accuracy. From the fits according to Eq. (1) the relaxation rate $1/\tau_{max}$ can be deduced, which is given at the frequency of maximum dielectric loss ε'' for a certain temperature. A second way to interpret the data is the use of a relaxation time distribution $L(\tau)$ of Debye relaxators with relaxation times τ. The imaginary part of the dielectric function is expressed by Eq. (2)

$$\varepsilon'' = (\varepsilon_s - \varepsilon_\infty) \int \frac{L(\tau)}{1 + \omega^2\tau^2} \mathrm{d}\tau \tag{2}$$

where ε_s and ε_∞ denote the low- and high-frequency limit of the permittivity. $L(\tau)$ can be extracted numerically from the data [19] or calculated (analytically) from the fit with HN functions [17,18]. To characterize the temperature dependence of the relaxation behavior, the averaged logarithmic relaxation time log τ_{med} is calculated (Eq. 3)

$$\log \tau_{med} = \langle \log \tau \rangle = \left(\int_{-\infty}^{+\infty} \log \tau \, L(\tau) \mathrm{d}\tau \right) / \left(\int_{-\infty}^{+\infty} L(\tau) \mathrm{d}\tau \right) \tag{3}$$

log τ_{med} is equal to log τ_{max} if the peak of a relaxation process is symmetrically broadened. The log τ_{med} can only be calculated with high accuracy if the relaxation time distribution function is known over a broad range. Hence, log τ_{max} is determined for molecules confined in zeolites (where the frequency range is limited), and log τ_{med} for nanoporous sol–gel glasses as host. Figure 3 shows the dielectric spectra for ethylene glycol (EG) confined in different zeolitic host systems at 160 K. The relaxation rates $\tau_{max}{}^{-1}$ for EG in the zeolitic host systems differ by up to six orders of magnitude: In zeolites with smaller pores (silicalite and sodalite) the relaxation rates of EG are significantly higher than in zeolite beta and $AlPO_4$-5. Especially for EG in sodalite, the relaxation strength is comparatively low. This is due to EG molecules which are immobilized by interaction with the zeolitic host matrix. Figure 4 shows the relaxation rate as a function of reciprocal temperature for EG as bulk liquid and confined in zeolites. EG in zeolite beta (solid triangles) and in $AlPO_4$-5 (open triangles) has a relaxation rate like that of the bulk liquid

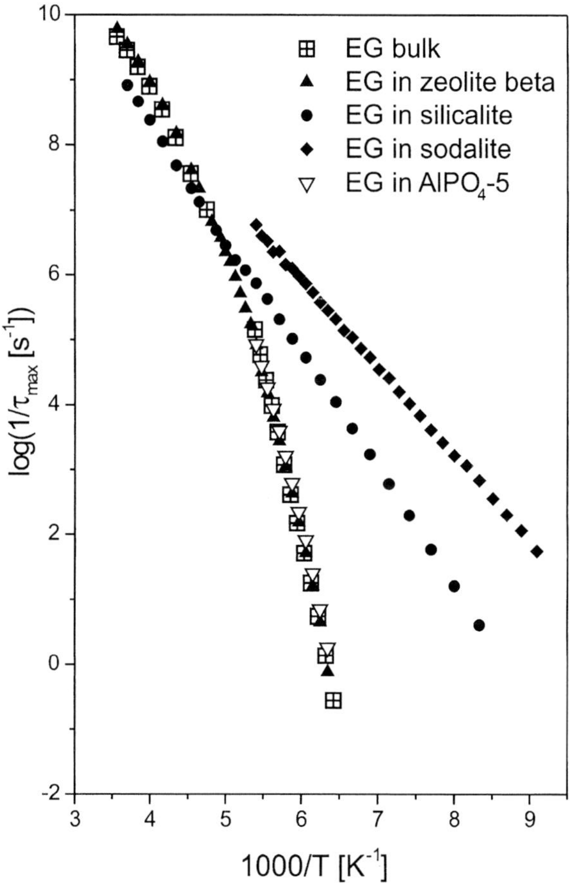

Fig. 4. Relaxation rate versus reciprocal temperature for ethylene glycol confined to different zeolitic host systems. The errors are smaller than the size of the symbols. Taken from Ref. [1] with permission.

(squares) and follows the temperature dependence according to the empirical Vogel–Fulcher–Tammann (VFT) equation (Eq. 4) [20–22]

$$\frac{1}{\tau} = A \exp\left(\frac{-DT_0}{T - T_0}\right) \tag{4}$$

where A is a prefactor, D is the fragility parameter, and T_0 is the Vogel temperature. The relaxation rates of EG in silicalite and sodalite show an Arrhenius-type temperature dependence.

The single-molecule relaxation of EG in sodalite at $T \approx 155$ K is about six orders of magnitude faster compared to the bulk liquid. Its activation energy is 26 ± 1 kJ mol^{-1} and corresponds to the value for bulk EG at high relaxation rates (29 ± 2 kJ mol^{-1}) [23]. The relaxation process of EG in silicalite has a larger activation energy (35 ± 2 kJ mol^{-1}) which is still smaller than the apparent activation energy (tangent to the VFT temperature dependence) of the bulk liquid close to T_g. Its Arrhenius-like temperature dependence resembles that of the single-molecule relaxation of EG in sodalite.

To study the molecular arrangement of the molecules in confined space, the molecular simulation program Cerius^2 was used on a Silicon Graphics workstation to model a finite zeolite crystal with four unit cells surrounded by vacuum. By "filling" the pores with EG a completely loaded nanoporous host–guest system can be simulated, and structural parameters such as distance between molecules, density, and length of hydrogen bonds can be determined. The simulations were carried out by using three different force fields: the Dreiding force field [24], the force field burchart-universal [25,26], and the consistent force field [26]. The three force fields provide the same results within the uncertainty for the quantities listed in Tab. 1.

The computer simulations of EG in zeolitic host systems show that in silicalite the molecules are aligned almost single-file-like along the channels and that in zeolite beta and in $AlPO_4$-5 two EG molecules are located side by side in the channels. However, for the distance between molecules, the average length of hy-

Tab. 1. Distance between molecules, average length of hydrogen-bonds (O–H···O bonds with a length up to 0.3 nm), and density, calculated from the molecular simulations for ethylene glycol confined in zeolite beta and silicalite and for the bulk liquid. For simulation of the bulk liquid a limited volume (6.64 nm^3) was filled with EG molecules until the bulk density of 1.113 g cm^{-3} was reached. In contrast, the densities of EG confined in zeolites are results of the simulation. The error is mainly caused by the uncertainty in calculating the accessible volume of the zeolitic channels.

	Distance between molecules (nm)	*Average length of H-bonds (nm)*	*Density (g cm^{-3})*
Bulk liquid	0.42 ± 0.01	0.23 ± 0.02	1.113
Zeolite beta	0.41 ± 0.01	0.25 ± 0.02	1.0 ± 0.1
Silicalite	0.42 ± 0.01	0.24 ± 0.02	1.0 ± 0.1

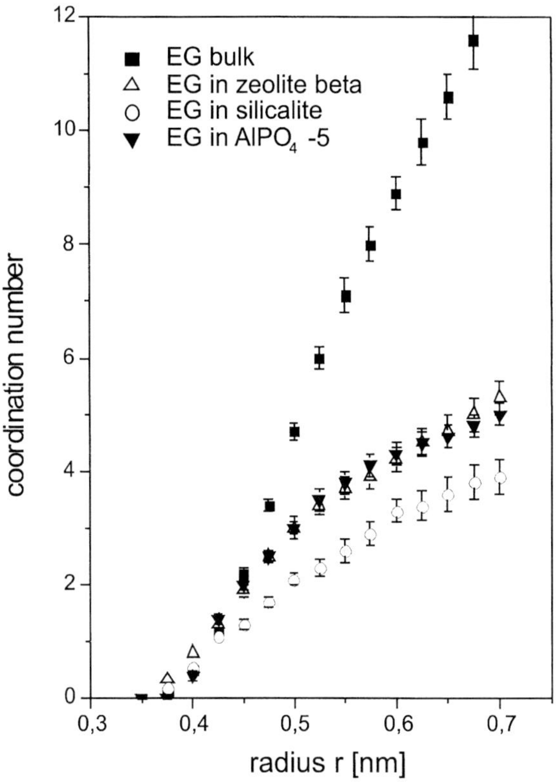

Fig. 5. Average number of neighboring molecules (coordination number) as a function of the radius of a surrounding sphere, calculated from the simulations for EG bulk liquid (squares), EG confined in zeolite beta (triangles), in silicalite (circles), and in $AlPO_4$-5 (solid triangles). Taken from Ref. [1] with permission.

drogen bonds, and the density, no significant change is found between the bulk liquid and the molecules in the restricted geometry (Tab. 1). However, for the number of neighboring molecules (coordination number) a pronounced difference is observed (Fig. 5): The coordination number of 11 corresponds to the maximum value in the case of the random close-packing model [27] and is found for the bulk liquid within a radius of $r = 0.66$ nm, for which EG in zeolite beta and in $AlPO_4$-5 has only five neighboring molecules. As $AlPO_4$-5 has nonintersecting one-dimensional channels, in contrast to zeolite beta the dimensionality of the host system seems to play only a minor role for the dynamics of hydrogen-bonded guest molecules. Further reduction in the channel size (as in the case of silicalite) decreases the average number of neighboring molecules by about 1. This results in a sharp transition from liquidlike dynamics to those of single molecules. In $AlPO_4$-5 only two molecules are located side by side in the one-dimensional channels; hence, the interactions are dominated by the nearest-neighbor molecules, and an ensemble as small as six EG molecules is sufficient to show liquidlike dynamics.

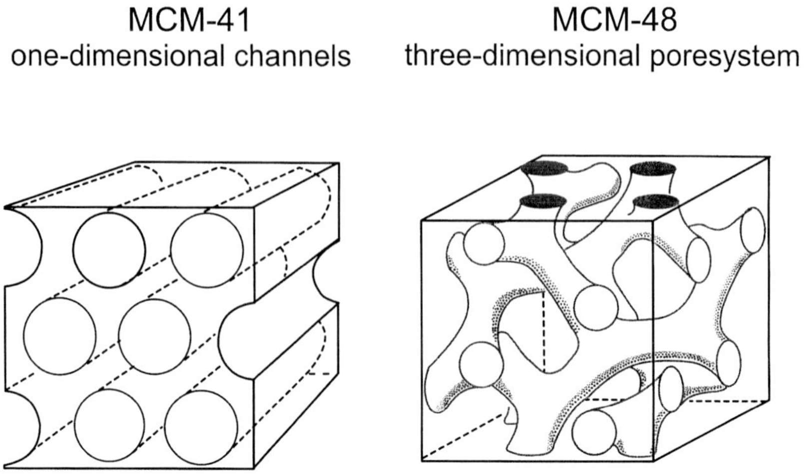

Fig. 6. Structures of mesoporous MCM-41 and MCM-48.

2.3 Propylene Glycol in Mesoporous MCMs

The dielectric measurements were carried out with MCM-41 and MCM-48, both having a pore diameter of 2.7 nm (Fig. 6). The spectra of propylene glycol in MCM-materials show one molecular relaxation processes which is assigned to the dynamic glass transition (α-relaxation) in the mesoporous environment [28].

In the uncoated pores of MCM-41 and MCM-48 propylene glycol shows a surface effect (Fig. 7). The molecular dynamics are shifted to lower values compared to the bulk liquid due to the formation of hydrogen bonds between the propylene glycol molecules and the hydrophilic silica pore walls of the MCM materials. After hydrophobization of the silica walls the formation of hydrogen bonds with the pore surface is hindered (see Fig. 8). The suppression of the surface effect results in a relaxation rate of propylene glycol in the coated mesoporous hosts which is comparable to that of the bulk liquid over the whole temperature range for MCM-41 and MCM-48 samples. From the size of the pores one has to conclude that the molecular rearrangements of PG take place on a length scale of ≤ 2 nm, in accordance with previous studies [9].

2.4 Poly(Vinyl Ether) in Mesoporous MCMs

It is possible to synthesise poly(vinyl ether)s directly in the channels of nanoporous zeolites and mesoporous MCM materials [29–33]. MCMs have pores with diameters in the range from 2 to 8 nm with a narrow pore size distribution. The framework of these porous materials is an amorphous aluminosilicate. To study the molecular dynamics of polymers in confined space the following host systems (Fig. 6 and Tab. 2) were used: MCM-41 with one-dimensional channels and a pore

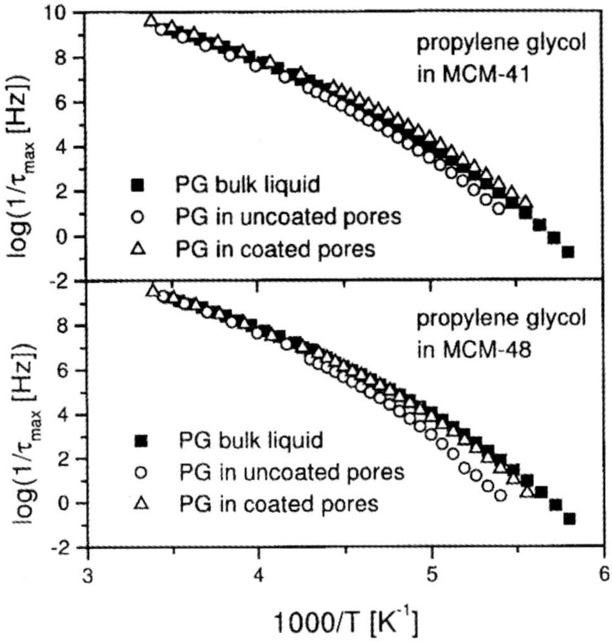

Fig. 7. Mean relaxation rate versus reciprocal temperature of propylene glycol as bulk liquid (solid squares) and confined in uncoated (open circles) and silanized pores (triangles) of MCM-41 (top) and MCM-48 (bottom) having a pore diameter of 2.7 nm. The errors are smaller that the size of the symbols. The Maxwell–Wagner polarization is omitted.

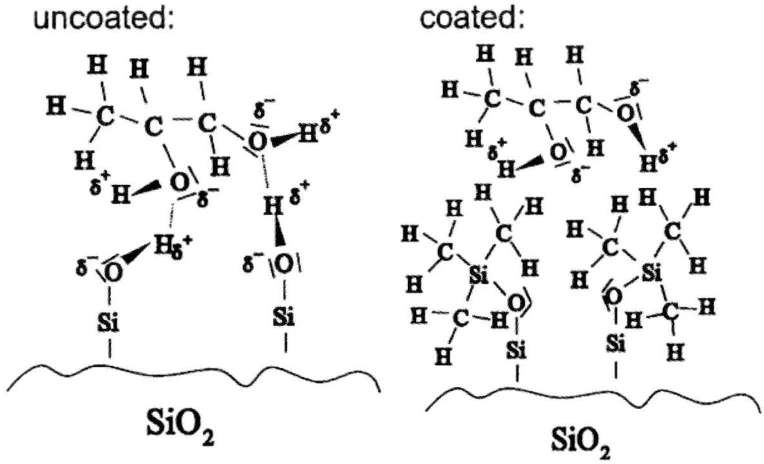

Fig. 8. Schematic diagram of propylene glycol in the neighborhood of an uncoated SiO_2 surface (left) and a silanized SiO_2 surface (right).

Tab. 2. Pore diameter, inner surface, specific pore volume, and specific channel length for the MCM-41 and MCM-48 materials in which poly(isobutyl vinyl ether) was synthesised.

	MCM-41	*MCM-48*
Pore diameter (nm)	3.6	2.5
Inner surface ($m^2\ g^{-1}$)	770	1840
Specific pore volume ($cm^3\ g^{-1}$)	1.25	1.34
Specific channel length (1010 m g^{-1})	11.6	27.3

diameter of 3.6 nm, and MCM-48 having pores with a pore size of 2.5 nm and a cubic structure [34,35] (The MCM materials used in these studies were not identical to those employed for the measurements with confined PG).

The polymerization was started by an initiator or by silanol groups at the inner surface of the host system. Figure 9 shows how surface-induced polymerization

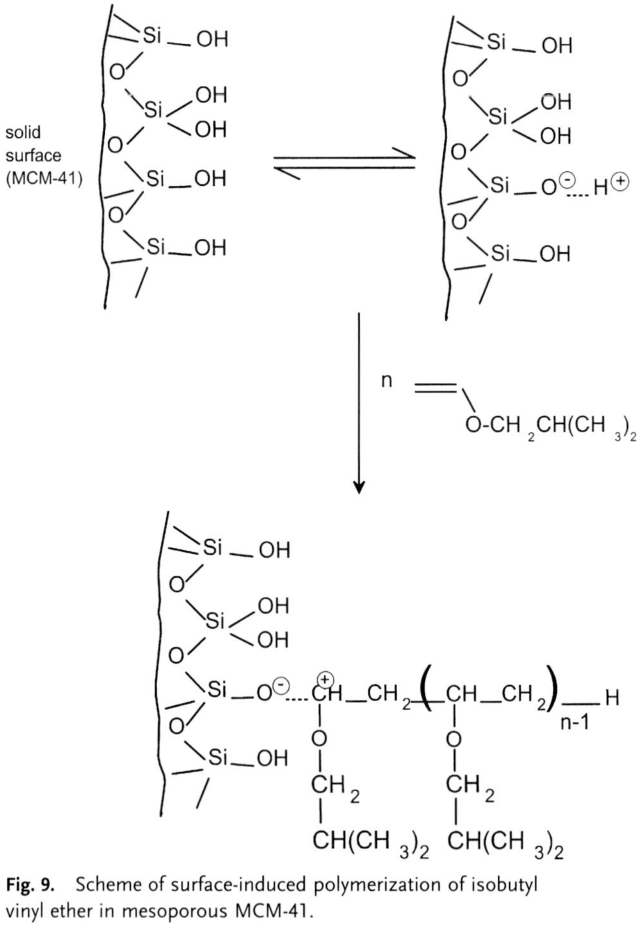

Fig. 9. Scheme of surface-induced polymerization of isobutyl vinyl ether in mesoporous MCM-41.

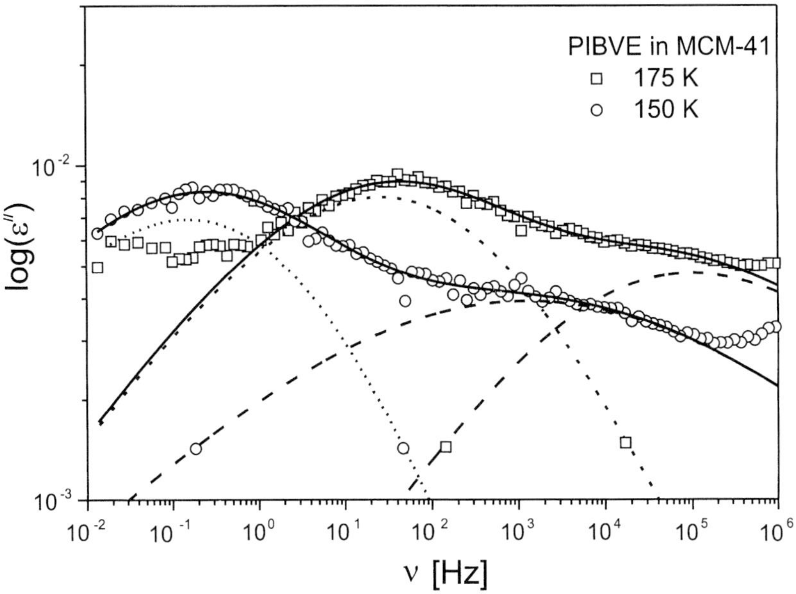

Fig. 10. Dielectric spectra of PIBVE confined in MCM-41 at 175 K (squares) and 150 K (circles). The solid line is a superposition of the HN fits for the α- (dotted line) and β-relaxation (dashed line) at the indicated temperatures.

works. In both cases the pores are filled only partially by the polymer. For the polymer in MCM materials the filling ratio is comparatively large (up to 43 vol %), and the dielectric spectra show distinct relaxation processes. Figure 10 shows the dielectric spectra of poly(isobutyl vinyl ether) (PIBVE) in MCM-41 for two different temperatures. Figure 11 shows the relaxation rates and the dielectric strength for PIBVE in the bulk and in the confining spaces of MCM-41. One process has a relaxation rate similar to the β-relaxation of the bulk polymer with an Arrhenius-like temperature dependence. It corresponds to the β-relaxation of the confined polymer and is assigned to fluctuations of the ether group [35]. The second process has also an Arrhenius-type temperature dependence. It is much faster than the α-relaxation of the bulk polymer. After annealing the sample, its relaxation rate slows down and approaches the dynamic glass transition of the bulk sample. This process is assigned to the α-relaxation of the confined polymer. The rate of the β-relaxation is almost uninfluenced by confinement and thermal treatment. The relaxation strength of both processes decreases after annealing. Comparing the molecular dynamics of PIBVE in MCM-41 and in MCM-48, faster relaxation rates are observed (Fig. 10) in the smaller pores of MCM-48. Hence, similarly to EG in zeolites, PIBVE in MCMs shows a confinement effect.

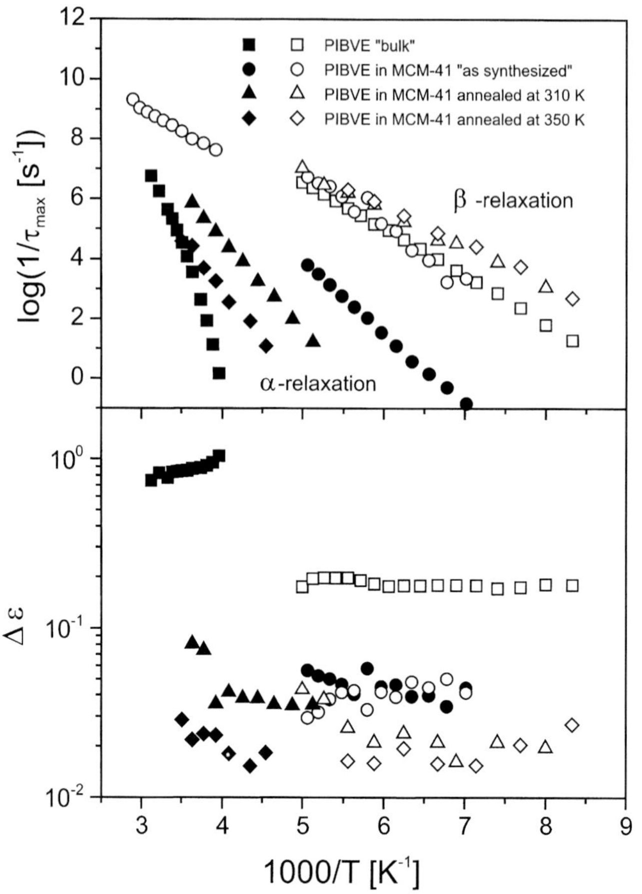

Fig. 11. Relaxation rate $1/\tau_{max}$ and dielectric strength $\Delta\varepsilon$ versus reciprocal temperature for PIBVE in the bulk and in MCM-41. The solid symbols correspond to the α-relaxation, and the open symbols to the β-relaxation. The thermal treatment of the samples is indicated in the figure.

2.5 Conclusions

Molecular dynamics in confined space is determined by the interplay between confinement and surface effects. This is demonstrated in this article for the following systems:

- Ethylene glycol (EG) in zeolitic host systems shows a pronounced confinement effect. Beyond a threshold channel size, the liquid character is lost, as indicated by a dramatically increased relaxation rate and an Arrhenius-like temperature dependence. Computer simulations of the molecular arrangement in the confin-

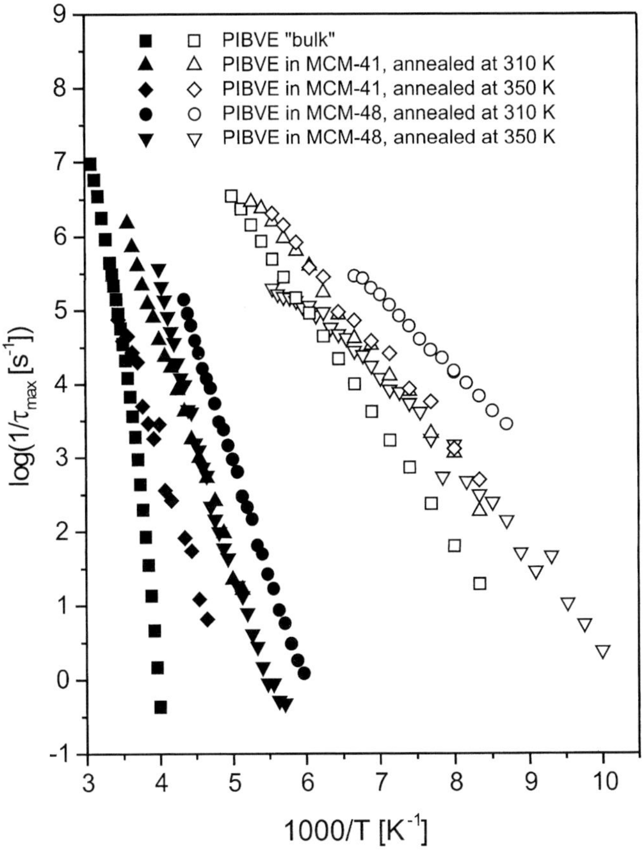

Fig. 12. Relaxation rate versus reciprocal temperature for PIBVE in the bulk (squares), in MCM-41 (upward triangles and diamonds) and in MCM-48 (circles and downward triangles). The solid symbols correspond to the α-relaxation, and the open symbols to the β-relaxation.

ing space prove that an ensemble as small as six EG molecules is sufficient to exhibit the dynamics of a bulk liquid.

- Propylene glycol (PG) shows for untreated native pores a surface effect, which results in an overall decrease of the molecular dynamics. This can be fully removed by making the boundary layer between the guest molecules and the solid host system hydrophobic. The molecular dynamics in the confined system is then faster than in the bulk liquid.
- Cationic host–guest polymerization enables the synthesis of poly (isobutyl vinyl ether) in nanoporous channels of MCMs with different topology. In full accord with the results for low molecular weight systems, a confinement effect is observed. Residual solvent acts as a plasticizer which enhances the molecular dynamics.

References

1 F. Kremer, A. Huwe, M. Arndt, P. Behrens, W. Schwieger, *J. Phys. Conens. Matter* **1999**, *11*, A175.

2 G. Adam, J.H. Gibbs, *J. Chem. Phys.* **1965**, *43*, 139.

3 E. Donth, *Glasübergang*, Akademie Verlag, Berlin, **1981**.

4 E. Donth, *Relaxation and Thermodynamics in Polymers, Glass Transition*, Akademie Verlag, Berlin, **1992**.

5 E.W. Fischer, E. Donth, W. Steffen, *Phys. Rev Lett.* **1992**, *68*, 2344.

6 E.W. Fischer, *Physica A* **1993**, *201*, 183.

7 D. Sappelt, J. Jäckle, *J. Phys. A* **1993**, *26*, 7325.

8 A. Huwe, F. Kremer, P. Behrens, W., Schwieger, *Phys. Rev. Lett.* **1999**, *82*, 2338.

9 W. Gorbatschow, M. Arndt, R. Stannarius, F. Kremer, *Europhys. Lett.* **1996**, *35*, 719.

10 M. Arndt, R. Stannarius, W. Gorbatschow, F. Kremer, *Phys. Rev. E* **1996**, *54*, 5377.

11 a) R. Stannarius, F. Kremer, M. Arndt, *Phys. Rev. Lett.* **1995**, *75*, 4698. b) M. Arndt, R. Stannarius, H. Groothues, E. Hempel, F. Kremer, *Phys. Rev. Lett.* **1997**, *79*, 2077.

12 W.M. Meier, D.H. Olson, C. Baerlocher, *Atlas of Zeolite Structure Types*, Elsevier, Amsterdam, **1996**.

13 J. Kärger, D.M. Ruthven, *Diffusion in Zeolites and Other Microporous Solids*, Wiley, New York, **1992**.

14 M. Bibby, M.P. Dale, *Nature* **1985**, *317*, 157.

15 C.M. Braunbarth, P. Behrens, J. Felsche, G. van de Goor. *Solid State Ionics* **1997**, *101–103*, 1273.

16 J.M. Newsam, M.M.J. Treacy, W.T. Koetsier, C.B. de Gruyter, *Proc. Roy. Soc. (London)* **1988**, *420*, 375.

17 S. Havriliak, S. Negami, *J. Polym. Sci. Part C* **1966**, *14*, 99.

18 S. Havriliak, S. Negami, *Polymer* **1967**, *8*, 161.

19 H. Schäfer, E. Sternin, R. Stannarius, M. Arndt, F. Kremer, *Phys. Rev. Lett.* **1996**, *76*, 2177.

20 H. Vogel, *Phys. Zeit.* **1921**, *22*, 645.

21 G.S. Fulcher, *J. Am. Ceram. Soc.* **1925**, *8*, 339.

22 G. Tammann, G. Hesse, *Z. Anorg. Allgem. Chem.* **1926**, *156*, 245.

23 B.P. Jordan, R.J. Sheppard, S. Szwarnowski, *J. Phys. D* **1978**, *11*, 695.

24 S.L. Mayo, B.D. Olafson, W.A. Goddard III, *J. Phys. Chem.* **1990**, *94*, 8897.

25 A.K. Rappe, C.J. Casewit, K.S. Colwell, W.A. Goddard III, W.M. Skiff, *J. Am. Chem. Soc.* **1992** *114*, 10 024.

26 E. Burchart, Thesis, Technische Universiteit Delft, **1992**.

27 N.E. Cusack, *The Physics of Structurally Disordered Matter*, Adam Hilger, Bristol, **1987**.

28 A. Huwe, F. Kremer, *ACS Symposium Series* 820 (**2002**) Chap. 20, 268–283.

29 A. Gräser, S. Spange, *Chem. Mater.* **1998**, *10*, 1814.

30 S. Spange, Y. Zimmermann, A. Gräser, *Chem. Mater.* **1999**, *11*, 3245.

31 S. Spange, A. Gräser, H. Müller, Y. Zimmermann, P. Rehak, C. Jäger, H. Fuess, C. Baetitz, *Chem. Mater.* **2001**, *13*, 3698.

32 S. Spange, A. Gräser, P. Rehak, C. Jäger, M. Schulze, *Macromol. Rap. Commun.* **2000**, *21*, 146.

33 S. Spange, A. Gräser, A. Huwe, F. Kremer, C. Tintemann, P. Behrens, *Chem. Eur. J.* **2001**, *7*, 3722.

34 C.T. Kresge, M.E. Leonowicz, W.J. Roth, J.C. Vartuli, J.S. Bech, *Nature* **1992**, *359*, 710.

35 J.S. Beck, J.C. Vartuli, W.J. Roth, M.E. Leonovicz, C.T. Kresge, K.D. Schitt, C.T.W. Chu, D.H. Olson, E.W. Shappard, S.B. McCullen, J.B. Higgins, J.L. Schlenker, *J. Am. Chem. Soc.* **1992**, *114*, 10834.

36 R. Casalini, D. Fioretto, A. Livi, M. Lucchesi, P.A. Rolla. *Phys. Rev. B* **1997**, *56*, 3016.

3
Conductive Structures in Mesoporous Materials

*Nikolay Petkov and Thomas Bein**

3.1
Introduction

3.1.1
Molecular Electronics

The miniaturization of electronic components and devices produced by the semiconductor industry in the highly successful "top-down" approach using lithography proceeds at a breathtaking pace. However, at some point one will have to face fundamental limits to this development [1], and the alternative "bottom-up" approach is expected to gain in importance. In this context, the concept of "molecular electronics" has been explored in which the basic electronic devices (wires and transistors) will be replaced by functional molecules created by bottom-up technology [2]. There are, of course, many challenging problems to be overcome before the first molecular electronic devices can be produced. Among them, the addressability of the conductive structures, the arrangement of the conductive molecules in a three-dimensional (3D) array, and the encapsulation of the conductive structures so that they are easily accessible and isolated from each other are of special interest. Another challenge is to develop a method for rapid screening and electrical characterization of new candidates for molecular electronic devices. This requires special techniques that can link the nanoscale world of molecular devices to the macroscopic world of conventional electronic circuits.

An assembly of molecular wires, switches, or connections could be addressed by constructing an ordered channel matrix of insulating material (e.g., metal oxide) in which these structures are encapsulated and therefore isolated from each other. On the other hand, solid supports with ordered porosity can offer a template matrix for guided growth of a variety of conductive structures, giving patterned arrays of aligned nanowires and connections. Such solid supports with ordered porosity ranging in diameter from sub-nanometer to several nanometers are, for example, the silica-based micro- or mesoporous materials, also called molecular sieves. Sol-gel chemistry provides different routes for the preparation of such structures in a variety of special morphologies ranging from bulk materials to thin films and nanofibers. The synthesis and applications of molecular sieves, layers, and membranes have been reviewed [3]. Recently the subject of encapsulation of conducting polymers in different hosts ranging from micro- and mesoporous materials, pillared clays, and layered materials to organic macrocycles possessing accessible cavities have been discussed [4,5]. The state of the art in molecular electronics with an emphasis on the synthesis and testing of components has also been reviewed [2].

In this chapter we focus on the utilization of mesoporous solids showing different morphologies and symmetries as hosts for the encapsulation of conductive materials at the nanometer scale. We focus on recent reports regarding the encapsulation of carbon filaments and nanotubes, metal nanowires and nanoarrays, and semiconducting nanoparticles and wires in ordered mesoporous hosts. Different morphologies including thin films hosting conductive guests are also covered.

3.1.2 Mesoporous Materials

Since the discovery of the so-called M41S mesoporous silica materials in the early 1990s, considerable synthetic and characterization effort has been devoted to these materials resulting in more than 2500 papers and numerous reviews [6,7]. This effort has led to the formation of a new class of porous solids with uniform pore sizes in the range 3–30 nm and extremely high surface areas that can be easily tuned by choosing the proper surfactant and co-surfactant agents. At present, periodic mesoporous silica materials may be readily synthesized in a wide range of pH, at temperatures ranging from ambient to approximately 150 °C, using a variety of surfactants and polymers as structure directing agents, thus leading to different mesophase structures and morphologies. Features such as structure, composition, pore diameter, surface area, and morphology of the products can be controlled by the inorganic source material, molar composition, template type, and the condensation/hydrolysis process. It is the intriguing interaction of liquid-crystal templating and the cooperative self-assembly of inorganic and organic (micellar) species that is responsible for the formation of these materials [8–12]. The mechanistic pathways for the formation of mesoporous materials are still controversial, but in any case the resultant inorganic material mimics the liquid-crystal mesophase [4,13]. As a result mesoporous materials with different mesophase structure have been obtained: hexagonal MCM-41, SBA-15; cubic MCM-48, SBA-1, SBA-11, SBA-16; FDU-1; 3D hexagonal SBA-2; lamellar MCM-50; disordered mesophase structures KIT-1, MSU-X, and others [4,14–19]. This class of inorganic solids is not limited to silicates or aluminosilicates, but mesoporous forms of a variety of other metal oxides were also obtained [20,21]. Recently, mesoporous carbons (CMK) were also prepared as replicas of the silica-based mesoporous materials [22–24]. The walls of the pore-structure of most of these materials are essentially amorphous; this often leads to lower hydrothermal stability compared to many zeolites and limits some of the desirable applications. Recently a new promising class of hybrid inorganic/organic materials possessing crystallinity in the mesoporous walls has been discovered, thus opening prospects for novel applications [25]. The option of introducing different molecular functionality by co-condensation or post-synthetic reactions inside the porous system of these materials makes them promising candidates for a variety of technological applications ranging form heterogeneous catalysis to the creation of sensor or optical devices [26–30].

Mesostructured materials with specific morphologies (thin films, fibers, monoliths, and free-standing membranes) have also been prepared using appropriate synthesis conditions [31–39]. This has opened new avenues to many advanced nanotechnological applications such as the preparation of optical and electronic devices at the nanometer scale and the creation of special strategies for addressing them.

3.1.3
General Synthetic Methods for Nanowires

Continuous micrometer long metal nanowires can be prepared by electrochemical template replication, electrochemical step-edge metal decoration, and electroless plating [40–49]. Metal salt impregnation and metalorganic chemical vapor infiltration of solid porous substrates followed by metal reduction are also considered to be effective routes towards such systems [50–57]. The electrochemical template synthesis of metal nanowires (established by Martin, Moskovits, and Searson [40–42] involves electrochemical deposition of metal into cylindrical pores of an inert, nonconductive host material: usually porous alumina or polycarbonate membranes with pore sizes ranging from tens of nanometers to several hundred nanometers [40–42,49]. This approach results in metal wires that are nanoscopic in diameter but macroscopic in length. Electrochemical step-edge metal decoration employs selective deposition of a metal at atomic step edges of a single crystal surface [44–47]. The wire thickness in the direction normal to the substrate is only several atomic layers. Electroless reduction methods use metal catalyst nanoparticles to nucleate and guide the growth of continuous metal nanowires. It has been shown that chiral lipid tubules can be metallized by electroless reduction giving nanometer thin metal wires [48]. Metallized DNA molecules were also subjected to electroless reduction, resulting in a metal nanowire between two metal electrodes [58,59]. Mo_3Se_3 nanowires were used as reducing and sacrificial templates for the preparation of micrometer-long metal nanowires (Au, Ag, Pt) [60]. Metal nanowires were also prepared using carbon nanotubes as templates [61].

3.2
Metal Nanowires and Nanoarrays in Mesoporous Hosts

Here we discuss in more detail different approaches used for the preparation and characterization of various metal nanowires in mesoporous solids. The templating role of the mesoporous matrix is clarified.

Metal infiltration through wet impregnation or metalorganic chemical vapor deposition (MOCVD) in the hollow channels of different porous supports are usually used to load porous materials with metallic particles for different catalytic applications [62–64]. This approach has been expanded to the preparation of continuous metal wires. Thus, metal salt impregnation and metalorganic chemical vapor infiltration followed by hydrogen reduction so far are the basic synthetic strategies used to prepare metal nanoparticles and nanowires in different mesoporous solids having hexagonal (MCM-41, SBA-15), cubic (MCM-48), or disordered (KIT-1) structure. In an early article, Ryoo et al. described a technique to probe the local channel arrangement of mesoporous silica by preparing Pt-loaded mesoporous materials, in conjunction with transmission electron microscopy (TEM) (Fig. 1) [50]. MCM-41, MCM-48, and KIT-1 mesoporous materials were loaded with tetraammine platinum (II) nitrate $Pt(NH_3)_4(NO_3)_2$ by an incipient wetness impregnation technique, followed by reduction of the Pt precursor at 400 °C in a flow of hydrogen. To obtain a higher degree of Pt loading, the impregnation was repeated several times. In a more recent paper, the authors describe a technique for the preparation of Pt networks in hexagonal MCM-41, SBA-15, and cubic MCM-48 materials in which

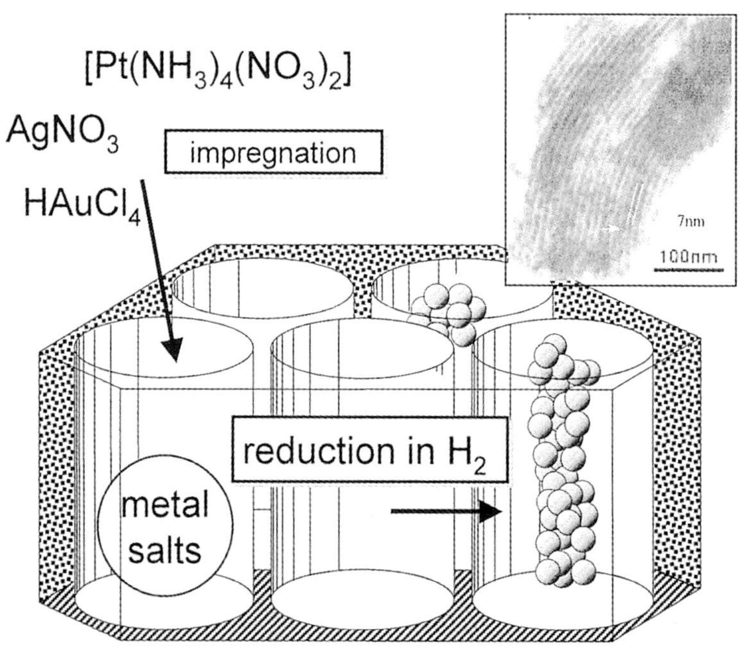

Fig. 1. A strategy for the incorporation of noble metal nanowires in mesoporous hosts through incipient wetness impregnation of salts followed by reduction [50–54]. Inset: Pt nanowires prepared through repeated wet impregnation of SBA-15 with $Pt(NH_3)_4(NO_3)_2$ followed by reduction [52].

the Pt loading approaches 70 wt.-% [51]. Template-free Pt nanowires were obtained by dissolving the mesoporous silica matrix in HF. The TEM investigations show the resulting template-free Pt nanowires and Pt networks with diameters similar to the channel dimensions of the mesoporous host. X-ray diffraction data confirmed the crystallinity of the Pt nanowires and networks. Recently, Terasaki and Ryoo used the same synthetic approach together with high-resolution electron microscopy (HRTEM) to probe the local structure of SBA materials and to demonstrate the microporosity existing in the channel walls of these materials [52]. A comparison between Pt wires prepared in hexagonal MCM-41 and in SBA-15 materials shows that isolated wires can be obtained after removing the MCM-41 host, whereas in the case of SBA-15 material the wires are interconnected because of the microporosity in the channel walls. The microporosity of the SBA materials is explained by the specific interaction of the ethylene oxide groups of the copolymers used as organic templates for the preparation of this class of mesoporous solids.

Following a similar impregnation approach, ordered silver nanocrystal arrays were prepared in cubic MCM-48 mesoporous silica [53]. Thermal treatment at 300 °C led to the formation of metallic Ag as confirmed by high-angle XRD. Stucky et al. demonstrate a general synthetic route for the preparation of noble metal nanowires (Au, Ag, Pt) encapsulated in SBA-15 materials [54]. In order to incorporate the metal precursors inside the mesoporous channels, calcined mesoporous SBA-15 was impregnated with aqueous solutions containing noble metal salts such as $Pt(NH_3)_4(NO_3)_2$, $HAuCl_4$, and $AgNO_3$. TEM and EDAX investigations showed the metal loading to be 5–15 wt.-%. The average length of a single nanowire is 500 nm with a uniform diameter of 7 nm, consistent with the channel diameter of the mesoporous host. The dimensions of the nanowires can be controlled by changing the loading, annealing temperature, and annealing time.

Chemical vapor infiltration is another approach used to prepare metal nanowires inside the mesoporous host matrix. Cheon et al. demonstrated the synthesis of Pd nanoballs and nanowires in MCM-48, MCM-41, and SBA-15 materials, by chemical vapor infiltration with the organometallic precursor $Pd(hfac)_2$ (hfac = 1,1,1,5,5,5-hexafluoroacetylacetonate) (Fig. 2) [55,56]. The organometallic precursor was sublimed into the degassed pores of the mesoporous materials at 55 °C and decomposed at 150 °C in a flowing H_2/N_2 mixture. It was shown that Pd nanoball domains of 35–40 nm dimension consisting of 3D interconnected Pd networks whose shapes and pores are a replication of the MCM-48 template can be obtained after careful dissolution of the silica framework. In the case of hexagonal MCM-41 and SBA-15 materials, it was demonstrated that adjustable 100–150 nm Pd nanowires can be prepared with diameters determined by the channel dimensions of the mesoporous host. The thermal behavior of the Pd nanowires was studied *in situ* by TEM. A significant decrease in the melting point of these Pd wires was attributed to their small dimensions. In a recent communication, the *in situ* formation of gold nanoparticles within functionalized mesoporous silica via an organometallic "chimie douce" approach was reported (Fig. 3) [65]. The selective growth of Au nanoparticles could be achieved in thiol-functionalized mesoporous silica.

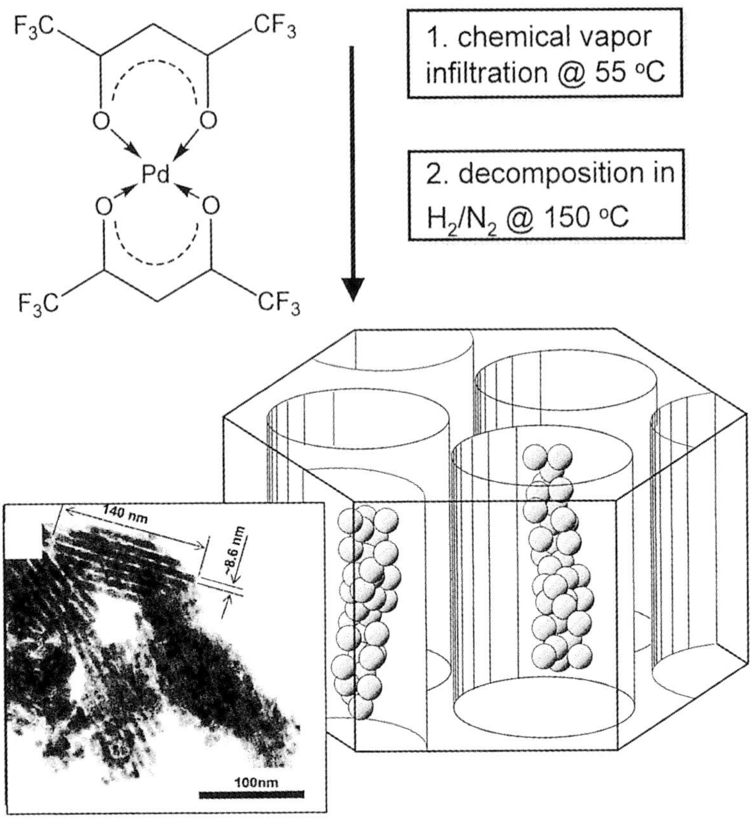

Fig. 2. Organometallic chemical vapor infiltration and decomposition to crystalline Pd nanowires at elevated temperatures [55,56]. Inset: Pd nanowires in SBA-15 [55].

The anchored Au precursor was subsequently treated with reducing agents such as sodium citrate and $NaBH_4$, resulting in nanometer-sized Au particles, as confirmed by XRD, TEM, and UV/vis measurements. Ag_xAu_{1-x} alloy nanocrystals were synthesized within the pores of mesoporous silica by repeated wet impregnation of gold and silver salts and subsequent annealing in air at 500 °C [66].

Unfortunately, electrical conductivity measurements were not performed in any of the studies discussed here, but potential applications of the metal nanostructures as conductive devices suitable for molecular electronics are occasionally mentioned. Recently, the preparation of RuO_2 nanowires in disordered mesoporous silica aerogels by cryogenic decomposition of RuO_4 was reported [67]. Two-point probe conductivity tests between various points in the center and on the outer surfaces of the loaded aerogel show metallic conductivity after annealing the composite material at 150 °C.

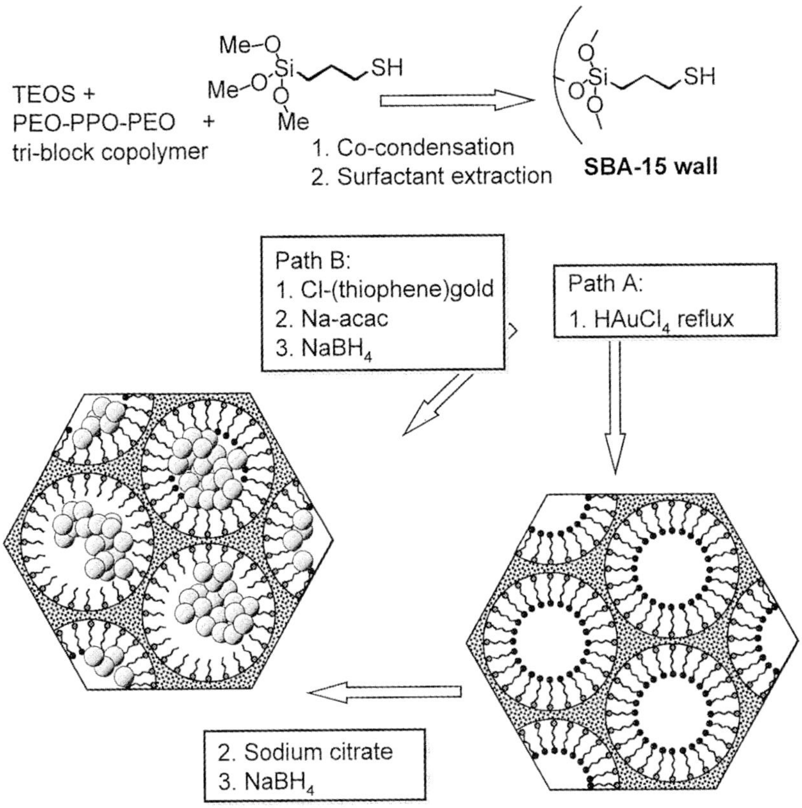

Fig. 3. Reaction pathway towards the preparation of gold nanoparticles in SBA-15 through an organometallic "chimie douce" approach [65].

3.3
Semiconductor Nanoparticles and Nanoarrays in Mesoporous Hosts

Semiconductor nanotechnology could lead to major breakthroughs in the design of electronic systems owing to the appealing perspectives offered by size-quantization effects [68]. The production of two-dimensional (2D) quantum wells is already a mature technology. The confinement of semiconductors in less than 3D is well established; it is commonly done by laser-assisted CVD or molecular beam epitaxy deposition techniques [69,70]. Nevertheless, the preparation of uniform 2D-confined quantum wires and 3D-confined quantum dots is still a challenge. The pore systems of mesoporous materials offer the potential for synthesizing 3D semiconductor heterostructures separated by insulating silica barriers in optically transparent hosts. The dimensions and the arrangement of the incorporated materials can be defined by the size, shape, and structural order of the pores of the selected host template [71–73]. Size quantization in 3D space controlled by a meso-

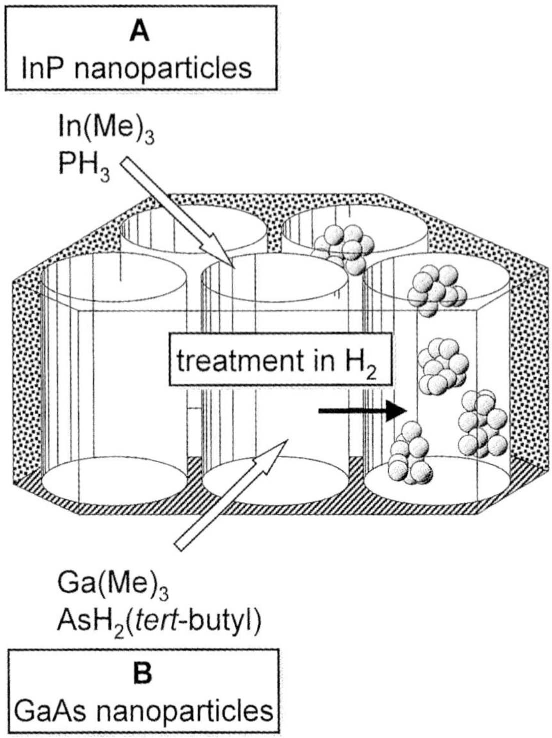

Fig. 4. Metalorganic chemical vapor deposition (MOCVD) of (A) InP and (B) GaAs semiconductor nanoparticles in MCM-41 hosts [75,76].

porous host system may lead to different optoelectronic effects such as shifting of the semiconductor band gap or enhanced nonlinear optical properties. The preparation of these nanoscale composite structures in the form of thin films on appropriate substrates is expected to offer added functionality [74].

Two main synthetic strategies toward the incorporation of semiconductors into mesoporous materials have been explored: metalorganic chemical vapor deposition, and wet chemical loading followed by treatment with a gaseous group V or VI hydride. For example, InP and GaAs direct gap semiconductors were prepared in MCM-41 material from trimethylindium and phosphine, and trimethylgallium and tert-butylarsine respectively in continuous flow MOCVD reactors (Fig. 4) [75,76]. In both cases a broad size-distribution of semiconductor particles grown on the internal and external surfaces of the MCM-41 material was obtained. The InP/MCM-41 and GaAs/MCM-41 heterostructures show blue-shifted UV/vis absorption and broad visible photoluminesence at room temperature consistent with the expected quantum-size effects. Recently several different wet chemical techniques were applied for the preparation of semiconductor heterostructures in mesoporous hosts. GaN, a wide band gap semiconductor, was synthesized in

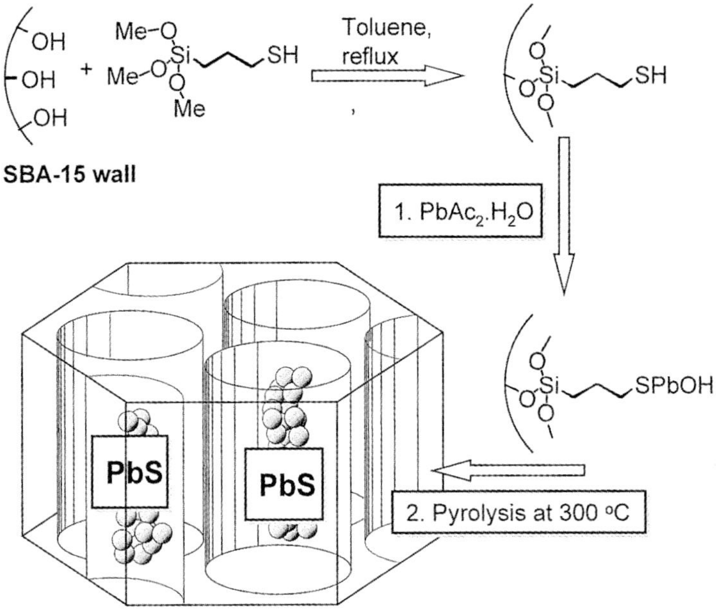

Fig. 5. Inclusion chemistry for the generation of PbS nanowires in a thiol-modified SBA-15 host [78].

a boron-doped MCM-41 mesoporous host by first impregnating the support with triazido(trimethylamine)gallium followed by heating in ammonia at 500 °C [77]. Excitation data and TEM images show the existence of quantum-confined GaN inside the pores of MCM-41 material together with larger GaN particles on the outer surface. PbS nanowires with a uniform diameter of 6 nm were prepared in a thiol-functionalized SBA-15 host (Fig. 5). A structural transition from nanocrystals to nanowires was achieved by increasing the loading of the inorganic compound in the host through –SH functionalization of the walls of the mesoporous material. TEM images show that PbS nanowires are incorporated preferentially inside the channel system of the SBA-15 mesoporous host [78].

CdSe quantum dots (attractive candidates for the fabrication of tunable light absorbers and emitters in LEDs) were confined in an MCM-41 host by wet impregnation of dimethyl cadmium and Se dissolved in tributylphosphine under vacuum, followed by heating at 325 °C in trioctylphosphine oxide to initiate growth of CdSe nanoparticles (Fig. 6) [79]. TEM, XRD, XPS, and optical measurements demonstrate the confinement of CdSe nanoparticles in the mesoporous channel matrix and the deposition of larger semiconductor particles on the outer surface of the mesoporous material. CdS semiconductor nanoparticles were incorporated into a thiol-modified MCM-41 matrix by immersion of the mesoporous support in a reactive solution that gave nanosized CdS particles (Fig. 7) [80]. Optical absorption spectra provide evidence for the nanoscale encapsulation of the CdS particles inside the mesoporous channels.

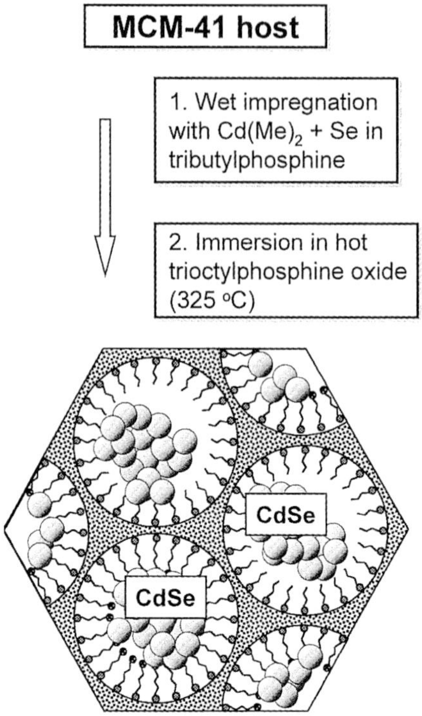

Fig. 6. Incorporation of CdSe nanoparticles in an MCM-41 host through reaction of precursors in hot trioctylphosphine oxide [79].

The wide-band semiconductor ZnO was prepared in ethylenediamine-functionalized MCM-41 by wet impregnation-complexing of Zn(II) to the ethylenediamine groups of the mesoporous support (Fig. 8) [81]. Subsequent calcination led to ZnO nanoparticles encapsulated in the mesoporous host, the former showing a blue shift in their absorption spectra. Very recently, diluted magnetic semiconductor quantum wires $Cd_{1-x}Mn_xS$ were prepared in MCM-41 hosts by wet impregnation with Cd and Mn acetates, followed by reaction with H_2S at 70 °C [82]. XRD, TEM, sorption, and IR measurements indicate the incorporation of the $Cd_{1-x}Mn_xS$ structure in the mesoporous channels of MCM-41 material. Additional photoluminescence, photoluminescence excitation spectroscopy and EPR investigations reveal the effect of quantum confinement, leading to an increase of about 200 meV in the direct band gap.

The confinement of CdS quantum dots in the hexagonal channel structure of mesoporous silica films prepared by spin-coating was reported for the first time. The incorporation of the CdS nanoparticles was achieved by repeated wet impregnation with Cd salt, followed by reaction with gaseous H_2S (Fig. 9). TEM images and UV/vis absorption measurements demonstrate that the size confinement and

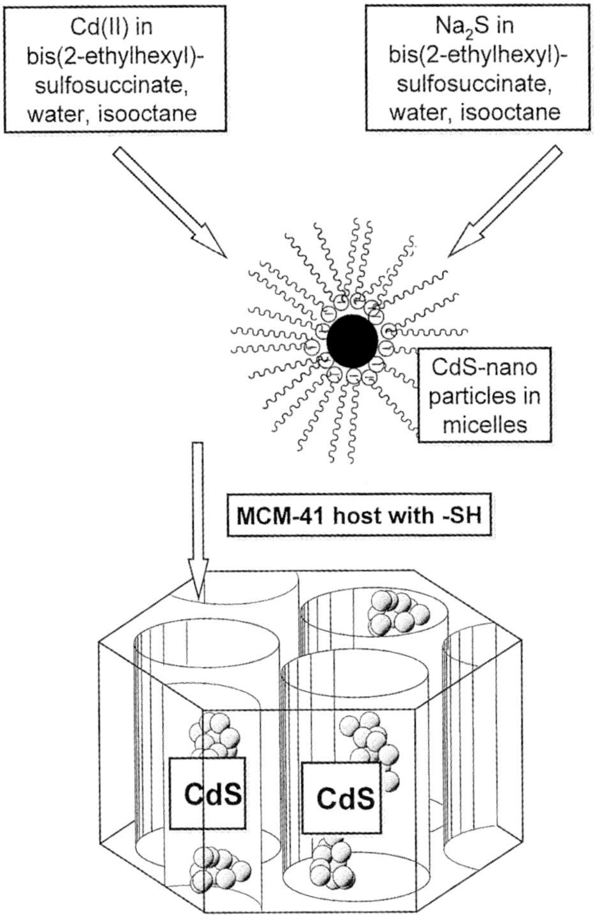

Fig. 7. Growth of CdS nanoparticles in reverse micellar solutions and incorporation of the semiconductor nanocrystals in a thiol-modified MCM-41 host [80].

the 3D arrangement of the ordered porous structure (prepared as a 300 nm film) controls the growth of a CdS superlattice [83].

3.4
Carbon Nanotubes and Graphitic Filaments in Host Materials

During the last decade carbon nanotubes have attracted considerable attention owing to their unique electronic properties [84–86]. As far as theoretical calculations are concerned, a carbon nanotube is assumed to be an infinitely long cylinder with a monolayer of hexagonally ordered carbon atoms in the tube wall

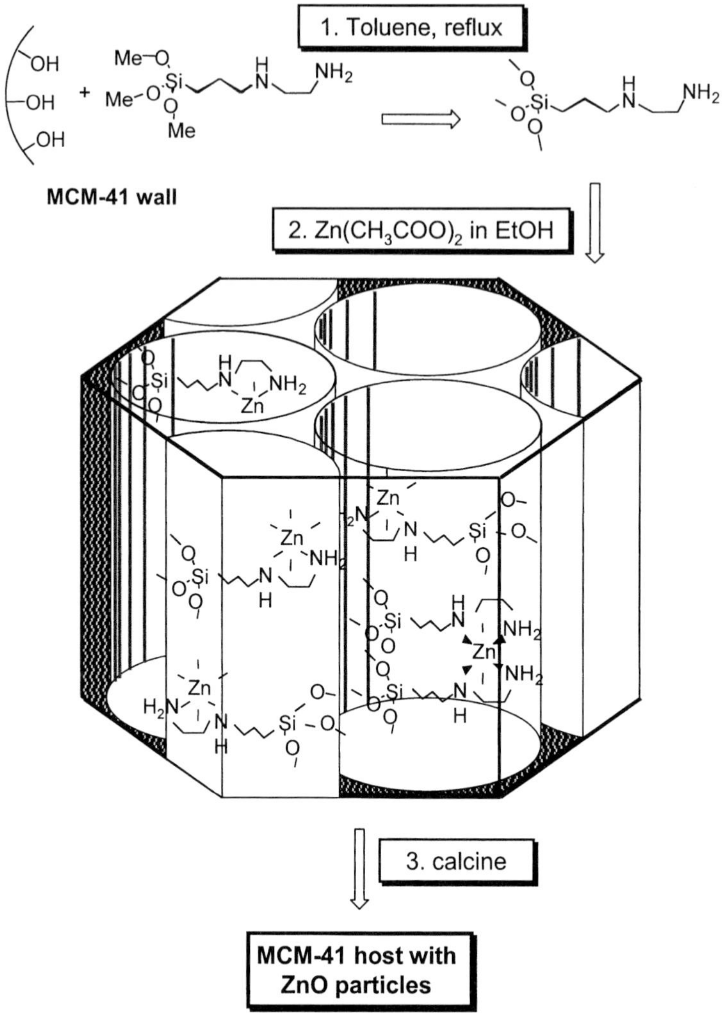

Fig. 8. Generation of ZnO in an MCM-41 host through complexation of Zn(II) ions with grafted amine ligands followed by calcination [81].

[85,87]. Calculations predict that carbon nanotubes are either semiconducting or metallic, depending on their diameter and the orientation of the carbon hexagons of the honeycomb structure with respect to the nanotube axes (helicity) [86]. Nanotubes of carbon are synthesized in two categories: single-walled carbon nanotubes (SWCNT) and multiwalled carbon nanotubes (MWCNT). The latter consist of concentric cylinders placed around a hollow center, with spacing between the layers close to that in graphite. Carbon nanotubes are prepared by arc-discharge, catalytic decomposition of hydrocarbons, or laser-assisted methods, and present

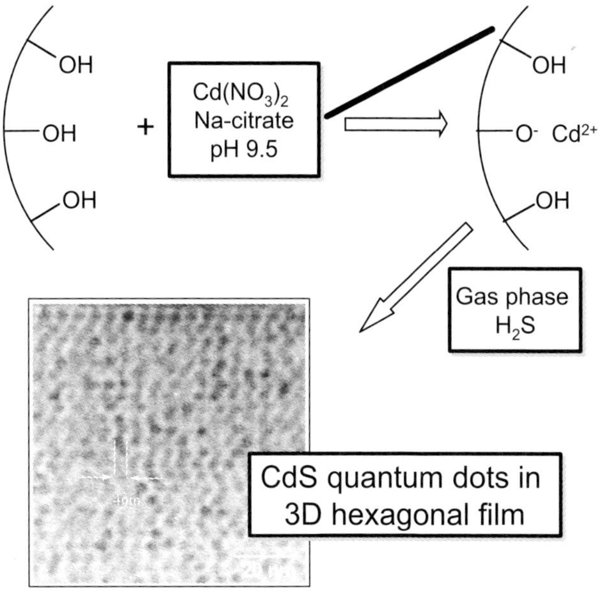

Fig. 9. Introduction of CdS nanoparticles into a 3D hexagonal mesoporous film. Inset: TEM image of a cross section through the film [83].

synthetic strategies permit the preparation of 3D organized arrays of carbon nanotubes [86,88–92]. Certainly, the incorporation of these structures with their remarkable 1D electronic properties in an insulating matrix can open new opportunities for investigation and for the construction of novel composite materials.

Transition-metal loaded molecular sieves were used for the large-scale synthesis of carbon nanotubes by catalytic decomposition of hydrocarbons [93–95]. Depending on the type of catalytic metal center, carbon source, and molecular sieve used as a catalytic support, carbon nanotubes with different diameter and nanotube alignment were prepared. Quasi-aligned MWCNT were prepared on Fe-containing mesoporous silica by decomposition of acetylene at 700 °C [96]. In this system, the templating role of the porous system is not yet understood in detail. Recently, aligned carbon nanotube patterns were prepared on cubic mesoporous films by the decomposition of acetylene [97]. In this study, patterned mesoporous films were deposited using the micromolding-in-capillaries technique (MIMIC) with preformed Fe-containing solutions.

Single crystals of microporous aluminophosphate molecular sieve $AlPO_4$-5 with long 1D channels of 0.73 nm diameter were used to prepare the smallest possible 0.4 nm SWCNT *inside* [98–102]. The SWCNT were prepared by pyrolysis of tetrapropylamine (TPA) molecules that had served as templates for the synthesis of the $AlPO_4$-5 crystals, in a vacuum at 500–800 °C. Polarized optical microscopy, polarized Raman spectroscopy, TEM, and X-ray scattering were used to demonstrate the encapsulation of the 0.4 nm carbon nanotubes in the $AlPO_4$-5 channel structure. DC conductivity measurements (*I–V* curves) suggest that the confined carbon nanotubes are intrinsic semiconductors with interesting electrical properties.

3.5 Conclusions

Based on the numerous different examples for the use of host–guest chemistry aimed at the synthesis and stabilization of nanoscale conductive structures, it becomes increasingly clear that this type of inclusion chemistry within mesoporous materials offers a vast range of opportunities for the generation of wires and particles of metallic and semiconducting materials. It is also anticipated that further research devoted to this family of nanostructures will enhance synthetic control regarding the generation of novel molecular electronic systems, and increase our knowledge about their unique physical properties.

References

1 P. Packan, *Science* **1999**, *285*, 2079.
2 J. M. Tour, *Acc. Chem. Res.* **2000**, *33*, 791.
3 T. Bein, *Chem. Mater.* **1996**, *8*, 1636.
4 T. Bein, *Stud. Surf. Sci. Catal.* **1996**, *102*, 295.
5 D. J. Cardin, *Adv. Mater.* **2002**, *14*, 553.
6 J. S. Beck, J. C. Vartuli, W. J. Roth, M. E. Leonowicz, C. T. Kresge, K. D. Scmitt, C. T-W. Chu, D. H. Olson, E. W Sheppard, S. B. McCullen, J. B. Higgins, J. L. Schlenker, *J. Am. Chem. Soc.* **1992**, *114*, 10 834.
7 C. T. Kresge, *Adv. Mater.* **1996**, *8*, 181.
8 F. Schüth, *Stud. Surf. Sci. Catal.* **2001**, *135*, 1.
9 A. Monnier, F. Schüth, Q. Huo, D. Kumar, D. Margolese, R. S. Maxwell, G. D. Stucky, M. Krishnamurty, P. Petroff, A. Firouzi, M. Janicke, B. F. Chmelka, *Science* **1993**, *261*, 1299.
10 Q. Huo, D. I. Margolese, U. Ciesla, D. G. Demuth, P. Feng, T. E. Gier, P. Sieger, A. Firouzi, B. F. Chmelka, F. Schüth, G. D. Stucky, *Chem. Mater.* **1994**, *6*, 1176.
11 G. S. Attard, J. C. Glyde, C. G. Göltner, *Nature* **1995**, *378*, 366.
12 Q. Huo, D. I. Margolese, U. Ciesla, P. Feng, T. E. Gier, P. Sieger, R. Leon, P. M. Petroff, F. Schüth, G. D. Stucky *Nature* **1994**, *368*, 317.
13 J.C. Vartuli, K. D. Schmitt, C. T. Kresge, W. J. Roth, M. E. Leonowicz, S. B. McCullen, S. D. Hellring, J. S. Beck, J. L. Schlenker, D. H. Olson, E. W. Sheppard, *Chem. Mater.* **1994**, *6*, 2317.
14 Q. Huo, D. I. Margolese, G. D. Stucky, *Chem. Mater.* **1996**, *8*, 1147.
15 D. Zhao, J. Feng, Q. Huo, N. Melosh, G. H. Fredrickson, B. F. Chmelka, G. D. Stucky, *Science* **1998**, *279*, 548.
16 C. Yu, Y. Yu, D. Zhao, *Chem. Commun.* **2000**, 575.
17 P. Tanev, T. J. Pinnavaia, *Science* **1995**, *267*, 865.
18 M. Kruk, M. Jaroniec, R. Ryoo, J. M. Kim, *Microporous Materials*, **1997**, *12*, 93.
19 S. A. Bagshaw, E. Prouzet, T. J. Pinnavaia, *Science* **1995**, *269*, 1242.
20 P. Yang, D. Zhao, D. I. Margolese, B. F. Chmelka, G. D. Stucky, *Nature* **1998**, *396*, 152.
21 F. Schüth, *Chem. Mater.* **2001**, *13*, 3184.
22 S. Jun, S. H. Joo, R. Ryoo, M Kruk, M. Jaroniec, Z. Liu, T. Oshuma, O.

Teresaki, J. Am. Chem. Soc. **2000**, *122*, 10 712.

23 R. Ryoo, S. H. Joo, M. Kruk, M. Jaroniec, *Adv. Mater.* **2001**, *13*, 3184.

24 S.-S. Kim, T. J. Pinnavaia, *Chem. Commun.* **2001**, 2418.

25 S. Inagaki, S. Guan, T. Ohsuna, O. Terasaki, *Nature* **2002**, *416*, 304.

26 K. Moller, T. Bein, *Chem. Mater.* **1998**, *10*, 2950.

27 A. Stein, B. J. Melde, R. C. Schroden, *Adv. Mater.* **2000**, *12*, 1403.

28 T. Asefa, C. Yoshina-Ishii, M. J. MacLachlan, G. A. Ozin, *J. Mater. Chem.* **2000**, *10*, 1751.

29 A. Sayari, S. Hamoudi, *Chem. Mater.* **2001**, *13*, 3151.

30 S. J. Brian, G. Wirnsberger, G. D. Stucky, *Chem. Mater.* **2001**, *13*, 3140.

31 Y. Lu, R. Ganguli, C. A. Drewien, M. T. Anderson, C. J. Brinker, W. Gong, Y. Guo, H. Soyez, B. Dunn, M. H. Huang, J. I. Zink, *Nature* **1997**, *389*, 364.

32 C. J. Brinker, Y. Lu, A. Sellinger, H. Fan, *Adv. Mater.* **1999**, *11*, 579.

33 D. Zhao, P. Yang, N. Melosh, J. Feng, B. F. Chmelka, G. D. Stucky, *Adv. Mater.* **1998**, *10*, 1380.

34 R. C. Hayward, P. Alberius-Henning, B. F. Chmelka, G. D. Stucky, *Micropor. Mesopor. Mater.* **2001**, *44-45*, 619.

35 G. Wirnsberger, G. D. Stucky, *Chem. Mater.* **2000**, *12*, 2525.

36 B. Lebeau, C. E. Fowler, S. R. Hall, S. Mann, *J. Mater. Chem.* **1999**, *9*, 2279.

37 S. Pevzner, O. Regev, R. Yerushalmi-Rozen, *Curr. Opin. Colloid Interface Sci.* **2000**, *4*, 420.

38 F. Marlow, M.D. McGeene, D. Zhao, B. F. Chmelka, G. D. Stucky, *Adv. Mater.* **1999**, *11*, 632.

39 F. Kleitz, F. Marlow, G. D. Stucky, F. Schüth, *Chem. Mater.* **2001**, *13*, 3587.

40 C. A. Foss, M. J. Tierney, C. R. Martin, *J. Phys. Chem.* **1992**, *96*, 9001.

41 C. K. Preston, M. Moskovits, *J. Phys. Chem.* **1993**, *97*, 8495.

42 T. M. Whitney, J. S. Jiang, P.C. Searson, C. L. Chien, *Science* **1993**, *261*, 1316.

43 R. E. Benfield, D. Grandjean, J. C. Dore, Z. Wu, M. Kröll, T. Sawitowski, G. Schmid, *Eur. Phys. J. D* **2001**, *16*, 399.

44 T. Jung, R. Schlitter, J. K. Gimzewski, F. J. Himpsel, *Appl. Phys. A* **1995**, *61*, 467.

45 P. Gambardella, M. Blanc, H. Brune, K. Kuhnke, K. Kern, *Phys. Rev. B* **2000**, *61*, 2254.

46 S. Morin, A. Lachenwitzer, O. M. Magnussen, R. J. Behm, *Phys. Rev. B* **1999**, *83*, 5066.

47 M. P. Zach, K. H. Ng, R. M. Penner, *Science* **2000**, *290*, 2120.

48 M. S. Spector, R. R. Price, J. M. Schnur, *Adv. Mater.* **1999**, *11*, 337.

49 J. K. N. Mbindyo, T. E. Mallouk, J. B. Mattzela, I. Kratochvilova, B. Razavi, T. N. Jackson, T. S. Mayer, *J. Am. Chem. Soc.* **2002**, *124*, 4020.

50 C. H. Ko, R. Ryoo, *Chem. Commun.* **1996**, 2467.

51 H. J. Shin, C. H. Ko, R. Ryoo, *J. Mater. Chem.* **2001**, *11*, 260.

52 Z. Liu, O. Terasaki, T. Ohsuna, K. Hiraga, H. J. Shin, R. Ryoo, *Chem. Phys. Chem.* **2001**, *4*, 229.

53 L.-Z. Wang, J.-L. Shi, W.-H. Zhang, M.-L. Ruan, J. Yu, D.-S. Yan, *Chem. Mater.* **1999**, *11*, 3015.

54 Y.-J. Han, J. M. Kim, G. D. Stucky, *Chem. Mater.* **2000**, *12*, 2068.

55 H. Kang, Y. W. Jun, J. I. Park, K.-B. Lee, J. Cheon, *Chem Mater.* **2000**, *12*, 3530.

56 K.-B. Lee, S.-M. Lee, J. Cheon, *Adv. Mater.* **2001**, *13*, 517.

57 C. Yang, H. Sheu, K. Chao, *Adv. Funct. Mater.* **2002**, *12*, 143.

58 E. Braun, Y. Eichen, U. Sivan, G. Ben-Yoseph, *Nature* **1998**, *391*, 775.

59 J. K. Mbindyo, B. D. Reiss, B. R. Martin, C. D. Keating, M. J. Natan, T. E. Mallouk, *Adv. Mater.* **2001**, *13*, 249.

60 J. H. Song, Y. Wu, B. Messer, H. Kind, P. Yang, *J. Am. Chem. Soc.* **2001**, *123*, 10 397.

61 Y. Zhang, H. Dai, *Appl. Phys. Lett.* **2000**, *77*, 3015.

62 R. Kohn, M. Froba, *Catal. Today* **2001**, *68*, 227.

63 F. Schuth, A. Wingen, J. Sauer, *Micropor. Mesopor. Mater.* **2001**, *44–45*, 465.

64 C. P. Mehnert, D. W. Weaver, J. Y. Ying, *J. Am. Chem. Soc.* **1998**, *120*, 12 289.

65 Y. Guari, C. Thieuleux, A. Mehdi, C. Reye, R. J. P. Corriu, S. Gomez-Gallardo, K. Philippot, B. Chaudret, R. Dutartre, *Chem.Commun.* **2001**, 1374.

66 H. Shi, L. Zhang, W. Cai, *J. Appl. Phys.* **2000**, *87*, 1572.

67 J. V. Ryan, A. D. Berri, M. L. Anderson, J. W. Long, R. M. Stroud, V. M. Cepak, V. M. Browning, D. R. Rolison, C. I. Merzbacher, *Nature* **2000**, *406*, 169.

68 A. P. Alivisatos *J. Phys. Chem.* **1996**, *100*, 13 226.

69 R.F. Xiao, H. B. Liao, N. Cue, X. W. Sun, H. S. J. Kwok, *Appl. Phys. Lett*, **1998**, *73*, 1104.

70 C. A. Tran, A. Osinski, R. F. Karlicek, I. Berishev, *Appl. Phys. Lett*, **1999**, *72*, 350.

71 G. A. Ozin, A. Kuperman, A. Stein, *Angew. Chem. Int. Ed. Engl.* **1989**, *28*, 359

72 G. D. Stucky, J. E. Mac Dougall, *Science* **1990**, *247*, 669.

73 M. J. Kelly, *Adv. Mater.* **1997**, *9*, 857.

74 H. Fan, S. Reed, T. Bear, R. Schunk, G. P. Lopez, C. J. Brinker, *Micropor. Mesopor. Mater.* **2001**, *44–45*, 625.

75 J. R. Agger, M. W. Anderson, M. E. Pemble, O. Terasaki, Y. Nozue, *J. Phys. Chem. B* **1998**, *102*, 3345.

76 V. I. Srdanov, I. Alxneit, G. D. Stucky, C. M. Reaves, S. P. DenBaars, *J. Phys. Chem. B* **1998**, *102*, 3341.

77 H. Winkler, A. Birkner, V. Hagen, I. Wolf, R. Schmechel, H. von Seggern, R. A. Fischer, *Adv. Mater.* **1999**, *11*, 1444.

78 F. Gao, Q. Lu, X. Liu, Y. Yan, D. Zhao, *NanoLett.* **2001**, *12*, 743.

79 H. Parala, H. Winkler, M. Kolbe, A. Wohlfart, R. A. Fischer, R. Schmechel, H. von Seggern, *Adv. Mater.* **2000**, *12*, 1050.

80 T. Hirai, H. Okubo, I. Komasawa, *J. Phys. Chem. B* **1999**, *103*, 4228.

81 W.-H. Zhang, Y.-L. Shi, L.-Z. Wang, D.-S., Yan, *Chem. Mater.* **2000**, *12*, 1408.

82 F. J. Brieler, M. Fröba, L. Chen, P. J. Klar, W. Heimbrodt, H-A. K. von Nidda, A. Loidl, *Chem. Eur. J.* **2002**, *8*, 185.

83 S. Besson, T. Gacoin, C. Ricolleau, C. Jacquiod, J.-P. Boilot, *NanoLett.* **2002**, *2*, 409.

84 M. S. Dresselhaus, G. Dresselhaus, M. Pimenta, *Eur. Phys, J. D* **1999**, *9*, 69.

85 P.C. Eklund, J. M. Holden, R. A. Jishi, *Carbon* **1995**, *33*, 959.

86 P. M. Ajayan, *Chem. Rev.* **1999**, *99*, 1787.

87 R. A. Jishi, M. S. Verkatarman, M. S. Dresselhaus, G. Dresselhaus, *Chem. Phys. Lett.* **1993**, *77*, 209.

88 S. Iijima, *Nature* **1991**, *354*, 56.

89 R. Andrews, D. Jacques, A. M. Rao, F. Derbyshire, D. Quan, X. Fan, E. C. Dickey, J. Chen, *Chem. Phys. Lett.* **1999**, *303*, 467.

90 L.-C. Qin, X. Zhao, K. Hirahara, Y. Miyamoto, Y. Ando, S. Iijima, *Nature* **2000**, *408*, 50.

91 B. Q. Wei, R. Vajtai, Y. Jung, J. Ward, R. Zhang, G. Ramanath, P. M. Ajayan *Nature* **2002**, *416*, 495.

92 V. Derycke, R. Martel, J. Appenzeller, Ph. Avouris, *NanoLett.* **2001**, *1*, 453.

93 K. Mukhopadhyay, A. Koshio, T. Sugai, N. Tanaka, H. Shinohara, Z. Konya, J. B. Nagy, *Chem. Phys. Lett.* **1999**, *303*, 117.

94 A. Zhang, C. Li, S. Bao, Q. Xu, *Micropor. Mesopor. Mater.* **1999**, *29*, 383.

95 N. He, Y. Kuang, Q. Dai, Y. Miao, A. Zhang, X. Wang, K. Song, Z. Lu, C. Yuan, *Mater. Sci. Eng. C* **1999**, *8–9*, 151.

96 W. Z. Li, S. S. Xie, L. X. Qian, B. H. Chang, B. S. Zou, W. Y. Zhou, R. A. Zhao, G. Wang, *Science* **1996**, *274*, 1701.

97 G. Zheng, H. Zhu, Q. Luo, Y. Zhou, D. Zhao, *Chem. Mater* **2001**, *13*, 2240.

98 Z. K. Tang, H. D. Sun, J. Wang, J. Chen, G. Li, *Appl. Phys. Lett.* **1998**, *73*, 2287.

99 Z. K. Tang, H. D. Sun, J. Wang, *Physica B* **2000**, *279*, 200.

100 N. Wang, G. D. Li, Z. K. Tang, *Chem. Phys. Lett.* **2001**, *339*, 47.

101 N. Wang, Z. K. Tang, G. D. Li, J. S. Chen, *Nature* **2000**, *408*, 50.

102 P. Launois, R. Moret, D. Le Bolloch, P. A. Albouy, Z. K. Tang, G. Li, J. Chen, *Solid State Commun.* **2000**, *116*, 99.

4
Density Functional Studies of Host–Guest Interactions in Sodalites

Joachim Sauer and René Windiks

4.1
Introduction

Nanostructuring of materials often results in novel electronic, magnetic, or optical properties. Use can be made of the nanoporous cavity structure of aluminosilicates such as zeolites to control the arrangement of clusters on the nanoscale. It has been shown that the presence of paramagnetic guest clusters with single electrons in zeolite hosts gives rise to materials with special electronic, magnetic, and optical properties [1–3]. Due to their simple framework structures sodalites are particularly suited for fundamental studies of such host–guest interactions.

Sodalites have nanoporous aluminosilicate frameworks built from alternating corner-sharing SiO_4 and AlO_4^- tetrahedra. The framework structure can be described as a space-filling body-centered cubic (bcc) lattice of $[4^6 6^8]$ polyhedra of T atoms, also known as β-cages or sodalite units. In sodium sodalite $(Na^+)_3[SiO_2(AlO_2^-)]_3$ each of the sodalite cages contains 3 sodium cations (Fig. 1, left). When this material is treated with sodium vapor each sodalite cage can take up an additional sodium atom, and a paramagnetic $(Na_4)^{3+}$ cluster is formed [4]. In this cluster the single electron is shared by all four sodium cations, like an F center in ionic solids (Fig. 1, right). Formally the electron replaces the anion present in the cages of many sodalites and the name sodium electro sodalite (SES) is used for the material with the composition $(Na)_4[SiO_2(AlO_2^-)]_3$, in accord with the common nomenclature for sodalites. The unpaired electrons form a regular bcc lattice with nearest-neighbor electron–electron distances of 7.67–7.69 Å, i.e., half the length of the body diagonal of the sodalite unit cell, $(a\sqrt{3})/2$. The distance between two next nearest-neighbor unpaired electrons is identical with the cell constant, $a = 8.86$–8.88 Å.

Depending on the fraction of cages with an additional sodium atom, SES changes its color from white to blue, purple, and finally black [5]. The last-named is also known as "black sodalite". If all cages are filled with $(Na_4)^{3+}$ clusters the density of electron spins is 3×10^{21} spins·cm^{-3}, much higher than the F center concentration reached in other materials so far, and close to the Mott criterion for

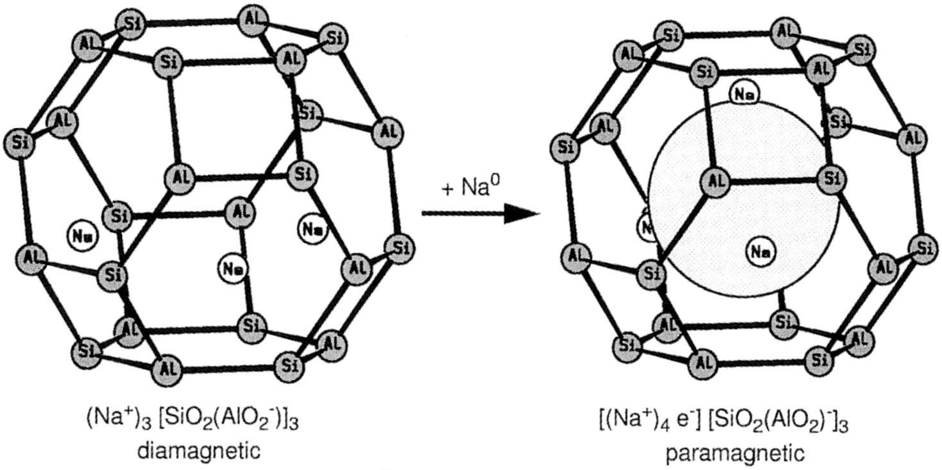

Fig. 1. Formation of a paramagnetic $(Na_4)^{3+}$ cluster in a sodalite cage by sodium doping.

the insulator/metal transition [6]. From their isotropic g tensors it was concluded that the unpaired electrons in the $(Na_4)^{3+}$ sites have predominantly s character [7].

Engelhardt et al. [8] showed by ^{29}Si and ^{27}Al MAS-NMR spectroscopy that the aluminosilicate framework is not an inert matrix but interacts with the paramagnetic clusters. In contrast to the ^{29}Si and ^{27}Al NMR spectra of unloaded sodium sodalite, which show only one line (all framework Si and Al atoms are crystallographically equivalent), the spectra of SES with 73% Na loading showed several lines shifted to low fields by up to 70 ppm (Fig. 2). Lines in these shift ranges had never been observed before for any sodalite or other aluminosilicate. They are due to substantial paramagnetic NMR shifts which originate from the interaction of the Si and Al framework nuclear spins with the electron spins of the paramagnetic $(Na_4)^{3+}$ clusters. When almost all sodalite cages are filled with an additional Na atom (96% Na loading) only the lines with maximum shift to low fields persist (line 7 in Fig. 2). Engelhardt et al. explained the observed MAS-NMR spectra by the following model [8]: In the sodalite structure each T atom (T=Si, Al) is surrounded by four β-cages. In a partially Na loaded SES the size of the paramagnetic shift depends on the number of β-cages containing paramagnetic $(Na_4)^{3+}$ clusters (A-type

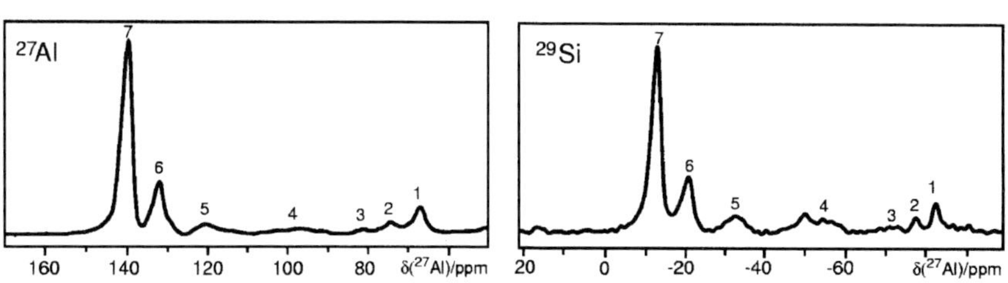

Fig. 2. ^{27}Al and ^{29}Si MAS NMR spectra of SES at 295K with 73% Na loading [8].

cages) and diamagnetic $(Na_3)^{3+}$ units (B-type cages). Hence, a T atom can have 5 different magnetic environments of type $[n\mathrm{A}(4-n)\mathrm{B}]$, where $n = 0$–4. Because of the high concentration of paramagnetic A cages in the highly loaded sample (73%), the strong line at lowest field (line 7) is assigned to the [4A0B] case. The resonance at highest field (line 1) is assigned to the completely diamagnetic environment [0A4B]. Lines 4–6 are assigned to intermediate cases [1A3B], [2A2B], and [3A1B], respectively. Weak lines close to line 1 in the partially loaded sample are assigned to T atoms with [0A4B] environment with one (line 2) or two (line 3) next-nearest-neighbor A cages.

In the absence of any dipole pseudocontact interactions the paramagnetic NMR shift, $-(B-B_0)/B_0$, is given by the Fermi contact interaction (Eq. 1)

$$-(B-B_0)/B_0 = A_N \chi_e(T)/N\mu_0 g_e \mu_B g_N \mu_N \tag{1}$$

where N is the number of nuclei, μ_0 the permeability of vacuum, g_e and g_N are the electronic and nuclear g values, and μ_B and μ_N the Bohr and nuclear magneton. The paramagnetic NMR shift is proportional to the hyperfine coupling constant (Eq. 2)

$$A_N = (2\mu_0/3) g_e \mu_B g_N \mu_N \rho(r_N), \tag{2}$$

which in turn depends on the electron spin density $\rho(r_N)$ at the T nucleus at r_N. Hence, the electron spin density at the T nuclei can be derived from the observed paramagnetic shifts.

The temperature dependence of the paramagnetic shift is given by the dectronic contribution to the magnetic susceptibility $\chi_e(T)$, as described by the Curie–Weiss law (Eq. 3)

$$\chi_e(T) = C/(T-\Theta), \tag{3}$$

where C is the Curie constant, T is the absolute temperature, and Θ is the Weiss temperature. When fitting the observed temperature dependence of the ^{27}Al NMR resonance at lowest field (line 7 in Fig. 2) to the Curie–Weiss law, Srdanov et al. obtained a negative Weiss temperature of -178 ± 8 K [9], indicative of a strong antiferromagnetic interaction below the critical (Néel) temperature. Antiferromagnetic ordering was confirmed by EPR measurements, and from susceptibility measurements the Néel temperature was estimated as $T_c = 48 \pm 2$ K. More recent measurements of the temperature dependence of NMR shifts of SES yielded $\Theta = -168 \pm 5$ K and $T_c = 54 \pm 2$K [10].

These findings raised the question whether the ideas of the atomic structure of SES can explain the peculiar electronic, magnetic, and optical properties, and induced quantum mechanical studies of SES (see Refs. [11–15] and references therein). This chapter focuses on our studies [13,14] which try to answer the question: Is there antiferromagnetic order in SES and how strong is the calculated antiferromagnetic interaction compared to parameters derived from observed

Weiss temperatures and critical temperatures? Using calculated electron spin densities, can we rationalize the Engelhardt model for explaining the different lines observed in the ^{29}Si and ^{27}Al MAS-NMR spectra of SES?

4.2 Theory

For systems with weakly interacting spins localized on different sites Heisenberg has suggested an effective Hamiltonian to describe the energy as a function of the different spin states. For a system with two spins localized on sites a and b the Heisenberg Hamiltonian is given by Eq. (4)

$$\mathscr{H} = -J_{ab}\mathscr{S}_a\mathscr{S}_b = -J_{ab}\tfrac{1}{2}(\mathscr{S}^2 - \mathscr{S}_a{}^2 - \mathscr{S}_b{}^2), \tag{4}$$

where $\mathscr{S}$, $\mathscr{S}_a$, and $\mathscr{S}_b$ are the operators for the total electron spin, the spin at site a, and the spin at site b; and J_{ab} is the magnetic coupling constant.

For the energies of the states with parallel spins (triplet = ferromagnetic coupling) and with antiparallel spins (singlet = antiferromagnetic coupling) this yields Eq. (5).

$$E_T - E_S = -\tfrac{1}{4}J_{ab} - \tfrac{3}{4}J_{ab} = -J_{ab} \tag{5}$$

The parameter J_{ab} can be calculated from first principles if the energies of the tiplet and singlet states are calculated as eigenvalues of the electronic Hamiltonian. The valence bond treatment of two weakly coupled electrons described by localized and orthogonal orbitals, ϕ_a and ϕ_b (cf. two weakly interacting H atoms) yields Eq. (6)

$$E_T - E_S = -2\,K_{ab} \tag{6}$$

with the exchange integral (Eq. 7)

$$K_{ab} = \int \phi_a{}^*(1)\phi_b{}^*(2)(1/r_{12})\phi_a(2)\phi_b(1)\mathrm{d}r_1\mathrm{d}r_2 \tag{7}$$

Molecular orbital description with the effective Hamiltonian [16] (cf. Ref. [17])

$$H(1,2) = H_c(1) + H_c(2) + 1/r_{12}$$

uses one-determinant and two-determinant wavefunctions for the triplet and singlet states, respectively (Eq. 8).

$$T = |\psi_1\alpha\psi_2\alpha| \quad \text{and} \quad S = \frac{1}{\sqrt{2}}(|\psi_1\alpha\psi_1\beta| + |\psi_2\alpha\psi_2\beta|). \tag{8}$$

Transformation into localized and orthogonal orbitals ϕ_a and ϕ_b (Eq. 9)

$$\phi_a = \frac{1}{\sqrt{2}}(\psi_1 + \psi_2)$$
$$\phi_b = \frac{1}{\sqrt{2}}(\psi_1 - \psi_2) \tag{9}$$

yields Eq. (10)

$$E_T - E_S = -2\ K_{ab} + 4\ H_{ab}{}^2/U \tag{10}$$

The integral H_{ab} over the core Hamiltonian H_c is given by Eq. (11)

$$H_{ab} = \int \phi_a{}^*(1) H_c(1) \phi_b(1) dr_1 \tag{11}$$

and the integral U for the on-site electron repulsion is given by Eq. (12)

$$U = \int |\phi_a(1)|^2 (1/r_{12}) |\phi_a(2)|^2 dr_1 dr_2. \tag{12}$$

U and H_{ab} (also known as transfer integral) are parameters of the Hubbard Hamiltonian. Hence, the general expression for the Heisenberg coupling parameter as derived by Anderson [18] is

$$J_{ab} = 2\ K_{ab} - 4\ H_{ab}{}^2/U. \tag{13}$$

The first term is always positive, favors ferromagnetic interaction, and is known as potential exchange. The second contribution is always negative. It favors antiferromagnetic coupling and is known as kinetic exchange.

A periodic array of spins as in a crystal can be also described by the above Heisenberg Hamiltonian (Eq. 14)

$$\mathscr{H} = -\Sigma_{a<b} J_{ab} \mathscr{S}_a \mathscr{S}_b, \tag{14}$$

However, to find the solutions for the energies of the different spin eigenstates, further assumptions have to be made. As the exchange interactions are expected to fall off exponentially with distance, neglect of all J_{ab} beyond nearest-(n) and possibly next-nearest-(nn) neighbor interactions seems possible. For crystals with a single kind of magnetic site in which all magnetic lattice sites are crystallographically equivalent, this yields.

$$\mathscr{H} = -\mathscr{S}_a(\Sigma_n J_n \mathscr{S}_b + \Sigma_{nn} J_{nn} \mathscr{S}_b) \tag{15}$$

These approximations are known as mean field theory of magnetism [19,20] and yield for the Weiss temperature

$$\Theta = S(S+1)(|Z_n J_n| + |z_{nn} J_{nn}|)/3k \tag{16}$$

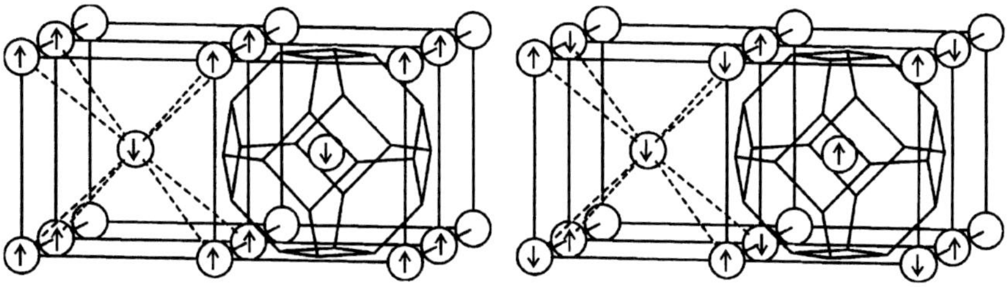

Fig. 3. Antiferromagnetic ordering of type I (left) and type II (right) for the bcc lattice of spins in SES. The dashed lines denote interactions between nearest neighbor $(Na_4)^{3+}$ clusters.

where S is the spin, and z_n and z_{nn} are the numbers of nearest and next-nearest neighbors in the lattice.

Above a critical temperature the thermal effects prevent the ordering of spins and the material is paramagnetic. Below the critical temperature the magnetic interactions outweigh thermal effects and the spin system assumes an ordered state. The type of magnetic ordering that actually exists below the critical temperature is that which corresponds to the highest value of T_c. Ferromagnetic ordering with all spins parallel is unique, but two kinds of antiferromagnetic ordering are possible. Figure 3 shows this for the bcc lattice. Antiferromagnetic (AF) ordering of type I orients all nearest-neighbor spins antiparallel, but all next-nearest-neighbor spins parallel. Ordering of type II has half the nearest-neighbor spins antiparallel and half parallel, while the orientations of all next-nearest-neighbor spins are antiparallel. Different expressions for calculating the critical temperature result for the two kinds of antiferromagnetic ordering [21] (Eqs. 17, 18).

$$T_{c,I} = S(S+1)(|z_n J_n| - |z_{nn} J_{nn}|)/3k \tag{17}$$

$$T_{c,II} = S(S+1)|z_{nn} J_{nn}|/3k. \tag{18}$$

If both T_c and Θ are known from experiments J_n and J_{nn} can be estimated using the above equations.

Which kind of ordering exists depends on sign and relative size of J_n and J_{nn}. For single spins arranged in a bcc lattice ($S = \frac{1}{2}$, $z_n = 8$ and $z_{nn} = 6$) AF ordering of type I will exist if

$$J_n < 0 \quad \text{and} \quad J_{nn} > 0 \quad \text{or} \quad |\tfrac{3}{2} J_{nn}| < |J_n|,$$

and AF ordering of type II if

$$J_{nn} < 0 \quad \text{and} \quad |\tfrac{3}{2} J_{nn}| > |J_n| \quad (\text{for arbitrary values of } J_n).$$

Mean field theory puts limits on the theoretically possible ratio of Θ and T_c: $-3 < \Theta/T_c < 1$ for both AF-I and AF-II ordering. The limiting value $\Theta/T_c = -3$ is reached for $\frac{3}{2}J_{nn} = J_n$. If J_{nn} is zero or neglected (nearest neighbor approximation), then

$\Theta/T_c = -1,$

and J_n can be obtained from the Weiss temperature using Eq. (19)

$$|J_n| = 3k\Theta/z_n S(S+1) = 3k\Theta/2 \quad (19)$$

Note that there are other definitions of the Heisenberg Hamiltonian that use J with the opposite sign (Ref. [17]) or that use J with half the value (Refs. [9,10,20]) it has in the definition adopted here (Ref. [19,21]. Hence, J values reported in different papers may differ in sign or by a factor of 2, and comparison is only meaningful if the definition is known.

4.3 Magnetic Ordering and Heisenberg Coupling Constants

Table 1 summarizes observed Weiss temperatures for SES and the analogue potassium electro sodalite (PES). The $\Theta = 200 \pm 10$ value for SES is an average of results obtained by a least-squares fit of integrated EPR intensities to the Curie–Weiss law and from magnetic susceptibility measurements using the SQUID method [9]. The second value is from the fit of the paramagnetic shift of the ^{27}Al resonance as a function of the temperature to the Curie–Weiss law (cf. Eq. 3). Mean field theory with first-neighbor interaction only (Eq. 19) yield Heisenberg coupling constants between 7.2 and 8.6 meV. These values can be compared with results of density functional theory (DFT) for the SES crystal applying periodic boundary conditions. The calculations yield total energies for the ferromagnetic and antiferromagnetic states of SES at 0 K (only antiferromagnetic ordering of type I has been studied so far) and all show that the antiferromagnetic ordering is more stable. Because the eigenvalues of the Heisenberg Hamiltonian for a bcc lattice of

Tab. 1. Comparison of Heisenberg exchange parameter J_n, derived from observed Weiss temperatures (Θ, within the mean field theory considering first-neighbor coupling only, and quantum mechanical calculations of kinetic exchange using the Anderson model (cf. Eq. 13).

	Observed			*DFT calculation*				
	−Θ/K	*Ref.*	*J_n/meV^a*	*J_n/meV^b*	*H_{ab}/meV*	*U/eV*	*Method*	*Ref.*
SES	200 ± 10	[9]	8.6	8.6	100.5	4.72	U-PW91^c	[14]
	178 ± 8	[9]	7.7	7.6	94.9	4.71	U-PW91^d	[14]
	168 ± 5	[10]	7.2	5.2	87.0	5.80	LSDA^c	[11]
				3.6	61.2	4.18	U-PW86^c	[12]
				3.6	71.0	5.86	U-PBE^e	[15]
PES	400	[24]	17.2	4.7	81.0	5.63	U-PBE^e	[15]

^{a} Derived from observed Weiss temperature using Eq. (22).
^{b} Kinetic exchange derived from DFT calculations using Eq. (13).
^{c} Structure A – Srdanov, cf. Ref. [11].
^{d} Structure B [22].
^{e} Structure C [23].

spins are not known, the Heisenberg coupling parameter cannot directly be derived from these calculations. Therefore, in most DFT studies the calculated band structures are fit to a Hubbard Hamiltonian whose parameters are then used to estimate the kinetic exchange according to Eq. (13). Note that the DFT calculations have been made for three different observed SES crystal structures denoted A (cf. Ref. [11]), B [22], and C [23].

The DFT calculations listed in Table 1 all assume antiferromagnetic ordering of type I and yield lower energies for this state than for the ferromagnetic state. Blake and Metiu [12] limit their DFT calculations to the single electrons of the $(Na_4)^{3+}$ cluster and replace the sodium ion cores by effective potentials. The aluminosilicate framework is described semiempirically. These calculations predicted J_n of about 3.6 meV. Sankey et al. [11] considered the valence electrons on all atoms explicitly and replaced only core electrons by pseudopotentials. A minimal basis set of sp^3 hybrid orbitals was used for calculations within the local spin density approximation (LSDA). An estimate for J_n of about 5.2 meV was derived. Spin density functional calculations employing the full potential linearized augmented plane wave method (FLAPW) [25] by Windiks and Sauer [14] yield estimates of J_n between 6.4 and 9.5 meV, the most likely value being 8.1 ± 0.5 meV. Madsen et al. [15] use the same method (with slightly different parameters) and arrive at smaller J_n of about 3.6 meV. Given the approximations connected with the mean field theory, the agreement is good, and the DFT calculations support the picture of the magnetic interactions derived from experiments.

Madsen et al. also made calculations on PES [15]. In agreement with experiment, antiferromagnetic order is also predicted (the calculations assume type I ordering), but the calculated increase in J_n when Na^+ is replaced by K^+ is much smaller than observed.

The observed $-\Theta/T_c$ ratios of PES and SES (Tab. 2) are very far from the theoretical value of 1 that mean field theory predicts when next-nearest-neighbor inter-

Tab. 2. Heisenberg coupling parameters for nearest and next nearest neighbor interactions, J_n and J_{nn}, respectively, derived from observed transition temperatures, T_c, and Weiss temperatures, Θ, compared to DFT results for J_n and J_{nn}.

	Observed			DFT calculations	
	PES	SES	SES	Periodic b.c.	Molecular
Ref.	[24]	[10]	[9]	[14]	[13]
$-\Theta$/K	400	168 ± 5	178 ± 8		
T_c/K	70	54 ± 2	48 ± 2		
$-\Theta/T_c$	5.7	3.1 ± 0.2	3.7 ± 0.3		
J_n/meV	10.1–14.2[a]	4.8–4.9[a]	5.3–6.5	7.6–8.6	12.0–15.0
J_{nn}/meV	9.5–4.0[a]	3.3–3.1[a]	4.4–2.8	0.6	2.7–3.7
J_n/J_{nn}	1.1–3.5	1.46–1.58[a]	1.20–2.3[a]		4.0–4.7

[a] First and second number correspond to AF-I and AF-II ordering, respectively.

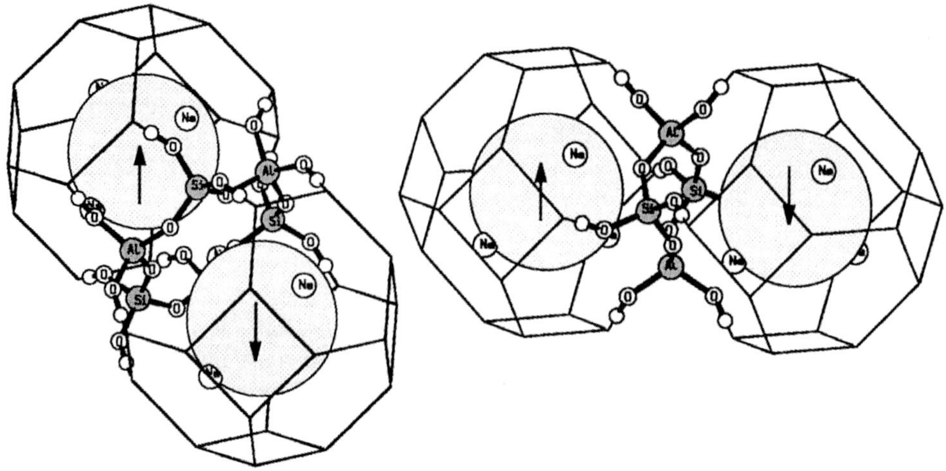

Fig. 4. Molecular models used for determining the Heisenberg coupling constants J_n (left) and J_{nn} (right) from the energy difference of the parallel and antiparallel spin orientations (triplet and singlet state, respectively) [13]. The arrows show the singlet orientation.

actions are neglected. It is also larger than the maximum theoretical value of 3 derived when next-nearest-neighbor interactions are introduced. However, the experimental estimates for SES, in particular the most recent one, $-\Theta/T_c = 3.1$ [10], are close to the maximum value at the border between orderings of type I and type II. Assuming either AF-I or AF-II ordering, mean field theory yields very similar values for J_n and J_{nn} of about 4.8–4.9 and 3.3–3.1 meV, respectively. While J_n still falls into the range of DFT estimates, J_{nn} is much larger than derived from the Anderson formula (Eq. 13) using fitted U and H_{ab} parameters.

However, Windiks and Sauer attempted to determine J_n and J_{nn} directly from DFT calculations without any fitting. They used two finite models of SES which include a nearest-neighbor or a next-nearest-neighbor pair of paramagnetic $(Na_4)^{3+}$ clusters (Fig. 4). These models neglect interactions with all other magnetic sites and describe the electronic structure only approximately. From the energies of the triplet and singlet states of these clusters, J can be obtained using Eq. (5). These models yield the proper value for J_{nn}, while J_n is 2.5 to 3 times larger than the value derived from Θ and T_c.

4.4 Spin Density Distribution

Figure 5 shows the spin density calculated for the antiferromagnetic state using unrestricted DFT and periodic boundary conditions [14]. A similar picture can be

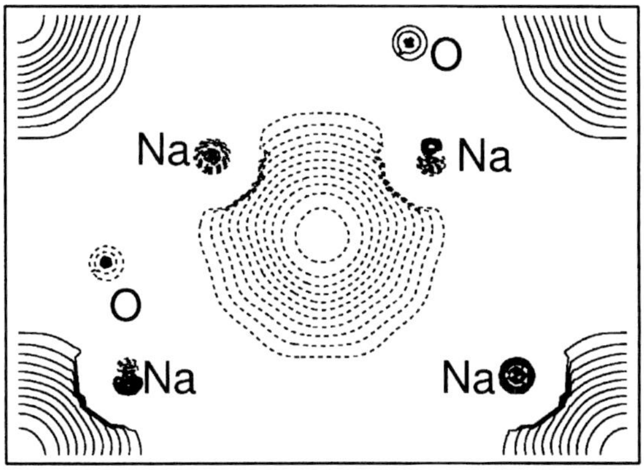

Fig. 5. Spin density distribution of the antiferromagnetic state obtained by DFT calculations for the SES crystal (periodic boundary conditions) after Ref. [14]. Shown is the (110) plane which contains the centers of two $(Na_4)^{3+}$ clusters at (0,0,0) and (1,1,1)(a/2). Note that the contour lines are the same as used in Fig. 8 of Ref. [15], but different from Fig. 6 of Ref. [14], which shows the same data.

found in Fig. 8 (top) of Ref. [15]. The most striking feature is that the maximum of the spin density is not at any of the nuclei, but at the center of the tetrahedron formed by the four Na^+ cores. Such a off-nuclear maximum has been found before in unrestricted HF calculations [23]. The spin density at the off-nuclear maximum is about 0.004 a.u., and that at the Na nuclei between 0.047 and 0.056 a.u. (for the ferromagnetic state 0.080–0.082 a.u.). Spin densities of the same order of magnitude (0.072–0.075 a.u.) have been observed for isolated $Na_4{}^{3+}$ clusters in sodalite [7,26,27].

The other noteworthy feature of the spin density is a substantial spin polarization on the framework atoms. This is evidence for electronic interactions between the aluminosilicate framework as the host system and the bcc lattice of paramagnetic $Na_4{}^{3+}$ clusters as guests. This interaction is also the origin of the observed paramagnetic shifts of the NMR signals for the framework Si and Al nuclei.

4.5 Paramagnetic NMR Shifts for ^{27}Al and ^{29}Si Framework Nuclei

The Engelhardt model relates the different lines observed in the ^{29}Si and ^{27}Al MAS-NMR spectra of SES (Fig. 2) to electron spin densities on Si and Al nuclei, respectively, surrounded by an increasing number of paramagnetic $Na_4{}^{3+}$ clusters. Density functional calculations of these spin densities to support this model meet the following difficulties. Periodic boundary conditions can only be applied to completely ordered spin states which exist below the critical temperature, but not for the temperatures at which paramagnetic shifts are observed. The spin densities on the T sites obtained for the antiferromagnetic state are zero, and those obtained

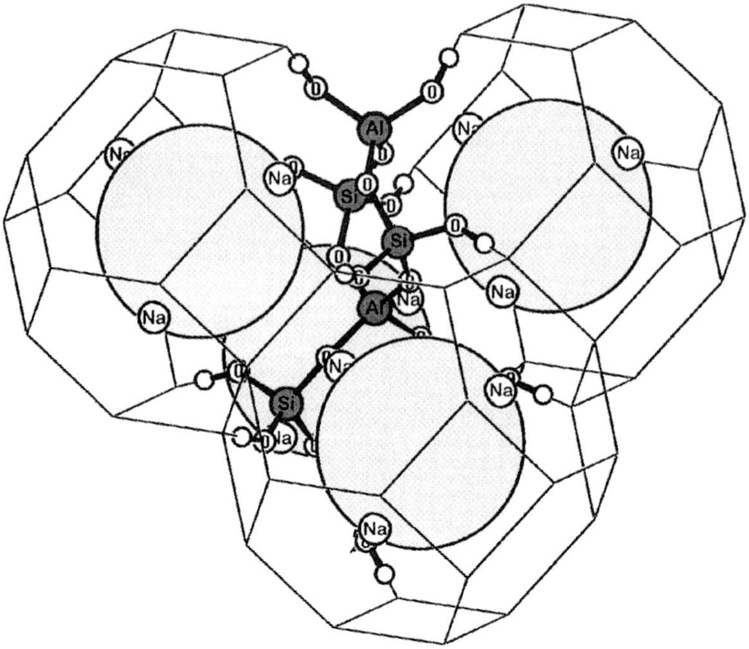

Fig. 6. Molecular model used in spin density calculations. $(Na_4{}^{3+})_n(Na_4{}^{3+})_{4-n}[(AlO_2{}^-)SiO_2]_5$, $n = 1–4$ [13].

for the ferromagnetic state are about 0.01–0.02 a.u., one order of magnitude larger than derived from the paramagnetic NMR shift (about 0.001 a.u.). Finite models, as shown in Fig. 6, are more adequate as they neglect all paramagnetic $Na_4{}^{3+}$ clusters but the immediate neighbors of the T site considered. This is equivalent to an arbitrary orientation of all other spins in the system. As far as the spins of the neighboring $Na_4{}^{3+}$ clusters included in the model are concerned, also states with all spins aligned will be populated in the paramagnetic state and will make the largest contribution to the paramagnetic shift. Therefore, the calculations on finite models with one to four paramagnetic $Na_4{}^{3+}$ clusters were always made for the states with all spins parallel.

The design of the finite cluster model used (Fig. 6) follows general principles [28]. It consists of two four-membered aluminosilicate rings having the central T site in common ("spiro" connection). The dangling bonds at the boundary of the cluster are saturated by H atoms. Depending on the $[n\text{A}(4–n)\text{B}]$ environment modeled, $4n$ sodium ions of the n paramagnetic $Na_4{}^{3+}$ clusters are explicitly considered, while the $(4–n)$ diamagnetic $(Na^+)_3$ clusters are represented by point charges. This cluster is embedded in a finite array of point charges representing neglected framework and sodium ions. This ensures that the whole model system is electrically neutral.

Figure 7 shows the results calculated for different $[n\text{A}(4–n)\text{B}]$ environments and compares them with spin densities derived from paramagnetic NMR shifts. For

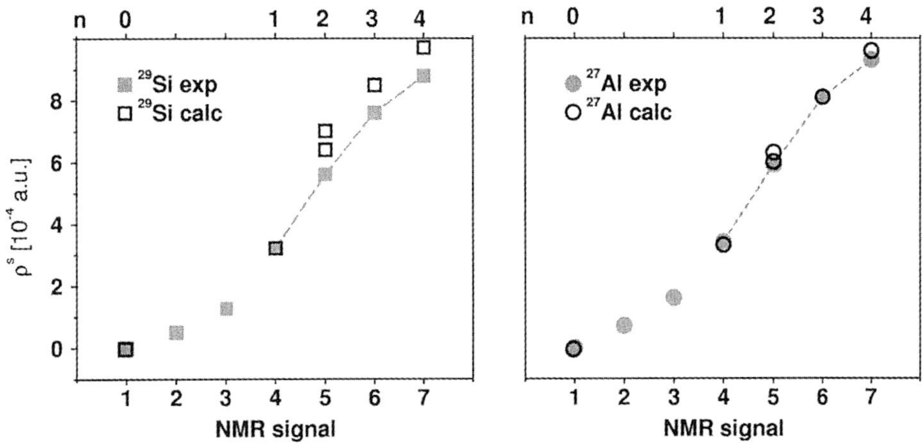

Fig. 7. Spin densities at Si and Al nuclei of the sodalite framework with different environments [nA(4−n)B], $n = 0$–4, derived from paramagnetic NMR shifts [2,8] and calculated by DFT using finite molecular models [13].

$n = 2$ two possible arrangements of the Na_4^{3+} clusters exist (they can occupy either the two cages connected by the four-membered ring or the two cages connected by the six-membered ring) which cannot be resolved in experiments. In general, the DFT calculations reproduce the stepwise increase in the spin densities on the Si and Al nuclei with increasing number of paramagnetic Na_4^{3+} clusters around the T site, including the fact that the steps become smaller for $n = 3$ and $n = 4$. The calculated spin densities provide convincing support for Engelhardt's assignment of the observed lines in the ^{29}Si and ^{27}Al MAS NMR spectra, even if their absolute values are known to vary when changing technical details of the calculation such as the specific functional adopted and the basis set used.

4.6 Concluding Comment

Not only does the computational approach provide an explanation for the peculiar electronic and magnetic properties of alkali electro sodalites, it can also be used to study hypothetical systems that are not accessible to experiments. If we seek an answer to the question how the properties of an bcc lattice of paramagnetic Na_4^{3+} clusters would change if the latter were not embedded in an aluminosilicate guest matrix but fixed in free space at the same positions, computations may provide an answer. When the same DFT calculations as made for SES are applied to an bcc lattice of bare Na_4^{3+} clusters, the result is a diamagnetic ground state even if the starting density was chosen such that the spin moments of the two Na_4^{3+} clusters in the unit cell were oppositely aligned. The valence electrons are completely delocalized, and the band structure indicates that the system is metallic, while SES is predicted to be a small-gap (about 0.1 eV) semiconductor. In contrast, in SES the aluminosilicate framework localizes the single valence electron in each of the Na_4^{3+} clusters. This is also supported by the Heisenberg coupling constants J_n and J_{nn} calculated for finite molecular models. They are one order of magnitude larger for pairs of bare Na_4^{3+} clusters than for the corresponding pairs in SES [13].

Acknowledgement

This work has profited from numerous discussions with Günter Engelhardt, who suggested these studies. We also thank him for comments on the manuscript.

References

1 G.D. Stucky, V.I. Srdanov, W.T.A. Harrison, T.E. Gier, N.L. Keder, K.L. Moran, K. Haug, H.I. Metiu in *Supramolecular Architecture, Synthetic Control in Thin Films and Solids*, T. Bein (Ed.), *ACS Symposium Series 499*, American Chemical Society, Washington, **1992**, p. 294.

2 G. Engelhardt in *Solid-State NMR Spectroscopy of Inorganic Materials*, J.J. Fitzgerald (Ed.), *ACS Symposium Series 717*, American Chemical Society, Washington, **1998**, p. 266.

3 N.P. Blake, H. Metiu, *Physics of Novel Materials, Vol. 10*, M.P. Das (Ed.), Canberra International Physics Summer Schools, **1997**, Canberra, Australia, World Scientific, (London), **1999**, p. 86.

4 R.M. Barrer, J.F. Cole, *J. Phys. Chem. Solids* **1968**, *29*, 1755.

5 V.I. Srdanov, K. Haug, H. Metiu, G.D. Stucky, *J. Phys. Chem.* **1992**, *96*, 9039.

6 F.N. Mott, *Metal-Insulator Transition*, Taylor & Francis, London, **1990**.

7 S.D. McLaughlan, D.J. Marshall, *Phys. Lett. A* **1970**, *32*, 343.

8 G. Engelhardt, M. Feuerstein, P. Sieger, D. Markgraber, G. Stucky, V. Srdanov, *J. Chem. Soc. Chem. Commun.* **1996**, 6, 729.

9 V.I. Srdanov, G.D. Stucky, E. Lippmaa, G. Engelhardt, *Phys. Rev. Lett.* **1998**, *80*, 2449.

10 I. Heinmaa, S. Vija, E. Lippmaa, *Chem. Phys. Lett.* **2000**, *327*, 131.

11 O.F. Sankey, A.A. Demkov, T. Lenosky, *Phys. Rev. B* **1998**, *57*, 15129.

12 N.P. Blake, H. Metiu, *J. Chem. Phys.* **1998**, *109*, 9977.

13 R. Windiks, J. Sauer, *Phys. Chem. Chem. Phys.* **1999**, *1*, 4505.

14 R. Windiks, J. Sauer, *J. Chem. Phys.* **2000**, *113*, 5466.

15 G.K.H. Madsen, B.B. Iversen, P. Blaha, K. Schwarz, *Phys. Rev. B* **2001**, *64*, 195102.

16 P.J. Hay, J.C. Thibeault, R. Hoffmann, *J. Am. Chem. Soc.* **1975**, *97*, 4884.

17 C.A. Daul, I. Ciofini, A. Bencini, in *Reviews of Modern Quantum Chemistry*, K.D. Sen (Ed.), World Scientific, **2002**, in press.

18 P.W. Anderson, *Phys. Rev.* **1959**, *115*, 2.

19 A.H. Morrish, *The Physical Principles of Magnetism*, Wiley, New York, **1965**, pp. 649.

20 J.S. Smart, *Effective Field Theories of Magnetism*, W. B. Saunders Company, Philadelphia, **1966**.

21 J.S. Smart, *Phys. Rev.* **1952**, *86*, 968.

22 N.P. Blake, V.I. Srdanov, G.D. Stucky, H. Metiu, *J. Phys. Chem.* **1996**, *104*, 8721.

23 G.K.H. Madsen, C. Gatti, B.B. Iversen, L. Damjanovic, G.D.

Stucky, V.I. Srdanov, *Phys. Rev. B* **1999**, *59*, 12359.

24 H. Tou, Y. Maniwa, K. Mizoguchi, L. Damjanovic, V.I. Srdanov, *J. Magn. Magn. Mater.* **2001**, *226–230*, 1098.

25 P. Blaha, K. Schwarz, J. Luitz, *WIEN97 - A Full Potential Linearized Augmented Plane Wave Package for Calculating Crystal Properties*, Technische Universität Wien, Wien, **1999**.

26 W.G. Hodgsen, J.S. Brinen, E.F. Williams, *J. Chem. Phys.* **1967**, *47*, 3719.

27 J.B.A.F. Smeulders, M.A. Hefni, A.A.K. Klaassen, E. deBoer, U. Westphal, G. Geismar, *Zeolites* **1987**, *7*, 347.

28 J. Sauer, *Chem. Rev.* **1989**, *89*, 199.

5
Electronic Structure of Zeolite-Stabilized Ions and Quantum Dots

Gion Calzaferri, Stephan Glaus, Claudia Leiggener, and Ken'Ichi Kuge*

5.1
Introduction

Zeolitic materials can act as hosts for supramolecular organization of molecules, ions, complexes, clusters, and quantum-sized particles. They allow the design of precise and reversible functionalities [1]. The possibility of arranging zeolite microcrystals of good quality and narrow size distribution as dense monolayers on different substrates can be used to realize distinct properties [2–7]. New electronic structures are accessible either by specific geometrical arrangements of guests, made possible by the structure of the host, and/or by explicitly involving the electronic properties of the host.

Three functionalities are of special importance in our research: intrazeolite ion transport, intrazeolite charge transport, and intrazeolite electronic excitation energy transport (energy migration). The zeolite host is not actively involved in these processes, but provides the necessary geometrical and chemical environment. It can also lead to greatly improved chemical stability of incorporated species by shielding them from chemicals with which they would otherwise react or by preventing intramolecular rearrangements by limiting the available free space. A number of methods have been developed for preparing zeolites containing the desired molecules, ions, complexes, or clusters. These are crystallization inclusion, ion exchange, incorporation from the gas phase, and *in situ* synthesis. Each of these methods has its advantages and disadvantages depending on the specific problem to be solved.

The most interesting phenomena seem to occur in micrometer- and nanometer-sized crystals. In spite of the very fast progress of nanoscience techniques, unambiguous structure determination, detection of intrazeolite charge transport, and interpretation of photophysical phenomena are still difficult and time-consuming. An improved understanding of the electronic structure of these host–guest materials is therefore of decisive importance.

Especially for the synthesis of quantum-confined semiconductor clusters such as CdS [8–12], CdSe [10], CdO [13], GaP [14], PbS [15], Se [16], Si [17,18], SnO_2 [19],

TiO_2 [13,20], ZnO [13,21,22], ZnS [10,11,21], and ZnSe [10], different types of zeolites are used, because their cavities determine to a great extent the size and shape of the clusters. In this chapter, we focus on the framework of zeolites A, Y, and L (Fig. 1). These zeolites are crystalline aluminosilicates with cavity and channel structures. Their lattices are enormous polyanions which contain cations for charge compensation.

In Section 5.2 the electronic structure of a zeolite framework, derived from $H_8Si_8O_{12}$ as the smallest cage molecule of relevant size and structure, is discussed. Section 5.3 describes a general concept of the electronic structure of Cu^+-, Ag^+-, and Au^+-loaded zeolites, while Ag^+-loaded zeolites are discussed in detail in Section 5.4. We discuss the synthesis and analysis of quantum-sized clusters in the cavities of zeolites in Section 5.5. These tunable semiconductor materials open a variety of fascinating phenomena. Section 5.6 deals with the controversial topic of intrazeolite charge transport in the channels of zeolites and the decisive role of the interface is discussed. A new idea for solving the interfacial problem is reported.

5.2 $H_8Si_8O_{12}$: A Model for the Vibrational and Electronic Structure of Zeolite A

The framework of zeolite A can be generated by placing cubic T_8O_{12} double four-rings (D4R) in the centers of the edges of a cube of 12.3 Å length and by connecting the D4R's by oxygen bridges. The center of the unit cell is a large cavity, also named α-cage, with a free diameter of 11.4 Å; 8-membered rings with a free diameter of 4.1 Å give access to the large cavity. The relation between the D4R and the zeolite framework is shown in Fig. 2 (right). The bridging oxygen atoms are omitted in the middle and upper part of this figure, as is usual in this kind of drawing. They have been added in the next step, which leads to the $H_8Si_8O_{12}$ molecule. We have shown that this molecule is an excellent model for studying properties not only of the D4R secondary building unit and thus of zeolite A, but also for advancing our understanding of aluminosilicate-based zeolites in general, because it is easy to correlate the vibrational and the electronic structure of O_h-$H_8Si_8O_{12}$ with that of the hypothetical O_h-$H_{24}Si_{24}O_{36}$ [23–27]. The latter has the structure of the sodalite cage, which can be used as a link to many zeolites.

Remarkably, among the many orbitals of O_h-$H_8Si_8O_{12}$ there is exactly one of A_{2g} symmetry [28]. This pure oxygen lone pair, which cannot interact with AOs from centers other than oxygen, is the HOMO; it is followed by a number of oxygen lone pairs between −10.75 and −11.7 eV which interact only slightly with the Si atoms. A comparison of the calculated one-electron energy levels in the HOMO region and the measured photoelectron spectrum is illustrated in Fig. 3. The calculated first ionization energy of 10.7 eV is low but in good agreement with the experimental observation. To get a feel for the consequences of this relatively high lying HOMO, we compare it with the first ionization energy of water, which is 12.6 eV and attributed to the energy of the p-type oxygen lone pair of the water molecule

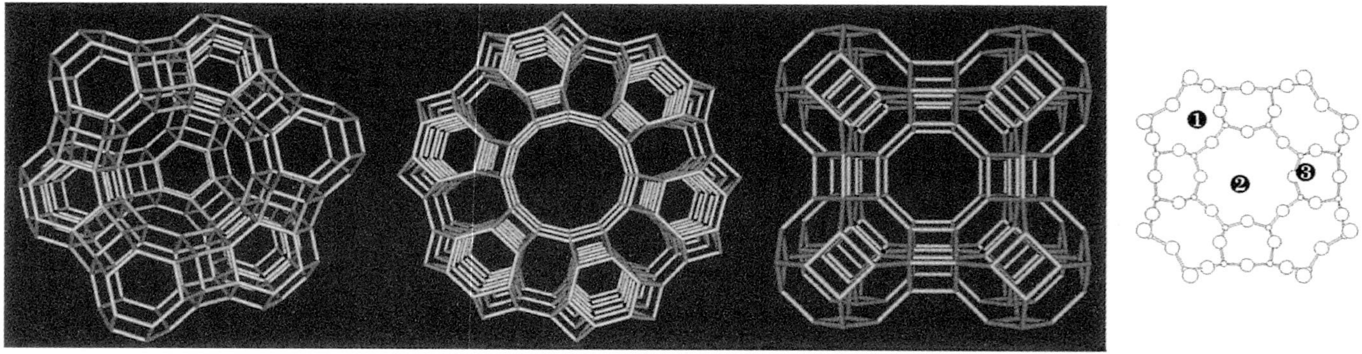

Fig. 1. Framework of zeolite Y (left), zeolite L (middle), and zeolite A (right). The bridging oxygen atoms are omitted. Cation positions in zeolite A – 6-ring (1), 8-ring (2), and 4-ring (3) – are shown on the right.

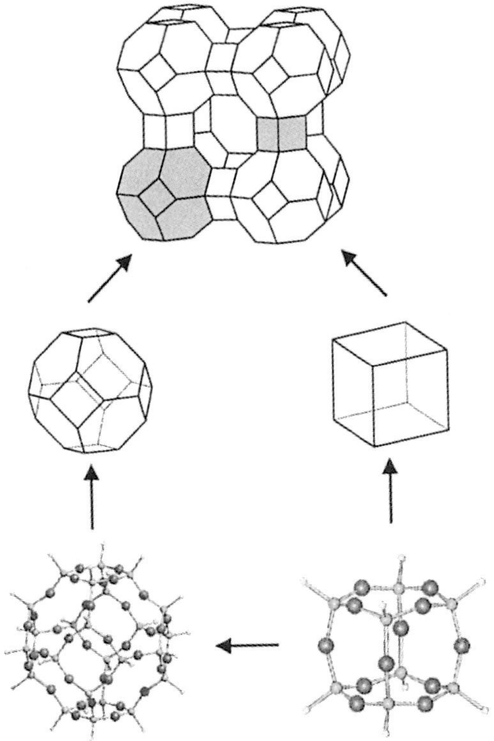

Fig. 2. Relation between the framework of zeolite A and $H_8Si_8O_{12}$ (right) and $H_{24}Si_{24}O_{36}$ (left).

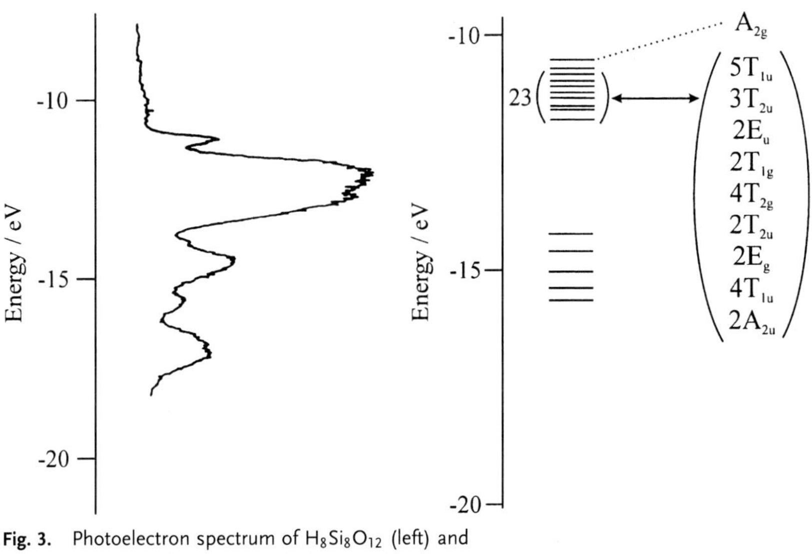

Fig. 3. Photoelectron spectrum of $H_8Si_8O_{12}$ (left) and calculated occupied electron levels (right) [28].

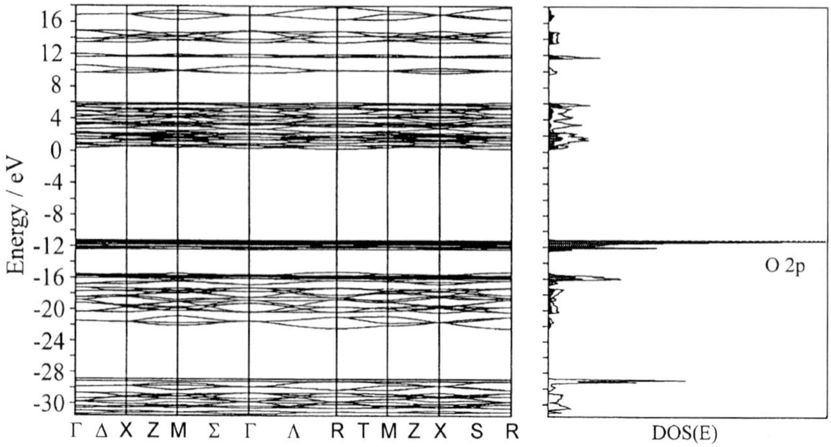

Fig. 4. Band structure (left) and DOS (right) of the silicon dioxide analogue of zeolite A. The oxygen 2p density is projected (shaded regions).

[29]. We note that the first ionization energy of α-quartz is 10.4 eV, as determined by valence-band spectroscopy [30].

The link between the electronic structure of $H_8Si_8O_{12}$ and that of zeolite A was discussed in Ref. [24]. The result is summarized in Fig. 4, which illustrates the band structure and density of states of the silicon dioxide analogue of zeolite A, which is especially simple because of the absence of co-cations. The flatness of the bands in the HOMO region indicates the presence of nonbonding states. Some bands below −14 eV are significantly bent and contribute to Si–O bonding. Further insight is gained from the density of states DOS(E), defined in such way that DOS(E)dE is the number of states in the interval E to $E + dE$. Since we are expressing the crystal orbitals as linear combination of atomic orbitals (LCAO) we can project specific atomic orbitals or linear combinations thereof. In Fig. 4 this is done by shading the oxygen 2p contributions and leaving the 2s oxygen and the silicon contributions blank. This shows that the HOMO region consists of nearly pure oxygen 2p lone pairs, which we denote as |O<.

5.3
Electronic Structure of Cu^+-, Ag^+-, and Au^+-Loaded Zeolites

The Al^{3+} centers in zeolites result in a negatively charged framework (AlO_2^-). This charge is compensated by exchangeable cations, which influence the band structure to some extent. It is, however, reasonable to assume that the HOMO and the LUMO regions are in general similar to those illustrated in Fig. 4. Provided this is correct, the HOMO–LUMO region of zeolites containing monovalent cations of the type Li^+, ..., Cs^+, Cu^+, Ag^+, Au^+ can be drawn qualitatively as illustrated in

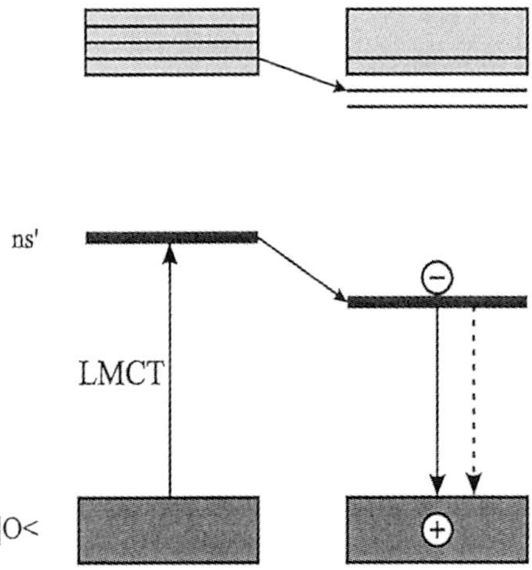

Fig. 5. Energy diagram of a metal cation in a zeolite framework. The HOMO region consists of many narrowly spaced localized states strongly concentrated on the oxygen atoms. We call this the lone pair region of the silicate and abbreviate it as |O<. Some of the np′ levels may reach into the LUMO region of the silicate. Three lines have been added to indicate this.

Fig. 5. It consists of the oxygen lone pair region denoted as |O<, the empty ns′ level of the metal cations M^+, and the LUMO region of the zeolite, which may be modified by np′ contributions of M^+. The ns′ and np′ levels are modified to some extent with respect to the ns and np levels of the free cations by their interaction with the surrounding [31–34]. This scheme suggests the occurrence of ligand to metal charge transfer (LMCT) transitions of the ns′ ← |O< type, with excitation of an oxygen lone pair electron to the metal cation coordinated to a zeolite oxygen atom. The energy ΔE_{CT} needed for this transition is equal to the difference of the ionization potential $IP_{|O<}$ of the oxygen lone pair and the first ionization potential IP_M of the metal M, plus a correction Δ which stands for the antibonding interaction of the empty ns′ level of the metal ion M^+ with the environment (Eq. 1).

$$\Delta E_{CT}(ns' \leftarrow |O<) = IP_{|O<} - IP_M + \Delta \qquad (1)$$

This simple relationship allows us to estimate the energy of the charge transfer band for different situations, as shown in Tab. 1 for cations in water and in a silicate with oxygen lone pairs lying at about −10.7 eV, as discussed in Section 5.2.

Charge transfer transitions of the type ns′ ← |O< have been observed in Cu^+-A, Cu^+-X and Ag^+-A zeolites in the region of 28 000 cm^{-1}. This means that Δ is on the order of 0.5 eV or 4000 cm^{-1} for these transitions. CT transitions with larger Δ are discussed in Section 5.4. In water Ag^+ absorbs light at about 225 nm [35], in agreement with Eq. (1). Table 1 shows that Au has the largest IP_M, which gives Au its noble character. In general Au^+ is unstable but can be stabilized by specific ligands. This makes it difficult to incorporate Au^+ by conventional ion-exchange

Tab. 1. Estimation of the ns′ ← |O< LMCT charge transfer transition energy ΔE_{CT} for the cations M^+ in water and in a silicate with $IP_{|O<} = 12.6$ eV and 10.7 eV, respectively.

Cations	*IP_M/eV*	*ΔE_{CT}(ns ← |O<)–Δ/cm^{-1}*	
		$IP_{\|O<} = 12.6$ eV	*$IP_{\|O<} = 10.7$ eV*
Li	5.3600	56 216	43 067
Na	5.1200	60 326	44 183
K	4.3200	66 778	51 454
Rb	4.1600	68 078	52 745
Cs	3.8700	70 407	55 083
Cu	7.6800	39 679	24 356
Ag	7.5400	40 808	25 485
Au	9.1800	27 582	12 258

methods. However, some researchers reported the synthesis of Au^+-loaded zeolites by means of ion exchange with gold complexes or by sublimation of $(AuCl_3)_2$ [36]. The electronic spectra of the resulting materials reported so far suggest that Au^+ was still coordinated by chloride, and that direct interaction with the zeolite could not be observed. Clear assignment of the nature of the observed electronic transitions has not been established. More work is needed to obtain electronic spectra of Au^+ coordinated to zeolite oxygen atoms. We should also add that it is still difficult to locate the ns levels of divalent cations such as Ca^{2+} ($IP_1 = 6.01$, $IP_2 = 11.82$ eV) and Mg^{2+} ($IP_1 = 7.61$, $IP_2 = 14.96$ eV) in zeolites. We must leave the position and properties of the ns levels open at present.

Luminescence of the Cu^+-A [31], Cu^+-X [32], Cu^+-ZSM-5 [37,38], and Ag^+-A zeolites after ns′ ← |O< excitation occurs at 400–700 nm, depending on the sample and the conditions. This means that the Stokes shift is about 8000 cm^{-1}. Excitation of an electron from the oxygen lone pair level |O< into the empty ns′ orbital of the metal cation causes a formal reduction of M^+ to M^0. The radius r of M^+ is significantly smaller than that of M^0 ($r(Cu^+) = 0.96$, $r(Cu^0) = 1.35$, $r(Ag^+) = 1.26$, $r(Ag^0) = 1.6$ Å). This means that the LMCT transition expands the metal center by 0.3–0.4 Å and causes a change in its position. As a consequence, the ns′ level relaxes to a state of lower energy, which we denote the $(|O<)^{\oplus}(ns')^1$ state. The latter relaxes to the ground state by emitting a photon with a large Stokes shift or by radiationless processes, as illustrated in Fig. 5 (right).

5.4
Electronic Structure of Ag^+-Zeolite A

Rálek et al. reported in 1962 that hydrated colorless zeolite $Ag^+{}_xNa^+{}_{12-x}A$ turns yellow to brick red on activation [39]. No explanation of this phenomenon was given at that time. Later it was believed that the color change was due to formation of silver clusters $Ag^0{}_n$ in the cavities of Ag-zeolite A. These neutral silver species were assumed to form at elevated temperatures by an auto-reduction process in which O_2 from the zeolite framework was released [40]. We studied the vibrational spectra of Ag^+-zeolite A materials in some detail [41,42], and we recently showed that activation at room temperature under high vacuum is already sufficient to produce the yellow form of $Ag^+{}_xNa^+{}_{12-x}A$. The fully reversible color change, which depends on the hydration state of the silver zeolite, was attributed to electronic charge transfer transitions from the oxygen lone pairs of the zeolite framework to the empty 5s orbital of the Ag^+ ions, denoted as $Ag^+(5s) \leftarrow O(n)$ [33]. Pure sodium ($Na^+{}_{12}A$) and calcium zeolite A ($Ca^{2+}{}_6A$) are colorless in both their hydrated and their activated (dehydrated) states. Silver-containing sodium zeolite A is colorless in its fully hydrated form. In activated silver zeolite A materials, the Ag^+ is forced to undergo coordination by zeolite oxygen atoms because an insufficient number of water molecules is available. The question remained whether specific coordination sites which act as yellow and/or red "color centers" can be identified. We answered this question by studying the UV/Vis spectra of $Ag^+{}_xNa^+{}_{12-x}A$ and of $Ag^+{}_xCa^{2+}{}_{6-0.5x}A$ materials in their fully hydrated, in high-vacuum (HV) room-temperature (RT) dehydrated, and in HV elevated-temperature (ET) dehydrated states. A comparison of such spectra is shown in Fig. 6. The marked site preference of the ions in $Ag^+{}_xCa^{2+}{}_{6-0.5x}A$, probed by gas adsorption experiments, offered the unique possibility of investigating different coordination sites of Ag^+ ions in zeolite A [34]. Pure sodium and calcium zeolite A do not absorb light within the spectral range of 50 000–10 000 cm^{-1} that we investigated. This means that any absorption bands or colors observed in Ag-zeolite A materials are due to the presence of silver ions.

We found that 6- and 8-ring-coordinated Ag^+ give rise to electronic transitions in the near-UV region. An absorption in the visible region, at 22 000 cm^{-1}, was only observed for materials in which 4-ring-coordinated Ag^+ was present, and only they showed the typical deep yellow color. We also observed that Ag^+ avoids the 4-ring sites as long as possible in $Ag^+{}_xCa^{2+}{}_{6-0.5x}A$, namely, as long as x is less than 10. In the case of $Ag^+{}_xNa^+{}_{12-x}A$ either a Na^+ or a Ag^+ ion is forced to undergo coordination by a 4-ring site, because all other sites are occupied. The fact that the 22 000 cm^{-1} absorption responsible for the yellow color is already present at $x < 0.2$ proves that isolated Ag^+ ions are sufficient to cause it and that 4-ring coordination of Ag^+ is significantly stronger than that of Na^+. The red color of ET-activated samples is caused by a strong absorption band at 19 000 cm^{-1}. We observed that samples which remained colorless after RT activation never turned red, that samples with lower silver content than one Ag^+ per α-cage never turn red, and that RT dehydration under our experimental conditions was not sufficient to produce red samples. These observations strongly indicate that only samples with 4-ring-coordinated Ag^+ can give rise to the 19 000 cm^{-1} band, and this only occurs if a second Ag^+ ion is not too far away at a 6-ring site, so that they can interact to develop a corresponding low-lying state.

Molecular orbital calculations on a sufficiently large zeolite fragment consisting of 1296 atoms allowed us to address questions about the nature of the HOMO and of the LUMO region, the contributions of the zeolite framework atoms to the elec-

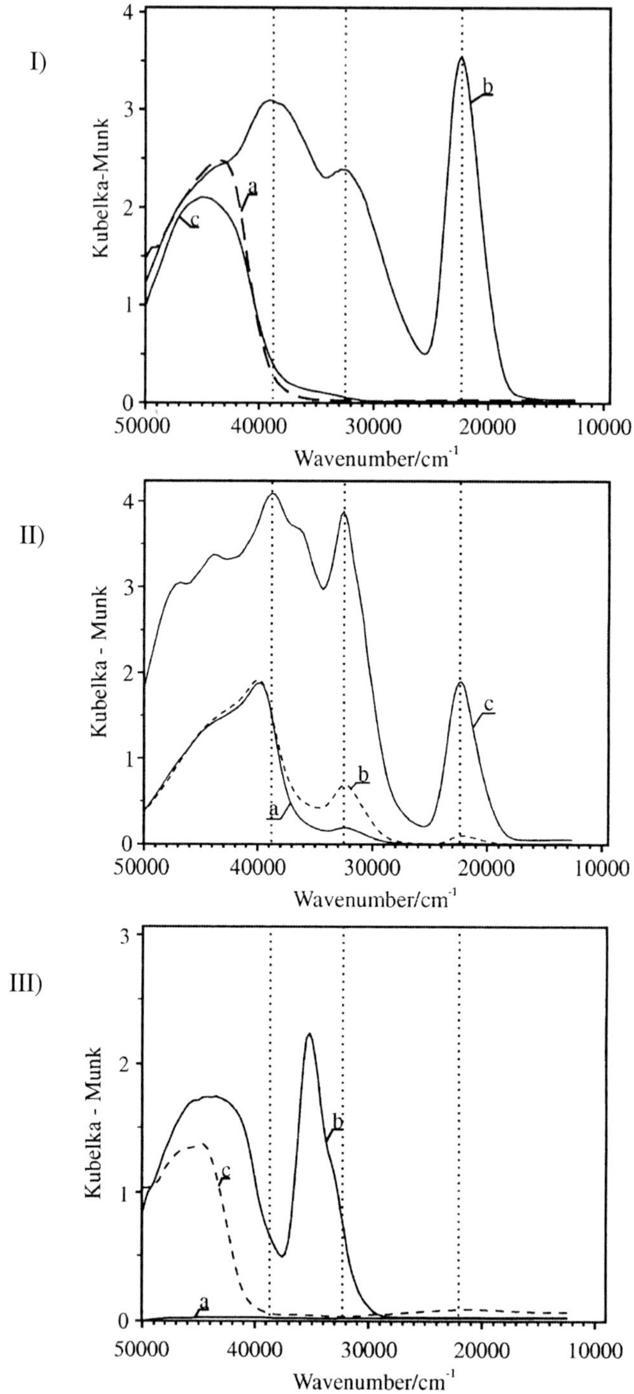
I)
Kubelka-Munk
Wavenumber/cm^{-1}
a
b
c
II)
Kubelka - Munk
Wavenumber/cm^{-1}
a
b
c
III)
Kubelka - Munk
Wavenumber/cm^{-1}
a
b
c

tronic transitions, the influence of the local symmetry of Ag^+ at 4- and at 6-ring sites, and the importance of Ag^+–Ag^+ interactions. To avoid geometries without experimental relevance, we restricted this study to Ag^+ at 4- and 6-ring sites known from X-ray measurements.

We found that the occupied frontier orbital region consists mainly of two bunches of levels: the HOMO region from about −11 to −12.6 eV and the next levels below −13.6 eV. The LUMO consists of a single level of mainly $Ag^+(5s)$ character. The LUMO+1 was energetically too high to be of relevance. Thus, the oscillator strength of transitions from the first 1244 levels to the LUMO were calculated. For 6-ring-coordinated Ag^+ all levels in the HOMO region are derived from mostly noninteracting oxygen lone pairs which we abbreviate as O(n). The LUMO is a rather pure $Ag^+(5s)$ orbital with some contribution from the three coordinating oxygen atoms. Thus, all electronic excitations in question are LMCT transitions from oxygen lone pairs to the silver ion. They are energetically located in the near-UV and depend only slightly on the polarization. The agreement in shape and position between the calculated and experimental spectra allowed us to conclude that 6-ring-coordinated Ag^+ gives rise to electronic transitions from zeolite oxygen lone pairs to the $Ag^+(5s)$ orbital in the near-UV. We denote such electronic transitions as $Ag^+(5s) \leftarrow O(n)$ [34].

The bands of 4-ring-coordinated Ag^+ are strongly polarized. Two almost degenerate low-energy absorption bands and a prominent high-energy band dominate the spectrum. The first one can be described as an $Ag^+(5s) \leftarrow O(n)$ LMCT transition. It is responsible for the yellow color. The near-UV band exhibits some $Ag^+(5s) \leftarrow \sigma$ character but it can still be regarded as an oxygen to silver LMCT transition. It is assigned to the 32 000 cm^{-1} band shown in Fig. 6 (I and II). Interestingly, only three of the 1244 electronic transitions from the frontier orbital region to the LUMO have significant intensity. This clearly demonstrates the paramount importance of oscillator strengths and how problematic an estimation of electronic spectra based on DOL arguments alone can be. We conclude that the calculated $Ag^+(5s) \leftarrow O(n)$ transition of 4-ring-coordinated Ag^+ is in agreement with the appearance of a deep yellow color in RT-activated Ag^+-zeolite A materials with occupied 4-ring positions in the sodalite cavity. The main difference between 4-, 6-, and 8-ring-coordinated Ag^+ is that the antibonding interaction of the $Ag^+(5s)$ orbital with the oxygen lone pair is weaker in the 4-ring position and larger in the others.

Six-ring-coordinated Ag^+ gives rise to electronic transitions in the near-UV, and 4-ring coordinated Ag^+ is responsible for the deep yellow color of the RT-activated material. This implies that similar $Ag^+(5s) \leftarrow O(n)$ LMCT transitions are to be ex-

Fig. 6. I) UV/Vis spectra of $Ag^+{}_6Na^+{}_6A$: a) freshly exchanged, never activated; b) activated at room temperature; c) activated at room temperature and exposed to pure water vapor before measurement. II) UV/Vis spectra of various silver-containing zeolites activated at room temperature: a) $Ag^+{}_{10}Ca^{2+}A$; b) $Ag^+{}_{11}Ca^{2+}{}_{0.5}A$; c) $Ag^+{}_{12}A$. III) UV/Vis spectra of zeolites Y: a) $Na^+{}_{69}Y$, b) $Ag^+{}_{69}Y$, activated at room temperature, c) the same as b) but after exposure to moisture [34].

pected in other Ag^+ exchanged zeolites. Ag^+-exchanged zeolite Y can be used as a test. We therefore show in Fig. 6 (III) UV/Vis spectra of pure $Na^+{}_{69}Y$ (a) and of RT/HV-dehydrated $Ag^+{}_{69}Y$ before (b) and after exposure to moisture (c). The main result is that an intense band at about 34 000 cm^{-1} appears on dehydration and vanishes on rehydration. By analogy one would also expect a similar type of LMCT transition in Cu^+ zeolite materials. $Cu^+(4s) \leftarrow O(n)$ LMCT transitions, reversible upon HV hydration/dehydration, have indeed been observed in Cu^+-zeolites A and X [31,32].

The luminescence properties of Ag^+-loaded zeolites depend on the amount of water which is available for coordination to the silver ions. $Ag^+{}_xCa^{2+}{}_{6-0.5x}A$ shows luminescence even in the fully hydrated state, while in the case of $Ag^+{}_xNa^+{}_{12-x}A$ luminescence can only be observed in partially hydrated states [42]. Because the electrostatic interaction of water with Ca^{2+} is stronger than those of water with monovalent cations, most of the water molecules in the hydrated zeolite are coordinated to Ca^{2+}, and only a few to Ag^+. If only monovalent cations are present, as in $Ag^+{}_xNa^+{}_{12-x}A$ materials, the number of water molecules which coordinate to Ag^+ is larger than in $Ag^+{}_xCa^{2+}{}_{6-0.5x}A$ samples. Thus the condition for luminescence of Ag^+-loaded zeolite A is that only a small number of water molecules coordinate to the silver ions. Figure 7 shows the absorption and luminescence spectra of fully hydrated $Ag^+Ca^{2+}{}_{5.5}A$. The luminescence intensity is enhanced by cooling the sample with liquid nitrogen.

Based on all the information which has been collected over the last few years, we can now draw the schematic state diagram in Fig. 8 for Ag^+-containing zeolites.

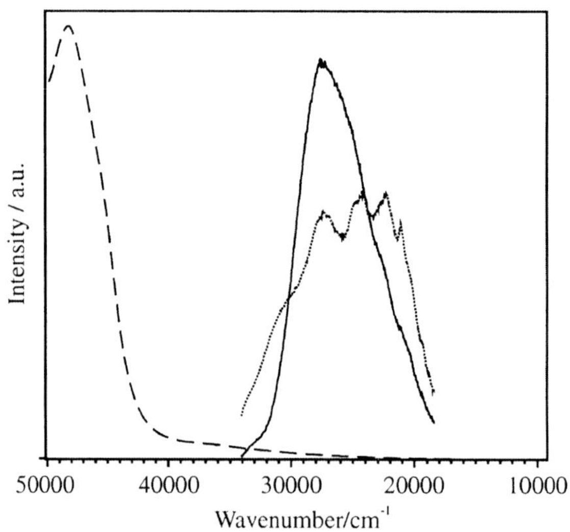

Fig. 7. Luminescence spectra at −195° C (solid line) and at room temperature (dotted line), excited at 250 nm, and diffuse reflectance spectrum (dashed line, Kubelka–Munk) of $Ag^+Ca^{2+}{}_{5.5}A$.

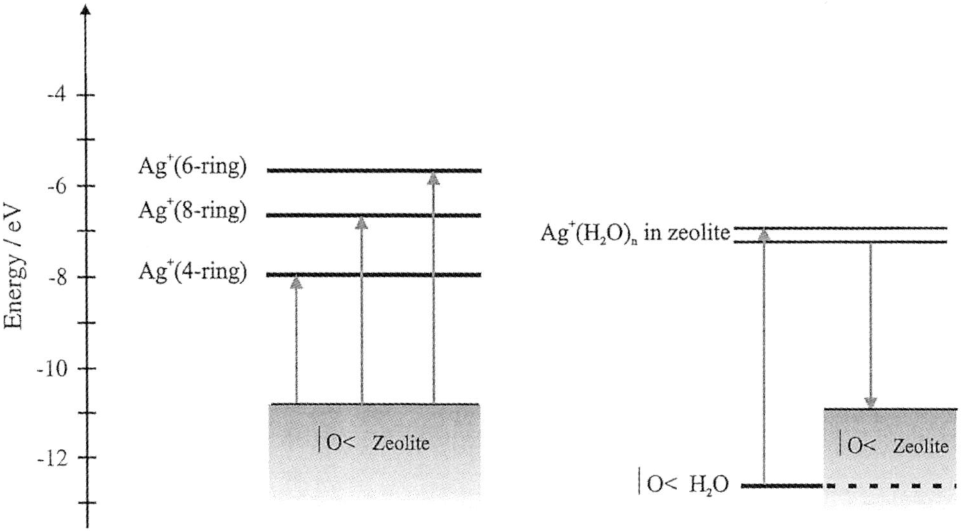

Fig. 8. Frontier orbital state diagram of Ag^+-loaded zeolite A. On the left side we show the levels observed in RT-activated zeolites in which all three sites are occupied by silver ions, while the scheme on the right side corresponds to situations typically observed in $Ag^+{}_xCa^{2+}{}_{6-0.5x}A$ materials containing some water.

5.5 Quantum-Sized Silver Sulfide Clusters in Zeolite A

The synthesis and the properties of semiconductor particles in the size regime of a few up to hundreds of angstroms continues to attract considerable interest [43]. Significant quantum confinement effects can be observed in clusters of II-VI and IV-VI compounds such as CdS [44], CdSe [45], ZnO [46], ZnS [47], and PbS [48]. While excellent progress has been made in the preparation and characterization of these materials, little is known about the properties of small Ag_2S species. This can be partly attributed to the fact that silver sulfide clusters show a strong tendency to aggregate into bulk material, which complicates their synthesis considerably. The well-defined cavities of zeolites provide a convenient environment for preparing clusters with a narrow size distribution or even cluster arrays [9]. We have shown that the frameworks of zeolites A and ZK4 prevent the silver sulfide clusters from aggregating, and we reported the synthesis and the optical absorption and emission spectra of silver sulfide/zeolite A composites. The preparation method is based on the observation discussed in Section 5.4 that Ag^+-loaded zeolite A can be reversibly activated at room temperature [49–52].

The low-temperature phase of bulk silver sulfide is stable up to approximately 177 °C and is usually denoted as α-Ag_2S. Historically, we can go as far back as 1833 when Michael Faraday made the remarkable discovery that silver sulfide behaves as an insulator at room temperature but exhibits high electrical conductivity at ele-

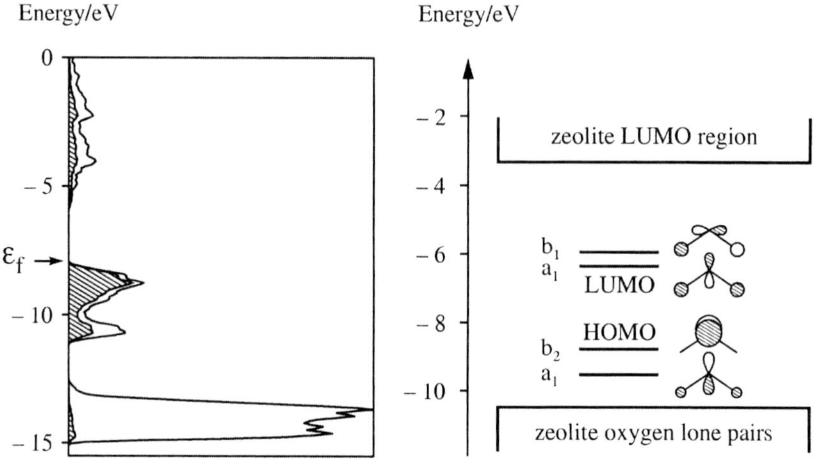

Fig. 9. Left: DOS plot of α-Ag_2S. The hatched region indicates the contribution of sulfur 3p states. The Fermi level ε_f is marked by an arrow. Right: HOMO–LUMO region of an Ag_2S molecule in comparison to the HOMO–LUMO region of zeolite A [49].

vated temperatures, leading him to the following conclusion [53]: "There is no other body with which I am acquainted, that, like sulphuret of silver, can compare with metals in conducting power for electricity of low tension when hot, but which, unlike them, during cooling, loses in power, whilst they, on the contrary, gain. Probably, however, many others may, when sought for, be found." Today it is well known that α-Ag_2S is a semiconductor with a monoclinic structure [54] and a band gap of approximately 1 eV at room temperature [46]. Figure 9 (left) shows the calculated density of states (DOS) of bulk α-Ag_2S. The electronic transition from valence band to conduction band is essentially a charge transfer from 3p(S) to 5s(Ag). This property is also observed in the Ag_2S molecule. The HOMO–LUMO region of such a molecule, and presumably also of larger silver sulfide clusters, fits well into the gap between the oxygen lone pairs of zeolite A and the zeolite A LUMO region (Fig. 9, right), and therefore a variety of electronic transitions arises [49].

Bulk silver sulfide has been considered for photoimaging and photodetection in the IR region [55], and small silver sulfide clusters are known to play an important role in photographic sensitivity [56–59]. It has been reported that silver sulfide clusters with sizes ranging from 23 to 76 Å can be synthesized in reverse micelles [60]. Another method utilizes the rapid expansion of an $AgNO_3$ solution in supercritical ammonia into an solution of Na_2S in ethanol. An average diameter of 73 Å was found after stabilizing the thus-formed particles with a suitable polymer [61]. Dosed addition of an $AgNO_3$ solution to a gelatin solution containing Na_2S was reported to yield silver sulfide clusters in the size regime between 30 and 100 Å [62]. Other methods use nylon thin films (cluster size ranging from 47 to 112 Å) [63], Nafion membranes (cluster size ranging from 50 to 150 Å) [64], or capping with cysteine/glutathione (average cluster size of 90 Å) [65] to stabilize the silver sulfide clusters. Some of the results obtained by the above-mentioned methods suggest the presence of a quantum-size effect for silver sulfide clusters with a diameter between 20 and 100 Å (see Refs. [61,62,65,66]), while other reports clearly negate the presence of such an effect (see, e.g., Ref. [63]).

The synthesis of silver sulfide particles in the cavities of zeolite A in the size regime below 15 Å can be divided into four steps: (1) loading of the zeolite with Ag^+, (2) activation of the Ag^+-loaded zeolite, (3) reaction with H_2S, and (4) rehydration. To explain the mechanism of cluster growth we start by examining the formation of silver sulfide clusters in zeolite samples with a low content of Ag^+, e.g., 0.05 Ag^+ per α-cage. Since the silver ions are evenly distributed among the α-cages, this implies that 5 % of them actually contain a silver ion in this case (see Fig. 3 in Ref. [33]). Activation of the Ag^+-loaded zeolite and subsequent adsorption of H_2S leads to reaction (2) inside an α-cage. The formation of protons can easily be observed.

$$Ag^+ + H_2S \rightleftharpoons AgSH + H^+ \tag{2}$$

It can be assumed that the silver atom of the AgSH molecule thus formed is at this stage still coordinated to zeolite framework oxygen atoms. Uptake of water during rehydration mobilizes the AgSH molecules through solvation. Encounter of two AgSH molecules leads to reaction (3).

$$AgSH + AgSH \rightleftharpoons Ag_2S + H_2S \tag{3}$$

H_2S is released from the zeolite. The equilibrium (3) is shifted to the right if the H_2S is removed. The reaction can be reversed in presence of H_2S. Further diffusion of the Ag_2S molecules has not been observed so far at room temperature [51,52].

In our first communication, we reported the luminescence spectra of Ag_2S–NaA samples [49]. Ag_2S–CaA samples were found to exhibit much stronger luminescence, usually readily visible at room temperature. Figure 10 shows the luminescence spectra of Ag_2S–CaA samples with varying silver sulfide content. The composition of the silver sulfide/zeolite A composites is $Ag_xS_{x/2}Na_{12-x}H_xSi_{12}Al_{12}O_{48}\cdot nH_2O$ for clusters in NaA and $Ag_xS_{x/2}Ca_{6-x/2}H_xSi_{12}Al_{12}O_{48}\cdot nH_2O$ for clusters in CaA (pseudo unit cell contents). We use the abbrevations Ag_2S–NaA-x and Ag_2S–CaA-x for the differently loaded samples, where x denotes the number of silver ions per α-cage of zeolite A.

Comparison of the spectra shown in Fig. 10 with the spectra of Ag_2S–NaA samples reported in Ref. [49] reveals the following similarities: A low silver sulfide content is characterized by a blue-green luminescence (480 nm for Ag_2S–NaA and 490 nm for Ag_2S–CaA) and a corresponding excitation spectrum with distinct and narrow bands. The Stokes shift is 1.3 eV. Given the distinct nature of the excitation bands and the low silver sulfide content, we conclude that the blue-green luminescence is caused by monomers of Ag_2S. Further evidence for this is reported in Ref. [52]. At higher loading levels, an orange-red luminescence becomes increasingly dominant.

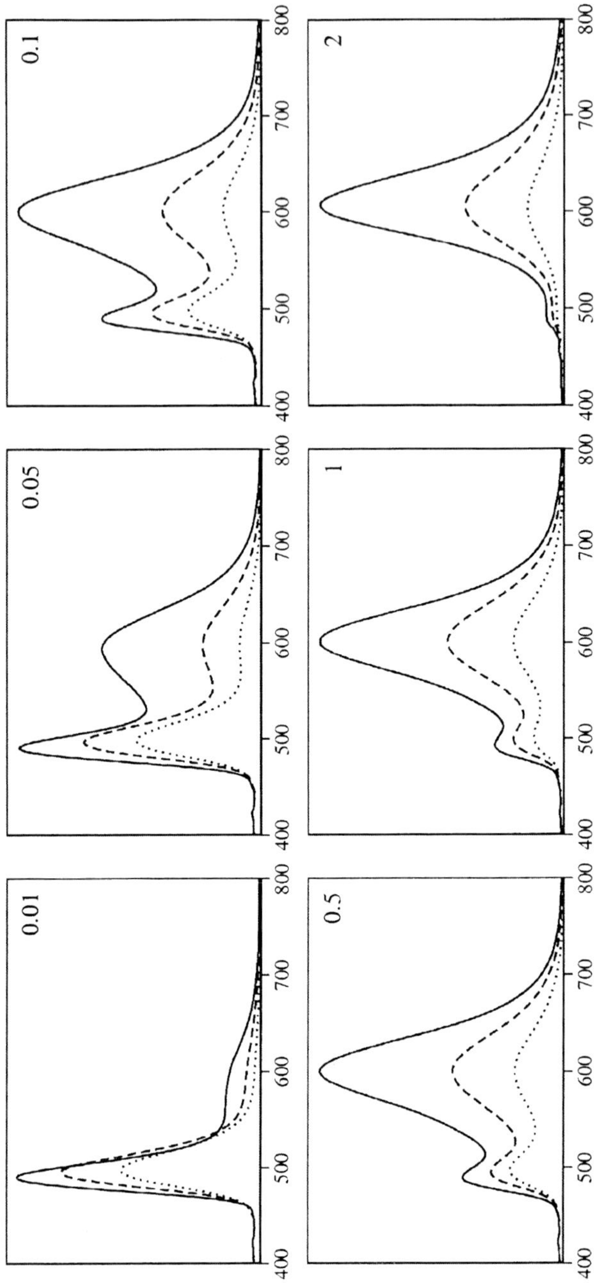

Fig. 10. Luminescence spectra of Ag_2S–CaA-x samples ($x = 0.01, 0.05, \ldots 2$) at −195 (solid lines), −100 (dashed lines), and −50 °C (dotted lines). The abscissa and the ordinate give the wavelength in nanometers and the emission intensity, respectively. Excitation was performed at 280 nm [51].

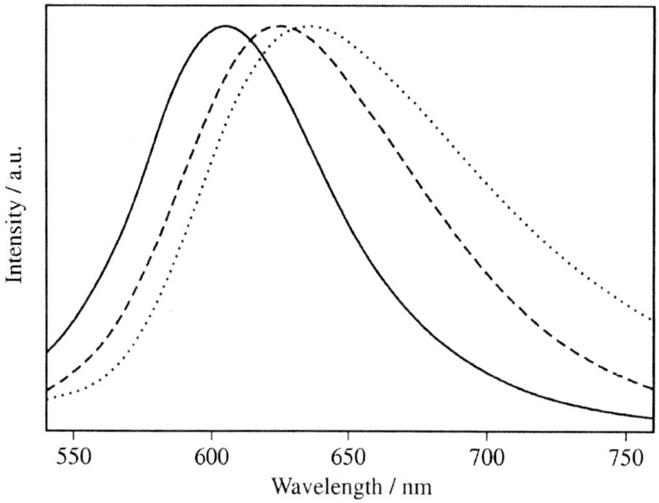

Fig. 11. Luminescence spectra of Ag_2S–CaA-2 (solid), Ag_2S–CaA-4 (dashed), and Ag_2S–CaA-6 (dotted) at −195 °C. Excitation was performed at 280 nm. The spectra are scaled to identical heights.

By using CaA as host material it is possible to produce luminescent samples with silver sulfide contents of up to at least Ag_2S–CaA-6. Figure 11 shows the low-temperature luminescence spectra of Ag_2S–CaA-2, Ag_2S–CaA-4, and Ag_2S–CaA-6. These samples only exhibit the long-wavelength emission, which is red-shifted with increasing silver sulfide content. This effect is accompanied by shortening of the luminescence lifetime. The following average decay times were measured for Ag_2S–CaA-x samples at −160 °C: 81 ($x = 2$), 49 ($x = 3$), 26 ($x = 4$), 9 ($x = 5$), and 2 μs ($x = 6$). The mechanism that causes this increased quenching at high loading levels is not yet understood.

The calculated electronic absorption spectrum of the Ag_2S monomer is essentially composed of four transitions between 300 and 400 nm, while the spectrum of AgSH features a single prominent transition at 276 nm. The NaSH monomer absorbs at even shorter wavelengths. This is in agreement with the assignment of the experimentally observed bands in the diffuse reflectance spectra of H_2S-loaded NaA and the Ag_2S–NaA samples. The HOMO of the Ag_2S, AgSH, and NaSH monomers is essentially a 3p(S) orbital. The electronic transition from this orbital to the corresponding LUMO generally has oscillator strengths of less than 0.05. We expect that these transitions are difficult to observe in diffuse reflectance spectra, but they are essential for the luminescence behavior of the composites. For the Ag_2S monomer the calculated HOMO–LUMO transition is at 505 nm. This corresponds well to the blue-green luminescence observed in Ag_2S–NaA-x, and Ag_2S–CaA-x samples of low silver sulfide content. The nature of the HOMO–LUMO transition is most likely maintained in larger silver sulfide clusters. This explains the significant Stokes shifts which can be observed in the excitation and emission spectra (see Figs. 5 and 6 in Ref. [51]) and the comparatively long luminescence lifetimes.

How does the electronic absorption spectrum of an Ag_2S monomer change on interaction with another Ag_2S monomer to form an Ag_4S_2 cluster? To answer this question, we first must address the characteristics of this interaction. The appropriate interaction geometry for the two monomers is not immediately evident, mainly because of the as-yet unknown structure of Ag_4S_2. Starting from the structures proposed by Bagatur'yants et al. [67], we selected the geometry of highest symmetry, namely, D_{2d}. The interaction was studied by reducing the distance between the two Ag_2S monomers. Figure 12 shows the corresponding correlation diagram. While the b_1 and b_2 levels of the monomers are little affected on interaction, splitting of the two a_1 levels (LUMO and HOMO–1) into levels of b_2 and a_1 symmetry is observed. The thus formed a_1 orbitals are stabilized with respect to the corresponding orbitals in the isolated monomers, while the b_2 orbital that is generated from the HOMO–1 is destabilized. The energy of the b_2 orbital which is formed on splitting of the LUMO is only slightly affected by variation of the Ag_2S–Ag_2S distance. The increasingly antibonding s–s and p–p interactions in this case are compensated by an increasingly bonding s–p interaction. The splitting of the a_1 levels on interaction leads to a red shift of the electronic absorption and luminescence bands (note the decrease in the HOMO–LUMO gap) [51].

When dealing with three-dimensional cluster arrays such as the Ag_2S–NaA-x and Ag_2S–CaA-x composites, the question arises whether the properties of these materials are due to crystal effects originating from interacting clusters. We evaluated the relevance of such effects by calculating the DOS of Ag_4S_2 clusters arranged in a cubic lattice. The MO diagram of the Ag_4S_2 cluster used for this purpose is depicted in Figure 12 (right) for an S–S distance of 5.6 Å. The development of the DOS upon variation of the lattice constant is shown in Fig. 13. Significant changes can be observed when the lattice constant is less than 10 Å. The band gap decreases at values below 9 Å and disappears at 7.5 Å. The distance between the centers of two α-cages in zeolite A is 12.3 Å [68]. The model therefore predicts that there is no through-space interaction between the clusters over the whole zeolite crystal up to a loading level of 4 Ag^+ per α-cage. The properties of Ag_2S–NaA-x and Ag_2S–CaA-x ($x \leq 4$) are therefore mainly determined by the presence of isolated clusters and by short-range interactions between these clusters. Such local interactions are likely to be between two clusters located close to a window connecting adjacent α-cages.

5.6 Intrazeolite Charge Transport

Intrazeolite charge transport is of great practical and fundamental importance but remains a controversial topic. We focus on "classical zeolites", the crystalline framework of which can be regarded as an enormous polyanionic system containing cations for charge compensation. The intrazeolite voids may contain solvent

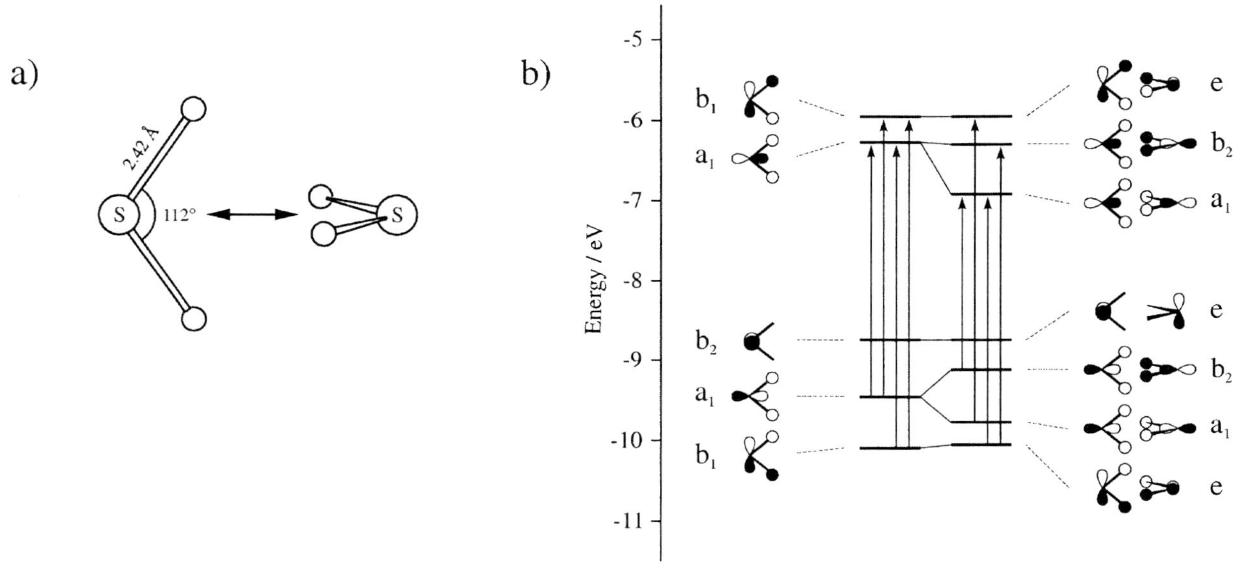

Fig. 12. a) Scheme of the structure used for Ag_4S_2. b) Frontier orbital correlation diagram of Ag_2S (left) and Ag_4S_2 (right). The contribution of the d orbitals to the molecular orbitals is marginal in this energy region and therefore not shown. Electronic transitions with an oscillator strength greater than 0.01 are indicated by arrows. The HOMO of Ag_2S and also of Ag_4S_2 is located at −8.7 eV.

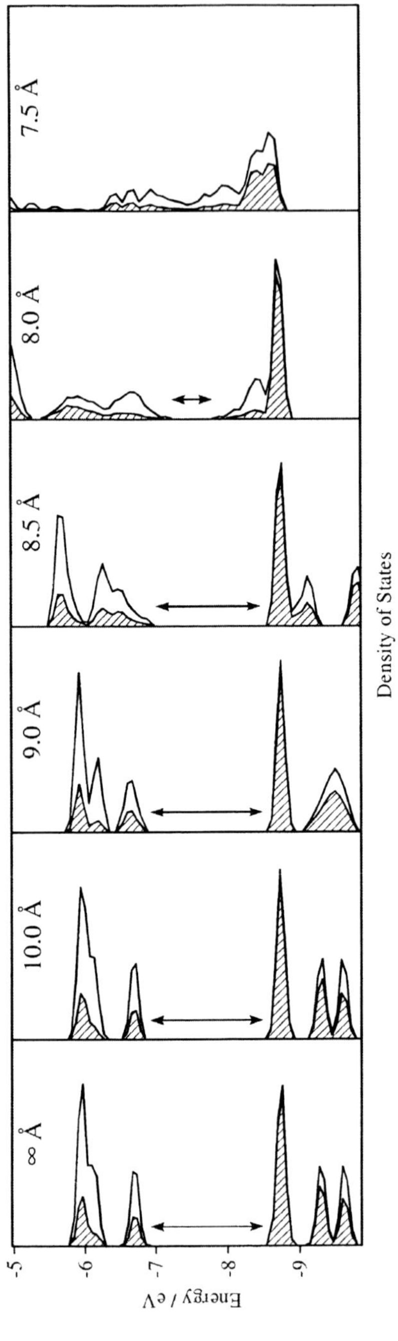

Fig. 13. DOS of Ag_4S_2 clusters in a cubic lattice with different lattice constants. The separated system is shown on the left. The band gap of each cluster array is marked by an arrow. The hatched region indicates the contribution of sulfur states.

molecules such as water, alcohol, or others. In these materials the framework acts as an insulating host. This means that any charge transport is due either to the charge-compensating cations or to guests, which can be molecules, ions, complexes, conducting polymers, clusters, and quantum-sized particles. Different types of intrazeolite charge transport should be distinguished, according to the transport mechanism. *Ion conductivity* is determined by the mobility of cations such as Na^+ or K^+ inside the channels [69]. We name charge transport governed by intrazeolite redox processes *redox conductivity.* In such a process the exchange of cations with the surroundings plays a role in order to maintain charge neutrality [70,71]. *Semiconductor conductivity* is to be expected in materials in which guests such as nanosized particles or conducting polymers form a sufficiently well developed band structure [51]. *Metal conductivity* can occur if the band gap becomes small enough [24]. Combinations of these mechanisms are possible, depending on the composition and on the specific arrangement of the guests within the zeolite cavities.

The most delicate region for obtaining unambiguous results in charge transport experiments on zeolite materials is located at the interface. We reported intrazeolite charge transport on Cu^{2+}-zeolite Y, on Ag^+-zeolite A, and on methylviologen-zeolite Y electrodes, in which zeolite microcrystals are deposited as monolayers on glassy carbon disk electrodes [71]. Baker et al. questioned the relevance of zeolite electrodes prepared as monolayers [72], a statement which we contradicted [71c]. He and others later focused mainly on conditions in which the electroactivity of intrazeolite species is suppressed [73]. In the meantime impressive progress on the preparation of zeolite monolayers has been made [2–7]. We feel that any further investigations on intrazeolite charge transport should concentrate on monolayers of the best possible quality. An other approach is to investigate single nanocrystals with, e.g., AFM, SEM, and confocal optical microscopy methods.

We now consider the challenge of charge transport in one-dimensional channels and we present a new idea for solving the interfacial problem. The framework of zeolite L shown in Fig. 14 serves as an example. The primitive vector $\boldsymbol{c}$ corresponds to the channel axis, while the primitive vectors $\boldsymbol{a}$ and $\boldsymbol{b}$ are perpendicular to it, enclosing an angle of 120°. Zeolite L crystals usually have cylindrical morphology. The number of parallel channels which coincide with the $\boldsymbol{c}$-axis of the hexagonal framework is equal to 1.07 ${r_{cyl}}^2$, where r_{cyl} is the radius of the crystal in nanometers. For example, this means that a cylinder 600 nm in diameter and 300 nm in length has about 100 000 parallel channels, each consisting of 400 unit cells. The channels have been filled with a large variety of molecules. This led to materials with exciting photophysical properties (see, e.g., Ref. [74]).

Among the many molecules which have already been inserted into zeolite L, we discuss only methyl viologen (MV^{2+}) as an example, because intrazeolite charge transport in MV^{2+}-Y zeolite electrodes has already been demonstrated [71d]. About 85 % of the unit cells can be filled with this molecule. On the basis of Rietfield refinement of X-ray data and molecular modeling, the model for the MV^{2+} location shown in Fig. 15 was derived. The MV^{2+} lies along the channel wall, and the angle between the main MV^{2+} axis and the c-axis of the zeolite is 27° [75]. We conclude

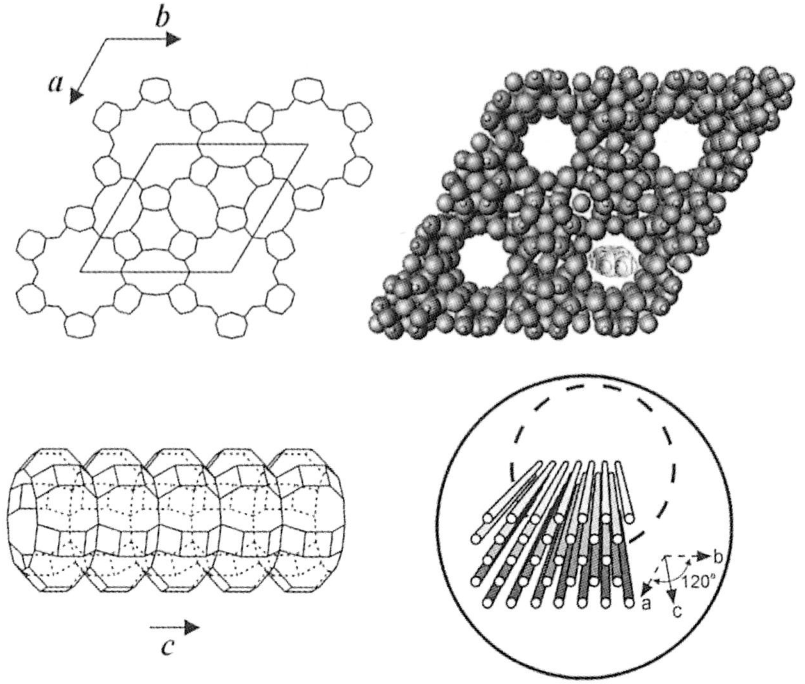

Fig. 14. Framework of zeolite L. Top: top view, perpendicular to the **c**-axis, displayed as stick (left) and as van der Waals (right) representations with a molecule entering the zeolite channel. Bottom: side view of a channel along the **c**-axis, without bridging oxygen atoms (left). Schematic view of some channels in a hexagonal zeolite crystal with cylindrical morphology (right).

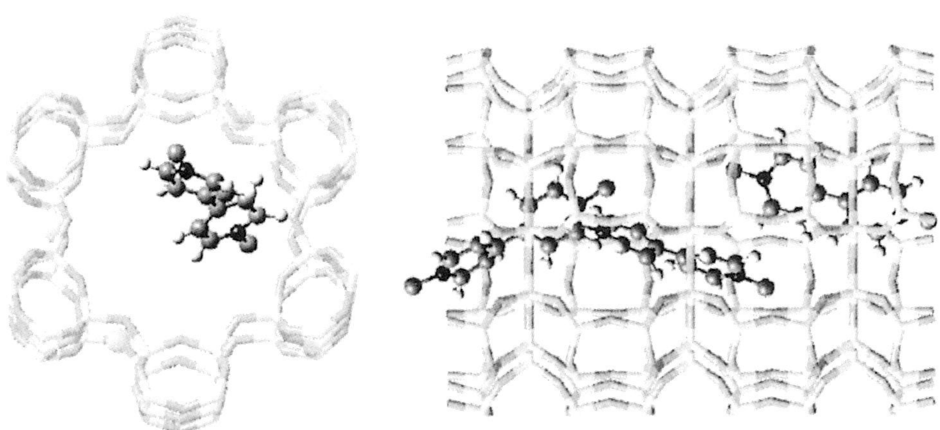

Fig. 15. Location of the MV^{2+} cation inside the channel of zeolite L. Left: side view of the channel depicting a likely arrangement of the molecules along the one-dimensional channel. Right: view along the channel axis showing a position and orientation of a molecule.

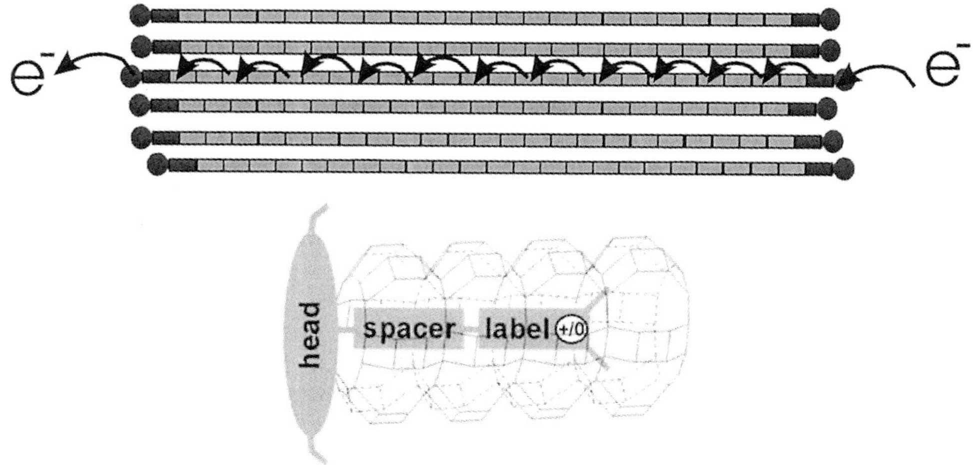

Fig. 16. Principle of the stopcock approach. The channels are filled with electron-conducting guests. An electron is injected on one side via the stopcock contact molecule. It travels along the channel to the other side if a corresponding voltage has been applied.

that in such materials intrazeolite charge transfer along channels of MV^{2+} molecules is to be expected. Two problems must be considered. Since MV^{2+} entered the zeolite channel by means of ion exchange, it can leave it by the same route. This must be prevented if a stable material should result. There are several ways to do this. We have shown for resorufin [76] that using a solvent which cannot enter the channels of zeolites prevents the exit of the intrazeolite molecules. This principle has been successfully used by us in many other cases. This means that using an appropriate solvent for the electrolyte, such as polycarbonates, or an ionic liquid can solve the problem. Another possibility is the use of a conductive polymer as a closure. The other problem lies at the electrode/zeolite interface, as stated above. We believe that it can be solved by applying the closure and stopcock approach illustrated in Fig. 16. It was discussed with respect to work on photonic antenna systems for light harvesting, transport, and trapping [1]. First experimental evidence for its functionality in photonic antenna materials was recently reported [77]. A stopcock generally consists of three components: a head, a spacer, and a label. The tail moiety (spacer + label) has a longitudinal extension of at least one unit cell along the *c*-axis. The head has a lateral extension that is larger than the channel width and prevents its penetrating into the channels. The channels are therefore terminated in a pluglike manner. Depending on the requirements, stopcocks can be either applied on both sides of the cylinders or only on one side. If redox conductivity is envisaged, care must be taken that the mobility of the cations, which is needed for charge compensation, is not hindered.

Figure 17 shows what an electrode for redox conductivity could look like. From this it is clear that the success of this approach depends mainly on the skill of

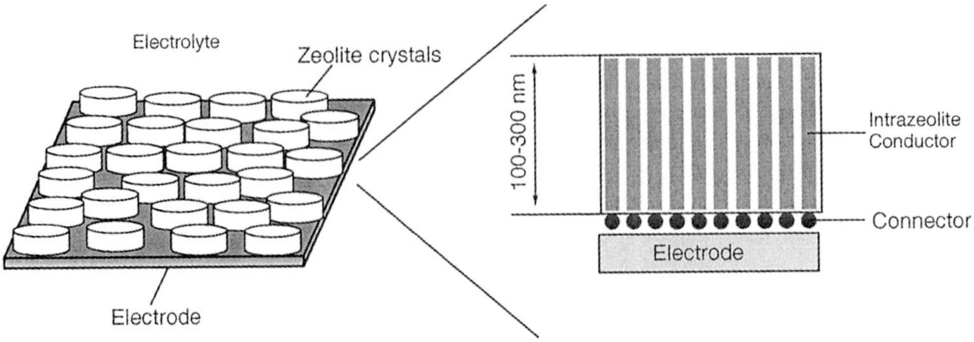

Fig. 17. Principle of the stopcock-electrodes described for hexagonal zeolite L crystals. The zeolite crystals are placed on an electrode such that contact to the electrode is made via the connector, which is the head of the stopcock. The electrode must be such that the resistance to the connector is minimized. It can, e.g., consist of a conducting polymer, gold, glassy carbon, or other materials. This principle applies similarly for a single nano- or microcrystal or for an ensemble. The connector on the electrolyte side, for which several possibilities can be envisaged, is not shown.

preparing zeolite crystals of the right morphology, of preparing monolayers on the electrode surface, and on applying the stopcock principle. It is a challenging approach and we look forward to how it will develop in the near future.

5.7 Conclusions

The electronic structure of a typical zeolite material can be regarded as a superposition of the electronic structures of the framework, the charge-compensating cations, the solvent molecules, and the guest species. The band gap and the HOMO position of the zeolite framework are similar to those of α-quartz, in spite of the fact that the material is less dense. The charge-compensating cations result in new electronic states, some of which lie within the band-gap region. Their interaction with each other depends on their nature and the mean distance between them. The presence of solvent molecules, usually water, influences their interaction with the zeolite framework. The electronic structures of the zeolite framework and the charge-compensating cations are not influenced by each other, if the solvation shell shields the cations from coordinating to the zeolite oxygens. Partial removal of the solvent leads to incomplete saturation of the coordination shell of the cations, which try to compensate this by being closer to the zeolite oxygen atoms. The states involved can be perturbed considerably, and this changes the optical properties. Silver zeolite is an excellent example for which this perturbation can be tuned reversibly. Equations (4) and (5) summarize and simplify the results reported in Section 5.4.

$$\text{Zeolite framework} + \text{Ag}^+_{\text{aq}} \underset{\text{rehydration}}{\overset{\text{dehydration}}{\rightleftharpoons}} \text{Zeolite} - \text{oxygen} \ldots \text{Ag}^+(\text{H}_2\text{O})_n \tag{4}$$

$$\text{Zeolite} - \text{oxygen} \ldots \text{Ag}^+(\text{H}_2\text{O})_n \underset{\text{partial rehydration}}{\overset{\text{further dehydration}}{\rightleftharpoons}} \begin{matrix} \text{Zeolite} - \text{oxygen} \ldots \text{Ag}^+(\text{H}_2\text{O})_{n-x} \\ \vdots \\ \text{Zeolite} - \text{oxygen} \ldots \text{Ag}^+(\text{H}_2\text{O})_{n-x} \end{matrix} \tag{5}$$

The dots symbolize electronic interactions between zeolite oxygen atoms and Ag^+ (horizontal) and between different $Ag^+(H_2O)_{n-x}$ clusters (vertical).

These changes can be monitored by observing the oxygen to silver charge transfer transitions, which can be turned on and off. They are also seen in the infrared spectra. It was even possible to identify the reversible coordination of Ag^+ to the 4-ring site of zeolite A, which causes the reversible deep yellow coloring of this otherwise colorless material (Eq. 6) [33,34].

$$\text{Zeolite framework} + \text{Ag}^+_{\text{aq}} \underset{\text{rehydration}}{\overset{\text{dehydration}}{\rightleftharpoons}} \text{Zeolite} - 4 - \text{ring} - \text{oxygen} \ldots \text{Ag}^+(\text{H}_2\text{O})_n \tag{6}$$

The extra framework cations protrude into the internal void space of zeolites. The adsorbed guest molecules are exposed to the considerable electric fields of these cations, especially in the absence of solvent molecules. As a result, otherwise infrared inactive molecules like H_2, N_2, and O_2 are polarized and show IR spectra when embedded in a zeolite [78]. An electric-field effect was used to explain the astonishing polarization of the electronic transition moments of oxonine and pyronine molecules in zeolite L, despite the fact that the electronic absorption and fluorescence spectra of these dyes are only slightly influenced by the host [79]. We do not expect that such effects are important in the Ag_2S–NaA and the Ag_2S–CaA host–guest materials, which consist of three-dimensional quantum dot lattices, as described in Section 5.5. From our observations we conclude that Ag_2S and Ag_4S_2 in zeolite A and ZK4 behave as individual species with well-defined properties [52]. It is remarkable that, e.g., in the case of Ag_4S_2 cluster materials the intercluster distances must be shorter than about 10 Å before the electronic coupling is sufficient to influence the electronic structure. This does not exclude, however, dipole–dipole and other long-range coupling. It can be expected that not only silver sulfide cluster arrays with different structures can be produced by applying the method described in Ref. [51], but also that it can be extended to other metal cations embedded in a zeolite. Comparison of different Ag_2S–zeolite composites will yield further insight into the specific interactions which govern the properties of such host–guest systems. The use of zeolite as host material opens possibilities for the assembly of highly organized structures. Well-defined, close-packed monolayers of high mechanical stability can be prepared on various substrates by using size-selected zeolite A crystals [2]. Furthermore, covalent linkage of zeolite crystals to glass facilates the assembly of oriented monolayers and micropatterned structures [3,4]. Such assemblies are of special interest for the exploitation of the unique optical and electronic properties of these materials. They are of similar importance for advancing our knowledge of intrazeolite charge transport.

References

1 a) G. Calzaferri, M. Pauchard, H. Maas, S. Huber, A. Khatyr, T. Schaafsma, *J. Mater. Chem.* **2002**, *12*, 1. b) G. Schulz-Ekloff, D. Wöhrle, B. van Duffel, R.A. Schoonheydt, *Microporous Mesoporous Mater.* **2002**, *51*, 91.

2 P. Lainé, R.Seifert, R. Giovanoli, G. Calzaferri, *New J. Chem.* **1997**, *21*, 453.

3 a) K. Ha, Y.-J. Lee, H.J. Lee, K.B. Yoon, *Adv. Mater.* **2000**, *12*, 1114. b) A. Kulak, Y.-J. Lee, Y. S. Park, K.B. Yoon, *Angew. Chem. Int. Ed.* **2000**, *39*, 950. c) K. Ha, Y.-J. Lee, Y.S. Park, G.S. Lee, and K.B. Yoon, *Adv. Mater.* **2001**, *13*, 594. d) K. Ha, Y.-J. Lee, D.-Y. Jung, J. H. Lee, K. B. Yoon, *Adv. Mater.* **2000**, *12*, 1614.

4 L. Huang, Z. Wang, J. Sun, L. Miao, Q. Li, Y. Yan, and D. Zhao, *J. Am. Chem. Soc.* **2000**, *122*, 3530.

5 S. Mintova, B. Schoeman, V. Valtchev, J. Sterte, S. Mo, and T. Bein, *Adv. Mater.* **1997**, *9*, 585.

6 S. Mintova, V. Valtchev, *Zeolites* **1996**, *16*, 31.

7 P. Yang, T. Deng, D. Zhao, P. Feng, D. Pine, B. F. Chmelka, G.M. Whitesides, G.D. Stucky, *Science* **1998**, *282*, 2244.

8 N. Herron, Y. Wang, M.M. Eddy, G.D. Stucky, D.E. Cox, K. Moller, T. Bein, *J. Am. Chem. Soc.* **1989**, *111*, 530.

9 G.D. Stucky, J.E. Mac Dougall, *Science* **1990**, *247*, 669.

10 G.A. Ozin, M.R. Steele, A.J. Holmes, *Chem. Mater.* **1994**, *6*, 999.

11 M.M. Garcia, H. Villavicencio, M. Hernandez-Velez, O. Sanchez, J.M. Martinez-Duart, *Mater. Sci. Eng.* **2001**, *C15*, 101.

12 J.R. Martin-Palma, M. Hernandez-Velez, I. Diaz, H. Villavicencio-Garcia, M.M. Garcia-Poza, J.M. Martinez-Duart, J. Perez-Pariente, *Mater. Sci. Eng.* **2001**, *C15*, 163.

13 X.S. Zhao, G.J. Millar, G.Q. Lu, *J. Porous Mater.* **1996**, *3*, 61.

14 J.E. Mac Dougall, H. Eckert, G.D. Stucky, N. Herron, Y. Wang, K. Moller, T. Bein, D. Cox, *J. Am. Chem. Soc.* **1989**, *111*, 8006.

15 Y. Wang, N. Herron, *J. Phys. Chem.* **1987**, *91*, 257.

16 A.A. Demkov, O.F. Sankey, *J. Phys. Condens. Matter* **2001**, *13*, 10433.

17 O. Dag, A. Kuperman, G.A. Ozin, *Adv. Mater.* **1995**, *7*, 72.

18 O. Dag, G.A. Ozin, H. Yang, C. Reber, G. Bussière, *Adv. Mater.* **1999**, *11*, 474.

19 H.-J. Schwenn, M. Wark, G. Schulz-Ekloff, H. Wiggers, U. Simon, *Coll. Polym. Sci.* **1997**, *275*, 91.

20 X. Liu, K.-K. Iu, J.K. Thomas, *Chem. Phys. Lett.* **1992**, *195(2-3)*, 163.

21 W.-H. Zhang, J.-L. Shi, H.-R. Chen, Z.-L. Hua, D.-S. Yan, *Chem. Mater.* **2001**, *13*, 648.

22 M. Wark, G. Schulz-Ekloff, N.I. Jaeger, *Bulg. Chem. Comm.* **1998**, *30(1-4)*, 129.

23 M. Bärtsch, P. Bornhauser, G. Calzaferri, R. Imhof, *J. Phys. Chem.* **1994**, *98*, 2817.

24 M. Brändle, G. Calzaferri, Res. Chem. Intermed. **1994**, *20*, 783.

25 P. Bornhauser, G. Calzaferri, *J. Phys. Chem.* **1996**, *100*, 2035.

26 G. Calzaferri in *Tailor-made Silicon-Oxygen Compounds, From Molecules to Materials*, R. Corriu, P. Jutzi (Eds.), Vieweg, Braunschweig, 1996, p.149.

27 C. Marcolli, P. Lainé, R. Bühler, G. Calzaferri, J. Tomkinson, *J. Phys. Chem. B* **1997**, *101*, 1171.

28 G. Calzaferri, R. Hoffmann, *J. Chem. Soc. Dalton Trans.* **1991**, *917*.

29 D.W. Turner, C. Baker, A.D. Baker, C.R. Brundle, *Molecular Photoelectron Spectroscopy*, Wiley-Interscience, London, **1970**.

30 D.L. Griscom, *J. Non-Cryst. Solids* **1977**, *24*, 155.

31 R. Beer, G. Calzaferri, I. Kamber, *J. Chem. Soc. Chem. Commun.* **1991**, 1489.

32 G. Calzaferri, R. Giovanoli, I. Kamber, V. Shklover, R. Nesper, *Res. Chem. Intermed.* **1993**, *19*, 31.

33 R. Seifert, A. Kunzmann, G. Calzaferri, *Angew. Chem.* **1998**, *110*, 1603; *Angew. Chem. Int. Ed.* **1998**, *37*, 1521.

34 R. Seifert, R. Rytz, G. Calzaferri, *J. Phys. Chem. A* **2000**, *104*, 7473.

35 S. Glaus, G. Calzaferri, *J. Phys. Chem. B* **1999**, *103*, 5622–5630.

36 a) Y.-M. Kang, B.-Z. Wan, *Catal. Today* **1995**, *26*, 59. b) T.M. Salama, S. Takafumi, R. Ohnishi, M. Ichikawa, *J. Phys. Chem.* **1996**, *100*, 3688.

37 S. Bordiga, G. Turnes Palomino, D. Arduino, C. Lamberti, A. Zecchina, C. Otero Arean, *J. Mol. Catal. A: Chem.* **1999**, *146*, 97.

38 G. Spoto, A. Zecchina, S. Bordiga, G. Ricchiardi, G. Martra, G. Leofanti, G. Petrini, *Appl. Catal. B: Environ.* **1994**, *3*, 151.

39 M. Rálek, P. Jíru, O. Grubner, H. Beyer, *Collect. Czech. Chem. Commun.* **1962**, *27*, 142.

40 T. Sun, K. Seff, *Chem. Rev.* **1994**, *94*, 857.

41 J. Baumann, R. Beer, G. Calzaferri, B. Waldeck, *J. Phys. Chem.* **1989**, *93*, 2292.

42 G. Calzaferri, W. Suter, B. Waldeck, *J. Chem. Soc. Chem. Commun.* **1990**, 485.

42 R. Seifert, Inauguraldissertation der philosophisch-naturwissenschaftlichen Fakultät der Universität Bern, **1999**.

43 Nanoscale Materials Special Issue, *Acc. Chem. Res.* **1999**, *32*, 387.

44 T. Vossmeyer, L. Katsikas, M. Giersig; I.G. Popovic, K. Diesner, A. Chemseddine, A. Eychmüller, H. Weller, *J. Phys. Chem.* **1994**, *98*, 7665.

45 C.B. Murray, D.J. Norris, M.G. Bawendi, *J. Am. Chem. Soc.* **1993**, *115*, 8706.

46 A. van Dijken, E.A. Meulenkamp, D. Vanmaekelbergh, A. Meijerink, *J. Lumin.* **2000**, *90*, 123.

47 J. Nanda, S. Sapra, D. D. Sarma, N. Chandrasekharan, G. Hodes, *Chem. Mater.* **2000**, *12*, 1018.

48 F. W. Wise, *Acc. Chem. Res.* **2000**, *33*, 773.

49 D. Brühwiler, R. Seifert, G. Calzaferri, *J. Phys. Chem. B* **1999**, *103*, 6397–6399.

50 G. Calzaferri, D. Brühwiler, S. Glaus, D. Schürch, A. Currao, C. Leiggener, *J. Imag. Sci. Technol.* **2001**, *45*, 331–339.

51 D. Brühwiler, C. Leiggener, S. Glaus, G. Calzaferri, *J. Phys. Chem. B* **2002**, *106*, 3770.

52 G. Calzaferri, C. Leiggener, S. Glaus, D. Schürch, K. Kuge, *Chem. Soc. Rev.*, **2003**, *32*, 29.

53 M. Faraday, *Philosoph. Transact.* **1833**, *507*.

54 R. Sadanaga, S. Sueno, *Mineral. J.* **1967**, *5*, 124.

55 S. Kitova, J. Eneva, A. Panov, H. Haefke, *J. Imaging Sci. Technol.* **1994**, *38*, 484.

56 T. Tani, *J. Imaging Sci. Technol.* **1995**, *39*, 386.

57 J.W. Mitchell, *J. Imaging Sci. Technol.* **1998**, *42*, 215.

58 E. Charlier, M. Van Doorselaer, R. Gijbels, R. De Keyzer, I. Geuens, *J. Imaging Sci. Technol.* **2000**, *44*, 235.

59 R.C. Baetzold, *J. Imaging Sci. Technol.* **1999**, *43*, 375.

60 L. Motte, M.P. Pileni, *J. Phys. Chem. B* **1998**, *102*, 4104.

61 Y.-P. Sun, J.E. Riggs, H.W. Rollins, R. Guduru, *J. Phys. Chem. B* **1999**, *103*, 77.

62 V.M. Belous, V.I. Tolstobrov, O.I. Sviridova, K.V. Chibisov, *Dokl. Phys. Chem. (Engl. Transl.)* **1982**, *262*, 75.

63 K. Akamatsu, S. Takei, M. Mizuhata, A. Kajinami, S. Deki, S. Takeoka, M. Fujii, S. Hayashi, K. Yamamoto, *Thin Solid Films* **2000**, *359*, 55.

64 H. W. Rollins, F. Lin, J. Johnson, J.-J. Ma, J.-T. Liu, M.-H. Tu, D.D.

DesMarteau, Y.-P. Sun, *Langmuir* **2000**, *16*, 8031.

65 M.C. Brelle, J.Z. Zhang, L. Nguyen, R.K. Mehra, *J. Phys. Chem. A* **1999**, *103*, 10194.

66 A. Henglein, M. Gutiérrez, H. Weller, A. Fojtik, J. Jirkovský, *Ber. Bunsenges. Phys. Chem.* **1989**, *93*, 593.

67 A.A. Bagatur'yants, A.A. Safonov, H. Stoll, H.-J. Werner, *J. Chem. Phys.* **1998**, *109*, 3096.

68 D.W. Breck, *Zeolite Molecular Sieves*, Wiley, New York, 1974.

69 a) A.K. Jonscher, *Universal Relaxation Law*, Chelsea Dielectrics, London, **1996**. b) R. Gerhardt, *J. Phys. Chem. Solids* **1994**, *55*, 1491. c) T. Ohgushi, S. Sato, *J. Solid State Chem.* **1990**, *87*, 95. d) E. K. Andersen, I.G.K. Andersen, E. Skou, *Chem. Solid State Mater.* **1992**, *2*, 210.

70 a) D.R. Rolison, *Chem. Rev.* **1990**, *90*, 867. b) D.R. Rolison in *Advanced Zeolite Science and Applications*, J.C. Jansen, M. Stöcker, H.G. Karge, J. Weitkamp (Eds.), *Stud. Surf. Sci. Catal.* **1994**, *85*, 543.

71 a) J.-W. Li, G. Calzaferri, *J. Electroanal. Chem.* **1994**, *377*, 163. b) J.-W. Li, K. Pfanner, G. Calzaferri, *J. Phys. Chem.* **1995**, *99*, 2819. c) J.-W. Li, K. Pfanner, G. Calzaferri, *J. Phys. Chem.* **1995**, *99*, 12368. (d) G. Calzaferri, M. Lanz, J.-W. Li, *J. Chem. Soc. Chem. Commun.* **1995**, 1313.

72 M.D. Baker, C. Senaratne, M. Mc Brien, *J. Phys. Chem.* **1995**, *99*, 12367.

73 C. Senaratne, J. Zhang, J. Fox, I. Burgess, M.D. Baker, *Microporous Mesoporous Mater.* **1999**, *33*, 281.

74 G. Calzaferri, H. Maas, M. Pauchard, M. Pfenniger, S. Megelski, A. Devaux in *Adv. Photochem.* D.C. Neckers (Ed.), Vol. 27, Wiley-VCH, Weinheim, 2002 pp. 1–50.

75 B. Hennessy, S. Megelski, C. Marcolli, V. Shklover, Ch. Bärlocher, G. Calzaferri, *J. Phys. Chem. B* **1999**, *103*, 3340.

76 D. Brühwiler, N. Gfeller, G. Calzaferri, *J. Phys. Chem.* **1998**, *102*, 2923.

77 H. Maas, G. Calzaferri, *Angew. Chem. Int. Ed.* **2002**, *41*, 2284.

78 a) H. Förster, M. Schuldt, *J. Chem. Phys.* **1977**, *66*, 5237. b) H. Böse, H. Förster, W. Frede, *Chem. Phys. Lett.* **1987**, *138*, 401. c) S. Bordiga, E. Garrone, C. Lamberti, A.F. Zecchina, C.O. Arean, V.B. Kazanski, L.M. Kustov, *J. Chem. Soc. Faraday Trans.* **1994**, *90*, 3367.

79 S. Megelski, A. Lieb, M. Pauchard, A. Drechsler, S. Glaus, Ch. Debus, A.J. Meixner, G. Calzaferri, *J. Phys. Chem. B* **2001**, *105*, 25.

6
Cetineites: Nanoporous Semiconductors with Zeolite-Like Channel Structure

Frank Starrost, Oliveo Tiedje, Wolfgang Schattke. Jörg Jockel, and Ulrich Simon

6.1
Introduction

The fabrication of nanostructured materials by chemical preparation techniques has become an important interdisciplinary field of research. On the one hand, various chemical routes have been developed to synthesise well-defined metal or semiconductor nanoparticles with sizes of a few up to tens of nanometers. In this size range the dimensionality determines the material properties [1–3]. On the other hand, great efforts are made to synthesise new compounds with a nanoporous open-framework structures. The most prominent compounds are zeolites and related microporous and mesoporous oxides, which are strongly bonded open-framework structures in which pores or channels of nanometer lateral extension are formed. The accessibility of these pores to various guest molecules makes these materials important for many applications, e.g., in catalysis, chemical sensing, water treatment, and separation processes [4]. Due to their chemical composition these nanoporous oxides have wide electronic band gaps. Most of them are electrically insulating, or their electrical conductivity results from the mobility of exchangeable cations [5].

Within the last few years, attempts have been made to synthesise microporous semiconductors. These new materials are of enormous interest from the scientific and the technical points of view, since many new physical properties may be expected from these solids. They might be applicable as chemical sensing devices, electro- or photocatalysts, and magnetoresistive devices [1–5]. Examples are $RbAg_5S_3$ and $CsAg_5S_3$, $A_2MP_2Se_6$ (A = K, Rb, Cs; M = Mn, Fe), KCu_4AsS_4, $(Me_3NH)_2[Sb_5S_9O_2]$, and $TMA_2Sb_3S_7$ (TMA = tetramethyamine) and related structures. Some of these compounds are stabilized by template molecules, and transform into thermodynamically stable compact (i.e., nonporous) phases when the template is removed. Little is known about the electrical and optical properties of most of these phases because of their their low chemical stability and the fact that they are only obtainable as microcrystalline powders instead of large crystals, which makes electrical characterization much more difficult.

In this paper we report on cetineite-type oxoselenoantimonates(III) with zeolite-like channel structure which show a pronounced photoconductivity [6–12]. These phases could be obtained as large single crystals as they are stable in air and free of template molecules. Here we describe the electronic and optical properties, as well as structure–property relationships, of cetineites from experimental and theoretical points of view. In Section 6.2 the history and chemical synthesis of the cetineites is presented. Section 6.3 discribes the experimental methods, and Section 6.4 gives an introduction to the AFC method (see also Ref. [13]) and lists the computational details. The experimental results are compared with the calculated properties in Section 6.5. Finally, a conclusion is given in Section 6.6.

6.2 Synthesis and Structure

The mineral cetineite was first described in 1987 by Sabelli and Vezzalini [14], who found crystals thereof in Le Cetine mine in Tuscany, Italy, a place famous among mineral collectors. In early the twentieth century the mine was used to extract antimony from stibnite (Sb_2S_3) ore. Slags from the extraction process were deposited in heaps where – among a number of other minerals – cetineite crystals formed in cavities as a result of the natural weathering of the residues of the mining process. The cetineite crystals found at the mine have the formula $(K,Na)_{3+x}(Sb_2O_3)_3(SbS_3)(OH)\cdot(2.82-x)H_2O$ with $x = 0.5$ [15]. They occur as tufts of acicular (needlelike) crystals in which the needles are parallel to [001]. The orange-red crystals (Fig. 1) have been found with lengths of up to 0.5 mm and diameters of

Fig. 1. A photograph of a cetineite mineral (copyright by Robert Vernet, Collection G. Favreau, used by kind permission).

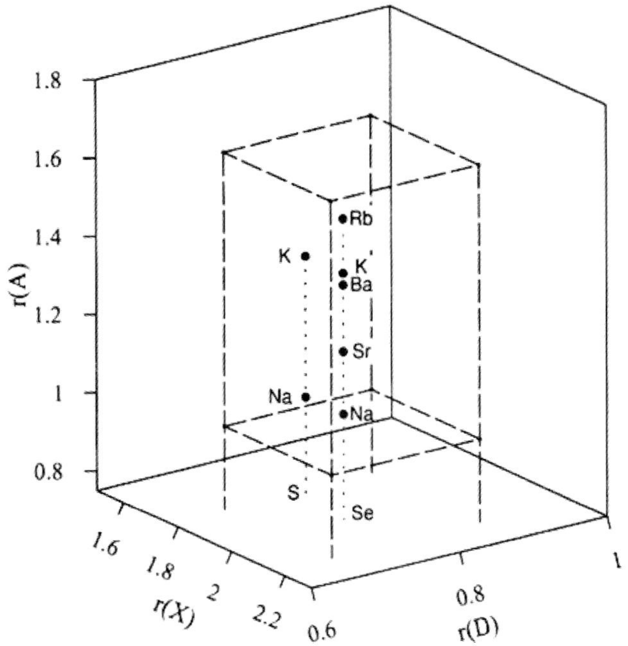

Fig. 2. Three-dimensional structure field map for cetineite phases [8]. Atomic radii are given in Å.

15 μm. A number of synthetic analogs of the mineral cetineite are known. The general formula is $A_6[Sb_{12}O_{18}][SbX_3]_2{\cdot}(6{-}mx{-}y)H_2O{\cdot}x[B^{m+}(OH^-)_m]{\cdot}y\check{T}$, where $A = Na^+, K^+, Rb^+$; $X = S^{2-}, Se^{2-}$; $B = Na^+, Sb^{3+}$; and $\check{T}$ stands for an unoccupied lattice site. The crystal formulas are abbreviated by (A;X).

Even before the natural mineral was discoverd, Graf and Schäfer described a synthetic analogue of (K;S) in 1975 [16]. The diffraction data, however, did not reveal the presence of the channel-filling guests. Since then several analogues have been synthesised [17,18]. Progress in the synthetic method has led to the preparation of single crystals up to a length of 2 mm and a hexagonal cross section of 3×10^{-2} mm^2 [8,12]. In the following the synthetic crystals are referred to as "cetineites".

According to Liebau, Fig. 2 shows the three-dimensional structure field map for cetineite phases. This clearly reflects the flexibility of the ionic components A and X with ionic radii of $r_A = 1.02$–1.52 Å and $r_X = 1.84$–1.98 Å, and also the rigidity of the covalently bonded Sb of the channels [17,18]. Therefore, new cetineite phases with variable K/Na ratio were synthesised by hydrothermal reaction of elemental Sb and S in an aqueous mixture of NaOH and KOH, following the experimental route as described in Refs. [6,7]. Orange-red single crystals with a maximum length of 1.2 mm and a hexagonal cross section of 9×10^{-2} mm^2 were obtained. Comparison of the potassium content of the cetineite crystals with the composition of the synthesis mixture shows preferred incorporation of potassium into the structure [9,19].

The two main structural characteristics of cetineite crystals (Fig. 3) are large infinite $Sb_{12}O_{18}$ tubes and isolated SbX_3 pyramids (Figs. 4 and 5). The tubes are made of interconnected SbO_3 pyramids and have a 6_3 axis. The tubes form a two-dimensional hexagonal lattice perpendicular to their axes, as is found in ordered arrangements of single-walled carbon nanotubes [20]. The SbX_3 pyramids are located in the tubular interstices; the plane in which the three X atoms lie is perpendicular to the tube axes. In a single interstice, infinite in the *c* direction, all pyramids have the same orientation; this orientation, however, varies randomly from one interstice to the next. When all interstitial pyramids point in the same direction the crystal space group is $P6_3$. The tube arrangement without the interstitial pyramids has an additional symmetry, namely, a mirror plane perpendicular to the axis in the plane containing O(1) and the A atom, i.e., the space group is $P6_3/m$. The hexagonal lattice constants for the cetineite class are $1.41 \leq a \leq 1.46$ nm and $c \approx 0.55$ nm; the voids within the tubes have a diameter of about 0.7 nm.

From chemical considerations it is assumed that there are strong Sb–O bonds that stabilize the $Sb_{12}O_{18}$ tubes. The binding among the tubes is thought to be via the SbX_3 pyramids, which are assumed to form ionic bonds to the A atoms on the inner wall of the tube [17] and additional weaker bonds to the Sb(2) atoms of the tube wall (numbers in parentheses following the atom name denote geometrically inequivalent atoms of the same species). The host lattice can (and readily does) accept a number of guest molecules. A common guest species is water in connection with the B^{m+} molecules, where B is an alkali metal, for example. Water molecules can be extracted from the tubes by applying vacuum and heating the crystals. However, the ease of extraction and the degree to which it is actually performed is dependent on the specific crystal. Defects in the tubes hinder the extraction. From well-formed crystals of (K;Se) the water can be reversibly expelled at room temperature under high vacuum within a few seconds [7].

6.3
Experimental Setups

Three experimental methods were used to determine the crystal band gaps: (1) UV/Vis transmission spectroscopy, (2) wavelength-dependent measurement of the photoconductivity, and (3) measurement of the thermal activation energy of conductivity [10,21]. With the exception of the determination of the thermal acivation energy of (Na;S) and (K;S), all measurements were performed under ambient conditions on single crystals of each cetineite phase, which were synthesised by hydrothermal reaction of elementary antimony and sulfur or selenium in aqueous NaOH or KOH [8]. Since the *ab* faces of the crystals in some cases are capped by an amorphous oxide layer the crystal ends were broken off. Conductivity measurements were performed by providing silver contacts to the clean faces as well as to the sides (see Fig. 11). Except for (K;Se), in which the tubes are occupied only by removable water molecules, all other compounds contain traces of Sb^{3+}, Na^+, or K^+ ions in addition to water and OH^-. These ions could not be extracted, pre-

(a)

(b)
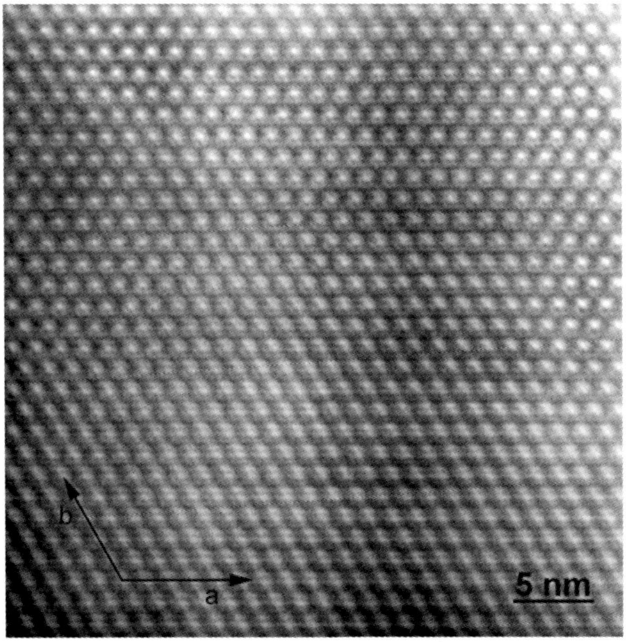

Fig. 3. Top: SEM image of a hexagonal (Na;Se) single crystal. Bottom: electron micrograph of (K;Se) showing the nanoporous structure.

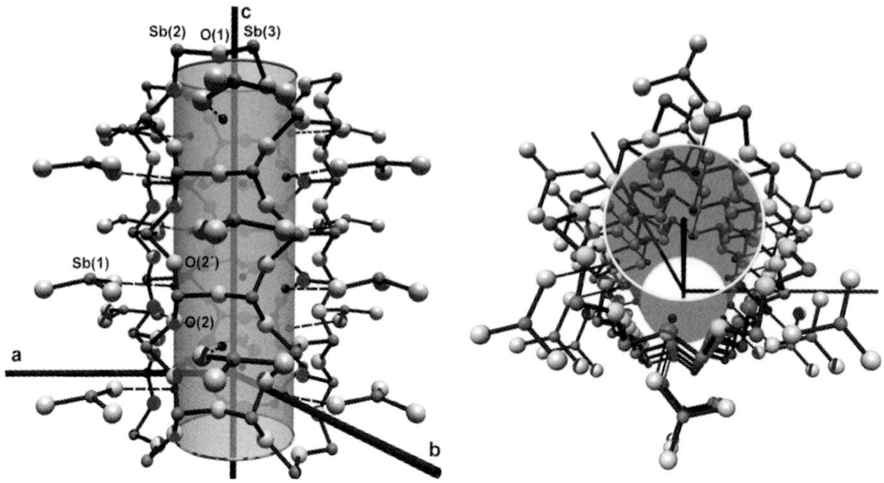

Fig. 4. A perspective view of the cetineite (Na;Se). The shaded tube is included only as a guide for the eye.

sumably due to structural defects. Using the first two methods, band gaps in the region of about 2.0–2.4 eV were measured, whereas the temperature-dependent measurements yielded activation energies of about 0.5 eV, reflecting a fundamental gap of about 1.0 eV. Photoemission spectroscopy experiments were performed on samples of a number of randomly oriented crystals with a photon energy of

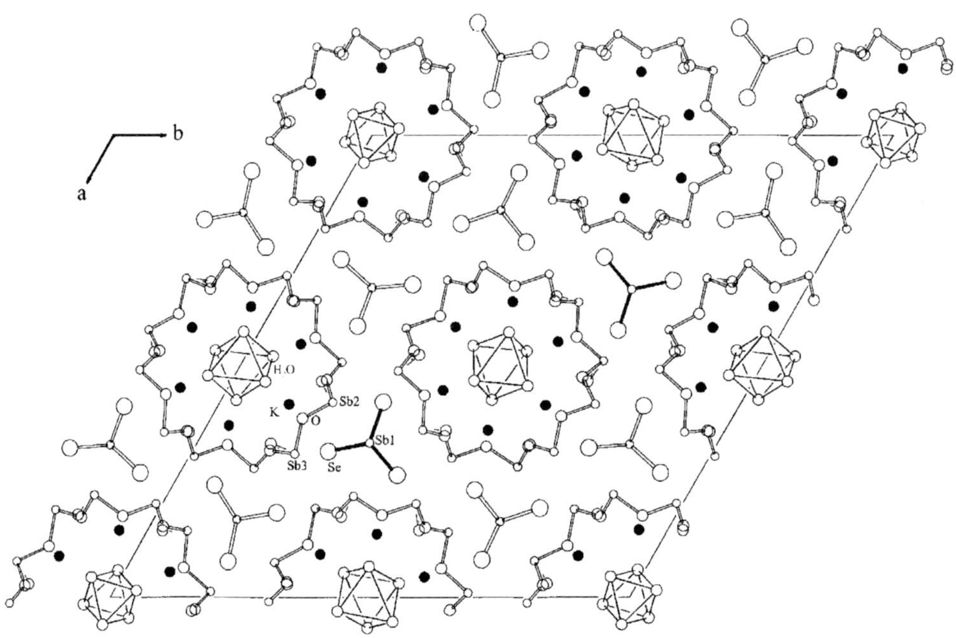

Fig. 5. The cetineite structure of (K;Se) viewed from the top.

21.22 eV from a helium Iα discharge lamp. The overall energy resolution was better than 100 meV. Random orientation was viewed as an angle integration to compare the photocurrent with the calculated density of states.

6.4 The Augmented Fourier Component Method: Computational Details

Linear augmented plane-wave (LAPW) calculations yield electronic band structure and wavefunctions of very high accuracy [22]. The muffin-tin (MT) approach to the crystal density and potential, which adopts spherical averaging of these two functions in the MT spheres around the atoms and constant quantities in the interstitial region, yields very good results for close-packed materials, such as elemental metals [23]. However, in open structures, where the interstitial region has a volume equal to or greater than the volume covered by the MT spheres, the approximation is hard to justify. In the semiconductor silicon, for example, which crystallizes in the diamond structure where only 34 % of the unit cell volume is covered by touching MT spheres, the MT approximation does not yield a band gap in the electronic band structure. A first step to remedy this situation is the introduction of "empty spheres", i.e., additional spheres without atomic nuclei, in which the charge density and potential are allowed to adopt values different from the rest of the interstitial volume, thus adding a primitive corrugation of the functions there. A better solution is a potential that can take on a general shape, as is implemented in the well-known full-potential LAPW (FLAPW) method [29]. The FLAPW method has shortcomings, however, particularly in its convergence properties, which is the reason why we developed the augmented Fourier component (AFC) method [13]. The main difference between the two methods is that in the FLAPW method the APW representation is retained for construction of the charge density, while in the AFC method the density is separated into two parts: a Fourier part $\rho^{F}(r)$, which represents the charge density of the valence electrons, and a narrow MT part $\rho^{MT}(r)$ describing the nucleus and the core electron distribution in a spherically symmetrical average. The Fourier charge density distribution is easily accessible in this method, because the ELAPW $\mathbf{k}{\cdot}\mathbf{p}$ method readily yields the Fourier transform of the valence wave functions. Additionally, the AFC method can be accelerated by transferring charge, whose distribution is given by the higher Fourier components of ρ^{F}, to the MT part. This combination of a significant gain of speed and a small but controllable loss of accuracy allows very fast LAPW calculations to be made with a general-shape potential. The extension of the ELAPW AFC method to structure optimization was recently implemented in the computer code. The perturbational $\mathbf{k}{\cdot}\mathbf{p}$ method is a fast approximate version of our $\mathbf{k}{\cdot}\mathbf{p}$ approach in which the lower solutions at a reference point $\mathbf{k}_0$ multiplied by a phase factor $\exp[i(\mathbf{k}-\mathbf{k}_0)r]$ are used as basis functions. Using this approach we significantly reduce the size of the Hamiltonian matrix, and the overlap matrix is the unity matrix. Since the AFC method provides the Fourier transform of the wave functions and the momentum operator is diagonal in a plane-wave basis, calculation of the momentum matrix elements is simplified, and the order of scaling in the construction of the Hamiltonian matrix is reduced. This approach has generally been used for calculations requiring a larger number of **k** points. For the calculations a simplified version of the structure described in Section 6.2 was used: The contents of the inner tubes were omitted, so that the formula is $A_6[Sb_{12}O_{18}][SbX_3]_2$. Also, the orientation of the intertubular pyramid chains was assumed to be parallel in all cases. For (K;Se) the $2 \times 2 \times 1$ superstructure was not taken into account. The unit cell contains 44 atoms of 8 inequivalent types with the atomic positions and lattice constants taken from X-ray diffraction data [6]. These experimental parameters have been established to a high degree of confidence. Calculations with slightly differing positions from earlier experiments (concerning mainly the positions of the chalcogen atoms) showed that the main features of these results are stable with respect to small changes of the geometry. The guest ions B^{m+} and the water molecules or hydroxyl groups OH^- were left out of the calculations because they are considered to be strongly disordered [25] and the presented discussion neither needs nor justifies the additional computational burden at this stage. Ionic conductivity carried by the water and/or the B^{m+} ions was excluded by impedance measurements [7,19,21]. The lattice constants are given in Tab. 1.

All the results presented here were calculated using the AFC ELAPW $\mathbf{k}\cdot\mathbf{p}$ method. The calculation of all observables is based on self-consistent potentials determined by the AFC method. The basis size of the decisive calculations was 3000 functions, or about 70 APWs per atom. This ratio has been found to be sufficient for other complex structures, such as perovskites [26]. For Ru_2Si_3, Wolf et al. [27] found that the electronic structure is well converged with 75 LAPWs per atom. The parameter $|G_{max}|S_{min}$ was about 4.5, where S_{min} is the smallest sphere radius. For the self-consistent calculations, the densities were computed at the **k** point Γ, which is justified by the large real-space dimensions of the unit cell translating into a very small Brillouin zone. Earlier calculations [10,11] were carried out within the MT approximation. Aspherical corrections were introduced by adding an empty sphere within the tubes. The new results using the AFC method presented here offer a much better description of the physics. The results show that the former calculations describe the basic properties of the cetineites; however, new and important effects are discovered by this highly accurate calculation.

Tab. 1. The lattice constants used for the calculations of the four cetineites (in nm).

	(K;Se)	*(Na;Se)*	*(K;S)*	*(Na;S)*
a	1.4630	1.4423	1.4318	1.4152
c	0.5616	0.5565	0.5633	0.5576

6.5 Results

For the four cetineites (K;Se), (Na;Se), (K;S), and (Na;S) a number of electronic and optical properties were calculated, which are compared here with the experimental results where available.

6.5.1 Density of States

To show that our approach is viable the density of states (DOS) of (Na;Se) was calculated in three different ways, all based on the AFC ELAPW **k**·**p** method (see Fig. 6). One of the two curves calculated with the exact ELAPW **k**·**p** method is the sum of Lorentzians centered at Γ point band energies, while the other is derived from a calculation with 8 **k** points in the irreducible part of the Brillouin zone (IBZ). The third curve was derived by using the perturbational **k**·**p** method for 64 **k** vectors (the IBZ has one-twelth of the volume of the full Brillouin zone). The comparison in Fig. 6 shows that all three methods give the same result to a very high degree. This not only means that the single-**k**-point method is justified, but also that the

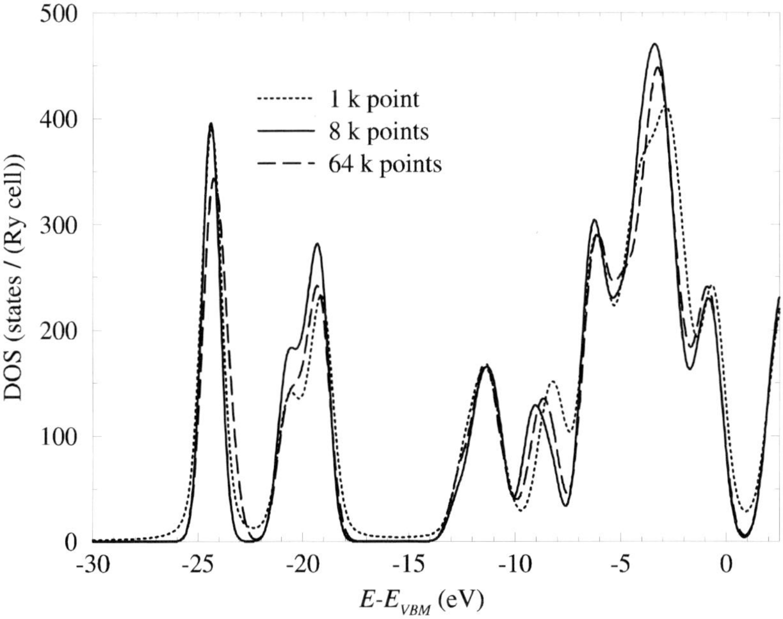

Fig. 6. DOS of the cetineite (Na;Se) computed by three different methods: with a single **k** point and 8 **k** points using the AFC ELAPW **k**·**p** method, and with 64 **k** points using the perturbational AFC ELAPW **k**·**p** method. The curves were convoluted with Gaussian curves with a full width at half maximum (FWHM) of 1 eV.

perturbational **k**·**p** method describes the electronic structure of the valence band region with very small differences from the exact **k**·**p** method. Shifts in the peaks of the three DOS curves due to dispersion of the band structure are minor. The worst example is the peak at about −9 eV, which in the single-**k**-point approximation is shifted by about 0.5 eV to higher energy. In all four cetineites, the valence DOS has one major band extending from the Fermi energy to about −6 eV. This peak can be roughly subdivided into two manifolds, covering approximately the range from 0 to −4 eV and from −4 to −6 eV. To a large degree the states in the first manifold are hybridized O 2p and chalcogen p states, in which the chalcogen p character dominates near the highest occupied state. The lower energy manifold is almost exclusively made up of oxygen p with a small contribution from Sb p states. The states in the energy interval from −6 to −8 eV are mainly of Sb 5s and O 2p character. So the binding of O and Sb in the wall is carried out by the states in these two energy intervals. The lower unoccupied states are to a large degree hybridized p orbitals of all the atoms in the crystal, except for the alkali metal. In the cetineites containing sulfur, at **k** point Γ the lowest unoccupied state is strongly delocalized in the *c* direction and is confined entirely to the interior of the Sb_2O_3 channels, see Fig. 20. Similar free-electron-like states have been observed in calculations on boron nitride nanotubes [28]. It will be difficult to experimentally observe this state in samples whose tubes are not completely evacuated, since we expect it to be destroyed by impurities. In the following discussions of gap sizes this state is therefore treated separately. The DOS for the four substances show a gap of between 1.5 and 1.8 eV (see Fig. 7). In the crystals containing sulfur, (K;S) and (Na;S), within this gap the DOS of the above-mentioned single free-electron-like state appears, which narrows the gap to 0.1 and 1.0 eV, respectively. This single state has a very high dispersion. These states are also visible, with some distortions indicating hybridization, at higher energies among the conduction bands of the selenium compounds. In the energetic interval extending a few electron volts around the gap, this band is easily identified, and it displays significant changes from compound to compound, whereas the other bands appear largely unchanged. The earlier calculations using the MT approximation of the potential yielded fundamental gaps of about 0.5 eV [11,12]. The MT potential approximation leads to a narrowing of the gap, an effect that also appears in the semiconductor silicon, for example, where the self-consistent MT potential yields a semimetal [29]. The gap size calculated by the AFC method still is slightly too small, which appears to be an effect of the treatment of many-particle interactions in the local density approximation (LDA). Photoconductivity measurements and UV/Vis spectroscopy gave a gap size of slightly greater than 2 eV at room temperature. The values are given in Tab. 2.

By scanning the temperature and monitoring the conductivity the electronic activation energies were determined, i.e., the minimum thermal energies that are necessary to excite electrons from the Fermi edge into the conduction band so that significant conductivity can be registered. For the (Na;Se) crystals the temperature interval was from 326.5 to 238.5 K and an activation energy of 0.43 eV was found. From 294 to 229 K the activation energy in (K;Se) was 0.49 eV, from 240 to 310 K in (K;S) yielded 0.5 eV, and from 268 to 327 K in (Na;S) gave 0.55 eV. This must be

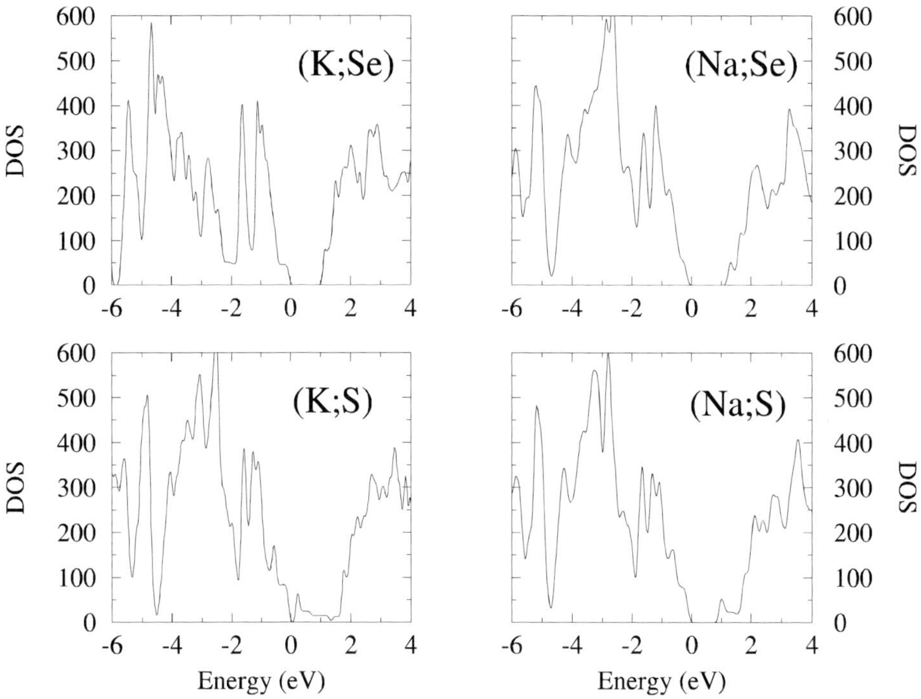

Fig. 7. DOS for the four cetineites (K;Se), (Na;Se), (K;S) and (Na;S) for the valence region, calculated with the perturbational **k**·**p** method for 64 **k** points in the IBZ. The (Na;Se) DOS shown in Fig. 4 is the one displayed here, convoluted with a Gaussian.

compared with one-half of the values of the indirect gaps of the band structure (see Tab. 2). The calculated DOS is compared to an angle-integrated photoemission spectrum of (Na;Se) taken at 21.22 eV photon energy. Figure 8 shows that the oxygen 2p-derived peak at about −3 eV and the antimony 5s-derived peak at about −12 eV are reproduced. Their relative intensities agree with those suggested by the

Tab. 2. Gaps of the cetineites, determined by measuring the absorption threshold $E_{gap,\,trans}$, the onset energy of photoconductivity $E_{gap,\,photo}$, and the thermal energies, which correspond to twice the electronic activation energies E_A. Further, the calculated first maximum of the squared plasma frequency $E_{gap,\,theor}$ and the indirect band gap $E_{indirect}$, taken from the band structure neglecting the gap state, are given.

Crystal	*(K;Se)*	*(Na;Se)*	*(K;S)*	*(Na;S)*
$E_{gap,\,trans}$	2.03 eV	2.12 eV	2.29 eV	2.38 eV
$E_{gap,\,photo}$	2.10 eV	2.15 eV	2.25 eV	2.40 eV
$E_{thermal} = 2\ E_A$	0.98 eV	0.86 eV	1.0 eV	1.1 eV
$E_{gap,\,theor}$	2.2 eV	2.4 eV	2.5 eV	2.6 eV
$E_{indirect}$	1.0 eV	1.3 eV	1.6 eV	1.7 eV

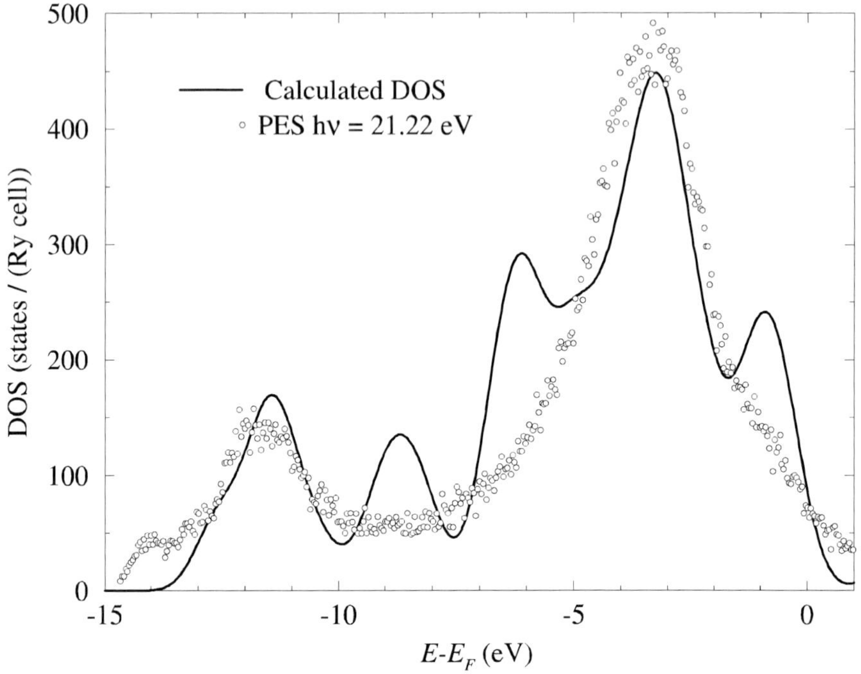

Fig. 8. Comparison of the photoemission spectrum for (Na;Se) (given in arbitrary units) and the DOS. The DOS was calculated using the perturbational **k**·**p** method for 64 **k** points in the IBZ and was convoluted with a Gaussian peak of 1 eV FWHM.

DOS curve. Two structures in the DOS at −6 and −9 eV are not resolved. This may possibly be due to the selection rules of the photoemission process, which have not been taken into account here or may be because only an incomplete angle integration was achieved experimentally by averaging the emissions from a number of differently oriented crystals.

6.5.2 Band Structure

The band structure of (Na;Se) is shown in Fig. 9. The cetineites have very similar band structures in which corresponding bands can easily be identified even though they sometimes are shifted energetically with respect to each other. The most striking feature of cetineite band structures is the stronger dispersion along the *c*-axis, i.e., in the ΓA direction. This dispersion indicates that the binding of the atoms making up the tubes is stronger than the binding among the tubes (this well-known effect is analogous to the transition-metal layered crystals, where the dispersion along the layer direction is much higher than that along the layer nor-

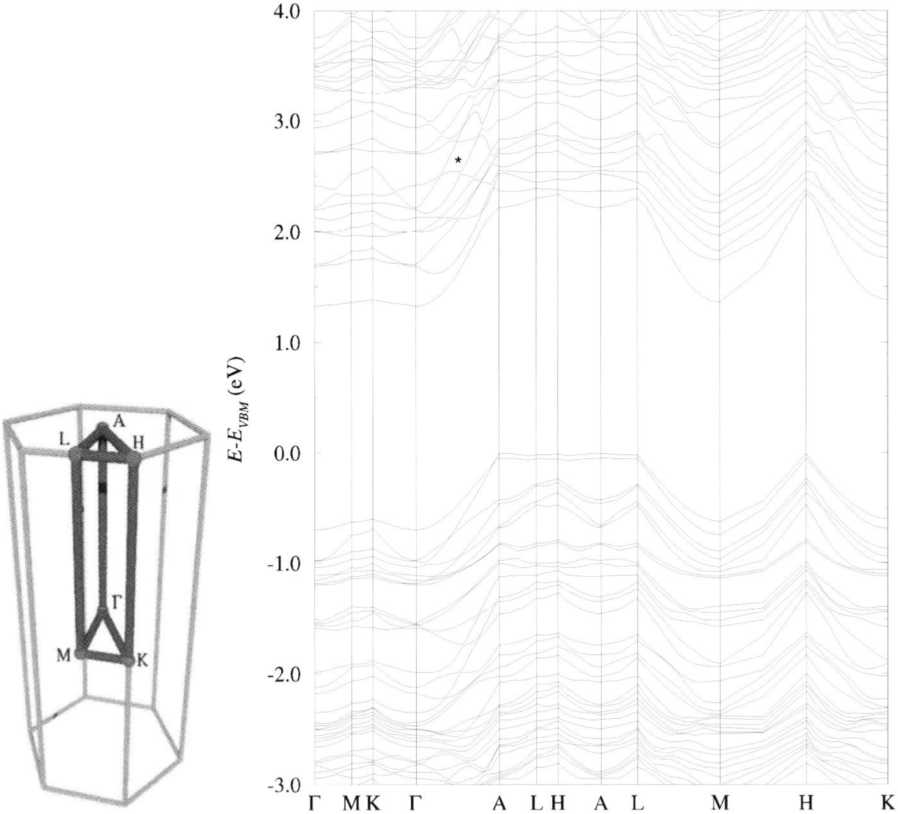

Fig. 9. Band structure of the (Na;Se) crystal in the hexagonal Brillouin zone, which is indicated on the left. An asterisk is used to denote the two dispersive bands noted in the text.

mal) [30]. In the sulfur compounds, the single state with very high dispersion discussed above is found inside the gap. The strong Se character near the top of the valence band (see Section 6.5.1) suggests that the selenium of the pyramids is responsible for a large part of the photoconductivity. Electrons from there will be the first to be excited to the conduction band. Since the lowest conduction bands have strong contributions from orbitals of nearly all the atoms in the crystal, no particular pathway for the conductivity can be deduced from their orbital composition. The two bands in the band structure marked by an asterisk in Fig. 9 are highly dispersive and prime candidates for the conduction of electricity. At Γ the states are composed of selenium, wall antimony, and oxygen orbitals, and this suggests a zigzag electron path between the selenium of the pyramids and the wall atoms. The *c*-axis dispersion of all bands has increased significantly with respect to the earlier MT calculations. However, similar bands are also present in the MT band structure, in particular the highly dispersive ones marked by the asterisk. The valence band maximum appears on the top of the Brillouin zone with hardly any energetic difference among the states there. The highest energy is found at the A point, the conduction band minimum 1.3 eV above at the Γ point, at almost the same level with the ΓMK plane. The band structures of the three other substances are displayed in Fig. 10.

6.5.3
The Dielectric Function

The complex dielectric function was calculated for all four crystals considered using the perturbational approach. The imaginary part is shown in Fig. 11 for **E** parallel to the crystal axis (the dependence of the function upon the polarization of the light is very small). Disregarding the (A;S) gap state, the sizes of the fundamental gaps are slightly larger than 2 eV, which is in excellent agreement with the experimental values for the gaps taken from transmission measurements. The energy onset of optical absorption, as represented by the imaginary part of the dielectric function, is significantly smaller for (K;Se) than for the three other cetineites, which approximately have a common onset energy. (K;Se) differs in two respects from the other compounds: K 3p and Se 4s are strongly hybridized here, and the optical properties are different. Experimentally, it is found that the substance crystallises in a superstructure. While the experimental chemical trend given in Tab. 6.2 is clearly visible for the cetineites containing selenium, it is less pronounced for those containing sulfur. For (K;S), a maximum appears at a photon energy of about 0.8 eV which is due to transitions to the highly delocalized state in the gap. This peak is not visible in the experiment. The comparison of the calculated absorption (see Fig. 12) with the measured transmission spectra shows that the chemical trend of the onsets of transmission in both theory and experiment agrees when cetineites of the same chalcogen and of the same alkali metal are compaired pairwise. We also have agreement in that the (Na;S) compound has the lowest onset and (K;Se) the highest. However, in the theoretical curves it is difficult to determine the relationship between (K;S) and (Na;Se).

6.5.4
Anisotropy of the Electrical Conductivity

Experimentally a strong anisotropy in the conductivity of cetineites has been measured [21] (see Tab. 3). The conductivity has been measured in white light parallel and perpendicular to the *c* axis by applying silver contacts to (Na;Se) crystals, as shown in Fig. 13. Two of those crystals measured at room temperature yielded the results given in Tab. 3, the typical variation with the sample preparation being visible in the difference of the results. In (Na;Se) the conductivity along the tubes is about 30 times higher than perpendicular to it. For a theoretical comparison, the squared plasma frequencies ${\omega^2}_{\mathrm{p},\mu\mu}$ were determined from the perturbational $\mathbf{k}{\cdot}\mathbf{p}$ results for 64 **k** points. The quantity ${\omega^2}_{\mathrm{p},\mu\mu}$ characterizes both the DOS and the averaged electron velocity at a given energy (see Eq. 1). While the conduction band

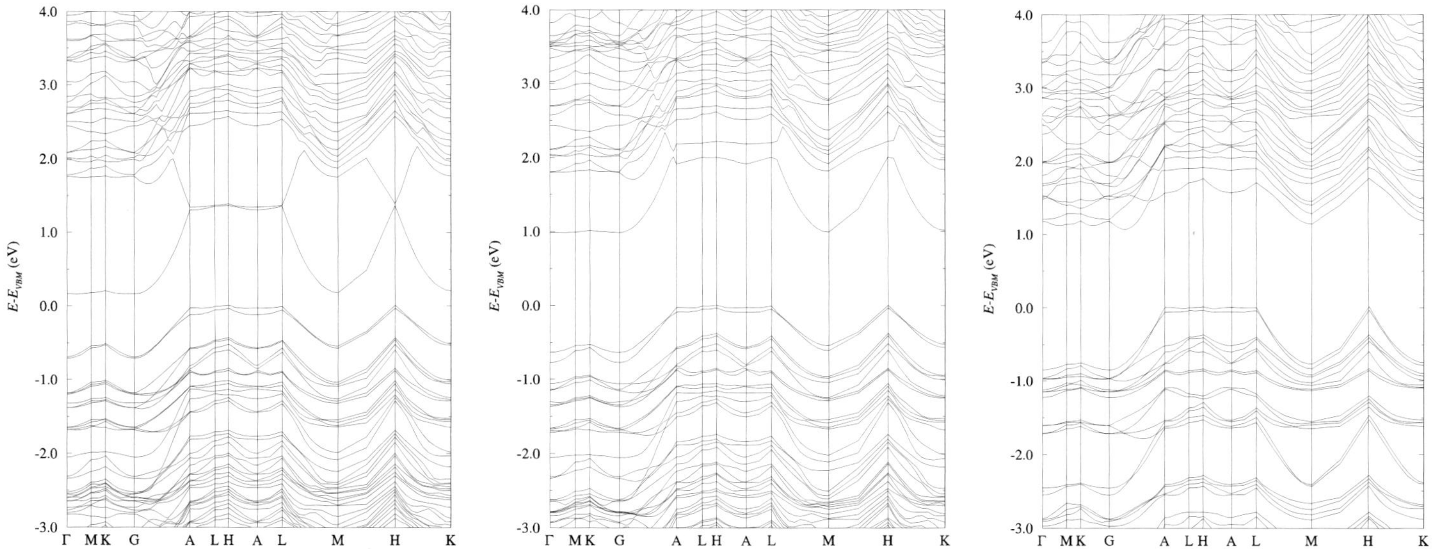

Fig. 10. Band structures of (K;S) (left), (Na;S) (middle), and (K;Se) (right), calculated with the ELAPW AFC **k p** method, middle and left figure are erroneously interchanged in Ref. 12.

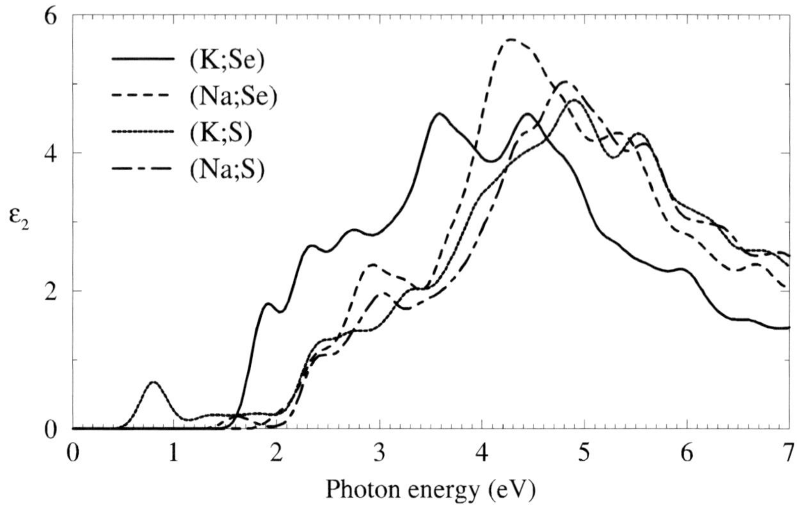

Fig. 11. The imaginary parts of the dielectric functions of the cetineites (K;Se), (Na;Se), (K;S), and (Na;S). The curves were convoluted with a Gaussian with a FWHM of 0.25 eV. The electric field vector **E** is parallel to the crystal axis *c*.

DOS provides the final states for the photoexcitation process, the averaged velocity determines the electron mobility and thus the magnitude of the photocurrent, which is measured to yield the conductivity in a given Cartesian direction μ. The $\omega i^2{}_{\mathrm{p},\mu\mu}(\mathrm{E})$ curves are displayed in Fig. 14. The squared plasma frequencies are defined by Eq. (1)

$$\omega^2{}_{\mathrm{p},\mu\mu}(E_{\mathrm{F}}) = \frac{1}{\pi^2}\sum_{\lambda}\int \mathrm{d}S_{\mathrm{F}}\frac{v_{\lambda}^2(k)_{\mu}}{|\mathbf{v}_{\lambda}(k)|} \tag{1}$$

where the integral is over the Fermi surface (generalized below to a **k** surface of constant energy), λ enumerates the energy bands, μ designates the real-space coordinates, and $v_{\lambda}(\mathbf{k})_{\mu} = \delta E_{\lambda}(\mathbf{k})/\delta k_{\mu}$. In a metal, the conductivity is proportional to the squared plasma frequency (Eq. 2)

$$\sigma_{\mu\mu}^{\mathrm{metal}} = \frac{\tau}{4\pi}\omega^2_{\mathrm{p},\mu\mu}(E_{\mathrm{F}}) \tag{2}$$

The values $\omega^2{}_{\mathrm{p},\mu\mu}(E)$ shown in Fig. 14 can be interpreted for semiconductors as the conductivity electrons encounter when excited to states of a certain energy E, as in the photoconductivity experiments. Clearly, the conductivity is considerably higher along the *c*-axis, that is, parallel to the tubes.

In a semiconductor, the conductivity tensor is given by an integral (Eq. 3)

$$\sigma_{\mu\mu}^{\mathrm{nonmetal}} = \frac{\tau}{4\pi}\int \mathrm{d}E\omega^2_{\mathrm{p},\mu\mu}(E)\left(-\frac{\partial f_0}{\partial E}\right) \tag{3}$$

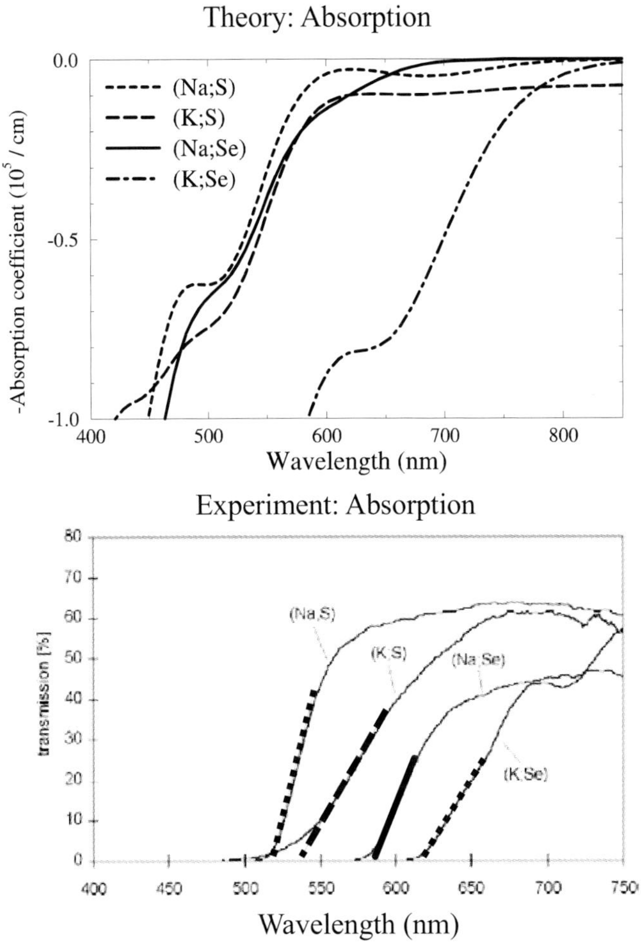

Fig. 12. The calculated absorption coefficient is compared with the measured transmission and the onset of photoconductivity.

Tab. 3. The conductivity of two (Na;Se) crystals at room temperature, parallel and perpendicular to the crystal axes [10,27].

	Crystal 1	*Crystal 2*
Conductivity, parallel $\sigma_{\parallel}$	5.4×10^{-7} S m^{-1}	6.8×10^{-7} S m^{-1}
Conductivity, perpendicular $\sigma_{\perp}$	1.96×10^{-8} S m^{-1}	2.23×10^{-8} S m^{-1}
Factor $(\sigma_{\parallel}-\sigma_{\perp})/\sigma_{\perp}$	26.6	29.5

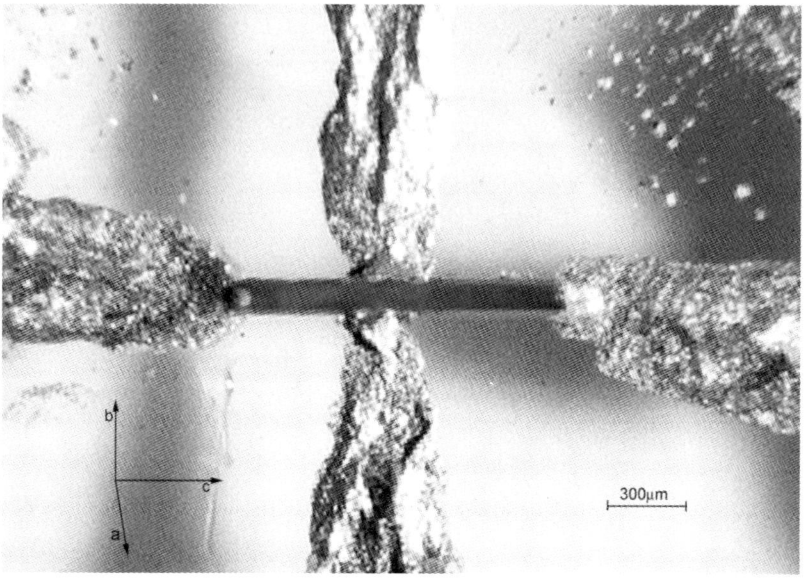

Fig. 13. A cetineite crystal contacted for measurement of the conductivity parallel and perpendicular to the crystal axis.

where f_0 is the electron distribution function and τ the relaxation time. The Fermi distribution will not be strongly distorted by the temperature, but the illumination will yield a nonzero distribution of carriers in the low conduction bands. Since the function f_0 applies to both spatial directions equally, it is a good approximation to assume that the ratio of the plasma frequencies will translate to the conductivity.

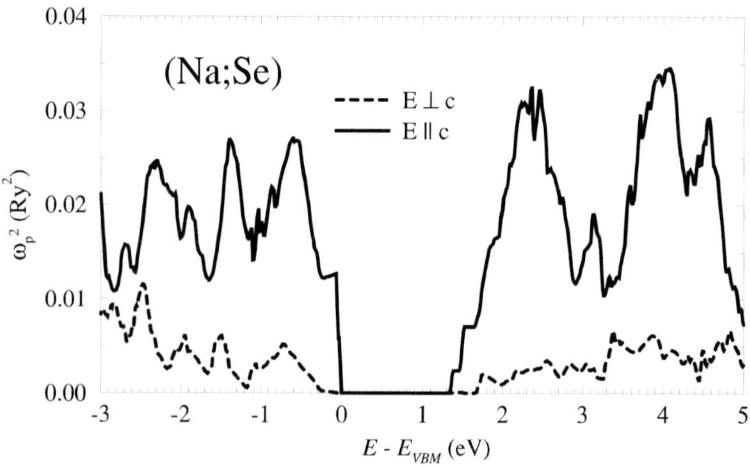

Fig. 14. Squared plasma frequencies for (Na;Se) calculated by the perturbational AFC ELAPW **k·p** technique.

Thus, the conductivity is much higher in the direction parallel to c than perpendicular to it. For the first peak in the conduction band, ranging from 1.5 to 3 eV, at the maximum the ratio is $\omega_{p\|}^2 : \omega_{p\perp}^2 = 13:1$.

This result can be easily seen in the band structure (Fig. 9) where the stronger dispersion parallel to ΓA translates into larger magnitudes of the derivatives parallel to this axis, which are integrated for the plasma frequencies. In the energy interval of the first conduction band peak a number of bands with a particularly high dispersion along ΓA can be observed, whereas around 3 eV the dispersion is reduced, which is reflected in the plasma frequencies. These results show that the order of magnitude of the anisotropy in the conductivity can be explained as a band-structure effect. The energy dependence of the plasma frequencies can also be used to extract the onset energies of the photoconductivity by assuming that it will set in with the first extended maximum in the conduction-band region. The mechanism assumed here is that valence band electrons are excited into conduction bands that show an appreciable conductivity, as expressed by the plasma frequencies. We assume that the DOS near the valence band maximum is high enough to support a sufficient population of photoexcited carriers to yield the current measured in the photoconduction experiments. To estimate the lowest excitation energy for an experimentally significant photocurrent we look at the first maximum to make sure that the resulting current can really be measured and to exclude a contribution from the misleading structure of the delocalized states in the sulfur compounds. We expect these free-electron-like states to be absent in the samples due to guest atoms present in the tubes. Since the absolute size of the gaps is somewhat ambiguous due to LDA effects, our main aim here is to establish whether we can reproduce the trend in the gap sizes measured in the photoconduction experiments. The photoconductivity mechanism is different from that leading to the optical spectra, where the transition probabilities between valence and conduction bands connected by the photon energy play a major role. When we look at the first extended maxima of the plasma frequency in the conduction band of the considered compounds (see Tab. 2), we find that they follow the experimental gap sizes.

6.5.5 Electron Density

The electronic density distribution in the cetineite (Na;Se) for the plane including the tube axis and parallel to the crystal a-axis is shown in Fig. 15, and a plane perpendicular to the c-axis is shown in Fig. 16. The atomic spheres around the nuclei are left empty so that the atoms are easier to identify. The charge density was calculated from the Γ point states, which is justified because of the very small Brillouin zone.

From the figures it is apparent that the tubes are nearly free of electronic density. The quite high charge density between the atoms of the tube walls indicates that the binding among the wall atoms is strong. Figure 14 shows that there is hardly any density surrounding the Na atoms, although some of it will be covered by the

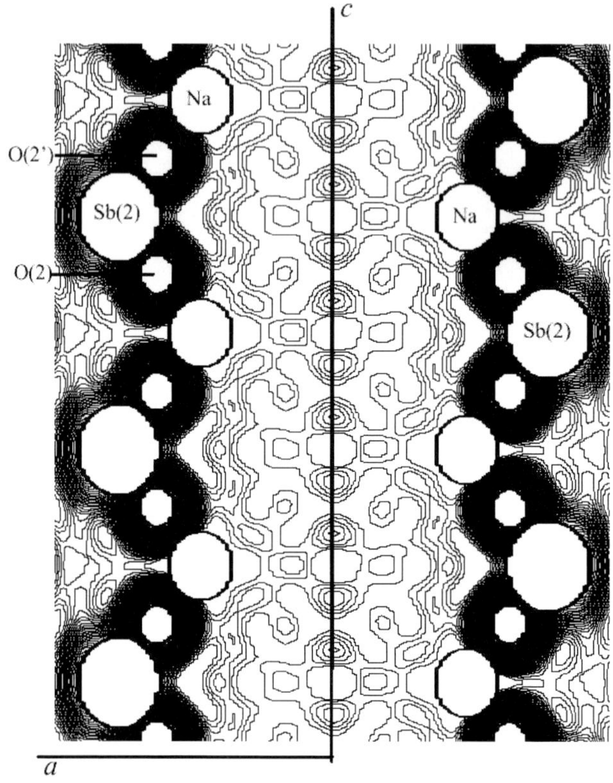

Fig. 15. The charge density of the cetineite (Na;Se) shown as a section through the crystal that includes the tube axis (shown by the central line labeled *c*) and is parallel to the *a*-axis. Three unit cells are shown, i.e., the height of the graph is 3*c* and the width *a*.

large sphere. Nevertheless, in the planes at $c = 0.7$ and 0.8 (not shown here) it is particularly visible that the Na atom is rather free of surrounding density in the opening in the wall made up of the oxygen atoms and Sb(2) and Sb(3). This confirms that in (Na;Se) the binding of the alkali metal to the pyramids, which is assumed to be responsible for holding the tubes together, is ionic in nature. This ionic binding was already suggested by the DOS: the Na and Se states do not interact. In particular the oxygen atoms are surrounded by a rather high electronic density. We have already mentioned the sizable contribution oxygen makes to the higher valence band, where the states are made up of O 2p hybridized with Sb 5p or Se 4p. Since the density distribution does not include the excited states these figures cannot directly help interpreting the channels of conduction. However, the almost vanishing ground-state density within the tubes, together with the nearly free-electron-like excited density localized there, may indicate an unencumbered,

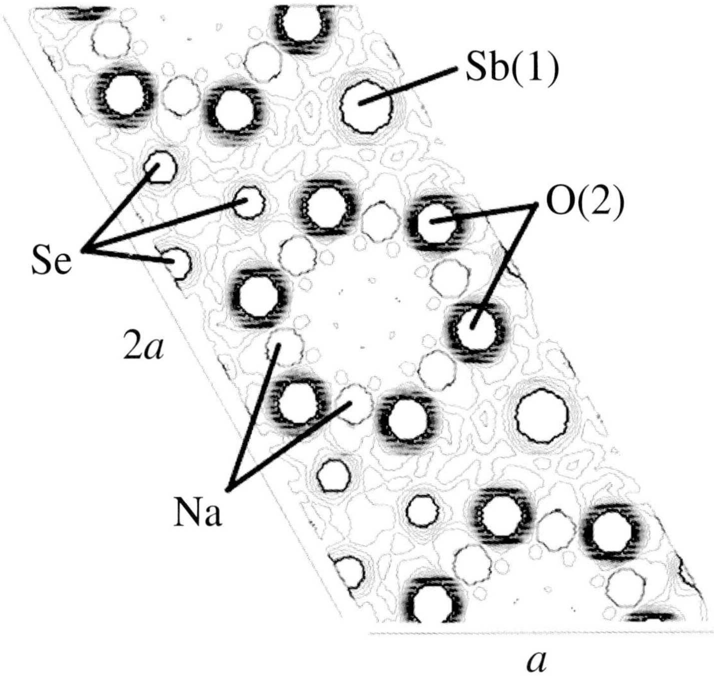

Fig. 16. Charge density of the cetineite (Na;Se) shown in a plane through the crystal perpendicular to the *c*-axis. Two unit cells are shown. The plane is for $c = 0.5$.

ballistic, electron path when a voltage is applied to the ends of the tubes under light incidence or other excitational processes.

6.5.6
Cetineite Mixed Phases

The investigations on the optical gaps of single-phase cetineites have shown that in comparison to the cetineites with X = Se the sulfur phases reveal a larger band gap. Changing the cation from $A = Na^+$ to $A = K^+$ leads to a red shift in the optical absorption. This was a first attempt to tune the optical properties by varying the chemical composition.

The optical gap in mixed phases with A = Na and K and X = S monotonously follows the chemical composition as the lattice constants. Figure 17 (top) shows the dependence of the lattice constants on the chemical composition, and Fig. 17 (bottom) the dependence of the optical band gap on the lattice constants of the crystals for the mixed phase (Na,K;S). Thus, the optical gap is directly determined by the chemical composition of the reaction mixture, and, in a narrow range, the properties may be tuned with respect to photoconductivity and optical absorption.

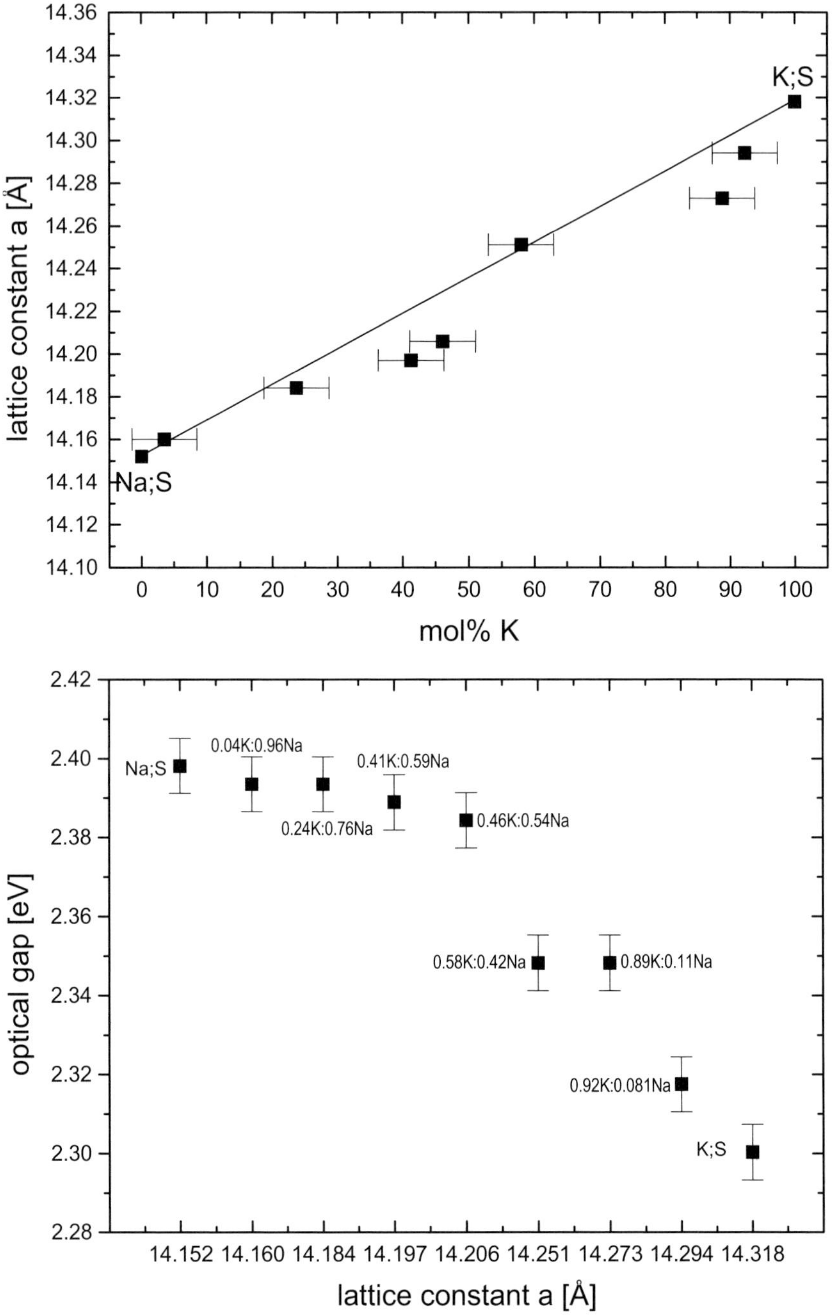

Fig. 17. Mixed phase (Na,K;S).Top: dependence of the lattice constants on the chemical composition. Bottom: dependence of the optical band gap on the lattice constants of the crystals.

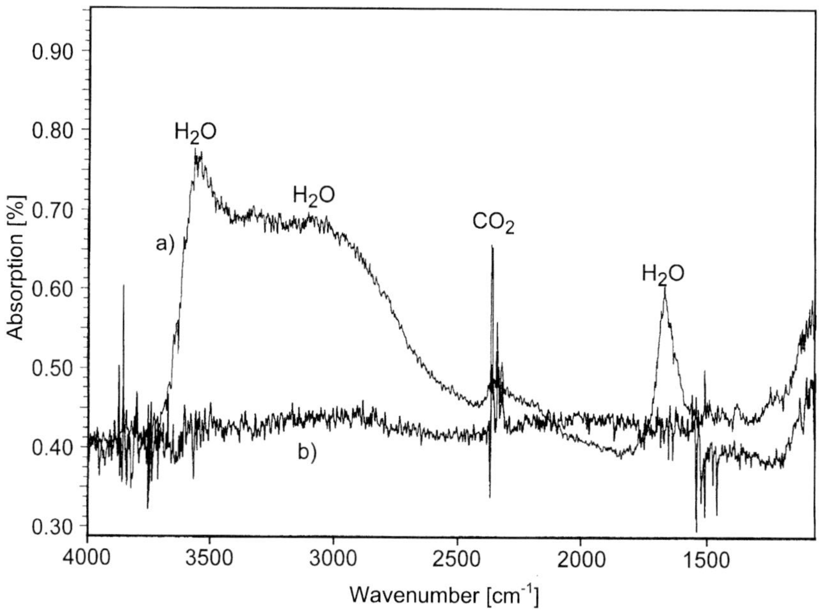

Fig. 18. IR spectra of (K;Se). a) At standard pressure. b) Under low pressure

6.5.7 Host/Guest-Interaction of (K;Se)

IR measurements on (K;Se) have shown that the water molecules inside the channels can be removed at room temperature under low vacuum. Figure 18 shows the IR spectra of (K;Se) with the channel filling water molecules under standard pressure (a) and under low vacuum (1×10^{-2} Pa; b).

The IR spectra in (a) show the characteristic regions of the OH vibrations at 3600–3000 cm^{-1} and at 1730–1600 cm^{-1}. This corresponds to the symmetric and antisymmetric stretching modes and the bending mode of the free water. To verify the assignment of the signals, the water was removed. Because the IR signals which appear under ambient conditions vanished under vacuum in all measured cases, they can ambiguously be identified as originating from the guests, i.e., from water.

For any use of (K;Se) as a molecular sieve or gas sensor, the interior of the empty channels must be accessible for guest molecules. Therefore sorption/desorption experiments in different atmospheres were performed. The resistance of a (K;Se) pressed pellet, contacted with silver paint on one side, was measured during sorption/desoption cycles. Figure 19 shows the resistance of (K;Se) in synthetic air, nitrogen, carbon dioxide, argon, and helium atmospheres. Between the sorption phases the host was evacuated.

In all sorption cycles the resistance of (K;Se) increases significantly in the presence of the above-mentioned gas molecules. Sorption experiments with lager mol-

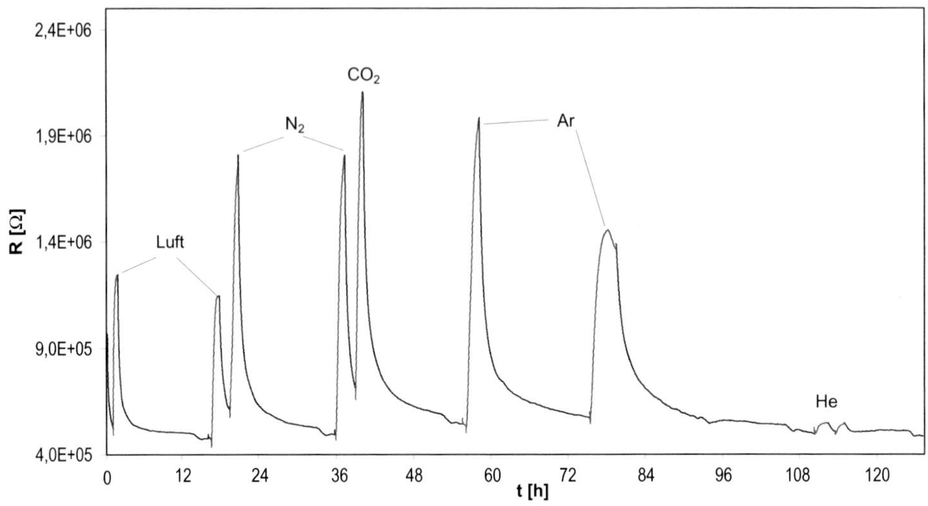

Fig. 19. Sorption/desorption cycles of a (K;Se) pressed pellet.

ecules such as 3-methylpentane und 2,2-dimethylbutane showed that a change in resistance can only be observed up to a molecular cross section of 0.49 nm. This result corresponds to the cross section of 0.5 nm determined by transmission electron microscopy [19].

The presence of guest atoms within the tubes of the Cetineite crystals is an important general topic with respect to preparation, properties, and application, and hence theoretical ab initio investigations are indispensable. The central question is why (K;Se) is the only compound that reversibly absorbs guests, whereas the others could not be even prepared without them. A first guess based on some theoretical reasoning was mentioned above.

However, for an understanding of stability and reaction paths total energy calculations are necessary. As our program code which we used for these materials initially was not capable of computing the total energy, we enlarged the code [31].

First calculations were devoted to the diffusion of the noble gas atoms He, Ne, Ar, and Xe for their energy barriers and diffusion times for the motion along the tube axes. The values obtained for the barriers are 120, 380, 250 and 470 meV, respectively. Details of the potential surfaces will be published elsewhere. The listed energies refer to one guest atom per unit cell. It can be assumed that the influence of guest atoms of neighboring cells in the tube direction are negligible in this first approach. Thus, the energies describe the diffusion of a single atom in an otherwise hollow tube yielding the difference in potential between the most preferred and the most avoided position of the guest along the symmetry axis, i.e., on the tube axis. The stable position differs for the light gases He and Ne and the heavier Ar and Xe, for which a stronger interaction with the wall is expected because of the larger atomic radius. Thus, their energy minimum is found near a plane perpendicular to the tubes and containing the potassium atoms, which polarize the guest electrons and themselves are polarized by them. In contrast, the light noble gases

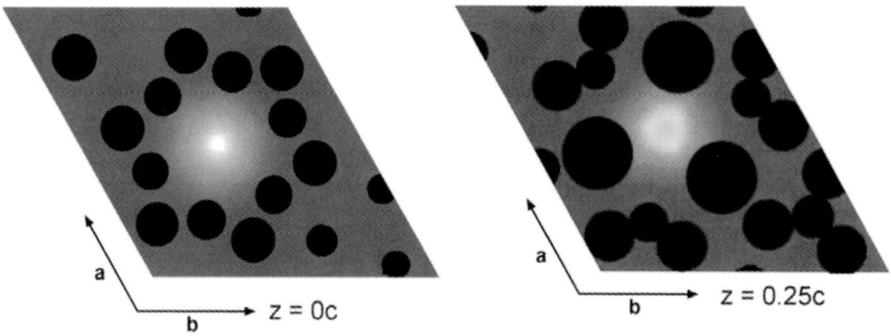

Fig. 20. Delocalized electron density in a channel of K;S; left for $z = 0c$ and right $z = 0.25c$ in the *ab*-plane, bright regions high density, calculated for Γ point with energy fixed at first excited eigen state.

have their energy minimum in a plane near the middle between two alkali planes. Their electronic configuration was found to be rather inert and not influenced by the tube-wall atoms. Only multipole contributions from the fixed nucleus and electron shells of the guest interact electrostatically with the wall-atom charges.

Further investigations on guests other than noble gases had to consider water and its hydrogen bonding because most of the impurities in the tubes of other compounds than (K;Se) contained water which could not be removed after preparation. We had to ensure that the code correctly describes such systems and calibrated by computing the vibrational modes of pure H_2O, for which much data can be found in the literature. The results compared well with other ab initio calculations and experiments with similar variance.

In the course of these investigations an interesting proposition could be discussed, namely, whether separation of the hydrogen isotopes can be achieved through different absorption rates because of different spatial extension of their zero-point oscillations. In fact, not hydrogen but water complexes showed to be possible candidates for isotopic separation, as revealed by the shape of the potential of some restricted configurations. A full geometric optimization was not possible with the available computing power. The configurations of main concern coincide with the experimentally found guest complexes, i.e., clusters of six octahedrally arranged water molecules. The positions of the hydrogen atoms and their distances from the wall are unknown and must be determined. From two model systems which we investigated it could be inferred that the possibility of isotopic separation by such guest complexes exists, but for quantitative values the hydrogen–wall distance should be determined experimentally or fully ab initio.

6.6 Conclusions

The properties of the electronic structure of cetineites have been investigated. Experimental and theoretical evidence is given that the crystals are nanoporous semiconductors – a very exiting property in the class of nanoporous materials with zeolite-like structure. The strong anisotropy found in the conductivity of the crystals parallel and perpendicular to the *c*-axis is directly reflected in the band structure, where the dispersion along lines in reciprocal space parallel to ΓA is much stronger than that along those perpendicular to this line. The calculations show that the conductivity along the axis is higher by about a factor of 10 than that perpendicular to it. The measurements found an anisotropy factor of 30, reflecting a quasi-one-dimensional conductivity. The atoms participating in the binding of the crystal were revealed by determining the composition of the valence-band density of states and the real-space charge density distribution. Particularly the oxygen atoms appear to be responsible for the stability of the tubes, which are bound ionically to each other by the SbX_3 pyramids and the A atoms. Only in (K;Se) does hybridization between the pyramids and potassium appear at semicore energies. As a further singular property the optical gap is smaller by about 0.5 eV in this compound than in the other cetineites. (K;Se) is also the only cetineite that is known to crystallize in a superstructure. The effective hybridization of binding states may indicate stronger bonding of potassium to the host structure in (K;Se). It is thus less likely to interfere with guest molecules. This could explain why (K;Se) can most easily be filled with water molecules, for example, and why these molecules are more easily extracted than from other cetineites. The generally strong dispersion in the ΓA direction of all the cetineites reflects strong binding within the tubes but rather weak interaction among them. However, despite the wealth of information on the properties of cetineites gained so far a number of questions remain to be resolved: What is the influence of the water or other guest molecules? What would be the result if the interstitial pyramid chains were not all oriented in the same direction? What is the exact nature of the delocalized state in the fundamental gap of the (A;S) cetineites, and how can we obtain experimental evidence for it? Work on these questions is on-going.

Acknowledgments

This work was supported by the Deutsche Forschungsgemeinschaft under Contracts Nos. SI 609/2-1 and SCHA 360/14-1.

References

1 G.A. Ozin, *Adv. Mater.* **1992**, *4*, 612.
2 G. Schön and U. Simon, *Colloid Polym. Sci.* **1995**, *273*, 101; **1995**, *273*, 202.
3 U. Simon, G. Schön, in H. S. Nalva (Ed.): Handbook of Nanostructured Materials and Nanotechnonoly, Vol. 3, Academic Press **1999**, pp. 131–175.
4 J. V. Smith, *Chem. Rev.* **1988**, 88, 149; S. L. Suib, *Chem. Rev* **1993**, *93*, 803.
5 U. Simon and M. E. Franke, *Microporous Mesoporous Mater.* **2000**, *41*, 1.
6 X. Wang, F. Liebau, *Eur. J. Solid State Inorg. Chem.* **1998**, *35*, 27; X. Wang, F. Liebau, *Z. Kristallogr.* **1999**, *214*, 820.

7 U. Simon, F. Schüth, S. Schunk, F. Liebau, X. Wang, *Angew. Chem. Int. Ed. Engl.* **1997**, *36*, 1121.
8 U. Simon, J. Jockel, F. Starrost, E.E. Krasovskii, W. Schattke, *Nanostruct. Mater.* **1999**, *12*, 447.
9 U. Simon, J. Jockel, F. Starrost, E.E. Krasovskii, W. Schattke, B. Marler, S. Schunk, M. Wark, H. Wellmann, *Stud. Surf. Sci. Catal.* **2000**, *129*, 683.
10 F. Starrost, E. E. Krasovskii, W. Schattke, X. Wang, F. Liebau, J. Jockel and U. Simon, *Z. Kristallog. Suppl.* **1998**, *13*, 90.
11 F. Starrost, E.E. Krasovskii, W. Schattke, J. Jockel, U. Simon, X. Wang, F. Liebau, *Phys. Rev. Lett.* **1998**, *80*, 3316.
12 F. Starrost, E.E. Krasovskii, Schattke, J. Jockel, U.Simon, R. Adelung, C. Kipp, *Phys. Rev. B* **2000**, 61, *23*, 15677.
13 E. E. Krasovskii, F. Starrost, W. Schattke, *Phys. Rev. B* **1999**, *59*, 10, 504.
14 C. Sabelli and G. Vezzalini, N. *Jb. Miner. Mh.* **1987**, *9*, 419.
15 C. Sabelli, I. Nakai, S. Katsura, *Am. Mineral.* **1988**, *73*, 398.
16 H. A. Graf and H. Schäfer, *Z. Anorg. Allg. Chem.* **1975**, *414*, 220.
17 F. Liebau and X. Wang, *Annual Meeting Abstract Book* (American Crystallographic Association, Pittsburgh 128 (**1992**).
18 X. Wang, *Z. Kristallogr.* **1995**, *210*, 693.
19 J. Jockel, Ph.D. thesis, Fachbereich Chemie der Universität Essen (**2000**).
20 A. Thess, R. Lee, P. Nikolaev, H. Dai, P. Petit, J. Robert, C. Xu, Y. Hee Lee, S. Gon Kim, A.G. Rinzler, D.T. Colbert, G.E. Scuseria, D. Toma'nek, J.E. Fisher, R.E. Smalley, *Science* **1996**, *273*, 483.
21 U. Simon, Habilitation thesis, Universität Essen (**1998**).
22 E.E. Krasovskii, *Phys. Rev. B* **1997**, *56*, 12, 866.
23 E.E. Krasovskii, A.N. Yaresko, V.N. Antonov, *J. Electron Spectrosc. Relat. Phenom.* **1994**, *68*, 157.
24 E. Wimmer, H. Krakauer, M. Weinert, A. J. Freeman, *Phys. Rev. B* **1981**, *24*, 864.
25 X. Wang, Ph.D. thesis, Kiel University (**1993**).
26 E.E. Krasovskii, O.V. Krasovska, W. Schattke, *J. Electron Spectrosc. Relat. Phenom.* **1997**, *83*, 121.
27 W. Wolf, G. Bihlmayer, S. Blügel, *Phys. Rev. B* **1997**, *55*, 6918.
28 X. Blase, A. Rubio, S. G. Louie, M.L. Cohen, *Europhys. Lett.* **1994**, *28*, 335.
29 F. Starrost, E. E. Krasovskii, W. Schattke, *Verhandl. DPG (VI)* **1998**, *33*, 741.
30 Leventi-Peetz, E. E. Krasovskii, W. Schattke, *Phys. Rev. B* **1995**, *51*, 17, 965.
31 O. Tiedje, Ph.D. thesis, Math.–Nat. Fakultät, Universität Kiel (**2000**).

Part 4
Optical Properties of Molecular Sieve Compounds

Introduction to Part 4

Franco Laeri

The periodic, crystallographically defined system of pores that forms the framework of molecular sieves is the source of inspiration on which many ideas for new compound materials grow. One idea consists of taking advantage of the pore matrix to arrange and fix guest ions and molecules in regular spatial positions and orientations. To realize this, nanoporous materials with certain symmetry properties of their pore framework must be on hand; the synthetic aspects of these are discussed in Part 1.

In an optical context the possibility of imposing a certain spatial symmetry on the arrangement of optically functional molecules is especially attractive. For example, to obtain a material with a large second-order polarizability a noncentrosymmetric ordering of the arrangement of molecular constituents is required. If the single molecules themselves exhibit a large nonlinear polarizability, but crystallize with centrosymmetric order, their noncentrosymmetric arrangement in nanoporous zeolitic hosts will preserve their function as second-harmonic generators [1]. An other important example in which imposing spatial order on a molecular ensemble can be used to nurture an optical effect, consists of arranging dyes side by side, one behind the other, in channels of a zeolite. This arrangement opens the way to long-distance nonradiative energy transfer and to the realization of light-harvesting photonic antennas [2].

These two examples show that it is possible to exploit the ordering principle of the nanoporous framework to conceive novel optical functional materials. Thus, the question arose whether, by combining the proper sieve structure with suitable molecular guests, one may compose material properties as it were at the drawing board. The answer must of course be documented with practical examples. However, we suspect that what can be achieved in reality will differ from the initial desideratum. The necessary investigations will certainly reveal new information about processes, which, properly considered, may lead to realistic ideas for modeling functional composites. The contributions in this part illustrate to what degree material properties can be designed "at the drawing board" in the present state of the art.

The focus in this part is on composite-material properties that are related to the interaction of the material with optical fields, i.e., its intrinsic optical properties, as well as on extrinsic optical properties, i.e., processes by which certain nonoptical properties of the composite are controlled by external optical fields. The chapters in this part thus start by laying out the respective initial idea, and then juxtapose this with a thorough discussion of the experimental findings.

In the first chapter K. Weh and M. Noack present zeolitic membranes in which the gas permeability can be switched by light-controlled nanovalves. The nanovalves consist of azobenzene molecules which reside in the pores of the zeolitic membrane material, either pentasils or faujasites. When irradiated with UV light of 360 nm a trans–cis isomerization reaction of the encaged azobenzene molecules is induced, whereas 436 nm blue radiation prompts the inverse reaction. Coupled with this isomer transition is a change in the molecular volume and, as a consequence, the permeable pore cross section of the material hosting the azobenzenes is controlled.

In a similar way, the following chapter by K. Hoffmann et al. addresses optical material properties such as absorption, refractive index, and birefringence, which are switched by external optical fields. The effects are based on composite materials consisting of a molecular sieve crystal matrix hosting azobenzene dyes. After the crystals were synthesized and the template was removed by a calcination process, azobenzene was infiltrated into the pores of NaX, $AlPO_4$-5, ZSM-5, or Silicalite-1 from the vapor phase or from solution. The dye molecules in the pores of these crystals are nearly perfectly aligned. Similar to the preceding chapter, the light-controlled trans–cis isomerization of the included dye guests alters the optical properties of the composite. Apart from changes in the absorption spectra, optical properties which depend on the orientation of the optical transition moment of the dye molecule ensemble are especially affected by the isomerization reaction. For example, the magnitude of the refractive index itself changes only marginally. However, the birefringence is drastically modified, as the orientation of the transition dipole moment in the trans and cis configurations does not coincide. As a consequence, photorefractive effects were observed.

When an optical field interacts with zeolitic composite materials the homogeneity of their properties plays an important role. Therefore, methods to characterize, for example, the spatial distribution of the composition, the guest concentration, or the guest molecule orientation, must be available. A method which allows the spectroscopic properties to be investigated with three-dimensional spatial resolution is scanning confocal microscopy. C. Seebacher et al. applied this method to characterize the influence of calcination on the spatial defect distribution, to reveal diffusion of guest molecules in the pore system of the host, to detect the spatial distribution of dye guests, and to specify the dynamics of the fluorescence of single molecules.

In the following chapter D. Sendor and U. Kynast discuss the conversion of UV light into visible light with lanthanide ions as luminophores which are exchanged into zeolite hosts. Usually the absorption of UV light in lanthanide ions is low, as the involved transitions are forbidden. Because of this the resulting light-conversion efficiency is unpractically low. As a way out, the authors investigate different schemes of sensitization. Together with the lanthanide ions, organic ligands are introduced to form complexes in which the ligands play the role of effective UV absorbers. In the complex the absorbed energy is then nonradiatively transferred to the luminescing lanthanide ion. The resulting luminescence properties and conversion efficiencies of the composite are discussed, as well as additional aspects of material preparation and properties.

The particular electronic structure of lanthanide ions, which results in long lifetimes of certain excited states, makes them suitable for lasing. The relevant transitions are in the near-infrared; in neodymium, for example, the resulting emission occurs at a wavelength around 1.06 μm. As is referred to in their name, zeolites (from Greek *zein*, to boil) contain water. Unfortunately, the laser emission lines of neodymium coincide with higher harmonics of water vibrations, which quench the lasing neodymium transition in zeolites. This is the reason why up to now lasing of lanthanides in zeolites was not reported. As an alternative to lanthanide ions, dyes encaged in zeolitic hosts can show laser emission [3]. The following three chapters are centered on lasing dyes in molecular sieve hosts.

An important concern when realizing a laser is the optical quality of the material. Absorption and scattering will eliminate light from the lasing path, thus causing losses, which must be compensated by the available amplification through stimulated emission processes. Absorption at the lasing wavelength and scattering losses must thus be avoided, as they increase the laser threshold. In their article Ö. Weiß et al. discuss the fabrication of laser-grade $AlPO_4$-5 crystals enclosing Pyridine 2 and DCM as laser dyes.

$AlPO_4$-5 composites are characterized by the channel structure of the host pores which align the included dye molecules. The optical transition dipole moment, which is responsible for the absorption of the pump light, as well as the laser emission in Pyridine 2 and DCM, is parallel to the molecular axis. The alignment of the molecules by the channels of the zeolitic host thus results in a dye system with nearly perfect optical order, a property which was not realized before in conventional dye lasers, and which leads to a series of new material properties, which are discussed in the article by L. Benmohammadi et al. One of the consequences is that the photostability of some molecules which are inclosed in regular pores is increased, i.e., photobleaching is reversible. Another important consequence of dye alignment in these systems is that optical emission along the alignment direction, which coincides with the hexagonal crystal axis, is suppressed. Lasing in these structures occurs in ring resonators which result naturally from total internal reflection at the crystal faces. It turns out that hexagonal ring lasers have peculiar optical wave properties. In fact, the wave equation in a hexagonal boundary can not be solved by the usual separation of variables ansatz. In this sense a hexagon forms an nonintegrable structure; a property which it shares with some chaotic billiards [4].

The pore dimension of zeolitic crystals like $AlPO_4$-5 is smaller than many laser dye molecules. J. Loerke and F. Marlow present mesoporous MCM-41-like materials which can host large molecules, such as Rhodamine 6G dyes. They discuss the preparation of these materials in the form of fibers. The fibers act as optical waveguides in which light amplification by stimulated emission processes is promoted.

The dimensions of single-crystal molecular sieve composites is usually less than one millimeter and most often around 10 μm. On the other hand, many practical applications require sample sizes of several millimeters. To increase the sample size J. Schneider et al. studied means to embed the molecular sieve grains in polymers. They discuss polymer mixtures exhibiting a refractive index which matches the composites, allowing a large number of composite grains to be combined in a solid, optically homogenous piece.

References

1 S. D. Cox, T. E. Gier, G. D. Stucky, J. Bierlein, *J. Am. Chem. Soc.* (**1988**), *110*, 2986.

2 G. Calzaferri, M. Pauchard, H. Maas, S. Huber, A. Khatyr, T. Schaafsma, *J. Mater. Chem.* (**2002**), *12*, 1.

3 U. Vietze, O. Krauß, F. Laeri, G. Ihlein, F. Schüth, B. Limburg, M. Abraham, *Phys. Rev. Lett.* (**1998**), *81*, 4628.

4 P. J. Richens, M. V. Berry, *Physica* (**1981**), *2D*, 495.

1
Modification of Gas Permeation by Optical Switching of Molecular Sieve–Azobenzene Membranes

Kornelia Weh and Manfred Noack

1.1
Introduction

The selectivity of molecular sieve membranes is generally affected by size and structure of the zeolite frameworks and of the permeant molecules. But it also depends on the polarity of both and on the interaction of adsorption and diffusion processes. In particular, the properties of zeolite membranes are determined by the type of zeolite, by the leak rate, pore size, and by the selection of the operation conditions during their use [1]. Zeolite hosts can be utilized to control the photophysical and photochemical reactions of encapsulated guest molecules. For example, the relative size of the host cavities in comparison to the size of the guest molecules is important for selectivity in photoreactions [2]. When the guest molecule azobenzene (AZB) absorbs light, its molecular size (C_4–C_4' distance) changes from 0.90 to 0.55 nm, and its geometry from planar to spherical. Furthermore, its dipole moment increases from 0.5 to 3.1 D by photoinduced isomerization from the thermodynamically stable trans to the metastable cis isomer [3]. If AZB is adsorbed in zeolite-type molecular sieve membranes with suitable pore structures, such as pentasils (MFI of type ZSM-5 or silicalite-1) with medium pore-opening diameters of 0.55 nm or faujasites (FAU of type NaY or NaX) with larger pore-opening diameters of 0.75 nm, modification of the transport and separation properties of such host–guest membranes for permeating gases by photoswitching is expected. The combination of the nanopore system of the zeolite membrane with the photoinduced trans–cis switching of the incorporated AZB leads to new collective properties of these MFI–AZB and FAU–AZB host–guest membranes.

1.2
Switchable Natural and Technical Membranes

Membranes used so far in industry have invariable permeation properties. In contrast, biological membranes in living organisms can change their permeation properties by the interplay of the type of ions, concentration, pH, UV radiation, or redox potential [4]. It is difficult to reproduce biomembranes. An important step towards switchable membranes was the development of a photoswitchable zeolite–AZB membrane that can be applied in technical gas-separation processes. Due to the optical switching of mass transport it could act as a microvalve in microreactors.

1.2.1
Realized Switchable Membrane Systems

There are attempts to produce switchable membranes for special applications. In most cases photoinduced trans–cis isomerization of AZB is applied for switching of different properties such as electrical resistance or conductivity [5,7], ion permeability [6,8,12], or gas permeance [9,10,13–15]. These membrane systems and their switchable effects and applications are listed in Tab. 1.

We already discussed the preparation of polymethacrylate (PM) membranes in which AZB is chemically bound to the polymer matrix [9,10]. In such membranes we could first observe the reversible change of gas permeance as a consequence of the trans–cis photoswitching of AZB. Similar experiments with a porous glass tube membrane which was modified by an organic AZB derivative are presented in Ref.

Tab. 1. Switchable membrane systems and their applications.

Switchable membrane system	*Switchable effect/application*	*Ref.*
Membrane consisting of AZB-containing water/oil microemulsions	variation of the electrical resistance	5
Bilayer membranes containing AZB	change of ion permeability	6
AZB in Langmuir–Blodgett films	change of electric conductivity	7
PVC membrane containing AZB-modified crown ethers	change of the electric potential and the ion permeability	8
Membrane consisting of supported polymethacrylate with AZB chemically bound to the polymer	reversible change of the gas permeance	9, 10
Thermosensitive membranes consisting of a porous substrate covered with a thermosensitive *N*-isopropylacrylamide grafted polymer	macromolecular separation by change of the hydrophobicity as a consequence of temperature change	11
Membranes consisting of a porous polymeric support with gold nanotubules	reversibly switchable permselectivity for cations or anions by change of the pH of the contacting solution	12
Porous glass membrane surface modified by an organic AZB derivative	variation of gas flow	13
Supported molecular sieve membranes consisting of MFI- or FAU-encapsulated AZB	reversible change of single-gas permeance and of gas-mixture permeance	14, 15

[13]. It was reported that the glass tube has pores with diameters of 3.4 nm and that the AZB derivative is mainly located on the outer glass surface. The influence of photoswitching on the gas flow is very low.

In this review we report on AZB adsorbed in molecular sieve membranes, in well-defined pore systems with pore diameters <1 nm [14,15]. For these host–guest membranes a higher influence of AZB trans–cis photoisomerization on the transport behavior of gases is expected because of the characteristic changes of its molecular size and dipole moment. The fundamentals of this novel concept of photoswitchable gas permeance across molecular sieve–AZB membranes are shown schematically in Fig. 1.

1.2.2 Requirements for Photoswitchable Molecular Sieve–AZB Membranes

The requirements for an effective mode of action of the photoswitchable host–guest membranes are: (1) The zeolitic host must be transparent and its pore system three-dimensional, which is the case for MFI and FAU; (2) densely intergrown MFI or FAU membrane layers must be synthesized, so that gas transport occurs exclusively through the regular zeolite pore system; (3) AZB must be adsorbed within the pores of MFI and FAU; (4) the incorporated *trans*-AZB must be reversibly photoswitchable to the cis configuration, which must have a sufficient lifetime for practical permeation measurements; and (5) measurement of the influence of the photoswitching on the transport properties of the molecular sieve–AZB membranes requires coupling of the permeation apparatus with a suitable irradiation device.

1.3 Characterization of Used Host–Guest Systems

Force-field-based Monte Carlo simulations were carried out to obtain information on a possible influence of gas transport by the trans–cis switching of the MFI–AZB or FAU–AZB membranes before making practical permeation measurements.

K. Hoffmann *et al.* investigated the photosensitive trans–cis isomerization of AZB incorporated in zeolite single crystals by UV/Vis spectroscopy at room temperature [16]. The spectra were taken in the wavelength region from 300 to 800 nm using a Microscope Spectral Photometer UMSP 80 (Carl Zeiss, Oberkochen) with a 75 W xenon lamp and a grating monochromator (band width 5 nm). The xenon lamp was also used for irradiating the sample; the diameter of the measurement spot on the microcrystals was 5 μm. This type of lamp produced an energy density of approximately 100 mW cm^{-2} behind the monochromator (band width 25 nm). After determination of the irradiation wavelengths for trans-to-cis (360 nm) and cis-to-trans switching (436 nm) and an exposure time of 3 min the photoinduced trans–cis isomerization of the zeolite-encapsulated AZB could be proven by UV/

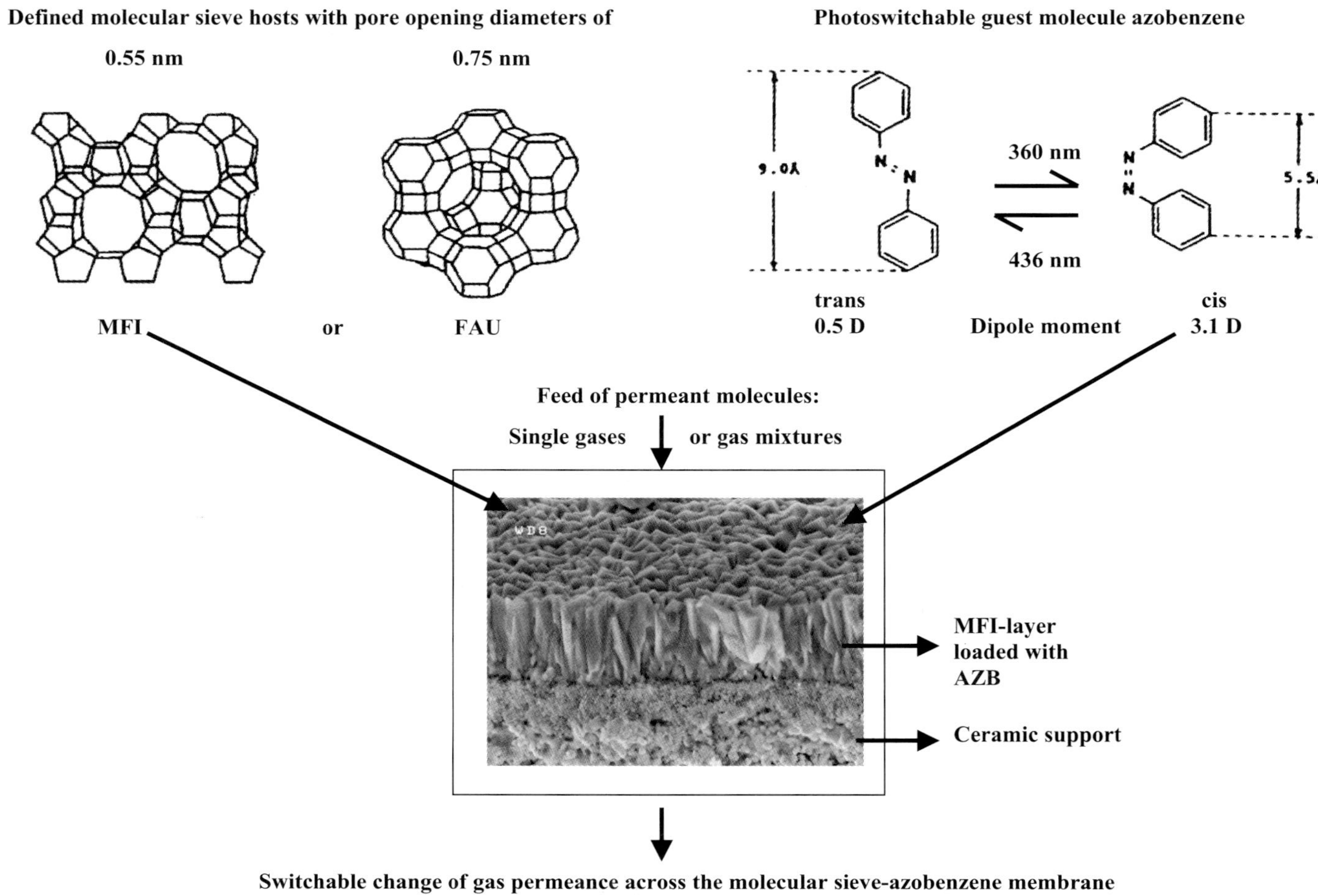

Fig. 1. Functional principle of photoswitchable molecular sieve–azobenzene membranes.

Vis spectroscopy. The large difference of 76 nm in the irradiation wavelengths allows easy photochemical switching of AZB. Furthermore, under these conditions, no byproducts of the photochemical treatment were detected. Based on these results we chose AZB as switchable guest molecule for molecular sieve host membranes.

1.3.1 Monte Carlo Simulations of the Free Pore Volume in the Host–Guest Systems MFI–AZB and FAU–AZB

For the host–guest system MFI–AZB energy-minimized structures assuming two different concentrations of AZB, namely, 1 and 4 molecules per unit cell (u.c.), were calculated with the aim of predicting the free volume for the unhindered diffusion of propane. AZB can influence the permeation of hydrocarbon molecules in different ways. First, it can be expected that AZB molecules occupy different volumes in their trans and cis configurations and thus alter the free volume for the diffusion of the hydrocarbon molecules. Second, *trans*- and *cis*-AZB may occupy different positions in the MFI framework: channel crossings and channel pore segments. Depending on the position, the efficiency of the AZB molecules as diffusion obstacles for permeant molecules can be significantly different [17,18]. The decrease in the self-diffusion coefficient in zeolites with increasing loading, i.e., decreasing free volume, was studied by NMR methods [19,20]. When four AZB molecules are adsorbed per u.c. of MFI, in the trans state, two of them are found in the straight channels, and two in the sinusoidal channels. In contrast, it can be predicted from energy-minimization calculations that *cis*-AZB molecules occupy all four channel intersections of the MFI u.c. In other words, both the number and the location of the AZB molecules contribute to switching the permeation of hydrocarbon molecules.

In a similar simulation, four molecules of AZB were assumed to reside in one u.c. of the host FAU. In this case, all AZB molecules are located inside the large supercages of the FAU structure, both in the trans and in the cis form. The free volume accessible for permeating hydrocarbons depends on the degree of pore filling with AZB and on whether AZB is present in the trans or cis configuration (Tab. 2). Furthermore, reduction of the free volume by the permeant itself (28.7×10^{-3} nm^3 for methane and 238.9×10^{-3} nm^3 for mesitylene) must be considered.

From the data of Tab. 2 two predictions can be made: (1) For both methane and mesitylene the free volume for diffusion in the presence of the trans form of AZB is about 6 % higher than that in the presence of the cis form. Therefore, we can predict a higher single-gas permeance in the trans state of the corresponding host–guest membrane than in the cis state. (2) In the trans form of encapsulated AZB, the difference in free volume for the permeant molecules methane and mesitylene is 408×10^{-3} nm^3. This difference is about 22 % higher in comparison with the difference of 317×10^{-3} nm^3 for the cis form of encapsulated AZB. From this finding one can expect better separation of binary gas mixtures in the trans state

Tab. 2. Accessible pore volume (10^{-3} nm^3) per u.c. for permeant molecules in the host–guest system FAU–AZB for the loading of 4 *trans*- or 4 *cis*-AZB per u.c. from Monte Carlo simulation. The loading with the permeant molecules methane and mesitylene, respectively, was one molecule per u.c.

Permeant	*FAU–trans-AZB*	*FAU–cis-AZB*
Methane	5880	5505
Mesitylene	5472	5183

than in the cis state of the FAU–AZB membrane. However, such predictions could only be made for the hypothetical cation-free SiO_2 framework of FAU, because calculations on the real FAU lattice were numerically too involved.

Kirschhock and Fuess [21] reported on different arrangements of *m*-dinitrobenzene in the polar NaY by use of diffraction and simulation methods. They came to the following conclusions: (1) Polar interactions are decisive. The cations tend to occupy positions in such a way that interaction of the highly polar nitro groups with as many cations as possible is ensured; and (2) steric factors play a role as well: in simultaneous adsorption of dinitrobenzene and aniline, an arrangement is achieved which minimizes steric hindrance between the framework and the guest molecules. Based on these models one can assume similar locations, especially of the more polar *cis*-AZB, in the framework of NaX.

NaX is more polar than NaY [26], and polar interactions are important for gas permeation. The polar *cis*-AZB and more polar permeant gases like the quadrupolar CO_2 [22] can undergo electrostatic interactions with the Na^+ ions at the sites SII and SIII inside the NaX cavities in addition to the steric factors that should lead to characteristic influences on the gas transport across the molecular sieve-AZB membranes.

The heats of adsorption of *trans*- and *cis*-AZB on MFI and FAU (frameworks assumed to consist of pure SiO_2) were determined by Monte Carlo simulation (Tab. 3).

It follows from the heats of adsorption in Table 3 that AZB is more strongly adsorbed in the narrow pores of MFI than in the wider pores of FAU, and that in both zeolites *cis*-AZB is more strongly adsorbed than *trans*-AZB. In comparison to these values the measured heats of adsorption of the permeant gases on silicalite-1 and NaX are about one order of magnitude lower (Tab. 4). In practical permeation

Tab. 3. Heats of adsorption (kJ mol^{-1}) of *trans*- and *cis*-AZB on MFI and FAU estimated by a Monte Carlo simulation assuming frameworks of pure SiO_2. Loading: 4 AZB per u.c.

Host–guest system	*Trans state*	*Cis state*
MFI–AZB	−218	−231
FAU–AZB	−139	−142

Tab. 4. Measured heats of adsorption at 300 K on silicalite-1 and NaX powders [23–26] and kinetic diameter of gas molecules [27].

Gas	*Heat of adsorption (kJ mol^{-1})*		*Kinetic diameter (nm)*
	Silicalite-1	*NaX*	
H_2		−6.8	0.289
CO_2	−27.2	−49.1	0.330
O_2	−16.3	−15.0	0.346
N_2	−17.6	−19.9	0.364
CH_4	−20.9	−19.2	0.380
n-C_4H_{10}	−50.3	−41.4	0.430
SF_6	−34.4	−28.2	0.550

measurements no desorption of AZB by the permeating gases was observed at room temperature.

CO_2 in NaX has the highest affinity of the gases listed in Tab. 4 because of strong ion–quadrupole interactions [28]. For nonpolar gases with larger kinetic diameters such as n-C_4H_{10} and SF_6 the heat of adsorption in the narrow pores of silicalite-1 increases. This is attributable to an increased dispersion energy due to the smaller pore size of the MFI framework.

1.3.2 Reversible Photoinduced Azobenzene Isomerization in the Host–Guest Systems MFI–AZB and FAU–AZB

In the pores of MFI molecular sieves two configurations of AZB exist that can be photochemically converted into each other on irradiation with light of different wavelengths [3]. Short-wavelength irradiation of the thermodynamically stable *trans*-AZB with light of $\lambda = 360$ nm at room temperature leads to the cis isomer, whereas long-wavelength irradiation with light of $\lambda = 436$ nm or heat regenerates the trans isomer. The process is associated with spectral changes in the UV/Vis region. After irradiation at 360 nm, the π–π^* absorbance around 300–320 nm is blue-shifted and decreases, whereas the n–π^* absorbance at 410 nm increases [29,30]. This reversible photochemical behavior has been found for AZB incorporated in NaX as well. Using AZB-loaded zeolite crystals as a model system, characteristic light-induced changes of the absorption spectra were detected by means of UV/Vis microscope spectroscopy on NaX single crystals, as shown in Fig. 2.

After short-wavelength irradiation of the incorporated *trans*-AZB trans-to-cis isomerization takes place on a timescale of a few minutes. Long-wavelength irradiation or a dark reaction without irradiation of the photostationary cis state regenerates the thermodynamically favored trans isomer. The cis-to-trans thermal relaxation of the metastable state was monitored by UV/Vis spectroscopy of the n–π^* absorption as a function of time. During microscope-spectroscopic investigations the thermal relaxation is complicated by the disturbance of the dark re-

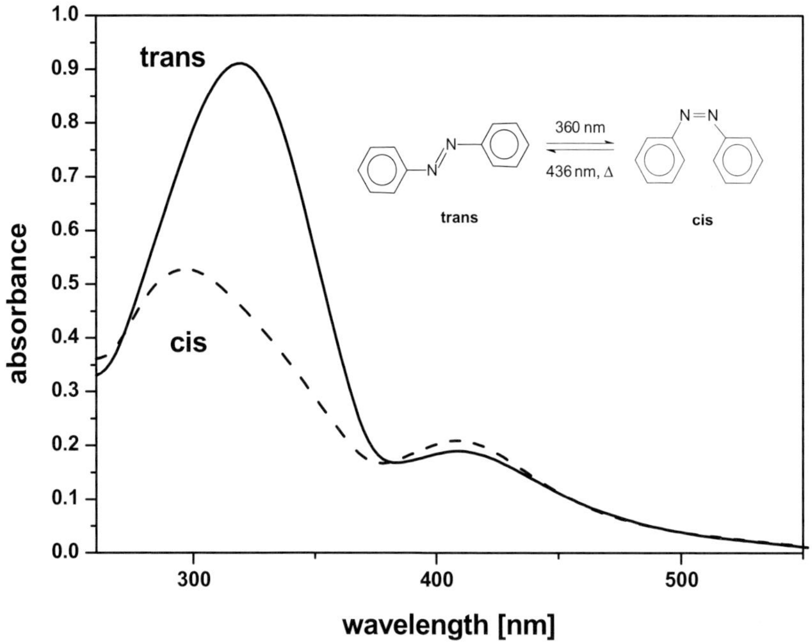

Fig. 2. UV/Vis spectrum of azobenzene incorporated in NaX after irradiation at 436 nm (solid line) and after irradiation at 360 nm (broken line) at room temperature.

action by the observation light [30]. Taking into consideration this influence on the thermal relaxation rate, the lifetime τ of metastable *cis*-AZB in NaX is found to be about 2 h [31]. In MFI it is about 9 h [30]. The reversibility of the photoisomerization of AZB within the pores of NaX follows from reversible photoinduced changes of the maximum absorption after alternating irradiation with light of 360 and 436 nm, as shown in Fig. 3.

The reversibility of photoswitching of MFI-encapsulated AZB [30] and of AZB chemically bound to polymethacrylate [9,10] has also been confirmed.

1.3.3
Preparation and Irradiation of FAU-AZB and MFI-AZB Membranes

FAU membranes of type NaX were prepared on α-Al_2O_3 supports with 200 nm pore diameter in the finest layer, coated with FAU nanocrystals (seeds). The following hydrothermal synthesis took place for 12 h at 363 or 373 K in a conventional oven or for 2 h at 393 K in a microwave oven [32]. The FAU membrane preparation and characterization by X-ray diffraction (XRD), field-emission scanning electron microscopy (FE-SEM), and energy dispersive X-ray analysis (EDX) are described in more detail in Ref. [33]. The α-Al_2O_3-supported MFI membrane was synthesized by a two-step mechanism using seed-coated α-Al_2O_3 supports, too.

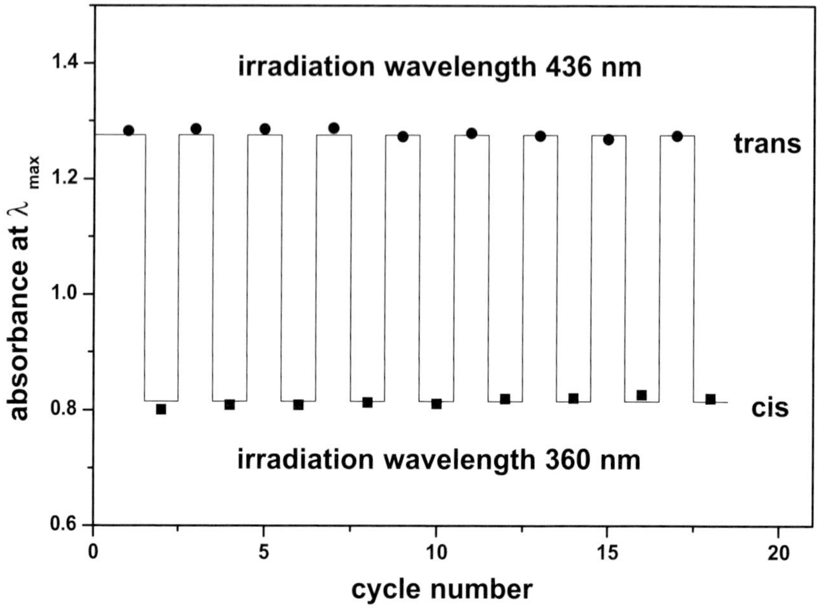

Fig. 3. Reversibility of the switching process of azobenzene in NaX under alternating short-wavelength (360 nm; squares) and long-wavelength (436 nm; circles) irradiation, monitored at λ_{max}(trans) = 320 nm and λ_{max}(cis) = 300 nm by UV/Vis spectroscopy at room temperature.

The hydrothermal synthesis was performed at 443 K. More details on the preparation and characterization of the MFI membranes are given in Refs. [34,35]. The membranes consisting of thin layers of dense intergrown zeolite crystals of type silicalite-1 (Si/Al = 90, layer thickness 10 μm) or of type NaX (Si/Al = 1.5, layer thickness 1.5 μm) on ceramic supports were used as starting materials for preparing photoswitchable molecular sieve–AZB membranes. The AZB was adsorbed into the zeolite pores by gas-phase diffusion, as described in Ref. [15]. A first indication of the successful incorporation of AZB into the zeolitic cavities was a change in the color of the membranes from white to yellow-orange. The adsorption of the AZB was quantitatively determined by thermal gravimetric analysis (TGA) of MFI and FAU crystals in powder form, loaded with AZB under the same conditions as the membranes. The loading of MFI with AZB for an adsorption time of 16 h at 393 K led to an AZB content of 3.5 ± 0.5 wt %, which corresponds to about 1–1.3 AZB per u.c.. Loading of FAU at 343 K for 3 h resulted in an AZB content of 10 ± 0.8 wt %, which corresponds to about eight molecules of AZB per u.c. Immediately after AZB loading, the membranes were mounted into a full-metal permeation cell and sealed with Viton rings. The cell allowed the *in situ* irradiation of the switchable host–guest composite membranes before and during the permeation measurements. To prevent noticeable desorption of AZB and to minimize

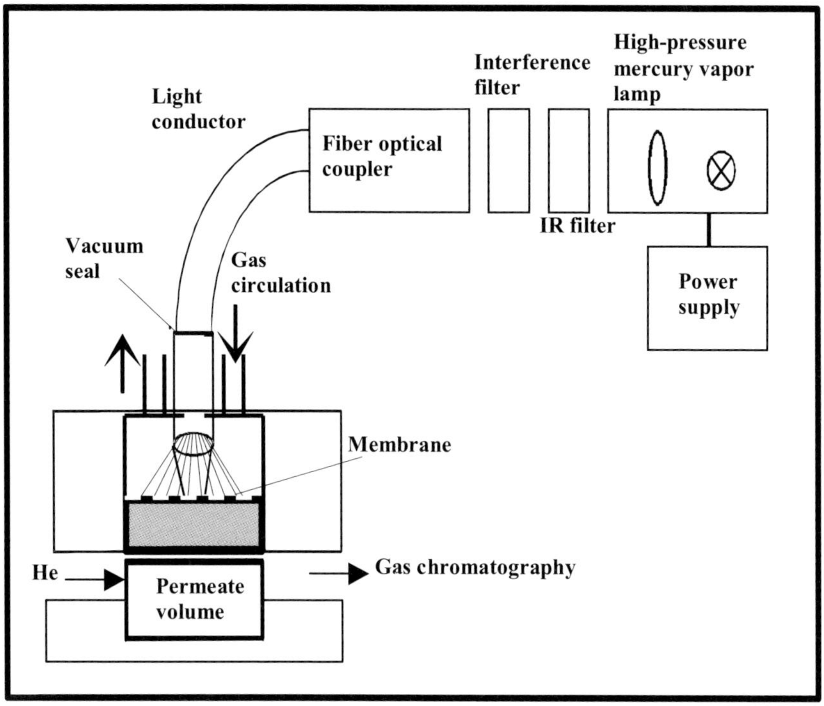

Fig. 4. Setup of the irradiation device linked to a permeation apparatus.

thermal relaxation of the metastable *cis*-AZB the permeation measurements were carried out at room temperature. The membranes were irradiated with a high-pressure mercury vapor lamp (HBO 100 W), and the required irradiation wavelengths of 360 and 436 nm were selected with two exchangeable interference filters. The illumination source was connected to the permeation cell by flexible fiber optics. The irradiated membrane area was 19.6 mm^2, which corresponded to a circular spot of 5 mm diameter. Due to optical and spectral losses a light intensity of 20 mW was available for this area. This resulted in an irradiation duration of 10–25 min before and during the permeation measurements. The arrangement of the irradiation device linked to the permeation apparatus is shown in Fig. 4.

1.4 Results and Discussion

Essential properties of molecular sieve membranes are the permeances and the selectivities. The experimental conditions of permeation measurements on single gases and binary gas mixtures are described in Refs. [15,33]. The permeances of single gases through the MFI–AZB membrane in the *trans* state follow the sequence $H_2 > CO_2 > O_2 > N_2 > CH_4 > n\text{-}C_4H_{10}$, whereas in the FAU–AZB membrane the sequence is $H_2 > CH_4 > N_2 > O_2 > CO_2 > n\text{-}C_4H_{10} > SF_6$. Both adsorption and diffusion processes control the net mass transport through the membranes, depending on the relative size of permeant molecules, the mean free volume in the pores, and the polarities of the permeant molecules and the framework. Since the silicalite-1 membrane is nonpolar, the permeance sequence is determined solely by the kinetic molecular diameter of the gases. The NaX membrane has an electrostatic potential, and the accessible SII Na^+ ions located near the six-rings interact with polar molecules. On the other hand, CO_2 has a quadrupole moment [22] and can form adsorption complexes with the SII Na^+ ions like other polar molecules [36]. Thus, although CO_2 has a smaller diameter its diffusion is hindered by electrostatic interactions. These interactions also result in a higher heat of adsorption of CO_2 in NaX compared to those of the nonpolar gases (see Tab. 4). The separation factors of equimolar binary gas mixtures in the AZB-free NaX membrane are a further indication for electrostatic interactions. Whereas the separation factor α_{N_2/CH_4} of the mixture of unpolar gases N_2 and CH_4 is 3.3, for the mixture of the nonpolar N_2 with quadrupolar CO_2, α_{N_2/CO_2} is higher, namely, 8.4 [33].

1.4.1
Switchable Single-Gas Permeance Across MFI–AZB and FAU–AZB Membranes

Figure 5a shows that the transport of the smallest gas molecule H_2 was practically not influenced by the photochemical trans–cis switching of the silicalite-1-AZB membrane. The other gases, with higher kinetic molecular diameter, were more strongly inhibited in their transport by cis switching. After switching back into the thermodynamically stable trans state, the gas permeances increased to the former values in all the cases. In the silicalite-1–AZB membrane the increase of the switching effect parallels the increasing kinetic gas diameters. The differences between the permeances on trans–cis switching were highest for CH_4 and $n\text{-}C_4H_{10}$.

Across the NaX–AZB membrane the small H_2 molecules were also practically not influenced by trans–cis switching (Fig. 5b). As expected, the other gas permeances were higher in the trans state. However, because of the heteropolar centers within the framework of the NaX–AZB membrane, electrostatic forces act in addition to dispersion forces. Therefore, the switching effect of this membrane did not correlate with the kinetic molecular diameters of the permeant gases. The quadrupolar CO_2 and polarizable gases like CH_4, $n\text{-}C_4H_{10}$, and SF_6 can electrostatically interact with the Na^+ ions inside the supercages of NaX. The polar *cis*-AZB is also expected to interact with the accessible Na^+ ions. Furthermore, dipole–dipole interactions of *cis*-AZB can further increase the diffusion barrier inhibiting gas transport through the NaX pores in additional to the fact that *cis*-AZB is bulkier than *trans*-AZB. Therefore, this membrane had the highest trans–cis selectivities. The calculated trans–cis selectivities $S_{trans/cis}$, that is, the ratio of the permeances in the trans and cis states of the photoswitchable host–guest membranes are listed in Tab. 5.

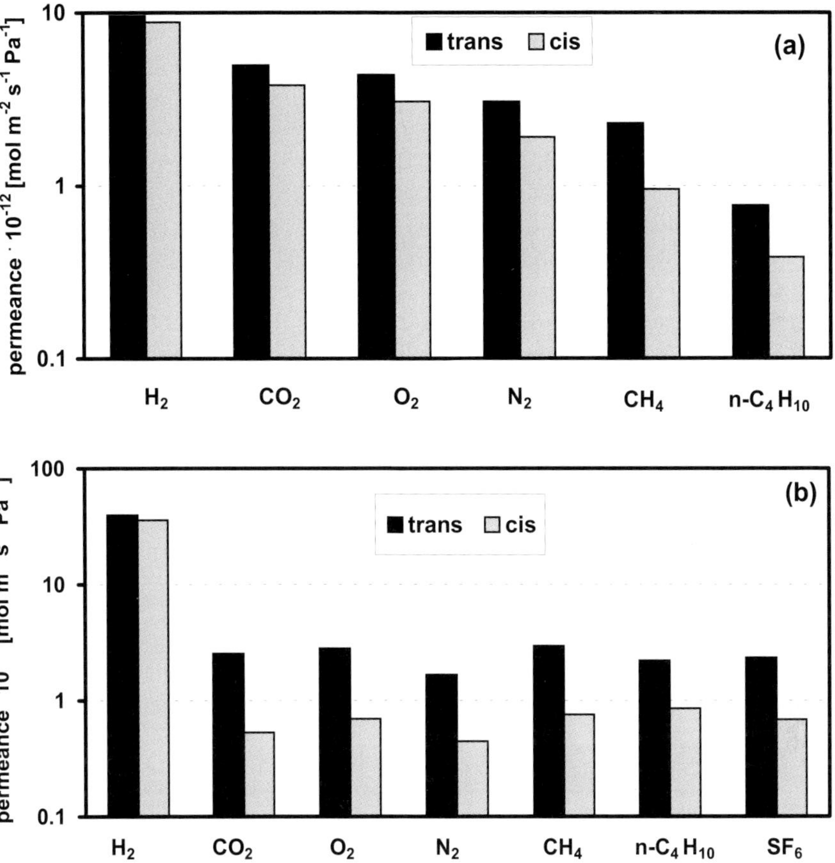

Fig. 5. Single-gas permeances on trans–cis photoswitching at room temperature and $\Delta p = 1$ bar across the silicalite-1-AZB membrane (a) and the NaX–AZB membrane (b).

Tab. 5. Calculated selectivities $S_{trans/cis}$ as ratio permeance$_{trans}$/permeance$_{cis}$ at room temperature of the photoswitchable MFI–AZB and FAU–AZB membranes in comparison to the PM–AZB membrane [9,10] and glass–AZB membrane [13].

Permeant gas	*MFI–AZB*	*FAU–AZB*	*PM–AZB*	*Glass–AZB*
He				1.022
H_2	1.1	1.1	1.1	
CO_2	1.3	4.8		
O_2	1.4	4.1		
N_2	1.6	3.7		1.015
CH_4	2.4	3.9	1.7	
n-C_4H_{10}	2.0	2.6	1.3	
SF_6		3.4	4.0	

In the nonpolar MFI–AZB membrane the trans–cis selectivity increased with increasing gas molecular diameter, but in the polar FAU–AZB membrane the electrostatic interactions are predominant. Thus, the FAU–AZB membrane has the highest trans–cis selectivity of 4.8 for the quadrupolar CO_2, and the trans–cis selectivities were higher overall, except for H_2. For comparison, the PM–AZB membrane, in which AZB is chemically bond to the polymer matrix [9,10], had lower trans–cis selectivities for CH_4 and n-C_4H_{10} than both the molecular sieve–AZB membranes. The glass membrane modified with the organic AZB derivative 11-[4-{(4′-hexylphenyl)azo}phenoxy]undecanoic acid [13] had only a minimal switching effect on the gas flow. The AZB derivative was mainly located on the outer surface of the porous glass tube and not in the pores of 3.4 nm diameter. Considering this fact and the interplay between the molecular size of the azo derivative and the pore size of the glass, gas flow only could be influenced very weakly by photoswitching.

The trans–cis photoswitching of the molecular sieve–AZB membranes and the associated changes in the gas permeances was reversible over at least 60 switching cycles.

As an example, Fig. 6 shows the reversible change in CH_4 permeance during trans–cis photoswitching of the NaX–AZB membrane, in a manner similar to that confirmed by UV/Vis investigations (cf. Fig. 3).

The switching effect of the zeolite–AZB membranes is mainly caused by the differences in size and polarity of the two AZB isomers. The guest molecule *cis*-AZB is bulkier and more polar than *trans*-AZB. The higher permeances of the

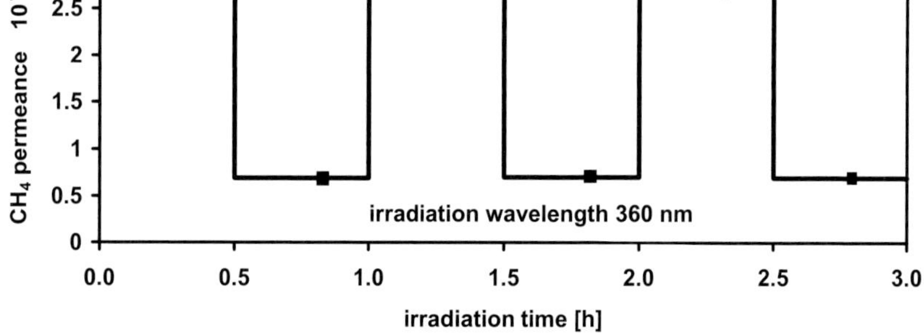

Fig. 6. Reversibility of the permeance of CH_4 under alternating trans (436 nm; circles) and cis photoswitching (360 nm; squares), monitored by single-gas permeance measurements at room temperature and $\Delta p = 1$ bar across the NaX–AZB membrane.

permeant gases in the trans state of the membranes due to a larger free volume were predicted by the Monte Carlo simulation.

1.4.2 Switchable Gas-Mixture Permeance across the NaX Membrane

In the single-gas permeance measurements it was found that trans–cis photoswitching of zeolite-encapsulated AZB can modulate the permeance of a permeating gas by a factor ≤ 4.8 (cf. Tab. 5). Separation of the equimolar gas mixtures CH_4/CO_2 and N_2/CO_2 across the NaX–AZB membrane was measured in the same combined irradiation/permeation apparatus at room temperature. The gas mixtures were fed over the membrane while repeated trans–cis switching cycles were carried out. The irradiation was carried out for 1–2 h either with light of 360 nm for shifting the photochemical balance in the direction of the cis form or with light of 436 nm in direction of the trans form of the NaX-hosted AZB. The measurements were carried out in the same way as described in Refs. [15,33] by using an on-line capillary gas chromaograph (GC) equipped with a thermal conductivity detector (TCD). The measured separation factors α (real selectivities) were compared with the corresponding permselectivities PS (ideal selectivities) that were calculated as the ratios of the corresponding single-gas permeances in the trans and cis states of the FAU–AZB membrane (Tab. 6).

The separation factors α were calculated as $\alpha_{(1/2)} = (x_1/x_2)_{\text{permeate}}/(x_1/x_2)_{\text{retentate}}$, where x is the mole fraction of gases 1 and 2.

On the NaX–AZB membrane, the separation factors of the equimolar gas mixture CH_4/CO_2 are generally lower than those of N_2/CO_2. CH_4 can be polarized by, e.g., Na^+ ions, *cis*-AZB, or traces of H_2O, and its diffusion could be inhibited like that of the quadrupolar CO_2. This, however, is not the case for the nonpolar N_2. Therefore, trans–cis photoswitching was more effective for the separation of the equimolar mixture N_2/CO_2. In the trans state the separation factor α_{N_2/CO_2} was 49, and decreased to 32 in the cis state. The polar *cis*-AZB ($\mu = 3.1$ D) can intensify its interaction with the Na^+ ions accessible in the supercages of NaX. In addition, *cis*-AZB is bulkier and blocks more pore volume. Therefore, as seen from the single-gas permeances, the transport of all permeant gases, including N_2 and CH_4, was more strongly inhibited by *cis*-AZB. Thus, for both gas mixtures under study the separation factors α were lower after cis switching, and better gas separation took place after trans switching of the NaX–AZB membrane. This was also predicted by the Monte Carlo simulation.

Tab. 6. Separation factors α of equimolar gas mixtures across the NaX–AZB membrane with trans–cis photoswitching at room temperature and $\Delta p = 0$ bar and the corresponding permselectivities PS.

NaX–AZB	PS_{N_2/CO_2}	α_{N_2/CO_2}	PS_{CH_4/CO_2}	α_{CH_4/CO_2}
Trans	1.33	49	1.74	3.30
Cis	1.17	32	1.70	1.90

1.5
Summary

The strongest reversible changes in gas permeance by photoswitching reported in the literature up to now were measured across the selected host–guest membranes of type MFI–AZB and, especially, of type FAU–AZB. The most important conditions for the functionality of the photoswitchable zeolite–AZB membranes were: (1) The synthesis of densely intergrown, permselective zeolite layers of types silicalite-1 and NaX; (2) the adsorption of AZB within the three-dimensional zeolite pore systems; (3) reversible trans–cis photoswitchability of the adsorbed AZB and optimal irradiation conditions and (4) sufficient lifetime of the metastable *cis*-AZB in the photoswitchable zeolite–AZB membranes.

All these conditions could be fulfilled. Photoswitchable host–guest membranes of types silicalite-1–AZB and NaX–AZB allowed reversible changes of single-gas permeances and of separation factors for equimolar binary gas mixtures. Photoswitching is performed by alternating irradiation with light of two suitable wavelengths (360 and 436 nm) in a combined permeation/irradiation apparatus. UV/Vis spectroscopic investigations confirmed the reversible trans–cis photoswitching of FAU- and MFI-encapsulated AZB over numerous switching cycles. A Monte Carlo simulation predicted different accessible pore space volumes for the trans and cis states of AZB. In the case of *trans*-AZB a higher free volume was available for the permeating gas molecules compared with the cis isomer. This theoretical forecast was in good agreement with the permeation measurements, which showed an increase in the single-gas permeances in the trans state of the zeolite–AZB membranes in comparison with the cis state. The trans–cis switching and the associated change in the gas permeance was reversible over many cycles. In contrast to the silicalite-1–AZB membrane, the photoinduced switching effect was more effective for the NaX–AZB membrane because of the higher AZB content and the heteropolar surfaces of the FAU framework that allow adsorptive interactions with both AZB and the permeant molecules. The Monte Carlo simulation predicted a higher separation factor for a binary gas mixture in the trans form of FAU-hosted AZB than in the cis form. This was also confirmed by the permeation measurements. Equimolar mixtures of N_2/CO_2 and CH_4/CO_2 had higher separation factors on trans switching of the NaX–AZB membrane than on cis switching. The measured real selectivity for the mixture of the nonpolar N_2 and the quadrupolar CO_2 is 36 times higher in the trans and 27 times higher in the cis state than the corresponding calculated permselectivities, obtained as the ratio of the single-gas permeances of N_2 and CO_2. This is a proof of the strong interaction of this gas mixture with the photoswitchable NaX–AZB membrane. The change in permeation properties by photoinduced trans–cis switching of the molecular sieve–AZB membranes was more evident for permeant gases with higher kinetic diameter like CH_4 and n-C_4H_{10} in the silicalite-1-AZB membrane. In the NaX–AZB membrane adsorption-controlled separation dominated. Therefore, in this membrane the change in the permeation properties by photoinduced trans–cis switching was more evident for quadrupolar and polarizable permeant gases.

Acknowledgements

The authors acknowledge financial support by the DFG (No 322/1-5). We thank J. Caro for helpful discussions, K. Hoffmann and F. Marlow for their cooperation in spectroscopic investigations, J. Sauer and K.-P. Schröder for enabling performance of the Monte Carlo simulation, and I. Sieber and P. Toussaint for analytical and technical assistance.

References

1 F. Mizukami in I. Kiricsi, G. Pal-Borbely, J.B. Nagy, H.G. Karge (Eds.), Porous Materials in Environmentally Friendly Processes, Studies in Surface Science and Catalysis, Vol. 125, Elsevier Science, Amsterdam, **1999**, pp. 1–12.

2 V. Ramamurthy, Chimia 46, **1992**, 359.

3 H. Rau in J.F. Rabek (Ed.), Photochemistry and Photophysics, Vol. 4, CRC Press, Boca Raton, Florida, **1990**, Chap. 4, pp. 120–141.

4 M. Höfer, Transport durch biologische Membranen, Verlag Chemie, Weinheim, **1977**, pp. 1–50.

5 D. Balasubramanian, S. Subramani, C. Kumar, *Nature* **1975**, *254*, 252.

6 Y. Okahata, H. Lim, S. Hachiya, *J. Chem. Soc. Perkin Trans. 2* **1984**, 989.

7 H. Tachibana, T. Nakamura, M. Matsumoto, H. Komizu, E. Manda, H. Niino, A. Yabe, Y. Kawabata, *J. Am. Chem Soc.* **1989**, *111*, 3080.

8 J. Anzai, T. Osa, *Tetrahedron* **1994**, *50*, 4039.

9 K. Weh, M. Noack, R. Ruhmann, K. Hoffmann, P. Toussaint, *Chem. Ing. Tech.* **1998**, *70*, 718.

10 K. Weh, M. Noack, R. Ruhmann, K. Hoffmann, P. Toussaint, J. Caro, *Chem. Eng. Technol.* **1998**, *21*, 408.

11 Y.J. Choi, T. Yamaguchi, S. Nakao, *Ind. Eng. Chem. Res.* **2000**, *39*, 2491.

12 C.R. Martin, M. Nishizawa, K. Jirage, M. Kang, S.B. Lee, *Adv. Mater.* **2001**, *13*, 1351.

13 T. Jin, A.H. Ali, T. Yazawa, *Chem. Commun.* **2001**, 99.

14 M. Noack, P. Kölsch, J. Caro, K. Weh, German Patent 19 952 725, granted 17.05.2001.

15 K. Weh, M. Noack, K. Hoffmann, K.-P. Schröder, J. Caro, *Microporous Mesoporous Mater.* **2002**, *54*, 15.

16 K. Hoffmann, F. Marlow, J. Caro, *Zeolites* **1996**, *16*, 281.

17 C. Förste, A. Germanus, J. Kärger, H. Pfeifer, J. Caro, W. Pilz, A. Zikanova, *J. Chem. Soc. Faraday Trans. 1* **1987**, *83*, 2301.

18 J. Caro, M. Bülow, J. Kärger, H. Pfeifer, *J. Catal.* **1988**, *114*, 186.

19 V. Kukla, J. Kornatowski, D. Demuth, I. Girnus, H. Pfeifer, L.V. Rees, S. Schunk, K.K. Unger, J. Kärger, *Science* **1996**, *272*, 702.

20 J. Kärger, D.M. Ruthven, Diffusion in Zeolites and Other Microporous Solids, Wiley, New York, **1992**, pp. 427–512.

21 C. Kirschhock, H. Fuess, *Zeolites* **1996**, *17*, 381.

22 J.E. Spice, Chemical Bonding and Structure, Leipzig (1971), Akademische Verlagsgesellschaft, Geest & Portig K.-G., Lizenz des Verlages Friedr. Vieweg & Sohn, Braunschweig, pp. 259–261.

23 J.A. Dunne, M. Rao, S. Sircar, R J. Gorte, A.L. Myers, *Langmuir* **1996**, *12*, 5896.

24 H. Stach, Promotion B: Experimental and Theoretical Investigations about the Adsorption Equilibrium of Unpolar and Polar Molecules on Zeolites of Type FAU, Academy of Science of GDR, Berlin, **1975**, pp. 70–72.

25 M.A. Levin, V.V. Serpinsy, T.S. Jakubov, A.A. Isirikyan, M.B. Gorbunov, Academy of Sciences of the GDR, Reprints of the Workshop

III, Adsorption in Microporous Adsorbents, **1987**, *2*, pp. 66–73.
26 W. Zhu, Proefschrift, Adsorption and Diffusion in Microporous Materials: An Experimental Study with the TEOM, Technische Universität Delft, **2001**, p. 35.
27 D. W. Breck, Zeolite Molecular Sieves, Wiley, New York, **1974**, pp. 64–351.
28 R. M Barrer, R. M. Gibbson, *Trans Faraday Soc.* **1965**, *61*, 948.
29 K. Hoffmann, F. Marlow, J. Caro, *Adv. Mater.* **1997**, *9*, 567.
30 K. Hoffmann, U. Resch-Genger, F. Marlow, *Microporous Mesoporous Mater.* **2000**, *41*, 99.
31 K. Hoffmann, U. Resch-Genger, F. Marlow, Chap. 2 of this part.
32 M. Noack, P. Kölsch, K. Weh, R. Schäfer, J. Caro, German Patent 100 27 685.7, assigned to Institut für Angewandte Chemie Berlin-Adlershof e. V., 26.05.2000.
33 K.Weh, M. Noack, I. Sieber, J. Caro, *Microporous Mesoporous Mater.* **2002**, *54*, 27.
34 M. Noack, P. Kölsch, J. Caro, M. Schneider, P. Toussaint, I. Sieber, *Microporous Mesoporous Mater.* **2000**, *35–36*, 253.
35 M. Noack, P. Kölsch, R. Schäfer, P. Toussaint, I. Sieber, J. Caro, *Microporous Mesoporous Mater.*, **2001**, *49*, 25.
36 P.J. O'Malley, C.J. Braithwaite, *Zeolites* **1995**, *15*, 198.

2
Photosensitive Optical Properties of Zeolitic Nanocomposites

*Katrin Hoffmann, Ute Resch-Genger, and Frank Marlow**

2.1
Introduction

Composites of molecular sieve crystals and organic guests have been widely investigated for advanced, especially optical, applications in the area of materials research. [1,2,3] The nanosized channel system of molecular sieves provides a size- and shape-selective matrix for organizing and orienting adsorbates on the nanometer scale. The constrained environment strongly influences optical and photochemical properties of the adsorbed guest molecules. The alignment of guest molecules results in dichroic and birefringent properties. The present study describes the postsynthetic encapsulation and microspectroscopic characterization of large molecular sieve-based composite crystals with a special focus on materials for photosensitive birefringence switching caused by intrazeolitic isomerization of azo-dye-containing nanocomposites. Investigations on the efficient reversible optical switching process are presented, and the influence of host–guest composition on switching parameters is described in detail.

In advanced optical nanodevices, the requirements on molecular sieve crystals as hosts for functional guest molecules are usually quite different from those for classical applications, especially with regard to their macroscopic size and shape. There has been considerable research effort dedicated to the synthesis of large, optically perfect zeolite crystals. Large crystals of molecular sieves of IUPAC structure types [4] AFI [5], MFI [6], and FAU which are optically clear in the ultraviolet and visible spectral region were used for the present optical investigations on photosensitive nanocomposites. The octahedral crystals of the three-dimensional FAU network consist of 6.6 Å sodalite cages and supercages of 11.8 Å diameter. The MFI structure forms an anisotropic three-dimensional channel network with straight channels (about 5.5 Å) parallel to [010], crosslinked by sinusoidal channels extending in the [100] direction of the prismatic crystals. The straight pore system of the AFI structure with 7.3 Å circular pores extends parallel to the length axis of the hexagonal crystal.

In most cases, optical applications require the construction of guest–host arrangements by incorporating functional guest molecules. The modification of the crystals must be carried out without affecting the inorganic framework. Crystallization inclusion of functional guest species during synthesis is one way to obtain extraction-stable composite materials without performing any postsynthetic treatment. [7,8,9] Ion exchange [10,11], *in situ* guest synthesis [12], and adsorption from gas or liquid phase [13,14,15] also allow insertion of guests into the pores of large zeolite single crystals.

In the present study, composite systems were obtained by adsorption of organic molecules with dimensions which fit to the pores of molecular sieves. After calcination of the zeolitic material, necessary to remove structure-directing organic templates, guest molecules were introduced from the vapor phase (e.g., Ref. [16]) or from solution [17]. To prevent co-adsorption of the solvent during adsorption from the liquid phase, we used sterically demanding 1,3,5-triisopropylbenzene as solvent, which is too large to enter the pore systems itself. By this method, various polar and nonpolar functional molecules were incorporated, especially in $AlPO_4$-5 single crystals [18,19].

2.2 Characterization of Nanocomposites by Polarization-Dependent UV/Vis Spectroscopy

Developments in nonclassical applications of zeolites have raised the necessity of characterizing organic/inorganic hybrid materials. For this reason, UV/Vis spectroscopy has been adapted to studying the structure of large zeolitic host–guest composite crystals and intrazeolite chemistry [20,21].

For visual microscopic investigations on zeolitic composites [22], polarization-dependent microspectroscopy allows the alignment of guest molecules with respect to the direction of molecular sieve pores to be determined and the homogeneity of the guest molecule distribution to be probed. In addition, optical spectroscopy supplies valuable information about the content of guest molecules, the influence of the environment, and the properties of electronic states.

2.2.1 Alignment of Guest Molecules

The alignment of adsorbed organic molecules constrained by molecular sieve pores can be investigated by polarization-dependent UV/Vis spectroscopy on large individual molecular sieve crystals. Using linearly polarized light makes the anisotropic light absorption of composite crystals visible, as shown in Fig. 1 for an anisotropic arrangement of guest molecules in the restricted geometry of the host lattice. If the dipole moment of the electronic transition of incorporated rod-shaped molecules is parallel to the electric field vector of the polarized light, the crystals strongly absorb light, and a dark color results. For other orientations of the dipole moment, the crystals are colorless or only slightly colored, depending on the degree of alignment of the guest molecules. Anisotropic light absorption sensitively

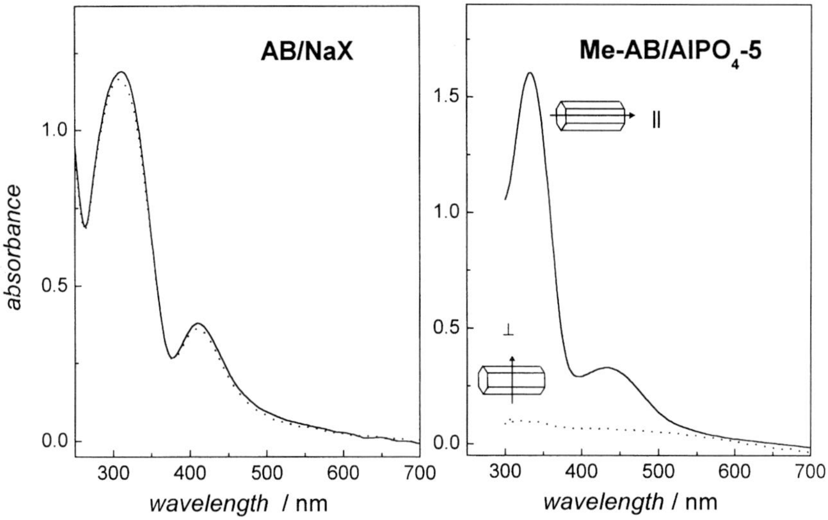

Fig. 1. UV/Vis absorption spectra of azo dyes encapsulated in isotropic and anisotropic molecular sieves, recorded with horizontally (solid line) and vertically polarized light (dotted line). In the case of $AlPO_4$-5 these directions correspond to polarizations parallel and perpendicular to the hexagonal *z*-axis (signs $\|$ and $\perp$, respectively). Left: isotropic absorbance of azobenzene (AB) in supercages of NaX. Right: anisotropic absorbance of 4-methylazobenzene (Me-AB) incorporated in the one-dimensional pore system of $AlPO_4$-5.

indicates guest incorporation within the host, in contrast to the adsorption of molecules on outer surfaces.

A small value of the dichroic ratio *d*, which is defined as the ratio of absorbencies perpendicular and parallel to the crystal's length axis (Eq. 1)

$$d = A_{\perp}/A_{\|} \tag{1}$$

indicates a pronounced optical anisotropy of composite crystals, with a strong absorbance parallel to the straight channels of $AlPO_4$-5 and weak absorbance perpendicular to this direction (Fig. 1). If, in the spectral range considered, only a single transition moment of the guest molecules with a polarization along the molecular long axis is relevant, the tilt angle φ of adsorbate molecules with respect to the one-dimensional molecular sieve pores can be calculated from the dichroic ratio (Eq. 2)

$$\tan^2 \varphi = 2d \tag{2}$$

This method was used to determine a mean tilt angle of $\varphi = 16^\circ$ for the system *p*-nitroaniline/$AlPO_4$-5 [20].

2.2.2 Guest Content of Nanocomposites

Information on the adsorbate content of zeolitic nanocomposites can usually be obtained from thermogravimetric investigations [23]. It can also be determined by spectrophotometric methods, either directly or indirectly. In the first case, the host crystal is dissolved in acidic solution and, after neutralization and centrifugation, the concentration of the dye is determined spectrophotometrically in the supernatant [24]. Alternatively, an indirect method relies on a gas-phase loading process carried out with a well-defined host–guest ratio and followed by the removal of unincorporated dye molecules by Soxhlet extraction. The dye concentration in the collected Soxhlet extracts, determined by UV/Vis spectroscopy, allow conclusions to made about the completeness of loading [25]. By all these methods, only the total loading of a batch can be obtained. The dissolution method is destructive and limited to organic compounds with sufficient stability to aqueous acids.

Nondestructive characterization of individual molecular sieve crystals is possible by using microspectroscopic techniques. The dye concentration c of guest molecules can be determined according to the the Lambert–Beer law (Eq. 3).

$$A_{\text{iso}} = \varepsilon \cdot c \cdot l \tag{3}$$

with some preconditions [20]. Here, ε denotes the molar absorptivity of the dye, l is the thickness of the molecular sieve crystal, and A_{iso} is the calculated absorbance in a hypothetical isotropic solution. The molecules in the channels of molecular sieves are, however, not isotropic, but aligned by the channels. Assuming that they are tilted with respect to the pore direction by a certain mean angle φ and isotropically distributed perpendicular to this direction, the isotropic absorbance can be calculated from polarization-dependent absorption spectra by using Eq. (4).

$$3A_{\text{iso}} = A_{\|} + 2A_{\perp}. \tag{4}$$

It turned out that the determination of guest molecules by microspectroscopy is especially suitable for well-shaped zeolite single crystals with low chromophore concentration [21].

2.2.3 Birefringence of Nanocomposites

Refractive index and birefringence are relevant parameters of materials for optical applications. To investigate refractive indices of molecular sieve nanocomposites, a microcrystal prism method has been developed. [17] For unloaded samples of $AlPO_4$-5 molecular sieves a very small difference in refractive indices for different polarizations was found (Eq. 5).

$$n' = n_{\text{e}} - n_{\text{o}} \tag{5}$$

The small birefringence (<0.005) of unloaded $AlPO_4$-5-crystals [17], however, is drastically enhanced in dye-loaded samples (up to 0.1).

The birefringent properties of molecular-sieve crystals can easily be visualized between two crossed polarizers under a microscope. The crystals appear in artificial colors due to the interference of the ordinary and the extraordinary light waves, which pass through the crystal with different phase velocities. Using this optical effect, one can quantitatively determine the birefringence of zeolitic host–guest materials by measuring the transmittance of the crystal in this setup [19]. The transmittance of a birefringent crystal depends on the phase shift and interference between ordinary and extraordinary light waves. From the oscillatory behavior of the wavelength-dependent transmittance of a nanocomposite crystal, the phase shift δ can be determined, which is connected to the birefringence n' by Eq. (6)

$$\delta = 2\pi \cdot l \cdot n'/\lambda \tag{6}$$

where l is the crystal thickness. This spectroscopic method allows the determination of the birefringence of dye-loaded molecular sieves with an accuracy of 10^{-3}. It has been applied to birefringent AFI- and MFI-based composites [19].

2.2.4 UV/Vis Spectroscopic Properties of Zeolite-Encapsulated Guest Molecules

For optical applications of host–guest composites both the orientation and the electronic states of adsorbed molecules are important. In addition to the molecular orientation, UV/Vis absorption bands can provide information on the nature of electronic states. Intermolecular interactions and the microenvironment of included functional molecules influence the position and intensity of absorption bands.

Microspectroscopic investigations on azo dyes in large crystals of $AlPO_4$-5 revealed a number of characteristic spectroscopic properties [18,19,26]. Zeolite-encapsulated azobenzene-type [27] molecules show an intense short-wavelength π–π^* band and a weak longer wavelength n–π^* transition, as shown in Fig. 2 (top). Aminoazobenzene-type [27] and push–pull-substituted pseudostilbene-type molecules [27] are characterized by a long-wavelength π–π^* bands, because the sequence of the n–π^* and π–π^* states on the energy scale is reversed in comparison to azobenzene-type molecules. The increased charge-transfer character of the band causes a strong dependence of the band position on the solvent polarity (Fig. 2, bottom). The bathochromic shift of the spectra of $AlPO_4$-5-encapsulated azo dyes compared to the absorption in hydrocarbon solution indicates a strongly polar environment of the dyes. Because of the hydrophilic character of the $AlPO_4$-5 host framework, the spectroscopic properties of the postsynthetically incorporated azo dye guest molecules seem to be dominated by the polar environment of coadsorbed water.

Since not only absorption bands but also fluorescence bands directly correspond to optical transitions between electronic states of adsorbed organic molecules, flu-

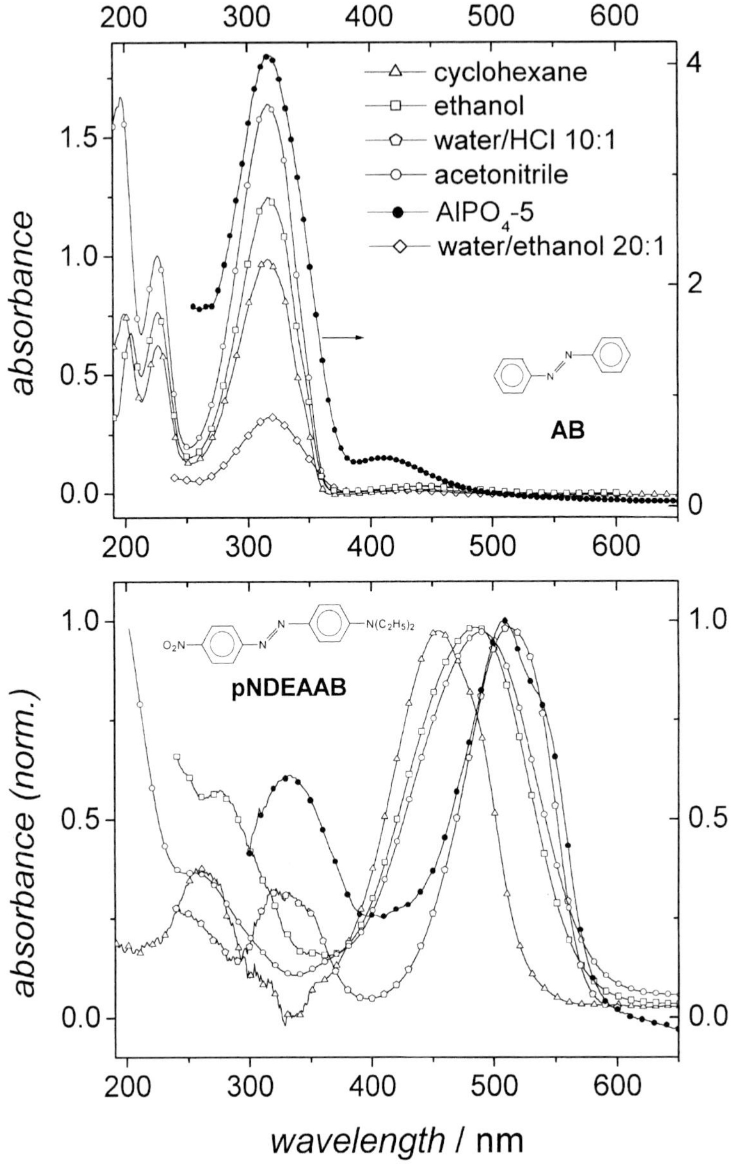

Fig. 2. Absorption spectra of azobenzene (top) and 4-diethylamino-4′-nitroazobenzene (pNDEAAB, bottom) in various solvents (spectra labeled with open symbols) and incorporated in large crystals of $AlPO_4$-5 (spectra labeled with solid circles).

orescence spectroscopy of microcrystals provides information on the nature of electronic states, intermolecular interactions between functional molecules, and the influence of the environment of zeolite-encapsulated molecules [14,28,29,30,31]. As an example, the fluorescent organic dye 2,2′-bipyridyl-3,3′-diol ($BP(OH)_2$) was investigated in pores of $AlPO_4$-5. $BP(OH)_2$ shows polarity- and proticity-dependent spectroscopic features, well-characterized in solution, which should allow the nature of the interior of $AlPO_4$-5 to be probed. The position of both the absorption and emission bands obtained by microscope-spectroscopic investigations on individual $BP(OH)_2$-containing $AlPO_4$-5 crystals, as well as the resulting Stokes shift of 6300 cm^{-1}, are similar to those of the dye in aqueous, weakly protic media. The spectroscopic features of incorporated $BP(OH)_2$ are likely influenced by host–guest interaction with a few relatively weak Brønsted acid sites of $AlPO_4$-5, but are mainly affected by co-adsorbed water molecules within the hydrophilic $AlPO_4$-5 framework [32].

2.3 Opto-Optical Switching of Azo Dye Guest/Zeolitic Host Materials

The encapsulation of molecules in zeolite channels is a promising possibility for the design of advanced optical materials, such as materials with laser [33], light-harvesting [15], sensor [34], nonlinear optical [35], and switching properties [19], because of the enhanced chemical and photochemical stability of matrix-incorporated organic compounds [15,36]. Here, investigations on materials for photosensitive birefringence switching are presented, and the influence of host–guest composition on the optical switching process is described in detail.

2.3.1 Photochromism

Organic photochromism is the reversible phototransformation of chemical species between two forms having different absorption spectra [37]. For photochromic processes, changes in molecular structure or charge distribution by different mechanisms are necessary. Intrazeolitic photoinduced processes have been found in faujasite supercages of NaY, in which 6-nitro-1′,3′,3′-trimethylspiro[2*H*-1]-benzopyrane-2,2′-indoline shows photochromic properties due to reversible bond cleavage [12,38]. In composite materials containing azo dyes, trans-cis isomerization which is reversible even in rigid matrices has been found [39]. Similar to various other media, in molecular sieve pores [19,40,41,42,43] two conformations of azobenzene with different spectroscopic properties [27,44] exist. Irradiation of *trans*-azobenzene with ultraviolet light leads to the metastable cis isomer. During this process, the π–π^* band is diminished and blue-shifted, whereas the intensity of the n–π^* band around 425 nm increases (Fig. 3). Irradiation with blue light or thermal treatment regenerates the thermodynamically stable trans isomer. Distinct photochromic effects have been found for several azo dyes incorporated in

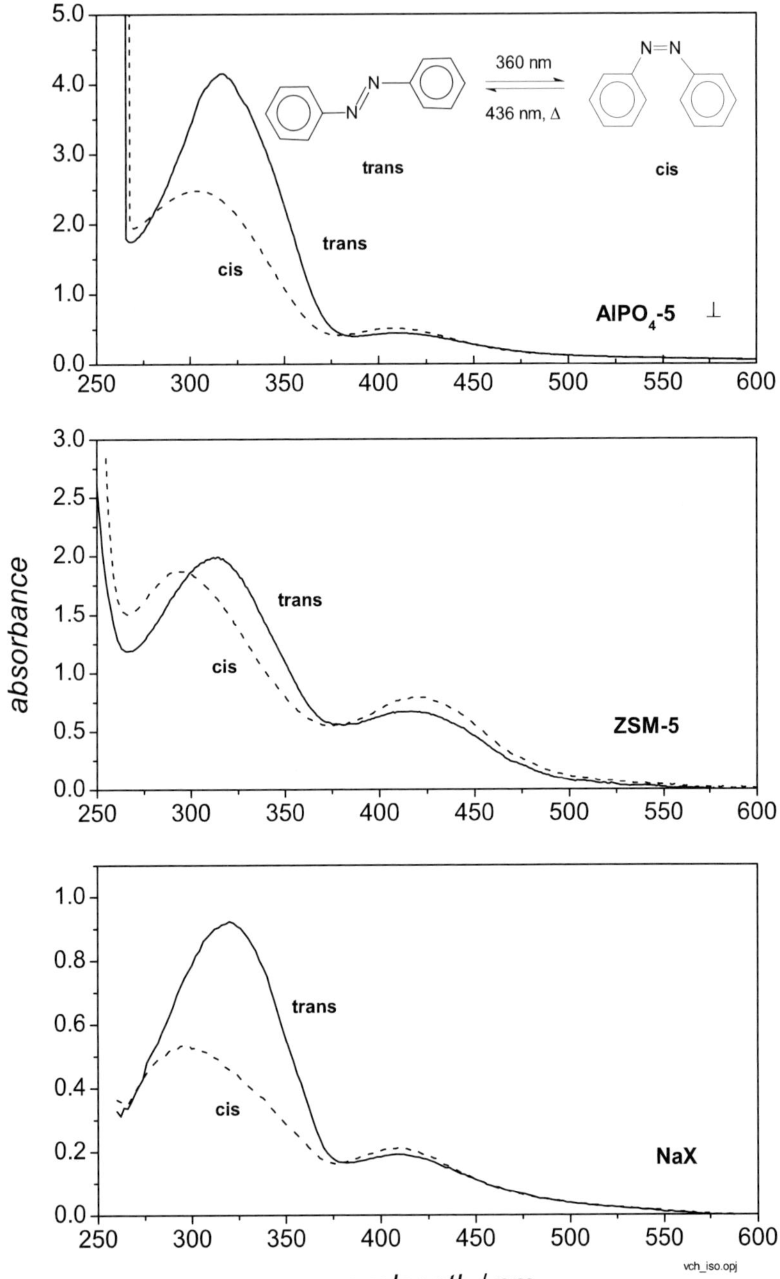

Fig. 3. Photochromism of azobenzene incorporated in molecular sieves. The absorption of dye-loaded zeolite microcrystals was measured after short-wavelength (broken line) and long-wavelength (solid line) irradiation. In the case of ZSM-5 and $AlPO_4$-5, the measurement direction was perpendicular to the straight channels. The azobenzene-loaded $AlPO_4$-5 crystals were measured with light polarized perpendicular to the crystal's length axis, because of high absorption of these crystals. NaX and ZSM-5 were measured with unpolarized light.

cages of NaX and in the channel structures of $AlPO_4$-5, ZSM-5, and Silicalite-1 (Tab. 1).

The ratio of trans to cis isomer in the photostationary states formed on irradiation is kinetically controlled by the rate constants of the trans-to-cis and the cis-to-trans conversions. The photostationary fraction of the cis isomer in nano-composites consisting of the isotropic zeolite host NaX and azobenzene can be estimated from the π–π^* absorption band by Eq. (7)

$$[\text{cis}]/[\text{trans}]_0 = (1 - A/A_0)/(1 - \varepsilon_{\text{cis}}/\varepsilon_{\text{trans}}). \quad (7)$$

Here, [cis] and [trans] are the concentrations of cis isomer after irradiation and the initial trans concentration, A is the absorbance after irradiation, A_0 the absorbance of the trans isomer, and $\varepsilon_{\text{cis}}/\varepsilon_{\text{trans}}$ the ratio of the molar absorption coefficients of the cis and trans isomers [45]. To estimate the photostationary cis content in NaX, absorption data besides the maximum π–π^* absorption and a value of $\varepsilon_{\text{cis}}/\varepsilon_{\text{trans}} = 0.05$, determined for azobenzene in polystyrene [46], were used. The $\varepsilon_{\text{cis}}/\varepsilon_{\text{trans}}$ ratios reported for azobenzene in several solvents and host systems [45,46] are similar. This rough estimate gives a photostationary fraction after irradiation of about 60% of *cis*-azobenzene in zeolite host NaX, which corresponds to typical values of photostationary cis fractions reported for various host systems [45,47,48].

2.3.2 Photosensitive Refractive Index Switching

The photochromic properties of azobenzene are not sufficient for direct applications in molecular optical microswitches intended for information processing. Photoinduced changes of the absorption spectra are, however, accompanied by differences in other physical properties. For example, dielectric properties are altered, and small changes in refractive index occur during photochromic reactions. The birefringence changes of composites are drastically enhanced if an ensemble of aligned guest molecules in molecular sieve channels is trans–cis switched by optical stimulation. This switching process becomes visible as different interference colors of birefringent molecular sieve crystals observed between crossed polarizers of a microscope.

The dye-loaded $AlPO_4$-5 crystal in Fig. 4 appears green before and red after short-wavelength irradiation, corresponding to different transmission spectra. The transmission spectra can be recalculated into birefringences as shown in Fig. 5 for a different system.

The large photoinduced birefringence changes are a direct consequence of altering the molecular alignment during intrazeolitic trans–cis photoisomerization of azobenzene [19,49,50]. Different shapes of the azobenzene isomers cause different orientations of the molecules in the microporous framework of the host materials. Molecular sieve crystals which mainly contain the better aligned rod-shaped *trans*-azobenzene show a higher birefringence than the cis-state composite encapsulating preferentially the bent *cis*-azobenzene molecules.

Tab. 1. Switching parameters of photosensitive nanocomposites. Maximum $\Delta n'_{max}$ and relative refractive index change $\Delta n'_{rel}$, photorefractive sensitivity $S_{n'2}$ of the trans-to-cis conversion, rate constants k_{therm} and lifetimes τ_{cis} of the metastable cis states are presented. Time intervals Δt in seconds, used for the kinetic measurements, are indicated in brackets. $\theta_{1/2}$ is a measure for the influence of the probe light. Absorption maxima λ_{max} or shoulders (sh) and optimum switching wavelengths λ_{irr}^{cis} are also given. Trans and cis data correspond to photostationary states obtained after short-wavelength and long-wavelength irradiation. Photochromic data of azobenzene in n-hexane are given for comparison [44].

Host	Guest	c/mass-%	$\Delta n'_{max}$	$\Delta n'_{rel}$	$S_{n'2}/cm^2\ kJ^{-1}$	k_{obs}/s^{-1}	k_{therm}/s^{-1}	τ_{cis}/h	$\theta_{1/2}/s$	λ_{irr}^{cis}/nm	$\lambda_{max}^{\pi\pi^*}/nm$ (trans)	$\lambda_{max}^{n\pi^*}/nm$ (trans)	$\lambda_{max}^{\pi\pi^*}/nm$ (cis)	$\lambda_{max}^{n\pi^*}/nm$ (cis)
Hexane [44]	AB										316	449	274	440
NaX	AB	15	(isotropic)	(isotropic)		2.5×10^{-3} (60 s) 1.0×10^{-3} (180 s) 3.6×10^{-3} (600 s)	1.3×10^{-4}	2	1079	350	320	408	297	409
ZSM-5	AB	3	0.008	0.2	−0.5	2.8×10^{-4} (60 s) 1.2×10^{-4} (180 s) 3.2×10^{-5} (3600 s)	3.2×10^{-5}	9	450	350	314	416	295	420
Silicalite-1	AB	<3	0.003	0.15							ca. 310	440	–	438
$AlPO_4$-5	AB	10	0.045	0.3	−4.1	2.4×10^{-5} (60 s) 1.2×10^{-5} (120 s) 2.5×10^{-6} (3600 s)	2.0×10^{-6}	140	630	345	318*	410	304	408
$AlPO_4$-5	Me-AB	<1	0.006	0.2	−2.1	3.9×10^{-5} (60 s) 1.4×10^{-6} (180 s) 6.7×10^{-6} (600 s)	3.6×10^{-6}	78	680	347	334	430	331	425
$AlPO_4$-5	4,4′-Me-AB	<1	0.005	0.2						340	341	462sh	340	462sh
$AlPO_4$-5	MeO-AB	ca. 1	0.003	0.08	−1.6	1.9×10^{-4} (60 s) 5.4×10^{-5} (180 s) 7.5×10^{-5} (600 s)	3.6×10^{-5}	8	254	360	349	474	347	472

* From measurements with light polarized perpendicular to the crystal's length axis.

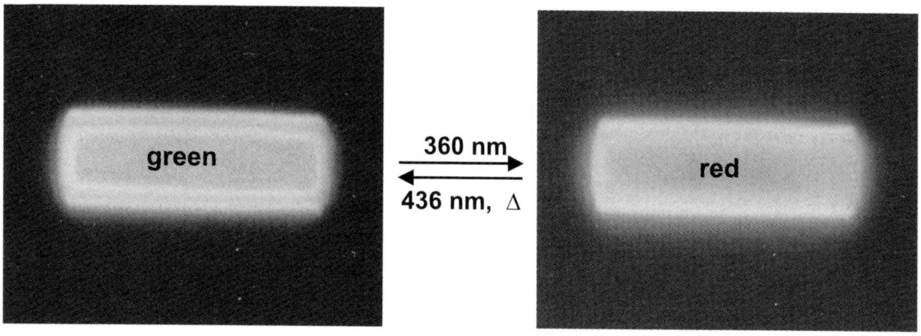

Fig. 4. Photosensitive birefringence of an azobenzene-loaded $AlPO_4$-5 crystal observed between two crossed polarizers.

The azobenzene/$AlPO_4$-5-based opto-optical birefringence switch shows a high contrast ratio. It is realized on the micrometer scale, which is determined by the size of the nanocomposite crystals.

2.3.3 Switching Parameters of Zeolite-Based Photosensitive Materials

Applications of organic/inorganic materials with photosensitive refractive index changes require the determination of switching parameters such as the dynamic range of the photosensitive effect, the stability of both switching states, the photorefractive sensitivity $S_{n'2}$, response time, nondestructive read-out capability, and reversibility of the effect [51]. The influence of guest molecules and host materials on several switching parameters has been investigated with regard to the photoswitchable properties of azo-dye-loaded molecular sieve composites.

2.3.3.1 Influence of the Host on Stability of Switching States, Dynamic Range, Sensitivity, and Reversibility

The type of molecular sieve influences the magnitude and the characteristics of the switching process. This indicates an important role of the host–guest interactions in the isomerization process [19].

Since the stability of the switching states determines the field of possible applications of photosensitive nanocomposites, the lifetimes τ_{cis} of metastable *cis*-azobenzene in different molecular sieve hosts were investigated. A lifetime τ_{cis} of 140 h was found for $AlPO_4$-5 composites, whereas the cis isomer in ZSM-5 and NaX has considerably shorter lifetimes (see Tab. 1).

The host systems of AFI, MFI, and FAU types differ in polarity, acidity, geometry, and size of internal voids. Because of the interaction of guest molecules with the host framework, acidic sites may play an important role in the stability of the metastable states. The complex influence of environmental polarity and acidity on

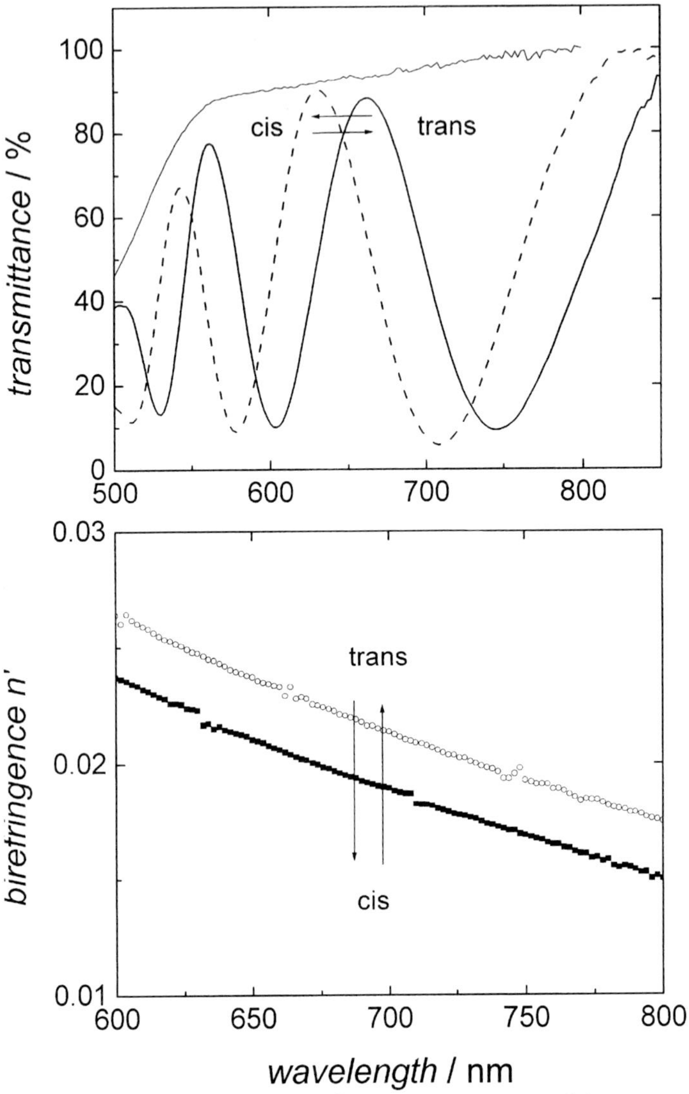

Fig. 5. Photosensitive birefringence of an azobenzene-loaded silicalite-1 micro-crystal. Top: oscillatory behavior of the transmittance of the crystal investigated between two parallel polarizers after irradiation with short- (broken line) and long-wavelength light (solid line). The transmittance of the same crystal observed with polarization parallel to the direction of the straight MFI channels is given for comparison (thin line). Bottom: birefringence changes of the microcrystal calculated from the transmission measurements depicted above.

the thermal relaxation of the cis isomer of azobenzene is a widely investigated phenomenon. The rate of thermal cis-to-trans conversion k_{therm} decreases with increasing polarity and polarizability of the solvent [52]. The polarity of the environment stabilizes the more polar cis isomer, whereas the acidity increases the relaxation rate of the cis isomer. The acid sites in ZSM-5 and NaX may also be responsible for the faster decomposition of photogenerated *cis*-azobenzene in these zeolitic hosts compared to $AlPO_4$-5. Considering the relaxation rate constants, which are related to the lifetime by $\tau_{cis} = 1/k_{therm}$, of the metastable states with respect to the requirements for the design of devices, one can state that both the thermodynamically stable trans state and the metastable cis state may be regarded as having long-term stability in the dark.

Microspectroscopic investigations of thermal relaxation are complicated by the influence of the measurement or the probe light [55]. To eliminate this effect, the rate constants k_{therm} were extrapolated from the linear dependence of the observed rate constants k_{obs} on the observation frequency ν_{obs} (the reciprocal time interval Δt between two measurements) according to Eq. (8).

$$k_{obs} = k_{therm}(1 + \theta_{1/2}\nu_{obs}) \tag{8}$$

The measurement specific constant $\theta_{1/2}$ characterizes the magnitude of the influence of the probe light on k_{obs} [55]. Contrary to kinetic investigations on $AlPO_4$-5 and ZSM-5-nanocomposites, which are based on birefringence measurements beyond the absorption bands of the dyes, in case of nonbirefringent NaX composites, kinetic data are derived from measurements of the time development of the absorbance of incorporated azobenzene. Here, the strong influence of the probe light during microspectroscopic investigations becomes evident, as indicated by the parameter $\theta_{1/2}$. From the high value $\theta_{1/2}$ for azobenzene in NaX (Tab. 1) it follows, that already a time interval of about $\Delta t = 1000$ s between two absorption measurements doubles the rate constant of the cis-to-trans relaxation. A much shorter value of Δt caused the same effect in the case of $AlPO_4$-5 and ZSM-5.

The maximum change in refractive index $\Delta n'_{max}$, called the dynamic range of the photosensitive effect, is one of the main characteristics of photosensitive materials. This parameter is also strongly influenced by the zeolitic host. For $AlPO_4$-5 composites, photosensitive refractive index changes $\Delta n'_{max}$ of up to 0.045 have been found. Molecular sieve-dependent fluctuations of $\Delta n'_{max}$ between 0.03 and 0.045 are mainly attributed to different dye contents, because the concentration-independent relative index change $\Delta n'_{rel}$ (Eq. 9) remains nearly constant.

$$\Delta n'_{rel} = \Delta n'_{max}/n'_{trans} \tag{9}$$

The magnitude of the photosensitive effect in azobenzene-loaded $AlPO_4$-5 is about ten times higher than in polymeric hosts [53]. It is among the largest ever reported (e.g., in photochromic sol–gel materials [39]) and offers the possibility of realizing zeolite-based optical switches on the micrometer scale.

To determine the energy efficiency of the trans-to-cis conversion of an azobenzene/$AlPO_4$-5 composite, the photorefractive sensitivity $S_{n'2}$ [54] of the process was determined. The parameter $S_{n'2}$, which describes the refractive index change per unit incident energy density, was obtained by observing the kinetics of the change in refractive index on irradiation. Among the zeolitic nanocomposites, AB–

$AlPO_4$-5 crystals show the highest values of $S_{n'2}$ [49], comparable to those reported for inorganic photorefractive materials like lithium niobate [54].

The usual problems of destructive read-out of optical switchable materials are minimal for azo dye/molecular sieve composites because of the low spectral overlap between the absorption bands used by the switching light ($\lambda < 450$ nm) to generate the different states of the bistable material, and the read-out light in the wavelength region of $\lambda > 600$ nm. Nevertheless, an influence of the probe light was found, as indicated by the dependence of the rate constants k_{obs} on the observation frequency ν_{obs} (see Tab. 1).

The reversibility of the photoswitching process was checked by alternating irradiation of a composite crystal with light of the wavelengths 360 and 436 nm to generate the cis and trans isomers of azobenzene, respectively.

For NaX nanocomposites, a cyclability number Z_{50} of about 48 was found, where Z_{50} is the number of cycles required to reduce the initial absorbance by 50% at a specific wavelength [37]. The difference in birefringence in the cis and trans states of $AlPO_4$-5 and ZSM-5 composites changes only slightly with the number of cycles, and the cyclability is some orders of magnitude larger. It depends strongly on the preparation of the samples. The reversible photosensitive change in the refractive index of the composite material over many cycles is shown in Fig. 6 [55].

2.3.3.2 Influence of the Guest on Optimum Excitation Wavelength, Stability of Switching States, and Dynamic Range

The photoinduced refractive index change $\Delta n'_{max}$ is strongly influenced by the zeolitic host, as mentioned above, and also by the incorporated guest molecules. Different guest molecules, of course, require matched irradiation wavelengths for switching. To find the optimal switching wavelengths, the cis/trans composition in the photostationary state at different irradiation wavelength λ_{irr} was investigated. The cis/trans composition of the nanocomposites is reflected in their birefringence, which is dependent on the irradiation wavelength, as shown in Fig. 7.

Minimum birefringence corresponds to composite crystals with a maximum content of cis isomer, which was obtained by irradiation with the corresponding optimum switching wavelengths $\lambda_{irr}{}^{cis}$. The results are summarized in Tab. 1. The type of guest molecules clearly influences the wavelength $\lambda_{irr}{}^{cis}$ that generates a maximum of cis molecules. The influence of the molecular sieve host plays a minor role, as can be seen from the nearly unchanged $\lambda_{irr}{}^{cis}$ observed for azobenzene in different zeolites. The photostationary state (i.e., the ratio of trans and cis isomers formed upon irradiation) depends on the respective absorbance of the isomers at the irradiation wavelength. Therefore, different optimum switching wavelengths $\lambda_{irr}{}^{cis}$ correspond to different absorption maxima $\lambda_{max}{}^{\pi\pi^*}$ and $\lambda_{max}{}^{n\pi^*}$ of the incorporated guest molecules. (Tab. 1) This behavior offers a way of tuning the switching wavelengths of zeolite-based optical switches by varying the photochromic guest molecules.

Lifetimes τ_{cis} of the metastable cis isomers have been determined for different guest molecules (see Tab. 1). The fastest thermal relaxation was found for the derivative *p*-OMe-AB incorporated in $AlPO_4$-5 and the slowest for AB in $AlPO_4$-5.

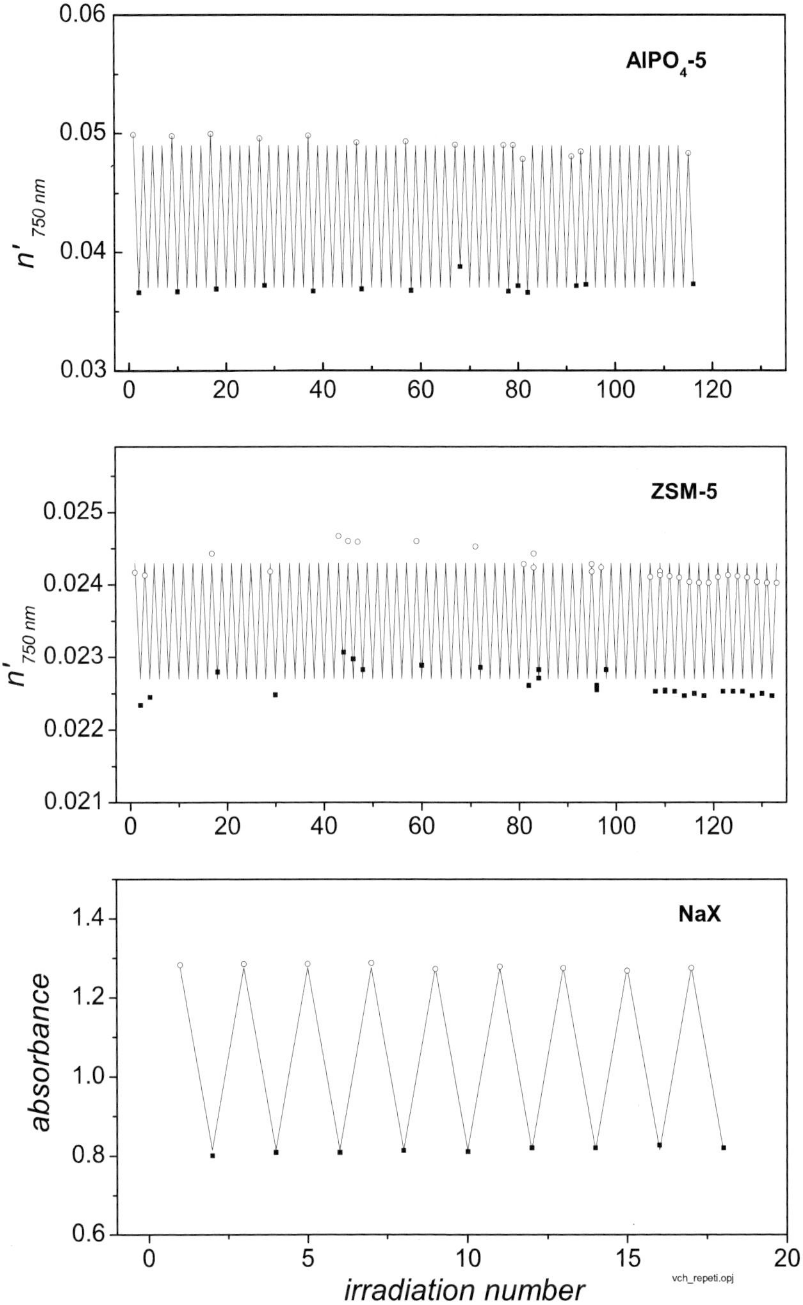

Fig. 6. Reversibility of switching processes of azobenzene-loaded $AlPO_4$-5, ZSM-5, and NaX composites under alternating short-wavelength (360 nm, 2 min; solid squares) and long-wavelength irradiation (436 nm, 2 min; open circles) with 16, 27, and 5 mW cm^{-2}, respectively.

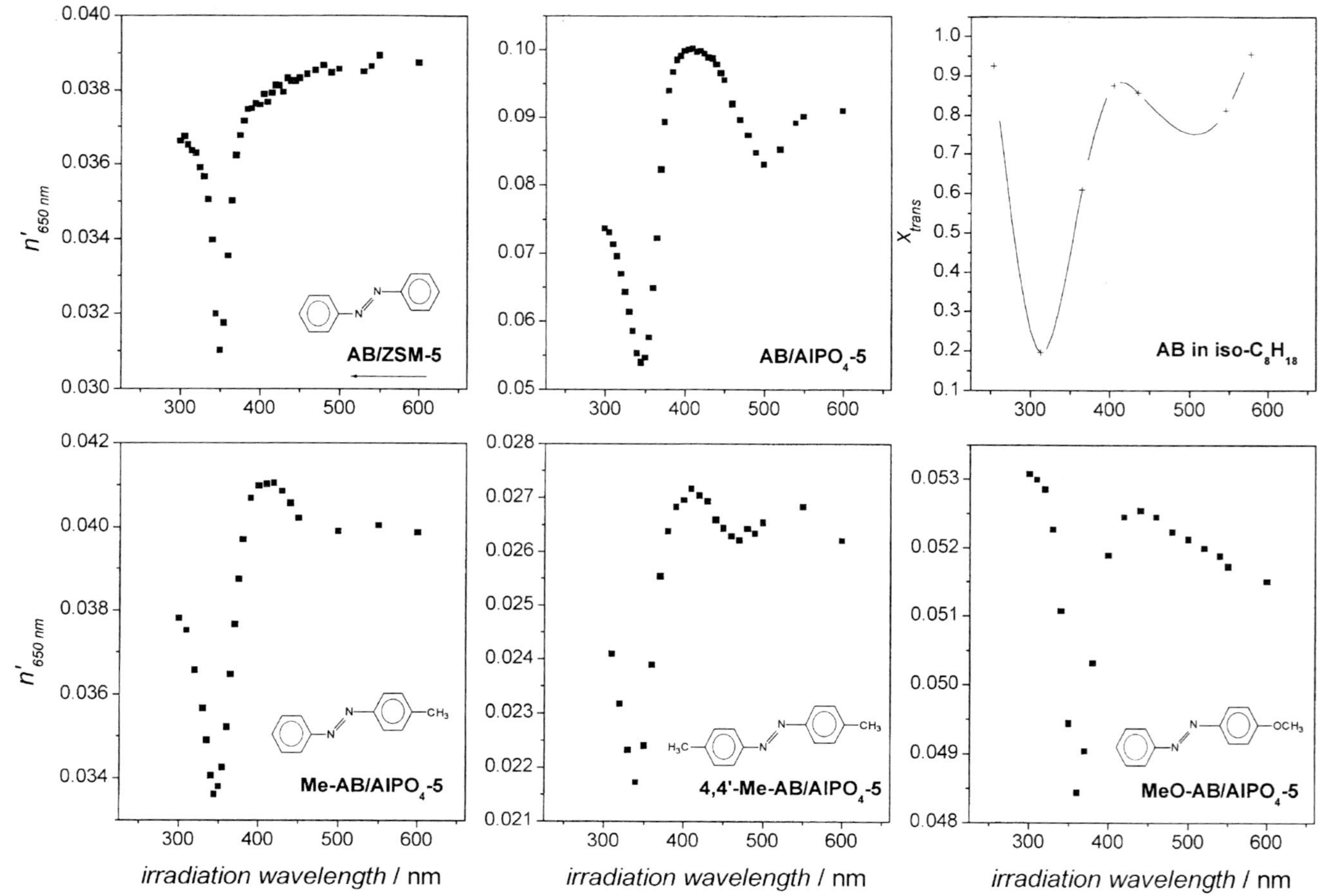

Fig. 7. Birefringence n'_{650nm} of dye-loaded molecular sieve crystals as a function of irradiation wavelength. Before each birefringence measurement, the crystal was irradiated with light of wavelength λ_{irr}. The duration of each irradiation was 5 min to ensure that a photostationary state was nearly reached. The measurements were carried out in decreasing order of λ_{irr} from 600 to 280 nm. The dependence of the photostationary fractions of trans isomer x_{trans} of azobenzene in isooctane on the irradiation wavelength is shown for comparison (from Ref. [48]).

These findings correspond to data for these dyes in solution, where all *p*-substituted azobenzenes thermally isomerize faster than the parent compound, regardless of the nature of the substituent [56].

Like the lifetime of the cis state, the dynamic range of the photosensitive effect is strongly influenced by adsorbate molecules. The maximum birefringence changes of nanocomposites depend on the intrazeolitic concentration of photochromic guest molecules and on the cis/trans ratio in the photostationary state. The faster thermal relaxation of the metastable cis isomer of substituted azobenzenes has an effect on the phototationary states and seems to be a major cause of the observed decrease in dynamic range of the photosensitive effect relative to azobenzene.

2.4 Summary

Photosensitivity is one of the interesting optical properties which can be achieved in zeolite-based host–guest composites. The modular composition of the host–guest materials and the strong organizing influence of the zeolite channels is exploited to tune and enhance the photosensitve effects in these materials.

The photosensitivity of the composites is realized by the incorporation of photochromic guests aligned by the zeolitic host. This geometrical alignment does not restrict the molecular isomerization reactions that are the basis for the photosensitive effects. However, the ordering of guest molecules leads to a drastic enhancement of the photosensitivity of alignment-dependent optical properties, since the arrangement of the guest molecules is also controlled by molecular isomerization. Therefore, the effect of photoisomerization on refractive indices is about ten times higher for azobenzene in the molecular sieve host $AlPO_4$-5 than in polymeric hosts. To explore the relationship between dye loading, alignment, and refractive index changes, polarization-dependent UV/Vis microspectroscopic studies of the host–guest systems were performed. It was shown that the loading level, the degree of alignment, and the birefringence of these composites can be efficiently determined by spectroscopic measurements on individual composite crystals.

Azobenzene/$AlPO_4$-5 composites display large refractive index changes of up to $\Delta n'_{650\text{nm}} = 0.045$, which is much larger than corresponding values found for inorganic or polymeric photorefractive materials. Moreover, the observed refractive index changes have similar energy efficiency $S_{n'2}$ to those of widely used inorganic materials (e.g., $LiNbO_3$), the switching states are stable for at least hours, and the switching process is reversible. The switching material has the capability for nondestructive readout, and its properties can be tuned by the selection of host–guest composition. All these features make switchable zeolitic host–guest systems promising materials for applications in all-optical microswitches.

Acknowledgements

The authors would like to thank the German Science Foundation (DFG) for financial support (RE 1203/2, Ma 1745/3). The helpful cooperation with H. Baumgärtel (Free University Berlin), E. Biller, and the ACA Berlin-Adlershof during the course of this work is gratefully acknowledged.

References

1 G.A. Ozin, *Adv. Mater.* **1992**, *4*, 10.

2 P. Behrens, G. D. Stucky in *Comprehensive Supramolecular Chemistry*, J. L. Atwood, J.E.D. Davies, D.D. MacNicol, F. Vögtle, J. M. Lehn (Eds.), Pergamon Press, Oxford, **1996**, p. 722.

3 F. Marlow, W. Dong, K. Hoffmann, J. Loerke in *Handbook of Porous Solids*, F. Schüth, K. S. W. Sing, J. Weitkamp (Eds.), VCH-Wiley, Weinheim, **2002**, p. 3029.

4 W.M. Meier, D.H. Olson, Ch. Baerlocher, *Atlas of Zeolite Structure Types*, 4th rev. ed., Butterworth-Heinemann, New York, **1996**, pp. 26, 146.

5 Synthesis: I. Girnus, see: a) I. Girnus, M.M. Pohl, J. Richter-Mendau, M. Schneider, M. Noack, D. Venske, J. Caro, *Adv. Mater.* **1995**, *7*, 711; b) I. Girnus, K. Jancke, R. Vetter, J. Richter-Mendau, J. Caro, *Zeolites*, **1995**, *13*, 33.

6 Synthesis: J. Kornatowski/G. Finger, see: J. Kornatowski, *Zeolites* **1988**, *8*, 77.

7 M. Bockstette, D. Wöhrle, I. Braun, G. Schulz-Ekloff, *Microporous Mesoporous Mater.* **1998**, *23*, 83.

8 I. Braun, G. Schulz-Ekloff, M. Bockstette, D. Wöhrle, *Zeolites* **1997**, *19*, 128.

9 a) G. Ihlein, F. Schüth, O. Krauß, U. Vietze, F. Laeri, *Adv. Mater.* **1998**, *10*, 1117. b) R. Hoppe, G. Schulz-Ekloff, D. Wöhrle, C. Kirschhock, H. Fuess, L. Uytterhoeven, R. Schoonheydt, *Adv. Mater.* **1995**, *7*, 61. c) M. Ehrl, F. W. Deeg, C. Bräuchle, O. Franke, A. Sobbi, G. Schulz-Ekloff, D. Wöhrle, *J. Phys. Chem.* **1994**, *98*, 47. d) D. Wöhrle, G. Schulz-Ekloff, *Adv. Mater.* **1994**, *6*, 875.

10 R. Hoppe, G. Schulz-Ekloff, D. Wöhrle, E.S. Shpiro, O.P. Tkachenko, *Zeolites* **1993**, *13*, 222.

11 N. Gfeller, S. Megelski, G. Calzaferri, *J. Phys. Chem. B* **1999**, *103*, 1250.

12 C. Schomburg, M. Wark, Y. Rohlfing, G. Schulz-Ekloff, D. Wöhrle, *J. Mater. Chem.* **2001**, *11*, 2014.

13 S.D. Cox, T.E. Gier, G.D. Stucky, J. Bierlein. *J. Am. Chem. Soc.* **1988**, *110*, 2986.

14 V. Ramamurthy in *Photochemistry in Organized and Constrained Media*, V. Ramamurthy (Ed.), VCH-Publishers, New York, **1991**, p. 429.

15 M. Pauchard, A. Devaux, G. Calzaferri, *Chem. Eur. J.* **2000**, *6*, 3456.

16 L. Werner, J. Caro, G. Finger, J. Kornatowski, *Zeolites* **1992**, *12*, 658.

17 C. Striebel, K. Hoffmann, F. Marlow, *Microporous Mater.* **1997**, *9*, 43.

18 K. Hoffmann, D. Prescher, F. Marlow, *J. Inf. Rec.* **1998**, *24*, 191.

19 K. Hoffmann, F. Marlow, J. Caro, *Adv. Mater.* **1997**, *9*, 567.

20 K. Hoffmann, F. Marlow, J. Caro, *Zeolites* **1996**, *16*, 281.

21 F. Marlow, K. Hoffmann, G. G. Lindner, I. Girnus, G. van de Goor, J. Kornatowski, J. Caro, *Microporous Mater.* **1996**, *6*, 43.

22 F. Marlow, J. Caro, *Zeolites* **1992**, *12*, 433.

23 F. Marlow, D. Demuth, G. Stucky, F. Schüth, *J. Phys. Chem.* **1995**, *99*, 1306.

24 S. Megelski, G. Calzaferri, *Adv. Funct. Mater.* **2001**, *11*, 277.

25 Z. Lei, A. Vaidyalingam, P.K. Dutta, *J. Phys. Chem. B* **1998**, *102*, 8557.

26 F. Marlow, K. Hoffmann in *MRS Proceedings of the 12th International Zeolite Conference*, Vol. III, M. M. J. Treacy, B. K. Marcus, M. E. Bisher, J. B. Higgins (Eds.), Materials Research Society, Warrendale, PA, **1999**, p. 2121.

27 H. Rau in *Photochemistry and Photophysics*, J. F. Rabek (Ed.), Vol. 4, CRC Press Inc., Boca Raton, Florida, **1990**, p. 120.

28 G. Calzaferri, D. Brühwiler, S. Megelski, M. Pfenninger, M. Pauchard, B. Hennessy, H. Maas, A. Devaux, U. Graf, *Solid State Sci.* **2000**, *2*, 421.

29 S. Megelski, A. Lieb, M. Pauchard, A. Drechsler, S. Glaus, C. Debus, A.J. Meixner, G. Calzaferri. *J. Phys. Chem. B* **2001**, *105*, 25.

30 K. Hoffmann, F. Marlow, J. Caro, *J. Fluoresc.* **1994**, *4*, 75.

31 K. Hoffmann, F. Marlow, J. Caro, S. Dähne, *Zeolites* **1996**, *16*, 138.

32 K. Rurack, K. Hoffmann, W. Al-Soufi, U. Resch-Genger, unpublished results.

33 I. Braun, G. Ihlein, F. Laeri, J.U. Nöckel, G. Schulz-Ekloff, F. Schüth, U. Vietze, Ö. Weiss, D. Wöhrle, *Appl. Phys. B* **2000**, *70*, 335.

34 a) J.L. Meinershagen, T. Bein, *J. Am. Chem. Soc.* **1999**, *121*, 448, b) S. Mintova, S.Y. Mo, T. Bein, *Chem. Mater.* **2001**, *13*, 901.

35 F. Marlow, J. Caro, L. Werner, J. Kornatowski, S. Dähne, *J. Phys. Chem.* **1993**, *97*, 11286.

36 D. Wöhrle, A.K. Sobbi, O. Franke, G. Schulz-Ekloff, *Zeolites* **1995**, *15*, 540.

37 H. Bouas-Laurent, H. Dürr, *Pure Appl. Chem.* **2001**, *73*, 639.

38 a) I. Casades, S. Constantine, D. Cardin, H. Garcia, A. Gilbert, F. Marquez, *Tetrahedron* **2000**, *56*, 6951; b) D. Wöhrle, G. Schulz-Ekloff, I. Braun, C. Schomburg, F. Laeri, U. Vietze, M. Ganschow, Y. Rohlfing, T. Bogdahn-Rai, *J. Inf. Rec.* **2000**, *25*, 87.

39 F. Chaput, J. Biteau, K. Lahlil, J.P. Boilot, B. Darracq, Y. Levy, J. Peretti, V.I. Safarov, G. Parent, A. Fernandez-Acebes, J. M. Lehn, *Mol. Cryst. Liq. Cryst.* **2000**, *344*, 77.

40 A. Corma, H. Garcia, S. Iborra, V. Marti, M.A. Miranda, J. Primo, *J. Am. Chem. Soc.* **1993**, *115*, 2177.

41 M. Kojima, T. Takagi, T. Goshima, *Mol. Cryst. Liq. Cryst.* **2000**, *344*, 179.

42 K. Weh, M. Noack, K. Hoffmann, K.P. Schröder, J. Caro, *Microporous Mesoporous Mater.* **2002**, *54*, 15.

43 Y. Kuriyama, S. Oishi, *Chem. Lett.* **1999**, 1045.

44 H. H. Perkampus, *UV-Vis Atlas of Organic Compounds*, 2. Ed., VCH Weinheim, **1992**. (*n*-Hexane: $\varepsilon_{\pi\pi^*}{}^{trans} = 22\,400$, $\varepsilon_{n\pi^*}{}^{trans} = 405$, $\varepsilon_{\pi\pi^*}{}^{cis} = 5000$, $\varepsilon_{n\pi^*}{}^{cis} = 1250$ L mol^{-1} cm^{-1}).

45 Y. Morishima, M. Tsuji, M. Kamachi, K. Hatada, *Macromolecules* **1992**, *25*, 4406.

46 J.G. Victor, J.M. Torkelson, *Macromolecules* **1987**, *20*, 2241.

47 A. Yabe, Y. Kawabata, H. Niino, M. Matsumoto, A. Ouchi, *Thin Solid Films* **1988**, *160*, 33.

48 G. Zimmermann, L. Chow, U. Paik, *J. Am. Chem. Soc.* **1958**, *80*, 3528.

49 F. Marlow, K. Hoffmann, *Ber. Bunsenges. Phys. Chem.* **1997**, *101*, 1731.

50 F. Marlow, K. Hoffmann, German patent pending 196 44 636.8; 17.10.96.

51 N. Tamai, H. Miyasaka, *Chem. Rev.* **2000**, *100*, 1875.

52 S.W. Afanasiev, L.W. Moiseeva, L.P. Salukaev, *Z. Fiz. Khim.* **1978**, *52*, 2507.

53 a) Z. Sekkat, D. Morichere, M. Dumont, R. Loucif-Saibi, J.A.

Delaire, *J. Appl. Phys.* **1992**, *71*, 1543. b) N. Tanio, M. Irie, *Jpn. J. Appl. Phys.* **1994**, *33*, 3942. c) J. A. Delaire, K. Nakatani, *Chem. Rev.* **2000**, *100*, 18717 and references therein.

54 P. Günther, J.P. Huignard, *Photorefractive Materials and their Applications*, Vols. 1 & 2, Springer, Berlin, **1988/1989**, pp. 1–53.

55 K. Hoffmann, U. Resch-Genger, F. Marlow, *Microporous Mesoporous Mater.* **2000**, *41*, 99.

56 E.R. Talaty, J.C. Fargo, *Chem. Commun.* **1967**, 65.

3
Confocal Microscopy and Spectroscopy for the Characterization of Host–Guest Materials

*Christian Seebacher, Christian Hellriegel, Fred-Walter Deeg, Christoph Bräuchle**

3.1
Introduction

In recent years the synthesis and characterization of functional materials that can be controlled on a nanometer scale have increasingly become the focus of scientific research. In this field, physics [1], chemistry [2–5], and biology [6,7] must work together closely to achieve this common goal. From a chemical perspective a solid material with an ordered structure in the nanometer range can act as a host environment for the incorporation of guest molecules. Here, the basic task is to organize the guest molecules to form a supramolecular structure within the host material. As a consequence the interplay of guest and host can result in a functional material [8]. Almost any molecule is conceivable as a guest molecule for such systems. Fluorescent dye molecules are of particular interest. They can be addressed and investigated by confocal optical microscopy and spectroscopy in a straightforward manner. Furthermore, confocal optical microscopy is not restricted to surfaces or electrically conducting materials, unlike other scanning probe techniques such as atomic force and tunneling microscopy, while still maintaining good spatial resolution. Fluorescent dye molecules as optically addressable moieties incorporated in nanometer-sized channels and cages are the mainstay of our investigations, and of the present article. Molecular devices on a single-molecule basis, such as optical switches, shutters or valves, and rotors, using dye molecules as guests in mesoporous hosts are some of the ideas pursued by our group (Fig. 1).

There is a huge variety of host materials, of which inorganic molecular sieves are a particularly interesting category. A large number of molecular sieves with well-defined pore systems have been synthesized [5]. One attractive feature of these materials is their availability with numerous pore sizes and shapes, variations of the chemical properties within the pore, and the topology of the entire pore system, ranging from one-dimensional channels to three-dimensional interconnected cage structures (Fig. 2). In a similar way the guest dye molecules are also available in different variations (Fig. 2). The combination of different host materials and guest molecules leads to a manifold of host–guest systems and enables the design

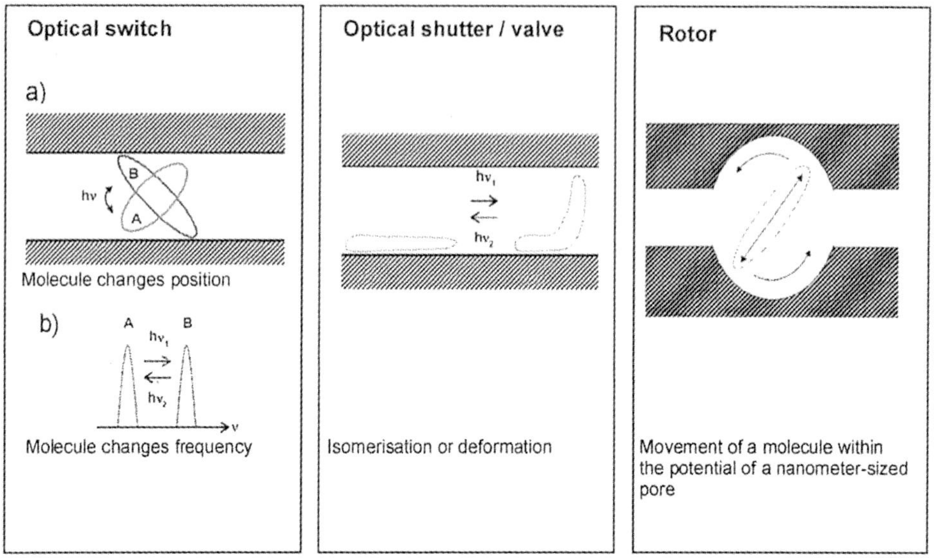

Fig. 1. Schematic illustration of the potential applications of optically addressable molecules incorporated into molecular sieves. Left: an optical switch based on orientational (a) or frequency (b) changes. Middle: a shutter/valve mechanism which functions by optically addressing a conformational change in a molecule. Right: a molecule rotating within the potential of a nanometer-sized pore illustrates the concept of a molecular rotor.

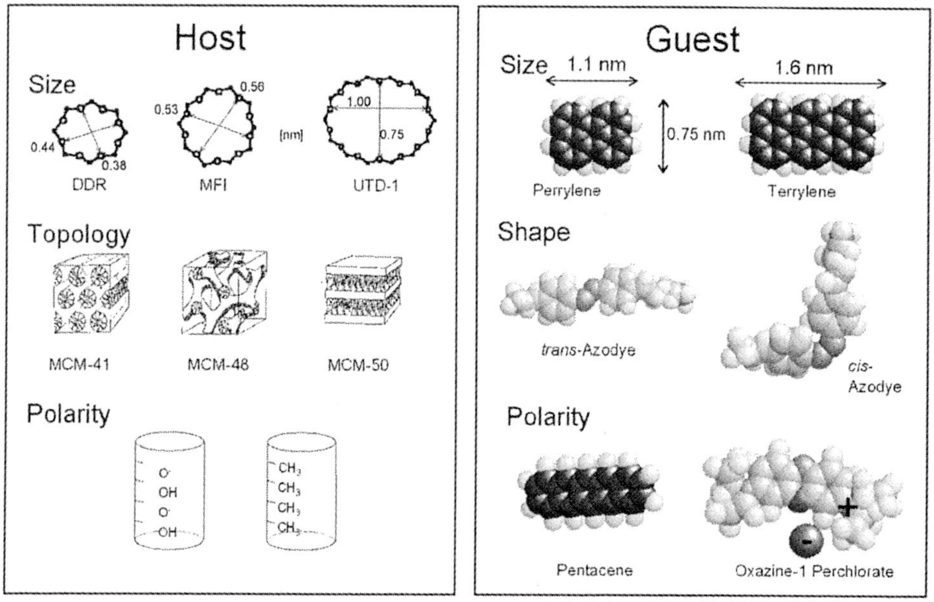

Fig. 2. A collection of substrates for host–guest interactions.

of materials with specific properties. Some examples with intriguing potential applications have been realized on this basis [2]. It was shown that the incorporation of *p*-nitroaniline molecules into the one-dimensional channel structure of $AlPO_4$-5 is characterized by a macroscopically well-defined orientation of the guest molecules. This allows efficient optical second-harmonic generation with this material [9]. Another example is the generation of molecular wires by the synthesis of isolated, conducting, one-dimensional polymer chains in the channels of mesoporous MCM-41 [10]. More recently, laser activity was demonstrated in a single micrometer-sized dye-loaded $AlPO_4$-5 crystal. In this case the resonator was the hexagonal morphology of the host crystal [11]. Another example is a synthetic light-harvesting complex based on dye molecules incorporated in zeolite L. In this material the absorbed energy can migrate within the crystal [12]. Different optical switches have been realized on the basis of host–guest materials in bulk [13a], as well as on a single-molecule level [13b]. As a last example, dye molecules immobilized in molecular sieves have been used as sensors for different chemical compounds [14–16].

3.2 Confocal Microscopy

Confocal optical microscopy allows the visualization of samples with high optical resolution in all three spatial dimensions [17,18]. The use of highly sensitive detectors further increases the capabilities of a confocal microscope, even enabling the investigation of single, spatially separated molecules [19].

The basic idea behind confocal microscopy [19b] is that – in contrast to common microscopy – only the spot that is detected is also illuminated. Technically, this is accomplished by illuminating the object with laser light (an idealized point source) and by placing a pinhole in the detection focal plane. This pinhole acts as a spatial filter that only lets light pass that stems from the focal point of the objective lens. As a consequence, the light scattered from objects lying outside the focal spot is suppressed. Mathematically, the confocal principle is formulated by a convolution of the detection and excitation point-spread functions. This leads to a reduction of the detection volume to about 0.1 fL. The reduction of the detection volume is mainly due to a significant improvement of the depth-resolution along the optical axis. The lateral resolution is improved by a factor of $\sqrt{2}$. This ingenious principle is best visualized as shown in Fig. 3. The size of the detection volume depends on the wavelength and on the numerical aperture (N.A.) of the objective used. The calculated detection volume [18] is shown in Fig. 3, with a full-width at half maximum of 194 nm in the focal plane and 505 nm along the optical axis. These values are obtained with an excitation wavelength of 633 nm and an objective with an N.A. = 1.3. In practice the resolution achieved is typically 300 nm in the focal plane and 900 nm along the optical axis. To obtain images the detection volume must be scanned across the sample in all three dimensions. Commonly, the *xy*-plane is used to acquire images in focal planes that are cross sections through the samples. For three-dimensional images these cross sections are then arranged in

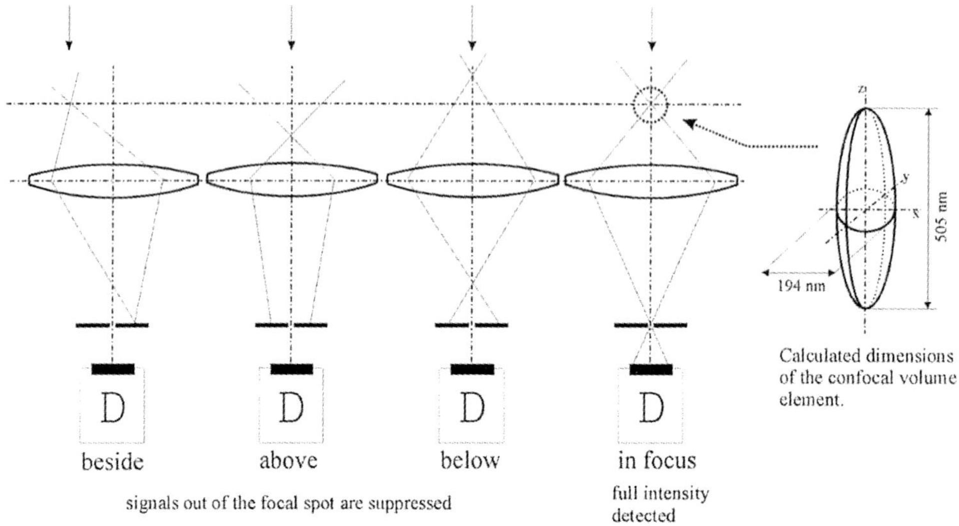

Fig. 3. Schematic illustration of the confocal principle. Only light originating from the focal plane can fully pass the detection pinhole (between the objective lens and the detector D). Light stemming from other areas is suppressed. On the right: calculated confocal volume element with a FWHM of 194 nm in the focal plane and 505 nm along the optical axis (oil-immersion objective, $n = 1.5$, N.A. $= 1.3$, and $\lambda_{ex} = 633$ nm).

stacks along the z-axis (optical axis). The scanning procedure is a limiting factor for the speed with which images can be recorded.

In summary, the main advantage of confocal microscopy compared to conventional microscopy is the significant improvement in spatial resolution.

The investigations presented in the following were conducted by using a modified inverted confocal laser scanning microscope (Zeiss LSM 410). The excitation light is provided by various laser systems. The laser beam is directed to the sample by a dichroic mirror and an objective. Oil-immersion objectives with high numerical aperture (N.A ≥ 1.3) were used to achieve a high spatial resolution and a high efficiency of light detection. The best results are obtained when the samples are embedded in a medium with a high refractive index, such as poly(methyl methacrylate), PMMA, or ethylene glycol. To modulate the polarization plane of the excitation light a $\lambda/2$ plate is placed just before the objective. The fluorescence light of the sample is collected by the same objective, passes the dichroic mirror and is spatially filtered by a pinhole in the manner explained above. Additional fluorescence filters eliminate backscattered excitation light. The fluorescence light is detected by a photomultiplier, an avalanche photodiode (for single-molecule sensitivity), or, for recording fluorescence spectra, by a prism/CCD camera arrangement. The last two variations are self-made extensions of the Zeiss LSM 410 microscope. Figure 4 depicts examples of the different measurement techniques conducted with the described confocal microscope. As mentioned before, images are recorded by scanning the focus across the sample. The complete three-dimensional infor-

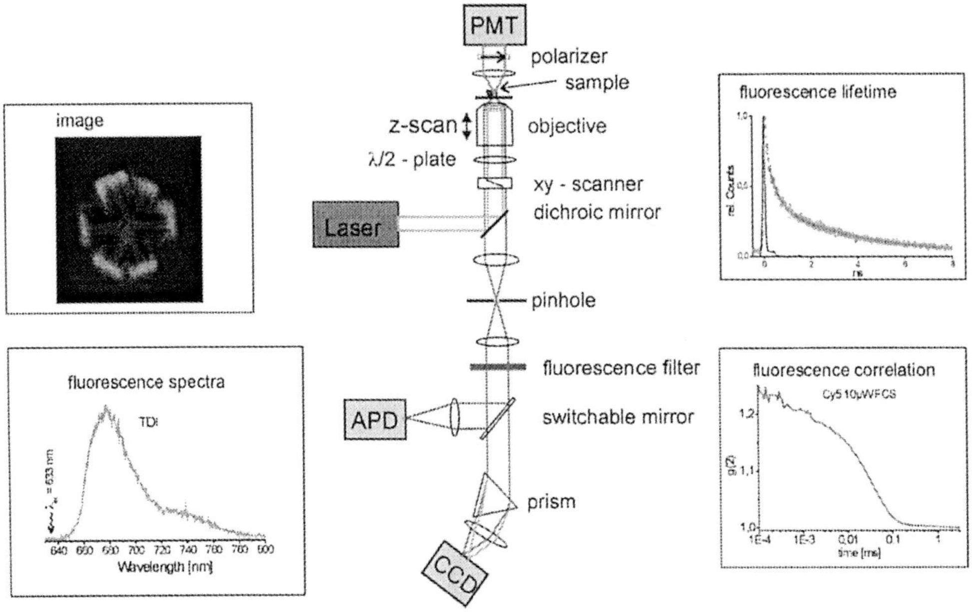

Fig. 4. Schematic overview of the confocal microscope used, and examples of different measurement possibilities. For the description of the individual measurement techniques, see text.

mation is acquired by scanning a stack of several pictures in the *xy*-plane with different *z* positions. To visualize the obtained three-dimensional information, a two-dimensional projection of the fluorescence intensity can be calculated for any viewing angle (Fig. 5). A three-dimensional impression is obtained by displaying a sequence of projections as a complete rotation.

Apart from images, also spectral, structural, and dynamic information can be obtained from different positions in the sample with the high spatial resolution of the confocal microscope. In other words, confocal microscopy enables spectroscopic analysis of very small probe volumes. By spectral information we mean fluorescence spectra, which can be recorded by various spectrographic dispersion methods. The prism/CCD arrangement mentioned above is the most sensitive, but at the cost of wavelength resolution. In principle, excitation spectra can also be recorded, but this method requires the availability of broadly tunable laser sources. Structural information, e.g., orientations of the guest molecules relative to the crystal axis, can be obtained by determining the orientation of the transition dipole moments of the fluorescing chromophores. The orientation can be measured in two ways. First, by recording the fluorescence in dependence on the excitation polarization planes; second, by analyzing the fluorescence light with polarization filters. The evaluated orientation of the transition dipole moments is determined with respect to the orientation of the crystal axis.

Dynamic phenomena taking place on a timescale smaller than 10 ns, such as

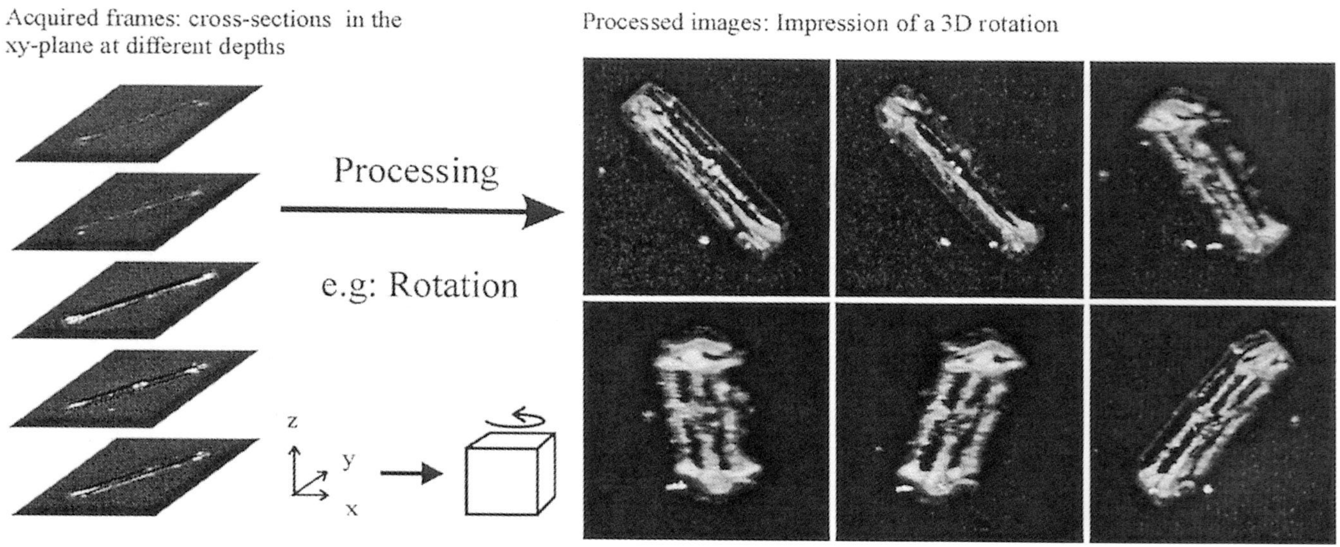

Fig. 5. Acquisition and visualization of a three-dimensional fluorescence image. On the left a stack of cross sections through the sample with different depths is shown. From this data six projections of the crystal at different angles were calculated, resulting in the pictures on the right and giving the visual impression of a rotation.

fast relaxation processes of excited states, are observed, e.g., by measuring the fluorescence decay after excitation with a pulsed laser. The most sensitive detection method is time-correlated single-photon counting (TCSPC) [21], which is a further implementation of the LSM 410 microscope. This method has a lower temporal resolution (ca. 50 ps) than less sensitive methods such as fluorescence up-conversion (ca. 50 fs) [22]. Dynamic phenomena taking place on a longer timescale, such as triplet-state dynamics or diffusion, can be investigated by recording the fluorescence autocorrelation curve [23], which is a further extension to the original LSM-410 microscope (in our case, consisting of an avalanche photodiode and a correlation card).

3.3 Results

In this section the observations made for the hosts and guests depicted in Fig. 6 will be presented. The chosen organic dyes show good fluorescence properties, and are well suited for confocal fluorescence microscopy. The short axis of the rectangular-shaped molecules range from ca. 0.7 nm for stilbene and tetracene to 1 nm for terrylenediimide (TDI) and 4-dicyanomethylene-2-methyl-6-(*p*-dimethylamino-styryl)-4*H*-pyran (DCM). Therefore, some molecules do not properly fit into the small-diameter pores of $AlPO_4$-5 (AFI, ca. 0.73 nm) and silicalite-1 (MFI, ca. 0.56 nm) crystals. The M41S-type materials (MCM) have pore widths of greater than 2 nm and can therefore accommodate even larger molecules.

3.3.1 Spatial Heterogeneities

In spite of the increasing number of applications encompassing devices based on host–guest materials, many basic properties are still unknown. One prerequisite for the realization of certain types of devices is the availability of large, defect-free host crystals. In addition, for host–guest systems, the well-defined spatial distribution of the molecular guests within the host is of great importance. However, in many cases a perfect crystal and a homogeneous distribution of the guest within a molecular sieve is just taken for granted and not justified by experiment. A fast and convenient technique for monitoring the quality of the host material and characterizing the guest distribution is therefore of considerable interest. We recently demonstrated that confocal fluorescence microscopy suits this purpose very well and represents a powerful and expeditious tool for visualizing the defect structure of molecular sieves and the guest distribution therein [24]. In the following, some examples will be presented which illustrate the potential of this method for the characterization of molecular sieves and host–guest interactions in these materials.

3.3.1.1 Staining Defect Structures in Silicalite-1 (MFI)

Silicalite-1 (pure-silica MFI-type structure) forms sizeable coffin-shaped crystals with a well-defined morphology, as shown in Fig. 7 (the structure is depicted in

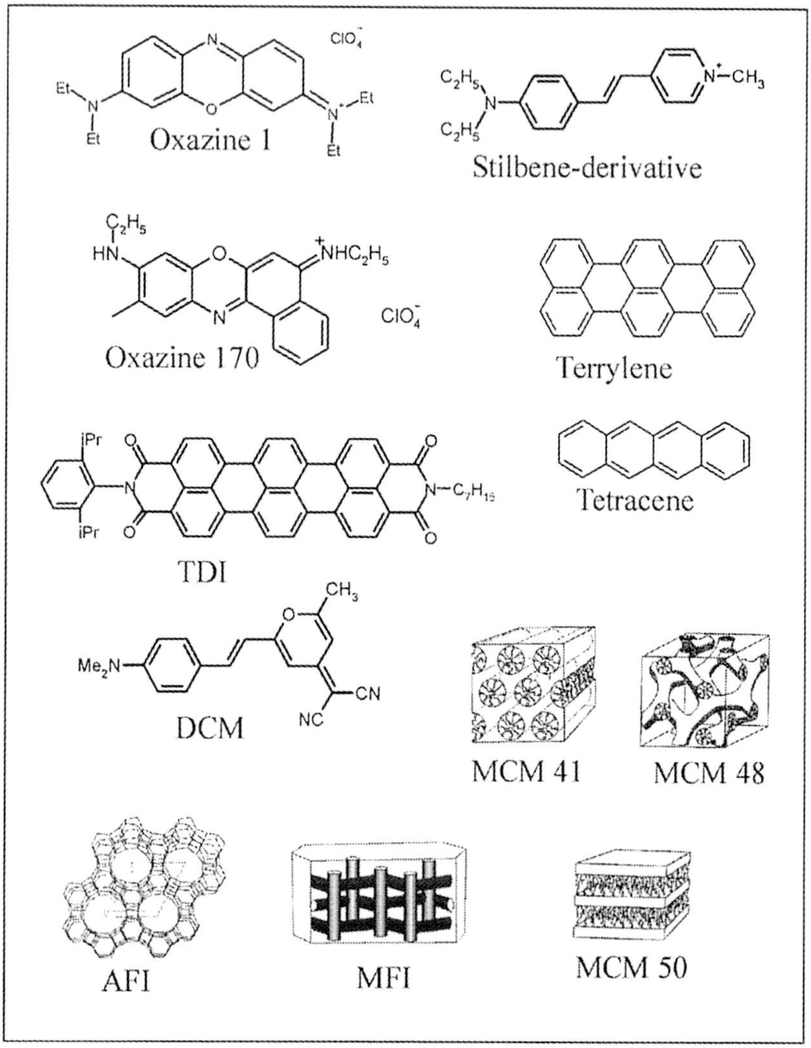

Fig. 6. Overview of the examined pore systems and dye molecules. The host structures are abbreviated according to the framework-type codes.

Fig. 6). Silicalite-1 was synthesized R. Jäger (P. Behrens group, Universität Hannover) from tetrapropylammonium bromide, hydrofluoric acid, ammonium fluoride, Cab-o-sil M5, and water in molar ratios of 0.12:0.10:0.11:1:37 [25]. Heating at 600 °C in air for 6 h yields the calcined material. The external morphology of the calcined MFI appears well defined, with indications of an additional defect structure in this transmission image. However, the full three-dimensional defect structure cannot be resolved from these images. To achieve this, the silicalite-1 crystals

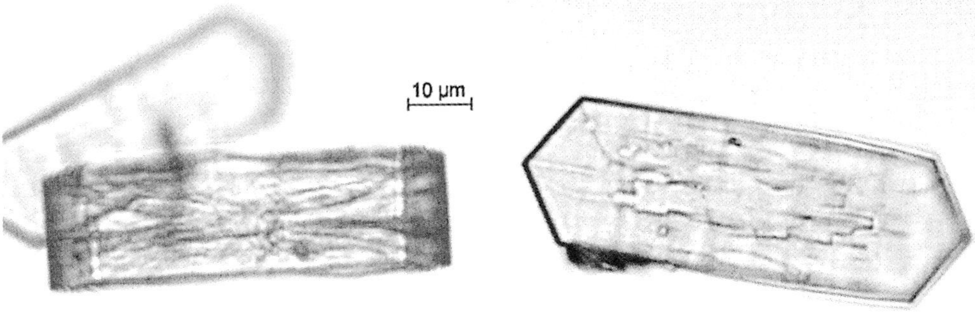

Fig. 7. Optical transmission images of a calcined silicalite-1 crystal viewed from the top (left) and from the side (right).

were immersed in a ca. 10^{-5} m solution of 4-(4-diethylaminostyryl)-1-methylpyridinium iodide (denoted stilbene dye) in ethylene glycol to stain this defect structure. The dye molecules diffuse from the liquid solution into the crystal, as shown by a confocal fluorescence image taken after about 3 h of immersion (Fig. 8). The image is a cross section through the middle of the crystal. This implies that the visible pattern comes from the inner part of the crystal. As is clearly evident from the image, the chromophores agglomerate in an irregular bandlike structure along the length of the crystal. Since the size of the stilbene dye molecules (∅ ca. 740 pm) exceeds that of the crystallographically defined pores of the MFI structure (∅ $<$ 560 pm), the recorded fluorescence must come from a defect structure.

Fig. 8. Left: confocal fluorescence image of a section through the center of a calcined MFI crystal after immersion in a solution of 4-(4-diethylaminostyryl)-1-methylpyridinium iodide (a stilbene dye) in ethylene glycol. The sample was excited at 442 nm and the fluorescence monitored with a 470 nm long-pass filter. Right: fluorescence image of the MFI crystal in Fig. 8 (left) after additional immersion in a solution of Oxazine 1 in ethylene glycol. The sample was excited at 633 nm, and the fluorescence monitored with a 650 nm long-pass filter to ensure that the luminescence depicted originates from Oxazine 1 molecules only.

To confirm this, the same experiment was carried out with a different dye, with an even larger molecular width than the stilbene dye. The same crystal was treated with a solution of 10^{-5} m Oxazine 1 in ethylene glycol and investigated under the microscope. Oxazine 1 and the stilbene dye can be monitored separately by the appropriate choice of excitation and detection wavelengths. Figure 8 (right) shows the fluorescence of Oxazine 1 (excited at 633 nm and monitored with a 650 nm long-pass filter) about 1 h after addition of the second solution. The oxazine fluorescence reveals the same pattern as is observed with the stilbene dye. First, this proves that the luminescent images highlight a defect pattern characteristic of the selected crystal. Second, this shows that the voids are large enough not to become blocked by the first dye. As emphasized earlier, confocal microscopy allows the three-dimensional imaging of samples. This feature can supply additional information and more meaningful visualizations of the mesostructure under investigation. Whereas Fig. 8 shows two-dimensional sections through the crystal, Fig. 9 (left) depicts the projection of the corresponding three-dimensional reconstruction of the stilbene fluorescence in the MFI crystal. (More projections of this crystal can be seen in Fig. 5). In the three-dimensional reconstruction an intricate network of interconnected defects is observed. The defect structures are typical for calcined MFI crystals. An analogous experiment with an uncalcined MFI crystal after immersion in a stilbene dye solution for 3 h (i.e., comparable to Fig. 8, left) was carried out (cf. Fig. 9, right). No fluorescence was detected from the interior of the crystal, and all luminescence observed originates from the outer surface. This means that the detected defect structure develops during the calcination process.

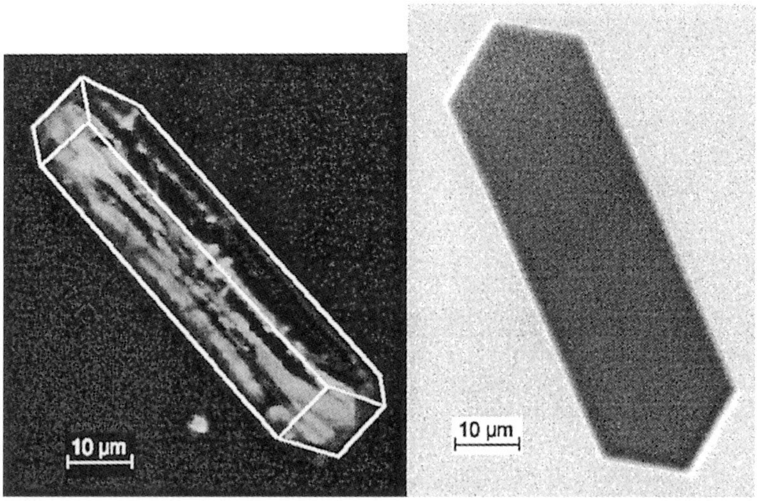

Fig. 9. Left: projection of the corresponding three-dimensional reconstruction of the stilbene fluorescence in the same MFI crystal as in Figs. 7 and 8. White lines have been added to outline the contours of the crystal. Right: fluorescence image of an uncalcined MFI crystal after immersion in a solution of the stilbene dye in ethylene glycol. The sample was excited at 442 nm and the fluorescence monitored with a 470 nm long-pass filter.

One particular effect concerning fluorescence quantum efficiencies must be taken into account. Although the frames depicted in Fig. 8 were recorded under very similar conditions, the background solution is strongly fluorescent in the case of the oxazine dye, but nearly dark for the stilbene dye. This is due to the fact that the stilbene chromophores have a low fluorescence quantum yield in solution (ca. 10%). However, as demonstrated in Fig. 8 (left), the same chromophore is strongly luminescent after diffusion into the solid specimen (ca. 35% on the inner surface). The loss of flexibility by adsorption onto the inner surface results in an increased fluorescence quantum yield. This allows enhancement of the contrast achieved in these images, but it also makes quantitative determination of chromophore content in the crystal more difficult.

3.3.1.2 Staining Defect Structures in $AlPO_4$-5 (AFI)

A molecular sieve which has generated particular interest as a host material is the aluminum phosphate $AlPO_4$-5 (AFI structure depicted in Fig. 6), because of its wide channels and the possibility to synthesize large crystals [26]. The large crystals used in our investigations were synthesized by Ö. Weiss (F. Schüth group, MPI für Kohleforschung, Mülheim).

An example of a uncalcined crystal with a length of several hundred micrometers is shown in the transmission micrograph in Fig. 10a. The crystal exhibits an almost perfect morphology with only two defects, indicated by the arrows in Fig. 10a. A fluorescence image of this crystal after immersion in a solution of 10^{-5} M Oxazine 1 in ethylene glycol (same procedure as described above for MFI) for a few hours is depicted in Fig. 10b. No fluorescence can be detected inside the crystal or at its surface; only the liquid solution around the crystal exhibits fluorescence. Obviously, no chromophores penetrate into the crystal or accumulate at its surface. A typical example for an $AlPO_4$-5 crystal after calcination can be seen in Fig. 10c. Although the morphology is unchanged, additional light-scattering structures become visible in the crystal. Again, we stained the sample by immersing it in a solution of 10^{-5} M Oxazine 1 in ethylene glycol, in this case for more than three months. After immersion, the crystals were rinsed with ethanol, and the Oxazine 1 fluorescence was recorded with the confocal microscope. As is evident from the cross section through the center of the crystal in Fig. 10d, the dye has partly penetrated into the interior of the crystal. The fluorescent spots with ellipsoidal shape in the middle of the crystal clearly outline internal defects. Here, we would like to reiterate that the figure in question is a cross section through the center of the crystal. Hence, the fluorescence observed is not a surface artifact. The large bright spots at both ends of the crystal are probably due to diffusion of the dye into the sample through the intact regular pores of $AlPO_4$-5. Even though the width of the Oxazine-1 dye molecule exceeds the pore diameter by approximately 0.1 nm, it may penetrate the pores very slowly, because of a certain flexibility inherent in both the dye and the crystal.

These investigations illustrate that calcined $AlPO_4$-5 crystals can acquire pronounced defect structures and that these defects can be visualized by staining the crystals with appropriate chromophores. The absence of defect structures in the uncalcined crystals suggest that they are of better quality.

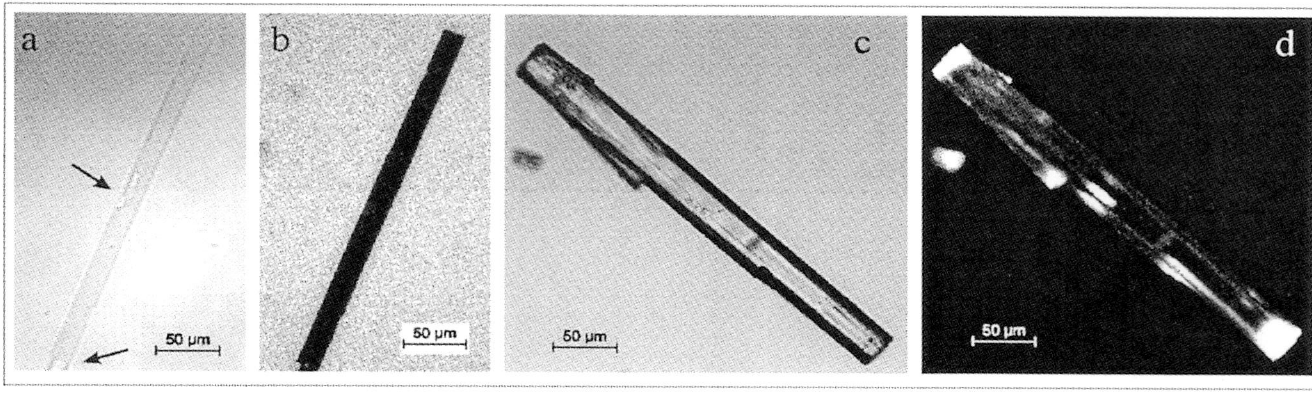

Fig. 10. Visualization of defects in $AlPO_4$-5 crystals. a) Optical transmission image of an uncalcined $AlPO_4$-5 crystal. b) Fluorescence image of the same crystal after immersion in a solution of Oxazine 1 in ethylene glycol. The sample was excited at 633 nm, and the fluorescence monitored with a 650 nm long-pass filter. c) Optical transmission image of a calcined $AlPO_4$-5 crystal. d) Fluorescence image of the crystal in c) after immersion in a solution of Oxazine 1 in ethylene glycol for three months and rinsing in ethanol (same recording conditions as in b).

3.3.1.3 Staining During Synthesis: DCM in $AlPO_4$-5 (AFI)

Apart from the method of loading the host by diffusion of the guest species after synthesis (and possibly after calcination) of a molecular sieve, there are alternative procedures available to incorporate chromophores into porous hosts, one being guest inclusion during synthesis [27]. In the following it will be shown that this technique does not guarantee a homogenous distribution of the guest. Again, confocal fluorescence microscopy allows monitoring of this distribution in a straightforward manner.

As mentioned in the introduction, one of the most exciting new applications of a host–guest systems based on a molecular sieve is the synthesis of a dye microlaser on the basis of $AlPO_4$-5 [11]. One of the samples used was an *in situ* synthesized composite of the well-known laser dye DCM with $AlPO_4$-5 (structures depicted in Fig. 6). These crystals were obtained by Ö. Weiss (F. Schüth group, MPI für Kohleforschung, Mülheim). They are exceptionally large due to the addition of HF during synthesis [26]. We used the confocal microscope to record a three-dimensional fluorescence image of such a DCM/$AlPO_4$-5 crystal. Transmission and fluorescence images of the whole specimen are shown in Fig. 11. The transmission image (Fig. 11a) shows a large $AlPO_4$-5 crystal with a well-defined morphology. In contrast, in the confocal fluorescence image (Fig. 11b) a heterogeneous distribution of the dye is apparent. A higher resolution image of the central part of another $AlPO_4$-5 crystal from same batch shows a similar pattern (Fig. 12, top). The dark area along the main axis of the crystal, which becomes broader from the center to the end of the crystal, forms a substructure. The images suggest that the dye molecules are excluded from this volume. Cross-section fluorescence images of the crystal enable a better visualization of these substructures. In Fig. 12 (bottom) three cross-sections are shown, with increasing distance to the center of the crystal

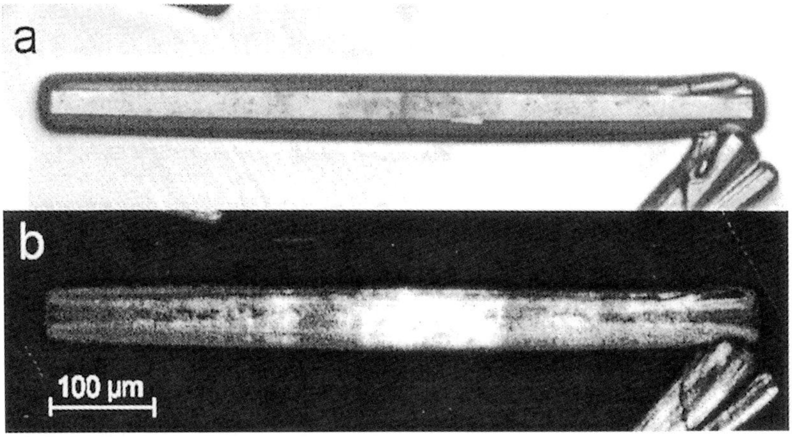

Fig. 11. Transmission (top) and fluorescence (bottom) image of a large $AlPO_4$-5 crystal stained with DCM dye during synthesis.

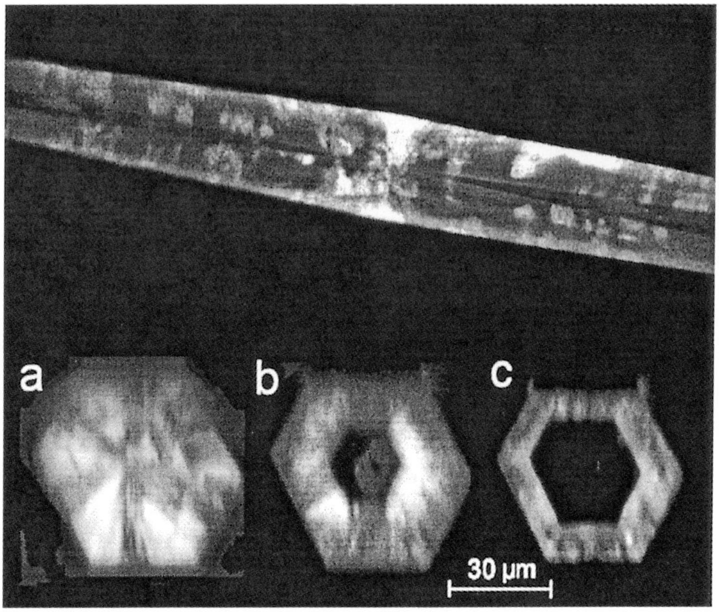

Fig. 12. Top: detailed fluorescence image of the central part of a large $AlPO_4$-5 crystal stained with DCM dye. Bottom: cross sections of the large $AlPO_4$-5 crystal stained with DCM. From left to right: increasing distance from the center.

along the main axis. The dark region has a hexagonal shape that increases in size from the center to the end of the crystal, whereas the irregularly stained region is confined to the peripheral part of the hexagon.

The origin of the spatial distribution of the guest cannot be inferred directly from these measurements. However, the visualization of these heterogeneities will certainly contribute to the development of better materials for lasing crystals.

3.3.2 Observation of Diffusion

Confocal microscopy is not limited to the static visualization of guest distributions. A series of pictures also allows a detailed insight into the dynamics of the diffusion processes taking place during the incorporation of guest chromophores. Figure 13 shows a monolithic fragment of an uncalcined mesostructured M41S sample. For the synthesis of the monolithic M41S-type samples, monododecyloctaethylene glycol ether (OMO) was used as structure-directing agent with tetramethoxysilane (TMOS) as silica source [28]. This material has hexagonally ordered pores as in the MCM-41 material (cf. Fig. 6). The samples were synthesized by S. Altmaier (P. Behrens group, Universität Hannover). A ca. 10^{-5} m solution of the stilbene

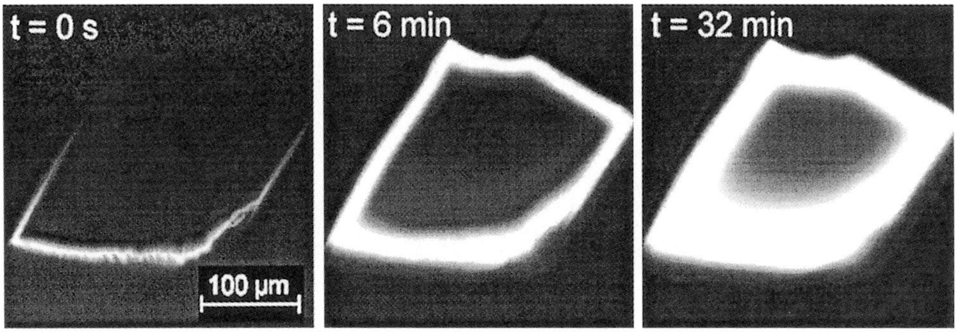

Fig. 13. Time-resolved imaging of the penetration of stilbene dye into a monolithic fragment of an uncalcined mesostructured M41S. Fluorescence image obtained at the beginning of the dye uptake from ethylene glycol solution (a), after 6 min (b), and after 30 min (c).

dye (cf. Fig. 6) in ethylene glycol was added to the sample, and the stilbene dye fluorescence was monitored with the microscope. The three images in Fig. 13 were recorded at different times after immersion in the chromophore solution. The confocal microscope records the images in a scanning fashion, starting at the top, with a total scan time of about 1 min for each image. The left image shows the fluorescence at the beginning of the chromophore incorporation. Due to the finite recording time, chromophores have already penetrated the lower part of the fragment by the time the lower part of the image was scanned, whereas the upper part is still dark. Due to the above-mentioned low fluorescence quantum yield in solution, the liquid surrounding the fragment is dark and the contrast in the picture is drastically enhanced. Figure 13 middle (after 6 min) and right (after 30 min) depict the progress of dye diffusion into the molecular sieve with time. As there is no preferential direction in the observed dye uptake no unidimensional order in the pore structure is observed on an optically limited length scale. In this case the diffusion is much faster than, for instance, in the case of Oxazine 1 in $AlPO_4$-5 (see Fig. 10 d). The timescale of guest diffusion from the solution into a molecular sieve obviously depends on the size of the molecule and the extensions of the pores, besides many other factors. Time-resolved micrographs allow a straightforward estimation of the diffusion coefficient. For the specific example depicted here a value of $D = 1.7 \times 10^{-10}$ cm^2 s^{-1} was found. (cf. Fig. 14 left).

To put this into a meaningful context we have performed similar measurements with different dyes and solvents, for the same host material (Fig. 14, right). Surprisingly, diffusion becomes slower in calcined materials, and ionic dyes in polar solvents diffuse more slowly than nonionic dyes in apolar solvents. The size effect, i.e., that small molecules (tetracene) diffuse faster than large molecules (TDI), is as expected from theory.

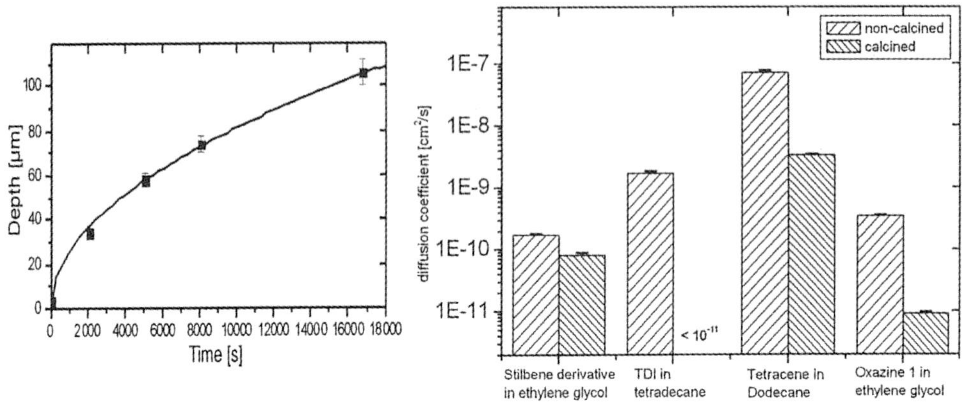

Fig. 14. Left: analysis of the time-dependent penetration depth by diffusion. A diffusion coefficient of $D = 1.7 \times 10^{-10}$ cm^2 s^{-1} was derived from the data. The data points were fitted with the diffusion relation $d = (2t)^{1/2}$. Right: diffusion coefficients for the uptake of different dye/solvent systems by the calcined and uncalcined M41S material.

3.3.3
Stilbene Derivative in $AlPO_4$-5 (AFI)

Homogeneous inclusion of dye was observed when $AlPO_4$-5 crystals were synthesized in the presence of the stilbene dye (structure depicted in Fig. 6). The crystals were synthesized by A. Glaue (P. Behrens group, Universität Hannover) using heated ethylene glycol. The obtained crystals are just large enough to allow confocal measurements (see Fig. 15a). The confocal fluorescence images show a uniform distribution of the dye. It can be shown that the intensity of the fluorescence is strongly dependent on the polarization of the excitation. The fluorescence almost vanishes when the polarization plane is perpendicular to the long crystal axis (Fig.

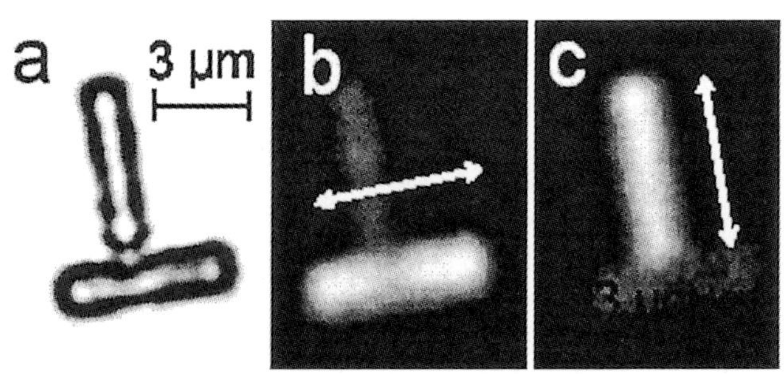

Fig. 15. Images of two orthogonally oriented stilbene-dye-loaded $AlPO_4$-5 crystals. a) Transmission image; b) and c) fluorescence images of the same two crystals. The arrows indicate the polarization plane of the excitation light.

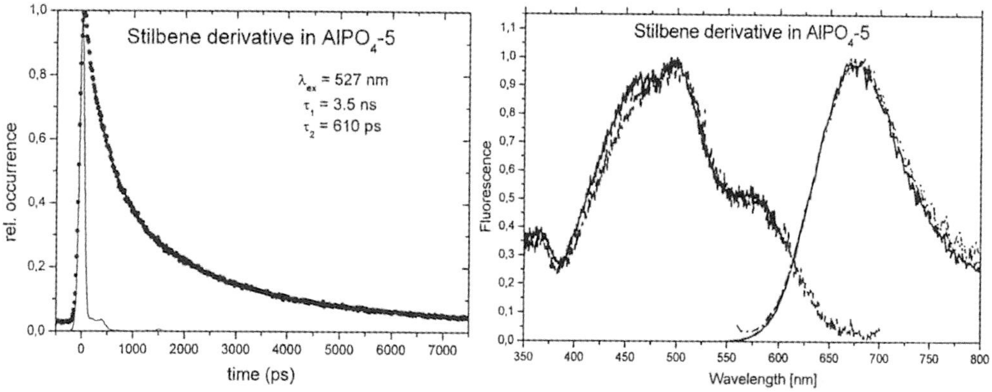

Fig. 16. a) Fluorescence lifetime measurements of stilbene dye in $AlPO_4$-5; the thin line corresponds to the instrument response function. The decay was determined to be biexponential with decay times of 610 ps and 3.5 ns. b) Fluorescence excitation spectra and fluorescence emission spectra of stilbene dye in $AlPO_4$-5. The lines correspond to different excitation (400 and 420 nm) or detection (580 and 620 nm) wavelengths. The spectra are not significantly affected by this choice of wavelengths.

15b and c). From this we conclude that the transition dipole moment of the incorporated dye, and hence the dye itself, is orientated along the pores.

Besides the structural information, the dynamics of the fluorescence decay were examined. The dyes were excited with a pulsed Nd-YLF laser (70 ps, $\lambda = 527$ nm), and the fluorescence photons detected with a fast single-photon detector (microchannel plate). The fluorescence of the dye molecules shows a biexponential decay (Fig. 16a), which can be explained by the presence of two distinct populations or by the presence of two distinct relaxation processes. However, none of these heterogeneities significantly affect the fluorescence excitation and the fluorescence emission spectra of this sample (see Fig. 16b)

3.3.4
Terrylene in MCM-48 and MCM-50

Different mesostructured M41S materials were prepared with incorporated terrylene by standard template-assisted synthesis by S. Altmeier (P. Behrens group, Universität Hannover; the structures are shown in Fig. 6). The material is available as a powder consisting of crystal agglomerates with a crystal size below the optical resolution limit (Fig. 17). Therefore, it is not possible to determine from the image whether the dye is located on the surface or in the interior of the crystal. However, due to the high hydrophobicity of terrylene, we assume that it remains located within the hydrophobic part of the template used for the synthesis, and is thus located in the inner part of the crystal. The chemical environment of the molecules can be assumed to be very similar in both host materials, MCM-48 and MCM-50.

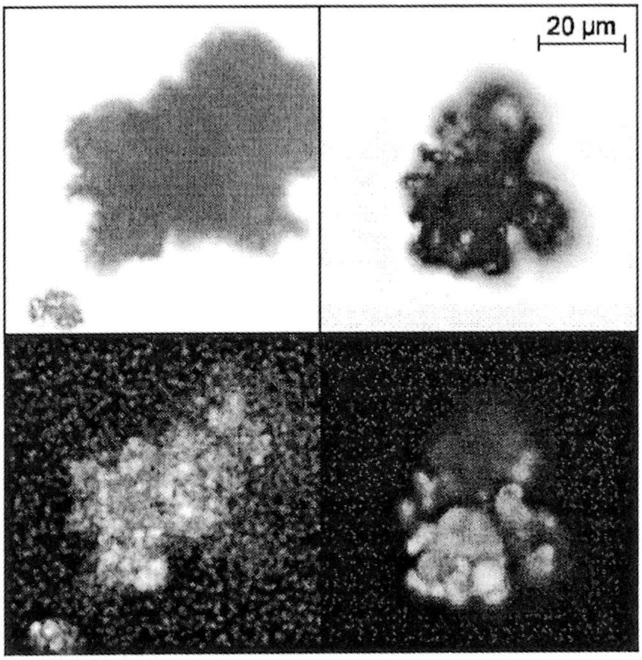

Fig. 17. Top row: transmission images; bottom row: corresponding fluorescence images of terrylene in MCM48 (left) and in MCM50 (right).

However, fluorescence lifetime measurements show an interesting difference in the relaxation behavior of the terrylene dye in the two hosts (Fig. 18). The most pronounced difference between these systems is the appearance of an additional fast fluorescence relaxation in the time range of 400 ps which may be displayed by some of the dye molecules in the MCM-48 sample. This difference must be attributed to the different topology of the pore structure.

3.3.5 Single Molecules: Perspectives

The development of various techniques for the detection and characterization of individual molecules has been one of the most important breakthroughs in the area of optical spectroscopy within the last decade [19]. Single-molecule measurements have shown that the properties measured are normally inhomogeneously distributed. This means that each molecule has its individual value for a specific property. Thus, single-molecule measurements will not only yield the mean value of such a distribution but also disclose the distribution itself. Furthermore, the recording of the fluorescence of an individual chromophore as a function of time can reveal dynamic phenomena, which are otherwise hidden in bulk spectroscopy due

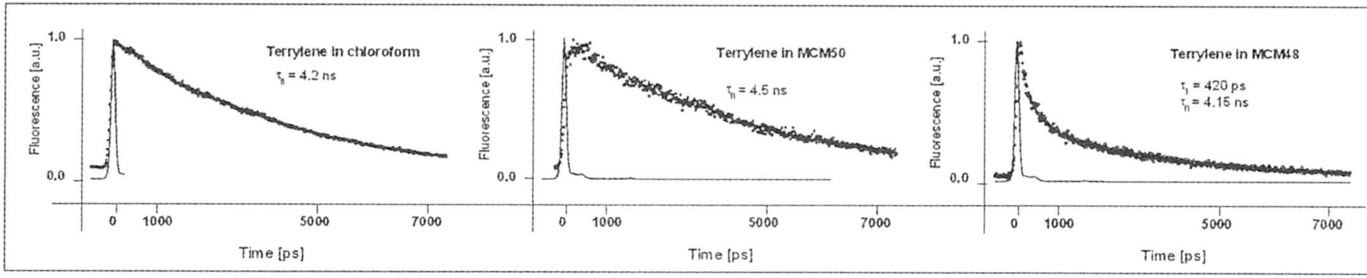

Fig. 18. Fluorescence lifetime measurements of terrylene in chloroform (left), in MCM50 (middle), and MCM48 (right). The curves were fitted with mono- or biexponential decay functions; the thin line is the instrument response function.

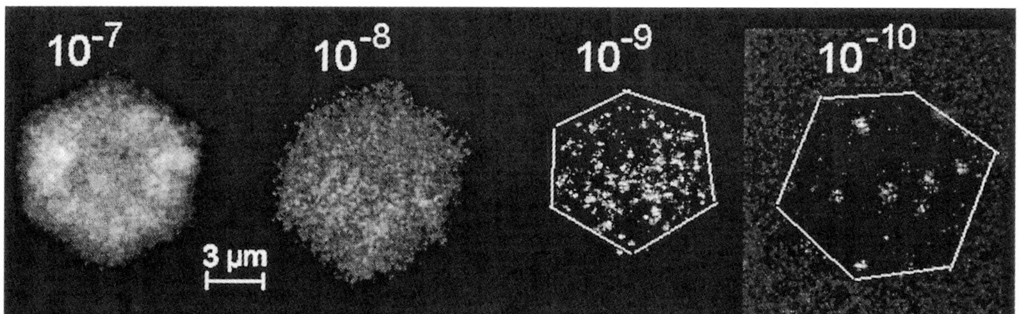

Fig. 19. Fluorescence images of $AlPO_4$-5 crystals loaded with Oxazine 170 dye molecules. The numbers denote the concentration of Oxazine 170 in mmol g^{-1} used during the synthesis of the crystals. The lines in the highly diluted samples represent the shape of the crystals as taken from the respective transmission images. The fluorescent spots on the most diluted sample are the diffraction-limited images of the fluorescence of single Oxazine 170 dye molecules.

to the averaging effect of the ensemble. For instance the jumping of a molecule between the singlet and triplet manifold can be observed in real-time in a single-molecule experiment [29]. In a bulk experiment this jumping would not be observable, because of the nonsynchronous behavior of the individual molecules and thus would be averaged out. The characterization of host–guest systems, in particular with regard to the distribution of properties and to the sensing of dynamic phenomena, can become significantly more detailed by the use of single-molecule spectroscopy as an evaluation technique.

We were recently able to detect single chromophores in molecular sieves. A series of samples of oxazine dyes incorporated into $AlPO_4$-5 (AFI) crystals was investigated. The $AlPO_4$-5 crystals loaded with Oxazine 170 (structure depicted in Fig. 6) were synthesized by M. Ganschow (D. Wöhrle group, Universität Bremen) using a microwave oven to reduce the reaction time and thereby avoid the degradation of the dye molecules [30]. The samples were examined for series of dye concentrations, and the corresponding fluorescence images were recorded. At a dye concentration above 10^{-8} mmol g^{-1} the images show a uniform fluorescence signal pattern. Below this concentration distinct fluorescing spots can be observed. The spots have a diameter of 300 nm, which corresponds to the diffraction limit of the microscope (see Fig. 19).

A simple verification of the presence of single molecules is given by the "digital" on/off bleaching behavior that can be observed by recording a time trace of the fluorescence intensity. In a photobleaching experiment the samples with high dye concentrations show an exponential decrease of fluorescence intensity with time (Fig. 20, first graph). Then, at a lower dye concentration (Fig. 20, second graph), the decrease in fluorescence is no longer smooth, and this indicates the presence of a small ensemble of molecules in which the individual bleaching behavior of each molecule becomes apparent. The third graph shows a two-step bleaching fluorescence "decay". This clearly indicates the presence of two single molecules with individual bleaching times. Single-step bleaching, as expected for single molecules, was found for samples at a dye concentration below 10^{-8} mmol g^{-1}. This confirms that the observed dots in Fig. 19 (right) are the diffraction-limited patterns of the fluorescence from single Oxazine 170 molecules. Another interesting feature visible in the last graph of Fig. 20 is the presence of some on/off steps, termed "blinking" [31]. Various effects can give rise to this behavior, e.g., triplet formation, transient quenching, and transient charge transfer. Again, it is worth noting that these dynamic processes are also present in the bulk sample, but are hidden by the averaging effect of the ensemble. The sum over all of the individual on/off dynamic traces of the single molecules present in an ensemble yields the observed exponential decay.

A further application of single-molecule spectroscopy is measuring the orientational distribution of guest molecules incorporated in molecular sieves. Here again, we can not only measure the mean orientation angle of the molecules, as in ensemble experiments (cf. Section 3.3.3, Fig. 15), but also the explicit distribution, i.e., the orientation of each individual molecule [32]. In future experiments we will

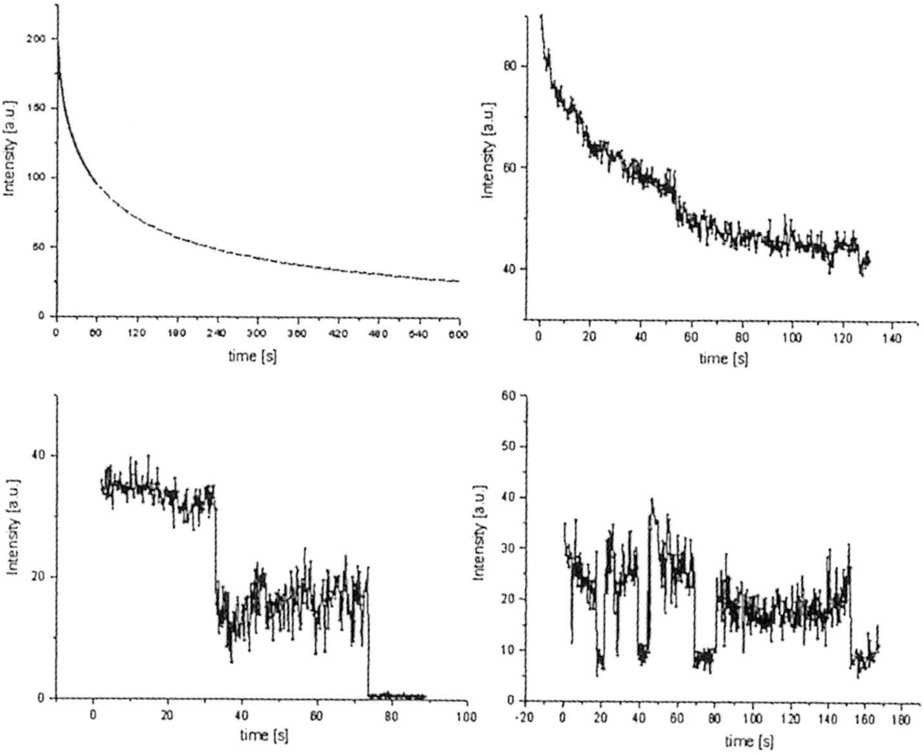

Fig. 20. Fluorescence intensity versus time plots of Oxazine 1 incorporated in $AlPO_4$-5 crystals. The concentration decreases from left to right, and the appearance of single molecules is indicated. The count rate can be estimated by multiplying the arbitrary units by a factor of 100 for the left curve and by 3 for the remaining curves). For details, see text.

continue the examination of single molecules incorporated in molecular sieves to reveal the degrees of freedom of such molecules in nanometer-sized environments. This should finally lead to single-molecule-based functional devices, as described in the introduction (see Fig. 1).

3.4 Conclusion

In conclusion, the main advantage of optical confocal microscopy is the capability to detect a very small volume element on the order of 0.1 fL. Three-dimensional images can be reconstructed from the scanned cross sections through the sample. By this method it is possible to spatially resolve defect structures in porous molecular sieves, and to show the distribution of incorporated fluorescent guest molecules in diverse host materials. In addition, it is also possible to observe the uptake process of dyes into host materials and to evaluate the respective diffusion coefficients. Furthermore, the orientational, spectral, and dynamic properties of molecules in a tiny volume element of the sample can be characterized. Finally, by improving the sensitivity of the detectors, investigations on the level of single molecules can be achieved and additional information accessed. Altogether, these capabilities permit a straightforward characterization of host–guest composites, a promising new class of supramolecular materials.

References

1 J.K. Gimzewski, C. Joachim, *Science* **1999**, *283*, 1683.

2 P. Behrens, G.D. Stucky, *Novel Materials Based on Zeolites* in *Comprehensive Supramolecular Chemistry* (Eds. J.L. Atwood, D.D. MacNicol, J.E.D. Davies, F. Vögtle), Vol. 7 (Eds. G. Alberti, T. Bein), Pergamon Press, Oxford, **1996**, p. 721.

3 K. Moller, T. Bein, *Chem. Mater.* **1998**, *10*, 2950.

4 G.A. Ozin A. Kuperman, A. Stein, *Angew. Chem.* **1989**, *101*, 373; *Angew. Chem. Int. Ed. Engl.* **1989**, *28.*

5 F. Schüth, *Chem. unserer Zeit* **1995**, *29*, 42.

6 Y. Ishii, T. Yanagida, *Single Mol.* **2000**, *1*, 5.

7 S. Weiss, *Science* **1999**, *283*, 1676

8 P.J. Langley, J. Hulliger, *Chem. Soc. Rev.* **1999**, *28*, 279.

9 J. Caro, F. Marlow, M. Wübbenhorst, *Adv. Mater.* **1994**, *6*, 413.

10 C.G. Wu, T. Bein, *Science* **1994**, *265*, 1757.

11 G. Ihlein, F. Schüth, O. Krauß, U. Vietze, F. Laeri, *Adv. Mater.* **1998**, *10*, 1117.

12 N. Gfeller, S. Megelski, G. Calzaferri, *J. Phys. Chem. B* **1998**, *102*, 2434.

13 a) D. Wöhrle, G. Schultz-Eckloff, I. Braun, C. Schomburg, F. Laeri, U. Vietze, M. Ganschow, Y. Rohlfing, T. Bogdahn-Rai, *J. Inf. Rec.* **2000**, *25*, 87; b) F. Kulzer, S. Kummer, R. Matzke, C. Bräuchle, T. Basché, *Nature* **1997**, *387*, 688.

14 J.L. Meinershagen, T Bein, *J. Am. Chem. Soc.* **1999**, *121*, 448.

15 M. Wark, G. Grubert, M. Warnken, G. Schultz-Eckloff, M. Ganschow, Y. Rohlfing, T Bogdahn-Rai, D. Wöhrle, *Applied Mineralogy.* Ed: Rammimair et al., Balkerna, Rotterdam, **2000**, p. 253.

16 R.W.J. Scott, S.M. Yang, G. Chabanis, N. Coombs, D.E. Williams, G. Ozin, *Adv. Mater.* **2001**, *13*, 1468.

17 J.B. Pawley (Ed.), *Handbook of Biological Confocal Microscopy*, Plenum Press, New York, **1995**.

18 T.R. Corle, G.S. Kino, *Confocal Scanning Optical Microscopy and Related Imaging Systems*, Academic Press, San Diego, **1996**.

19 T. Basché, W. E. Moerner, M. Orrit, U. P. Wild (Eds.), *Single-Molecule Optical Detection, Imaging and Spectroscopy*, VCH, Weinheim, **1996**.

19b T. Wilson, C. Sheppard, *Theory and Practice of Scanning Optical Microscopy*, Academic Press, London, **1984**.

21 W. Becker, H. Hickl, C. Zander, K.H. Drexhage, M. Sauer, S. Siebert, J. Wolfrum, *Rev. Sci. Instr.* **1999**, *70, 1835.*

22 M. Maroncelli, G.R. Fleming, *J. Chem. Phys.* **1997**, *86*, 6221.

23 W.W. Webb, *Appl. Opt.* **2001**, *40*, 3969.

24 C. Seebacher, J.P. Rau, F.W. Deeg, C. Bräuchle, S. Altmaier, R. Jäger, P. Behrens, *Adv. Mater.* **2001**, *13*, 1374.

25 J.L. Guth, H. Kessler, R. Wey, *Stud. Surf. Sci. Catal.* **1986**, *28*, 121.

26 I. Girnus, M. Poll, J. Richter-Mendau, M. Schneider, M. Noack, D. Venzke, J. Caro, *Adv. Mater.* **1995**, *7*, 711.

27 R. Hoppe, G. Schulz-Ekloff, D. Wöhrle, E.S. Shpiro, P.P. Tkachenko, *Zeolites* **1993**, *13*, 222.

28 G.S. Attard, J.C. Glyde, C.G. Göltner *Nature* **1995**, *378*, 366.

29 T. Basché, S. Kummer; C. Bräuchle, *Nature* **1995**, *373*, 132.

30 M. Ganschow, G. Schulz-Ekloff, M. Wark, M. Wendschuh-Josties, D. Wöhrle, *J. Chem. Mater.* **2001**, *11*, 1823

31 W. Göhde, U.C. Fischer, H. Fuchs, J. Tittel, T. Basché, C. Bräuchle, A. Hermann, K. Müllen, *J. Phys. Chem. A*, **1998**, *102*, 9109.

32 C. Seebacher, C. Hellriegel, C. Bräuchle, M. Ganschow, D. Wöhrle, unpublished results.

4
New Microlasers Based on Molecular Sieve/Laser Dye Composite Materials

Özlem Weiß, Ferdi Schüth, Justus Loerke, Frank Marlow, Lhoucine Benmohammadi, Franco Laeri, Christian Seebacher, Christian Hellriegel, Fred-Walter Deeg, and Christoph Bräuchle*

4.1
Introduction

One of the objectives in the search for novel materials for optical applications is to develop smaller high-efficiency devices made of pure substances or composites. This ambition can be pursued by either "top down" or "bottom up" strategies. The former approach is realized by reducing the size of devices made of given materials in order to modify and improve optical properties; the latter starts from small atomic units assembled to clusters and nanostructures which are ordered in two or three dimensions on a length scale in the submicrometer domain.

A third approach consists of using structures already ordered in three dimensions as host and modifying the host properties by including optically active species as guests. For this purpose molecular sieves can serve as versatile host materials. Their microporous three-dimensional structures with channel or cavity architectures are ideally suited as ordering and stabilizing frameworks. Introducing optically active molecules, clusters, or ions into the pore system results in materials with new properties. The properties of the composites can be tuned by modification of the host crystal or the guest molecule.

4.2
Host–Guest Composites based on Molecular Sieves

Over the past ten years several novel host–guest composites based on molecular sieves with new functions have been proposed or were realized. The different host systems that have been used range from zeolites and aluminophosphates to ordered mesoporous materials [1–4]. To realize new luminophores, optically efficient inorganic ions, for example, Ce^{3+} [5] or Eu^{3+} [6,7], were introduced by ion exchange into zeolite X and Y, which are suitable hosts due to their optical transparency [8]. The advantage of such systems is that the optically relevant centers are located at crystallographic sites in the pore system, thus allowing further modification by additional ligands. For example zeolite X:Eu^{3+} composites show a quantum efficiency (QE) of 7% due to quenching effects by OH^- ions [7,9]. However, after introduction of molybdate ions as sensitizers and calcination to obtain water-free sodalite [3], the QE was increased to 55%.

In another example the creation of new data-storage devices based on spectral hole burning was intended by exchanging cationic dye molecules into the pores of zeolitic host materials. The channel diameter of the host material must be large enough to accommodate the guest molecule. Thionine [10] and methylene blue [11] were incorporated in zeolite L and NaY, respectively. The fact that the guests are incorporated as monomers inspired further efforts in this field, but the exchange rates turned out to still be very low.

Larger molecules, such as laser dyes, porphyrins, and other chromophores, can be incorporated by crystallization inclusion, in which the guest molecules are simply added to the synthesis gel. The successful incorporation and, compared to NaY, increased uptake of methylene blue in $AlPO_4$s [12] directed attention towards aluminophosphates as host materials. The stabilizing function of the solid host matrix was shown with $AlPO_4$-5/porphyrin composites [13]. Bulky cationic porphyrins were incorporated in defect sites of $AlPO_4$-5. The incorporated dyes exhibit an enhanced photostability in comparison to their photostability in solution. This observation stimulated the search for new pigments which could be stabilized against bleaching. Various photo-unstable coumarines and azo dyes [14] could be introduced into $AlPO_4$-5 without impairment, due to the reduced crystallization times made possible by microwave synthesis. For the inclusion of large rhodamine derivatives [15] it was shown that the inclusion rate is increased when interactions of the guest molecule with the $AlPO_4$-5 framework are comparable to the framework–template interactions. This was realized with positively charged rhodamine dyes having template-analogous functional groups.

Another approach to produce host–guest composites is the adsorption of the guest from the gas or the liquid phase. This method is suitable for small organic molecules which would decompose under the synthesis conditions of microporous materials. For example, a new photoswitch was realized with azobenzene in $AlPO_4$-5 and ZSM-5 [16,17]. Introduction of various small molecules with a high dipole moment, such as *para*-nitroaniline or dimethylaminobenzonitrile, into $AlPO_4$-5, $AlPO_4$-11, VPI-5, and ZSM-5 yielded materials with high second-order optical polarizability. These composites show anisotropic light absorption and second harmonic generation (SHG) [18–22]. Raman spectroscopy proved the straight arrangement of the molecules to form dipole chains in the channels [23], and studies of pyroelectric effects clarified the loading mechanisms [24].

4.3 Microporous Aluminophosphates

Since aluminophosphates are the most frequently used family of host materials, they are discussed in more detail in the following. The huge family of molecular sieves was expanded in 1982 by the microporous aluminophosphates [25]

($AlPO_4s$). In analogy to zeolites, in $AlPO_4s$ the lattice sites are occupied by tetrahedrally coordinated Al^{3+} and P^{5+} ions connected by oxygen atoms to give a neutral framework, in contrast to zeolites, whose framework is normally negatively charged. This class of material was modified first by substitution of the framework atoms by silicon (SAPO) [26], and later also by further elements, e.g., metal-substituted $AlPO_4s$ (MeAPOs). $AlPO_4s$ have a defined framework of regularly arranged pores with diameters up to 1.3 nm. The possibility of substitution with optically active ions or inclusion of guest molecules in the channel system make these materials interesting for designing new optical materials.

A particular member of this class is $AlPO_4$-5 (AFI structure), which crystallizes in rodlike hexagonal prisms and has a one-dimensional channel system with a pore size of 0.73 nm. It is possible to synthesize $AlPO_4$-5 crystals with good optical transparency and low internal scattering losses. Large $AlPO_4$-5 crystals have been synthesized in various dimensions by different methods. The crystal size ranges from 100 μm in microwave synthesis [27] to 500–750 μm for crystals grown under conventional hydrothermal conditions [28] with triethylamine (TEA) as template. It was also reported that $CrAPO_4$-5 and MgAPO-5 crystals were achieved with lengths of up to 1 mm and 1.7 mm, respectively, when synthesized from concentrated gels with HF, whereby these crystals showed distorted morphologies [29,30].

4.3.1
Synthesis of Large, Perfect $AlPO_4$-5 Crystals

The successful synthesis of large perfect $AlPO_4$-5 crystals is influenced by several parameters. Amongst the most crucial ones is the aluminum source. It was shown that an aluminum oxide hydrate sol, AlO(OH), yields high quality AFI crystals with length of up to 200 μm, whereby the source of the commercially available supply turned out to be very important. This is a consequence of the different preparation methods employed by different suppliers. With an amorphous $Al(OH)_3$ precipitate which transforms into hydrargillite by aging we developed a suitable aluminum source for obtaining optically clear AFI crystals with perfect hexagonal morphology [31,32]. Alternatively, working at higher dilution with tripropylamine (TPA) as template and addition of alcohol to the synthesis mixture resulted in the formation of very large $AlPO_4$-5 crystals as well [33]. The obtained crystals are hexagonally shaped with a maximum length of up to 1.3 mm (Fig. 1a) and exhibit perfect faces (Fig. 1b). The insolubility of the template in the absence of alcohol induces deep pits in the surface of the crystals, though the length is retained and reaches up to 1.4 mm (Fig. 1c). During these experiments a further byproduct of $AlPO_4$-36 syntheses was identified and structurally classified as $AlPO_4$-36 (Fig. 1d), in contrast to previously used synthesis systems with TEA, in which only dense aluminum phosphate phases were identified as byproducts. In general it can be stated that it is possible to synthesize $AlPO_4$-5 with good optical quality, but the synthesis is very susceptible to small deviations in parameters (see also Part 1, Chap. 4).

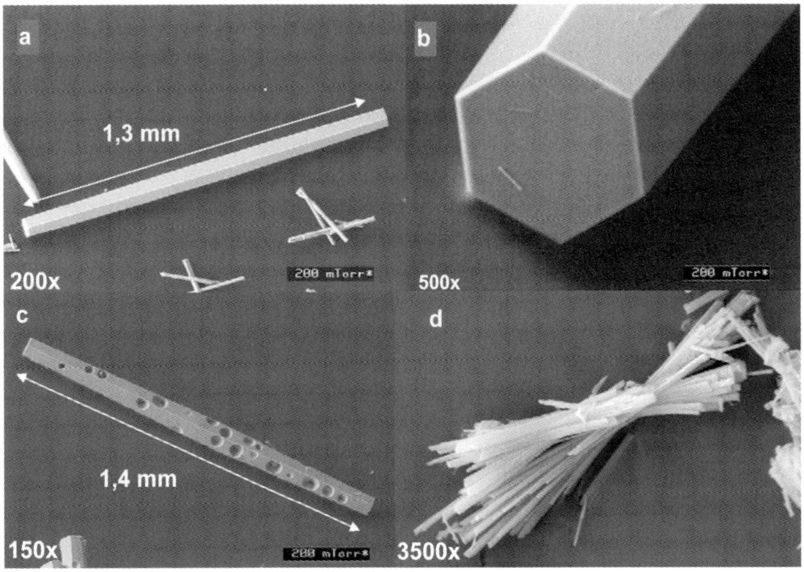

Fig. 1. Scanning electron micrographs of a) perfect hexagonal $AlPO_4$-5, b) view of the (001) plane of an $AlPO_4$-5 crystal, c) 1.4 mm sized crystal with holes, d) $AlPO_4$-36 crystals.

4.4 Single-Crystal Microlasers

A new addition to the field of laser materials are composites based on molecular sieves. The introduction of laser-active ions and commercially available laser dyes into the well-defined pore structure of zeolites and zeolitic materials opens new perspectives. By varying the laser-active guest species a large number of microlaser systems could be realized. The new composite materials can also give further information on the host or the guest. For example, host effects on the electronic nature of a laser dye and influence of the guest on the crystallization mechanism were observed. The fundamental advantage of using $AlPO_4$-5 as laser material is its morphology. Plane parallel end faces are favorable prerequisites for a Fabry–Perot resonator, and with the hexagonal morphology a ring resonator can be generated. With the ability to control the crystal size, the optical properties of the resonator can be modified, and the low roughness of the crystal faces makes polishing of the surface unnecessary. This is mandatory if microlasers are to be prepared, since for micrometer-sized objects additional processing steps, such as polishing, are difficult and costly.

In first attempts to realize new laser materials it was considered that, like in the ruby laser, Cr^{3+} ions in SAPOs could show laser activity, but it turned out that although the luminescence of CrAPSOs is similar to that of ruby, the emission is inhomogeneously broadened [34], and no laser emission could be generated. Furthermore, an attempt was made to produce a four-level laser analogous to the Nd:YAG laser by introducing Nd^{3+} ions into SAPOs [32]. However, the emission intensities of this material were also low compared to Nd:YAG [35] owing to quenching processes by OH^- ions [36], and again it was not possible to generate laser light. The route based on simple ions as laser active species was thus considered to be unpromising, and attention focused on a different pathway.

The approach which proved to be successful in designing new laser materials relies on crystallization inclusion of organic laser dye molecules in $AlPO_4$-5 crystals [37–39] and MCM-41 fibers [40]. The advantage of the $AlPO_4$-5/laser dye material is, besides the given ring resonator geometry, the alignment of the molecules in the one-dimensional channels and the thermally and chemically stable crystalline matrix. An important point for optimizing such systems is to understand inclusion and crystallization mechanisms in order to open modification pathways, and detailed optical characterization is needed to quantify these new materials. In the following the characteristics of $AlPO_4$-5/laser dye materials will be described in detail.

4.4.1 Morphology of $AlPO_4$-5/Laser Dye Crystals

The synthesis of $AlPO_4$-5 crystals with a well-developed morphology is the basis for designing a laser-active material. With the given morphology a flexible microlaser system with frequencies spanning the whole UV-vis region can be generated by varying the included laser dye. By means of crystallization inclusion different dyes have been incorporated, ranging from different coumarine dyes absorbing in the UV [41,42], Pyridine 2 [37] and DCM [43] (Fig. 2) absorbing in the visible region, to bulky derivatives of rhodamine [15].

In synthesizing new microporous laser materials it is very important that the presence of the dye does not disturb the crystallization, which would lead to distorted morphologies. Depending on the synthesis method, different types of crystal morphologies result. With conventional hydrothermal synthesis large, perfect

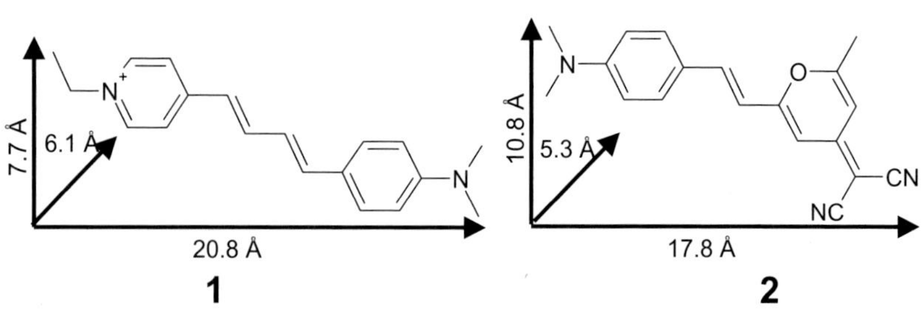

Fig. 2. Van der Waals dimensions and molecular structures of the dyes 1-ethyl-4-(4-(*p*-dimethylaminophenyl)-1,3-butadienyl)pyridinium perchlorate (**1**, trade name: Pyridine 2) and 4-dicyanomethylene-2-methyl-6-(*p*-dimethylaminostyryl)-4*H*-pyran (**2**, trade name: DCM).

rodlike crystals are obtained [42], whereas microwave synthesis results in small barrel-shaped specimens [15]. In the following, the focus is on the optical properties of $AlPO_4$-5 crystals loaded with Pyridine 2 and DCM. By comparing the inclusion behavior of these two dyes, information on degrees of loading and the location of the dye molecules in dependence on the molecular structure and molecular dimensions of the guest molecules was obtained.

We observed that in conventional hydrothermal synthesis the cationic dye Pyridine 2 interacts strongly with the growing crystal environment and thus leads to intergrown crystals. With increasing dye content the presence of Pyridine 2 resulted in a bundlelike morphology of $AlPO_4$-5 [37]. This fact is attributed to the interaction of template-analogous functional groups of the dye with the surrounding framework, similar to the template–framework interactions. In the presence of TPA, hydroxide ions are required as counterions. They coordinate framework aluminum atoms and result in a distorted fivefold coordination [44,45]. Judging from the structure of Pyridine 2 and given the position of the pyridinium moiety, protonation of the aminophenyl group will lead to a bifunctionally active molecule that influences crystal growth. A high fraction of distorted coordination environments of framework atoms will result, and these could be the origin of defect positions leading to the observed intergrowth. With uncharged molecules these interactions are minimized and crystal growth is less disturbed. Defects of the pore structure can also be induced by the inclusion of molecules with formation of internal mesoporous cavities. The calcination behavior of $AlPO_4$-5/DCM crystals is an indication that such defects form. The brown coloration that develops after calcination at 873 K in air, which is due to blocking of the channels by carbonaceous residues, is much less pronounced at higher degrees of dye incorporation [42], and this suggests the presence of more mesopores which facilitate mass transfer.

4.4.2
Optical Properties of Laser Dyes in $AlPO_4$-5

A well-defined spatial distribution of the dye molecules is of extreme importance for host–guest systems. With powder samples information must be gained from reflection spectra. Then it must be ensured that byproducts are not present which could influence inclusion levels and spectroscopic properties. Single-crystal characterization gives direct information on the degree of dye loading and the location of the molecules. The knowledge about inhomogeneity and possible limits of the dye concentration are important for further modifications. If the qualitative and quantitative characteristics of a new material are known, direct comparison with commonly used materials is possible.

Absorption spectra of single $AlPO_4$-5/Pyridine 2 (Fig. 3a) and $AlPO_4$-5/DCM (Fig. 3b) crystals show that the maximum absorption is in the same spectral region as when the molecule is dissolved in an organic solvent [46], that is, the organic molecule is still intact. With the obtained spectra not only the spectroscopic properties of the dye molecules are characterized, but solvent effects on absorption and fluorescence spectra can be used to provide information on solute–solvent inter-

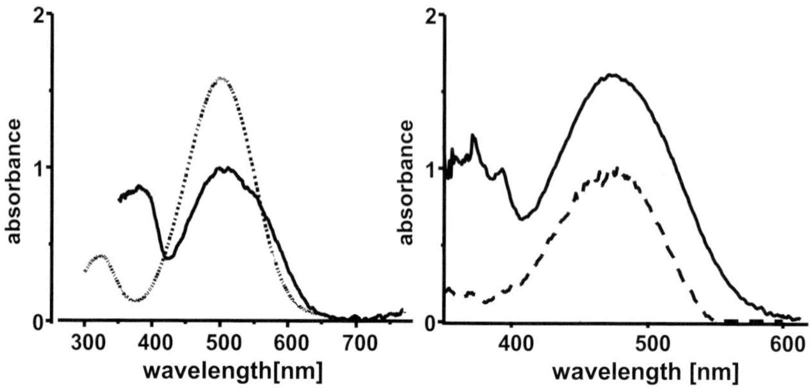

Fig. 3. Absorption spectra of single crystals (solid lines) of $AlPO_4$-5/Pyridine 2 (left) and $AlPO_4$-5/DCM crystals and corresponding spectra in ethanol (dashed lines).

actions [47] and thus the host–guest system. DCM is a highly solvatochromic molecule that will sensitively reveal dipole–dipole interactions of the $AlPO_4$-5 lattice with the dye molecule as spectral shifts. The measured absorption maximum at 472 nm corresponds to spectra in ethanol, which suggests that the dipole–dipole interactions of the $AlPO_4$-5 framework with DCM are similar to those of ethanol with DCM. The main absorption band of Pyridine 2 is also not shifted significantly. The shift of the second absorption band to around 380 nm, compared to about 320 nm in solution, can be attributed to an increase in the dipole moment of the excited state, which would thus be lowered in energy by the host lattice.

Proof that the dye is located in – or at least oriented by – the pore system is given by analyzing the crystals with polarized light (Fig. 4). The crystals only absorb light polarized parallel to the length axis of the crystal. The transition dipole moment of

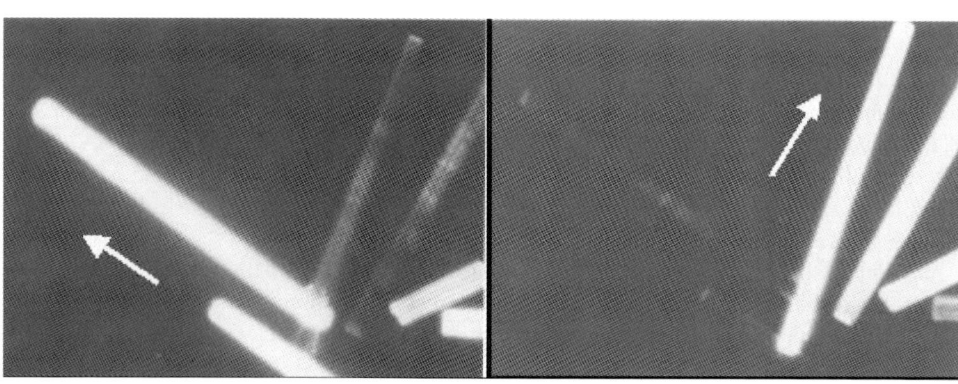

Fig. 4. Photographs of fluorescing $AlPO_4$-5 crystals. The arrows indicate the polarization plane of the excitation light.

these dyes is basically parallel to the long axis of the molecules. The polarized absorption parallel to the *c*-axis of the crystal, i.e., in the direction of the channels, that is, the dye molecules are aligned in the channel direction. This would not be the case for dyes adsorbed on the external surface or in large mesopores. As DCM exceeds the pore size of $AlPO_4$-5, adsorption in smaller mesopores having the same orientational relationship to the channels explains the polarized absorption observed in $AlPO_4$-5/DCM crystals.

4.4.3 Dye-Loading Profiles

The macroscopically anisotropic character of the dye-loaded $AlPO_4$-5 crystals is evident from polarization-dependent UV-vis studies. Polarized absorption spectra give information about alignment of the molecules and allow very low dye contents to be detected. By means of calculations based on the Beer–Lambert law it is possible to quantify degrees of loading [48]. Due to the size of the single crystals it is possible to "map" the dye content quantitatively by recording spectra along the *c*-axis of the crystal. The dye concentration in DCM-loaded crystals is high in the middle of the crystal and decreases towards either end (Fig. 5). Similar profiles are detected for $AlPO_4$-5/Pyridine 2 crystals.

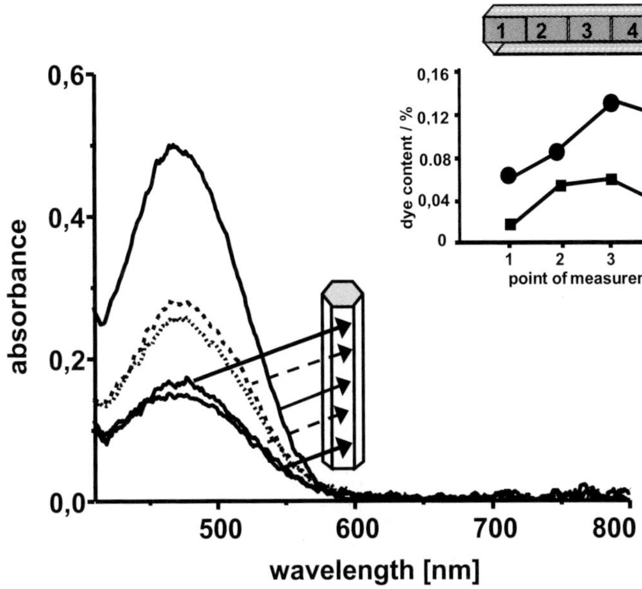

Fig. 5. Absorption spectra of DCM in a single $AlPO_4$-5 crystal. The arrows indicate the points of measurement (inset: comparison of DCM (squares) and Pyridine 2 (circles) degrees of loading along the crystal *c*-axis).

The higher degrees of loading of Pyridine 2 compared to DCM (Fig. 5, inset) reveal that the size of the Pyridine 2 molecule is more favorable with respect to incorporation in the channels. The loading profile indicates that incorporation rates are very high in the early stages of crystal growth, since these crystals start to grow from the center to both sides.

A powerful tool for gaining insight into the spatial distribution of fluorescent species in composite materials is confocal microscopy [49,50]. Here, a point light source is employed which produces a small and geometrically well defined focus spot in the sample plane. Fluorescence light which is emitted from the excitation spot is sent to the detector. With an additional small pinhole in the detection path, only light originating from the excitation spot can be detected, and fluorescence from out of focus locations is thus blocked. By moving the focus through the sample, a depth profile of the dye concentration can be obtained. By stacking the images of consecutive optical sections, each with a thickness of less than 0.5 μm, it is possible to construct a three-dimensional representation of the object. With this technique it is possible to visualize mesopores in molecular sieves by imaging areas in which molecules diffuse, although the nominal pore size of a given structure is too small for the accommodated guest [51]. Consequently, it is possible to characterize quality and homogeneity of crystal samples which are considered for optical applications. Furthermore, with the fluorescent species acting as probe molecules, internal defect structures can be revealed, which provides not only information about the suitability of the material as a host, but can also elucidate some crystallization mechanisms.

Dye concentration profiles of $AlPO_4$-5 crystals loaded with Pyridine 2 and DCM visualized by confocal microscopy give information on the dye distribution and the internal structure of large $AlPO_4$-5 crystals [52]. In both cases, spatial organization of the dye is observed. Sections along the c-axis of the crystal reveal that $AlPO_4$-5/Pyridine 2 crystals show a higher loading in the inner part of the crystal (Fig. 6,

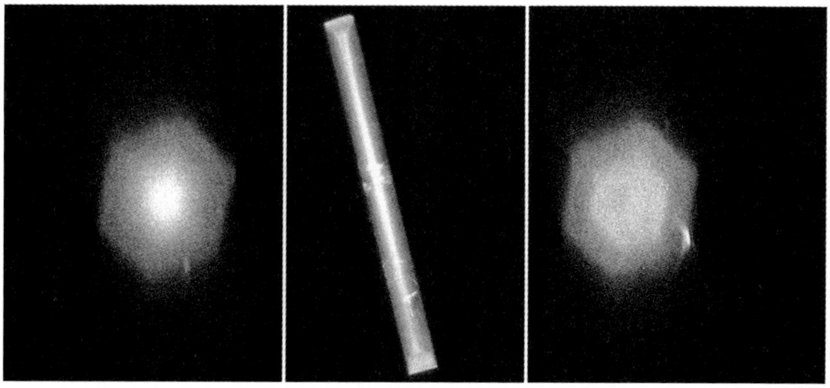

Fig. 6. Fluorescence image section through the length axis of an $AlPO_4$-5/Pyridine 2 crystal (middle) and corresponding perpendicular sections in the middle (left) and at the end (right) of the crystal.

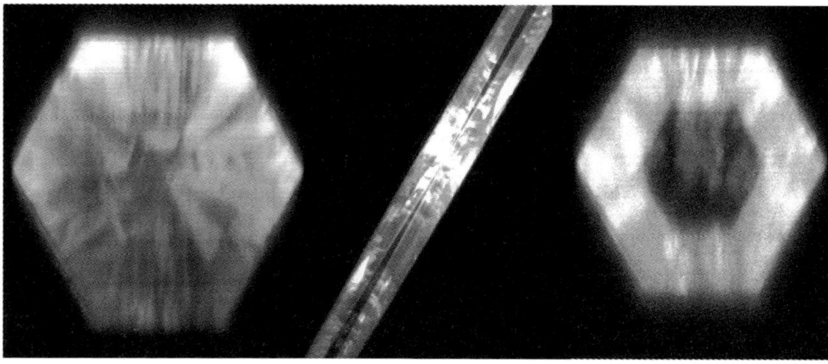

Fig. 7. Fluorescence image section through the length axis of an $AlPO_4$-5/DCM crystal (middle) and corresponding perpendicular sections in the middle (left) and at the end (right) of the crystal.

middle). In both materials the crystals exhibit two distinct regions: internal and external hexagonal domains. In $AlPO_4$-5/Pyridine 2 crystals the dye is preferentially located in the core of the crystal (Fig. 6, left), and the loaded domain becomes larger towards the ends of the crystal (Fig. 6, right). Thus the more highly loaded regions form an hourglass structure. In contrast to the loading profile of Pyridine 2, $AlPO_4$-5/DCM crystals exhibit a concentration distribution with an inverse-hourglass structure (Fig. 7, middle). In the middle part of the crystal almost no sectioning is observed (Fig. 7, left), whereas towards the ends of the crystal, DCM is concentrated in the outer hexagonal domain, and almost no DCM molecules are in the inner hexagonal core (Fig. 7, right). These images reveal the internal domain structure of $AlPO_4$-5. The crystals have a less disturbed internal core in which smaller molecules can be incorporated. The outer shell is a more defect-rich region with mesopores in which molecules can be incorporated although they exceed the pore size. The incorporation mechanism is determined by the crystallization mechanism of the host material and directs the guest molecules to their corresponding regions. At the same time, the structure of the outer shell is influenced additionally by the presence of larger molecules. However, the details of the growth process and dye incorporation are not fully comprehended yet.

4.4.4
Laser Activity in $AlPO_4$-5/Dye Crystals

Lasing from a modified molecular sieve was first demonstrated with $AlPO_4$-5/Pyridine 2 crystals [37]. When excited with the second harmonic of a Nd:YAG laser (523 nm) and when the pump irradiation exceeds a certain threshold, spectrally narrow spikes appear in the fluorescence spectra (Fig. 8, right). The morphology of the $AlPO_4$-5/Pyridine 2 microlaser is characterized by a perfect center, but defective crystal ends (Fig. 8, left) [39]. The laser action from the central part is characterized

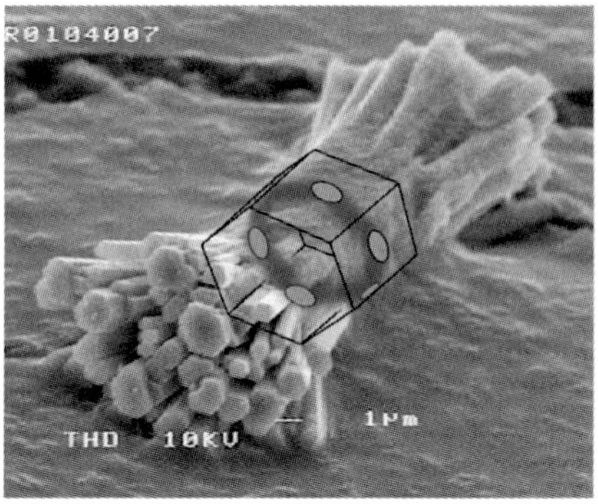

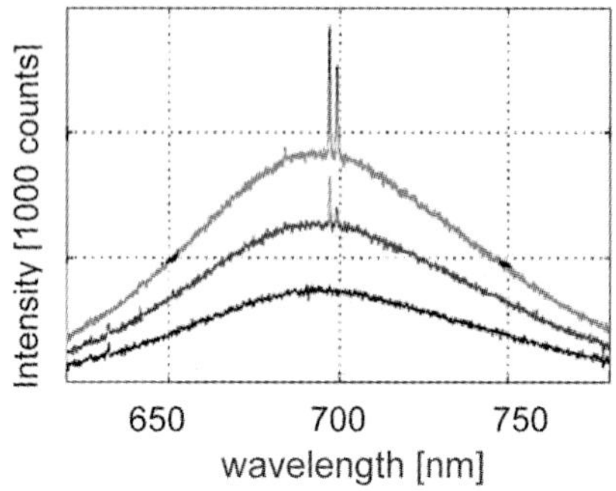

Fig. 8. Morphology of a typical $AlPO_4$-5/Pyridine 2 crystal with lasing properties (left; the whispering-gallery mode inside a hexagonal resonator is indicated by the sketch) [39]. Corresponding fluorescence spectra (right) with increasing pump intensities. Laser spikes appear when the pump intensity reaches the gain [37].

by a whispering-gallery mode due to total internal reflection at the hexagonal sides. Laser emission could not be detected from perfect $AlPO_4$-5/Pyridine 2 crystals as the necessary high degrees of dye loading imply disturbed morphology [see also F. Laeri's contribution].

However, large $AlPO_4$-5/DCM crystals do show laser activity even if they are perfect hexagonal prisms. Spectral spikes are detected in the region of 605–635 nm with a discrete line spacing indicating that an ensemble of modes propagates in the hexagonal cavity (Fig. 9). The expected wavelength spacing of the modes reflects the size of the resonator. The size of the studied crystal can be characterized by the distance d between opposite sides. According to calculations as described in Ref. [39] the observed line spacing of 2.26 nm corresponds to a distance of $d_{calc} = 38.7$ μm, which is in the range of the microscopically obtained distance of $d_{exptl} = 28.4$ μm. Independent of the laser-active molecule, modification of the resonator can be realized by changing crystal size and structure type. A smaller crystal means smaller resonator dimensions and hence lower lasing thresholds [39].

4.5 Outlook

To generate new laser materials based on molecular sieves, two opposing treads have to be considered. On the one hand, high degrees of loading for laser operation are possible by using charged molecules, but on the other hand adjusting dye concentration is difficult, since these dyes interact strongly with the growing crystal, and the resonator geometry is distorted. This problem can be solved by using uncharged molecules.

Technical application is first of all hindered by the fact that the synthesis yields very heterogeneous crystal sizes and morphologies. Homogeneous product quality is the basic prerequisite for integrating the materials into actual devices, which

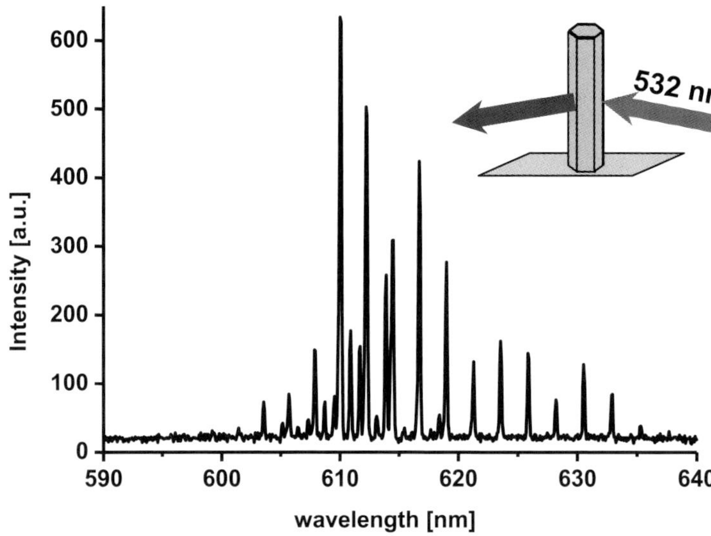

Fig. 9. Laser emission spectra of a single $AlPO_4$-5/DCM crystal (inset: principle of single-crystal measurement).

could be realized by alignment in an electric field [53] or directed growth on a surface. Another problem is the observed loading profile, which is due to the internal structure of $AlPO_4$-5. Thus, further studies using other molecular sieves, such as MFI crystals or smaller AFI crystals, could yield new resonator types or homogeneous loadings, respectively. Smaller crystal sizes would also result in a lower laser threshold and higher laser efficiencies, which could lead to use as lasing pigments in different applications such as paints for the automobile industry, cosmetics, and security markers.

These studies also show that the well-known characterization techniques and spectroscopy methods have reached their limits, and new approaches must be found to characterize optically functional molecular sieves. Since fluorescent species in the channels and pores of the host structures can act as probes providing information on the internal structure of host material, this could open new avenues to understanding the intriguing growth behavior of $AlPO_4$-5 and other molecular sieves.

References

1 G.D. Stucky, I. McDougall, *Science* **1990**, *247*, 669.

2 G.A.Ozin, *Adv. Mater.* **1992**, 4, 612–649.

3 J. Sauer, F. Marlow, F. Schüth, *Handbook of Advanced Electronic and Photonic Materials and Devices*, H.S. Nalwa (Ed.), Vol. 6, Academic, London, **2001**.

4 R.C. Hayward, P. Alberius-Henning, B.F. Chmelka, G.D Stucky, *Microporous Mesoporous Mater.* **2001**, *44*, 619–624.

5 M. Bredol, U. Kynast, C. Ronda, *Adv. Mater.* **1991**, *3*, 361.

6 U. Kynast, V. Weiler, *Adv. Mater.* **1994**, 6, 937.

7 C. Borgmann, J. Sauer, U. Kynast,

T. Juestel, F. Schüth, *MRS Conference Proceedings, Proceedings of the 12th International Zeolite Conference*, 2241 (1998).
8 S. Engel, U. Kynast, K.K. Unger, F. Schüth, *Stud. Surf. Sci. Catal.* **1994**, *84*, 477.
9 C. Borgmann, J. Sauer, T. Jüstel, U. Kynast, F. Schüth, *Adv. Mater.* **1999**, *11*, 45.
10 G. Calzaferri, N. Gfeller, *J. Phys. Chem.* **1992**, *96*, 3428.
11 R. Hoppe, G. Schulz-Ekloff, D. Wöhrle, E. Shpiro, O.P. Tkachenko, *Zeolites* **1993**, *12*, 222.
12 S. Wohlrab, R. Hoppe, G. Schulz-Ekloff, D. Wöhrle, *Zeolites* **1992**, *12*, 862.
13 D. Wöhrle, A.K. Sobbi, O. Franke, G. Schulz-Ekloff, *Zeolites* **1995**, *15*, 540.
14 I. Braun, G. Schulz-Ekloff, M. Bockstette, D. Wöhrle, *Zeolites* **1997**, *19*, 128.
15 M. Bockstette, D. Wöhrle, I. Braun, G. Schulz-Ekloff, *Microporous Mesoporous Mater.* **1998**, *23*, 83.
16 K. Hoffman, F. Marlow, J. Caro, *Adv. Mater.* **1997**, *9*, 567.
17 K. Hoffmann, U. Resch-Genger, F. Marlow, *Microporous Mesoporous Mater.* **2000**, *41*, 99.
18 S.D. Cox, T.E. Gier, G.D. Stucky, J. Bierlein, *J. Am. Chem. Soc.* **1988**, *110*, 2286.
19 S.D. Cox, T.E. Gier, G.D. Stucky, *Chem. Mater.* **1990**, *2*, 609.
20 L. Werner, J. Caro, G. Finger, J. Kornatowski, *Zeolites*, **1992**, *12*, 658.
21 J. Caro, G. Finger, J. Kornatowski, J. Richter-Mendau, L. Werner, B. Zibrowius, *Adv. Mater.* **1992**, *4*, 273.
22 F. Marlow, J. Caro, L. Werner, J Kornatowski, S. Dähne, *J. Phys. Chem. A* **1993**, *97*, 11286.
23 F. Marlow, W. Hill, J. Caro, G. Finger, *J. Raman Spectrosc.* **1993**, *24*, 603.
24 F. Marlow, M. Wübbenhorst, J. Caro, *J. Phys. Chem. A* **1994**, *98*, 12 315.
25 S.T. Wilson B.M. Lok, C.A. Messina, T. Cannan, E.M. Flanigen *J. Am. Chem. Soc.* **1982**, *104*, 1146.
26 B.M. Lok, C.A. Messina, R.L. Patton, T.R. Cannan, E.M. Flanigen, *J. Am. Chem. Soc.* **1984**, *106*, 6092.
27 I. Girnus, K. Jahncke, R. Vetter, J. Richter-Mendau, J. Caro, *Zeolites* **1995**, *15*, 33.
28 G. Finger, J. Richter-Mendau, M. Bülow, J. Kornatowski, *Zeolites* **1991**, *11*, 443.
29 S. Thiele, K. Hoffmann, R. Vetter, F. Marlow, S. Radaev, *Zeolites* **1997**, *19*, 190.
30 R.A. Rakoczy, S. Ernst, M. Hartmann, Y. Traa, J. Weitkamp, *Catal. Today* **1999**, *49*, 261.
31 D. Demuth, G.D. Stucky, K.K. Unger, F. Schüth, *Microporous Mater.* **1995**, *3*, 473.
32 S. Schunk, D.G. Demuth, B. Schulz-Dobrick, K.K. Unger, F. Schüth, *Microporous Mater.* **1996**, *6*, 273.
33 Ö. Weiß, G. Ihlein, F. Schüth, *Microporous Mesoporous Mater.* **2000**, *35–36*, 617.
34 D. Demuth, K.K. Unger, F. Schüth, G.D. Stucky, V.I. Srdanov, *Adv. Mater.* **1994**, *6*, 931.
35 S. Schunk, diploma thesis, Mainz, **1994**.
36 J.F. Tanguay, S.L. Suib, *Catal. Rev. Sci. Eng* **1987**, *29*(1), 1.
37 G. Ihlein, F. Schüth, O. Krauß, U. Vietze, F. Laeri, *Adv. Mater.* **1998**, *10*, 1117.
38 U. Vietze, O. Krauß, F. Laeri, G. Ihlein, F. Schüth, B. Limburg, *Phys. Rev. Lett.* **1998**, *81*, 4628.
39 I. Braun, G. Ihlein, F. Laeri, J. Nöckel, G. Schulz-Ekloff, F. Schüth, U. Vietze, Ö. Weiß, D. Wöhrle, *Appl. Phys. B* **2000**, *70*, 335.
40 F. Marlow, M.D. McGhee, D.Y. Zhao, B.F. Chmelka, G.D. Stucky, *Adv. Mater.* **1999**, *11*, 632.
41 I. Braun, G. Schulz-Ekloff, M. Bockstette, D. Wöhrle, *Zeolites* **1997**, *19*, 128.
42 Ö. Weiß, U. Wüstefeld, J. Loerke, F. Marlow, F. Schüth, *J. Solid State Sci.*, **2002**, *167*, 302.
43 Ö.Weiß, F. Schüth, L. Benmohammadi, F. Laeri, *Stud. Surf. Sci. Catal* **2001**, *135*, 161.

44 K.H. Schnabel, G. Finger, J. Kornatowski, E. Loeffler, C. Peuker, W. Pilz, *Microporous Mater.* **1997**, *11*, 293.

45 S. Popescu, S. Thomson, R.F. Howe, *Phys. Chem. Chem. Phys.* **2001**, *3*, 111.

46 U. Brackmann, *Lambdachrome Laser Dyes*, Lambda Physik GmbH, Göttingen, **2001**.

47 C. Reichardt, *Solvents and Solvent Effects in Organic Chemistry*, VCH, Weinheim, **1988**.

48 K. Hoffmann, F. Marlow, J. Caro, *Zeolites* **1995**, *16*, 281.

49 M. Minsky, Microscopy Apparatus, U.S. Patent No. 301467, Dec. **1961**.

50 T.R. Corle, G.S. Kino, *Confocal Scanning Optical Microscopy and Related Imaging Systems*, Academic Press, San Diego, **1996**.

51 C. Seebacher, J. Rau, F.-W. Deeg, C. Bräuchle, S. Altmeier, R. Jäger, P. Behrens, *Adv. Mater.* **2001**, *13*, 1374.

52 Ö. Weiß, F. Schüth, C. Seebacher, C. Hellriegel, F.-W. Deeg, J. Loerke, F. Marlow, in preparation.

53 J. Caro, G. Finger, J. Kornatowski, J. Richter-Medau, L. Werner, B. Zibrowius, *Adv. Mater.* **1992**, *4*, 273.

5
Luminescence of Lanthanide Organometallic Complexes

*Dorota Sendor and Ulrich Kynast**

5.1
Introduction, Motivation, and Scope

Due to the particular structure of zeolites, which exhibit well-organized cavities or channels in the nanometer regime, they are of immediate interest for optical phenomena that rely on a controlled three-dimensional arrangement of active atomic, ionic, molecular, or even oligomeric species. Being strongly dependant both on topology and the chemical nature of the environment, luminescence may be viewed as a typical example for such behavior.

In combination with a large band gap of typically greater than 5 eV (e.g., zeolite X [1]), the spacious zeolite voids provide appreciable degrees of freedom with regard to the accessible spectral range and the choice of the luminescent centers to be incorporated or manipulated within. In the following, the terms "doping" refer to the substitution of Na^+ ions by ion exchange and "loading" to the encapsulation of organic entities.

Supramolecular organization in zeolite hosts has been utilized extensively for organic guest species exhibiting a variety of optophysical effects, some highlights of which are the subject of reviews in this book. Furthermore, the optical properties of zeolites have been investigated in the context of quantum dots by the incorporation of II-VI and III-V semiconductors [2–8] and transition metallates [9–11]. The significant influence of energy and electron transfer on optical response [12–14] and photochemical behavior [15,16] have also been the subject of intense studies.

Emphasis in the luminescence of ion-exchanged zeolites has been on usage as an analytical tool to characterize ion locations with regard to photochemical and catalytic properties rather than to render them optically functional as luminophores [17–23]. The latter is somewhat surprising, as the lanthanide ions in particular offer a broad range of specific advantages with respect to luminescent phenomena, which have led to applications ranging from, e.g., phosphors, lasers, image-storage and after-glow materials [24] to sensors [25] and immunoassays [26–28]. Inherently, many lanthanides concentrate their efficient emission in very narrow lines, which may be highly advantageous, for example, in matching the sensitivity function of an instrument or the human eye [29]. Furthermore, the luminescence mechanism allows apparent Stokes shifts of several electron volts, so that excitation and emission may be separated by margins of several electron volts, which allows completely white or transparent materials with no absorption in the energetic vicinity of the emission lines to be obtained. Some potential uses of lanthanide-functionalized zeolites as phosphors [1,30–32], latent-image storage materials [33], and sensors [34] have been outlined.

However, the use of lanthanide-exchanged zeolites as efficient photostimulated emitters may be hampered severely by insufficient optical absorptivity of several of the ions of interest due to the forbidden nature of the f–f transitions involved. This also holds true for red-emitting Eu^{3+} and green-emitting Tb^{3+} ions, for which appropriate sensitization schemes must be applied, unless the excitation wavelength is lower than 230 nm, where transitions to d states become accessible. Sensitization of, e.g., Tb^{3+} can be accomplished with co-exchanged Ce^{3+}, which absorbs via allowed f–d transitions and itself is an efficient emitter in zeolites [30], although the overall optical performance remains limited, since an optimum balance between Ce^{3+} and Tb^{3+} concentrations cannot be established [1]. Another method to sensitize Tb^{3+} emission involves the incorporation of transition metallates such as molybdates and tungstates [35], although efficiency here remains too low for phosphor applications due to thermal quenching and undesired energy back-transfer to the metallate.

In the case of Eu^{3+}, numerous luminescent materials benefit from excitation into an electric dipole allowed $O \rightarrow Eu^{3+}$ charge transfer (CT) transition. Among them is the very popular $Y_2O_3:Eu^{3+}$. Unfortunately, as is evident from reflectance spectra of Eu^{3+}-doped zeolites, the less rigid oxygen coordination causes the same CT to extend to low energies (see Fig. 4a), so that radiationless decay into the ground states is favored. Additionally, imbibed water can coordinate the Eu^{3+} ions and cause nonradiative quenching via high-frequency OH vibrations [36]. Sensitization with molybdates was shown to increase the overall Eu^{3+} light output, but a true gain in quantum efficiency was obtained only after conversion to the anhydrous sodalite structure [37].

A broadly applicable and successful measure to overcome too weak an optical absorptivity of luminescent ions at the desired wavelengths is their complexation with organic ligands, which can act as strong absorbers and transfer the collected excitation energy to the emitting ion [38–42]. Eventually, this pathway enabled the use of organolanthanide complexes in applications such as electroluminescence [43–45] and immunoassays [26].

The widely accepted luminescence mechanism in these complexes has been known since 1942 [38], and was the theme of various papers [46,47]. It involves a singlet–singlet absorption of the organic ligand, followed by intersystem crossing to a characteristic ligand triplet state and intramolecular energy transfer to the lanthanide metal ion, which eventually relaxes to the ground state by its characteristic emission.

Investigations on the optical properties of zeolite-encapsulated organometallic complexes were described as early as 1980 ([Ru(bipy)-Y] [48], [Rh{*N*,*N*′-bis(salicylaldehyde)}-X] [49]) and have been under study ever since [50–54]. Complexes of lanthanides in zeolites are discussed in [55–58], and recently their inclusion in MCM-41 was investigated as well [59]. In addition to enhanced optical absorption, the method also provides means to alter or supress unfavorable CT transitions and to screen especially the Eu^{3+} ion from water coordination. The hybrid materials eventually obtained in this manner have the charm of constituting convenient and manipulable solids, and at the same time they preserve some favorable molecular optical properties of the complexes, such as the efficient intramolecular energy transfer channels.

Zeolite X of the faujasite family and aromatic carboxylate and β-diketonate complexes as luminescent centers form the core of our investigations. Apart from their high efficiencies in the "free" state, the aromatic carboxylate and diketonate classes of complexes were chosen for geometrical reasons (cavity size of zeolite X), ease of preparation, and the option to screen the effects of substitution of the aromatic ring and the $-CO-CH_2-CO-$ diketonate backbone on the luminescence of the encapsulated species. Last, but not least, for several of the "free" complexes single-crystal data were available or could be generated. From comparisons with these, insight into the structure and the identification of mechanisms governing the efficiency in zeolites was acquired, and generalizations towards structure–property relationships were found.

5.2 Synopsis

The complexes under consideration can be subdivided into poly-, amino-, and hydroxocarboxylates; pyridine derivatives; and co-coordinated complexes. Two diketonates and one co-ligated derivative were included in the present work. $[Eu^{III}(ttfa)_3]$ (ttfa = tris-1-thenyl-4,4,4-trifluorobutane-1,3-dionato) was chosen for its reasonable efficiency: for solid $[Eu(ttfa)_3(H_2O)_2]$ quantum yields of 23 [60] to 48% [61] have been reported (in DMF and acetone solution, 38 and 15%, respectively, have been measured [62]). Other available diketonates were reported to have at best similar or lower solution quantum yields (tris(pentane-2,4-dionato)Tb^{III} in DMF: 27%; tris(1,1,1,5,5,5-hexafluoropentane-2,4-dionato)Eu^{III} in DMF: 38%; tris(1-phenylbutan-1,3-dionato)Eu^{III}: 9% in DMF [62]), or revealed lower efficiencies in our investigations and were therefore discarded at an early stage.

The emission intensity data given in Tabs. 1–3 represent the best samples obtained from our screening program. The data given for the zeolites containing complexes of tere, btca, tetra, and phba were obtained by loading with excess ligand and are from single experiments, as are the results on co-cordination of [Tb-X] with phen and bipy; for each of the other zeolite complexes, series of parameters were checked (lanthanide concentration vs. amount of ligand vs. amount of zeolite). References given in the tables do not necessarily refer to spectral data.

Since the determination of accurate quantum yields for powder materials is a very tedious and time-consuming procedure requiring a dedicated setup [60,62],

Tab. 1. Emission intensities of Ln^{3+} carboxylates relative to $NH_4[Tb(pic)_4]$ excited at 285 nm ($\Theta_{254} = 74\%$) and $[Eu(ttfa)_3]\cdot 2\ H_2O$ excited at 300 nm ($\Theta = 23\%$) [60]. Zeolite polycarboxylate complexes were prepared with excess ligand and $[Ln_{16}\text{-}X]$; for others, see text. Compositions of zeolite complexes: $[Tb_{20}(sal)_4\text{-}X]$, $[Tb_{20}(mhba)_6\text{-}X]$, $[Tb_8(phba)_8\text{-}X]$, $[Tb_8(oaba)_{16}\text{-}X]$, $[Tb_8(maba)_8\text{-}X]$, $[Tb_{12}(paba)_6\text{-}X]$. (*: this work.)

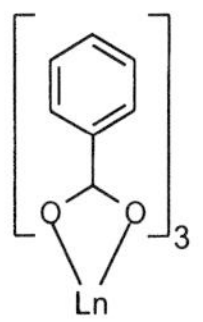

Benzoates, 'benz'

'free'	nr	I_{rel}	λ_{exc}	Ref.
Eu	(1)	0.5	280	47 73
Tb	(2)	0.4	280	*
in X	nr	I_{rel}	λ_{exc}	Ref.
Eu	(3)	<0.1	254	*
Tb	(4)	0.3		

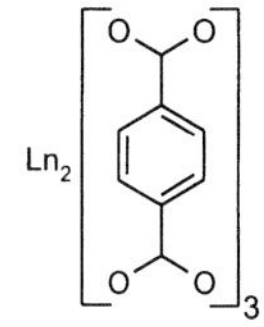

Terephtalates, 'tere'

'free'	nr	I_{rel}	λ_{exc}	Ref.
Eu	(5)	<0.1	280	74 *
Tb	(6)	0.2	280	
in X	nr	I_{rel}	λ_{exc}	Ref.
Eu	-	-	-	*
Tb	(7)	0.2	254	

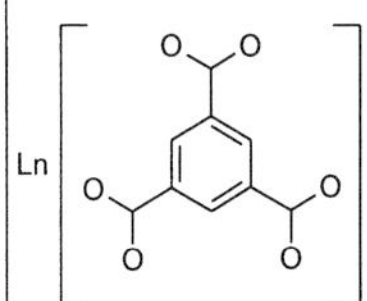

Benzene tricarboxylates, 'btca'

'free'	nr	I_{rel}	λ_{exc}	Ref.
Eu	(8)	0.5	280	74 75
Tb	(9)	0.4	280	*
in X	nr	I_{rel}	λ_{exc}	Ref.
Eu	(10)	<0.1	254	*
Tb	(11)	0.3	254	

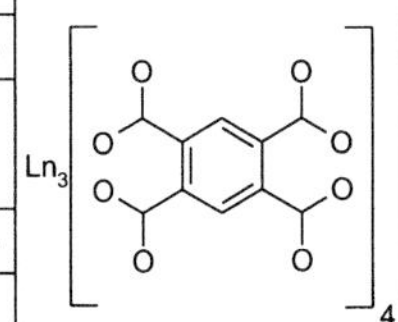

Benzene tetracarboxylates, 'tetra'

'free'	nr	I_{rel}	λ_{exc}	Ref.
Eu	-	-	-	74 *
Tb	(12)	0.2	254	
in X	nr	I_{rel}	λ_{exc}	Ref.
Eu	-	-	-	*
Tb	(13)	<0.1	254	

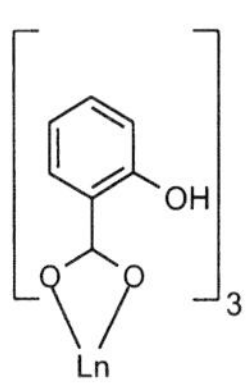

Salicylates, 'sal'

'free'	nr	I_{rel}	λ_{exc}	Ref.
Eu	(14)	≈0	-	77 *
Tb	(15)	0.4	284	
in X	nr	I_{rel}	λ_{exc}	Ref.
Eu	(16)	≈0	-	*
Tb	(17)	0.2	310	

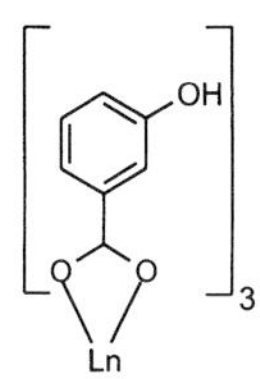

m-Hydroxy benzoates, 'mhba'

'free'	nr	I_{rel}	λ_{exc}	Ref.
Eu	(18)	≈0	-	77 78
Tb	(19)	0.5	328	*
in X	nr	I_{rel}	λ_{exc}	Ref.
Eu	(20)	≈0	-	*
Tb	(21)	0.4	328	

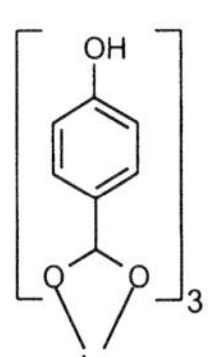

p-Hydroxy benzoates, 'phba'

'free'	nr	I_{rel}	λ_{exc}	Ref.
Eu	(22)	≈0	-	77 *
Tb	(23)	0.4	288	
in X	nr	I_{rel}	λ_{exc}	Ref.
Eu	(24)	≈0	-	*
Tb	(25)	<0.1	300	

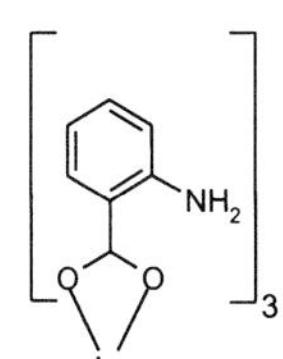

o-Amino benzoates 'oaba'

'free'	nr	I_{rel}	λ_{exc}	Ref.
Eu	(26)	≈0	-	79 *
Tb	(27)	0.4	380	
in X	nr	I_{rel}	λ_{exc}	Ref.
Eu	(28)	≈0	-	*
Tb	(29	0.1	307	

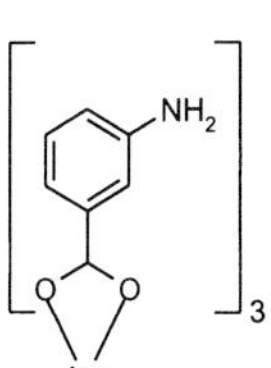

m-Aminobenzoates, 'maba'

'free'	nr	I_{rel}	λ_{exc}	Ref.
Eu	(30)	≈0	-	*
Tb	(31)	0.3	320	
in X	nr	I_{rel}	λ_{exc}	Ref.
Eu	(32)	≈0	-	*
Tb	(33)	<0.1	300	

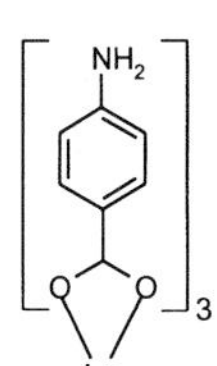

p-Amino benzoates, 'paba'

'free'	nr	I_{rel}	λ_{exc}	Ref.
Eu	(34)	≈0	-	87 *
Tb	(35)	0.5	310	
in X	nr	I_{rel}	λ_{exc}	Ref.
Eu	(36)	≈0	-	*
Tb	(37)	0.4	285	

Tab. 2. Emission intensities of Ln^{3+} picolinates and β-diketonates relative to $NH_4[Tb(pic)_4]$ excited at 285 nm ($\Theta_{254} = 74\%$) and $[Eu(ttfa)_3]\cdot 2\ H_2O$ excited at 300 nm ($\Theta = 23\%$) [60]. For preparation, see text. Compositions of zeolite complexes: $[Ln_{16}(pic)_{28}\text{-}X]$, $[Ln_{12}(dipic)_{28}\text{-}X]$, $[Eu(ttfa)_3\text{-}X]$, $[Eu_{12}(dpac)_4\text{-}X]$, $[Tb_{12}(dpac)_5\text{-}X]$. (*: this work.)

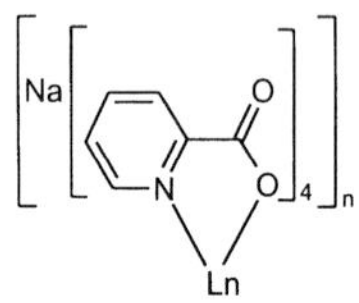

Picolinates, 'pic'				
'free'	nr	I_{rel}	λ_{exc}	Ref.
Eu	(38)	1.3	280	80 *
Tb	(39)	**1.0**	**285**	
in X	nr	I_{rel}	λ_{exc}	Ref.
Eu	(40)	1.2	300	*
Tb	(41)	0.3	273	

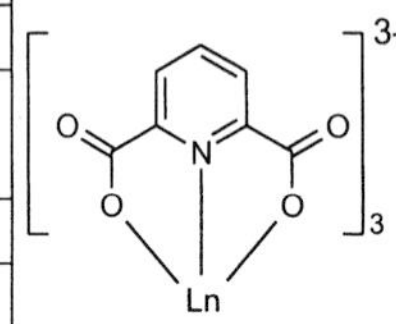

Dipicolinates, 'dipic'				
'free'	nr	I_{rel}	λ_{exc}	Ref.
Eu	(42)	1.9	278	68, 81-84,*
Tb	(43)	0.2	278	
in X	nr	I_{rel}	λ_{exc}	Ref.
Eu	(44)	0.1	278	57 58 *
Tb	(45)	0.3	278	

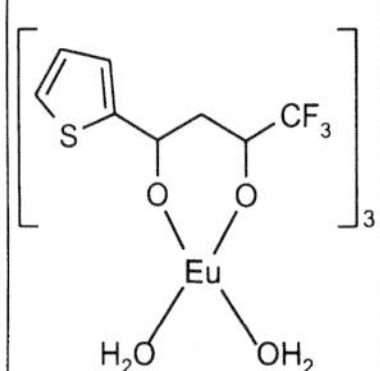

Thenylacetylacetonates, 'ttfa'				
'free'	nr	I_{rel}	λ_{exc}	Ref.
Eu	(46)	**1.0**	**300**	60 85 86,*
Tb	-	-	-	
in X	nr	I_{rel}	λ_{exc}	Ref.
Eu	(47)	2.2	330	*
Tb	-	-	-	

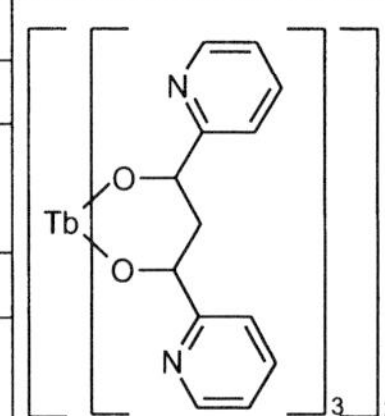

Dipyridylacetylacetonates, 'dpac'				
'free'	nr	I_{rel}	λ_{exc}	Ref.
Eu	(48)	0.03	320	107 *
Tb	(49)	≈0	-	
in X	nr	I_{rel}	λ_{exc}	Ref.
Eu	(50)	0.2	266	*
Tb	(51)	see text		

Tab. 3. Emission intensities of co-coordinated Ln^{3+} complexes relative to $NH_4[Tb(pic)_4]$ excited at 285 nm ($\Theta_{254} = 74\%$) and $[Eu(ttfa)_3]\cdot 2\ H_2O$ excited at 300 nm ($\Theta = 23\%$) [60]. Compositions of zeolite complexes: $[Eu(ttfa)_3\cdot(phen)\text{-}X]$, $[Tb_3(benz)_3(phen)\text{-}X]$, $[Tb_{16}(benz)_{16}(bipy)_8\text{-}X]$, $[Tb_{16}(sal)_4(phen)_4\text{-}X]$, $[Tb_{16}(sal)_6(bipy)_3\text{-}X]$, $[Tb_1(paba)_3\cdot(phen)\text{-}X]$, $[Tb_{12}(paba)_6(bipy)_3\text{-}X]$. (*: this work.)

Eu(ttfa)₃ · (phen)

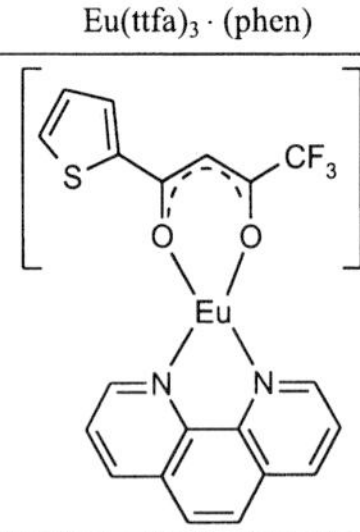

Tb(benz)₃ · (bipy), (phen)

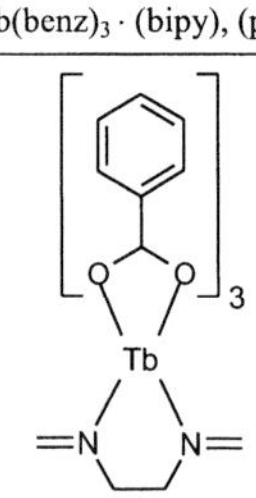

Tb(sal)₃ · (bipy), (phen)

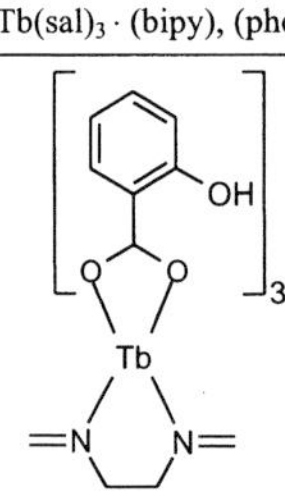

Tb(paba)₃ · (bipy), (phen)

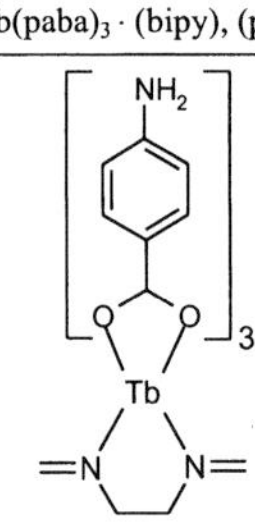

nr	I_{rel}	λ_{exc}	Ref.
(52)	2.9	320	42 *
(53)	2.4	330	

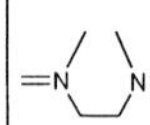

bipy, phen

phen	Tb(benz)₃ nr	I_{rel}	λ_{exc}	Ref.	Tb(sal)₃ nr	I_{rel}	λ_{exc}	Ref.	Tb(paba)₃ nr	I_{rel}	λ_{exc}	Ref.
free	-	-	-	*	(57)	0.5	325	*	(61)	0.2	320	87 *
in X	(54)	<0.1	325		(58)	0.3	310		(62)	≈0	-	
bipy	nr	I_{rel}	λ_{exc}	Ref.	nr	I_{rel}	λ_{exc}	Ref.	nr	I_{rel}	λ_{exc}	Ref.
free	(55)	0.4	280	*	(59)	0.5	365	*	(63)	0.3	325	*
in X	(56)	0.2	320		(60)	<0.1	300		(64)	0.3	285	

Tab. 4. Lifetimes of selected free and encapsulated Eu^{3+} and Tb^{3+} ions and molecules. bb = broad-band excitation. (*: this work.)

Material	*τ/ms (λ_{exc}/nm)*	*Lit.*	*Material*	*τ/ms (λ_{exc}/nm)*	*Lit.*
$[Eu(H_2O)_{8.9}]^{3+}$, H_2O	0.11	58	Y_2SiO_5:Tb	2.0 (310)	*
Eu^{3+}-X,Y	0.14-0,42 (395)	58	$Tb(acac)_3 \cdot 2H_2O$	1.0 (310, 490)	*
	0.15 / 0.6 (394)	63	$Tb(acac)_3$, H_2O	0.610	62
Eu^{3+}/Sol Gel	0.216	64	Tb^{3+}/sol gel	0.495 (310)	64
$Eu(acac)_3$/Sol Gel	0.385	65	$Tb(ohba)_3(CH_2)_6N_4$	1.80	66
$Eu(ttfa)_3 \cdot 2H_2O$	0.265 (bb)	67,42	$Na_3[Tb(dipic)_3]$	2.0	68
$Eu(ttfa)_3$phen	0.865 (bb)	42	Tb(phen)	0.805 (330)	69
Eu(phen)	0.908 (332)	69	$Tb(paba)_3phen \cdot 2H_2O$	1.30 (342)	71
$Na[Eu(EDTA)(H_2O)_4]$	0.317	70	$Na[Tbpic_4]$	1.639 (272)	*
$Na[Eu(pic)_4]$	1,603 (285)	*	$Tb(benz)_3$	2.391 (280)	*
$Na_3[Eu(dipic)_3]$	1.57	58	$[Tb_{16}(pic)_{28}$-X]	1.121 (272)	*
$Eu(paba)_3phen \cdot 2H_2O$	0.955 (342)	71	$[Tb_2Eu_1(pic)_6$-X]	1.335 (272)	*
$[Eu(bipy)_2$-Y]	0,17 / 0,43 (319)	56	$[Tb_{16}(benz)_{32}X]$	1.328 (280)	*
[Eu(dipic)-Y]	0,26 (275)	57,58	$Na[Tb(pic)_4]$/Sol Gel	1.684 (272)	*
[Eu(btfa)-Y]	0,288	57	$Tb(paba)_3(phen)$/Sol Gel	1.69 (342)	71
$Eu(paba)_3$phen/Sol Gel	1.161 (342)	71	$[Tb(benz)_3]$/Sol Gel	0.487 (280)	*,72

they have been determined for selected examples only. Some correlation of the efficiencies given with literature data may also be gained from the decay times provided in Tab. 4.

The integral emission intensity of all Tb complexes is evaluated vs. solid $(NH_4)[Tb(pic)_4]$, whose quantum yield of 74% at 254 nm excitation can be extrapolated to near unity at 285 nm excitation; for red-emitting Eu complexes the data are given relative to solid $[Eu(ttfa)_3(H_2O)_2]$ excited at 309 nm. Samples with less than 10% emission intensity compared to the reference materials are not given quantitatively, because the errors are too large. Data assigned as "≈ 0" indicate signal to noise ratios approaching unity. An exception was made for the dpac complexe to demonstrate the beneficial influence of the zeolite on its emission.

Thus, the figures given in Tabs. 1–3 do not reflect true quantum yields, but are merely light outputs I_{out}, i.e., the integrated emission intensity (400–700 nm) multiplied by the ratio of excitation intensity and lamp intensity at the excitation wavelength (Eq. 1).

$$I_{out} = \frac{I_{exc}(\lambda)}{I_{lamp}(\lambda)} \cdot \int_{400}^{700} I \, d\lambda \qquad (1)$$

The emission intensity of the various complexes and their X-encapsulated analogues thus obtained was additionally corrected for long-term lamp drift by comparison with the emission intensity of commercial $BaMgAl_{10}O_{17}$:Eu (BAM, Philips). Differences in scatter between two samples of the same chemical composition, but with different morphology and slightly different positioning of the sample, as well as unaccountable drifts of the excitation source may contribute to the overall error. In unfavorable cases, the relative error between two measurements on one and the same sample amounted to $\pm 10\%$.

In general, the emission spectra for the Tb^{3+} complexes correspond to Fig. 2a. Occasionally, the intensity ratios of the various $^5D_4 \rightarrow {}^7F_J$ transitions were correlated to symmetry and ligand field, as reviewed in, e.g., Ref. [87]. However, the changes found were too small to justify conlusions other than the fact that cross-relaxation between Tb^{3+} ions is not required for the $^5D_4$ emissions, because ligand vibrations relax the $^5D_3$ state into $^5D_4$, and consequently $^5D_3$ emissions are never observed; see, e.g., the emission spectra of $[Tb(pic)_4]^-$ and its derivatives (Fig. 5a).

The $^5D_0 \rightarrow {}^7F_J$ emissions of Eu^{3+} complexes are usually more informative [88,89], but we shall restrict ourselves here to noting that no centrosymmetric co-ordination was found, as is apparent from the low intensities of the $^5D_0 \rightarrow {}^7F_1$ lines (Figs. 4a, 5a). For Eu^{3+}, too, emissions from states higher in energy than $^5D_0$ are practically absent.

As is evident from the tables, with the exception of the β-diketonates (Tab. 2) and possibly the dipicolinates, generally no improvement in light output is observed for zeolite-encapsulated complexes over free complexes.

A noticable improvement, however, is obtained for $[Eu(ttfa)_3(H_2O)_2]$ in comparison with its co-coordination product with phen ($[Eu(ttfa)_3(phen)]$, Tab. 3). This behavior may also be related to a somewhat increased absorptivity of the complexes, although with Eu^{3+} it is predominantly a consequence of the replacement of coordinated water. In the case of $[Eu(ttfa)_3(phen)]$, albeit somewhat weaker, the enhancing effect of the phen co-ligand is also maintained in the zeolite complex $[Eu(ttfa)_3(phen)\text{-}X]$. The co-coordinated complexes also give rise to questions regarding steric limitations, as their molecular size approaches the dimensions of the zeolite cavity (see Section 5.3.5).

Some of the general principles guiding the performance of lanthanide complexes in zeolites are discussed on the basis of selected examples in the following sections.

5.3 Examples

5.3.1 Preparative Aspects

Doping with lanthanide (Ln) ions was performed by the procedure described in Ref. [1]. The incorporation of the organic ligand was performed by two different methods. In method A, the doped zeolite was first predried at 250–280 °C in vacuo at 10^{-5} mbar for several hours, followed by thoroughly mixing the solid components with ligand in large excess in comparison to the molar amount of zeolite. Then a static vacuum was applied to the mixture at 100 °C for 24 h. Excess ligand was then removed by washing with ethanol or by pumping at 100 °C for several

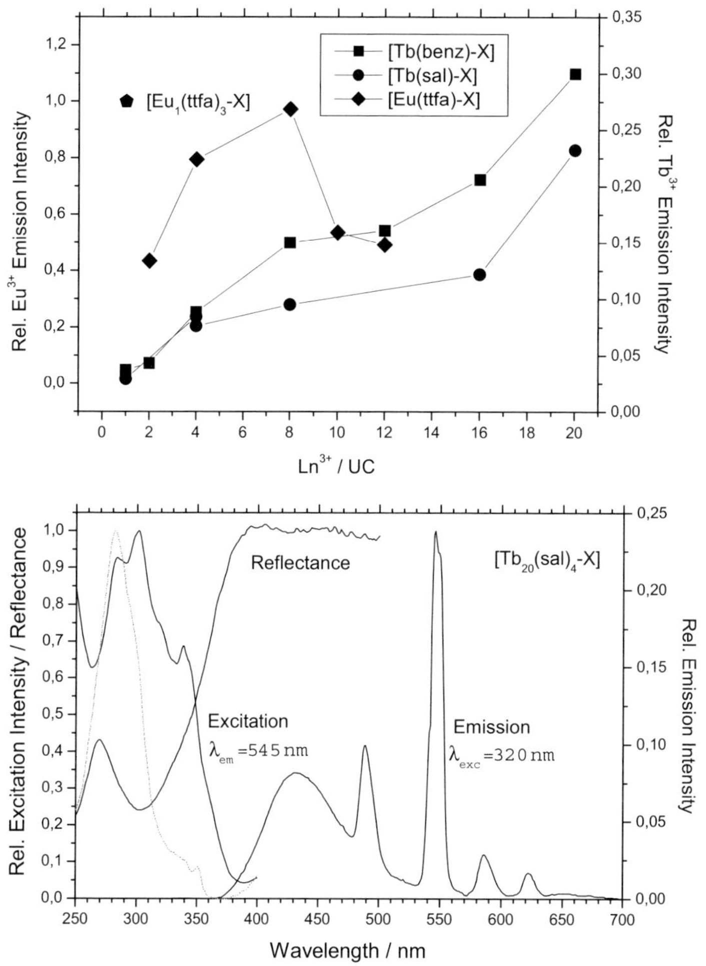

Fig. 1. Left: emission intensities of complexes vs. Ln^{3+} doping levels in zeolites. Benzoate series obtained by excess loading (method A, see text); sal and ttfa series by method B; sal emission intensity includes ligand phosphorescence. Right: optical spectra of $[Tb_{20}(sal)_4\text{-X}]$. Dotted curve obtained by monitoring the excitation of the ligand at 420 nm.

hours. Method A was chosen if "saturation" loading with ligand was to be ensured; e.g., in the case of the benzoates depicted in Fig. 1a an average of 32 benz molecules per unit cell (UC) were incorporated. However, for the dipic ligand it was shown [57,58] that the absolute loading can additionally depend on the number of Ln^{3+} ions present.

For several investigations, it was necessary to prepare series of materials with controlled amounts of ligand, or with controlled lanthanide/ligand ratios. For such cases method B was preferred, in which the doped zeolites were partially dehydrated in an ampoule, after which the desired amount of ligand was added, and the ampoule sealed under high vaccum and heated. The temperature chosen depended on the ligand (sal, pic: 140 °C; paba, maba, oaba: 100–130 °C; benz: 100 °C; ttfa: 70 °C; mhba, phba: 130 °C). After breaking the ampoule and washing with ethanol, the samples were measured and subjected to further treatments.

Similar procedures have also been described in the literature [48], but other methods, such as free diffusion of the ligand or the free complex into the zeolite from solution have also been applied [56–58]. Generally, we have refrained from using these, as we encountered problems with leaching of the complexes at high ligand loadings, especially with strongly complexing ligands (e.g. β-diketonates). Also, incorporation of large amounts of ligand could not be accomplished.

We assume that the presence of small amounts of water promotes intrazeolite complex formation, which is a particularly important factor in the formation of the β-diketonate complexes. But, on the other hand, the removal of water from the zeolite cages to provide sufficient space is also essential. As a consequence, when predrying steps are omitted, an up to tenfold decrease in efficiency has been found.

5.3.2
Effects of Doping Levels and Location in the Zeolite

One of the first factors governing the light output of the samples is the dependence on the lanthanide doping level. For Tb^{3+}-containing systems an increase in the Tb^{3+} emission intensity is systematically found at "saturation" loadings when the Tb^{3+} content is increased (method A). The increase in intensity is almost linear at lower Tb^{3+} doping levels and additionally dependent on the details of preparation, e.g., removal of excess ligand by vacuum sublimation or washing. With respect to emissivity and Tb^{3+} concentration, purely Tb^{3+}-doped zeolites X behave similarly [1].

In Fig. 1a the role of the Tb^{3+} content on efficiency is demonstrated for series of [Tb(benz)-X] zeolites in comparison with [Tb(sal)-X]. The same dependence of the emission intensity on Tb^{3+} concentration is generally also found for other carboxylates, but not in the same fashion for Eu^{3+}, as established for the series [Eu(ttfa)-X] (Fig.1a), [Eu(pic)-X], and, in accordance with the literature [57,58], [Eu(dipic)-X]. A series of zeolites with varying Eu^{3+} and dipicolinic acid contents clearly revealed that doping with Eu^{3+} in excess of the amounts needed for the formation of the actual complexes $[Eu(dipic)_3]^{3-}$ within the supercages reduces the efficiencies and lifetimes. In contrast, in the analogous Tb^{3+} series an almost linear increase in emission intensity is found at compositions containing up to 16 Tb^{3+}/UC.

In the literature, EXAFS studies [90,91], a Rietveld analysis [92], and spectroscopic data [30] have previously been reported on zeolites X containing more than 8 Ln^{3+}/UC, which had been calcined at up to 600 °C after ion exchange. It was concluded that in these, Ln^{3+} ions in excess of 8/UC lead to double occupation of sodalite cages, in which each Ln^{3+} ion is coordinated by three H_2O molecules after rehydration. Since in our preparation, a calcination step at 600 °C of zeolite X after primary ion exchange is performed as well, we assume that all materials investigated host the 1–16 lanthanide ions in the sodalite cages. Ions in excess of 16/UC are expected to lie within the supercage. However, complexation with ligands may mobilize the lanthanide ions, which additionally requires some water residing in the pores (see also below).

Therefore, in plotting emission intensity versus Eu^{3+} concentration, the point of decreasing or constant efficiency marks the onset of "free" Eu^{3+} ions, i.e., those that are not complexed by organic ligands. The free ions will most likely remain located in sodalite cages with water molecules attached ($Eu^{3+}{}_{sodalite}$). At this stage possible energy transfer channels must be considered, which are discussed in more detail in Section 4.5.3, but some conclusions are presented here: free ions, $Eu^{3+}{}_{sodalite}$, are the main contributors to the depreciation of the comparably efficient Eu^{3+} complexes in the supercages, which at the same time represent the centers of excitation. Complexed Eu^{3+} can now be deactivated by energy transfer to inefficient Eu^{3+} ions in the sodalite cages. In summary, the emissive $^5D_0$ state of complexed Eu^{3+} ions is quenched by the energy transfer $[Eu^{3+}(^5D_0)]_{complexed} \rightarrow [Eu^{3+}(^5D_0)]_{sodalite}$, eventually reducing the emission intensity of the whole system. However, in addition to this energy transfer by exchange, energy transfer from excited ligand triplets of complexes within supercages to $[Eu^{3+}(^5D_0)]_{sodalite}$ may also contribute to the observed losses.

As opposed to Eu^{3+}, Tb^{3+} is known to exhibit high intrinsic quantum yields in sodalite cages [1]. Therefore, "excess" doping and energy transfer mechanisms, unequivocally involved, will have little or no depreciating effect on the efficiency in the Tb^{3+} systems, and an increase in the Tb^{3+} loading beyond the complexation limit can still constructively contribute to the overall efficiency.

5.3.3 Nature of Encapsulated Complexes

5.3.3.1 Salicylates

The results presented for benzoates in Fig. 1a are based on syntheses in an excess of ligand, which may lead to free excess ligand in the zeolite. Optical effects resulting from such "overloading" are difficult to unravel, unless the free ligand can be detected by its phosphorescence. Yet another difficulty would appear if the absence of ligand → ligand energy transfer cannot be guaranteed, because in this case the excitation leading to phosphorescence of the free ligand could also originate directly from ligands bonded to lanthanide ions. However, such energy transfer due to Coulomb interaction between the states of the ligands was neither detected in the zeolitic matrix nor in the free complexes (Section 5.3.4).

Examples for the presence of free ligand and its phoshorescence were found in zeolite-inserted complexes of, e.g., sal, oaba, and maba. Salicylic acid is particularly suited for investigation with regard to the optical effects resulting from the presence of free ligand, as both Na(sal) and H(sal) have a constant fluorescence quantum yield of $\Theta = 50\,\%$ in the spectral excitation region between 250 and 350 nm [93]. Furthermore, the blue sal phosphorescence is completely quenched in pure complexes with Tb^{3+}, resulting in efficient green emission. In the optical spectra, irradiation at 300 nm or below yields a strong ligand emission band at 420 nm accompanied by some Tb^{3+} emission at 550 nm, whereas excitation at at 320 nm or above predominantly leads to Tb^{3+} emission.

A comparison with absorption, excitation, and emission spectra in solutions of H(sal), Na(sal), and $Tb(sal)_3$ showed that (1) short wavelengths (<300 nm) excite free ligand, which is still capable of energy transfer to Tb^{3+} to some degree; and (2) the maximum of the lower energy excitation at 344 nm corresponds to sal coordinated to Tb^{3+}. The same excitation maximum is observed in the free complex, so that the 344 nm excitation can also be attributed to Tb(sal) species in the zeolite.

In the zeolite series referred to above, method B was used to prepare the materials. The spectra of [$Tb_{20}(sal)_4$-X] are reproduced as an example in Fig. 1b, where the ligand fluorescence is still clearly visible at 420 nm despite the high Tb^{3+} content. This material also displayed the highest overall emission intensity in the [Tb(sal)-X] series. The ligand fluorescence intensity is almost exclusively dependent on the amount of Tb^{3+} present, and the emission ratios (ligand emission/Tb^{3+} $^5D_4$) decrease nearly exponentially with increasing Tb^{3+} doping, but the ligand luminescence is not quenched completely, even at the highest Tb^{3+} doping levels. As depicted in Fig. 1a, the plot of emission vs. Tb^{3+} doping level for [Tb(sal)-X] is also strongly reminescent of the curve obtained for the analogously prepared [Tb(benz)-X].

Note that the materials obtained from the above syntheses may not immediately represent equilibrium with regard to positioning of the various species; we detected a strong spectral change in [$Tb_4(sal)_4$-X] and [$Tb_8(sal)_8$X] after serendipitous exposure to ambient laboratory conditions for several months. In these samples, previously recorded ligand luminescence had vanished completely or dropped to a very low level, which we explain in terms of rearrangement of the ligand within the zeolite, possibly promoted by moisture. The more highly loaded analogue [$Tb_4(sal)_{12}$X], retained the ligand luminescence, although at somewhat lower intensity.

A thorough analysis of the data complies with the following picture:

- Low Tb^{3+} doping levels and low Tb^{3+}/ligand ratios (e.g., [$Tb_4(sal)_{12}$-X]) initially lead to the inclusion of free sal and cationic $Tb(sal)^{2+}$ species rather than the complex $Tb(sal)_3$. In this state, excited sal does not "find" Tb^{3+}. The Tb^{3+} emission signal at this composition must therefore be due to either $Tb(sal)^{2+}$ or to the energy transfer sal $\rightarrow$ $Tb^{3+}{}_{sodalite}$.
- Between 8 and 16 Tb^{3+}/UC, the course of the ligand/Tb^{3+} emission ratio indicates the formation of a new species $Tb(sal)_2{}^+$, because the ligand phosphorescence is strongly suppressed. Above 16 Tb^{3+}/UC, isolated complexes $Tb(sal)_3$ may also form within the supercage.
- Allowing sufficient time for rearrangement leads to the removal of ligand luminescence in, e.g., [$Tb_8(sal)_8$-X]. On the basis of stoichiometric arguments or the available Tb^{3+}/ligand ratios, the eventual formation of one $Tb(sal)^{2+}$ species per supercage is assumed.
- In [$Tb_4(sal)_{12}$-X], in which the formation of $Tb(sal)_2{}^+$ in half of the supercages still leaves uncomplexed sal, phosphorescence is maintained even after equilibration.

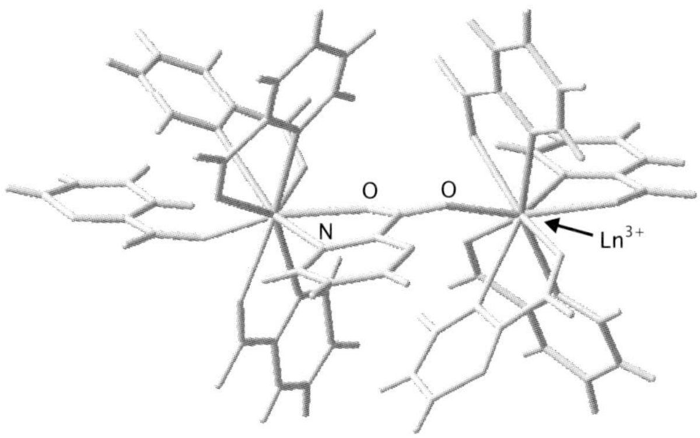

Fig. 2. Section of the $-[Ln(pic)_4]^- -[Ln(pic)_4]^- -$ anion chains found in the structures of $NH_4[Tb(pic)_4]$ and $NH_4[Eu(pic)_4]$.

This description given for salicylates may also be used to understand the behavior of the benzoates depicted in Fig. 1a, and holds equally true for the other substituted benzoates listed in Tab. 1.

5.3.3.2 **Picolinates**

As is evident from Tab. 2, outstanding performance as free complexes makes the picolinates another ligand system deserving attention. For both $Na[Tb(pic)_4]$ and, e.g., $[Tb_{16}(pic)_{28}\text{-}X]$, quantum yields near unity and 50%, respectively, could be established and confirmed by decay measurements (see Tab. 4).

Free picolinate complexes of Ho and Nd had previously been elucidated structurally [94,95], and investigations in solution [96] and on $Eu(pic)_3$ in a sol–gel matrix [97] have been published.

Our own structure determination revealed the same polymeric chains of $M[Ln(pic)_4]$ (M=Na, NH_4; Ln=Tb,Eu) units as in $[(CH_2)_2N_2H_5]^+[Ho(pic)_4]^-$ [95], bridged by the carboxylate group of one pic ligand, with a coordination number of nine for the Ln^{3+} ion (Fig. 2 [104]). The Ln–Ln distances are 633 pm within, and 1055 and 1257 pm respectively, between chains. As outlined for the benzoates, substituted benzoates, and dipicolinates, the Tb^{3+} emission intensities of zeolite complexes here, too, increase with the Tb^{3+} content and, additionally, with ligand loading. Up to 28 pic can be incorporated at 20 Tb^{3+}/UC, and this material is also the most efficient sample prepared. As expected, comparable efficiencies could not quite be achieved with Eu^{3+} materials. The compositional data already suggest the preferred formation of a $[Tb(pic)_4]^-$ anion within the supercage. In analogy to zeolitic hosts, very similar results with regard to efficiency and the composition of the luminescent entity have been reported for sol–gel matrixes [97] and are supported by sol–gel experiments of our own.

Further evidence for the presence of $[Ln(pic)_4]^-$ anions is obtained from comparison and analysis of the excitation and absorption spectra and the chemical behavior in solution. It should be noted that the $M[Ln(pic)_4]$ compounds are readily water soluble; thus, the polymeric structure of the solid is broken up and the coordination of water is assumed. In solution only one optical absorption at 270 nm is measured at pH values of 5–7 for $[Tb(pic)_4]^-$, representative of the fully coordinated ligand, in which both the carboxylate and the pyridine N atom are bound to Tb^{3+} or Eu^{3+}. At pH values above 7, an additional absorption band occurs at 320 nm, which we ascribe to pic ligands that exhibit a nonbonded N atom as a result of the increasing coordination by hydroxo groups. Whether these coordinate in a terminal or bridging fashion is not known. This 320 nm absorption band of the Tb^{3+} complex also gives rise to a corresponding excitation at this wavelength; hence, ligands without N coordination nevertheless remain coordinated to the metal ion through the carboxylate group.

The crystal structure of $M[Tb(pic)_4]$ exhibits a structural similarity with respect to N-coordination in that it displays ligands with significantly varying Tb–N bond lengths ranging from 257 to 270 pm. Correspondingly, the optical spectra of free solid $M[Tb(pic)_4]$ has an excitation band ranging from 260 to well over 320 nm, which reflects the various degrees of Tb–N bonding (Fig. 3a).

In the zeolite, excitation bands due to pic molecules with dangling or weakly coordinating N atoms appear above 300 nm. They are encountered if the formation of $[Tb(pic)_4]^-$ ions is hampered because too few ligands are present per Tb^{3+} ion. In, e.g., $[Tb_{16}(pic)_8\text{-}X]$, the excitation intensity of this 320 nm excitation amounts to about 30% of that of the 270 nm excitation, because at an average of 1 pic and 2 Tb^{3+} per supercage, the formation of $Tb(pic)^{2+}$ species involving carboxylate and lattice coordination is favored over full complexation via the pyridyl N atom. Doubling the amount of ligand reduces the 320 nm excitation intensity to about 15% of that of the 270 nm band, and for $[Tb_{16}(pic)_{28}\text{-}X]$ the excitation band is located exclusively at 270 nm. At the same time the emission intensity increases by a factor of 2.1 and 2.6, respectively. The composition of $[Tb_{16}(pic)_{16}\text{-}X]$ may thus be rewritten as $[(Tb_{sodalite})_9\{Tb(pic)_4\}_7\text{-}X]$ or $[(Tb_{sodalite})_8\{Tb(pic)_{3.5}\}_8\text{-}X]$, implying that up to 8 $[Tb(pic)_4]^-$ will form per UC, corresponding to one anion per supercage. Again, as was discussed for the benzoates and salicylates before, we thus assume surplus Tb^{3+} to reside in sodalite cages; by the same token given for the [Tb(benz)-X] and [Tb(sal)-X] series, surplus $Tb^{3+}{}_{sodalite}$ also contributes to the overall efficiency of the picolinates. The description given above is in good aggreement with the observations and interpretations given for solutions of Tb(pic) species and H(pic) in the literature [96].

5.3.3.3 **Thenyltrifluoroacteylacetonates**

For optical applications, β-diketonates are of interest because, apart from their efficiencies, they can be excited at longer wavelengths, so that, e.g., applications in combination with UV-emitting LEDs are conceivable (see Fig. 5a).

One of the β-diketonate complexes well known for its luminescence is $[Eu(ttfa)_3(H_2O)_2]$, whose molecular dimensions of at least 1000 pm are in the same range as those of the zeolite X supercage; hence, the complex can not enter as a whole. By the same token, impregnation with complexes from solution, as described in Refs. [56–58] require dissociation prior to entry into the zeolite.

The effect of tffa on Eu^{3+}-doped zeolite X is spectacular. It can be demonstrated

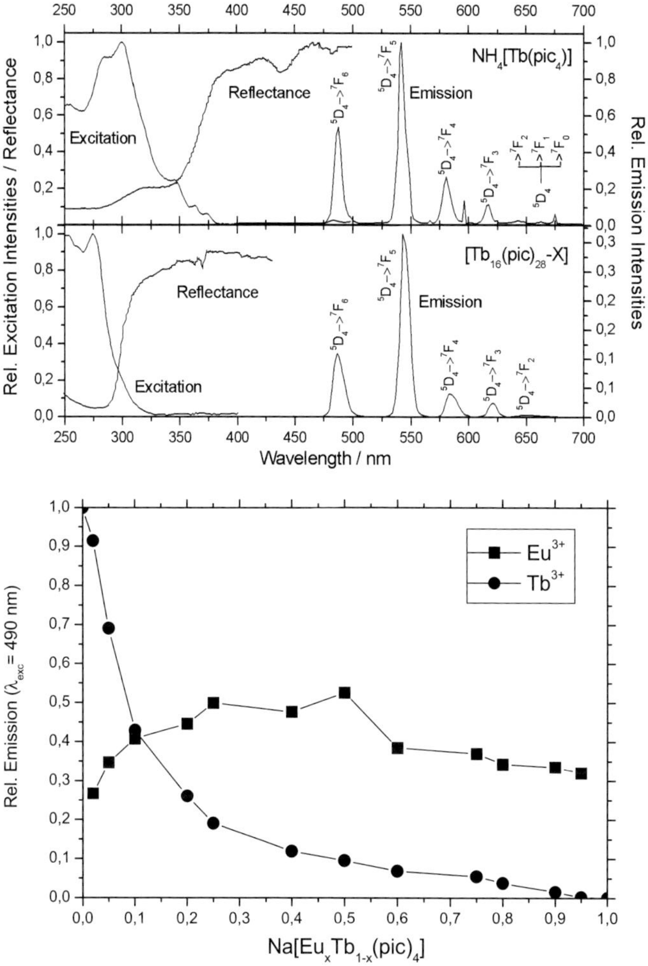

Fig. 3. Left: optical spectra of $NH_4[Tb(pic)_4]$ and $[Tb_{16}(picl)_{28}$-X]. Right: emission intensities of Tb^{3+} and Eu^{3+} in the substitution series $NH_4[Tb,Eu(pic)_4]$.

simply by mixing, e.g., $[Eu_8$-X] with H(ttfa) even in the solid state. A few minutes after mixing the components, irradiating the sample with a 366 nm UV source leads to bright red emission, whereas pure $[Eu_8$-X] shows no luminescence at this wavelength.

Considering the respective sizes of complex and zeolite cavity, synthesis by method B was chosen. In a series of different compositions, as well as in the presence of excess H(ttfa), the sample $[Eu_1(ttfa)_3$-X] showed the highest efficiency, and formation of a stoichiometric complex may be assumed (represented as an isolated dot in Fig. 1a). Next to $[Eu(ttfa)_3$-X] a maximum is obtained for $[Eu_8(ttfa)_{13}$-X], the

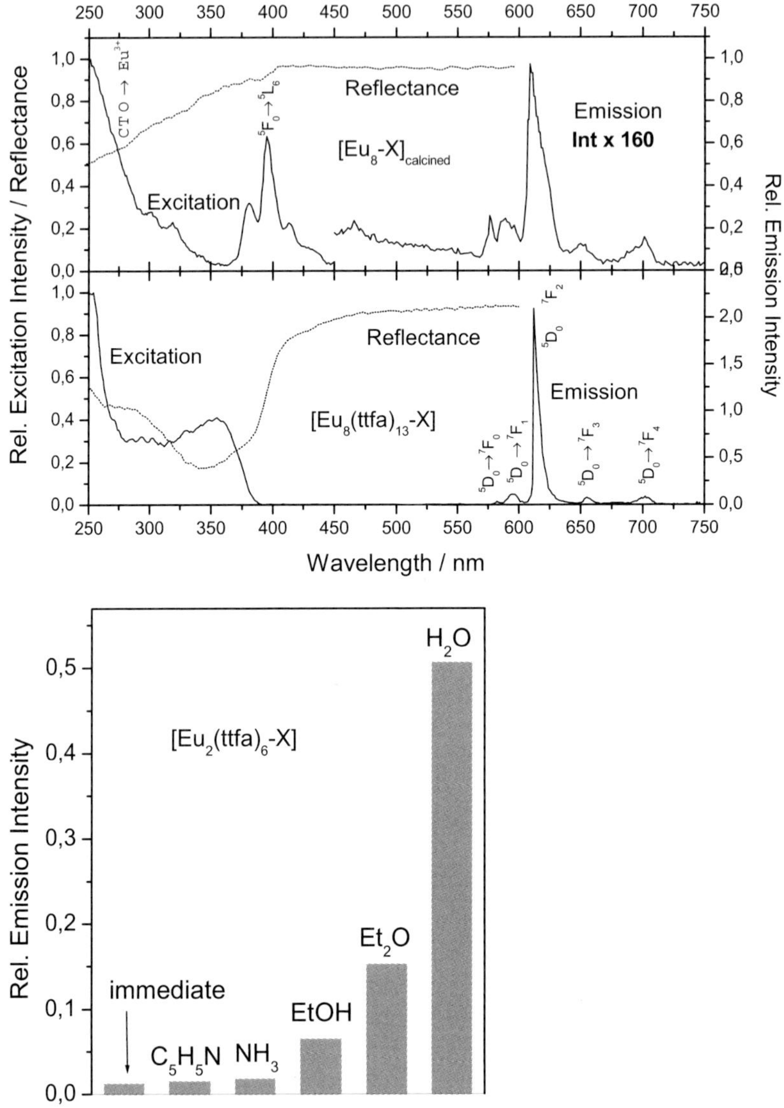

Fig. 4. Left: optical spectra obtained from freshly calcined [Eu_8-X] and rehydrated [Eu_8(ttfa)$_{13}$-X]. Right: effect of exposure to atmospheres saturated with various agentsfor 24 h.

spectra of which are reproduced in comparison with [Eu_8-X] in Fig. 4a. In this sample, the increase of the emission intensity amounts to a respectable factor of 350 after complex formation in comparison to [Eu_8-X]. Apart from [Eu(ttfa)$_3$-X], the behavior of the emission intensity vs. Eu^{3+} loading in this system again seems to correlate with the formation of up to 8 cationic Eu(ttfa)$^{2+}$ species per UC. Only [Eu(ttfa)$_3$-X] assumes an exceptional position in the series. Possibly, the ligands of the Eu(ttfa)$_3$ entity extend into neighboring supercages and prevent formation of the fully ligated species therein. In accord with the arguments presented before, higher doping then leads to an energy transfer $Eu(ttfa)^{2+} \rightarrow Eu^{3+}{}_{sodalite}$ that causes losses in the emission intensity due to the low efficiency of Eu^{3+} ions populating sodalite cages.

In several cases, a strong influence of intermissions required for equilibration and aftertreatments has been pointed out. A particularly remarkable example of this was found for [Eu$_n$(ttfa)$_{3n}$-X]. Freshly prepared samples were recorded and remeasured after exposure to various atmospheres, which caused the dramatic changes in the emission intensities depicted in Fig. 4b for, e.g., [Eu$_2$(ttfa)$_6$-X], in which a fiftyfold increase results from overnight rehydration. For [Eu$_1$(ttfa)$_3$-X] it was 216-fold and could not be depicted in the graph. This observation may seem surprising at first sight, as Eu^{3+} should show quenching on water contact. The explanation certainly has to be sought in the weak acidity of the H(ttfa) ligand, which cannot protonate the zeolite lattice spontaneously, and is probably related to the initial inclusion of the Eu^{3+} in six-ring windows of sodalite cages. Release of Eu^{3+} is mediated by water present after rehydration; at the same time, complexation of the ligand is promoted by its deprotonation, which is necessary for complexation to eventually take place. The efficiencies of good samples obtained in this manner already surpass free [Eu(ttfa)$_3$(H_2O)$_2$] by a factor of two, but the coordination of some water is still suspected.

5.3.3.4 Comparison of Ligands

The formation of cationic species, as deduced from the dependence of emission intensities on particular compositions in a variety of samples, as well as from the phosphorescence of the ligand in the sal system, is in contrast to the interpretations for bipy, dipic, pic, and ttfa complexes in zeolites, for which the presence of $[Ln(dipic)_3]^{3-}$, $[Ln(pic)_4]^{-}$, and [Eu(ttfa)$_3$] within the supercages must be assumed, at least at lower doping and loading levels.

It can be expected here that the nature of the complex species formed not only depends on size, but also on the affinity of the ligand for the lanthanide ion, which is reflected in the complex-formation constant. Typically, the complexes of pyridine carboxylates and diketonates are several orders of magnitude more stable than those of substituted benzoates (e.g., $La(dipic)^{+}$: $\beta_1 = 7.9$, $La(ttfa)^{+}$: $\beta_1 = 9.53$, $La(sal)^{2+}$: $\beta_1 = 2.8$ [98]), which in turn is a consequence of the nonchelating bonding mode of, e.g., the hydroxyl group of the sal ligand, as observed in the solid-state free complexes [99]. Note that none of the free complexes with substituted aromatic carboxylates under consideration involves coordination via the additional substituent (paba [100], mhba [101], phba [102], Eu(oaba)$_3$bipy [103]). For paba we investigated this in detail and solved single-crystal structures for numerous coordination modes, among which were materials in which excess H(paba) is embedded in the crystal unit cell. In no instance could coordination of the amino group be found [104]. Further support for the presence of cationic complex ions is gained from the elucidation of energy transfer, as discussed in the next section.

5.3.4 Energy Transfer

5.3.4.1 Energy Transfer between Free and Complexing Ligands ($L_\gamma \rightarrow L_{Ln3+}$)

The phosphorescence of the sal ligand can also yield information on existing energy transfer channels. Surprisingly, intermolecular energy transfer between uncomplexed molecules ($L\gamma$) and complexing ligands (L_{Ln3+}), which should be of the dipole–dipole type and thus extend over the dimensions of one unit cell and allow the excitation to probe large volumes for the presence of an Ln^{3+} ion, seems to be of only secondary significance. The arguments leading to this conclusion are laid out in the following for the [Tb(sal)-X] system, but similar results are found in, e.g., paba- and pic-doped zeolites:

- At low Tb^{3+} doping levels, as in samples ranging from $[Tb_4(sal)_4\text{-}X]$ to $[Tb_4(sal)_{12}\text{-}X]$, strong ligand phosphoresecence is observed, at least before the above-mentioned equilibrium distribution of the components is achieved. In $[Tb_4(sal)_4\text{-}X]$ the initial phosphorescence is understood in terms of a large spacial separation between sal and Tb^{3+}, which prevents energy transfer. After equilibration, formation of the complex $Tb(sal)^{2+}$ results in Tb^{3+} emission rather than ligand phosphorescence. In $[Tb_4(sal)_{12}\text{-}X]$, $Tb(sal)_2{}^+$ in half of the supercages still leaves the other half for the accomodation of uncomplexed sal; therefore, phosphorescence is maintained even after equilibration. Up to this point, ligand → ligand interactions need not enter the line of arguments.
- Increasing the ligand content from $[Tb_4(sal)_4\text{-}X]$ to $[Tb_4(sal)_{12}\text{-}X]$ increases the overall absorption, but leaves the Tb^{3+} emission intensity constant. If ligand → ligand energy transfer were to take place, at least some increase in the Tb^{3+} emission intensity would be expected. With respect to efficacy, it should be noted that an increase of absorption, which is not paralleled by an increase in emission intensity, corresponds to a reduction of the luminescence quantum yield.
- At high loading *and* doping, (e.g., in $[Tb_{16}(sal)_{16}\text{-}X]$) ligand phosphorescence is still observed. In the presence of ligand → ligand energy transfer, this phosphorescence should be absent, because all ligand excitation would inevitably be transferred to Tb^{3+} and contribute to the characteristic emission of the latter.

In Tb^{3+}-doped zeolites with paba and pic loading, in which phosphorescence does not contribute to the overall emission between 400 and 700 nm, doubling the ligand content at constant Tb^{3+} doping systematically reduces the quantum yield by a factor of roughly 2. Furthermore, the same reduction factor is found regardless of the absolute Tb^{3+} content.

Therefore, here too ligand → ligand energy transfer is assumed to be absent. In further support of this, as will be discussed in more detail below, results on the pure complexes $Na[Tb,Eu(pic)_4]$, $Na[La,Tb(pic)_4]$, $[Tb,Eu(benz)_3]$, and $[La,Tb(benz)_3]$ also give no indication of this type of energy transfer.

5.3.4.2 Free ligand → Free Ln^{3+} Energy Transfer ($L_\gamma \rightarrow Ln^{3+}{}_{sodalite}$)

As opposed to direct energy transfer between ligands themselves, transfer appears to be possible from ligands within the supercage (L_γ) to lanthanide ions located in the sodalite cages. This transfer is not expected to be of high efficiency, as it involves broad-band-emitting ligand states and lanthanide excitations, which are $f \rightarrow f$ in nature. Such processes involving transfer from a band emitter to a line absorber are disadvantageous with respect to spectral overlap and usually require high Ln^{3+} concentrations to afford short-range transfer by exchange interaction [105]. However, in the zeolite, relatively short distances of less than 500 pm between ligands in supercages and Ln^{3+} ions hosted in the sodalite cage can be realized, so that this transfer is observed. In the case of Tb^{3+}, it should additionally be noted that an overlap contribution of Tb^{3+} d states as potential band absorbers cannot be expected, as they are located at too high an energy corresponding to about 230 nm [1].

Experimental evidence for $L_\gamma \rightarrow Ln^{3+}{}_{sodalite}$ transfer could be obtained in the case of Tb^{3+}, if the Tb^{3+} emission intensity could be increased by a surplus of ligand, while the available free Tb^{3+} ions are located in sodalite cages. Several samples prepared within the series of [$Tb_n(benz)_{32}$-X] zeolites fulfill these conditions. In these, free ligand is unambiguously present, and at $n > 8$, Tb^{3+} is assumed to be located in sodalite cages; hence, the steep increase in emission intensity between $n = 8$ and $n = 20$, is ascribed to a $benz_\gamma \rightarrow Tb^{3+}{}_{sodalite}$ energy transfer (see Fig. 1a). Note that a conceivable energy transfer by exchange between Tb^{3+}-centered states of complexed and sodalite-hosted ions, $[Tb^{3+}(^5D_4)]_{benz} \rightarrow [Tb^{3+}(^5D_4)]_{sodalite}$, which is discussed below, would not yield the increase in emission intensity observed in the [$Tb_n(benz)_{32}$-X] series.

The constructive cooperation between the states of free ligands and sodalite-encaged ions could clearly be traced in several others of the samples prepared as well. Examples were found in [Tb(dipic)-X] (prepared with excess ligand, method A) and in the [Tb(pic)-X] series, e.g., the sample [$Tb_{16}(pic)_{28}$-X] (=[$(Tb_{sodalite})_9(Tb(pic)_4)_7$-X]; see Section 5.3.3.2).

Destructive transfer $L_\gamma \rightarrow Ln^{3+}{}_{sodalite}$, leading to quenching of Ln^{3+}, is more difficult to assess. In, e.g., [Eu_n(dipic)-X] with approximately 30 dipic/UC, the emission intensity remains practically constant on increasing the Eu^{3+} doping levels in the range above 8 Eu/UC. This would comply with $dipic_\gamma \rightarrow Eu^{3+}{}_{sodalite}$ transfer, because $Eu^{3+}{}_{sodalite}$ is extremely inefficient. Unfortunately, the presumed energy transfer $[Eu^{3+}(^5D_0)]_{dipic} \rightarrow [Eu^{3+}(^5D_0)]_{sodalite}$ by exchange (see below), which would also cause a decrease of the emission intensity, obscures the situation.

5.3.4.3 $Ln^{3+} \rightarrow Ln^{3+}$ and Energy Transfer between Complexing Ligands ($L_{Ln3+} \rightarrow L_{Ln3+}$)

The next question arising is concerned with the existence of energy transfer by direct exchange interaction of lanthanide ion states $[^{2S+1}Ln_J(1)] \rightarrow [^{2S+1}Ln_J(2)]$, which decreases exponentially with the distance between the ions and has a critical transfer distance of about 500 pm. Since the transfer $L_\gamma \rightarrow Ln^{3+}{}_{sodalite}$ is restricted to short distances and dipole–dipole (ligand → ligand) transfer is absent, communication between supercages will not be possible. Bearing in mind, however, that free ligand in supercages can transfer excitation energy to Tb^{3+} in neighboring sodalite cages, as discussed in the Section 5.3.4.2, then transfer between Tb^{3+} ion states ($[Tb^{3+}(^{2S+1}Ln_J)]_{complexed} \rightarrow [Tb^{3+}(^{2S+1}Ln_J)]_{sodalite}$; $[Tb^{3+}(^{2S+1}Ln_J)]_{complexed} \rightarrow [Tb^{3+}(^{2S+1}Ln_J)]_{complexed}$) and ultimately, e.g., $Tb^{3+} \rightarrow Eu^{3+}$ transfer should also be possible. Depending on the actual position of the lanthanide ions within the sodalite cages, $Tb^{3+}{}_{sodalite} \rightarrow Tb^{3+}{}_{sodalite}$ transfer between neighboring sodalite cages themselves is also conceivable, if we consider that the separation is just over 300 pm.

With regard to $Ln^{3+} \rightarrow Ln^{3+}$ energy transfer, the free complexes and zeolite analogues of the picolinates were studied in more detail. Stepwise substitution of Tb^{3+} by Eu^{3+} in this system is a suitable monitor, because the free complexes and zeolitic materials of both Tb^{3+} and Eu^{3+} show efficient luminescence. Furthermore, free $(NH_4)[Eu(pic)_4]$ and the derived $(NH_4)[Tb_nEu_{1-n}(pic)_4]$ mixtures have identical structures, as is evident from their IR spectra. Additionally, the crystal structures of the isostructural $(NH_4)[Ln(pic)_4]$ and $Na[Ln(pic)_4]$ reveal that no water is coordinated to the lanthanide ions (see also Section 5.3.3.2). For Eu^{3+} this is of particular significance, since obscuration of deactivation due to high-frequency vibrations can thus be excluded. Finally, the assumption seems reasonable that in the zeolitic materials, too, water in the immediate coordination sphere of the lanthanide ions, which are present as $[Ln(pic)_4]^-$ anions, is absent.

First, the mixed $(NH_4)[Tb,Eu(pic)_4]$ materials were excited at 490 nm, corresponding to Tb^{3+} ($^7F_6$–$^5D_4$) excitation. The excited Tb^{3+} ($^5D_4$) state can then transfer its energy into the Eu ($^5D_1$) level. After an internal relaxation step, Eu^{3+} ($^5D_0$–$^7F_2$) emission can be monitored at 612 nm, as the Eu^{3+} ($^5D_0$) state is efficiently emissive in the picolinates. At the same time, the intensity of the Tb^{3+} ($^5D_4 \rightarrow {}^7F_5$) emission at 545 nm will be reduced if transfer takes place. The intensity ratios between the Tb^{3+} and Eu^{3+} emissions at different Tb^{3+}/Eu^{3+} contents thus reflect the energy-transfer efficiency. For 50% substitution of Tb^{3+} against Eu^{3+} a decline of the Tb^{3+} emission intensity to 50% will be monitored in the absence of energy transfer, whereas in the presence of transfer, the Tb^{3+} emission intensity is expected to be lowered to less than 50% to an extent that depends on the transfer efficiency. As depicted in Fig. 2b, the drop of the Tb^{3+} emission to 50% is already attained at a surprisingly low degree of substitution of less than 10%, and this clearly indicates strong exchange interaction Tb^{3+} ($^5D_4$) ↔ Eu ($^5D_1$). With regard to the anionic chain structure of $[Ln(pic)_4]^-$ (Fig. 2), with Ln^{3+}–Ln^{3+} distances of 633 pm relevant to this type of energy transfer, one can additionally estimate that on average at least five $[Tb^{3+}(^5D_4)] \rightarrow [Tb^{3+}(^5D_4)]$ transfer steps will have been traversed, or a distance of roughly 3 nm covered, before a Eu^{3+} ion is reached.

It may be of interest that excitation at 280 nm, which corresponds to ligand excitation, results in exactly the same dependence, in agreement with the hypothesis that all energy transfer observed occurs through the $[Tb^{3+}(^5D_4)] \rightarrow [Eu^{3+}(^5D_1)]$ channel rather than $[pic(on\ Tb^{3+})] \rightarrow [pic(on\ Eu^{3+})] \rightarrow [Eu^{3+}(^5D_1)]$ or [pic(on

$Tb^{3+})] \rightarrow [Eu^{3+}(^5D_1)]$. To prove this, a similar sample series was prepared with $[Tb,La(pic)_4]^-$, as here an emission signal of, e.g., >50% for $Na[La_{0.5}Tb_{0.5}(pic)_4]$ unequivocally indicates interaction between neighboring ligands. At the composition corresponding to, e.g., 50% substitution of Tb^{3+} by La^{3+}, we found the Tb^{3+} intensity to be only slightly greater than 50% of that of $Na[Tb(pic)_4]$. The results from this series therefore seem to contradict the presence of energy transfer channels between ligands.

Unfortunately, $(NH_4)[La,Tb(pic)_4]$, according to IR spectroscopic and powder-diffraction data, has a structure different from $(NH_4)[Tb(pic)_4]$. In comparison to $(NH_4)[Tb,Eu(pic)_4]$, $(NH_4)[La,Tb(pic)_4]$ is almost insoluble in water, which suggests a different linkage, i.e., structure of the polymer. Therefore, materials based on $[Tb_nLa_{1-n}(pic)_4]^-$ mixed complexes could not yet be crystallized in sufficient quality, and a conclusive correlation with the structure remains an open task. However, the conclusions regarding the decrease of the Tb^{3+} emission intensity upon Eu^{3+} and La^{3+} substitution were supported by a series of $[Tb,Eu(benz)_3]$ and $[La,Tb(benz)_3]$ samples, respectively. As before, increasing $[Tb(^5D_4)] \rightarrow [Eu(^5D_1)]$ energy transfer is found upon stepwise Eu^{3+} substitution. With increasing La^{3+} substitution, the decrease in the Tb^{3+} emission is almost linear, which again provides evidence for the absence of energy transfer between the neighboring benzoate ligands. Here too, structural evidence would be desirable to arrive at yet more quantitative transfer data. The absence of energy transfer between complexing ligands ($L_{Ln}{}^{3+} \rightarrow L_{Ln}{}^{3+}$) complements the findings already established for the transfer from free ligand to ligands in, e.g., the benzoates (Section 5.3.4.2).

In the zeolites, $Ln^{3+} \rightarrow Ln^{3+}$ transfer is not so easy to assess because the interaction is dependent not only on the doping ratio Eu^{3+}/Tb^{3+}, but also on the overall doping level ($Tb^{3+} + Eu^{3+}$)/UC), the amount of ligands and their coordination mode, the location in the lattice, and individual quantum efficiencies. Nevertheless, the presence of $Tb^{3+} \rightarrow Eu^{3+}$ transfer at least may be concluded from Tb^{3+} vs. Eu^{3+} substitutions in $[Tb,Eu(pic)_8\text{-}X]$: at low doping levels, the Tb^{3+}/Eu^{3+} emission ratio shows little correlation with the $Tb^{3+} + Eu^{3+}$ content. Instead, the emission ratios are close to the values expected in the absence of energy transfer. $[Tb_2Eu_1(pic)_8\text{-}X]$, for example, yielded a Tb^{3+} emission 2.7 times higher than that Eu^{3+}, which only reflects the lower efficiency of Eu^{3+}. In contrast, $Tb^{3+} \rightarrow Eu^{3+}$ energy transfer is obviously present at high lanthanide content. A sample of composition $[Tb_{16}Eu_4(pic)_8\text{-}X]$ only yielded 50% of the Tb^{3+} emission intensity measured in $[Tb_{16}(pic)_8\text{-}X]$. Using the previously justifed assumption of 8 possible $[Ln(pic)_4]^-$ anions within supercages (Section 5.3.3.2) and 12 Ln^{3+} cations in sodalite cages, the reduction in the Tb^{3+} emission is due to an energy transfer from complexed Tb^{3+} to Eu^{3+} in sodalite cages ($[Tb^{3+}{}_{picolinate}(^5D_4)] \rightarrow [Eu^{3+}{}_{sodalite}(^5D_0)]$). This type of energy transfer has been described recently in Ref. [33] and was proposed as part of a mechanism for optical data storage. Once more, analogous findings on benzoates provide additional support for the results on $Ln^{3+} \rightarrow Ln^{3+}$ energy transfer outlined above.

To this end, apart from the intramolecular ligand $\rightarrow Ln^{3+}$ energy transfer described in the introduction, an additional four conceivable energy transfer channels have been considered in zeolites. The experimental results on their occurence are summarized as follows:

1. Energy transfer between free and complexing ligands ($L_\gamma \rightarrow {L_{Ln}}^{3+}$): not observed.
2. Free ligand $\rightarrow$ free Ln^{3+} energy transfer ($L_\gamma \rightarrow {Ln^{3+}}_{sodalite}$): found in several instances.
3. $Ln^{3+} \rightarrow Ln^{3+}$ energy transfer: found in several instances.
4. Energy transfer between complexing ligands (${L_{Ln}}^{3+} \rightarrow {L_{Ln}}^{3+}$, see Section 5.3.3.4) : not observed.

Although investigated on selected examples only, a quite general validity of the picture thus outlined is expected with respect to energy transfer in zeolites.

5.3.5 Size

Another important general aspect in the design of optically functional zeolite hybrid materials are the rigid geometric constraints due to cavity and channel sizes. The efficiency of $[Eu(ttfa)_3(H_2O)_2]$ can be improved dramatically by co-coordinating ligands, most prominently phen [42], which affords the molecular complex [Eu(ttfa)$_3$(phen)] with quantum yields in excess of 80%. The increase in efficiency over $[Eu(ttfa)_3(H_2O)_2]$ was already related to the absence of coordinated water above. It seemed worthwhile to analogously displace the water suspected to be present in [Eu$_1$(ttfa)$_3$-X], too. To accomplish this, attempts to co-coordinate Eu(ttfa)$_3$ with phen were performed, although the desired complex was expected to be at the limit of the containment capability of the supercage. We were thus surprised to observe that [Eu(ttfa)$_3$(phen)-X] formed readily.

The optical characteristics of the complex formed within the zeolite resemble those of the free complex almost perfectly, as is evident from the comparison presented in Fig. 5a. Free and zeolite-enclosed [Eu(ttfa)$_3$(phen)] also have very similar IR spectra. Slight shifts of the absorption bands were also observed after incorporation of [Eu(ttfa$_3$)(phen)] in sol–gel and PMMA matrixes [106], and these can in turn be used as an indication for the actual degree of incorporation. Furthermore, similar attempts to form co-ligated complexes in a series of other systems listed in Tab. 3 led to phosphorescence of the phen ligand at low doping and loading levels. Its phosphorescence may thus be used as a monitor that indicates free phen and non-co-coordinated complex species in isolated supercages rather than co-ligated complexes. Significantly, no phen phosphorescence is observed in [Eu(ttfa)$_3$(phen)-X] and, in accordance with the above information, formation of the co-coordinated complex is concluded.

The emission intensity of [Eu(ttfa)$_3$(phen)-X] is reduced to some 80% compared to free [Eu(ttfa)$_3$(phen)]. With regard to quantum yield, however, the reduction is compensated by a lower absorption at the excitation wavelengths of 320 and 330 nm, respectively, which were used for this measurement. Within experimental

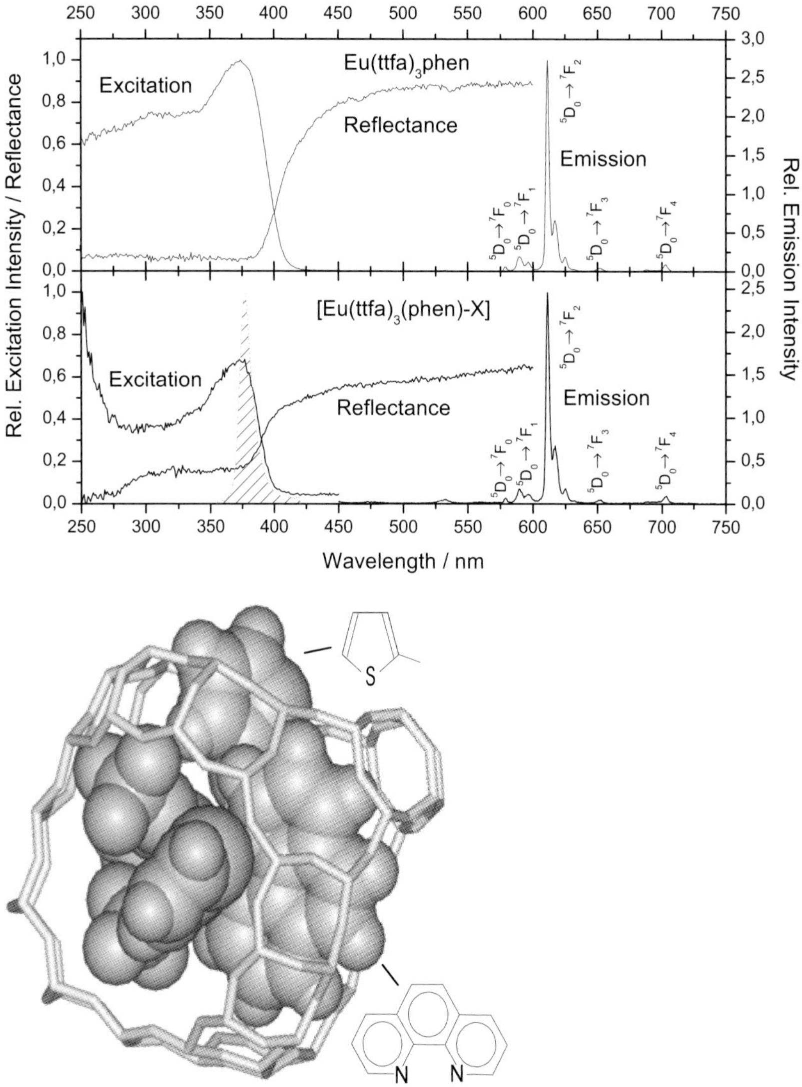

Fig. 5. Left: optical spectra of free [Eu(ttfa)$_3$(phen)] and [Eu(ttfa)$_3$(phen)-X]. The hatched area in the bottom part depicts the emission spectrum of a commercial UV LED for comparison. Right: structural model of [Eu(ttfa)$_3$(phen)] within a supercage (Hyperchem, Spartan).

error, the two systems therefore have nearly the same efficiency of approximately 80%.

Because the structure of [Eu(ttfa)$_3$(phen)-X] is unknown, the results from molecular modeling are depicted in Fig. 5b. In this model, merely the distances between lattice oxygen atoms and fluorine atoms appear to be critical with respect to repulsion. Only in one case do the predicted O and F contacts fall short of the sum of Van der Waals radii of 280 pm, by 20 pm.

5.3.6 Surface Efficiency

Motivated by the observation that yet another β-diketonate complex, namely, $Tb(dpac)_3$, which is nonluminescent as the free complex $[\{Tb(dpac)_3\}_2]$ [107], started luminescing on contact with solid zeolite (see also [Eu(ttfa)-X] in Section 5.3.3.3), the effect of surface versus bulk complex formation was studied.

Materials obtained from a 1:10 mixture of solid $[Tb(dpac)_3]_2$ and solid zeolite X, after some aging under ambient atmosphere and washing with EtOH, showed an unexpected Tb^{3+} emission amounting to 20% of the intensity obtained for $(NH_4)[Tb(pic)_4]$.

In the free complex the absence of emission, i.e., quenching, is caused by too low an excited triplet state of the ligand. The low-lying triplet, in turn, is presumably a result of the peculiar structure of the free complex, with bidentate, asymmetrically bridging ligands [107]. Hence we ascribe the occurrence of emission in the mixture to the coordination of zeolite surface oxygen atoms to the complex. The formerly low-lying triplet of the latter is raised in energy, as the need for bridging ligands is eliminated by the presence of zeolite oxygen atoms in the coordination sphere of Tb^{3+}, and a strong ligand excitation band appears at 270 nm, next to which no Tb^{3+} excitation exists.

By using preparative method B, it was also possible to incorporate the ligand H(dpac) into zeolites doped with various amounts of Tb^{3+} to yield "bulk" loaded complexes. At 1 Tb^{3+}/UC, offering three ligand molecules, only small amounts of ligand are inserted (0.03 dpac/UC). In this material, f–d excitation due to Tb^{3+} in sodalite cages can be recorded at 225 nm. It has six times the intensity of an excitation band at 270 nm, which originates from the minute amounts of complex concurrently formed within the supercages. The total Tb^{3+} emission intensity obtained from this material amounts to $I_{rel} = 0.01$. However, using a zeolite doped with 12 Tb^{3+}/UC and a large excess of ligand led to incorporation of 5.4 ligand molecules per UC. As a result, a fivefold intensity increase of the ligand excitation at 270 nm is obtained. At the same time, the Tb^{3+} f–d excitation signal is only doubled, despite the much higher Tb^{3+} content, and is a consequence of the lower efficiency of the complex in comparison to Tb^{3+} in sodalite cages. These data unambiguously reflect the formation of intrazeolite complexes, whose degree of formation can be judged by the intensity of the respective excitations of ligand and Tb^{3+}. An increase in the emission to $I_{rel} = 0.06$ could be measured for the "bulk" complex $[Tb_{12}(dpac)_{5.4}\text{-}X]$. It is also noteworthy that diffusion of the complex from solution to the zeolite voids does not yield noticeable loading. Also, formation of surface complexes is not observed in corresponding experiments with aerosils.

In conlusion, despite the considerable size of the ligand, formation of the complex $Tb(dpac)_3$ takes place within the zeolite, although the surface complexes discussed above remain the more efficient species by a factor of thirty. Nevertheless, the findings represent a case in which the luminescence quenching of a free complex due to particular structural features is relieved by interaction with the zeolite matrix.

5.4 Concluding Remarks

The data basis acquired in the recent past on the optical performance of zeolites, not just for incorporated organometallic complexes , gives rise to optimism with respect to applications as optically functional hybrid materials in various fields. Some understanding of the sensitive preparative procedures could be gained to direct product compositions and exploit the undoubtedly available efficient luminescence mechanisms. Specific energy transfer channels, reminescent of classical phosphors and luminophores, may now be selected and exploited. There is still a lack of conclusive structural evidence, which is highly desirable to optimize towards high efficiencies. The issue of photostability under prolonged irradiation remains to be addressed to judge the true competitiveness of these novel materials.

References

1 T. Jüstel, D.U. Wiechert, C. Lau, D. Sendor, U. Kynast, *Adv. Funct. Mater.* **2001**, *11*, 105.
2 J.E. Mac Dougall, H. Eckert, G.D. Stucky, N. Herron, Y. Wang, K. Moller, T. Bein, D. Cox, *J. Am. Chem. Soc.* **1989**, *111*, 8006.
3 G.D. Stucky, J.E. MacDougall, *Science* **1990**, *247*, 669.
4 G.A. Ozin, C.L. Bowes, M.R. Steele, *Mat. Res. Symp. Proc.* **1992**, *277*, 105.
5 M. Wark, G. Schulz-Ekloff, N.I. Jaeger, *Catal. Today* **1991**, *8*, 467.
6 A.D. Yoffe, *Adv. Phys.* **1993**, *42*, 173.
7 G. Tel'biz, I. Blonskij, S. Shevel, V. Voznyi, in *Novel Materials*, H. Chon, S.-K. Ihm, Y.S. Uh (Eds.), *Sudies in Surface Science and Catalysis*, Vol. 105, Elsevier, Amsterdam, **1997**, p. 293.
8 A. Aparisi, V. Fornes, F. Marquez, R. Moreno, C. Lopez, F. Meseguer, *Solid State Electron.*, **1996**, *40*, 641.
9 G.A. Ozin, S. Özkar, *Adv. Mater.* **1992**, *4*, 11.
10 K. Möller, T. Bein, S. Özkar, G.A. Ozin, *J. Phys. Chem.* **1991**, *95*, 5276.
11 C. Lau, H.-J. Mai, U. Kynast, *Microporous Mesoporous Mater.* **2001**, *47*, 339.
12 M. Pauchard, S. Huber, R. Méallet-Renault, M. Maas, R. Pansu, G. Calzaferri, *Angew. Chem. Int. Ed.* **2001**, *40*, 2839.
13 G. Calzaferri, *Chimia* **1998**, *52*, 525.
14 G. Calzaferri, *Catal. Today* **1997**, *39*, 145.
15 V. Ramamurthy, P. Lakshminarasimhan, C.P. Grey CP, L. Johnston, *J. Chem. Soc. Chem. Commun.* **1998**, 2411.
16 J.C. Scaiano, H. Garcia, *Acc. Chem. Res.* **1999**, *32*, 783.
17 J. Dedecek , B. Wichterlova, P. Kubat, P, *Microporous Mesoporous Mater.* **1999**, *32*, 63.
18 J. Dedecek , B. Wichterlova, *Phys. Chem. Chem. Phys.*, **1999**, *1*, 629.
19 T. Arakawa, T. Takata, G.-Y. Adachi, J. Shiokawa, *J. Chem. Soc. Chem. Commun* **1979**, *453*.

20 S.L. Suib, R.P. Zerger, G.D. Stucky, T.I. Morrison, G.K. Shenoy, *J. Chem. Phys.* **1984**, *80*, 2203.

21 J. Bartlett, R. Cooney, R. Kydd, *J. Catal.* **1988**, *114*, 58.

22 J. Tanguay, S.L. Suib, *Catal. Rev. Sci. Eng.* **1987**, *29*, 1.

23 M.D. Baker, M.M. Olken, G.A. Ozin, *J. Am. Chem. Soc.* **1988**, *110*, 5709.

24 S. Shionoya, W.M. Yen, *Phosphor Handbook*, CRC Press, Boca Raton, **1999**, p 655.

25 D. Parker, P. Kanthi-Senanayake, J. A. Williams, *J. Chem. Soc. Perkin Trans.* **1998**, *2* 2129.

26 F.S. Richardson, *Chem. Rev.* **1982**, *82*, 541.

27 J.-C.G. Bünzli in J.C.G. Bünzli, G.R. Choppin (Eds.), *Lanthanide Probes in Life, Chemical and Earth Sciences*, Elsevier, Amsterdam, **1989**, Chap 7.

28 I. Hemilä, *Application of Fluorescence in Immunoassays*, Wiley Interscience, New York, **1991**.

29 M. Koedam, J.J. Opstelten, *Lighting Res. Technol.* **1971**, *3*, 205.

30 U. Kynast, V. Weiler, *Adv. Mater.* **1994**, *6*, 937.

31 C.R. Ronda, *J. Alloys Compd.* **1995**, *225*, 534.

32 U. Kynast, V. Weiler, EP 96 200 282, **1996**.

33 W. Chen, R. Sammynaiken, Y. Huang, *J. Appl. Phys.* **2000**, *88*, 1424.

34 M. Warnken, K. Lazar, M. Wark, *Phys. Chem. Chem. Phys.* **2001**, *3*, 187.

35 U. Kynast, V. Weiler, EP 96 200 281, **1996**.

36 G. Stein, E. Würzberg, *J. Chem. Phys.* **1975**, *62*, 208 (OH quenching Eu).

37 C. Borgmann, J. Sauer, T. Jüstel, U. Kynast, F. Schüth, *Adv. Mater.* **1999**, *11*, 45.

38 S.I. Weissmann, *J. Chem. Phys.* **1942**, *10*, 214.

39 G. Crosby, R. Whan, R. Alire, *J. Chem. Phys.* **1961**, *34*, 743.

40 W. Sager, N. Filipescu, F. Serafin, *J. Phys. Chem.* **1965**, *69*, 1092.

41 S. Sato, M. Wada, *Bull. Chem. Soc. Jpn.* **1970**, *43*, 1955.

42 E.V. Melent'eva, L.L. Kononenko, E.G. Koltunova, N.S. Poluektov, *Zh. Prikl. Spectrosk.* **1966**, *5*, 245.

43 J. Kido, H. Nagai, Y. Okamoto, T. Skotheim, *Chem. Lett.* **1991**, 1267.

44 H. Boerner, U. Kynast, W. Busselt, M. Haase, DE 44 28 450, **1996**.

45 M.A. Baldo, M.E. Thompson, S.R. Forrest, *Nature* **2000**, *403*, 750.

46 M. Bredol, U. Kynast and C.R. Ronda, *Adv. Mater.* **1991**, *3*, 361.

47 A.M.V. de Andrade, N.B. da Costa Jr., O.L. Malta, R.L. Longo, A.M. Simas, G.F. de Sa, *J. Alloys Compd.* **1997**, *250*, 412.

48 W. DeWilde, G. Peeters, J. Lunsford, *J. Phys. Chem.* **1980**, *84*, 2306.

49 K. Balkus, A. Welch, B. Gnade, *Zeolites* **1990**, *10*, 272.

50 Y.I. Kim, T. Mallouk, *J. Phys. Chem.* **1992**, *96*, 2879.

51 W. Turbeville, D.S. Robbins, P.K. Dutta, *J. Phys. Chem.* **1992**, *96*, 5024.

52 E.S. Brigham, P. Snowden, Y.I. Kim, T. Mallouk, *J. Phys. Chem* **1993**, *97*, 8650.

53 P. Lainé, M. Lanz, G. Calzaferri, *Inorg. Chem.* **1996**, *35*, 3514.

54 M. Sykora, J.R. Kincaid, P. Dutta, N. Castagnola, *J. Phys. Chem. B* **1999**, *103*, 309.

55 M. Bredol, U. Kynast, C.R. Ronda, T. Welker, DE 41 22 009, **1993**.

56 I.L.V. Rosa, O.A. Serra, E.J. Nassar, *J. Luimin..* **1997**, *72–74*, 532.

57 M. Alvaro, V. Fornés, S. Garcia, J.C. Scaiano, *J. Phys. Chem. B* **1998**, *102*, 8744.

58 J. Dexpert-Ghys, C. Picard, A. Taurines, *J. Incusl. Phenom. Macrocycl. Chem.* **2001**, *39*, 261.

59 Q. Xu, L. Li, B. Li, J. Yu, R. Xu, *Microporous Mesoporous Mater.* **2000**, *38*, 351.

60 E.O. Malta, H.F. Brito, J.F.S. Menezes, F.R.G. de Silva, C. de Mello Donega, S. Alves Jr., *Chem. Phys. Lett.* **1998**, *282*, 233.

61 K. Yamazoe, A. Yoshino, Y. Yoshiharu, I. Iwami, K. Kawasaki, DE 3050703 C2, **1983**.

62 M.L. Bhaumik, C.L. Telk, *J. Opt. Soc. Am.* **1964**, *54*, 1211.

63 S. Lee, H. Hwang , P. Kim , D.J. Jang, *Catalysis Letters* **1999**, *57*, 221.

64 M. Wang, G. Qian, *Mater. Sci. Eng. B* **1998**, *55*, 119.

65 V.C. Costa W.L. Vasconcelos, *J. Sol-Gel Sci. Technol.* **1998**, *13*, 605.

66 F. Lianshe M. Qingguo, *J. Phys. Chem. Solids* **2000**, *61*, 1877.

67 A.V. Hayes, H.G. Drickamer, *J. Chem. Phys.* **1982**, *76*, 114.

68 D. Albrecht, PhD thesis, **1999**, TU Graz.

69 H.J. Zhang, L.S. Fu, *Mater. Lett.* **1999**, *38*, 260.

70 H.-Y. D. Ke, E. Birnbaum, *J. Lumin.* **1995**, *63*, 917.

71 B. Yan, H. Zhang, *Mater. Res. Bull.* **1998**, *10*, 1517.

72 M. Bredol, T. Jüstel, S. Gutzov, *Opt. Mater.* **2001**, *18*, 337.

73 W. Li, G. Yu., S. Huang, *J. Alloys Compd.* **1993**, *194*, 19.

74 Z.-M. Wang, L.J. van de Burgt, G.R. Choppin, *Inorg. Chim. Acta* **1999**, *293*, 167.

75 B.S. Panigrahi, S. Peter, K.S. Vishwanathan, C.K. Methews, *Anal. Chim. Acta* **1993**, *282*, 117.

76 N.V. Petrochenkova, A.G. Mirochnik, A.P. Kulikov, V.E. Karasev, *Russ. J. Coord. Chem.* **1997**, *23*, 869.

77 H. Shou, J. Ye, Q. Yu, *J. Lumin.* **1988**, *42*, 29.

78 Y. Koizumi, H. Sawase, Y. Suzuki, T. Takeuchi, M. Shimoi, A. Ouchi, *Bull. Chem. Soc. Jpn.* **1984**, *57*, 1809.

79 B. Yan, H. Zhang, S. Wang, J. Ni, *Monatsh. Chem.* **1988**, *129*, 151.

80 H. Brittain, *J. Am. Chem. Soc.* **1978**, *17*, 2762.

81 K. Binnemanns, K.V. Herck, *Chem. Phys. Lett.* **1997**, *266*, 297.

82 L. Van Meervelt, K. Binnemans, K. Van Herck, C. Görller-Walrand, *Bull. Soc. Chim. Belges* **1997**, *106*, 25.

83 J.G. Kim, K. Yoon, *J. Alloys Compd.* **1998**, *274*, 1.

84 C.G. Sanny, J.A. Price, *Bioconjugate Chem.* **1999**, *10*, 141.

85 M.L. Bhaumik, P.C. Fletcher, L.J. Nugent, *J. Phys. Chem.* **1964**, *68*, 1490.

86 W. Dawson, J.L. Kropp, M. Windors, *J. Chem. Phys.* **1966**, *45*, 2410.

87 N. Sabbatini, S. Dellonte, G. Blasse, *Chem Phys. Lett.* **1986**, *129*, 541.

88 F.S. Richardson, *Chem. Rev.* **1982**, *82*, 541.

89 J. Bartlett, R. Cooney, R. Kydd, *J. Catal.* **1988**, *114*, 58.

90 F.J. Berry, M. Carbucicchio, A. Chiari, C. Johnson, E.A. Moore, M. Mortimer, F.F.F. Vetel, *J. Mater. Chem.* **2000**, *10*, 2131.

91 F.J. Berry, J. Marco, A.T. Steel, *J. Alloys Compd.* **1993**, *194*, 167.

92 J. Sauer, Diploma thesis, **1997**, J.-W.-Goethe-Univ., Frankfurt.

93 H.G. Küttner H.L. Selzle, E.W. Schlag, *Z. Naturforsch. A* **1974**, *29*, 224.

94 P. Starynowicz, *Acta Cryst Sect. C* **1991**, *47*, 32.

95 P. Starynowicz, *Acta Cryst Sect. C* **1993**, *49*, 1895.

96 H.G. Brittain, *Inorg. Chem.* **1978** *17*, 2762.

97 Y. Zhang, M. Wang, J. Xu, *Mater. Sci. Eng. B* **1997**, *47*, 23.

98 *Lange's Handbook of Chemistry*, MacGraw-Hill, 15th ed., **1998**.

99 M. Jianfang, J. Zhongsheng, N. Jiazuan, *Chin. J. Inorg. Chem.* **1993**, *9*, 160.

100 M.S. Khiyalov, I.R. Amiraclanov, E.M. Movsumov, K.S. Mamedov, *Strukt. Khim.* **1982**, *23*, 119.

101 A.E. Koziol, W. Brzeska, *Russ. J. Coord.Chem.* **1990**, *21*, 183.

102 M.S. Khiyalov, I.R. Amiraclanov, K.S. Mamedov, E.M. Movsumov, *Koord. Khim.* **1981**, *7*, 445.

103 L. Jin, S. Lu, *Polyhedron* **1996**, *15*, 4069.

104 U. Kynast, P. Junk, unpublished results.

105 G. Blasse, B.C. Grabmeyer, *Luminescent Materials*, Springer, Berlin, **1994**, p. 94.

106 P.A. Tanner, B.Y. Hongjie Zhang, *J. Mater. Sci.* **2000**, *35*, 4325.

107 S. Brück, M. Hilder, P.C. Junk, U. Kynast, *Inorg. Chem. Commun.* **2000**, *482*, 688.

6
Microscopic Lasers Based on the Molecular Sieve $AlPO_4$-5

Lhoucine Benmohammadi, A. Erodabasi, K. Koch, Franco Laeri, N. Owschimikow, U. Vietze, G. Ihlein, Ferdi Schüth, Özlem Weiß, Ingo Braun, Matthias Ganschow, Günter Schulz-Eckloff, Dieter Wöhrle, J. Wiersig, and J. U. Nöckel*

6.1
Introduction

The regular, crystallographically defined nanometer-size pores and orifices in the structure of molecular sieves frame a location in which even larger guest molecules can be inserted. Depending on the characteristics of the inserted molecules, the physical and chemical properties of the molecular sieve will be altered. Thus, one may conceive of a molecular sieve as a medium in which selected molecules are inserted to design a material with desired properties. In this way a composite material is assembled which exhibits properties neither the host nor the guest alone has. For example, many organic molecules show a large nonlinear optical hyperpolarizability. For practical applications one could wish to transform this molecular property into a corresponding property of the bulk material. More often than not this takes an ugly turn, as especially the most promising molecules happen to form crystals with centrosymmetry. However, by hosting such molecules in the pores of a noncentrosymmetrical molecular sieve, the nonlinear optical properties of the guest can be passed onto the bulk material [1].

Since stimulated emission from a dye solution was demonstrated in 1966 [2], a wide variety of dyes which meet the requirements for lasing have been reported and commercialized [3]. The broadening of the electronic states by molecular vibrations and rotations results in a fluorescence with a bandwidth of several tens of nanometers. As a consequence, the laser emission of a typical dye laser can be tuned over a range of more than ± 25 nm from its central wavelength. But the molecular environment also plays a significant role. The emission spectrum and the laser efficiency are affected by the polarity, the acid–base equilibrium, the concentration of the solvent, etc. The laser transition occurs between vibrational excitations of the electronic ground state and the first electronically excited singlet state. After pumping the laser transition for several microseconds a considerable part of the dye population will have crossed to the triplet ground state, so that the concentration of lasing molecules in the singlet system is reduced. Thus, system crossing processes will remove molecules from the laser transition. As a result, the stimulated transition rate is reduced, which lowers the optical gain. Soon the gain can no longer compensate for the optical resonator losses, and the laser emission stops. This is the reason why it is not possible to attain continuous-wave laser emission with organic dyes, unless one replaces the excited triplet molecules with unexcited ones, as in dye lasers, in which the dye solution flows rapidly through the excitation volume.

A wide variety of organic dyes [3] can be used as laser medium, but, as already mentioned, in solution [4]. For obvious practical reasons a liquid dye solution is not the laser medium of choice, however good its laser properties might be. Therefore, solid-state dye lasers attracted considerable interest [5–8].

An alternative way to realize solid-state dye lasers is by including the laser dyes in the pores of molecular sieves [9,10]. Technologically, two issues have to be considered: The first concerns the size of the sieve pores, which determine the sizes of molecules that fit into the pores, and the second is related to the morphological quality of the molecular sieve crystal. Any crystal imperfections will cause light rays to be scattered off their path and thus introduce losses which are detrimental for the lasing process. We studied various combinations of $AlPO_4$-5 and SAPO-5 molecular sieves in which lanthanide and transition metal ions as well as organic dyes were inserted. It turned out that certain organic laser dyes can exhibit the necessary properties for lasing when they are hosted by $AlPO_4$-5 and SAPO-5 molecular sieves. The size with which $AlPO_4$-5 and SAPO-5 crystals can be synthesized with good morphology is limited to ≤ 1 mm [11]. On the other hand, hydrothermal synthesis on the laboratory scale already delivers thousands of microcrystals. It would not make much technological sense to try to apply mirror coatings on each microcrystal. Fortunately, laser mirrors are already built into the crystals as synthesized: $AlPO_4$-5 crystallizes in hexagonal prisms. In a dielectric hexagonal prism light rays can be confined by repeated total internal reflection to form a ring resonator, and the resulting light distribution that circulates around the hexagonal prism is known as a whispering-gallery mode (Fig. 1).

6.2 The Structure of the $AlPO_4$-5–Dye Compounds

6.2.1 Organic Dyes as Laser Gain Medium

Up to now, we observed lasing with three kind of dyes in two kinds of molecular sieves: 1-ethyl-4-(4-(*p*-dimethylaminophenyl)-1,3-butadienyl)-pyridinium perchlorate (Pyridine 2, Fig. 2a) in $AlPO_4$-5, ethanaminium-*N*-[6-(diethylamino)-9-[2-(*N*,*N*-

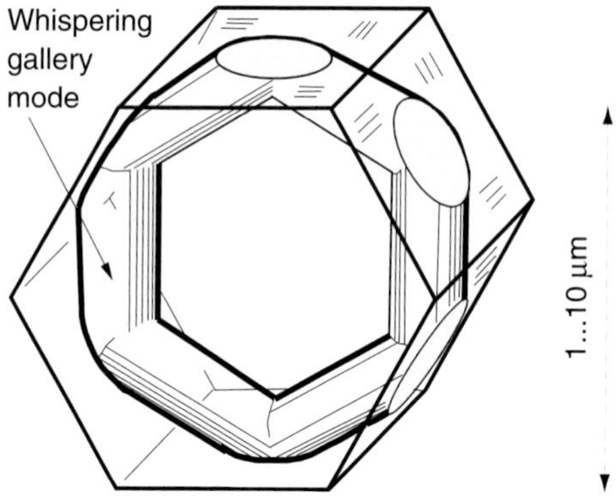

Fig. 1. Ray picture of a light bundle confined in a hexagonal prism by total internal reflection. The situation becomes more complex when the size of the prism approaches the wavelength; see Fig. 20 below.

dimethyl-3-amino-1-propoxycarbonyl)phenyl]-3*H*-xanthen-3-ylidene]-*N*-ethyl chloride (rhodamine BE50, Fig. 2b) in $AlPO_4$-5 [12–14] and recently in SAPO-5 as well, and 4-(dicyanomethylene)-2-methyl-6-(*p*-dimethylaminostyryl)-4*H*-pyran (DCM, Fig. 2c) in $AlPO_4$-5 [15].

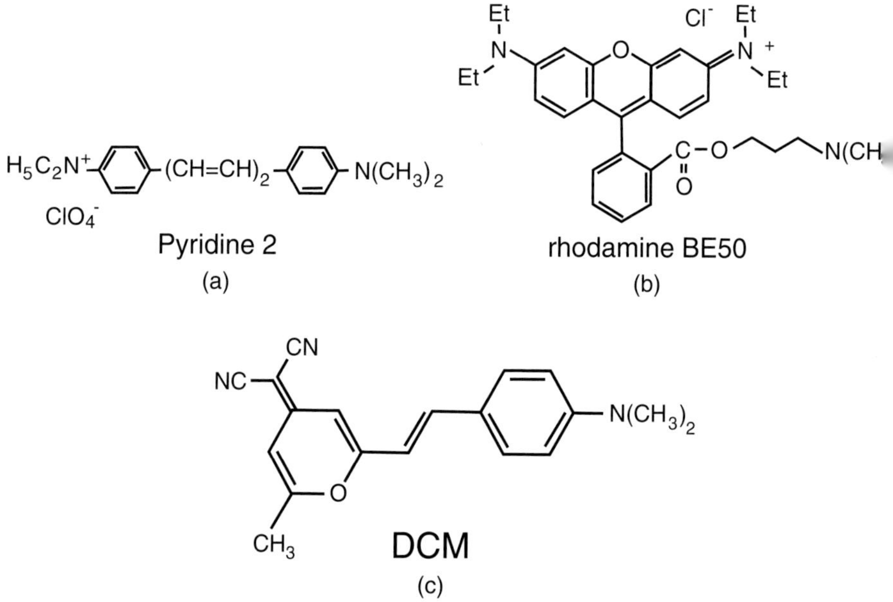

Fig. 2. Structure formulae of used laser dyes: (a) Pyridine 2 (trade name, cf. Ref. [3]), (b) the rhodamine B derivative BE50 [12], (c) DCM.

6.2.2
Synthesis of the Molecular Sieve/Dye Compounds

$AlPO_4$-5 and SAPO-5 crystals are grown from aqueous or alcoholic solutions under hydrothermal conditions, with the addition of an organic structuring agent, or template [16–19]. The template is necessary to direct the synthesis towards the desired structure. Pure $AlPO_4$-5 and SAPO-5 crystals are optically transparent from below 400 to above 800 nm ($n_{500nm} = 1.466$). X-ray patterns revealed systematic absences which are consistent with space group $P6cc$, and for $AlPO_4$-5 with $P\frac{6}{m}cc$ as well. The synthesis parameters of the laser samples are discussed in the contributions in Part 1 of this book.

6.2.3
Crystal Morphology

Corresponding to their hexagonal crystal structure $AlPO_4$-5 and SAPO-5 normally grow as hexagonal prisms (cf. Fig. 3). Not only the growth conditions, but also the size and electronic structure of the enclosed dye molecules can have a considerable influence on the crystal morphology. Pyridine 2 is enclosed without the ClO_4^- counter ion and thus exhibits a static dipole moment. Crystals containing small amounts of Pyridine 2 (concentration determined qualitatively under the microscope from the sample color depth) grow as regular hexagonal prisms (Fig. 4a). However, if the concentration is high, crystal growth is perturbed by the interaction of the molecular dipole with the dipoles in the pore walls (Al^{3+}, P^+). The perturbation manifests itself at the ends of the prism, where the crystal breaks into divergent needles to form a structure resembling a shaving brush (Fig. 4b). Lasing action was only observed in such brushlike crystals [10]. The regular prisms with their low Pyridine 2 concentration did not provide enough optical gain to overcome the losses. Whereas the Pyridine 2 molecules fit into the pores, rhodamine dyes are larger than the 0.73 nm pore diameter. By modifying the structure of rhodamine B

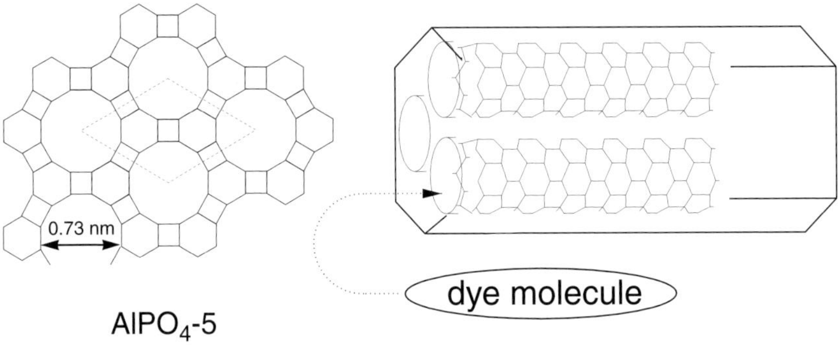

Fig. 3. Idealized crystal architecture of $AlPO_4$-5 with enclosed dye molecules. In the schematic structure representation the polygon corners represent aluminum (Al^{3+}) alternating with phosphorus (P^+), while the connecting lines stand for oxygen.

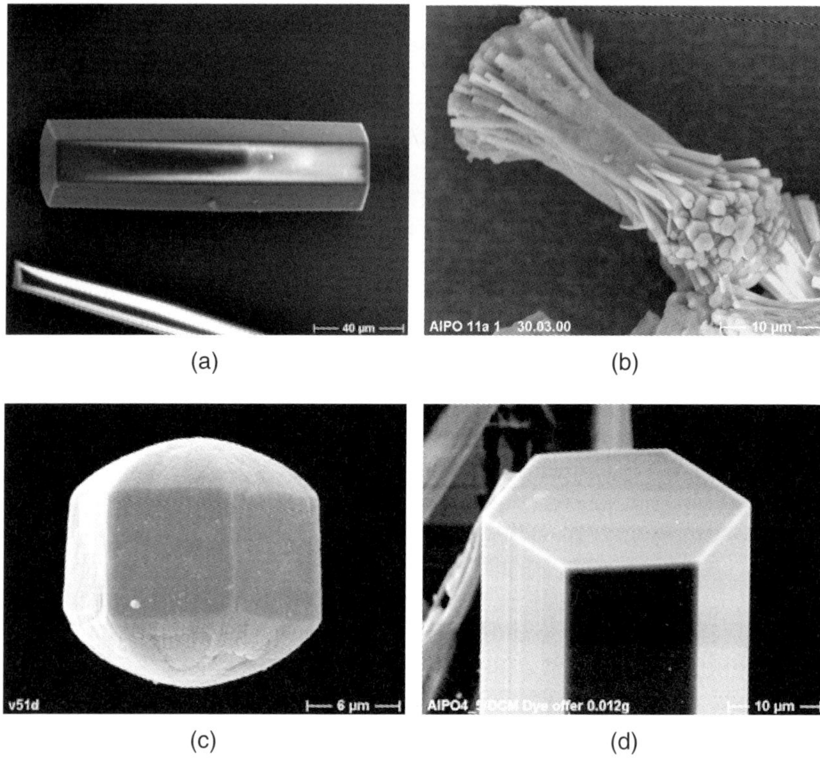

Fig. 4. Morphology types of $AlPO_4$-5 crystals loaded with the following dyes: (a) Pyridine 2, low concentration; (b) Pyridine 2, high concentration; (c) rhodamine BE50; (d) DCM. Lasing action was observed in types (b) to (d).

to mimic the electronic structure of a template molecule, it was possible to obtain a dye (rhodamine BE50), which was accepted by $AlPO_4$-5 [12] and SAPO-5 hosts. Figure 4c shows that the inclusion of these large molecules does not spoil the morphological quality of the crystal. We assume that the dye molecules reside in cavelike crystal defects in which the dye dipoles and the pore-wall dipoles are electrically equilibrated so that the lattice disturbances heal over a distance of a few lattice units. DCM molecules do not bear a significant static dipole moment and as a consequence do not interact with the wall charges as strongly as Pyridine 2. This is probably the reason why crystals containing a high concentration of DCM can be obtained with an undisturbed morphology (Fig. 4d).

6.2.4
Dye Molecule Alignment and Pyroelectric Material Properties

We already mentioned that the rhodamine BE50 molecules are too large to fit into the regular pores of the $AlPO_4$-5 structure and that they reside in defect pores. As

Tab. 1. Properties of dye/molecular sieve compounds

Enclosed dye	*Dimensions*	*Resides in*	*Static dipole*	*Aligned in pores*	*Oriented*
Pyridine 2	0.6 nm	pores	yes	yes	yes
Rhodamine BE50	(1.6×0.85) nm	defect pores	yes	no	no
DCM	ca. 0.6 nm	pores	yes	yes	unknown

will be shown in Section 6.3.1 their net alignment with the crystal axis is only weak. We assume that in the defect pores several nearly equiprobable orientations exist in which they can settle. On the other hand, Pyridine 2 and DCM are practically perfectly aligned with the crystal *c*-axis (cf. Tab. 1).

When the enclosed guest molecules exhibit a static electric dipole then the question arises whether neighboring molecular moments are oriented in the same direction or antiparallel. Whereas antiparallel dipoles cancel each other, parallel orientation of the molecular dipoles will result in a macroscopic dipole moment which endows the composite with pyroelectric properties. By periodically heating a Pyridine 2-loaded $AlPO_4$-5 crystal with a modulated laser beam we could induce pyroelectric charges at the ends of the crystal *c*-axis well above that of the empty host, and this proves that the dye guests exhibit an average net orientation. Figure 5 shows that when the heating position is changed from the top of the crystal to the bottom, the polarity of the resulting electric field changes its value. This implies that the majority of the Pyridine 2 molecules in the upper half of the crystal are oriented opposite to the ones in the lower part.

6.3 Optical Properties

6.3.1 Absorption, Dichroism, and Birefringence

Above we have shown that size and dipole moment of dye molecules in the channel pores of the $AlPO_4$-5 molecular sieve can determine their orientation with respect to the crystal *c*-axis. As a consequence the alignment of the optical transition moments (absorption as well as emission) of the dye molecule will be fixed as well. If, for example, the absorption moment of all the enclosed dye molecules is oriented parallel to a given direction (e.g., the *c*-axis) then the molecules will interact only with the corresponding parallel electric field component. Such a material will be dichroic, that is, it appears colored only for light linearly polarized parallel to the absorption moment; otherwise, it is transparent. In Figure 6 shows micrographs in which light is transmitted through $AlPO_4$-5 with Pyridine 2 for polarization parallel und perpendicular to the crystal *c*-axis. Whereas the Pyridine 2 samples appear transparent for perpendicular polarization, the rhodamine BE50 samples reveal

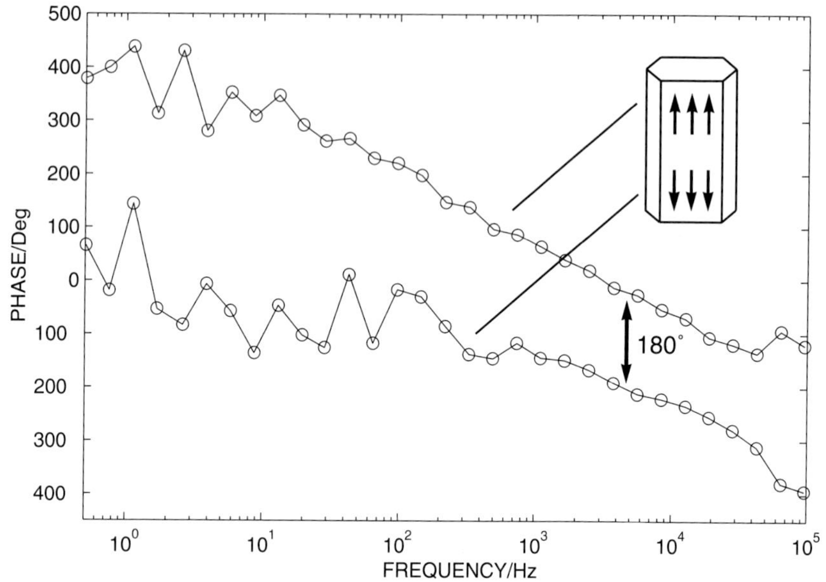

Fig. 5. Measurement of the current induced by the electric field of pyroelectrically generated charges in an $AlPO_4$-5 microcrystal. The phase of the current is shown as a function of the frequency of the temperature modulation. Below 100 Hz the current signal was too low (current < 10 fA) for the lock-in detector to synchronize. When the heating laser spot was moved from the upper part to the lower part of the crystal, the phase of the current signal changed by 180°. This implies that the majority of the dye molecules in the upper half of the crystal are aligned opposite to the ones in the lower part.

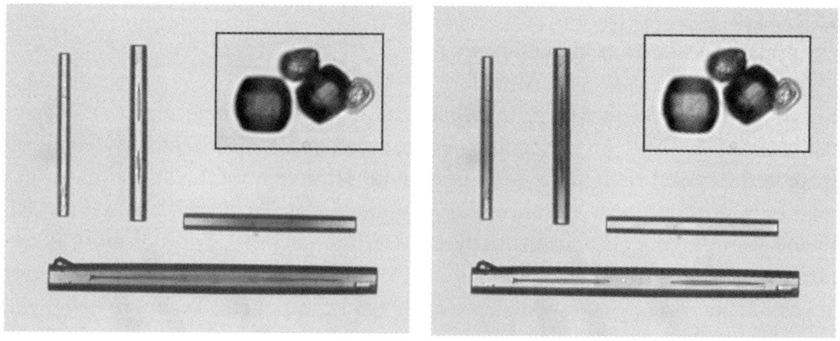

Fig. 6. Illustration of dichroism in $AlPO_4$-5 crystals loaded with Pyridine 2 (long prisms) and rhodamine BE50 (inset). The crystals are illuminated with linearly polarized light with horizontal (left), and vertical (right) polarization *P*. For *P* perpendicular to the crystal *c*-axis the Pyridine 2 composite is transparent, whereas the rhodamine BE50 compound shows little difference.

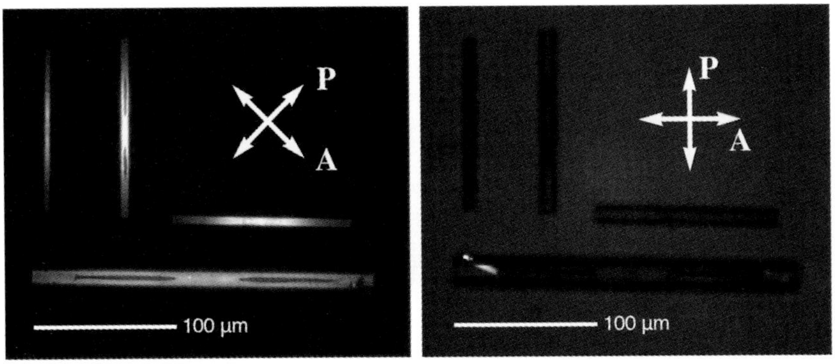

(a) (b)

Fig. 7. Micrograph illustrating the birefringence in $AlPO_4$-5 crystals loaded with Pyridine 2. The pictures are taken with crossed polarizer/analyzer. When the incident linear polarization direction coincides with a principal axis the polarization state is conserved (b). Otherwise the incident linear polarization is transformed into elliptically polarized light (a).

only a week modulation of the absorbance and appear colored for all polarization directions, which is a consequence of the low degree of orientational order. Similar to Pyridine 2 samples, crystals containing DCM also show strong dichroic properties, and this means that the DCM molecules in the channel pores are oriented as well. The presence of oriented dye molecules (Pyridine 2 and DCM, cf. Tab. 1) in the pores of the molecular sieve will also enhance the birefringence of the host (Fig. 7). This effect is less pronounced for rhodamine BE50 composites, and this again suggests that the dye is only weakly aligned. We assume that several nearly equiprobable orientations exist in which rhodamine BE50 settles. In Figs. 6 and 7 it can be seen that the corresponding effects increase towards the middle section of the crystal prisms, indicating a higher dye concentration in this part.

6.3.2
Fluorescence Emission and Decay Dynamics

6.3.2.1 Fluorescence Spectra

An obvious consequence of the aligned inclusion of dyes is the linearly polarized fluorescence emission in the case of Pyridine 2 and DCM composites. We observed polarization contrasts $(I_\perp \; I_\parallel)/(I_\perp + I_\parallel)$ of nearly 100 %. The emission of rhodamine BE50 samples shows a contrast of only 10–20 %, consistent with the observations above, according to which rhodamine BE50 seems to be only weakly oriented.

Figure 8 shows fluorescence spectra for Pyridine 2, rhodamine BE50, and DCM enclosed in $AlPO_4$-5 at varying concentrations. The Pyridine 2 and rhodamine BE50 spectra show a characteristic shift with concentration, whereas the DCM spectra seem not to be affected by concentration. One of the reasons for the spectral shift is re-absorption due to the overlap of the emission and absorption bands,

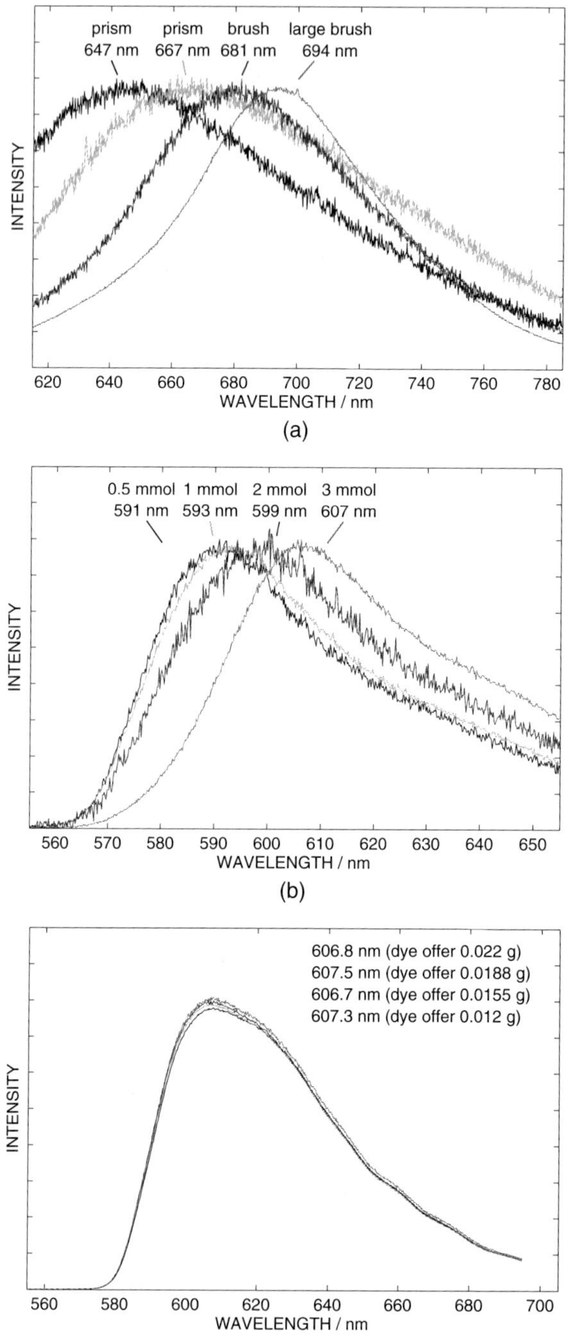

Fig. 8. Fluorescence spectra of typical samples and how they are affected by the dye concentration. $AlPO_4$-5 loaded with Pyridine 2 (a), rhodamine BE50 (b), DCM (c).

but concentration quenching effects can also contribute to the shift. Lasing action in Pyridine 2 samples was only observed in the deeply colored crystals with high concentration, whereas we found lasing DCM crystals with any of the shown concentrations, and for rhodamine BE50 all samples lased except the highly loaded ones (3 mmol), in which concentration quenching effects were too pronounced.

6.3.2.2 Spontaneous Emission Dynamics

The lifetime of the excited state of a dye molecule is an important parameter for modeling the laser [20, p. 118ff]. For this reason we characterized the decay dynamics of the relevant excited state by means of time-correlated single-photon counting methods (TCSPC) [21]. The excited state dynamics depends crucially on the interaction of the molecule with its environment. When the investigated laser dyes, i.e., Pyridine 2, rhodamine BE50, and DCM, are dissolved in their recommended solvent [3], their decay dynamics is perfectly described by a single exponential decay function $I(t) \propto \exp -t/\tau$. The same dyes enclosed in $AlPO_4$-5, however, reveal more complex decay dynamics. which formally can be modelled by a series of exponential functions (Eq. 1).

$$I(t) = \sum_{i=1}^{N} A_i \cdot \exp\left(-\frac{t}{\tau_i}\right). \tag{1}$$

We found that for the dyes which fit into the channel pores of the $AlPO_4$-5 host (Pyridine 2 and DCM) a series with two terms ($N = 2$) leads to stable results (a third term results in erratic parameters), whereas for rhodamine BE50, which resides in defect pores, the dynamics is more complex and requires four exponential terms ($N = 4$, cf. Fig. 9).

We also investigated the rotational dynamics of the dyes in their cages. Detection of the fluorescence emission for each polarization separately [22] reveals that Pyridine 2 and DCM do not rotate in the channel pores, whereas the orientation of the rhodamine BE50 molecules decorrelates on a timescale of 400 ps. The consequence is that Pyridine 2 and DCM molecules are under the influence of a quasistationary crystal-field environment, while rhodamine BE50 has enough dynamical degrees of freedom to sample many configurations of its entourage in a short time. In this regard the two exponential terms in the model dynamics of Pyridine 2 and DCM can be interpreted as an indication for two possible configurations of the molecules in the pores, one strongly interacting with the sieve walls (small τ), and the other more loosely bound to the sieve channel. This hypothesis is compatible with the findings on photostability discussed in Section 6.5. Extending this argument to rhodamine BE50 encaged in defect pores would imply four quasistable positions, but this is not compatible with its rotational mobility. In addition, we also can not expect that defect pores are so consistently alike. However, in contrast to rhodamine in solution, the encaged molecules are kept from diffusing, and also their mutual distance is fixed, i.e., nonfluctuating. Hence, it is conceivable that

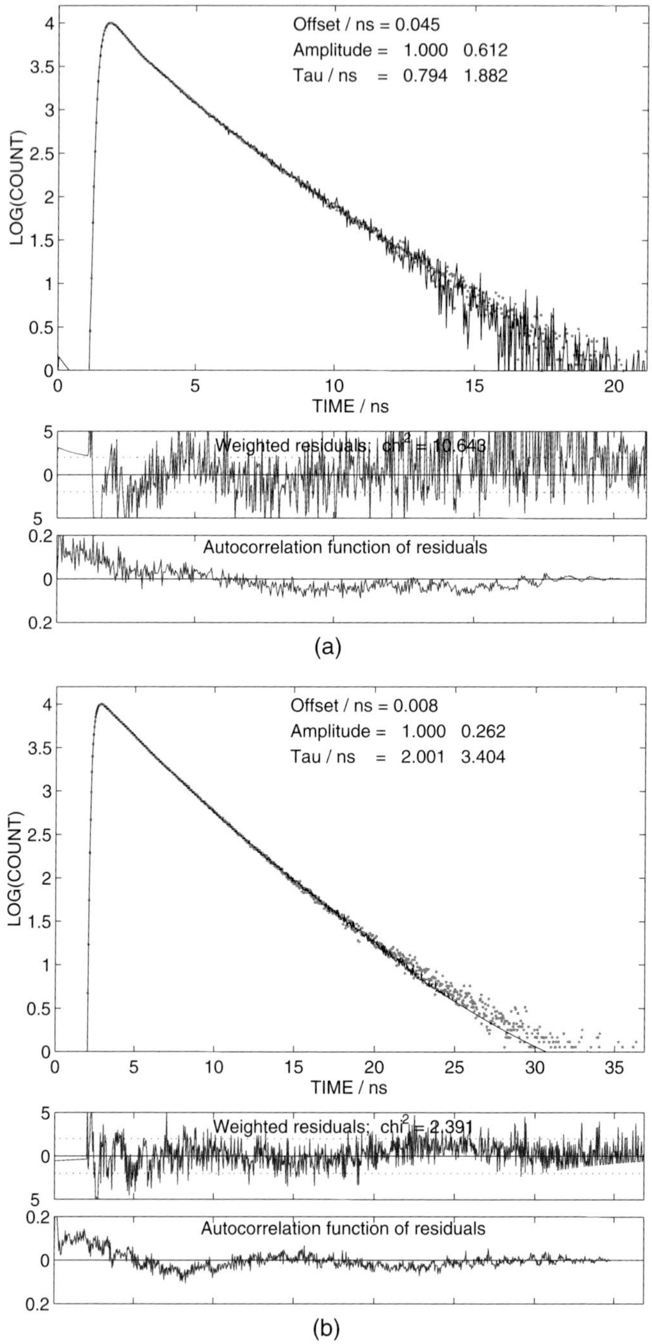
Offset / ns = 0.045
Amplitude = 1.000 0.612
Tau / ns = 0.794 1.882
LOG(COUNT)
TIME / ns
Weighted residuals; chi2 = 10.643
Autocorrelation function of residuals
(a)
Offset / ns = 0.008
Amplitude = 1.000 0.262
Tau / ns = 2.001 3.404
LOG(COUNT)
TIME / ns
Weighted residuals; chi2 = 2.391
Autocorrelation function of residuals
(b)

Fig. 9.

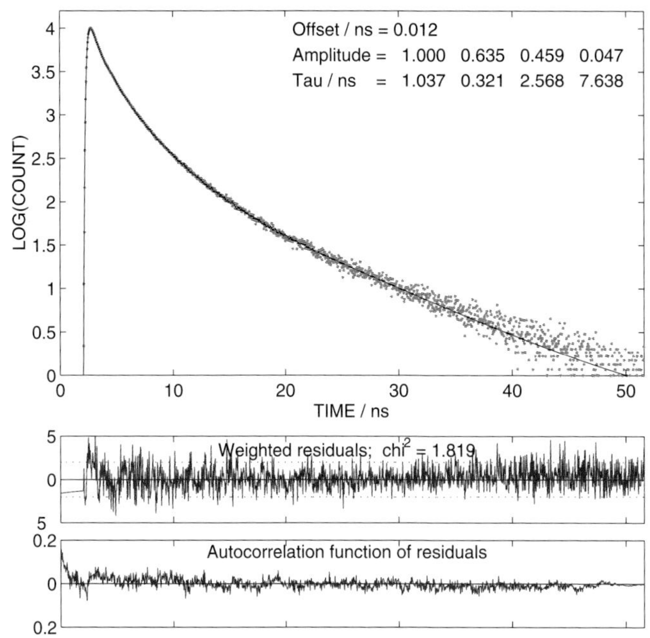

(c)

Fig. 9. Fluorescence decay dynamics of typical dye samples enclosed in $AlPO_4$-5: (a) Pyridine 2, (b) DCM, (c) rhodamine BE50, characterized with the time-correlated single-photon counting method (TCSPC). The dots represent the measured data and the solid line the fitted multi exponential decay model Eq. (1). The model parameters amplitude A_i and decay constant τ_i are given in the figures. Note also the different timescales. The residuals in (a) and (b) indicate a systematic discrepancy between the two term exponential model and the data. However, in both cases we could not determine statistically significant parameters for a three-term exponential model. On the other hand, the rhodamine BE50 decay can be perfectly fitted with a stable four-term exponential model.

compared to the solution case mutual interactions leading to deexcitation of excited molecules is reduced to a smaller set of now stationary alternatives. Their quantum mechanical interference then causes the observed deviation from single exponential decay. In fact, the decay dynamics of rhodamine BE50 is in good agreement with a decay process given by Eq. (2)

$$I(t) \propto \exp\left\{-A_1 \cdot \left[\ln\left(\frac{t}{A_2}\right)\right]^{A_3}\right\} \tag{2}$$

which is obtained for energy transfer in exchange interactions, or tunneling through energy barriers; cf. Eq. (2.15) in [23]. The fit of Eq. (2) is shown in Fig. 10c.

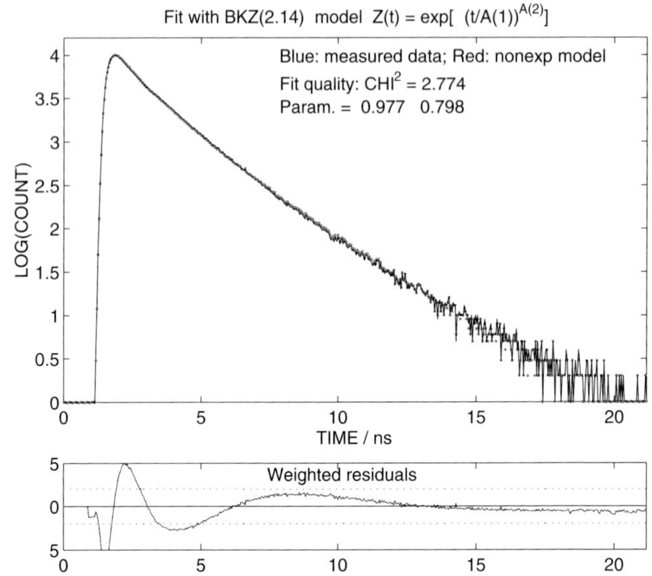

(a)

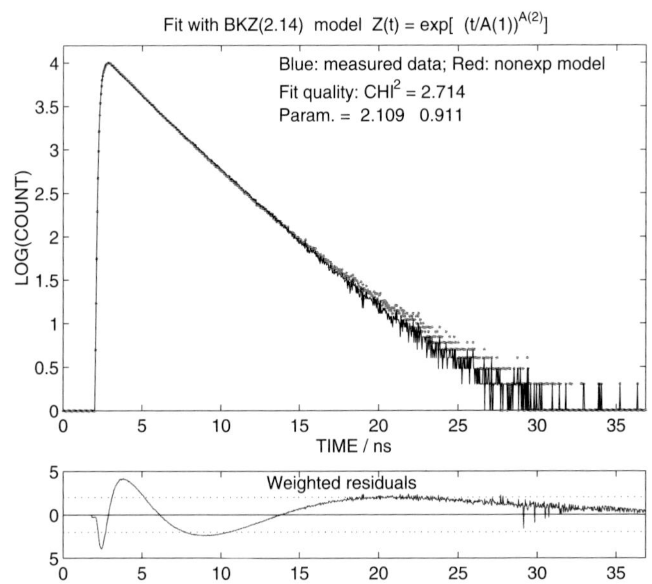

(b)

Fig. 10. Fluorescence decay dynamics of Pyridine 2 (a), DCM (b), and rhodamine BE50 (c) enclosed in $AlPO_4$-5, fitted with stretched exponential decay laws. (a) and (b) show the fit with Eq. (3), and (c) with Eq. (2).

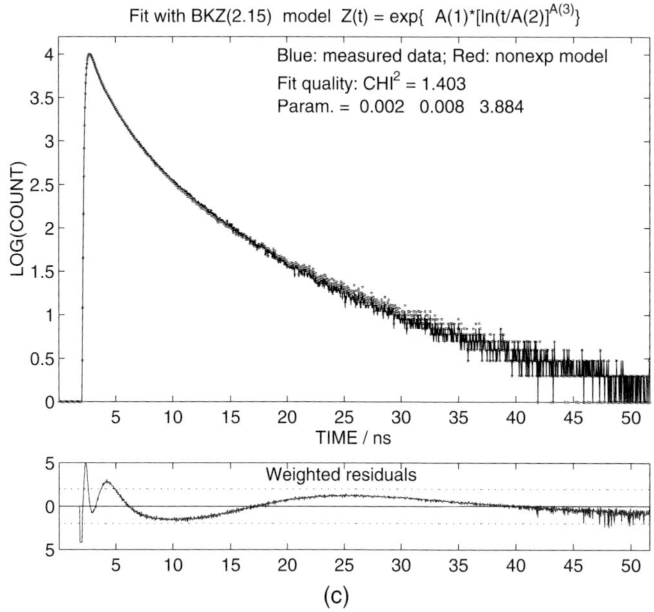

Fig. 10. *(continued)*

Alternatively to the two-position postulate for Pyridine 2 and DCM we checked for decay dynamics according to a generalized Förster law (Eq. 3), which is observed for nonradiative energy transfer of the excited molecule to a quenching center; cf. Eq. (2.14) in [23].

$$I(t) \propto \exp\left[-\left(\frac{t}{A_1}\right)^{A_2}\right]. \tag{3}$$

Figures 10a and b illustrate the fit of Eq. (3) with the measured decay. The fit with a regular Förster [24] law led to significantly larger misfits, characterized by χ^2 that exceeded the values given in Fig. 10 by more than a factor of two in the case of Pyridine 2 and DCM, and a factor of 16 for rhodamine BE50. Thus, at least for rhodamine BE50, we believe that Förster quenching processes can be ruled out as the cause of the nonexponential decay dynamics.

6.4 Laser Properties

Provided that microcrystals had a smooth part in the hexagonal section, and this part of the volume was free of scattering defects, then excitation above a certain pump threshold power was frequently accompanied by characteristic narrow peaks in the luminescence emission spectrum [9,10]. In fact, narrow spectral emission peaks in combination with a threshold of the excitation level are characteristic of laser action.

In this section we show how the ingredients which define a laser manifest themselves in the molecular sieve/dye composites. First we must identify the resonator required to enhance stimulated emission processes.

6.4.1 Structure of the Microresonator

We have shown that in the case of $AlPO_4$-5—Pyridine 2 and–DCM composites the dye molecules are enclosed with their optical transition dipole moments aligned parallel to the hexagonal *c*-axis of the host. On the other hand, emission and absorption of a dipole occurs mainly in directions normal to the dipole axis, but not parallel to the axis, and we mentioned above (section 3.2) that in fact, the luminescence is linearly polarized parallel to the crystal axis. As a consequence, light amplification in these samples can only occur for waves traveling in planes normal to the crystal axis. For these waves there exists a bundle of propagation directions normal to the *c*-axis which fulfill the conditions for total internal reflection (TIR) at the hexagonal prism faces. For this bundle of rays the faces of the prism form reflectors, which together shape an optical ring resonator in which the necessary feedback for laser action is provided (see Fig. 1). The optical mode oscillating in this ring resonator resembles a whispering gallery mode [25–29]. As the resonator dimensions correspond to few wavelengths, the structure of the optical field in the hexagonal boundaries does not match the intuitive picture shown in this figure. A discussion of the exact field distribution is presented in Section 6.4.5.

6.4.2 Temporal Coherence of the Laser Emission

When a cuvette filled with a laser dye solution is strongly pumped, narrowing of the emitted luminescence spectrum is observed. The reason is the increase in the probability of stimulated emission processes with increasing pump power. When the dye medium is located in an optical resonator which recycles emitted photons, the probability for stimulated emission processes is enhanced according to their lifetime. At the same time the emission spectrum narrows. As soon as the gain provided by stimulated emission processes overcomes the loss of resonator photons, the spectral width contracts to less than one nanometer. This fast width reduction marks the threshold of laser emission. According to Schawlow and Townes the observable spectral width $\Delta\nu$ is related to the net gain g and the linewidth of the empty resonator $\Delta\nu_e$ by Eq. (4) [20, p. 453].

$$\Delta\nu \approx \frac{2\Delta\nu_e}{\sqrt{g}} \tag{4}$$

The linewidth $\Delta\lambda_s$ which we can resolve in our experiments with the 1/4 meter spectrometer is ca. $\Delta\lambda_s = 0.3$ nm, corresponding to a frequency bandwidth of

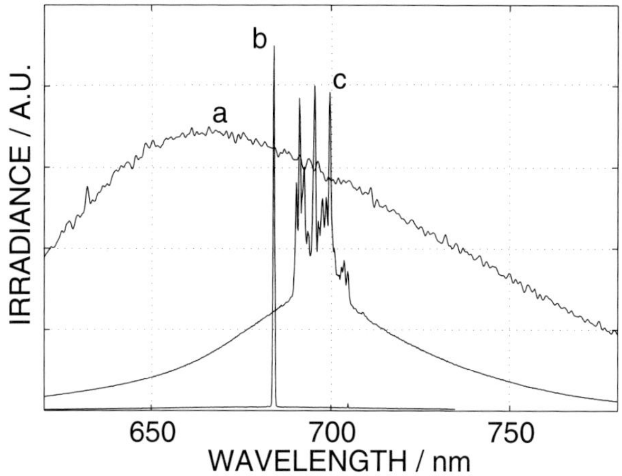

Fig. 11. Emission spectra of typical samples of Pyridine 2-loaded $AlPO_4$-5 samples. (a) Fluorescence emission of a regular hexagonal prism, (b) single-mode laser emission from a small (WOF = 4.5 μm) brush-shaped sample. (c) multimode emission of a large (WOF = ca. 22 μm) brush-shaped sample.

$\Delta\nu_s = (c/\lambda^2) \cdot \Delta\lambda = 250$ GHz. This bandwidth corresponds to a temporal coherence length $T_c = \Delta\nu_s$ of 4 ps or a coherence length $L_c = c \cdot T_c$ of $L_c = 1.2$ mm. Assuming a hexagonal crystal with a width over flats (WOF) of 10 μm, then L_c corresponds to ca. 30 resonator revolutions, i.e., a quality factor of 30. However, we believe (see arguments in Section 6.4.5) that the effective quality factor of the microresonators is higher. Therefore the effective coherence length of the observed laser emission is larger and the spectra narrower than we resolve experimentally.

The transition of the linewidth at the laser threshold was apparent as a sudden emergence of narrow peaks on the normal luminescence background which is generated by the nonlasing volume of the compound. Although single-line emission was observed from small crystal samples, molecular sieves with cross sections (WOF) around 10 μm showed a multiline emission spectrum, in which the line separation was consistent with the oscillation of multiple longitudinal laser modes. Some typical spectra are documented in Figs. 11–13.

6.4.3 Spatial Coherence of the Laser Emission

Laser emission is collimated and originates from an emission area with small extension. The consequence is that laser radiation can be focused to small spots, which in the case of single-mode emission is diffraction-limited. On the other hand, the emission from a small spot will necessarily diverge. If the emission spot size δ (beam waist) is related to the divergence angle α by $\alpha \approx \lambda/\delta$ then the emission is called spatially coherent. Thus, the spatially coherent emission from a small

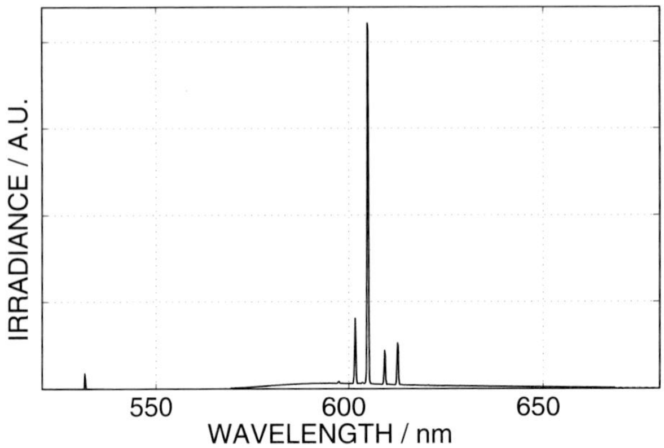

Fig. 12. Emission spectrum of a typical rhodamine BE50-loaded $AlPO_4$-5 sample WOF = ca. 7.5 μm).

spot on the order of a few wavelengths is not ideally collimated, but exhibits a certain divergence. Thus, when we consider spatial coherence of laser emission from the molecular sieve microcrystals, we have to reveal the difference of the emission geometry between omnidirectional fluorescence and laser emission. That the emission occurs in a cone is documented in Fig. 14. With an imaging spectrometer we also showed that small spots with a size of ca. 2 μm emit radiation with a spectrum that is consistent with a laser emission spectrum [9]. Figure 15 shows

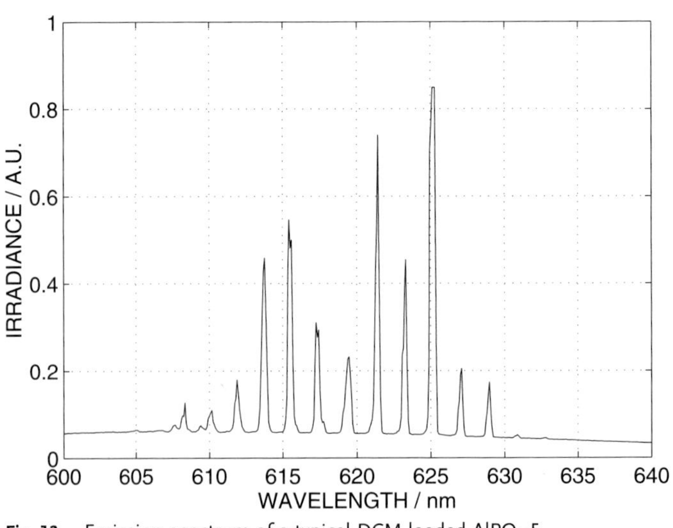

Fig. 13. Emission spectrum of a typical DCM loaded $AlPO_4$-5 sample (WOF = ca. 50 μm). The line spacing is consistent with the longitudinal mode spacing of a hexagonal resonator with WOF = 47.9 μm.

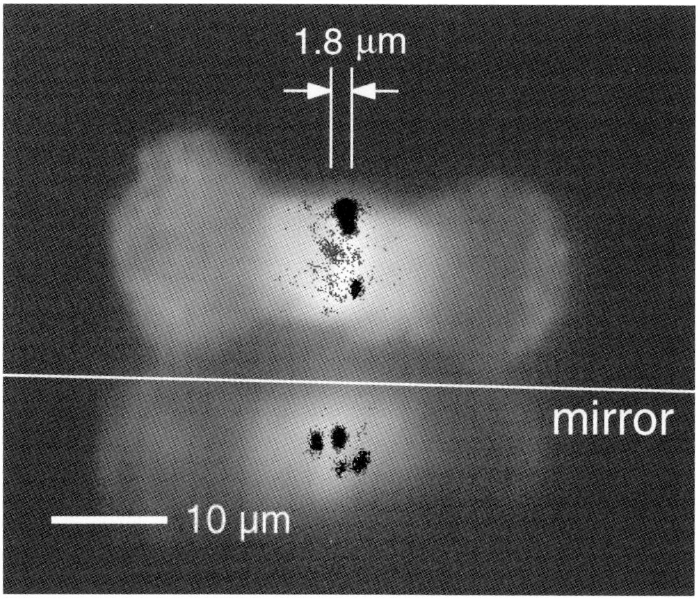

Fig. 14. Emission distribution of a whispering-gallery laser mode in a brushlike sample: The superposition of the omnidirectional fluorescence distribution just below the laser threshold is shown, and the black spots are the difference image of the luminescence just below and just above the laser threshold. Thus, the black spots mark the laser emission area. The sample lies on a mirror surface and the mirror image appears below the direct image. In the mirror image, which corresponds to the sample observed from a different angle, emission spots appear, which do not exist in the direct image, and vice versa. This proves that the emission from the black spots is directed into different nonoverlapping cones, which means that the emission exhibits spatial coherence properties which are consistent with small-area (1.8 µm) emitters.

that the spots from which the laser emissions originate are located along the edges of the prism sides. This is different from what one might expect from the situation represented in Fig. 1. This issue will be discussed in Section 4.5.

6.4.4 Laser Threshold and Differential Efficiency

To obtain Fig. 14 we exploited the fact that the geometry of the emission below and above the laser threshold changes drastically. The increase of spatial coherence can also be made visible as an increase of the emitted power evaluated in a fixed spatial angle (cone). In Fig. 16 the threshold is documented as the increase of emitted power per unit spatial angle as function of pump irradiance for the samples shown in Fig. 15. The slope of the curve is proportional to the differential efficiency of the laser. Plots are shown for a typical microlaser with WOF < 10 µm and one with WOF > 10 µm. Lasing threshold for the larger size samples was around 0.5 MW

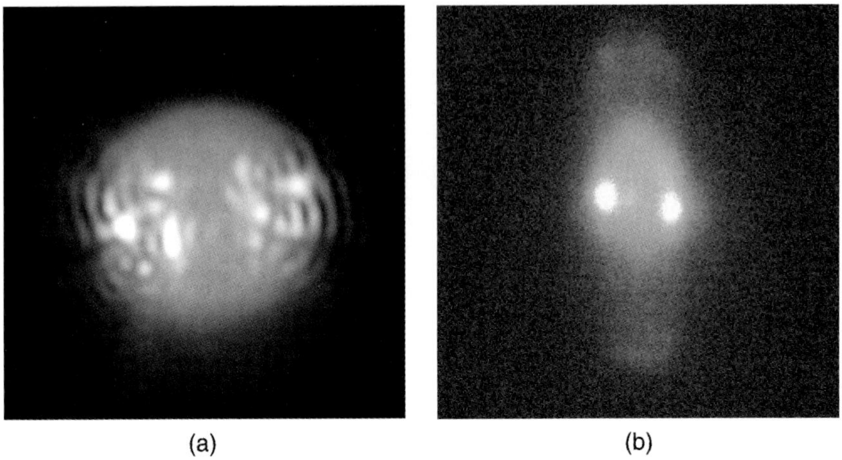

Fig. 15. Patterns of the laser emission show that the emission originates from regions along the edges of the side. (a) rhodamine BE50/$AlPO_4$-5 compound; WOF = 7.5 μm. (b) Pyridine 2/$AlPO_4$-5 compound; WOF = 4.5 μm.

cm^{-2}, regardless of the type of dye loading. On the other hand, crystals of smaller size (WOF = 4.5 μm) from the same synthesis batch had a considerably smaller threshold (0.12 MW cm^{-2}) and a factor of > 7 larger differential gain. Whether this is a consequence of quantum-size effects [30] is not clear at this moment.

It is informative to compare the threshold of molecular sieve microlasers with

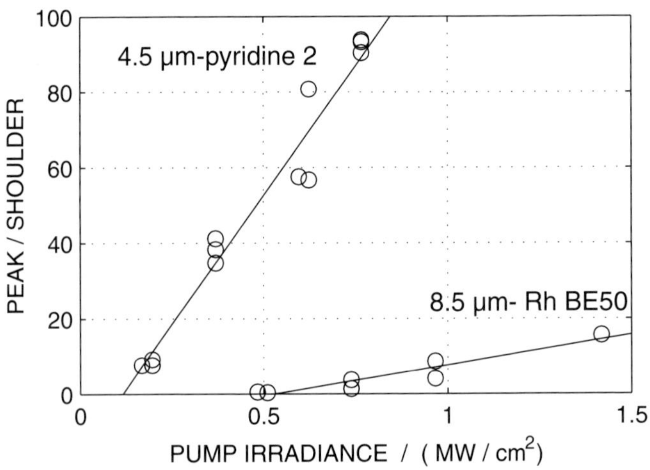

Fig. 16. Lasing threshold and differential efficiency of typical $AlPO_4$-5/dye compounds, showing the peak of the laser emission spectrum normalized by the fluorescence shoulder as a function of the pump power density for the samples shown in Fig. 15.

commercial semiconductor vertical cavity surface emitting lasers (VCSELs). VCSELs with a size comparable to the sample represented in Fig. 16 exhibit threshold currents of ca. 1 mA, which corresponds to 6.25×10^{15} s^{-1} electrons. On the other hand, the threshold power density of 0.12 MW cm^{-2} incident on the molecular sieve laser surface of 1×4.5 μm (cf. Fig. 15(b)) corresponds to a current of 1.4×10^{16} s^{-1} of 532 nm photons, or a factor of 2.24 greater than the electron rate. As the pump radiation is not polarized but the molecular ensemble of the considered laser is aligned, only half of the pump photons actually contribute to the inversion. Thus, in terms of elementary (quantum) pump processes needed to reach lasing threshold, the molecular sieve lasers are comparable to commercial VCSELs.

6.4.5
Field Distribution in the Hexagonal Ring Resonator

In Fig. 15 can be seen that the laser emission leaves the resonator at the edges of the hexagonal boundary. This is somewhat unexpected, as in the ray representation of the whispering-gallery mode in Fig. 1 there seems to be no field at the edges. On the other hand, it is known that as soon as a structure approaches dimensions of a few wavelengths the field distribution is no longer consistent with the picture of light rays, but must be worked out based on Maxwell's equations. We used a wave model which is based on Maxwell's wave equation [31,32]. But before we examine the wave model of a hexagonal resonator, let us say some words about the ray picture.

6.4.5.1 The Ray Picture of the Hexagonal Resonator

The main feature that distinguishes the hexagonal resonator from other common whispering-gallery-type cavities such as microdroplets [33] and semiconductor disk lasers [31,32], is that the latter do not exhibit sharp corners and flat sides. Portions of the boundary in convex resonators can act as focusing or defocusing elements, but the straight sides of a hexagon are neither one nor the other. The hexagon in fact constitutes a self-assembled realization of a *pseudointegrable* resonator [34]: For a polygon with precisely 120° angles between adjacent sides, any ray launched at some angle to the surface will go through only a finite number of different orientations [35], just as in the more familiar rectangular resonator, in which there are at most two nonparallel orientations for any ray path. In the hexagon, a ray encounters the interface with at most three different angles of incidence. Despite this apparent simplicity, there is no orthogonal coordinate system in which the wave equation for the hexagonal cavity can be solved by separation of variables. This property of nonintegrability is shared by wave equations whose classical (short-wavelength) limit exhibits chaos. However, ray paths in the hexagon display a degree of complexity that cannot be classified as chaotic, and hence the term pseudointegrability has been coined for these systems. The ray-wave duality in "billiards" of this type must be addressed in order to explain how they can support whispering-gallery modes that emit at the corners.

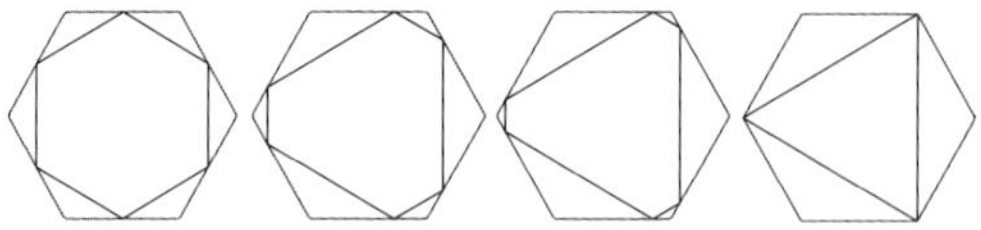

Fig. 17. All four orbits shown have the same length and the same angle of incidence with respect to the interface normal, $\chi = 60°$. They hence satisfy the condition for total internal reflection: $\sin\chi > 1/n = 0.69$. In the case of rays impinging on the corners (rightmost picture) the ray picture breaks down.

To characterize the sample size, one can specify either the diameter $2R$ of the hexagon measured over opposite corners or – more conveniently – the width over flats (WOF) satisfying the relation $R = \text{WOF}/\sqrt{3}$. The closed ray path underlying Fig. 1 is only one member of an infinite family of periodic orbits of the hexagon that all have the same length $L = 3 \times \text{WOF}$ as shown in Fig. 17. Long-lived cavity modes should be expected only if the corresponding rays satisfy the condition of total internal reflection (TIR) at the interface, $\sin\chi > 1/n$, where χ is the angle of incidence with respect to the surface normal. This is true for the orbits of Fig. 17. In a naive ray approach one would furthermore obtain the spectrum of modes by requiring an integer number of wavelengths to fit into L, leading to constructive interference after a round-trip. As we shall see shortly, this estimate is justified, even though a proper treatment of the ray-wave connection has to take into account that any given mode is in fact made up of a whole *family* of different ray paths. The ultimate breakdown of the ray model, however, is illustrated in Fig. 17 by the degenerate ray orbit hitting the corners, where the classical laws of refraction and reflection become undefined. One can imagine that as soon as a light beam hits a corner, the beam propagation direction becomes undefined. This is the reason why for hexagonal laser modes, clockwise or counterclockwise propagation is not a "good quantum number".

6.4.5.2 The Wave Picture

Spectral properties Because the hexagonal faces are neither focusing nor defocusing, there is no obvious way of determining the weight that should be given to individual members of a ray family to predict the spatial structure of the resulting mode. Full solutions of Maxwell's equations have therefore been carried out for the TM polarized modes of a dielectric hexagonal prism, using methods previously applied in [31,32].

In view of the experimental spectra presented in Section 6.4.2, the attention here is focused on three different sample sizes, with WOF of 4.5, 7.5, and 22 μm. The aim is to understand the observed laser line spacings and the emission directionality. Comparison of the calculated and observed linewidths will not be attempted because the simulations do not take gain narrowing into account.

First we verify that orbits of the type shown in Fig. 17 determine the characteristic mode spacing of these cavities, Fig. 18 shows light-scattering spectra for dif-

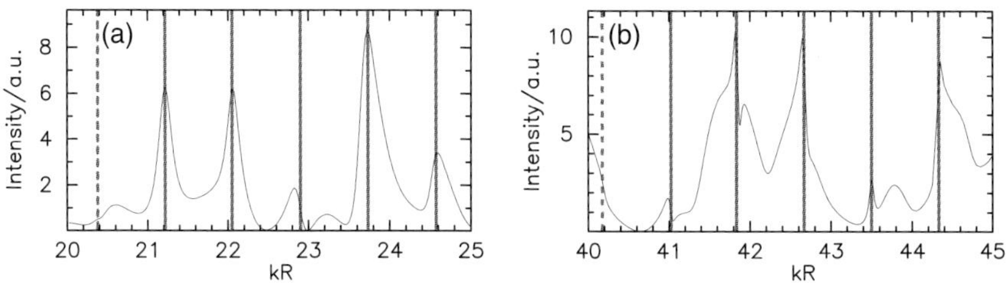

Fig. 18. Calculated scattering intensity spectra of a hexagonal cylinder for plane-wave incidence at 15° to a side face and detection at 60° from incidence. (a) corresponds to a spectral interval $\lambda \approx 653$–816 nm for WOF $= 4.5$ μm; (b) covers the interval $\lambda \approx 605$–680 nm for WOF $= 7.5$ μm. Vertical lines are guides to the eye, indicating narrow resonances. The spacing between resonances is $\Delta(kR) \approx 0.84$ in (a) and $\Delta(kR) \approx 0.83$ in (b), in good agreement with the characteristic mode spacing $\Delta(kR)_c \approx 0.83$ of a closed hexagonal orbit. Expected resonances not clearly seen in the above spectra are marked by dashed lines; they appear at other detection angles.

ferent sample sizes in the vicinity of the experimental wavelengths. Intensity is plotted versus dimensionless wavenumber kR, where $k = 2\pi/\lambda$. This is the natural scale for comparison with semiclassical predictions because modes differing by one node along a closed path will be equally spaced, with a characteristic separation $\Delta(kR)_c = 2\pi\, R/(n\, L) = 2\pi/(3\sqrt{3}n) = 0.825$ independent of the sample size. The expected wavelength spacing of the modes (free spectral range FSR) is $\Delta\lambda = \lambda^2 \times \Delta(kR)_c/(2\pi R) \approx 23$ nm in (Fig. 18a) and $\Delta\lambda \approx 11$ nm in (Fig. 18b). For WOF $= 22$ m, we obtain $\Delta\lambda \approx 4.9$ nm. Figure 18 indeed shows a series of resonant features with approximately the predicted wavevector spacing.

Each of the peaks marked in Fig. 18(b) is in fact a multiplet, which is not resolved because the splittings of the individual modes comprising the multiplet are smaller than their passive linewidths. There is evidence for this because several of the peaks are very asymmetric and, in particular, exhibit a steep slope on one side. For an isolated resonance, the most general lineshape that could arise is the Fano function (of which the Lorentzian is a special case), which, however, does not yield satisfactory fits here.

To reveal the multiplet structure, we modeled deviations from the ideal hexagonal shape, which could lead to narrower individual linewidths and increase the multiplet splitting. Shape perturbations were chosen that preserve the D_{6h} point group symmetry and hence remove only "accidental" quasidegeneracies. The actual perturbation that is present in the samples eluded experimental characterization, so that a model calculation can reproduce only generic features which are insensitive to the precise type of perturbation. One such feature is the *average* mode spacing after degeneracies have been lifted sufficiently.

Figure 19(a) shows the spectrum of a rounded hexagon in which the radius of curvature at the corners is $\rho \approx 0.9\lambda$ (assuming $\lambda \approx 610$ nm for definiteness). No qualitative difference to Fig. 18(b) is seen, except that the resonant features have become somewhat narrower, and thus we can identify two distinct series of modes

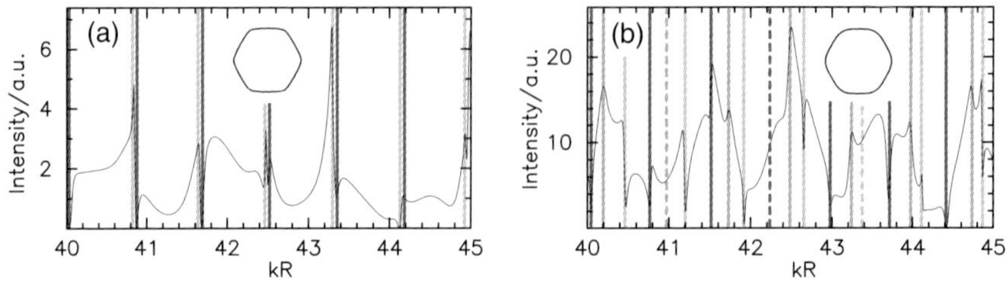

Fig. 19. Calculated scattering intensity spectra for slightly rounded hexagonal cavities (shapes depicted as insets). The incoming plane wave is at an angle of 15° to a facet in (a) and 30° in (b); detection occurs at 60° from incidence. Spacings between modes of the same color agree well with $\Delta(kR)_c \approx 0.83$ (cf. Fig. 18). All resonances in (a) appear as doublets. At $kR \approx 42.5$ the doublet structure is seen most clearly. In (b), stronger deviation from hexagonal shape leads to further lifting of degeneracies. Dashed lines mark expected resonances not seen at this observation angle.

with characteristic spacing $\Delta(kR)_c$. This indicates that departures from sharp corners are not resolved in the wave solution when their scale is smaller than λ. A qualitatively different spectrum is observed in Fig. 19(b) for $\rho \approx 3.7\lambda$. Here, the perturbation reveals three well-separated, interleaving combs of modes, again with period $\Delta(kR)_c$. There are 21 distinct resonances in the wavelength interval of Fig. 19(b), which translates to an average mode spacing of $\Delta\lambda \approx 3.6$ nm for a resonator with WOF = 7.5 μm, in good agreement with the experiment (cf. Fig. 12).

To verify that no further modes will be revealed by other choices of deformation, an independent estimate of the average density of modes can be made on the basis of semiclassical considerations (Eq. 5).

$$\left\langle \frac{\mathrm{d}N}{\mathrm{d}(kR)} \right\rangle = \frac{n^2 k\,R}{4} \times \left[1 - \frac{2}{\pi} \left(\arcsin\frac{1}{n} + \frac{1}{n}\sqrt{1 - \frac{1}{n^2}} \right) \right] \tag{5}$$

Here, $\mathrm{d}N$ is the number of modes in the interval $\mathrm{d}(kR)$. The result is $\left\langle \frac{\mathrm{d}N}{\mathrm{d}(kR)} \right\rangle \approx 4.6$, and hence we expect about 22 modes in the interval of Fig. 19(b), again in good agreement with the actual count.

Intensity profile In the ideal case one class of quasi-degeneracies is not removed by any of the perturbations in Fig. 19: their physical origin is time-reversal symmetry for the ray motion inside the cavity. Any of the periodic orbits in Fig. 17 can be traversed clockwise or counterclockwise. In the wave picture this can be interpreted as a standing wave. In Fig. 20 we show the field distribution of the standing wave field in hexagonal resonator with WOF = 4.5 μm as a superposition of clockwise and counter-clockwise propagating modes. In analogy to sine and cosine waves there exists a second, nearly degenerate, distribution with opposite par-

Fig. 20. Standing wave for WOF = 4.5 μm ($kR = 21.59$) in a false-color representation.

ity with respect to a hexagonal symmetry axis. A small frequency splitting exists because the nonintegrability of the ray motion implies that the propagation direction itself is not a "good quantum number", i.e., reversals of the sense of rotation are unlikely but not impossible in the wave equation. This corresponds to quantum tunneling and hence leads to exponentially small splittings on the scale of the individual resonance linewidths [36]. These multiplets have been counted as one resonance in Eq. (5). This small frequency splitting between clockwise (CW) and counterclockwise (CCW) modes exists, and is documented in the spectra of Fig. 21. In addition these measurements also indicate that the field dynamics of the CW and CCW modes is correlated in a manner similar to the amplitudes of two coupled pendulums.

High-intensity ridges inside the resonator form a whispering-gallery-like pattern that decays from the interface into the cavity center. The number of ridges in the radial direction (perpendicular to a side face) provides an approximate analogue of a transverse mode order, but on closer examination one sees that the number of ridges and nodal lines is not uniquely defined, especially along a diameter joining opposite corners. The modes can therefore not be properly labeled by "quantum

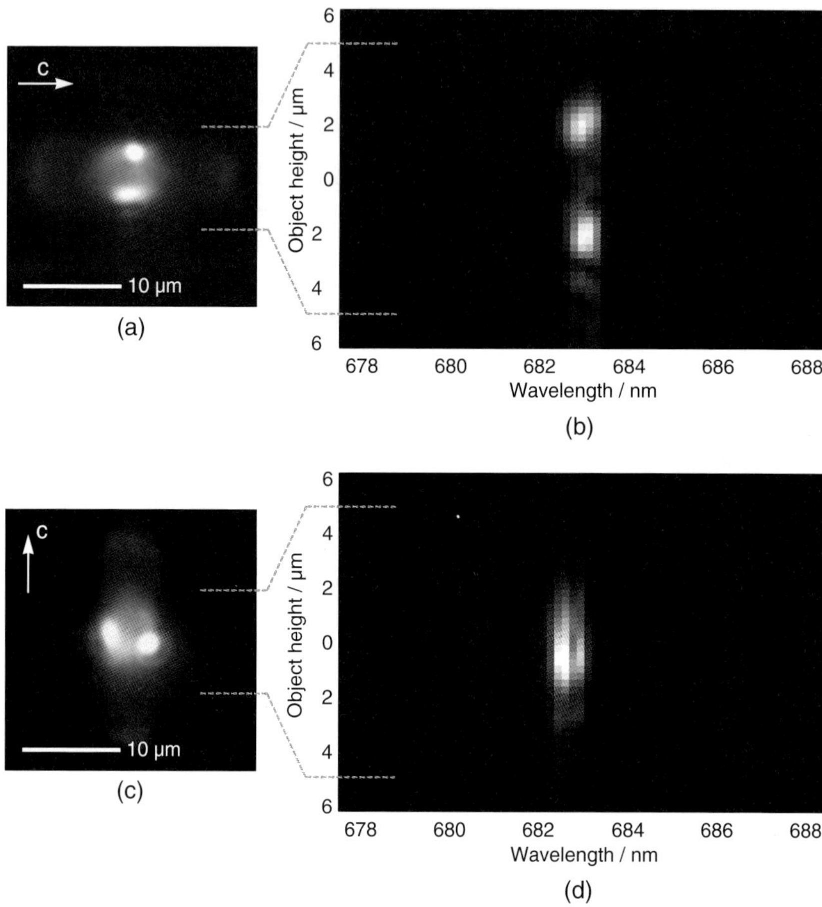

Fig. 21. Frequency splitting between (CW) and (CCW) modes of the sample shown in Fig. 15(b), made visible with an imaging spectrometer. The two emission spots in (a) correspond to CW and CCW laser modes. Their spectral distribution is shown in (b). The spectral separation of the lines (if visible) is less than the linewidth. For a second measurement the crystal was rotated by 90° (c). Now the emission of each spot arrives on the same detector pixels so that both spectra are coherently superposed. Clearly the spectrum in (d) is not compatible with an addition of the intensity of the two spectra in (b): In (d) a dark central line is visible which can be explained by a destructive interference of the CW and CCW fields. This, however, requires that the fields have slightly different frequencies and are (anti)correlated.

numbers" that characterize the number of radial and azimuthal nodes; this is a direct consequence of the nonintegrability of the problem. The most significant difference to the whispering-gallery modes of a circular cavity is clearly the anisotropic emission. High intensity is seen to emanate predominantly from the corners and is directed almost parallel to an adjacent crystal facet.

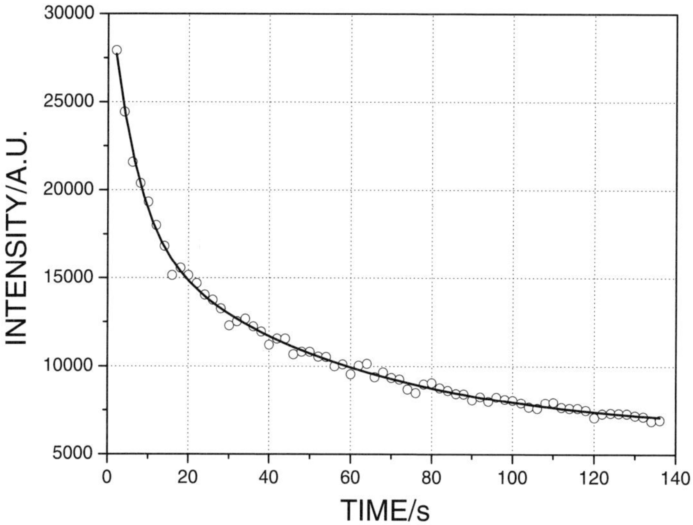

Fig. 22. Photostability of Pyridine 2: strong irradiation bleaches the fluorescence. (Irradiation: train of Q-switched Nd: YAG laser pulses at 532 nm with a power density of 5 MW cm^{-2}, repetition rate 10 Hz, and pulse width 15 ns.) The decrease of shown fluorescence intensity is fitted with the function $Y(t) = A_0 + A_1 \exp(-t/T_1) + A_2 \exp(-t/T_2)$, where the parameters are $A_0 = 6234$, $A_1 = 11761$, $T_1 = 51.9$ s, $A_2 = 13889$, $T_2 = 6.4$ s.

6.5 Photostability

The photostability was investigated for single microcrystals of $AlPO_4$-5 molecular sieves containing Pyridine 2 or DCM dye. The samples, regular hexagonal prisms as shown in Fig. 4(a) and (c), were irradiated with a constant train of strong laser pulses, and the corresponding fluorescence emission was monitored. As the fluorescence emission is proportional to the number of dye molecules, the detected fluorescence signal represents the size of the active, that is, the unbleached, dye population. The photostability history plots shown in Figs. 22 and 23 reveal that the fluorescence decrease due to the constant bleaching irradiation can be represented by a sum of two exponential functions (Eq. 6).

$$Y(t) = A_0 + A_1 \exp(-t/T_1) + A_2 \exp(-t/T_2) \tag{6}$$

In Section 6.3.2.2, where we discussed the spontaneous emission decay dynamics, we found strong evidence that in the $AlPO_4$-5 channels the Pyridine 2 and DCM dye molecules reside in two different characteristic host–guest arrangements (which here we label as arrangements I and II). The photostability investigation documented here in Figs. 22 and 23 bolsters this assumption. In addition, the photostability kinetics also reveal that one of the dye host–guest arrangements is

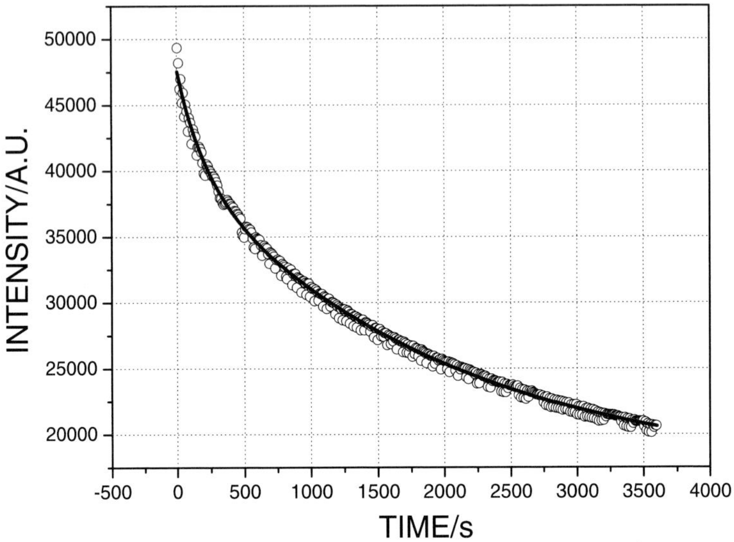

Fig. 23. Photostability of DCM: strong irradiation bleaches the fluorescence. (Irradiation: train of Q-switched Nd: YAG laser pulses at 532 nm with a power density of 0.11 MW cm^{-2}, repetition rate 30 Hz, and pulse width 50 ns.) The decrease of shown fluorescence intensity is fitted with the function $Y(t) = A_0 + A_1 \exp(-t/T_1) + A_2 \exp(-t/T_2)$, where the parameters are $A_0 = 16732$, $A_1 = 7461$, $T_1 = 198$ s, $A_2 = 23498$, $T_2 = 1999$ s.

photochemically more stable than the other, and that one recovers faster than the other.

If the dye molecules had been dissolved in a liquid, and the fluorescence had been bleached by strong irradiation, the bleaching would have been irreversible. However, this is not the case when the dyes Pyridine 2 and DCM are encaged in the pores of $AlPO_4$-5 molecular sieves. For encaged molecules, fluorescence bleaching is reversible. We characterized the bleaching and recovery kinetics, and in Fig. 24 we show a typical recovery history for DCM. For Pyridine 2 molecules encaged in $AlPO_4$-5 molecular sieves the recovery kinetics follow a similar pattern.

6.5.1 Model of the Photostability Kinetics

The figures in which the decrease of the fluorescence emission is plotted can thus be interpreted as resulting from a superposition of the kinetics in the molecular arrangements I and II, where each is represented by the fluorescence emission of the respective arrangement $\eta_{\mathrm{I}}(t)$ and $\eta_{\mathrm{II}}(t)$. If we assume that the kinetics of one arrangement is independent from the other, we can write Eq. (7) for the observed fluorescence intensity $\eta(t)$

$$\eta(t) = a_{\mathrm{I}}\eta_{\mathrm{I}}(t) + a_{\mathrm{II}}\eta_{\mathrm{II}}(t) \tag{7}$$

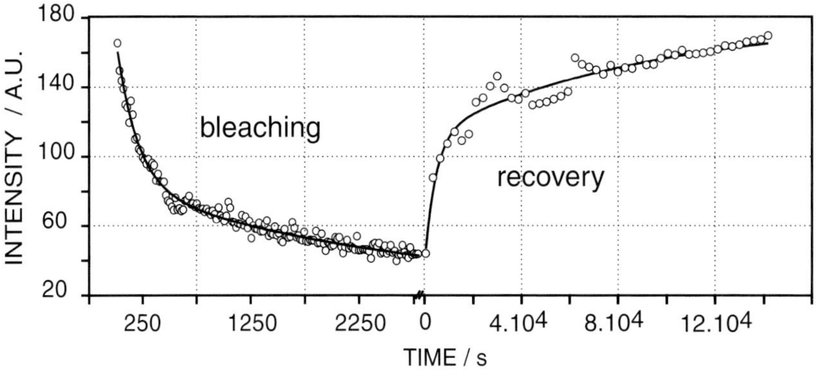

Fig. 24. Typical history of the bleaching and recovery kinetics of DCM dyes loaded inside the channel pores of $AlPO_4$-5 molecular sieves, characterized by their fluorescence intensity. For the given optical conditions the fluorescence decay can be fitted with the function $Y(t) = A_0 + A_1 \exp(-t/T_1) + A_2 \exp(-t/T_2)$ ($A_0 = 24$, $A_1 = 83$, $T_1 = 192$ s, $A_2 = 62$, $T_2 = 2384$ s), whereas the recovery is fitted with $Y(t) = B_0 - B_1 \exp(-t/T_S) - B_2 \exp(-t/T_L)$ ($B_0 = 178$, $B_1 = 63$, $T_S = 3703$ s, $B_2 = 64$, $T_L = 77202$ s). After ca. 12 h of storage in the dark at room temperature, the initial fluorescence level is recovered. Note the scaling of the recovery time interval is different from the bleaching time interval.

where the weights a_{I} and a_{II} represent the relative efficiency with which a pump photon is converted into a detected fluorescence photon. Note that as the coupling of the dye molecules to the optical field can be different for the molecular arrangements I and II, the weights a_i are not necessarily proportional to the respective fractional concentrations.

Immediately after a dye molecule emits a fluorescence photon the energy excess, i.e., the energy difference between exciting end emitted photon, remains stored in the vibrational degrees of freedom of the dye ground state. The photostability of the molecule is determined by the rate with which the molecule can dissipate this excess energy into its environment. If, compared to the excitation rate, the excess energy dissipation rate is too low, then the vibrational energy does not relax fast enough and accumulates in the vibrational degrees of freedom until some nonreversible reaction occurs, such as breaking of some molecular bonds. When enclosed in channel pores the dye molecules interact strongly with the host lattice. This allows fast transfer of the excess energy from the dye to the host lattice, which acts as an efficient heat bath, so limiting the energy accumulation to nondamaging levels.

On the other hand, for many dye molecules the transition from the ground to the excited singlet state is accompanied by an increase of the static dipole moment due to intramolecular charge transfer [37]. If the molecule is encaged, its electric coupling with the molecular sieve environment thus increases when excited [38]. As a consequence, the excited molecule, now more strongly interacting with the molecular sieve lattice, can access configurations which otherwise it could not attain in the ground state. We speculate that with the help of this process, encaged

DCM, which in the ground state normally exists as a trans isomer, can decay to the cis isomer ground state after excitation. However, in contrast to DCM dissolved in a liquid [39], and although the ground state interaction is weaker, the host–guest interaction can trigger spontaneous relaxations from the cis back to the trans ground state. This picture is compatible with a report according which the fluorescence efficiency of the DCM cis isomer is strongly reduced [40]. Together these assumptions shape a consistent hypothesis for the observed fluorescence bleaching and subsequent recovery of the fluorescence.

The bleaching measurements documented in Figs. 22–24 (as well as the fluorescence decay dynamics discussed in Section 6.3.2.2) reveal kinetics which are not compatible with the assumption of a homogenous dye population. Although we do not have detailed information on the physical or chemical nature of the arrangement of the dye guests inside the host channels, the measurements of the Pyridine 2 and DCM bleaching and recovery kinetics suggest the existence of two distinct sites where the dye molecule resides. In the following we refer to these possibilities as arrangements I and II. We also assume that the molecules reside exclusively either in arrangement I or II, and that they do not spontaneously mutate from I to II. Further we assume that above-mentioned enhanced isomerization in the excited state is possible for molecules in arrangement I as well as II, but with different efficiency. In Eq. (7) we accounted for this with the weights a_I and a_{II}.

With these assumptions the photobleaching effects documented in Figs. 22 and 23 can formally be described with a rate equation model (Fig. 25). Note that we implicitly assumed that the kinetics of arrangements I and II are not correlated.

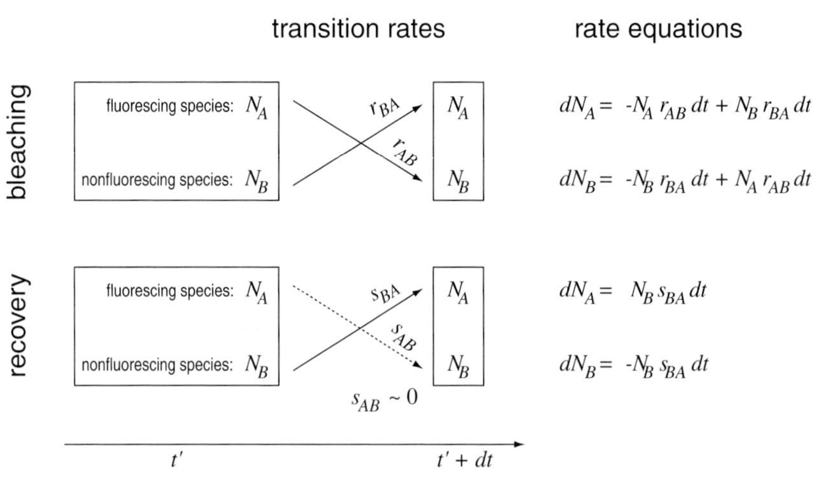

Fig. 25. Rate equation model for the bleach history which we assume to describe the processes in arrangement populations I and II. The transition rates r relate to purely laser induced processes, whereas the rates s are associated with purely non radiative processes. Specifically r_{AB} and r_{BA} characterize the bleaching effect of the laser radiation on the molecular system, while s_{BA} describes the recovery of the dye molecules. Thus, the bleach history reflected by the fluorescence results from both processes, whereas the recovery of the sample in the dark depends only on the s rates.

Therefore we can independently associate a corresponding bleaching model to each arrangement. In the following we introduce the single model, which we then associate with arrangements I and II.

We use A to denote the fluorescing state (e.g., the DCM trans isomer), and B for the state with quenched fluorescence (e.g., the cis isomer). Then N_A stands for the number of molecules in the A state. The sum $N_A(t) + N_B(t) = N_0$ corresponds to the total number of molecules, which is assumed to be constant. With this the transition rates shown in Fig. 25 can be collected into a differential equation which describes bleaching and concurrent recovery (Eq. 8).

$$\dot{N}_A + (r_{AB} + r_{BA} + s_{BA})N_A(t) = N_0(r_{BA} + s_{BA}) \tag{8}$$

This equation is solved with the standard ansatz, which results in Eq. (9).

$$N_A(t) = N_0\left(\frac{r_{AB}}{r_{AB} + r_{BA} + s_{BA}}\right)\exp[-(r_{AB} + r_{BA} + s_{BA})t] + N_0\frac{r_{BA} + s_{BA}}{r_{AB} + r_{BA} + s_{BA}} \tag{9}$$

The experimental evidence allows us to assume that transitions from the non-fluorescing B to the fluorescing A species are not induced by the bleaching irradiation, therefore we can assume $r_{BA} \ll r_{AB}$. With this Eq. (9) simplifies to

$$N_A(t) = N_0\left(\frac{r_{AB}}{r_{AB} + s_{BA}}\right)\exp[-(r_{AB} + s_{BA})t] + N_0\frac{s_{BA}}{r_{AB} + s_{BA}} \tag{10}$$

The observable fluorescence $\eta(t)$ (Eq. 7) results from the superposition of the emission of arrangements I and II, either of which is described by an equation corresponding to (Eq. 10). Relabeling r_{AB} with r_I or r_{II} and s_{AB} with s_I or s_{II} we obtain Eq. (11).

$$\eta(t) = \left(\frac{a_I s_I}{r_I + s_I} + \frac{a_{II} s_{II}}{r_{II} + s_{II}}\right) + \frac{a_I s_I}{r_I + s_I}\exp(-(r_I + s_I)t) + \frac{a_{II} s_{II}}{r_{II} + s_{II}}\exp(-(r_{II} + s_{II})t) \tag{11}$$

On the other hand, the observable fluorescence emission kinetics shown in Figs. 22 and 23 can be fitted with the Eq. (6). To compare the model function (Eq. 11) with the data fit (Eq. 6) we normalized both functions by their constant term (Eqs. 12 and 13).

$$\tilde{Y}_B(t) = 1 + \frac{A_1}{A_0}\exp(-t/T_1) + \frac{A_2}{A_0}\exp(-t/T_2) \tag{12}$$

$$\tilde{\eta}_B(t) = 1 + \left(1 + \frac{a_{II} s_{II}(r_I + s_I)}{a_I s_I(r_{II} + s_{II}}\right)^{-1}\exp(-(r_I + s_I)t)$$

$$+ \left(1 + \frac{a_I s_I(r_{II} + s_{II})}{a_{II} s_{II}(r_I + s_I)}\right)^{-1}\exp(-(r_{II} + s_{II})t) \tag{13}$$

Then we can identify the following relations between the model parameters and the fitted parameters (Eqs. 14–16).

$$r_{\mathrm{I}} + s_{\mathrm{I}} = \frac{1}{T_1} \tag{14}$$

$$r_{\mathrm{II}} + s_{\mathrm{II}} = \frac{1}{T_2} \tag{15}$$

$$\frac{a_{\mathrm{II}} s_{\mathrm{II}}}{a_{\mathrm{I}} s_{\mathrm{I}}} = \frac{A_1 T_1}{A_2 T_2} \tag{16}$$

We see that the model parameters can not be uniquely determined by the data fit. The spontaneous recovery rates s_i can be studied by probing the fluorescence activity while the sample recovers in the dark. A typical example is shown in Fig. 24. The dark recovery process is modeled according to the "recovery" part of Fig. 25 alone, which results in Eq. (17)

$$N_{\mathrm{A}}(t) = (\bar{N} - N_0)\exp(-s_{\mathrm{BA}} t) + N_0 \tag{17}$$

where $\bar{N}$ denotes the number of fluorescing molecules after bleaching $[\bar{N} = N_{\mathrm{A}}(t = 0)]$ and N_0 the total amount of molecules of respective arrangement.

Again the observable fluorescence $\eta(t)$ is given as the superposition of the emission of molecules of arrangements I and II, which with the above relabeling can be written as

$$\eta(t) = (a_{\mathrm{I}} + a_{\mathrm{II}}) + a_{\mathrm{I}}\left[\left(\frac{\bar{N}}{N_0}\right)_{\mathrm{I}} - 1\right]\exp(-s_{\mathrm{I}} t) + a_{\mathrm{II}}\left[\left(\frac{\bar{N}}{N_0}\right)_{\mathrm{II}} - 1\right]\exp(-s_{\mathrm{II}} t) \tag{18}$$

The function fitted to the measured data has the form (cf. Fig. 24) of Eq. (19)

$$Y(t) = B_0 - B_1 \exp(-t/T_{\mathrm{S}}) - B_2 \exp(-t/T_{\mathrm{L}}) \tag{19}$$

Again, it is convenient to compare normalized functions (Eqs. 20 and 21).

$$\tilde{Y}_R(t) = 1 - \frac{B_1}{B_0}\exp(-t/T_{\mathrm{S}}) - \frac{B_2}{B_0}\exp(-t/T_{\mathrm{L}}) \tag{20}$$

$$\begin{aligned} \tilde{\eta}_R(t) = 1 &+ \frac{a_{\mathrm{I}}}{a_{\mathrm{I}} + a_{\mathrm{II}}}\left[\left(\frac{\bar{N}}{N_0}\right)_{\mathrm{I}} - 1\right]\exp(-s_{\mathrm{I}} t) \\ &+ \frac{a_{\mathrm{II}}}{a_{\mathrm{I}} + a_{\mathrm{II}}}\left[\left(\frac{\bar{N}}{N_0}\right)_{\mathrm{II}} - 1\right]\exp(-s_{\mathrm{II}} t) \end{aligned} \tag{21}$$

Comparing the model function with the fit to the measured data we can identify the relations of Eqs. (22)–(24).

$$s_{\mathrm{I}} = \frac{1}{T_{\mathrm{S}}} \tag{22}$$

$$s_{\mathrm{II}} = \frac{1}{T_{\mathrm{L}}} \tag{23}$$

$$\frac{a_{\mathrm{I}}\left[\left(\frac{\bar{N}}{N_0}\right)_{\mathrm{I}} - 1\right]}{a_{\mathrm{II}}\left[\left(\frac{\bar{N}}{N_0}\right)_{\mathrm{II}} - 1\right]} = \frac{B_1}{B_2} \tag{24}$$

Equations (22)–(24), together with (14)–(16), allow us to determine the as-yet unknown parameters of the model. Thus, together with (22) and (23) we find (Eqs. 25–28)

$$r_{\mathrm{I}} = \frac{1}{T_1} - \frac{1}{T_{\mathrm{S}}} \tag{25}$$

$$r_{\mathrm{II}} = \frac{1}{T_2} - \frac{1}{T_{\mathrm{L}}} \tag{26}$$

$$\frac{a_{\mathrm{I}}}{a_{\mathrm{II}}} = \frac{A_2 T_2 T_{\mathrm{S}}}{A_1 T_1 T_{\mathrm{L}}} \tag{27}$$

$$\frac{\left(\frac{\bar{N}}{N_0}\right)_{\mathrm{I}} - 1}{\left(\frac{\bar{N}}{N_0}\right)_{\mathrm{II}} - 1} = \frac{B_1 A_1 T_1 T_{\mathrm{L}}}{B_2 A_2 T_2 T_{\mathrm{S}}} \tag{28}$$

Tab. 2. Model parameters of DCM sample shown in Fig. 24

Fit parameter	***Value***	***Model parameter***	***Value***
A_1	83	$\frac{a_{\mathrm{I}}}{a_{\mathrm{II}}}$	0.44
A_2	61	r_{I}	$4938 \times 10^{-6}\ \mathrm{s}^{-1}$
T_1	192 s	r_{II}	$407 \times 10^{-6}\ \mathrm{s}^{-1}$
T_2	2384 s	$\frac{1}{r_{\mathrm{I}}}$	202.5 s
B_1	63	$\frac{1}{r_{\mathrm{II}}}$	2460 s
B_2	64	s_{I}	$270 \times 10^{-6}\ \mathrm{s}^{-1}$
T_{S}	3703 s	s_{II}	$13 \times 10^{-6}\ \mathrm{s}^{-1}$
T_{L}	77 202	$\frac{\left(\frac{\bar{N}}{N_0}\right)_{\mathrm{I}} - 1}{\left(\frac{\bar{N}}{N_0}\right)_{\mathrm{II}} - 1}$	2.25

For the DCM sample shown in Fig. 24 the resulting model parameters are given in Tab. 2.

Up to now the photostability of lasing $AlPO_4$-5 molecular sieve samples containing the dyes Pyridine 2, DCM, and the rhodamine derivative rhodamine BE50 were investigated. Fluorescence recovery, however, was observed only with the samples containing Pyridine 2 and DCM. Whether the recovery is a consequence of the specific sample architecture described by Ö. Weiss et al. in Chapter 4 of this part, is an open question.

References

1. S.D. Cox, T.E. Gier, G.D. Stucky, J. Bierlein, *J. Am. Chem. Soc. 110*, 2986 (1988); S.D. Cox, T.E. Gier, G.D. Stucky, *Chem. Mater.* (**1990**), *2*, 609.
2. P.P. Sorokin, J.R. Lankard, *IBM J. Res. Develop.* (**1966**), *10*, 162.
3. U. Brackmann, *Lambdachrome Laser Dyes*, Lambda Physik, Göttingen, **1994**.
4. F.P. Schäfer, Ed., *Dye Lasers*, Springer, Berlin, **1977**.
5. M. Rifani, Y.Y. Yin, D.S. Elliot, M.J. Jay, S.H. Jang, M.P. Kelley, L. Bastin, B. Kahr, *J. Am. Chem. Soc.* (**1995**), *117*, 7572.
6. B. Kahr, S.H. Jang, J.A. Subramony, M.P. Kelley, L. Bastin, *Adv. Mater.* (**1996**), *8*, 941.
7. F.J. Duarte, *Opt. Commun.* (**1995**), *117*, 480.
8. C. Kallinger, M. Hilmer, A. Haugeneder, M. Perner, W. Spirkl, U. Lemmer, J. Feldmann, U. Scherf, K. Müllen, A. Gombert, V. Wittwer, (**1998**), *Adv. Mater. 10*, 920.
9. U. Vietze, O. Krauß, F. Laeri, G. Ihlein, F. Schüth, B. Limburg, M. Abraham, *Phys. Rev. Lett.* (**1998**), *81*, 4628.
10. I. Braun, G. Ihlein, F. Laeri, J. Nöckel, G. Schulz-Ekloff, F. Schüth, U. Vietze, Ö. Weiss, D. Wöhrle, *Appl. Phys. B* (**2000**), *70*, 335.
11. Ö. Weiß, G. Ihlein. F. Schüth, *Microporous Mesoporous Mater.* (**2000**), *35–36*, 617.
12. M. Bockstette, D. Wöhrle, I. Braun, G. Schulz-Ekloff, *Microporous Mesoporous Mater.* (**1998**), *23*, 83.
13. M. Wark, M. Ganschow, Y. Rohlfing, G. Schulz-Eckloff, D. Wöhrle, *Stud. Surf. Sci. Catal.* (**2001**), *135*, 160.
14. M. Ganschow, G. Schulz-Eckloff, M. Wark, M. Wendschuh-Josties, D. Wöhrle, *J. Mat. Chem.* (**2001**), *11*, 1823.
15. Ö. Weiß, F. Schüth, L. Benmohammadi, F. Laeri, *Stud. Surf. Sci. Catal.* (**2001**), *135*, 161.
16. S.T. Wilson in H. van Bekkum, E.M. Flanigen, J.C. Jansen (Eds.), *Introduction to Zeolite Science and Practice*, Elsevier, Amsterdam, **1991**, *Stud. Surf. Sci. Catal.* vol. 58, p. 137
17. I. Girnus, K. Jancke, R. Vetter, J. Richter-Mendau, *J. Caro, Zeolites* (**1995**), *15*, 33; I. Girnus, M. Poll, J. Richter-Mendau, M. Schneider, M. Noack, D. Venzke, J. Caro, *Adv. Mater.* (**1995**), *7*, 711.
18. S.A. Schunk, D.G. Demuth, B. Schulz-Dobrik, K.K. Unger, F. Schüth, *Microporous Mater.* (**1996**), *6*, 273; Ö. Akdogan, G. Ihlein, F. Schüth, *Micrporous Mesoporous Mater.*, in press.
19. H. Du, M. Fang, W. Xu, X. Meng, W. Pang, *J. Mater. Chem.* (**1997**), *7*, 551.
20. A.E. Siegman, *Lasers*, University Science, Mill Valley, **1986**.
21. The experimental method we applied to obtain the decay data as well as the fit approach is described in the URL: http://gaston.iap.physik.tu-darmstadt.de/lp/TCSPC/tcspc_deconv.pdf.
22. D.V. O'Connor and D. Phillips, *Time-correlated Single Photon Counting*, Academic Press, London, **1984**, p. 252.

23 A. Blumen, J. Klafter. G. Zumofen, in *Optical Spectroscopy of Glasses*, I. Zschokke (Ed.), Reidel, Dordrecht, **1986**.

24 T. Förster, *Fluoreszenz Organischer Verbindungen*, Vandenhoeck & Ruprecht, Göttingen, **1951**.

25 V.B. Braginsky, M.L. Gorodetsky, V.S. Ilchenko, *Phys. Lett. A* (**1989**), *137*, 393.

26 L. Collot, V. Lefèvre-Seguin, M. Brune, J.M. Raimond, S. Haroche, *Europhys. Lett.* (**1993**), *23*, 327.

27 P. Berman (Ed.), *Advances in Atomic Molecular and Optical Physics*, suppl. 2, Academic Press, New York, **1994**.

28 J. Rarity, C. Weisbuch (Eds.), *Microcavities and Photonic Bandgaps: Physics and Applications*, Kluwer, Dordrecht, **1996**.

29 M. Ducloy, D. Bloch (Eds.), *Quantum Optics of Confined Systems*, Kluwer, Dordrecht, **1996**.

30 Y. Yamamoto in M. Ducloy, D. Bloch (Eds.), *Quantum Optics of Confined Systems*, Kluwer, Dordrecht 1996, p. 241ff.

31 J.U. Nöckel, A.D. Stone, *Nature* (**1997**), *385*, 45.

32 C. Gmachl, F. Capasso, E.E. Narimanov, J.U. Nöckel, A.D. Stone, J. Faist, D.L. Sivco, A.Y. Cho, *Science* (**1998**), *280*, 1556.

33 A. Mekis, J.U. Nöckel, G. Chen, A.D. Stone, R.K. Chang, *Phys. Rev. Lett.* (**1995**), *75*, 2682.

34 P.J. Richens, M.V. Berry, *Physica* (**1981**), *2D*, 495.

35 A. Hobson, *J. Math. Phys.* (**1975**), *16*, 2210.

36 M.J. Davis, J.E. Heller, *J. Chem. Phys.* (**1981**), *75*, 246.

37 This effect is especially pronounced in DCM, for which the dipole moment increases from 6.6 to 20.2 D; S. Pommeret, T. Gustavsson, R. Naskrecki, G. Baldacchino, J.-C. Mialocq, *J. Mol. Liquids* (**1995**), *64*, 101.

38 A similar increase in the electric guest–host coupling is observed in dissolved DCM (solvation relaxation); see, e.g., P. van der Meulen, H. Zhang, A.M. Jonkman, M. Glasbeek, *J. Phys. Chem.* (**1996**), *100*, 5367.

39 T. Gustavsson, G. Baldacchino, J.-C. Mialocq, S. Pommeret, *Chem. Phys. Lett.* (**1995**), *236*, 587.

40 M. Meyer, J.-C. Mialocq, *Chem. Phys. Lett.* (**1988**), *150*, 484.

7
Laser Materials based on Mesostructured Systems

*Justus Loerke and Frank Marlow**

7.1
Introduction

Current developments in information and communication technologies create new demands on optically functional materials. These are of quantitative nature, to ensure the enormous growth rates in this field, as well as being connected with novel functionalities that enable drastic changes in the system architecture of communication networks. One of these changes is the replacement of electronic devices for information processing by optical devices. The ultimate goal of this replacement is the so-called transparent network, i.e., a network which can transport any optical information without converting it to electrical signals. Such a network can overcome the bottlenecks of current telecommunication systems.

For transparent networks, laser sources, amplifiers, and optical switches are needed for signal generation, regeneration, and routing during propagation through the network. The main focus in this field has long been exclusively on semiconductor devices, such as laser diodes, semiconductor waveguides, and directional couplers. However, there are also other emerging research fields based on alternative materials such as polymers or glasses for similar devices and applications. In this respect, much effort has been made in the search for solid-state dye lasers and other tunable microlasers. Applications in integrated optics demand devices which are small, cheap, simple to produce, and flexible in their spectral range. Systems investigated for these purposes include dye-doped polymer matrices [1], dye-doped glasses [2], organic molecular crystals such as tetracene [3], and porous materials [4].

Solid-state dye lasers, formed by doping a solid host with appropriate organic dye molecules, would offer wide variability in the emission wavelength and promise the realization of comparatively simple microlasers suited for integrated optical elements. Tuning of the emission region of a solid-state dye laser can easily be achieved by variation of the dopant dye molecules and tuning of the laser geometry. While most gas and solid-state lasers use a small number of distinct transitions for laser operation, organic dyes can operate over a quasicontinuum of lines over wide spectral ranges (spanning up to 100 nm) and could thus allow continuous tuning of the laser wavelength.

Mesoporous materials are among the host systems suited to the incorporation of very different dyes reaching from the UV to the IR region. For many MCM-41-like materials, the dye can be simply added during synthesis; the dye molecules are then incorporated in the mesopores of the structure by replacing a small fraction of the surfactant molecules in the structure-directing micelles. Because of the incorporation into the pores of the system the effects of the surrounding host material are strongly restricted. The interaction with the surrounding matrix or size effects of a rigid lattice that can influence or even quench emission of the dye are negligible for mesostructured host–guest composites.

Here we describe the efforts to construct solid-state dye lasers from mesoporous systems doped with laser dyes. The different approaches based on MCM-41-, SBA-15-, and SBA-3-like materials, which are used by different research groups, are reviewed. In detail, we present our own work on microlasers with controlled output, fabricated from dye-doped mesoporous fibers.

7.2 Synthesis of Mesoporous Materials for Optical Applications

The synthesis of mesoporous materials is highly versatile [5]. The mesoporous framework composition can be substituted by or doped with a wide range of compounds, and the micellar template allows many variations as well. The possibility of doping mesoporous materials with organic dyes is especially noteworthy, since it enables the incorporation of dye molecules without destroying the framework structure. Since optical materials are intended to generate, detect, and change the propagation of light, interaction of light with the electronic states of the material is required and must be optimized. Host–guest materials with a functional division between the optical properties of the matrix and the function of the included guest molecules, as well as the independent tuning possibilities connected with this separation, are an important step towards the realization of the complex functions required for optical materials [6].

7.2.1 Mesoporous Systems Useful for Optical Materials

Mesoporous materials have been investigated intensively with respect to their methods of formation, chemical properties, and catalytic applications [5,7,8]. A number of proposals have also been made to apply mesoporous systems for optical purposes [9,10]. Preliminary investigations of basic optical effects [11] or chemo-optical sensor applications have been pursued with powder materials, but these powder-based investigations must be considered exceptions. Since light should propagate in a defined way in optical devices, well-defined macroscopic forms such as monoliths, fibers, and films have attracted more attention. Well-defined objects can be produced by several methods, such as:

- Spinning of fibers [12]
- Drawing of fibers [13]

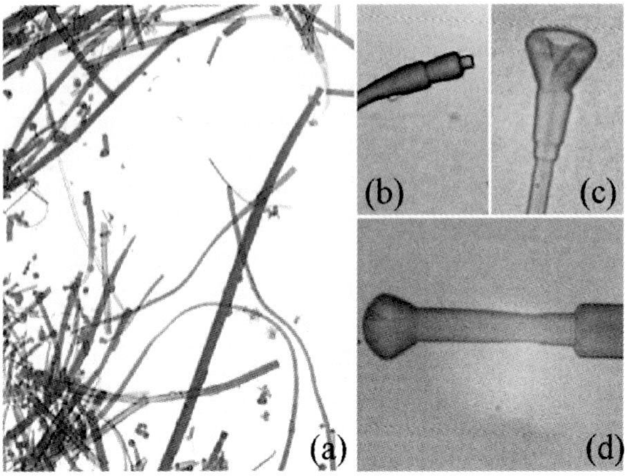

Fig. 1. Mesoporous fibers obtained by a slow spontaneous growth process. A typical mixture from a synthesis batch, consisting of fibers with thicknesses between 5 and 20 μm and small particles doped with Rh6G is shown in a). Normally the fibers are several millimeters long and their ends arise mainly from breaking during handling. The photographs b), c), and d) show special, naturally grown fiber ends. For details, see Section 7.2.3.

- Slow self-assembly of fibers [9,10]
- Dip- and spin-coating techniques for films [14,15]
- Slow evaporation techniques for crack-free monoliths [16]
- Microstructure formation by molding techniques [17]
- Self-assembly of complicated shapes from small particles [18–23]

All these techniques deliver optically useful structures on the micrometer scale. Examples of such structures are the mesoporous fibers shown in Fig. 1. The suitability of these fibers for optical applications has been demonstrated by their use as optical waveguides [9] and as a new type of laser material [10]. Rapid evaporation techniques in combination with dry spinning [12] might be especially useful for large-scale production of fibers. These fibers are also expected to show an ordered pore architecture within their well-defined morphology. The production of films, on the other hand, has a high potential in integrated optics, provided the lateral structuring problems of the films can be solved. The micromolding technique seems to be well suited for this aim.

7.2.2
Mesopore Environment

Mesopores represent an interesting environment for optically functional molecules from the materials science point of view, too. This is characterized by three specific properties, namely, pore definition, adsorbate organization, and accessibility of the pore system. The pores have a defined structure, similar to micropores, that provides a defined environment for the included species, which can be an organized adsorbate system within the mesopores. The well-defined host induces a well-defined composite. This is also exploited in mesoporous laser materials, in which the pore filling consists of an organized surfactant and optically functional molecules. The surfactant system is of special interest for optical applications since it avoids dye aggregation and stabilizes the dyes. Finally, the incorporation of optically functional molecules is possible in different ways, during or after mesopore synthesis. For the postsynthetic dye loading, regular pore openings, as are assumed to occur in the mesopore system of MCM-41, or micropores in the channel walls can be used. All these properties make mesostructures a variable and tunable host system.

The environment often strongly influences the optical and photochemical properties of molecules, which are decisive for optically functional host–guest materials. The absorption and fluorescence maxima as well as the dominant deactivation processes of exited states depend on molecular geometry and symmetry, which are strongly influenced by the surroundings of the molecules. Especially intersystem crossing rates, which have a strong influence on photochemical reactions, show a clear dependence on the molecular surrounding.

A specific advantage of the mesopore environment for emitting materials is the prevention of dye aggregation, as mentioned above. The fluorescent properties of laser dyes are distinctly changed by the formation of aggregates. It has been shown that dimerization almost completely quenches the fluorescence of Rhodamine 6G in solution [24]. In most dye lasers, this effect is successfully avoided by a low concentration of the dissolved lasing dye or the use of appropriate solvents. This aggregation seems to be naturally prevented during the inclusion in mesostructured solids. Most laser dyes are incorporated in monomeric form into the micelles filling the mesopores, so that the fluorescence is not affected by dimer quenching processes.

7.2.3 Fiber Synthesis

As mentioned in Section 7.2.1, there are numerous synthetic possibilities for obtaining mesoporous systems. Depending on the exact synthesis method, the structure and morphology of the obtained materials differ widely. Even the synthesis of one type of material, such as silica fibers, offers many ways of modifying the fiber morphology and the microenvironment of guest molecules. By means of morphology tuning, many sizes, shapes, and even complicated architectures can be obtained, in addition to straightforward modifications of the composition of the synthesis medium, e.g., variation of the included dye.

The fibers used for laser experiments up to now were prepared by a self-assembly process in a stirred aqueous system [10] with a typical molar composition of 100 H_2O:0.0312 $CH_3(CH_2)_{15}N(CH_3)_3Br$ (CTAB):3.69 HCl:0.00029 Rhodamine 6G. Then 0.038 mol of the silicon source tetrabutoxysilane (TBOS) was added to the solution without stirring to form a thin layer on top of the aqueous phase. After several days (typically between 5 and 14), spontaneously grown fibers were removed from the solution and dried in air. The fibers have hexagonally ordered mesostructures of cylindrical aggregates with a lattice constant of 4.7 nm [9]. The dye concentration in the fibers was on the order of 0.2 wt%, as determined by absorption measurements on individual fibers.

The fiber length exceeds several millimeters in each synthesis. Therefore, only fiber parts broken from whole fibers are used for the experiments. The ends of such fibers are smooth planes perpendicular to the axis corresponding to cleavage planes of the fiber structure (see Section 7.2.4). However, short unbroken fibers would also be of interest for optical applications, since this would allow the use of special shapes of the fiber endings controlled by the synthesis conditions. Fiber ends can vary between different shapes such as plane surfaces, 60° tips, and more complicated shapes, as shown in Fig. 1. These surfaces could be used for tuning of the reflectivity, output aperture, and mode selectivity.

Since the morphology of the fibers is decisive for optical functions in microdevices, methods for controlling the morphology have been explored. Interesting properties of the fibers are their thickness, surface roughness, fiber length, and the shape of the fiber end. The fiber thickness can be varied by means of the synthesis time or by addition of ethanol. The silicon source also has a strong influence [25] on the thickness, as well as on the surface roughness of the fibers [26]. Fibers best suited for optical waveguiding are obtained with the silicon source TBOS. In this case, the fibers are always smooth and of constant thickness.

7.2.4
Internal Structure

Although the use of mesoporous materials for many optical applications has been considered by a number of groups, until now their exact internal structure and formation mechanism are largely unknown. This is not always of direct importance for the optical functions, but the specific optical properties of the material can show a significant dependence on the mesoscale structure. For this reason, we have conducted a number of experiments to elucidate the internal structure of fibers grown in acidic media [27]. It was revealed that fibers grown by acidic synthesis form channels that wind around the central axis of the fiber and form a circular channel system. This architecture, shown in Fig. 2 in a micrograph of a microtomed fiber cross section, is completely unknown for inorganic systems.

The novel architecture, characterized by special circular symmetry instead of translational symmetry, was confirmed by X-ray diffraction (XRD) [28] and birefringence experiments [25,29]. The XRD experiments furthermore revealed a remarkable modification of scattering phenomena. For these circular structures, the Bragg law is valid only in a modified form, the diffraction peaks have a defined fine structure, and symmetry-related peaks show different intensities (for details, see Ref. [28]). Because of the modification of physical phenomena relative to materials with translational symmetry, we named these circular structures circulites instead of crystals. The XRD experiments additionally showed that the channels in the fibers are ordered in a helical form. From TEM investigations [26], it was con-

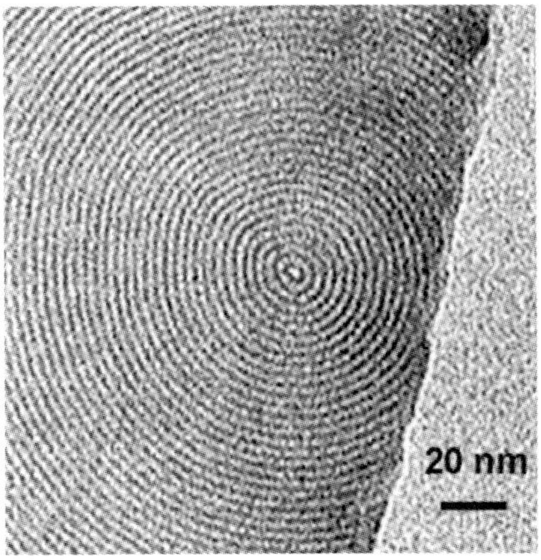

Fig. 2. TEM micrograph of the fiber cross section. The circular channels are wound around a central axis, visible in the middle of the picture. Channels are densely packed and form a coil [27].

cluded that these helices must have a very low pitch, which is very likely less than $2\pi\, d/\sqrt{3}$ ($\approx 3.6d$), where d is the channel distance [30].

The formation of the fibers is mainly controlled by the aggregation of micelles in the synthesis solution, where it is likely that rodlike micelles are formed. Under the influence of the silicon source, these rodlike micelles become unstable and tend to aggregate. However, at low micelle concentration, a flexible micelle is more likely to interact with itself instead of other micelles. This could lead to the formation of closed loops from flexible rodlike micelles, which would then tighten to reduce surface energy and grow into helical shapes, the seeds for fiber growth. This formation mechanism [25] would explain the formation of the circular symmetry of the fibers.

7.2.5 Morphology Control and Hierarchical Structures

Hierarchical structures are useful for optical functions, for which different length scales are important. While the interaction of light with the electronic states of the material is determined on the subnanometer level, the orientation and arrangement of molecules is determined on a nanometer scale. In addition, the propagation of light, involving scattering, diffraction, and waveguiding, is determined on the micrometer scale. To realize a function with a material, its structure must fulfill certain requirements on each length scale. Dye-doped mesoporous materials can be regarded as an example in which control on different length scales could be realized. The molecular scale is determined by special optically functional molecules. Spontaneous aggregation of micelles leads to the nanometer-scale structure,

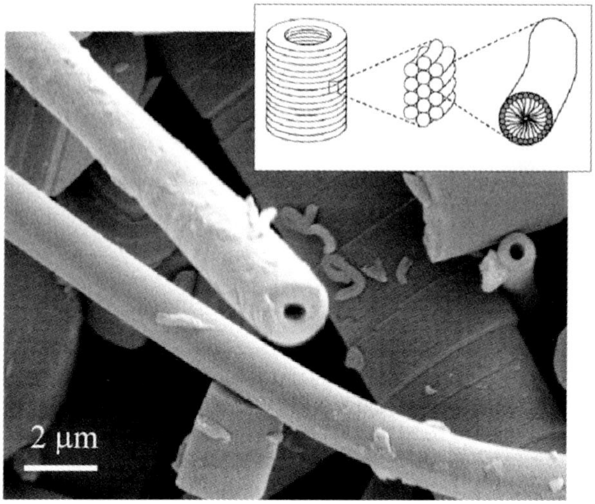

Fig. 3. SEM picture of hollow fibers. The inset shows the structural hierarchy realized in these mesoporous fibers schematically [31].

such as MCM-41. On the micrometer-scale, self-assembly processes lead to useful morphologies such as fibers. Alternatively, fiber spinning, dip coating, or microstructuring by molds can be used for making micrometer-scale structures [12–15].

Experiments on the growth kinetics and morphology of spontaneously grown fibers [26] have shown that the general internal architecture, consisting of a hexagonally ordered channel system wound around the central axis, is essentially unchanged by the use of different silicon sources, such as silicon alkoxides, and the addition of oil to the synthesis mixture. However, the general composition of the synthesis batch, consisting of circular particles and fibers, can be controlled through these parameters. In some syntheses, it is even possible to produce hollow fibers, as shown in Fig. 3 [31]. These fibers show central bores of several hundred nanometers up to several micrometers in diameter. This structure seems to be the result of fluctuations in the longitudinal growth process of the fiber. It is a nice example for a hierarchical structure consisting of an ordered surfactant system inside the pores (primary structure), the hexagonal near-ordering of the channels (secondary structure), and the circular fiber architecture with the central hole (tertiary structure).

Several parameters of this structure can be influenced by the synthetic conditions. For example, the formation of the central hole depends on the initial composition in the batch. Additionally, the internal structure can be fine-tuned by modification of the silicon source and the addition of oil. The wall thickness, the lattice constant of the channel ordering, and the pore size can be varied in this way, paving the way for optimization of the microenvironment of the included dye.

As mentioned above, the generation of hierarchical structures is interesting from a general point of view, since these structures would allow the tuning of optical properties and the construction of more complicated devices in the future. At this point of research, this is pure speculation, since one cannot yet speak of a controlled construction of these hierarchical structures. The spontaneous generation of special hierarchical structures can, however, be observed reproducibly in acidic mesopore syntheses. They show the interesting possibilities of this type of material and might be regarded as the first step towards the controlled generation of structure hierarchies.

7.3 Optically Amplifying Materials Based on Mesostructured Systems

The first step towards construction of devices such as microlasers or amplifiers is the synthesis of suitable materials (laser materials). The key property of a laser material is optical amplification. This property can conveniently be investigated by amplified spontaneous emission (ASE) experiments. The first mesoporous host–guest systems for which such measurements were carried out were mesoporous SBA-3-like fibers [10]. The fibers were doped with the laser dye Rhodamine 6G (Rh6G) during synthesis and were then excited by Nd:YAG laser pulses. The light emitted by the fiber shows a strong anisotropy due to the specific fiber form of the amplifying material. Fluorescence emitted perpendicular to the fiber axis is essentially unchanged when compared to the fluorescence of Rh6G in solution, even at high excitation densities. However, light emitted along the fiber axis shows a very different effect. While the spectrum is almost unchanged at very low excitation energies, the emission band shows significant narrowing above a threshold value of 15 kW cm^{-2}. The band width is reduced from about 50 nm, the band width of Rh6G in solution, to a minimum width of 7 nm. This so-called gain narrowing is caused by the amplification of spontaneously emitted fluorescence during propagation along the fiber axis. This amplified spontaneous emission (ASE) is a clear indication for a high-gain material. In addition to gain narrowing, the total intensity above the threshold rises faster than the spontaneous emission visible below the threshold. Furthermore, the output of the fibers shows only a small divergence angle, and a very narrow output beam of only a few degrees is formed (Fig. 4). The output of the excited fiber (visible on the right) can be seen as a bright spot on a screen to the left of the fiber.

Similar experiments were carried out with mesoporous SBA-15 waveguide structures formed by a micromolding technique [17]. These waveguides were also excited optically by Nd:YAG laser pulses. The emission along the waveguide axis also shows gain narrowing, with a minimal width of 7 nm, depending on excitation density and the size of the excitation region.

The experiments on fibers and micromolded waveguides have shown that mesostructured materials doped with a laser dye exhibit a high amplification during the propagation of light along the waveguide. This amplification shows the suitability of these materials for use as laser materials.

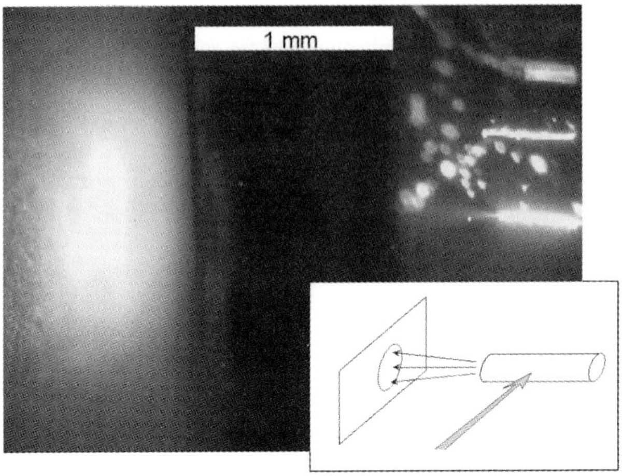

Fig. 4. Directed emission of a fiber. The fiber, visible by scattering in the fiber, is excited by laser pulses. The output beam is imaged onto a screen, as shown schematically in the inset. The opening cone of the beam is visible at the left end of the fiber [10].

7.4 Design of Microlasers

A laser is based on the interplay between amplification and feedback. Therefore, it is not enough to produce a laser material (i.e., an optically amplifying material), but one has also to show that there are realistic ways to combine this material with some feedback mechanism, which is normally provided by a laser resonator. This will then result in a microdevice showing laser emission which is coherent and spectrally narrow. These properties can be verified experimentally.

7.4.1 Priciples of Laser Design

There are a number of different approaches to microlaser design using integrated optical or micro-optical principles. Resonators range from comparatively simple lasers consisting of only a single crystal, in which reflections from the interfaces introducing feedback into the cavity, to quite complicated systems structured in all three dimensions. Resonator schemes that are often employed are the Fabry–Perot resonator, ring resonator, and distributed feedback, depending on the structuring possibilities of the material and the application field. These resonators are shown schematically in Fig. 5.

For fibers, waveguides and some crystals, the simplest resonator is formed by the amplifying object itself, with plane mirrors at the end surfaces forming a

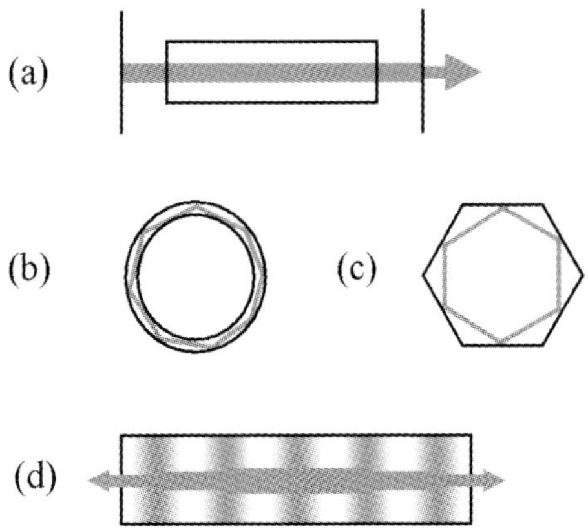

Fig. 5. Main principles for resonator construction. a) Fabry–Perot, b) ring, c) hexagonal whispering gallery, and d) distributed feedback. All these principles can be realized in microsystems such as microfibers or microcrystals.

Fabry–Perot-like resonator. Light is reflected at the interfaces of the amplifying structure and forms a standing wave. A fraction of this light wave is transmitted through a partially transparent output mirror and forms a well-defined laser beam. This controlled output is of crucial importance for any practical application of a laser. The Fabry–Perot principle allowed the first realization of a mesostructured laser with controlled output [32] and will be described in the next section.

For other systems, a closed ring resonator can be formed by the material by patterning procedures or the natural growth process, e.g., in hexagonal $AlPO_4$-5 crystals [4]. In this case, the path of light forms a closed curve, along which light waves travel in so-called whispering-gallery modes by internal reflection at the crystal surfaces. The output of this type of laser is observed at defects or at the corners of the hexagonal crystal, where a small part of the laser radiation is scattered out of the resonator. Mesoporous silica can also be used in a ring-resonator configuration [33]. Here, SBA-15 films doped with Rh6G were deposited on thin glass fibers by dip coating to form a microring resonator around the fiber, which acts as a support, and this allows ring modes to propagate around the circumference of the fiber by internal reflection. The output of this laser stems from scattering at the interfaces and is poorly defined.

Distributed feedback (DFB) lasers rely on a periodic spatial modulation of the material for the formation of Bragg reflections, for instance, by modulation of the refractive index of the active material or through spatial modulation of the pumping light. This approach can be realized especially well on the basis of mesoporous materials, since they can be structured easily by micromolding techniques. A periodic grating of dye-doped SBA-15 waveguides was created and illuminated perpendicular to the waveguide axis by laser light [34]. The periodic modulation of the structure forms a DFB laser and allows the generation and detection of laser light.

7.4.2
Realization of a Fabry–Perot Resonator

Individual microfibers doped with Rh6G by the synthetic method described in Section 7.2.3 were glued onto partially transparent mirrors obtained by the deposition of a 10 nm gold layer on a glass substrate. These films have a transmission of 20% at 590 nm. The fibers were fixed to the support with epoxy resin and coated with another 10 nm film of gold to give in a surface mirror with 20% transmission on the other fiber end. Some fibers were also embedded in silicon grease to reduce the refractive index mismatch between the fibers and the surrounding material and to decrease the acceptance angle for guided waves in the fibers. In this setup it was possible to pump the fibers optically with 10 ns pulses from a frequency-doubled Nd:YAG laser at 532 nm, corresponding to the absorption maximum of Rh6G, and to analyze the emitted light from the fibers with a monochromator (see the following sections).

7.4.3
Spectroscopic Properties

The effect of the above Fabry–Perot resonator on the emission can be seen in Fig. 6. After the fiber ends have been coated with a reflective layer to form a laser cavity the emission narrows drastically to a band with a width of about 2 nm. This band consists of many individual sharp lines with a very small linewidth of about 0.1 nm, which is the resolution limit of the used spectrograph. The laser cavity acts as a spectral filter that allows only certain modes to oscillate. These modes are characterized by a longitudinal mode number describing the number of nodes of the standing wave forming in the cavity. The mode structure of the laser resonator de-

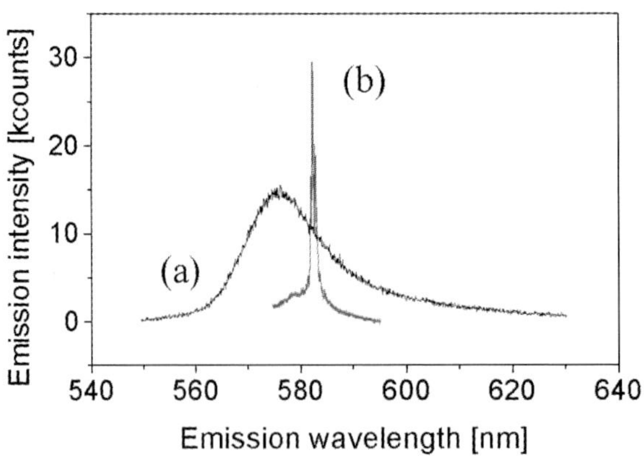

Fig. 6. Emission from fibers. Spectrum a) shows the gain-narrowed emission band of a fiber without mirrors, with a width (FWHM) of 18 nm. Spectrum b) shows the emission of a fiber after the deposition of mirrors, resulting in a line system with a total width of 2 nm [37]. The emission in b) is shifted with respect to a) due to the nonconstant reflectivity of the gold films.

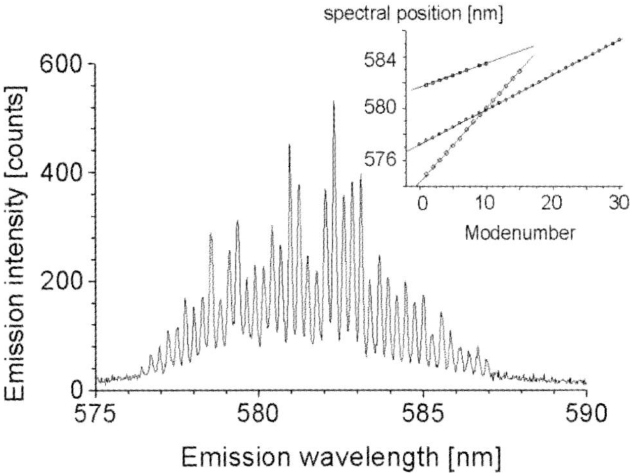

Fig. 7. Emission spectrum of a fiber of $L = 400$ μm at 700 kW cm^{-2}. The emission band consists of over 30 lines with an intermode spacing of 0.27 nm. The inset shows the spectral positions of the laser lines of three fibers of different length. The different resonator lengths result in intermode spacings of 0.18, 0.27, and 0.52 nm, visible in the slope of the lines [37].

pends on the cavity geometry, i.e., the geometry of the used fiber. Using the Fabry–Perot cavity as a first approximation, the intermode distance $\Delta\lambda$ between adjacent lines is then determined by the length L of the fiber, according to $\Delta\lambda = \lambda^2/2nL$. This formula is also valid for a waveguide, but with an effective refractive index $n_{\text{eff}} = \beta\lambda/2\pi$ characterizing the wave propagation along the fiber with the propagation constant β [35]. It can be shown [35] that n_{eff} must lie between the refractive index of the fiber n and the fiber surrounding. For our conditions one can moreover assume that n_{eff} is very close to n. Therefore, the mode separation increases when the cavity length is decreased, and fewer modes can oscillate in the fiber. A typical line spectrum can be seen in Fig. 7. Here, 37 equidistant lines are visible. The intermode distance of 0.27 nm corresponds well to the theoretical separation of 0.28 nm, calculated from the fiber length of $L = 400$ μm and the refractive index of $n = 1.46$. The mode separation is constant for all lines (with an error far below the spectrometer resolution), as predicted by theory and shown in the inset of Fig. 7.

The width of the emitted lines depends significantly on fiber parameters such as length, quality, and mirror reflectivity. Since laser dye molecules typically have a broad band of active transitions, the fiber laser operates on a quasicontinuum of laser-active transitions. The linewidth is then strongly determined by the cavity quality, i.e., the mirror reflectivities and the resonator losses. For a mirror reflectivity of $R = 0.8$, the theoretical linewidth for a passive Fabry–Perot resonator is $\delta\lambda = \lambda^2/2nLF$ [36], where the cavity finesse $F = \pi\sqrt{R}/(1 - R) = 14.1$. The theoretical Fabry–Perot linewidth (e.g., 0.05 nm calculated for a fiber length of 190 μm) is reached by the measured spectra if there are no scattering losses in the fiber. In the case of scattering losses in the fiber, the emission lines become broader. Additionally, the gain of the active medium has a narrowing effect on the linewidth, but this effect has up to now not been observed for the investigated fibers.

7.4.4
Threshold Behavior

The intensity of the fiber emission shows a nonproportional behavior with excitation density. In addition, there is a clear qualitative change in the emission spectrum. At low excitation densities, only low emission intensity is observed, which can be assigned to background fluorescence and ASE, forming a broad emission band. For excitation densities above a certain threshold, the laser line system appears in addition to the background emission, and the intensity of these lines rises much faster than the background fluorescence, which is characterized by the straight line at low intensities in Fig. 8. For excitation densities below $P_0 \approx 350$ kW cm^{-2}, the total intensity emitted from the fiber rises slowly and linearly, while no laser emission is visible. Above the threshold, laser lines are emitted, and the intensity of these lines rises approximately proportional to $P-P_0$. Below the threshold, resonator losses dominate, allowing only the emission of fluorescence. Above the threshold, the gain of the active material exceeds the total resonator losses, leading to an amplification of light and the formation of laser modes of high intensity in the resonator. Such threshold behavior is typical for lasers. Both the threshold behavior and the spectral properties show that the light emitted from the fiber resonators is indeed laser light.

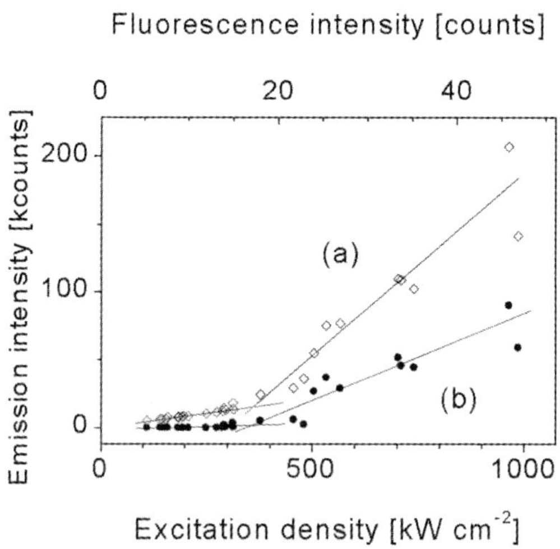

Fig. 8. Pump-intensity dependence of the emission of dye doped fibers. Total emission intensity (integrated intensity of spontaneous, amplified, and laser emission) is shown in a), and line b) shows the integrated intensity of the emission lines without spontaneous and amplified background. The threshold behavior is clearly visible [37].

7.5 Perspectives

The design of optically functional materials based on porous host systems filled with appropriate guests is a promising and open field. The possible applications in opto-electronics create a whole series of requirements for their functionality, especially in the field of switches, amplifiers, and lasers. The host–guest systems can potentially meet these functional requirements, since they can often be assigned to structural requirements on different length scales. It is on these length scales that structures or pores must be modified and appropriately filled. Mesoporous systems clearly show the possibility to design and combine construction elements of different length scales. These combinations could enable the realization of complex optical functions.

The laser materials based on mesoporous systems offer new possibilities for material construction which will be investigated in future projects. One of these new possibilities is the possibility of combining different dyes and organic guests. This offers the possibility of fine tuning, stabilization of laser dyes, realization of energy transfer cascades, and other specially designed photofunctional systems. The tuning and the enhancement of photostability will be one of the key issues of future projects on mesoporous fiber lasers. Another fascinating perspective is connected with the growth of fibers on supports, where the fibers could form arrays. This could lead to microlaser arrays or novel displays.

Finally, an interesting property of these materials is their variable internal architecture, which ranges from disordered, over crystal-like, to circular. Since these architectures are connected with specific anisotropies of the refractive index, they could be used for the control of the mode structure in resonators and fibers.

References

1 R. Sastre, A. Costela, *Adv. Mater.* **1995**, *7*, 198–202.
2 R. Reisfeld, E. Yariv, H. Minti, *Opt. Mater.* **1997**, *8*, 31–36.
3 J.H. Schön, C. Kloc, A. Dodabalapur, B. Batlogg, *Science* **2000**, *289*, 599–601.
4 I. Braun, G. Ihlein, F. Laeri, J.U. Nöckel, G. Schulz-Eckloff, F. Schüth, U. Vietze, Ö. Weiß, D. Wöhrle, *Appl. Phys. B* **2000**, *70*, 335–343.
5 U. Ciesla, F. Schüth, *Microporous Mesoporous Mater.* **1999**, *27*, 131–149.
6 F. Marlow, W.T. Dong, K. Hoffmann, J. Loerke in *Handbook of Porous Solids*, F. Schüth, K. Sing, J. Weitkamp (Eds.), Wiley-VCH, Weinheim, **2002**, 3000 pp.
7 C.T. Kresge, M.E. Leonowicz, W. J. Roth, J. C. Vartuli, J. S. Beck, *Nature* **1992**, *359*, 710–712.
8 S. Biz, M. L. Occelli, *Catal. Rev. Sci. Eng.* **1998**, *40*, 329–407.
9 Q. Huo, D. Zhao, J. Feng, K. Weston, S. K. Buratto, G. D. Stucky, S. Schacht, F. Schüth, *Adv. Mater.* **1997**, *9*, 974–978.
10 F. Marlow, M.D. McGehee, D. Zhao, B.F. Chmelka, G.D. Stucky, *Adv. Mater.* **1999**, *11*, 632–636.
11 I. Kinski, H. Gies, F. Marlow, *Zeolites* **1997**, *19*, 375–381.

12 P.J. Bruinsma, A.Y. Kim, J. Liu, S. Baskaran, *Chem. Mater.* **1997**, *9*, 2507–2512.

13 P. Yang, D. Zhao, B.F. Chmelka, G.D. Stucky, *Chem. Mater.* **1998**, *10*, 2033–2040.

14 M. Ogawa, *J. Am. Chem. Soc.* **1994**, *116*, 7941–7942.

15 S.H. Tolbert, T.E. Schäffer, J. Feng, P.K. Hansma, G.D. Stucky, *Chem. Mater.* **1997**, *9*, 1962–1967.

16 N. Melosh, P. Davidson, B. Chmelka, *J. Am. Chem. Soc.* **2000**, *122*, 823–829.

17 P.D. Yang, G. Wirnsberger, H.C. Huang, S.R. Cordero, M.D. McGehee, B. Scott, T. Deng, G.M. Whitesides, B.F. Chmelka, S.K. Buratto, G.D. Stucky, *Science* **2000**, *287*, 465–467.

18 H. Yang, N. Coombs, I. Sokolov, G.A. Ozin, *Nature* **1996**, *381*, 589–592.

19 I.A. Aksay, M. Trau, S. Manne, I. Honma, N. Yao, L. Zhou, P. Fenter, P.M. Eisenberger, S.M. Gruner *Science* **1996**, *273*, 892–898.

20 H.Yang, A. Kuperman, N. Coombs, S. Mamiche-Afara, G.A. Ozin, *Nature* **1996**, *379*, 703–705.

21 G. A. Ozin, H. Yang, I. Sokolov, N. Coombs, *Adv. Mater.* **1997**, *9*, 662–667.

22 S.M. Yang, H. Yang, N. Coombs, I. Sokolov, C.T. Kresge, G.A. Ozin, *Adv. Mater.* **1999**, *11*, 52–55.

23 I. Sokolov, H. Yang, G.A. Ozin, C.T. Kresge, *Adv. Mater.* **1999**, *11*, 636–642.

24 K.H. Drexhage, in *Dye Lasers*, F.P. Schäfer (Ed.), Springer Verlag, Berlin, **1973**, p. 159.

25 F. Marlow, F. Kleitz, *Microporous Mesoporous Mater.* **2000**, *44–45*, 671–677.

26 F. Kleitz, F. Marlow, G.D. Stucky, F. Schüth, *Chem. Mater.* **2001**, *13*, 3587–3595.

27 F. Marlow, B. Spliethoff, B. Tesche, D. Zhao, *Adv. Mater.* **2000**, *12*, 961–965.

28 F. Marlow, I. Leike, C. Weidenthaler, C. W. Lehmann, U. Wilczok, *Adv. Mater.* **2001**, *13*, 307–310.

29 F. Marlow, D. Zhao, G. D. Stucky, *Microporous Mesoporous Mater.* **2000**, *39*, 37–42.

30 Preliminary estimation. F. Marlow, unpublished results.

31 F. Kleitz, U. Wilczok, F. Schüth, F. Marlow, *Phys. Chem. Chem. Phys.* **2001**, *17*, 3486–3489.

32 J. Loerke, F. Marlow, *Proceedings of the 13th German Zeolite Conference*, Erlangen, March 7–9, **2001**.

33 G. Wirnsberger, G.D. Stucky, *Chem. Mater.* **2000**, *12*, 2525–2527.

34 B.J. Scott, G. Wirnsberger, M.D. McGehee, B.F. Chmelka, G.D. Stucky, *Adv. Mater.* **2001**, *13*, 1231–1234.

35 B.E.A. Saleh, M.C. Teich in *Fundamentals of Photonics*, Wiley & Sons, New York, **1991**, p. 278.

36 W. Demtröder, in *Laser Spectroscopy*, 2nd Ed., Springer Verlag, Berlin, **1998**, p. 247.

37 J. Loerke, F. Marlow, *Adv. Mater.* **2002**, *14*, 1745–1747.

8
Polymer-Embedded Host–Guest Systems

Juergen Schneider, Detlef Fanter, and Monika Bauer

Abstract

Zeolites loaded with organic chromophores and embedded in a polymer matrix are of interest with respect to novel optical applications. To avoid light scattering at the zeolite particle/polymer interface the refractive index of the matrix polymer was adjusted to that of the zeolite, and highly transparent composites were thus produced. For this purpose an immersion method for estimating refractive indices of zeolites was developed. Copolymers of methyl methacrylate and a fluorinated methacrylate were chosen as embedding polymer matrix. Faujasites HY loaded with Disperse Red were embedded in methacrylate copolymers. Wavelength-dependent optical transparency of the resulting composites was measured as a criterion for successful index matching.

8.1
Introduction

Chromophores embedded in inorganic molecular sieves are recently attracting increasing attention with respect to novel applications. Such materials were synthesized, e.g., at the University of Bremen. For the practical use of chromophores embedded in molecular sieves different options exist, e.g., (1) the use of micrometer-sized crystals [1,2], (2) a set of oriented crystals on a carrier [1,4–7], and (3) crystals dispersed in an organic polymer.

The last method would allow the preparation of (e.g., cuboidal) larger samples for optical applications due to coloring by the active chromophore-filled inorganic material. However, a fundamental prerequisite for dispersion in a polymer is optical homogeneity of the composite material polymer/molecular sieve. Since the size of the molecular sieve crystals in most cases is greater than 1 μm, light scattering occurs at the polymer/zeolite interface due to differences in refractive index between zeolite and polymer. Therefore, the refractive index of the polymer has to match that of the molecular sieve to minimize these scattering effects [8]. In addition, the optical transparency of samples prepared has to be considered as a criterion for successful index matching. To the best of our knowledge, no paper describes the preparation of molecular sieves in polymers with the aim of obtaining an optically homogeneous material. We investigated different zeolite materials, and for more detailed examinations faujasite HY and Disperse Red 1 (DR1) embedded in faujasite HY were selected.

Refractive indices $n(\lambda)$ of faujasites are mostly unknown but could be expected to be less than 1.49. The refractive index of conventional amorphous polymers is usually higher. Copolymerization of different monomers is a convenient way to adjust the refractive index of a polymer (see, e.g., Ref. [9]). In accordance with the Gladstone-Dale relation [10] partly fluorinated monomers allow the refractive index of polymers to be decreased by copolymerization with unfluorinated monomers, depending on the composition ratio used.

8.2 Experimental

8.2.1 Copolymers

8.2.1.1 Bulk Samples

Methyl methacrylate (MMA; Fluka, 99 %) and 2,2,2-trifluoroethyl methacrylate (TFM; ABCR, 99 %) were selected as suitable comonomers. The refractive indices of their copolymers can be adjusted between 1.44 and 1.49 depending on the monomer ratio used. Copolymers were synthesized by bulk polymerization with the initiator dibenzoyl peroxide (BPO; Fluka, 97 %). The homogeneous mixture of monomers and 1.5 wt % BPO relative to the monomer reacted in a round bottomed flask with magnetic stirrer and reflux condenser at 80 °C under nitrogen. When the mixture became viscous after about 45 min it was poured into a casting mold (Fig. 1) and the reaction was completed in an oven at 50 °C for 5 h. This

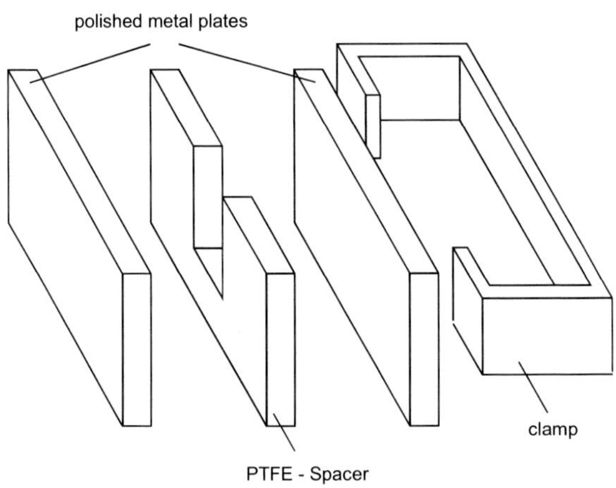

Fig. 1. Scheme of casting mold for preparation of composite samples

Tab. 1. Conditions of solution copolymerization

Parameter	*Copolymer 1*	*Copolymer 2*
Methyl methacrylate (MMA)	9.6 g	8 g
2,2,2-Trifluoroethyl methacrylate (TFM)	0.4 g	2 g
1,4-Dioxane (solvent)	90 g	90 g
Dibenzoyl peroxide (initiator)	200 mg	200 mg
Polymerization temperature	80 °C	80 °C
Polymerization time	3 h	3 h
Yield	82 %	80 %

molding process gave cuboidal samples having parallel, plane, and smooth surfaces, as are necessary for optical investigations.

8.2.1.2 Powder Material

For preparation of composite layers two copolymers were synthesized by free radical polymerization in solution in a stirred three-neck round-bottomed flask under nitrogen atmosphere with the conditions shown in Tab. 1.

The copolymers were isolated by pouring the solutions into an excess of water and filtering off the precipitated polymer. For further purification the polymer was dissolved in ethyl acetate and reprecipitated by adding an excess of ethanol. After collection by filtration and drying its molecular mass was determined by gel permeation chromatography (GPC), and it was used for layer formation.

8.2.2 Composite Preparation

8.2.2.1 Bulk Samples

The zeolite powder was added to the monomer mixture in amounts from 0.1 to 5 wt % relative to the monomers. Stirring the polymerizing system prevented sedimentation and yielded homogeneously dispersed zeolite particles in the resulting polymer matrix. Due to increasing viscosity of the polymerizing mixture sedimentation became increasingly impossible, and the viscous liquid was then poured into the casting mold for completion of the reaction, as described in Section 8.2.1.

8.2.2.2 Layers

For layer preparation copolymers (as described in Section 8.2.1.2) were dissolved in 2-ethoxyethyl acetate to form 35 wt % viscous solutions. To adjust the refractive index of the polymer phase equivalent amounts of solutions of both copolymers were mixed. The necessary amount of zeolite was added and dispersed. Due to the viscosity of the solution no settling of the inorganic particles was observed. Layers were prepared on 20 × 20 mm glass plates by spin coating for 10 s at 500 min^{-1} to equilibrate the solution on the plate and 15 s at 2000 min^{-1} to adjust the thickness of the wet layer, which was dried at 70 °C for 30 min and at 40 °C in vacuum for 30 min.

8.2.3
Optical Characterization of Materials

8.2.3.1 Refractive Indices of Zeolites

In order to select a suitable embedding polymer, the $n(\lambda)$ of zeolites were estimated in the wavelength range of interest ($\lambda = 350$–800 nm), where light scattering at the interface between the zeolite and the surrounding medium due to differences in refractive indices affects optical applications, i.e., where such a system becomes turbid. Photometric measurements (Spekol 11, Zeiss Jena) of suspended zeolites (without encaged dye molecules) in toluene/ethanol mixtures of different ratios resulted in a turbidity minimum at which the $n(\lambda)$ values of the zeolite and the solvent mixture correspond. The turbidity τ from the data obtained was calculated using Eq. (1)

$$\tau = 1/s \cdot \ln(I_0/I) \tag{1}$$

where s is the thickness of the transirradiated layer, I_0 the intensity of incoming light, and I the intensity of light measured after passing through the zeolite suspension.

A minimum turbidity indicates the coincidence of the $n(\lambda)$ values for the zeolite and the solvent mixture. Therefore, the $n(\lambda)$ of the solvent mixtures at this turbidity minimum were taken as an approximation for $n(\lambda)$ of the respective suspended zeolite and were measured by variable-angle spectroscopic reflection ellipsometry (VASE, SE 850 from Sentech Instruments). The basic principle of ellipsometry is that incoming polarized light undergoes a change in its polarization state due to its reflection at the sample surface. The Fresnel equation relates this change of polarization to optical constants of the sample material (e.g., refractive index) [11].

8.2.3.2 Refractive Indices of Copolymers

The dispersive complex refractive indices $n(\lambda)$ of copolymers of different composition were also estimated by the VASE method. To our knowledge such measurements on these copolymers used have not been published. The goal of this procedure was to obtain a set of dispersive refractive indices as a function of copolymer composition for matching with those of zeolites at a wavelength λ that is of interest for a particular application. Additionally, the dispersive behavior of the refractive index makes detection of artefacts and lateral surface inhomogeneities possible.

8.2.3.3 Transparency of Composites

The wavelength-dependent total transparency $T(\lambda)$ [12] was used as criterion for successful index matching. The measurements were carried out with an UV/Vis spectrometer (UV 2101PC, Shimadzu). The higher T is, the better is the match between the refractive indices of the composite components, polymer matrix and zeolite. Additionally, the dispersive behavior of the transparency characterizes the quality of homogeneity of the inner composite. For the pure copolymers (without zeolites) the transparency has to be very high in order to minimize influences of the copolymer matrix onto the composite to be prepared. Transparencies of bulk samples were measured directly relative to air, but for the layers on glass plates the transparency of uncoated plates was used as reference.

8.3 Results

8.3.1 Properties of Materials

8.3.1.1 Zeolites

Turbidity measurements for estimating the refractive indices of zeolites were carried out for different materials (NaX, NaY, HY faujasites, SAPO, and ALPO) as described in Section 8.2.3.1. An example for the dependence of the specific turbidity τ/c_{Zeo} (c_{Zeo} = zeolite concentration) of zeolite suspensions [13] as function of the mixture composition of the dispersing agent for different λ values is shown in Fig. 2. From these measurements refractive indices $n(\lambda)$ of zeolites were found to lie in the range between 1.45 and 1.49. The results of turbidity measurement are shown in Fig. 3.

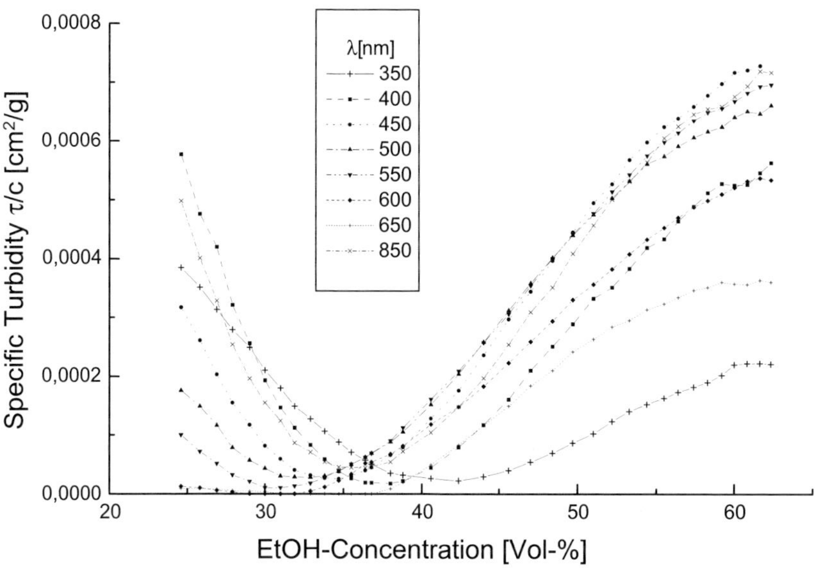

Fig. 2. Dependence of specific turbidity of a faujasite NaY suspension on ethanol content in a toluene/ethanol mixture.

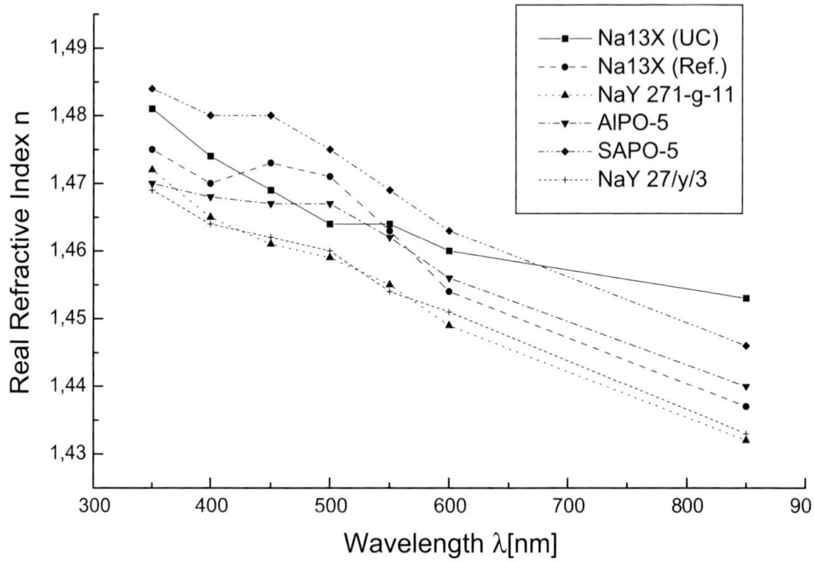

Fig. 3. Dispersive refraction index $n(\lambda)$ of different zeolites.

Only for AlPO-5 und SAPO-5 the estimated values could be compared with results of cooperating colleagues. Table 2 shows that agreement was quite good.

8.3.1.2 **Copolymers**

By copolymerization of MMA–TFM mixtures of different monomer ratios samples were obtained for measuring the dependence of refractive index of these copolymers on monomer composition. ^{1}H NMR investigation of selected copolymer samples of different composition showed that the monomer ratio used in the polymerization mixture and the resulting copolymer composition coincide acceptably. Thus, the monomer ratio can also describe the copolymer composition. Fig. 4 shows the ^{1}H NMR spectrum of a copolymer synthesized from a monomer 70:30 wt % TFM:MMA mixture.

The ratio of the area integrals for methylene hydrogen atoms of TFM and methyl hydrogen atoms of MMA gives contents of these monomer units of 71 wt % TFM and 29 wt % MMA. This ratio is in good agreement with the initial monomer composition.

Tab. 2. Comparison of results of different groups for estimation of refractive indices of zeolites at $\lambda = 670$ nm

	Marlow group, Berlin [14]	*Limburg group, Mainz [15]*	*Our results*
AlPO-5	1.456	1.452	1.454
SAPO-5	–	1.456	1.459

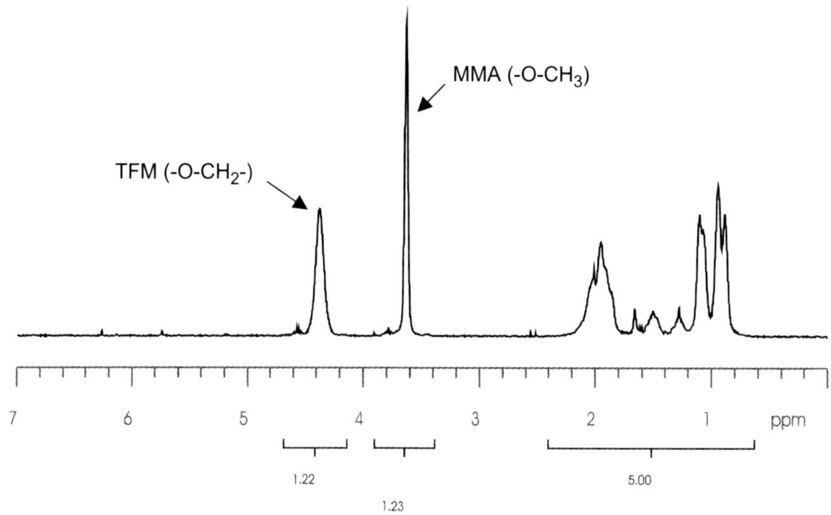

Fig. 4. ^{1}H NMR spectrum of a MMA–TFM copolymer.

The molecular weight of the copolymers was nearly independent of composition. Fig. 5 shows the molecular weight distribution of two copolymer samples.

Figure 6 shows the wavelength dependencies of refractive indices for MMA–TFM copolymers of different composition measured by VASE. The higher the wavelength of light used and the greater the amount of the fluorinated component, the lower the refractive index of the resulting copolymer. For glassy polymers like

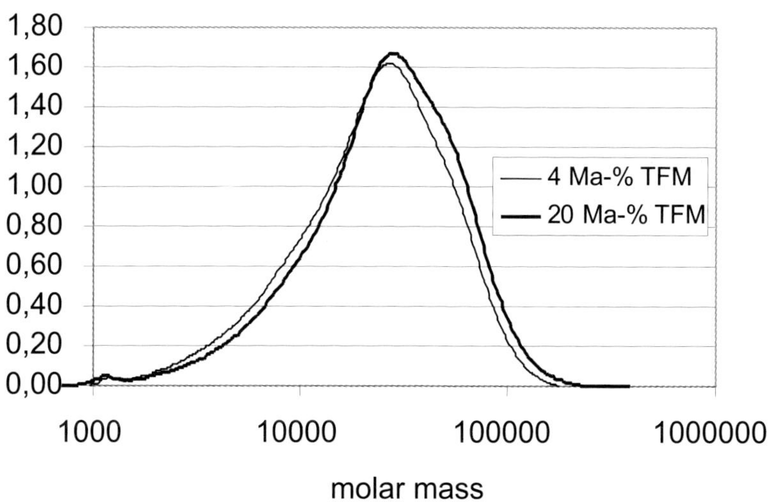

Fig. 5. Molar mass distribution of two embedding copolymers with different MMA/TFM ratios.

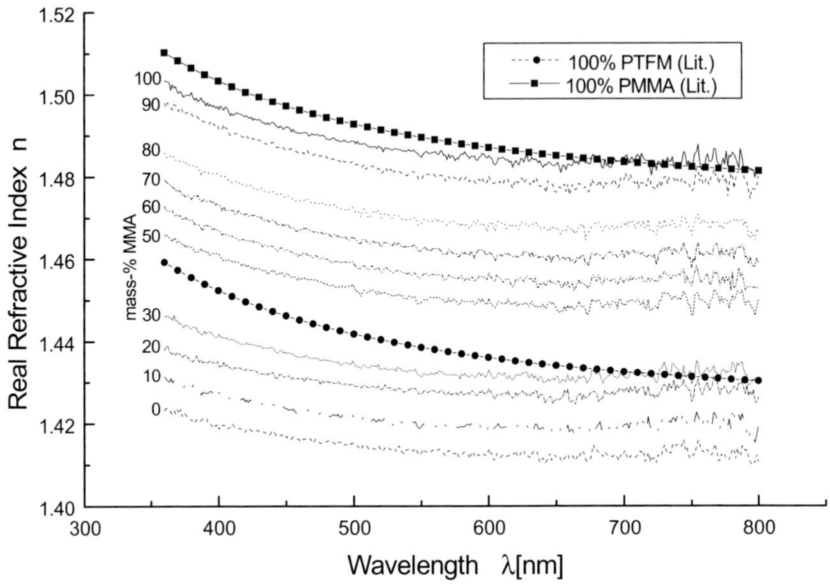

Fig. 6. Dispersive refraction index $n(\lambda)$ of TFM–MMA copolymers (compared with known values).

PMMA the Cauchy approximation [16] can be used to calculate the wavelength-dependent refractive indices (Eq. 2)

$$n(\lambda) = n_1 + n_2/\lambda^2 + n_3/\lambda^4 \tag{2}$$

For a sufficiently accurate approximation the n_3 term in Eq. (2) can be neglected [17], and for PMMA Eq. (2), using values from Ref. [18], follows.

$$n^{\mathrm{PMMA}}(\lambda) = 1.474 + 4700/\lambda^2 \tag{3}$$

Results were in a fairly good accordance with the data given in Fig. 5.

For PTFM the constant for the Cauchy equation is not available. Assuming the same value as for PMMA and using the known $n_D = 1.437$ [9] the dispersion of $n(\lambda)$ is described by Equation (4).

$$n^{\mathrm{PTFM}}(\lambda) = 1.423 + 4700/\lambda^2 \tag{4}$$

However, in this case we found essential differences between the estimated values from Eq. (3) and our experimental results (Fig. 6).

For our results a parameter fit gave Eqs. (5) and (6)

$$n^{\mathrm{PMMA}}(\lambda) = 1.477 + 3200/\lambda^2 \tag{5}$$

$$n^{\mathrm{PTFM}}(\lambda) = 1.408 + 1800/\lambda^2 \tag{6}$$

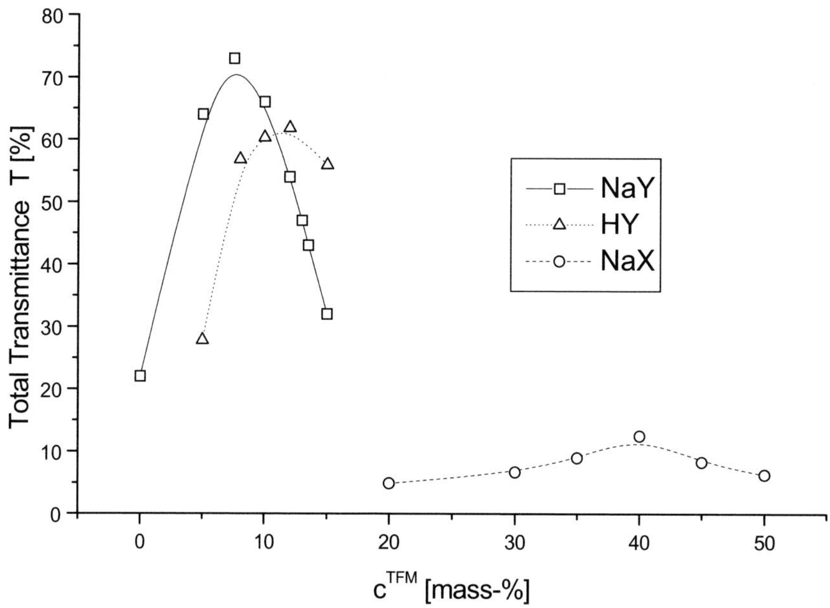

Fig. 7. Total transmittance T_D (at $\lambda_D = 589$ nm) of different TFM–MMA copolymers (expressed by c^{TFM}) filled with various faujasites.

8.3.1.3 Bulk Composites

The transmittance of light was chosen as criterion for successful index matching. For example, Fig. 7 shows the transparency of MMA–TFM copolymers of different compositions filled with various faujasites at a wavelength of $\lambda = 589$ nm. All curves pass through a maximum representing the point of optimum coincidence between the n_D values of the matrix polymer and the zeolite. A maximum transmission of about 75 % could be reached with an NaY faujasite. This material consists of particles with an average diameter of about $x_P = 3$ µm. Average particle sizes of HY and the NaX faujasite also investigated were 5 and 19 µm, respectively. Therefore, small particles (preferentially < 10 µm) improve the transparency of the composite materials. Table 3 lists the diameters of the zeolite particles.

The transmittance behavior of composites made from dye-doped zeolites is similar to that of unloaded ones. The main difference is that light absorption of the

Tab. 3. Mean particle diameter of zeolites investigated

	NaX zeolite	*NaY zeolite*	*HY zeolite*
Mean particle diameter x_P (µm)	19	3	5
Optimum transmission (%)	15	73	62

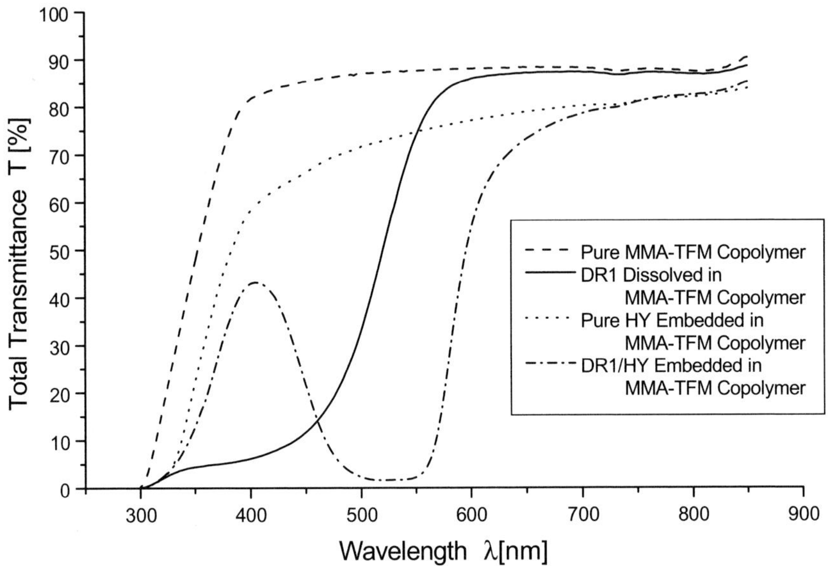

Fig. 8. Dispersive total transmittance $T(\lambda)$ of TFM–MMA copolymers, unfilled and filled with faujasite HY and/or dye.

dye occurs, as expected. Figure 8 compares the different transparency behavior of samples investigated. Pure MMA–TFM copolymer shows about 90% transmittance. After addition of 2 wt % unloaded zeolite an optimum transparency of 75–80% could be reached. Embedding of the azo-dye-loaded HY faujasite led to a similar light-transmittance behavior to the embedded unloaded zeolite, except for the absorption peak of DR1. It should be mentioned that during the polymerization reaction for the preparation of the composite samples no influence of the polymerization system on the loaded zeolites could be observed. The color remained red. But if the pure dye DR1 (without zeolite) was dissolved in the monomer mixture during polymerization the color changed to yellow. Figure 8 shows a blue shift of the absorption peak by about 100 nm. The reason for this effect is assumed to be a reaction between free radicals of the polymerizing system and the amino and/or nitro group of the dye. Further investigations are required to prove this assumption.

Such dependencies also were observed for composites loaded with different zeolite-embedded dyes. Scheme 1 shows the dyes used.

The absorption behavior of composites made from faujasites loaded with these dyes and MMA–TFM copolymers are shown in Fig 9.

It can be seen that there are no significant differences in transmission between DR1- and NDA-containing composites due to their similar chemical structure. The absorption maximum of the spiropyrane NBIPS lies below 400 nm. Due to the increasing absorption of the polymer matrix above 350 nm the minimum transparency of this dye could not be measured.

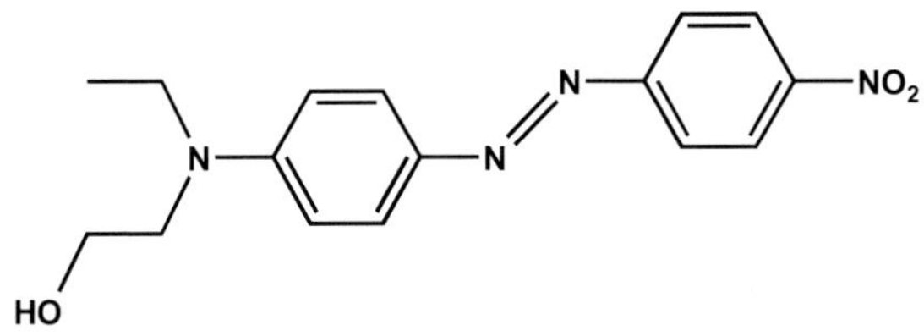

Disperse Red1: 4-ethyl-hydroxyethylamino-4′-nitroazobenzene (DR1)

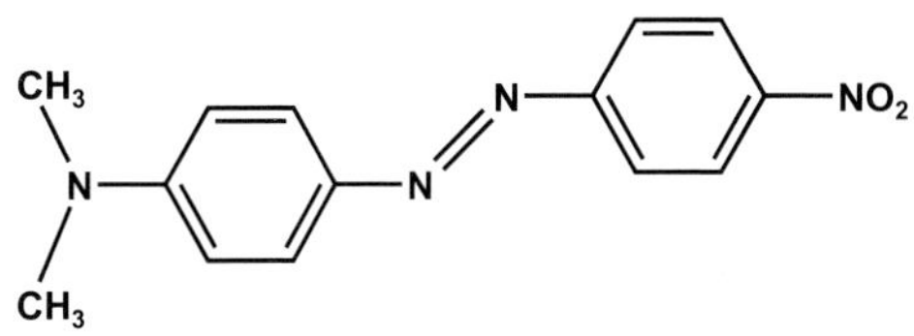

4-dimethylamino- 4'-nitroazobenzene (NDA)

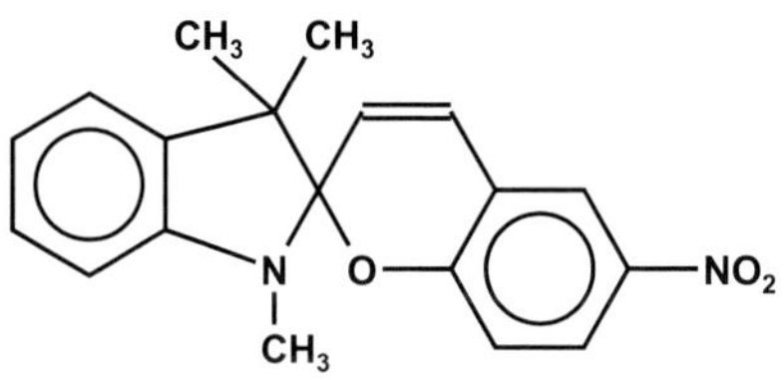

6-nitro-spiro[2H-1-benzopyran-2,2′-indoline] (NBIPS)

Scheme 1

If the polymer bulk is filled with increasing amounts of zeolite the absorption also increases. The reason is assumed to be the small optical anisotropy of zeolite particles resulting in scattering effects. Figure 10 shows the dependence of the transmission of the composites on zeolite concentration.

The evident result is that faujasite-encaged dyes do not participate in any reaction during polymerization process. Neither differences in polymerization rates between zeolite-containing monomer mixtures (with and without dye loading) and the pure monomer mixture, nor changes in the dye (identical absorption behavior) were observed.

8.3.1.4 **Composite Layers**

Layers were prepared from MMA–TFM copolymer solutions as described in Section 8.2.2.2. Results of transparency measurements of layers filled with different zeolite materials are shown in Fig 11.

Due to their small thickness pure polymer layers and such layers filled with small amounts of zeolite show practical 100% transparency. The absorption of

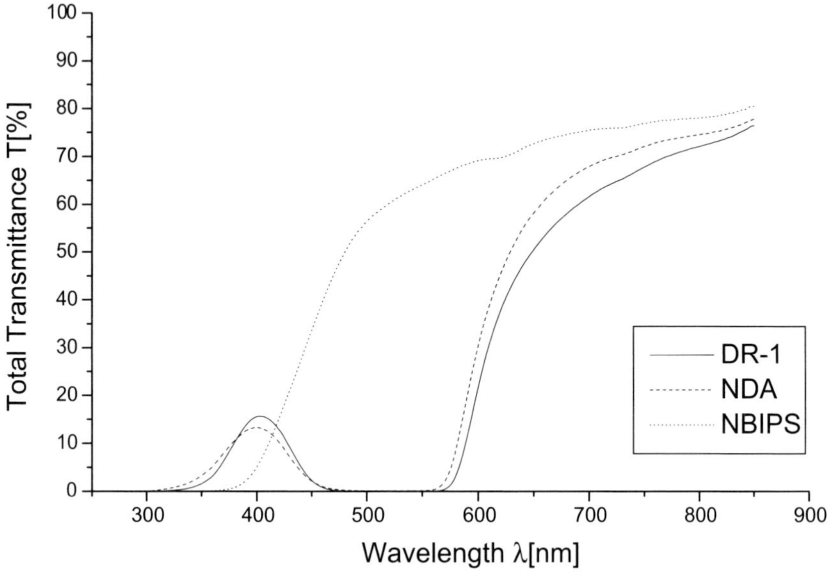

Fig. 9. Dependence of spectral transmission of composites of polymer and HY faujasites loaded with different dyes (dye content of loaded faujasite: 0.5 wt %)

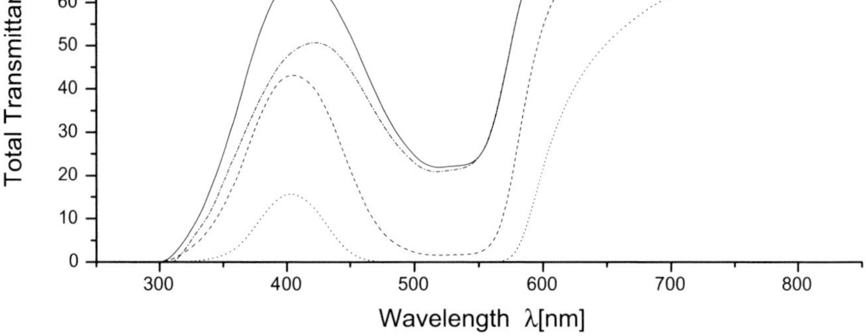

Fig. 10. Dependence of spectral transmission of different faujasite/DR1–polymer composites on concentration of loaded faujasites.

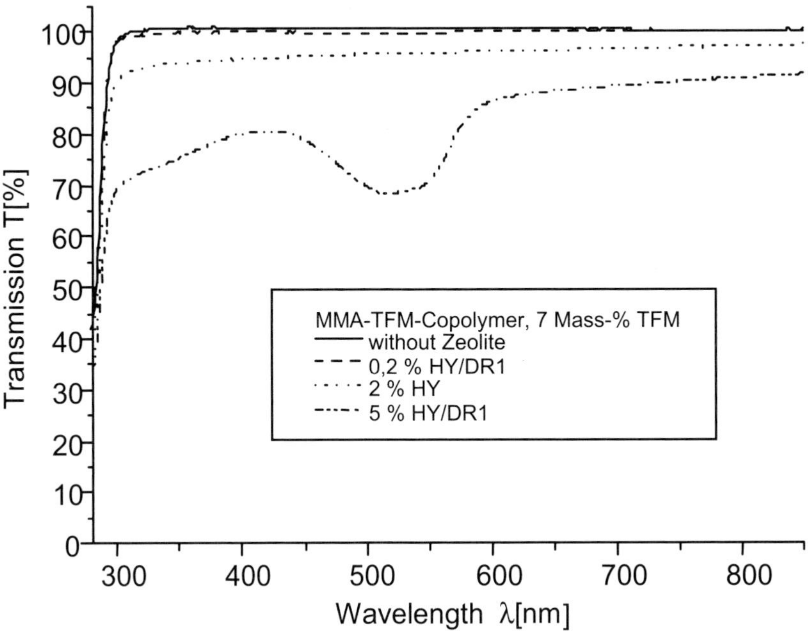

Fig. 11. Dependence of spectral transmission of layers of faujasite/DR1–polymer composites on the content of zeolite.

encaged dyes was not significant at these low concentrations (in the same range as for bulk samples). Only when zeolite amounts in the layers exceed 2 wt % a noticeable decrease in wavelength-dependent transparency was observed, and the characteristic absorption of the encaged dye also became significant.

8.4 Summary

8.4.1 Procedures

- From measurements of the turbidity of zeolite suspensions in different ethanol/toluene mixtures a specific ethanol/toluene mixture can be determined that shows a minimum of concentration-normalized turbidity for the zeolite. The refractive index of this solvent mixture can be measured easily and represents the refractive index of the suspended zeolite. By using this "immersion" method the wavelength-dependent refractive indices of different zeolites were measured and found to lie in the range of $n(\lambda) = 1.43$–1.49. These results were used to select an appropriate embedding polymer having the same refractive index.

- Free-radical copolymerization of methyl methacrylate (MMA) and 2,2,2-trifluoroethyl methacrylate (TFM) results in transparent, thermoplastic copolymers. Changing the ratio of MMA and TFM in the copolymer allows the refractive index to be adjusted in the range relevant to zeolites. Incorporating zeolites into these copolymers during bulk copolymerization led to composites of high optical transparency if the mean diameter of the zeolite particles did not exceed about 5 μm. Adjustment of the refractive indices of the copolymers could be performed by mixing copolymers having comonomer ratios. Phase separation of such mixtures in solution was not detected.
- By using variable-angle spectroscopic ellipsometry and UV/Vis measurements the wavelength-dependent refractive index and the transparency of base materials and composites could be measured.
- The spin-coating method was used to prepare thin layers from solutions of zeolite-filled copolymers.

8.4.2 Composite Properties

- As expected for total transparency of pure embedding polymers values could be measured in the order of $T = 92\,\%$. At wavelengths of less than 400 nm polymer absorption increases. Therefore, this copolymer is not suited for application in the wavelength region below 400 nm.
- Optimization of index matching was quantified measuring the transparency of composites. Results depend strongly on the particle size of the embedded zeolites. For an HY zeolite (mean particle diameter $x_T = 3$ μm) index matching led to a composite (2 wt % zeolite) having a high transparency of 73%. At the same concentration for a NaX zeolite (mean particle diameter $x_T = 19$ μm) the optimum transparency only reached 15%.
- Encaging of organic dyes in the zeolites examined had no influence on composite preparation and basic optical behavior. The main difference in optical properties between composites containing dye-loaded and unloaded zeolites is that wavelength-dependent transparency of the dye loaded composites reproduces the absorption behavior of the dye. The absorption maximum measured corresponds to that of the pure dye. Polymerization during composite preparation did not influence the dye in the zeolite cage. In contrast in the polymerizing monomer mixture dissolved Disperse Red 1 (DR1) changed in colour from red to yellow, indicating a reaction of free radicals with amino and/or nitro groups of the dye.

Acknowledgements

Financial support from Deutsche Forschungsgemeinschaft is gratefully acknowledged.

References

1 J. Caro, G. Finger, J. Kornatowski, J. Richter-Mendau, L. Werner, B. Zibrowins, *Adv. Mater.* **1992**, *4*, 273.

2 U. Vietze, O. Krauß, F. Laeri, G. Ihlein, F. Schüth, B. Limburg, M. Abraham, *Phys. Rev. Lett.* **1998**, *81*, 4628.

3 K. Hoffmann, F. Marlow, J. Caro, *Zeolites* **1996**, *16*, 281.

4 S. Feng, T. Bein in H. Chon, S.-K. Ihm, Y.S. Uh (Eds.), Progress in Zeolite and Microporous Materials, Studies in Surface Science and Catalysis, Vol. 105, Elsevier, Amsterdam, **1997**, p. 2147.

5 J. Caro, F. Marlow, K. Hoffmann, C. Striebel, J. Kornatowski, I. Girnus in H. Chon, S.-K. Ihm, Y.S. Uh (Eds.), Progress in Zeolite and Microporous Materials, Studies in Surface Science and Catalysis, Vol. 105, Elsevier, Amsterdam, **1997**, p. 2171.

6 T.-G. Tsai, K.-J. Chao, X.-J. Guo, S.-L. Sung, C.-N. Wu, Y.-L. Wang, H.-C. Shih, *Adv. Mater.* **1997**, *9*, 15.

7 I. Girnus, M.-M. Pohl, J. Richter-Mendau, M. Schneider, M. Noack, D. Venzke, J. Caro, *Adv. Mater.* **1998**, *7*, 711.

8 G. Carotenuto, L. Nicolais, *Appl. Compos. Mater.* **1995**, *2*, 385.

9 L.L. Beecroft, C.K. Ober, *J. Macromol. Sci. Pure Appl. Chem.* **1997**, *A34(4)*, 573.

10 J.C. Seferis in J. Brandrup, E.H. Immergut (Eds.), Polymer Handbook, Wiley, New York **1989**, p. VI/451.

11 D.E. Aspnes in E.D. Palik (Ed.), Handbook of Optical Constants of Solids, Academic Press, Orlando/London, **1985**, p. 89.

12 F.M. Willmouth in G.H. Meeten (Ed.), Optical Properties of Polymers, Elsevier, London/New York, **1986**, p. 278.

13 H. Lange, *Kolloid Z. Z. Polym.* **1969**, *223*, 24.

14 Ch. Striebel, K. Hoffmann, F. Marlow, The Microcrystal Prism Method for Refractive Index Measurements on Zeolite-based Nanocomposites, *Microporous Mater.* 9 (1997) 43.

15 B. Limburg, personal communication, **1997**.

16 H.G. Tompkins, A User s Guide to Ellipsometry, Academic Press, Boston **1993**, p. 5.

17 M. Born, E. Wolf, Principles of Optics, Pergamon Press **1993**, p. 95.

18 Werkstoffdatenbank POLYMAT, V. 5.2, Deutsches Kunststoff-Institut, Darmstadt **1998**.

Index

d

l

m

n